THIRD CANADIAN EDITION

NATURAL HAZARDS

Earth's Processes as Hazards, Disasters, and Catastrophes

EDWARD A. KELLER
University of California, Santa Barbara

DUANE E. DEVECCHIO
University of California, Santa Barbara

JOHN J. CLAGUE
Simon Fraser University

PEARSON
Toronto

Executive Acquisitions Editor: Cathleen Sullivan
Marketing Manager: Marlene Olsavsky
Program Manager: Darryl Kamo
Project Manager: Andrea Falkenberg
Developmental Editor: Patti Altridge
Production Services: Aptara®, Inc.
Permissions Project Manager: Marnie Lamb
Photo Permissions Research: Cordes Hoffman, Bill Smith Group
Text Permissions Research: Electronic Publishing Services
Cover Art Director: Jayne Conte
Cover Designer: Suzanne Behnke
Interior Art Director: Miguel Acevedo
Cover Image: Craig Hiltz

Credits and acknowledgments for material borrowed from other sources and reproduced, with permission, in this textbook appear on the appropriate page within the text.

10 9 8 7 6 5 4 3 2 1 [CKV]

Library and Archives Canada Cataloguing in Publication
Keller, Edward A., 1942 author
Natural hazards: earth's processes as hazards, disasters, and catastrophes/Edward A. Keller, Duane E. DeVecchio, John J. Clague.—Third Canadian edition.
Includes bibliographical references and index.
ISBN 978-0-13-307650-9 (pbk.)
1. Natural disasters—Textbooks. I. Clague, J. J. (John Joseph), 1946–, author II. DeVecchio, Duane E. (Duane Edward), 1970–, author III. Title.
GB5014.K44 2014 551 C2013-906532-6

10 9 8 7 6 5 4 3 2 1

PEARSON

ISBN: 978-0-13-307650-9

About the Authors

Edward A. Keller

Edward A. Keller is a professor, researcher, and writer, and, most importantly, a mentor and teacher to undergraduate and graduate students. Ed's students are currently working on earthquake hazards, how waves of sediment move through a river system following a disturbance, and on geologic controls on the habitat of endangered southern steelhead trout.

Ed was born and raised in California. He received bachelor's degrees in geology and mathematics from California State University at Fresno, and a master's degree in geology from the University of California at Davis. It was while pursuing his Ph.D. in geology from Purdue University in 1973 that Ed wrote the first edition of *Environmental Geology*, a text that became the foundation of an environmental geology curriculum in many colleges and universities. He joined the faculty of the University of California at Santa Barbara in 1976 and has been there ever since, serving multiple times as chair of both the Environmental Studies and Hydrologic Science programs. In that time he has authored more than 100 articles, including seminal works on fluvial processes and tectonic geomorphology.

Ed's academic honours include the Don J. Easterbrook Distinguished Scientist Award, Geological Society of America (2004); the Quatercentenary Fellowship from Cambridge University, England (2000); two Outstanding Alumnus Awards from Purdue University (1994, 1996); a Distinguished Alumnus Award from California State University at Fresno (1998); and the Outstanding Outreach Award from the Southern California Earthquake Center (1999).

Ed and his wife, Valery, who brings clarity to his writing, love walks on the beach at sunset and when the night herons guard moonlit sand at Arroyo Burro Beach in Santa Barbara.

Duane DeVecchio

Duane DeVecchio is currently a researcher and adjunct professor at the University of California at Santa Barbara, where he earned his Ph.D. in geology. Since starting his graduate education, Duane has devoted a significant amount of time to becoming an effective communicator of science. He is a passionate teacher and feels strongly that students need to develop critical thinking skills, which will enable them to evaluate for themselves data presented in graphs and tables from various sources. This is particularly important today, when the Internet and cable television offer accessibility to vast amounts of information, yet the validity of this information is often questionable or misleading. Integrating data from current and relevant research into core curriculum to illustrate the methodology and rigour of scientific investigations is essential to his teaching strategy.

Duane has a broad field-based background in the geological sciences and likes to share stories about his many months living alone in mobile trailers in the mountains and deserts so he could study rocks and landforms for his research. For his master's degree and post-master's research he conducted structural and stratigraphic analysis, as well as numerical dating of volcanic and volcaniclastic rocks in southeast Idaho and the central Mojave Desert of California, which record the Miocene depositional and extensional histories of these regions. His Ph.D. research was aimed at resolving fault slip rates and quantifying the earthquake hazard presented by several active fault-related folds growing beneath urbanized Southern California. Duane's current research interests focus on the timing and rates of change of Earth's surface due to depositional and erosional processes that result from climate change and tectonics.

When Duane is not teaching or conducting research, he enjoys whitewater rafting, rock climbing, snowboarding, and camping with his partner, Christy.

John J. Clague

John Clague is a professor of earth sciences at Simon Fraser University (SFU) in Burnaby, British Columbia, and an emeritus scientist with the Geological Survey of Canada. John was employed as a research scientist with the Geological Survey of Canada for 24 years, specializing in natural hazards, climate change, and ice age geology of western Canada. In 1998, he accepted a faculty position in the Department of Earth Sciences at SFU, where he is currently based. John became a Canada Research Chair in natural hazards research in 2003 and is currently director of the Centre for Natural Hazard Research. John is the author of 300 scientific papers on subjects as diverse as earthquakes, geochemistry, and archaeology. He has written popular books on Canadian earth science issues and on earthquakes and tsunamis in the Pacific Northwest. He is a fellow of the Royal Society of Canada, former president of the Geological Association of Canada, former president of the International Union for Quaternary Research, and a recipient of numerous awards and honours.

John is a strong proponent of the philosophy captured so eloquently by Margaret Mead: "Never doubt that a small group of committed people can change the world. Indeed, it is the only thing that has."

Brief Contents

1 Introduction to Natural Hazards 2

2 Internal Structure of Earth and Plate Tectonics 24

3 Earthquakes 46

4 Tsunamis 90

5 Volcanoes and Volcanic Eruptions 114

6 Landslides 156

7 Snow Avalanches 186

8 Subsidence and Soil Expansion and Contraction 208

9 River Flooding 234

10 Atmosphere and Severe Weather 272

11 Hurricanes and Extratropical Cyclones 306

12 Waves, Currents, and Coastlines 334

13 Wildfires 364

14 Climate Change 390

15 Impacts and Extinctions 422

Contents

Preface xxi

1 Introduction to Natural Hazards 2

Case History: Earthquake in Haiti, 2010: Lessons Learned 2

1.1 Why Studying Natural Hazards Is Important 5

Hazardous Natural Processes and Energy Sources 6

Hazard, Risk, Disaster, and Catastrophe 7

Death and Damage Caused by Natural Hazards 8

1.2 Magnitude and Frequency of Hazardous Events 9

1.3 Role of History in Understanding Hazards 10

1.4 Geologic Cycle 11

The Tectonic Cycle 11

The Rock Cycle 11

The Hydrologic Cycle 12

Biogeochemical Cycles 12

1.5 Fundamental Concepts for Understanding Natural Processes as Hazards 13

Concept 1: Hazards can be understood through scientific investigation and analysis. 13

Science and Natural Hazards 13

Hazardous Processes Are Natural 13

Prediction, Forecast, and Warning 14

Concept 2: An understanding of hazardous processes is vital to evaluating risk. 15

Concept 3: Hazards are commonly linked to each other and to the environment in which they occur. 16

Concept 4: Population growth and socio-economic changes increase the risk from natural hazards. 16

Concept 5: Damage and loss of life from natural disasters can be reduced. 17

Reactive Response: Recovery from Disasters 17

Proactive Response: Avoiding and Adjusting to Hazards 20

1.6 Many Hazards Provide a Natural Service Function 21

1.7 Climate Change and Natural Hazards 21

Summary 22

Key Terms 23

Did You Learn? 23

Critical Thinking Questions 23

MasteringGeology 23

2 Internal Structure of Earth and Plate Tectonics 24

Case History: Two Cities in Harm's Way 24

2.1 Internal Structure of Earth 26

Earth Is Layered and Dynamic 26

Continents and Ocean Basins Have Different Properties and History 28

2.2 How We Know about Earth's Internal Structure 28

2.3 Plate Tectonics 28

Movement of Lithospheric Plates 28

Types of Plate Boundaries 31

Rates of Plate Motion 37

Hot Spots 37

A Closer Look 2.1 Paleomagnetism and Seafloor Spreading 38

A Famous Canadian's Contribution to the Plate Tectonic Theory 40

2.4 Mechanisms That Move Plates 42

2.5 Plate Tectonics and Hazards 43

Revisiting the Fundamental Concepts Internal Structure of Earth and Plate Tectonics 43

Summary 44

Key Terms 44

Did You Learn? 44

Critical Thinking Questions 45

MasteringGeology 45

3 Earthquakes 46

Case History: The Toll of Earthquakes 46

3.1 Introduction to Earthquakes 49

Earthquake Magnitude 49

Earthquake Intensity 50

3.2 Earthquake Processes 54

Process of Faulting 54

Fault Types 54

Fault Activity 55

Tectonic Creep and Slow Earthquakes 55

Seismic Waves 55

3.3 Earthquake Shaking 57

Distance to the Epicentre and Focal Depth 57

Direction of Rupture 60

Local Soil and Rock Conditions 62

3.4 The Earthquake Cycle 63

3.5 Geographic Regions at Risk from Earthquakes 65

Plate-Boundary Earthquakes 65

Intraplate Earthquakes 66

3.6 Effects of Earthquakes and Linkages with Other Natural Hazards 68

Shaking and Ground Rupture 68

Liquefaction 68

Survivor Story Shaky Honeymoon 69

Land-Level Changes 70

Landslides 71

Fires 72

Disease 72

3.7 Natural Service Functions of Earthquakes 72

Groundwater and Energy Resources 72

Mineral Resources 72

Landform Development 72

3.8 Human Interaction with Earthquakes 73

Reservoirs 73

Deep Waste Disposal 73

Pumping of Oil or Gas 73

Hydraulic Fracturing 73

Nuclear Explosions 75

3.9 Minimizing the Earthquake Hazard 75

Earthquake Hazard Reduction Programs 75

Estimation of Seismic Risk 75

Case Study 3.1 The Denali Earthquake: Estimating Potential Ground Rupture Pays Off 76

Forecasts and Prediction 77

Status of Earthquake Prediction and Forecasting 79

Case Study 3.2 The 2009 L'Aquila Earthquake 80

Professional Profile Gail Atkinson, Seismologist 82

Earthquake Warning Systems 83

3.10 Perception of and Adjustment to the Earthquake Hazard 83

Perception of the Earthquake Hazard 83

Community Adjustments to the Earthquake Hazard 83

Personal Adjustments Before, During, and After an Earthquake 84

Revisiting the Fundamental Concepts Earthquakes 86

Summary 87

Key Terms 88

Did You Learn? 88

Critical Thinking Questions 88

MasteringGeology 89

4 Tsunamis 90

Case History: Giant Earthquake and Tsunami in Japan 90

4.1 Introduction to Tsunamis 94
Earthquake-Triggered Tsunamis 94
Case Study 4.1 Catastrophe in the Indian Ocean 96
Survivor Story Swept Away by a Tsunami 99
A Closer Look 4.1 Geologic Evidence for Tsunamis 100
Landslide-Triggered Tsunamis 100
Volcanic-Triggered Tsunamis 103

4.2 Regions at Risk 103

4.3 Effects of Tsunamis and Links with Other Natural Hazards 104

4.4 Minimizing the Tsunami Hazard 105
Detection and Warning 105
Structural Control 106
Tsunami Inundation Maps 107
Land Use 107
Probability Analysis 107
Education 108
Professional Profile Jose Borrero, Tsunami Scientist 110
Tsunami Readiness 111

4.5 Perception and Personal Adjustment to Tsunami Hazard 111
Revisiting the Fundamental Concepts Tsunamis 111

Summary 112
Key Terms 112
Did You Learn? 113
Critical Thinking Questions 113
MasteringGeology 113

5 Volcanoes and Volcanic Eruptions 114

Case History: Mount Unzen, 1991 114

5.1 Introduction to Volcanoes 114
How Magma Forms 114
Magma Properties 118
Volcano Types 120
Volcanic Features 124
Case Study 5.1 Planning for a Disaster in Naples 125
Volcanoes and Plate Tectonics 129

5.2 Geographic Regions with Active Volcanoes 131

5.3 Volcanic Hazards 133
Lava Flows 133
Pyroclastic Flows and Surges 135
Lateral Blasts 135
Ash Falls 135
Case Study 5.2 Icelandic Eruption Paralyzes Air Travel in Europe 137
Poisonous Gases 137
Edifice or Sector Collapse 139
Debris Flows and Other Mass Movements 139
A Closer Look 5.1 Mount St. Helens 1980–2008: From Lateral Blasts to Lava Flows 141

5.4 Links between Volcanoes and Other Natural Hazards 144
Professional Profile Catherine Hickson, Volcano Scientist 145

5.5 Natural Service Functions of Volcanoes 146
Volcanic Soils 146
Geothermal Power 146
Recreation 146
Creation of New Land 146

5.6 Minimizing the Volcanic Hazard 147
Forecasting 147
Volcanic Alert or Warning 150

5.7 Perception of and Adjustment to the Volcanic Hazard 150
Perception of Volcanic Risk 150
Survivor Story A Close Call with Mount St. Helens 151
Adjustments to Volcanic Hazards 151
Revisiting the Fundamental Concepts Volcanoes and Volcanic Eruptions 153

Summary 153
Key Terms 154
Did You Learn? 154
Critical Thinking Questions 154
MasteringGeology 155

6 Landslides 156

Case History: The Frank Slide 156

6.1 Introduction to Landslides 159
Types of Landslides 159
Forces on Slopes 163
A Closer Look 6.1 Estimating the Velocity of Landslides from Their Run-Up and Superelevation 164

6.2 Geographic Regions at Risk from Landslides 170

6.3 Effects of Landslides and Links with Other Natural Hazards 171

Effects of Landslides 171
Links between Landslides and Other Natural Hazards 171
Survivor Story Landslide 172
Case Study 6.1 Mount Meager Landslides and Consequent Flooding 173

6.4 Natural Service Functions of Landslides 175

6.5 Human Interaction with Landslides 175
Timber Harvesting 175
Urbanization 175

6.6 Minimizing Landslide Hazard and Risk 177
Identification of Potential Landslides 177
Professional Profile Matthias Jakob, Engineering Geologist 178
Prevention of Landslides 179
Landslide Warning Systems 182

6.7 Perception of and Adjustment to Landslide Hazards 182
Perception of Landslide Hazards 182
Adjustments to the Landslide Hazard 182
Personal Adjustments: What You Can Do to Minimize Your Landslide Risk 182
Revisiting the Fundamental Concepts Landslides 183

Summary 184
Key Terms 184
Did You Learn? 184
Critical Thinking Questions 185
MasteringGeology 185

7 Snow Avalanches 186

Case History: The Chilkoot Disaster 186

7.1 Introduction to Snow Avalanches 188
Snow Climatology 189
Avalanche Initiation 189
Weak Layers 190
Avalanche Motion 191
Avalanche Triggering 191
Terrain Factors 191

7.2 Geographic Regions at Risk of Avalanches 193

7.3 Impacts of Avalanches and Links with Other Natural Hazards 194
Impacts of Avalanches 194
Links between Avalanches and Other Natural Hazards 194

7.4 Natural Service Functions of Avalanches 195

7.5 Human Interaction with Avalanches 195

7.6 Minimizing Avalanche Risk 195
Location of Infrastructure 195
Structures in the Start Zone 196
Structures in the Track and Run-Out Zone 196
Control through the Use of Explosives 196
Case Study 7.1 Deadly Avalanche in Glacier National Park 197
Forecasting 198
Modelling 200

7.7 Avalanche Safety 201
Good Habits Minimize Risk 201
Professional Profile Grant Statham, Parks Canada 203

7.8 Avalanche Rescue and Survival 203
Avalanche Cords 204
Avalanche Transceivers 204
Probes 204
Shovels 204
Avalanche Dogs 204
Avalanche Survival 204
Revisiting the Fundamental Concepts Snow Avalanches 205

Summary 205
Key Terms 206
Did You Learn? 206
Critical Thinking Questions 206
MasteringGeology 207

8 Subsidence and Soil Expansion and Contraction 208

Case History: Venice Is Sinking 208

8.1 Introduction to Subsidence and Soil Expansion and Contraction 210
Karst 210
Survivor Story Sinkhole Drains Lake 213
Permafrost 213
Piping 215
Sediment Compaction 215
Expansive Soils 216
Earthquakes 216
Deflation of Magma Chambers 216

8.2 Regions at Risk from Subsidence and Soil Expansion and Contraction 218

8.3 Effects of Subsidence and Soil Expansion and Contraction 219
Sinkhole Formation 219

Groundwater Use and Contamination 219
Permafrost Thaw 219
Coastal Flooding and Loss of Wetlands 220
Soil Volume Changes 220
Case Study 8.1 Permafrost Thaw in Canada's North 221

8.4 Links between Subsidence, Soil Volume Changes, and Other Natural Hazards 222
Case Study 8.2 Loss of Wetlands on the Mississippi Delta 223

8.5 Natural Service Functions of Subsidence 225
Water Supply 225
Aesthetic and Scientific Resources 225
Unique Ecosystems 226

8.6 Human Interaction with Subsidence 226
Withdrawal of Fluids 226
Underground Mining 228
Permafrost Thaw 228
Restricting Deltaic Sedimentation 229
Draining Wetlands 229
Landscaping on Expansive Soils 229

8.7 Minimizing Subsidence Hazards 229
Restricting Fluid Withdrawal 229
Regulating Mining 229
Preventing Damage from Thawing Permafrost 229
Reducing Damage from Deltaic Subsidence 230
Stopping the Draining of Wetlands 230
Preventing Damage from Expansive Soils 230

8.8 Perception of and Adjustments to Subsidence and Soil Hazards 230
Perception of Subsidence and Soil Hazards 230
Adjustments to Subsidence and Soil Hazards 230
Revisiting the Fundamental Concepts Subsidence and Soil Expansion and Contraction 231

Summary 232
Key Terms 232
Did You Learn? 232
Critical Thinking Questions 233
MasteringGeology 233

9 River Flooding 234

Case History: The Alberta Floods of 2013 234

9.1 Introduction to Rivers 237
Earth Material Transported by Rivers 238
River Velocity, Discharge, Erosion, and Sediment Deposition 238

Channel Patterns and Floodplain Formation 238

9.2 Flooding 241

Case Study 9.1 Mississippi River Floods of 1973 and 1993 242

Flash Floods and Downstream Floods 246

Survivor Story Flooding from Hurricane Hazel 247

Case Study 9.2 The Saguenay Flood 250

Outburst Floods 250

A Closer Look 9.1 Magnitude and Frequency of Floods 251

9.3 Geographic Regions at Risk for Flooding 253

9.4 Effects of Flooding and Links between Floods and Other Hazards 255

9.5 Natural Service Functions 256

Fertile Land 256

Aquatic Ecosystems 256

Sediment Supply 256

9.6 Human Interaction with Flooding 256

Land-Use Changes 256

Dam Construction 257

Urbanization and Flooding 258

9.7 Minimizing the Flood Hazard 260

The Structural Approach 260

Channel Restoration: An Alternative to Channelization 263

Flood Forecasts and Advisories 264

9.8 Perception of and Adjustment to Flood Hazards 264

Perception of Flood Hazards 264

Adjustments to the Flood Hazard 265

Professional Profile Eve Gruntfest, Geographer 266

Relocating People from Floodplains 268

Personal Adjustments: What to Do and What Not to Do 269

Revisiting the Fundamental Concepts River Flooding 269

Summary 270

Key Terms 270

Did You Learn? 270

Critical Thinking Questions 271

MasteringGeology 271

10 Atmosphere and Severe Weather 272

Case History: The 1998 Ice Storm 272

10.1 Energy 274

Types of Energy 276

Heat Transfer 276

10.2 Energy at Earth's Surface 276
Electromagnetic Energy 277
Energy Behaviour 277

10.3 The Atmosphere 278
Composition of the Atmosphere 278
Structure of the Atmosphere 279

10.4 Weather Processes 280
Atmospheric Pressure 280
Vertical Stability of the Atmosphere 282
Fronts 282

10.5 Hazardous Weather 282
Thunderstorms 282
A Closer Look 10.1 Coriolis Effect 284
Case Study 10.1 Lightning 287
Tornadoes 289
Survivor Story Struck by Lightning 290
Case Study 10.2 The Edmonton Tornado 293
Blizzards, Extreme Cold, and Ice Storms 294
Fog 297
Drought 297
Dust and Sand Storms 299
Heat Waves 299

10.6 Human Interaction with Weather 301

10.7 Links with Other Hazards 301

10.8 Natural Service Functions of Severe Weather 302

10.9 Minimizing Severe Weather Hazards 302
Forecasting and Predicting Weather Hazards 302
Adjustment to Severe Weather Hazards 303
Revisiting the Fundamental Concepts Atmosphere and Severe Weather 303

Summary 304
Key Terms 305
Did You Learn? 305
Critical Thinking Questions 305
MasteringGeology 305

11 Hurricanes and Extratropical Cyclones 306

Case History: Hurricane Katrina 306
Survivor Story Hurricane Katrina 310

11.1 Introduction to Cyclones 311

Classification 311
Naming 311

11.2 Cyclone Development 312
Tropical Cyclones 312
Extratropical Cyclones 315

11.3 Geographic Regions at Risk for Cyclones 318
Case Study 11.1 Hurricane Juan 320

11.4 Effects of Cyclones 322
Storm Surge 322
High Winds 323
Case Study 11.2 Hurricane Sandy 324
Heavy Rains 326

11.5 Links between Cyclones and Other Natural Hazards 326

11.6 Natural Service Functions of Severe Weather 327

11.7 Human Interaction with Weather 327

11.8 Minimizing the Effects of Cyclones 328
Forecasts and Warnings 328

11.9 Perception of and Adjustment to Cyclones 331
Perception of Cyclones 331
Adjustments to Cyclones 331
Revisiting the Fundamental Concepts Hurricanes and Extratropical Cyclones 331

Summary 332
Key Terms 333
Did You Learn? 333
Critical Thinking Questions 333
MasteringGeology 333

12 Waves, Currents, and Coastlines 334

Case History: Harris Meisner's Farm by the Sea 334

12.1 Introduction to Coastal Hazards 337

12.2 Coastal Processes 337
Waves 337
Case Study 12.1 Rogue Waves 338
Beach Form and Processes 342

12.3 Sea-Level Change 343
Eustatic Sea-Level Change 344
Isostatic Sea-Level Change 344
Tectonic and Other Effects 344

12.4 Coastal Hazards 344
Rip Currents 344
Coastal Erosion 345
Professional Profile Phil Hill, Coastal Geologist 345
A Closer Look 12.1 Beach Budget 347
Sea-Level Rise 349
12.5 Links between Coastal Processes and Other Natural Hazards 350
12.6 Natural Service Functions of Coastal Processes 352
12.7 Human Interaction with Coastal Processes 352
12.8 Minimizing Damage from Coastal Hazards 356
Hard Stabilization 356
A Closer Look 12.2 E-Lines and E-Zones 357
Soft Solutions 358
12.9 Perception of Coastal Hazards 359
12.10 Future Coastal Zone Management 361
Revisiting the Fundamental Concepts Coastal Hazards 361
Summary 362
Key Terms 363
Did You Learn? 363
Critical Thinking Questions 363
MasteringGeology 363

13 Wildfires 364
Case History: Wildfires in British Columbia in 2003 364
13.1 Introduction to Wildfire 366
13.2 Wildfire as a Process 367
Fire Environment 368
13.3 Geographic Regions at Risk from Wildfires 371
Case Study 13.1 Wildfires in Canada 372
Case Study 13.2 The 2011 Slave Lake Wildfire 374
13.4 Effects of Wildfires and Links with Climate 376
Effects on the Geological Environment 376
Effects on the Atmosphere 376
Links with Climate 376
Professional Profile Bob Krans 377
13.5 Impacts of Wildfires on Plants and Animals 379
13.6 Natural Service Functions of Wildfires 379
Case Study 13.3 Yellowstone Fires of 1988 380

13.7 Fire Management 381
Scientific Research 381
Data Collection 381
Fire Suppression 381
13.8 Perception of and Adjustment to the Wildfire Hazard 383
Perception of Wildfire Hazard 383
Reducing Wildfire Risk 384
Survivor Story The Cedar Fire 386
Revisiting the Fundamental Concepts Wildfires 387
Summary 388
Key Terms 388
Did You Learn? 388
Critical Thinking Questions 388
MasteringGeology 389

14 Climate Change 390
Case History: Arctic Threatened by Climate Change 390
14.1 Global Change and Earth System Science: An Overview 394
14.2 Climate and Weather 394
14.3 The Atmosphere 394
Composition of the Atmosphere 394
Structure of the Atmosphere 397
Atmospheric Circulation 397
The Greenhouse Effect 398
14.4 How We Study Climate Change and Make Predictions of Future Climate 399
Tree Rings 399
Sediments 399
Ice Cores 400
Pollen 400
Global Climate Models 400
14.5 Climate Change on Long Timescales 402
Pleistocene Glaciation 402
14.6 Climate Change on Short Timescales 404
Evidence for Climate Change on Short Timescales 404
Causes of Climate Change on Short Timescales 406
14.7 Effects of Climate Change 409
Glacier Ice and Sea-Level Rise 410
Glacier Hazards 411
Thawing of Permafrost 411
Changes in Climate Patterns 411
Changes in the Biosphere 412

Desertification and Drought 414
Case Study 14.1 Palliser Triangle 414
Wildfires 415

14.8 Minimizing the Effects of Global Warming 415
International Agreements 416
Carbon Sequestration 417
Fossil Fuels and Future Climate Change 417
A Closer Look 14.1 Abrupt Climate Change 418

14.9 Adaptation 418
Revisiting the Fundamental Concepts Climate Change 419

Summary 420
Key Terms 420
Did You Learn? 421
Critical Thinking Questions 421
MasteringGeology 421

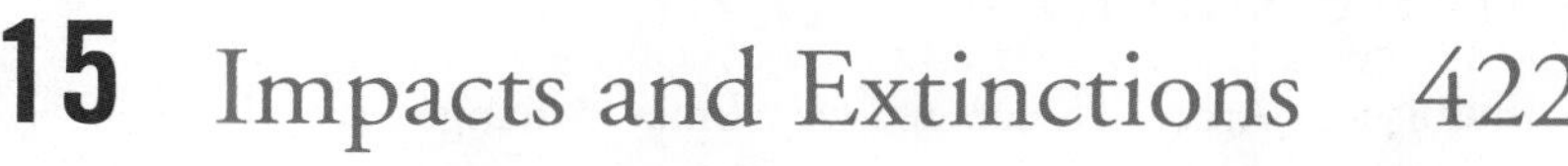

15 Impacts and Extinctions 422

Case History: The Tunguska Event 422

15.1 Earth's Place in Space 424
Asteroids, Meteoroids, and Comets 425

15.2 Airbursts and Impacts 425
Impact Craters 429
Survivor Story Meteorites in Illinois 430
Case Study 15.1 The Sudbury Impact Event 434
Case Study 15.2 Uniformitarianism, Gradualism, and Catastrophes 436

15.3 Mass Extinctions 437
K-T Boundary Mass Extinction 438

15.4 Links with Other Natural Hazards 440

15.5 Impact Hazards and Risk 440
Risk Related to Impacts 440
Managing the Impact Hazard 443
Revisiting the Fundamental Concepts Impacts and Extinctions 445

Summary 446
Key Terms 446
Did You Learn? 446
Critical Thinking Questions 446
MasteringGeology 447
Glossary 448
References 461
Index 478

Preface

Natural Hazards: Earth's Processes as Hazards, Disasters, and Catastrophes, Third Canadian Edition is an introductory university-level, non-technical survey of natural processes that have direct, often sudden, and violent impacts on humanity. The book integrates principles of geology, geography, hydrology, meteorology, climatology, oceanography, soil science, ecology, and solar system astronomy. It has been designed for a course for non-science majors and will help instructors guide students who may have little background in science through the geologic underpinnings and societal repercussions of hazardous Earth processes. It is also suitable for topical introductory courses in physical geology, physical geography, and Earth science.

In preparing the third edition of this book, we took advantage of the growing amount of information about natural hazards, disasters, and catastrophes. Since the second edition was published, many natural disasters and catastrophes have occurred. In 2010, drought, heat wave, and air pollution from wildfires killed several thousand people in Moscow; in 2011, great earthquakes and tsunamis struck Chile and Japan; and in 2012, a superstorm caused over $65 billion damage on the Atlantic coast of the United States. In 2013, destructive floods happened in southwest Alberta and Colorado, and wildfires destroyed more than 200 homes in New South Wales, Australia. These events are the result of enormous forces that are at work both inside and on the surface of our planet.

Climate change is causing glaciers to melt, permafrost to thaw, the atmosphere and oceans to warm, and sea level to rise more rapidly than had originally been forecast. These changes are caused in part by human activities, primarily the burning of fossil fuels, which releases vast amounts of carbon dioxide and other gases into the atmosphere. The interaction between humans and Earth processes has never been clearer, and the need for understanding these processes as hazards for our economy and society has never been greater.

This edition of *Natural Hazards* seeks to explain destructive Earth processes in an understandable way, illustrate how they affect us, and discuss how we can better prepare for, and adjust to, inevitable natural disasters.

An important point, central to both the text and course, is that Earth processes are not, in and of themselves "hazards," even though we describe them as such. Earthquakes, tsunamis, volcanic eruptions, landslides, floods, tornadoes, hurricanes, and wildfires are natural processes and have occurred for hundreds of millions of years. They are hazards only when they impact people. Ironically, human behaviour can turn hazardous events into disasters or, worse, catastrophes. The text strives to present these phenomena as natural geologic processes that have human impacts.

A course in natural hazards offers many benefits besides satisfying students' natural curiosity about such events. An informed citizenry is our best guarantee of a prosperous future. Armed with insights into the complex relations between people and the geologic environment, we will ask better questions and make better choices. On a local level, we will be better prepared to make decisions about where we live and how best to invest our time and resources. On national and global levels, we will be better able to advise our leaders on important issues related to natural hazards that impact our lives.

Distinguishing Features of the Third Edition

With these objectives in mind, we have incorporated into this edition of *Natural Hazards* a number of features designed to support the student and instructor.

A Balanced Approach

Many readers will naturally focus on natural hazards that threaten their community, province, or country, but globalization of our economy, near-universal information access, and the effects of humans on our planet require a broader, more balanced approach to the study of natural hazards and risk. A major earthquake in Tokyo or San Francisco would affect trade in the ports of Vancouver and Seattle, the economy of Silicon Valley in California, and the price of computers in Toronto and Halifax. The authors have tried to provide balanced coverage of natural hazards, with examples from Canada, the United States, and other countries. Topics covered in this edition include earthquakes and tsunamis on the British Columbia coast; the 1929 Grand Banks earthquake and tsunami; disastrous historic floods on the Fraser and Red rivers; a discussion of how Canada deals with floods; large landslides at Frank, Alberta, in 1903, near Hope, British Columbia, in 1965, and north of Pemberton, British Columbia, in 2010; the disastrous ice storm in Ontario and Quebec in 1998; the 1983 Edmonton tornado; Hurricanes Katrina, Hazel, and Juan; Superstorm Sandy in 2012; destructive wildfires in Kelowna, British Columbia, in 2003 and Slave Lake, Alberta, in 2011; recent killer earthquakes in southwest China, Haiti, New Zealand, and Japan; and an expanded treatment of climate change and its impacts in Canada. This edition also includes a new chapter on plate tectonics.

This book treats each topic as both a natural phenomenon and a human hazard. For example, the discussion of tsunamis includes a description not only of their characteristics, causes, global distribution, frequency, and effects, but also of engineering and nonstructural approaches that can be taken to reduce tsunami risk.

Five Fundamental Concepts

Five key concepts provide a conceptual framework of understanding that guide the reader:

1. **Hazards can be understood through scientific investigation and analysis.**

   Most hazardous events and processes can be monitored and mapped and their future occurrence and magnitude forecast based on the frequency of past events, patterns in their occurrence, and types of precursor events.

2. **An understanding of hazardous processes is vital to evaluating risk.**

   Hazardous processes are amenable to risk analysis because the probability and consequences of an event can be determined or estimated.

3. **Hazards are commonly linked to each other and to the environment in which they occur.**

   Hazardous processes are linked in many ways. For example, earthquakes can produce landslides and giant sea waves called tsunamis, and hurricanes cause flooding and coastal erosion.

4. **Population growth and socio-economic changes increase risk from natural hazards.**

   As a result of increasing human population, larger concentrations of economic wealth, and poor land-use practices, events that previously caused disasters are now causing catastrophes.

5. **Damage and loss of life from natural disasters can be reduced.**

   Minimizing the potential adverse effects of hazardous processes requires an integrated approach that includes scientific understanding, land-use planning and regulation, engineering, education, and proactive disaster preparedness.

These five concepts are introduced in Chapter 1 and revisited throughout the text. They provide a framework for understanding that can extend beyond the course into everyday life.

Survivor Stories and Professional Profiles

Many chapters contain the personal story of someone who has had a brush with disaster, as well as a profile of a scientist or other professional who has worked with a particular hazard. Although most of us will never experience a volcanic eruption, tsunami, or major hurricane, we are naturally curious about what we would see, hear, and feel if we did. For example, a scientific description of a volcanic eruption does not convey the amazement and terror that Canadian volcanologist Catherine Hickson experienced on the morning of May 18, 1980, when Mount St. Helens exploded (see her professional profile on page 145). Likewise, the story of the tsunami in the Indian Ocean in 2003 does not give us the real sense of what Christine Lang felt when she was caught up in and nearly killed in the tsunami (see Survivor Story in Chapter 4). To fully understand and appreciate natural hazards, we need both scientific knowledge and human experience. As you read the survivor stories, ask yourself what you would do in a similar situation. This knowledge could save your life someday, as it did for Tilly Smith and her family on the beach in Phuket, Thailand, in December 2003 (see Chapter 4).

People study and work with natural hazards for many reasons—curiosity, monetary reward, excitement, or the desire to help others prepare for events that might threaten their lives and property. As you read each professional profile, think about the person's motivation, the type of work he or she does, and how that work contributes to increasing human knowledge or reducing risk. For example, Grant Statham, who works for Parks Canada, has studied snow avalanches in western Canada for nearly three decades (see his professional profile in Chapter 7). For him, this study is both a vocation and an avocation. He has had a lifelong interest in avalanches and is striving to reduce avalanche injuries and deaths by promoting scientific study, educational programs, and heightened awareness of the hazard. Most of the survivor stories and professional profiles are based on interviews conducted exclusively for *Natural Hazards*.

Major New Material in the Third Edition

The third edition benefited greatly from feedback from instructors who used the previous edition, and many of the changes reflect their thoughtful reviews. New material for the third edition includes the following:

- NEW! A chapter on plate tectonics that reflects its overarching importance in Earth science.
- NEW! Revisiting the Fundamental Concepts, which reminds students of the unifying theme of the five fundamental concepts introduced in the first chapter. This feature identifies how the five fundamental concepts relate to each hazard.
- NEW! "A Closer Look" uses real-life events and data to enhance understanding and comprehension not only of the hazard, but also its mitigation.
- NEW! "Did You Learn" questions at the end of each chapter allow students to track their comprehension of the learning goals stated at the beginning of the chapter.
- Coverage of the most recent disasters on Earth, such as recent earthquakes in Haiti, Italy, New Zealand, Chile, Italy, and Japan; the Icelandic volcanic eruptions in 2010; wildfires in Alberta in 2011; the superstorm on the Atlantic coast of the United States in 2012; the flooding in Alberta in 2013; and more.
- Revised design throughout the book includes changes to the chapter openers, which feature new photos and content.
- Updated art with new figures, illustrations, and photos throughout the book. Each image has been reviewed for accuracy, relevance, and its educational impact.

Features of the Text

This text is sensitive to the study needs of students. Each chapter is clearly structured to help students understand the material and effectively review the major concepts. Each chapter is organized with the following study aids:

- Learning Objectives on the first page of each chapter clearly state what students should have achieved after completing their study of the topic.
- Selected features, including Case Studies, Closer Look boxes, Survivor Stories, and Professional Profiles, are added where appropriate to help students understand natural hazards through real-world examples.
- A discussion of the five fundamental concepts as they apply to the specific hazard at hand is followed by a summary. The summaries reinforce the major points of each chapter and help students focus on important concepts.
- References are included with each chapter to provide additional sources of information and give credit to the scholars who did the research reported in the chapter.
- Key terms are highlighted in **bold** where they are first introduced in detail in the text. These terms are listed at the end of each chapter. Other important technical terms that may be new to students are indicated in *italics*. All **key terms** and *useful terms* are defined in a glossary at the end of the book.
- Review questions at the end of each chapter will help students focus on the important subject matter.
- Critical thinking questions at the end of each chapter have been designed to stimulate students to think about important issues and to apply the information to both their lives and society.
- Appendices, which are included in the online resources, provide additional information useful for understanding some of the more applied aspects of geology that relate to natural hazards. This information may be helpful in supplementing laboratory and field exercises.

The Teaching and Learning Package

MasteringGeology™ with Pearson eText

MasteringGeology delivers engaging, dynamic learning opportunities that focus on course objectives and is responsive to each student's progress. It helps students absorb course material and understand difficult concepts.

- MasteringGeology provides a rich and flexible set of course materials to get instructors started quickly.
- MasteringGeology provides quick and easy access to information on student performance against learning outcomes. Instructors can add their own learning outcomes, to track student performance.
- The MasteringGeology gradebook and diagnostic tools capture the step-by-step work.

Assignable items in MasteringGeology include:

- Mathematics Review, Chemistry Review, Geography Review, and Test Bank questions.
- MasteringGeology includes a Study Area where students can access all their study resources. The Study Area includes geoscience animations, In the News RSS feeds, Self-Study Quizzes, Weblinks, Glossary, Flashcards, and an optional Pearson eText.

  See www.masteringgeology.com.

Hazard City: Assignments in Applied Geology is now part of MasteringGeology. In Hazard City, students will assume the role of a practising geologist and be assigned a variety of tasks: gathering and analyzing real data, evaluating risk, and making assessments and recommendations. Topics include:

Ground Water Contamination: Students use field and laboratory data to prepare a contour map of the water table, determine the direction of ground water flow, and map a contaminated area.

Volcanic Hazard Assessment: Students research volcanic hazards, collect field information, and use decision making to determine the potential impact of a volcanic eruption on different parts of Hazard City.

Landslide Hazard Assessment: Students research the factors that determine the landslide hazard at five construction sites and make recommendations for development.

Earthquake Damage Assessment: Students research the effects of earthquakes on buildings, explore Hazard City, and determine the number of people needing emergency housing given an earthquake of specific intensity.

Flood Insurance Rate Maps: Students estimate flood insurance premiums using a flood insurance rate map, insurance tables, and site characteristics.

Snowpack Monitoring: Students use climatic data to estimate variables that are key to flood control and water supply management.

Coal Property Evaluation: Students estimate what the potential value of a mineral property is by learning about mining and property evaluation and applying that knowledge in a resource calculation.

Landfill Siting: Students use maps and geological data to determine if any of five proposed sites meet the requirements of the State Administrative Code for landfill siting.

Shoreline Property Assessment: Students visit four related waterfront building sites—some developed and some not—and analyze the risk each faces due to shoreline erosion processes.

Tsunamis/Storm Surges: Students research the causes and effects of tsunamis and storm surges and then prepare a risk assessment report for a small oceanside village near Hazard City.

Map Reading: In order to identify the source of water contamination in a local stream, students learn how to read topographic maps and use their knowledge to plan their field work.

For the Instructor

Instruction Resource Centre

Everything you need, where you want it. Pearson Canada has assembled a first-rate package for *Natural Hazards*. The following instructor supplements are available for download from a password-protected section of Pearson Canada's online catalogue (www.pearsoncanada.ca/highered). Navigate to your book's catalogue page to view a list of available supplements. See your local sales representative for details and access.

- **Instructor's Manual.** The Instructor's Manual provides chapter outlines and objectives, classroom discussion topics, and answers to the end-of-chapter questions in the text.
- **Test Generator.** The TestGen provides more than 500 questions in multiple-choice, true/false, short-answer, and essay formats. It also tests recall, understanding, and application of the main data and concepts presented in the text.
- **PowerPoint® Presentation.** PowerPoint presentations are provided for each chapter and include all the figures and many photos from the text.
- **The Prentice Hall Geoscience Animation Library (ISBN 978-0-321-71684-2).** The Prentice Hall Geoscience Animation Library arose from a survey of instructors. We asked them to identify the concepts most difficult to teach using traditional, static resources. Then we animated them. Created through a unique collaboration of five of Prentice Hall's leading geoscience authors, the animations represent a significant advance in lecture presentation. Each animation is mapped to the corresponding chapter in *Natural Hazards* and is available on the Instructor's Resource site. They are provided as Flash files and, for convenience, are pre-loaded into PowerPoint slides.
- **CourseSmart for Instructors.** CourseSmart goes beyond traditional expectations—providing instant, online access to the textbooks and course materials you need at a lower cost for students. And even as students save money, you can save time and hassle with a digital eTextbook that allows you to search for the most relevant content at the moment you need it. Whether it's evaluating textbooks or creating lecture notes to help students with difficult concepts, CourseSmart can make life a little easier. See how when you visit www.coursesmart.com/instructors.
- **Pearson Custom Library.** For enrollments of at least 25 students, you can create your own textbook by choosing the chapters that best suit your own course needs. To begin building your custom text, visit www.pearsoncustomlibrary.com. You may also work with a dedicated Pearson Custom editor to create your ideal text—publishing your own original content or mixing and matching Pearson content. Contact your local Pearson representative to get started.
- **Technology Specialists.** Pearson's Technology Specialists work with faculty and campus course designers to ensure that Pearson technology products, assessment tools, and online course materials are tailored to meet your specific needs. This highly qualified team is dedicated to helping schools take full advantage of a wide range of educational resources, by assisting in the integration of a variety of instructional materials and media formats. Your local Pearson Education sales representative can provide you with more details on this service program.

For the Student

- **Pearson eText.** Pearson eText gives students access to the text whenever and wherever they have access to the Internet. eText pages look exactly like the printed text, offering powerful new functionality for students and instructors. Users can create notes, highlight text in different colours, create bookmarks, zoom, click hyperlinked words and phrases to view definitions, and view in single-page or two-page view. Pearson eText enables quick navigation to key parts of the eText using a table of contents, and provides full-text search. The eText may also offer links to associated media files, enabling users to access videos, animations, or other activities as they read the text.
- **CourseSmart for Students.** CourseSmart goes beyond traditional expectations—providing instant, online access to the textbooks and course materials you need at an average savings of 60%. With instant access from any computer and the ability to search your text, you'll find the content you need quickly, no matter where you are. And with online tools like highlighting and note-taking, you can save time and study efficiently. See all the benefits at www.coursesmart.com/students.

Acknowlegments

Many individuals, companies, and agencies provided information and images that we included in this book. In particular, we are indebted to the Geological Survey of Canada (Natural Resources Canada), Environment Canada, the U.S.

Geological Survey, and the National Oceanic and Atmospheric Administration for their excellent natural hazard programs and publications. We also extend our thanks to Tricouni Press for providing many figures and photos.

We appreciate and thank authors of papers cited in this book for their contributions. Without their work, this book could not have been written. We also thank the following scholars who dedicated their time reviewing chapters of this book:

John Gosse, *Dalhousie University*
Jeremy Hall, *Memorial University*
Norman Jones, *Bishop's University*
David McMullin, *Acadia University*
Mark Moscicki, *University of Guelph*
Catherine Pappas-Maenz, *Dawson College*
Rick Schneider, *University of PEI*
Elizabeth Sonnenburg, *McMaster University*
Maggie Squires, *Simon Fraser University*
Kim West, *University of Saskatchewan*

We also acknowledge our editors at Pearson Canada for their help and guidance in the preparation of the third Canadian edition. Our appreciation is extended to Kathleen McGill and Laura Armstrong, sponsoring editors; Patti Altridge, senior developmental editor; Andrea Falkenberg, project manager. Thanks are also due to Julie Fletcher, who copy-edited the manuscript, and Brett Gilley, who technically reviewed the material.

Edward A. Keller
Duane E. DeVecchio
Santa Barbara, California
John J. Clague
Burnaby, British Columbia

Monster on the Prairie (Saskatchewan)
Winner of Canadian Geographic's Whatever the Weather Photo Contest
Photographer Craig Hilts is a stock broker and storm chaser who gives talks to local schools on storm safety and preparedness. To see more of Craig's work visit: http://www.prairiefirephoto.com/

CHAPTER 1

▶ **CATASTROPHE IN HAITI** A building in Port-au-Prince, Haiti, destroyed by the devastating earthquake on January 12, 2010. Buildings such as this were not constructed to withstand the ground shaking from a strong earthquake. *(Getty Images, Design Pics /Rey old Mainse)*

Introduction to Natural Hazards

Learning Objectives

Natural processes, such as volcanic eruptions, earthquakes, landslides, tsunamis, floods, and hurricanes, threaten human life and property throughout the world. As the world's population continues to grow, disasters and catastrophes will become more common. An understanding of natural processes as hazards requires some basic knowledge of earth science. Your goals in reading this chapter should be to

- Recognize that natural disasters and catastrophes are high-energy events caused by natural Earth processes
- Understand that natural hazards have social, economic, and political dimensions that are just as important as the hazards themselves
- Understand the differences among hazard, risk, acceptable risk, disaster, and catastrophe
- Understand the concept that the magnitude of a hazardous event is inversely related to its frequency
- Recognize that many natural hazards are linked to one another
- Recognize that population growth, concentration of infrastructure and wealth in hazardous areas, and land-use decisions are increasing our vulnerability to natural disasters
- Understand that hazardous natural processes can also provide benefits

Earthquake in Haiti, 2010: Lessons Learned

One of the fundamental realities in the study of natural hazards is that people and governments are poorly prepared for rare natural disasters; they commonly behave as if a disaster will never happen. This unfortunate reality is well illustrated by five recent catastrophes: the tsunami in the Indian Ocean in December 2004, Hurricane Katrina on the U.S. Gulf Coast in August 2004, the Haiti earthquake in January 2010, the Tohoku (Japan) earthquake and tsunami in April 2011, and superstorm Sandy in October 2012. Each of these events provides hard lessons that can help us reduce the toll of future disasters. Here, we illustrate these lessons using the Haiti earthquake as an example.

The massive earthquake struck southern Haiti without warning in the late afternoon (local time) of January 12, 2010. The *epicentre* was near the town of Léogâne, about 25 km west of Port-au-Prince, Haiti's capital. After several tens of seconds of strong shaking, close to 240 000 people had lost their lives, an estimated 300 000 were injured, and more than a million had been rendered homeless (Figure 1.1).[1] Equally devastating was the loss of Haiti's infrastructure, including most of the significant buildings and other engineered structures in Port-au-Prince (Figure 1.2).

The earthquake had a *magnitude* of 7.0 and was much smaller than many recent catastrophic earthquakes, such as those in Sumatra, Indonesia, in 2004, China in 2008, Chile in 2010, and Japan in 2011. Yet it was one of the worst natural disasters in history, with a loss of life comparable to the quake that levelled the city of Tangshan, China, in 1976, killing more than 250 000 people. The loss of life for an earthquake of this size was appalling. In comparison, the magnitude 7.1 earthquake that struck the San Francisco area in 1989 caused only 63 deaths and about 4000 injuries.[2]

The built environments in Port-au-Prince, Jacmel, Jerémie, Les Cayes, and other urban areas in southern Haiti suffered grievous damage. The destruction and loss of life were exacerbated by poor building materials and construction practices stemming from a lack of official building codes and insufficient attention to planning.[3] Some of the problems included construction of heavy unsupported block walls; failure to use steel rods to reinforce concrete columns, walls, and floors; and use

◄ **FIGURE 1.1 EARTHQUAKE DEVASTATION IN HAITI** Unreinforced and poorly reinforced masonry and concrete slab buildings in Canapé Vert, a shanty town in the hills around Port-au-Prince, collapsed during the January 12, 2010, earthquake *(Eduardo Munoz/Landov)*

of marine sand in cement. Salt in the marine sand led to the corrosion and weakening of concrete. Buildings of all types failed—poured concrete, mortared and dry-stacked concrete blocks and stone, and scavenged wood and metal. Slums cover the slopes surrounding Port-au-Prince, which have expanded with little control in recent years due to migration of rural Haitians into the capital city. Most buildings on slopes lack proper foundations and, consequently, many slid down hillsides during the quake (Figure 1.1). Several tens of thousands of commercial buildings collapsed or were severely damaged, including Haiti's prized Cathédrale de Port-au-Prince, the National Assembly building, Palace of Justice (Supreme Court building), the headquarters of the United Nations Stabilization Mission, and several ministerial buildings. The prison civile de Port-au-Prince was also destroyed, allowing 4000 inmates to escape. The second floor of the presidential palace completely collapsed, leaving the third floor resting on the first (Figure 1.2).

The seaport ceased to function. Docks and piers slid into the sea, and cargo cranes fell from their footings. Damage was so extensive that vessels providing international relief were forced to dock along adjacent shores. Many roads were covered with rubble from collapsed buildings or were rendered impassable due to cracks caused by ground failure.

◄ **FIGURE 1.2 COLLAPSED BUILDING IN PORT-AU-PRINCE, HAITI** A front-end loader is ready for clean-up work in front of the destroyed National Palace in Port-au-Prince, the capital of Haiti. Poor construction of even important government buildings was largely responsible for the great loss of life from the earthquake in 2010. *(St. Felix Evans)*

Although it had horrific consequences, the Haiti earthquake is not unprecedented. In an average year, about 17 earthquakes with magnitudes equal to or larger than the Haiti quake occur on Earth. Several things made this earthquake different from most quakes of similar magnitude. First, it occurred in a heavily populated area—the population of Port-au-Prince preceding the earthquake was nearly 3 million. Second, buildings in the affected area were not constructed to withstand strong seismic shaking. Access to resources is limited in Haiti, and most of those scant resources are allocated to immediate needs—food and basic shelter—rather than less pressing concerns such as disaster mitigation. Resource availability to all but a small number of Haitians is particularly low. Haiti is the poorest country in the Northern Hemisphere, with an annual income per person of only $1300 in 2012, in comparison to Canada's per capita income of $41 100. This situation has been made worse by Haiti's government bodies, which have long made few resources available for governance issues, including the establishment and enforcement of building standards.

The Haiti earthquake carries a strong message for people living in earthquake zones, including much of the west coast of North America. Port-au-Prince experienced earthquakes even larger than the 2010 quake in 1751 and 1770.[1] However, a disaster that took place hundreds of years ago is commonly a forgotten one. Due to lack of experience, people and governments were complacent and could not conceive of such an event happening; they were thus completely unprepared, as reflected in the poor construction practices prevalent in Haiti. Scientists have argued that an earthquake as large as the 2010 Haiti event could strike Vancouver, Seattle, or Portland, although no one knows precisely when. These cities are much better prepared than Haiti was before January 12, 2010, but how will they cope when it's their turn?

What are the lessons of the Haiti earthquake? We must design buildings to meet the highest seismic standards: doing so clearly saves lives. We also must continue to provide adequate research funding to scientists who are seeking to better understand where earthquakes occur, how large they are likely to be, and when they are likely to happen. New technologies, including satellite-based sensors, offer opportunities to "measure the pulse" of Earth and provide clues that might someday allow us to more accurately forecast or even predict quakes. Communication is also important. We must review and upgrade communication infrastructure and chain-of-command protocols in earthquake-prone areas to ensure that emergency officials receive timely information and respond quickly after an earthquake. People living close to faults must know what to do in the event of an earthquake. Public education programs must teach people about earthquakes and provide instructions on how to prepare, how to act when the shaking starts, and what to do after it stops. One of the most important lessons of the Haiti catastrophe is that wealthy countries must help poorer ones prepare for earthquakes and other natural disasters. Canada and the United States must do more than just respond when disaster strikes—the standard approach to dealing with disasters in developing countries. A better strategy is a long-term proactive one aimed at helping poor countries develop and prepare for disasters before they occur.

The 2010 catastrophe in Haiti was largely human-caused. Had their buildings been constructed properly, the loss of life would have been much less. If Haiti were not impoverished, its buildings might have been more earthquake resistant and the country would have been better able to deal with the terrible aftermath of the earthquake, including an outbreak of cholera ten months after the earthquake.

1.1 Why Studying Natural Hazards Is Important

In the past few decades, earthquakes, floods, and hurricanes have killed several million people around the world; the average annual loss of life has been around 150 000 per year, with more than 300 000 deaths in 2005 alone. Financial loss from natural disasters now exceeds $50 billion per year and can be as high as $200 billion, as happened in 2005 (this figure represents direct property damage and does not include such expenses as loss of employment, mental anguish, and reduced productivity).

Four catastrophes—a cyclone accompanied by flooding in Bangladesh in 1970, earthquakes in China in 1976 and Haiti in 2010, and a tsunami in the Indian Ocean in 2004—claimed more than 230 000 lives each. A cyclone that struck Bangladesh in 1991 killed 145 000 people. Four years later in 1995, an earthquake in Kobe, Japan, claimed more than 5000 lives, destroyed many thousands of buildings, and caused more than $100 billion in property damage. An earthquake in northern Pakistan in 2005 killed 86 000 people (Figure 1.3). In the same year, Hurricane Katrina devastated New Orleans and killed over 1800 people. It was the most destructive natural disaster in United States history and the deadliest hurricane since one in Florida in 1928.[4] In 2011, the tsunami generated by a powerful earthquake off the coast of Japan claimed at least 16 000 lives. The World Bank estimated that the economic cost of this earthquake and tsunami was U.S.$235 billion, making it the most expensive natural disaster in world history. More recently, in October 2012, superstorm Sandy devastated the northeast Atlantic coast of the United States, with damage estimated at U.S.$75 billion.[5] Other notable disasters in the past 20 years include catastrophic flooding in Venezuela, Bangladesh, and central Europe; the strongest El Niño on record; deadly earthquakes

◄ **FIGURE 1.3 DEVASTATING EARTHQUAKE** People search for victims in the rubble of a 10-storey building in Islamabad, Pakistan, that collapsed during a large earthquake in 2005. About 86 000 people died and 3 million more were left homeless. *(Warrick Page/Getty Images)*

in Chile, India, Iran, Italy, New Zealand, and Turkey; a Category 5 hurricane in Central America; record-setting wildfires in British Columbia, Arizona, California, Colorado, and Utah; and a crippling ice storm in Ontario, Quebec, New Brunswick, and New England. During this period, Earth also experienced many of the warmest years of the past century—and probably even of the past millennium.

These events are the result of enormous forces that are at work both inside and on the surface of our planet. In this book, we explain these forces and their impacts on people and property. We also discuss how we can better prepare for natural disasters, thus minimizing their impact when they do occur.

All areas of Canada and the United States are at risk from at least one hazardous natural process.[6,7] Parts of western North America are prone to earthquakes and landslides and experience rare volcanic eruptions; the Pacific coast is vulnerable to tsunamis; the Atlantic and Gulf of Mexico coasts are threatened by hurricanes; forested areas of the continent are prone to wildfires; the mid-continent, from Texas to Ontario, is at risk from tornadoes and blizzards; and drought and flooding can occur almost anywhere. No area is considered hazard-free.

Loss of life from natural disasters in Canada has been decreasing in recent decades, although economic losses have increased. The current low loss of life is due, in part, to land-use planning, education, and high engineering standards. Canada also has been fortunate in having experienced few large natural disasters of the scope of Hurricane Katrina in the United States, the 2011 earthquake and tsunami in Japan, and the flood in Pakistan in 2010.

Hazardous Natural Processes and Energy Sources

In our discussion of natural hazards, we will use the word *process* to mean the ways in which events, such as volcanic eruptions, earthquakes, landslides, and floods, affect Earth's surface. All of these processes are driven by energy, and this energy is derived from three sources.

The first source of energy is Earth's internal heat, which produces slow convection in the solid but plastic mantle. The hazardous processes associated with this source of energy are earthquakes and volcanic eruptions. As we will see, earthquakes and volcanic eruptions are explained by the theory of *plate tectonics*, one of the basic unifying theories of science. Most earthquakes and active volcanoes occur at boundaries of tectonic plates, which are large blocks of Earth's crust.

The second source of energy is the sun. Energy from the sun warms Earth's atmosphere and surface, producing winds and causing evaporation of water. Circulation of the atmosphere and oceans and evaporation of water determine Earth's climate and drive the hydrologic cycle. These forces are, in turn, directly related to hazardous processes such as violent storms, floods, and coastal erosion. Solar energy is also ultimately responsible for wildfires and lightning.

The third source of energy is the gravitational attraction of Earth. Gravity is the force that attracts one body to another—in this case, the attraction of surface materials toward the centre of Earth. Because of gravitational attraction, rocks, soils, and snow on mountainsides and the water that falls as precipitation move downslope. Earth's gravitational field also attracts objects from space that can enter the atmosphere and explode or strike the surface of the planet.

The amount of energy released by natural processes differs greatly. The average tornado expends about 1000 times as much energy as a single lightning bolt, and Earth receives nearly a trillion times more solar energy each day than it receives from a lightning bolt. However, keep in mind that a lightning bolt's energy is focused at a point—a tree, for example—whereas solar energy is spread over the entire globe.

Events such as earthquakes, tsunamis, volcanic eruptions, floods, and fires are natural processes that have been occurring on Earth's surface for billions of years. They become hazardous only when they threaten human beings.

We use the terms *hazard, risk, disaster*, and *catastrophe* to describe our interaction with these natural processes.

Of course, not all hazards are natural. Many hazards are caused by people; examples include pandemics, warfare, and technological disasters such as regional power failures. The early 2009 outbreak in humans of a new strain of influenza that is endemic in pigs (H1N1 or "swine flu") is an example of a non-natural hazard. The virus rapidly spread around the world and was labelled a pandemic by the World Health Organization in June 2008. The Chernobyl nuclear accident in 1986, which resulted in the release of large quantities of radioactive particles into the atmosphere that spread over much of western Russia and Europe, is an example of a technological disaster.

Over the past century, the distinction between natural and human-induced hazards has become blurred, and technological disasters are increasing as the world's population grows and state economies become increasingly connected and interdependent. Social and technological hazards are important and interesting in their own right, but are beyond the scope of this book. Our focus is on hazardous solid Earth and atmospheric processes.

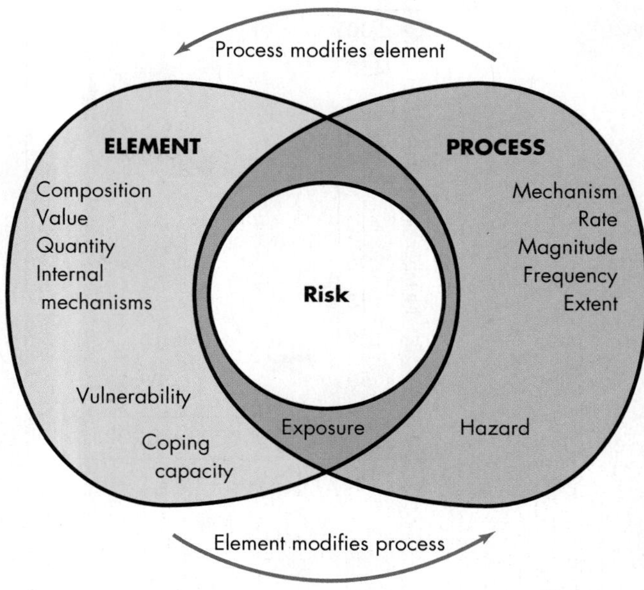

▲ **FIGURE 1.4 CONCEPTUAL MODEL OF RISK** Risk is a function of four factors—hazard, vulnerability, coping capacity, and exposure. *(From Clague, J. J., and N. J. Roberts. 2012. "Landslide hazard and risk." In* Landslides: Types, Mechanisms and Modeling, *eds. J. J. Clague and D. Stead, pp. 1–9. Cambridge, UK: Cambridge University Press. Reprinted with permission.)*

Hazard, Risk, Disaster, and Catastrophe

This book considers hazards within the human context: it focuses on the science of natural hazards and explores the social, economic, and political issues that hazards pose. The text recognizes that the human response to threats posed by natural hazards is just as important as hazard science itself.

Before discussing specific natural hazards, we need to define several key terms. **Hazard** is the probability that a specific damaging event will happen within a particular period of time.[8] This definition is common to both the natural and social sciences. **Risk**, on the other hand, is commonly a subject of social sciences because it is rooted not only in hazard but in vulnerability and coping capacity (Figure 1.4).[9,10,11] Definitions of risk are legion, but for our purposes, risk can be expressed by the following function:

$$\text{Risk} = f(\text{hazard, exposure, vulnerability, coping capacity})$$

Although risk can be conceptualized as a function of these four factors, their interrelationships cannot be described in mathematical terms or even fully understood. *Vulnerability* and *coping capacity* are latent states of the people and property that are at risk and are only manifested through the occurrence of a hazardous event. Vulnerability is the susceptibility of people and property to a hazardous event and is commonly thought of as having technological and human dimensions. Technological aspects are studied by engineers and geoscientists and include primary damage and life loss. Human aspects relate to a wide range of social issues, including loss of livelihood, physical displacement, and psychological and environmental impacts of hazardous events. Coping capacity is the ability of a population to respond to and reduce the negative effects of a hazardous event. *Exposure* is the overlap in space and time of a hazardous process and infrastructure or population.

The terms **disaster**, or *natural disaster*, and **catastrophe** refer to events that cause serious injury, loss of life, and property damage over a limited time and within a specific geographic area. Although the distinction between disaster and catastrophe is somewhat vague, the latter is more massive and affects a larger number of people and more infrastructure. Disasters may be regional or even national in scope, whereas catastrophes commonly have consequences far beyond the area that is directly affected and require huge expenditures of time and money for recovery. Examples of catastrophes are the Indian Ocean tsunami in December 2004, Hurricane Katrina in August 2004, and the Haiti earthquake in January 2010.

The United Nations (UN) designated the 1990s as the International Decade for Natural Hazards Reduction. The objectives of the UN program were to minimize loss of life and property damage from natural disasters, but the objectives were not met; rather, losses from disasters increased dramatically in the 1990s (Figure 1.5), along with similar economic losses (Figure 1.6). Achieving the UN objectives will require education and large expenditures of money to mitigate specific hazards and contain diseases that accompany disasters and catastrophes.[12]

The term **mitigation** is used by scientists, planners, and policymakers when describing efforts to prepare for disasters and to minimize their harmful effects. Mitigation can be defined as actions taken to reduce or eliminate the long-term risk to human life and property from natural hazards. For example, buildings in an earthquake-prone area can be retrofitted to better withstand seismic ground motions and thus reduce the risk to occupants during a future quake.

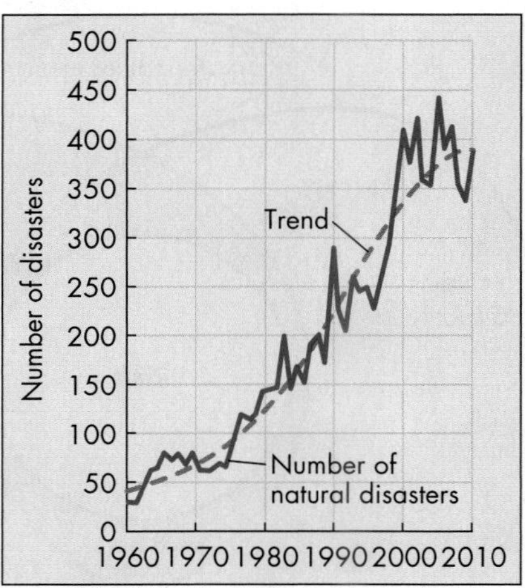

◄ FIGURE 1.5 INCREASE IN THE NUMBER OF NATURAL DISASTERS The number of natural disasters is increasing. Here, an event is considered a natural disaster if one of the following four criteria is met: 10 or more people are killed; 100 or more people are affected; a state of emergency is declared; or international assistance is requested. *(Based on Guha-Sapir, D., and P. Hoyola. 2012. Measuring the Human and Economic Impact of Disasters. UK: Government Office for Science)*

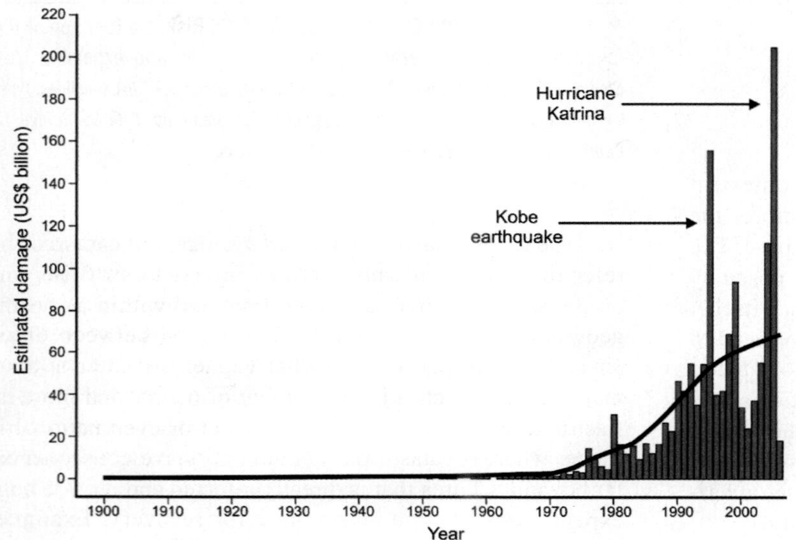

◄ FIGURE 1.6 THE RISING COST OF NATURAL DISASTERS Estimated damage (in billions of U.S. dollars) caused by natural disasters between 1900 and 2007. *(Data from EM-DAT, the OFDA/CRED International Disaster Database)*

Death and Damage Caused by Natural Hazards

Worldwide, flooding is the largest killer of people, followed by earthquakes, volcanic eruptions, and windstorms. Since 1974, about 2.5 million people have died in natural disasters. Most deaths were in developing countries. Asia has suffered the greatest losses, with about three-quarters of the total deaths and nearly one-half of the economic losses. Loss of life and economic loss could have even been worse were it not for improvements in warning systems, disaster preparedness, and emergency response.[13,14]

Natural hazards that cause the greatest loss of life in North America are not the same as those that cause the most property damage. Tornadoes and windstorms cause the largest number of deaths in most years, although lightning, floods, and hurricanes also take a heavy toll (Table 1.1). Loss of life from earthquakes in North America is surprisingly low, largely because buildings are constructed to a high standard. However a single large earthquake can cause tremendous property damage. For example, the Northridge earthquake in Los Angeles in 1994 caused U.S.$20–30 billion in property damage but killed only 60 people. The next great earthquake in a densely populated part of California could cause more than U.S.$100 billion in damage.[12]

Natural disasters cost Canada and the United States tens of billions of dollars annually. Because populations are increasing in high-risk areas of North America, we can expect losses to increase significantly in the future. Floods, landslides, expansive soils that shrink and swell, and frost each cause in excess of U.S.$1.5 billion in damage each year in the United States alone.

TABLE 1.1 Effects of Selected Hazards in Canada and the United States

Hazard	Deaths per Year	Catastrophe Potential
Flood	100	High
Earthquake[a]	>50	High
Landslide	30	Low
Snow avalanche	20	Low
Volcanic eruption[a]	<1	High
Coastal erosion	0	Low
Soil expansion	0	Very low
Hurricane	60	High
Tornado and windstorm	220	High
Lightning	125	Very low
Drought	0	Medium
Heat wave	100s	High
Wildfire[b]	<10	High
Freezing and frozen rain	<10	Medium
Permafrost thaw	0	High
Extraterrestrial impact	0	Very high

[a] Estimates based on recent or predicted loss over a 150-year period. Actual losses differ considerably from year to year and could be much greater in a given year.
[b] Most deaths are firefighters.

Source: Adapted from White, G. F., and J. E. Haas. 1975. Assessment of Research on Natural Hazards. *Cambridge, MA: MIT Press. Reprinted with permission.*

Note that the relations between loss of life and property damage discussed above apply only to the developed world—mainly North America, Europe, Japan, Australia, and New Zealand. Natural disasters in most developing countries claim far more lives than comparable events in North America. For example, the tsunami in the Indian Ocean in December 2004 killed nearly 230 000 people. In comparison, the tsunami in the North Pacific in 1964, although equal in size, killed 119 people. A notable characteristic of North American disasters, however, is their very large toll on the economy. Category 4 and 5 hurricanes typically cause billions of dollars in damage in southern U.S. states; the direct damage from Hurricane Katrina, the worst storm in U.S. history, was more than U.S.$80 billion, and indirect damage, including lost economic activity and employment, was several times that amount.[4]

Natural hazards differ greatly in their potential to create a catastrophe, mainly because of differences in the size of the area each affects (Table 1.1). Three processes—climate change, eruptions of super volcanoes, and large meteorite impacts—can have global repercussions. Large tsunamis, earthquakes, major volcanic eruptions, hurricanes, monsoon floods, and floods on large rivers have regional effects and may result in catastrophes. Landslides, snow avalanches, floods on small rivers, most wildfires, and tornadoes generally affect small areas and thus are rarely catastrophic. Coastal erosion, lightning, expansive soils, and permafrost thaw do not create catastrophes, but still cause much damage.[15]

Risks associated with natural hazards change with time because of changes in population and land use. As cities grow, neighbourhoods may extend onto hazardous land, such as steep hillsides and floodplains. Such expansion is a serious problem in many large, rapidly growing cities in developing nations. Urbanization alters drainage, increases the steepness of some slopes, and removes vegetation. Agriculture, forestry, and mining also remove natural vegetation and can increase erosion and sedimentation. Overall, damage from most hazardous natural processes in Canada is increasing, but the number of deaths from many hazards is decreasing because of better planning, forecasting, warning, and engineering.

1.2 Magnitude and Frequency of Hazardous Events

The *impact* of a hazardous event is partly a function of its magnitude—the size of an event or the amount of energy released—and partly a function of frequency. The *magnitude–frequency concept* asserts that an inverse exponential relationship exists between the magnitude of an event and its frequency (Figure 1.7). Large floods or earthquakes, for example, are infrequent, whereas small floods or earthquakes are common. The *recurrence interval*, also known as the return period, is the average time separating two events of the same magnitude. The magnitude–frequency relation for many natural phenomena can be approximated by an exponential equation of the type $M = Fe^{-x}$, where M is the magnitude of the event, F is the frequency, e is the base of the natural logarithm, and x is a constant.

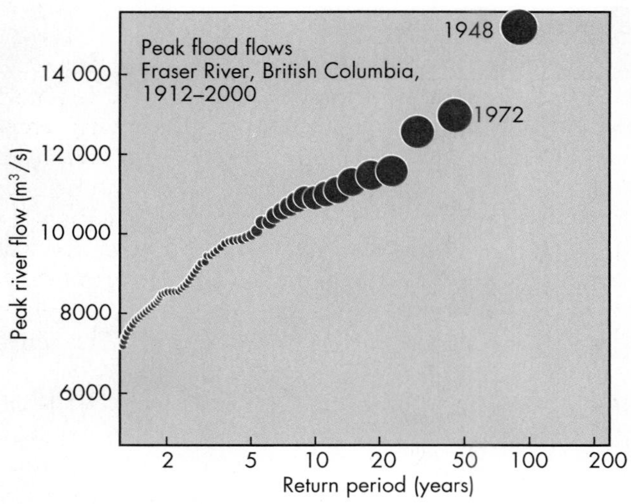

▲ **FIGURE 1.7 MAGNITUDE–FREQUENCY RELATION** The size and frequency of many physical processes are inversely related. The concept is illustrated here with this plot of average return periods for Fraser River (British Columbia) floods of different sizes. Large floods occur less frequently than small ones. Note that the horizontal scale (return period) is logarithmic, not linear. *(From Clague, J., and B. Turner. 2006.* Vancouver, City on the Edge; Living with a Geological Landscape. *Vancouver, BC: Tricouni Press. Reprinted with permission.)*

Impact is also influenced by many other factors, including climate, geology, vegetation, population, and land use. Land use can directly affect the magnitude and frequency of events. People have long tried to reduce the threat of floods, for example, by building *levees* along rivers. However, levees constrict the width of rivers, and the reduced width lessens the amount of water than can be transported during flood conditions. Ironically, our efforts to reduce floods may actually be causing larger, more frequent floods.

Four of the worst natural disasters in recent years were exacerbated by land-use practices—Hurricane Mitch in 1998; a flood on the Yangtze River in China, also in 1998; the tsunami in the Indian Ocean in 2004; and Hurricane Katrina in 2005. Hurricane Mitch devastated parts of Central America and claimed approximately 11 000 lives, and the Yangtze River flood killed nearly 4000 people. Land-use changes made the damage from these events particularly severe. For example, Honduras lost nearly one-half of its forests in the past century, and wildfires before Hurricane Mitch burned an area of 11 000 km^2. As a result of deforestation and the fire, hillside soils washed away, and with them went farms, homes, roads, and bridges. The story is much the same in China. About 85 percent of the forest in the Yangtze River basin has been removed through timber harvesting and conversion of land to agriculture. As a result of these changes, flooding on the Yangtze River is probably much more common and severe than it was previously.[16] The huge loss of life from the 2004 tsunami in the Indian Ocean was due in part to the increase in population along the shores of the Indian Ocean and in part to a growth in tourism in South Asia, especially in Thailand. Thousands of tourists were among the casualties (see Chapter 4). Hurricane Katrina severely damaged the coasts of Mississippi and Louisiana in 2005. Wetlands that might have buffered the hurricane's storm surge had been damaged or removed as a result of human development of the river and coastline (see Chapter 11).

These and other recent catastrophes may be warning signs of things to come. Human activities are likely increasing the severity of some natural disasters. China has heeded this lesson and banned timber harvesting in the upper Yangtze River basin, limited use of the Yangtze floodplain, and allocated several billion dollars for reforestation. If we want to minimize damage from natural disasters, we need to rehabilitate the land and strive for a more harmonious relationship with the processes that shape Earth's surface. An ancillary benefit of this approach is that future generations will have access to the resources that our planet offers.[16] Population growth in developing countries and reckless squandering of resources in the developed world, however, will make it difficult for humanity to achieve this goal.

1.3 Role of History in Understanding Hazards

An important principal emphasized in this book is that natural disasters are recurrent events; thus study of past events provides needed information for risk reduction. Whether we are studying floods, landslides, volcanic eruptions, or earthquakes, knowledge of historic events and the recent geologic history of an area is vital to understanding the hazard and evaluating its risk. For example, we can evaluate the risk of flooding along a particular river by identifying floods that have occurred in the recent past. Useful information can be obtained by studying aerial photographs and maps as far back as such records allow. We can extend the historic record by searching for evidence of past floods in stream deposits. Commonly, these deposits contain organic material such as wood or shells, which can be dated by the *carbon-14* method to provide a chronology of ancient flood events. This chronology can then be combined with the historic record of high flows to provide an overall perspective of the frequency and size of floods. Similarly, if we are studying landslides in a particular area, we must investigate both historic and prehistoric events to properly forecast the likelihood of future landslides. Geologists have the tools and training to "read the landscape" for evidence of past events and, by linking prehistoric and historic records, they extend our perspective of recurrent natural events far back in time.

But how far back in time do we need to go in order to understand natural hazards? The answer is: far enough to see the complete spectrum of events that can affect a region. In the case of floods and earthquakes, a geologist would like to reconstruct events over a period of hundreds or even thousands of years. In contrast, for rarer events such as meteorite impacts, the answer is millions or tens of millions of years. Although most of the hazardous processes considered in this book have existed since the birth of our planet, they are relevant only within the context of today. Geologic studies extend our perspective of recurrent events far back in time, providing insights on high-magnitude events that may not have occurred in historic time.

In summary, before we can fully appreciate a hazardous process, we must study in detail how it has operated in the past and the geologic features that it has produced or affected. The latter may be landforms, such as beaches or impact craters; structures, such as faults or folded rock; or Earth materials, such as lava flows or tsunami deposits. Hazard forecasts and warnings are more accurate if we integrate information about past events with an understanding of current conditions.

To fully understand natural hazards, you must have some knowledge of the geologic cycle—the processes that shape Earth's surface and modify Earth's materials. In the next few sections, we discuss the geologic cycle and then introduce five key concepts that are fundamental to understanding natural processes as hazards.

1.4 Geologic Cycle

Geology, topography, and climate govern the type, location, and intensity of natural processes. For example, hurricanes and cyclones form only over warm oceans and have different impacts, depending on the topography, and therefore geology, of the areas they strike.

Throughout much of the 4.6 billion years of Earth's history, the materials on or near the surface of the planet have been created and modified by numerous physical, chemical, and biological processes. These processes have produced the mineral resources, fuels, land, water, and atmosphere that we require for our survival. Collectively, these processes constitute the **geologic cycle**, which comprises a group of sub-cycles that includes

- the tectonic cycle
- the rock cycle
- the hydrologic cycle
- biogeochemical cycles

The Tectonic Cycle

The term *tectonic* refers to the large-scale geologic processes that deform Earth's crust and produce ocean basins, continents, and mountains. Tectonic processes are driven by forces deep within Earth. To describe these processes, we must use information about the composition and layering of Earth's interior and about the large blocks that form the outer shell of Earth, called *lithospheric plates*. The **tectonic cycle** involves the creation, movement, and destruction of lithospheric plates. It is responsible for rock and mineral resources that are essential to modern civilization, as well as hazards such as earthquakes and volcanoes. The tectonic cycle and its linkages to hazards are the subject of Chapter 2.

The Rock Cycle

Rocks are aggregates of one or more *minerals*. A mineral is a naturally occurring crystalline substance with a specific elemental composition and a narrow range of physical properties. The term **rock cycle** refers to worldwide recycling of three major groups of rocks, driven by Earth's internal heat and by energy from the sun. The rock cycle is linked to the other cycles; it depends on the tectonic cycle for heat and energy; the biogeochemical cycle for materials; and the hydrologic cycle for water, which plays a central role in weathering, erosion, transportation, deposition, and lithification of sediment.

Although rocks differ greatly in their composition and properties, they can be classified into three general types, or families, according to how they formed (Figure 1.8). *Crystallization* of molten rock produces *igneous rocks* beneath and on Earth's surface. Rocks at or near the surface break down chemically and physically by *weathering* to form particles known as *sediment*. These particles range in size from clay to very large boulders and blocks. Sediment formed by weathering is transported by wind, water, ice, and gravity to depositional sites such as lakes and oceans. When wind or flowing water slackens or evaporates, ice melts, or material moving under the influence of gravity reaches a flat surface, the sediment is *deposited*. During burial, the sediment is converted to *sedimentary rock* by a process called *lithification*. Lithification takes place by compaction and cementation of sediment during burial. With deep burial, sedimentary rocks may be *metamorphosed* by heat, pressure, and chemically active fluids into *metamorphic rock*. Metamorphic rocks may be buried to depths where pressure and temperature conditions cause them to melt, beginning the entire rock cycle again.

As with any of Earth's cycles, exceptions exist to the idealized sequence outlined above. For example, metamorphic

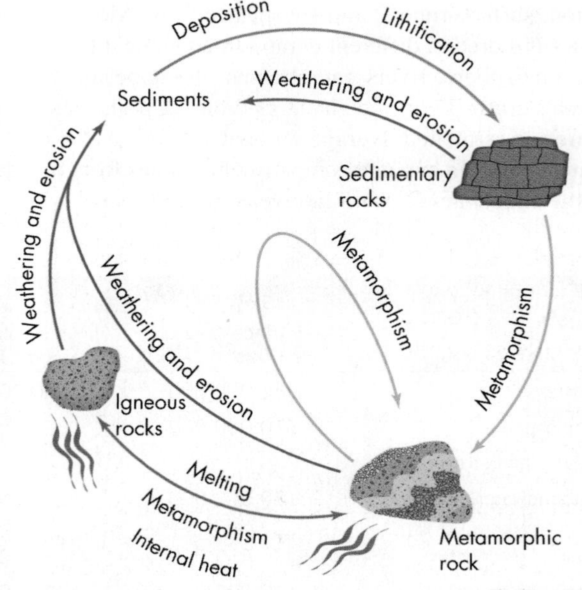

▲ **FIGURE 1.8 THE ROCK CYCLE** An idealized cycle showing the three families of rocks and important processes that form them.

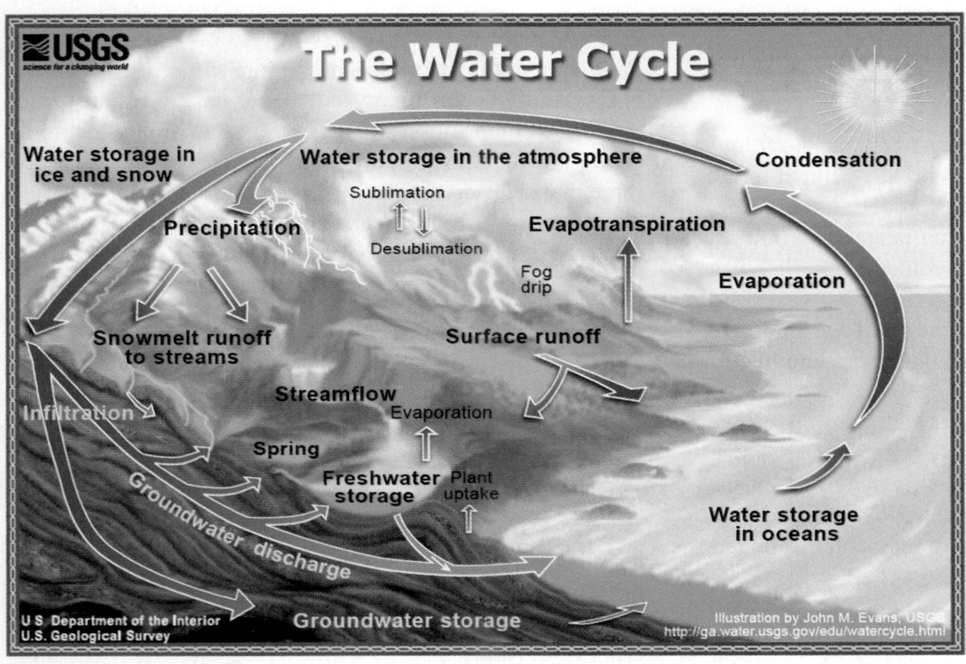

◄ **FIGURE 1.9 HYDROLOGIC CYCLE** Idealized diagram showing important processes and transfers that define the hydrologic cycle. *(Illustration by John M. Evans, USGS, http://ga.water.usgs.gov/edu/watercycle.html)*

rock might be changed into a new metamorphic rock without undergoing weathering or erosion (Figure 1.8), or sedimentary and metamorphic rocks might be uplifted and weathered before they continue on to the next stage in the cycle. Finally, some sediments have a biological or chemical origin, and some types of metamorphism do not involve deep burial.

The Hydrologic Cycle

The cycling of water from the oceans to the atmosphere, to continents and islands, and back again to the oceans is called the **hydrologic cycle** (Figure 1.9). This cycle is driven by solar energy and operates by way of evaporation, precipitation, surface runoff, and subsurface flow. Along the way, water is stored in different compartments, including oceans, the atmosphere, rivers and streams, the uppermost part of Earth's crust, lakes, and glaciers (Table 1.2). The **residence time**, or estimated average amount of time that a drop of water spends in any one compartment, ranges from just days in the atmosphere to a million years in ice sheets.

As you can see from Table 1.2, only a tiny percentage (0.3 percent) of the total water on Earth occurs in the atmosphere, rivers, lakes, and shallow subsurface sediments and rocks. This water, however, is tremendously important for life on Earth and for the rock and biogeochemical cycles. Surface and near-surface water moves chemical elements in solution, sculpts the landscape, weathers rocks, and transports and deposits sediments. It is also the source of the fresh water that makes life on land possible.

Biogeochemical Cycles

A **biogeochemical cycle** is the transfer or cycling of an element or elements through the atmosphere (the layer of gases surrounding Earth), lithosphere (Earth's outer rocky layer), hydrosphere (oceans, lakes, glaciers, rivers, and groundwater), and biosphere (organisms). It follows from this definition that biogeochemical cycles are intimately related to the tectonic, rock, and hydrologic cycles. The tectonic cycle provides water and gases from volcanic activity, as well as

TABLE 1.2 The World's Water Supply

Location	Surface Area (km^2)	Water Volume (km^3)	Percentage of Total Water	Estimated Average Residence Time
Oceans	361 000 000	1 230 000 000	97.2	Thousands of years
Atmosphere	510 000 000	12 700	0.001	9 days
Rivers and streams	—	1200	0.0001	2 weeks
Groundwater; shallow	130 000 000	4 000 000	0.31	Hundreds to many thousands of years to depth of 0.8 km
Lakes (freshwater)	855 000	123 000	0.009	Tens of years
Ice caps and glaciers	28 200 000	28 600 000	2.15	Hundreds of years to hundreds of thousands of years

Source: Data from U.S. Geological Survey.

heat and energy, all of which are required to transfer dissolved solids in gases, aerosols, and solutions. The rock and hydrologic cycles transfer and store chemical elements in water, soil, and rock.

Elements and chemical compounds are transferred via a series of storage compartments or reservoirs, which include air, soil, groundwater, and vegetation. For example, carbon is exhaled by animals, enters the atmosphere, and is taken up by plants by *photosynthesis*. When a biogeochemical cycle is well understood, the rate of transfer, or *flux*, among all of the compartments is known. However, determining these rates globally is a daunting task. The amounts of important elements like carbon, nitrogen, and phosphorus in each compartment, and their rates of transfer between compartments, are known only approximately.

1.5 Fundamental Concepts for Understanding Natural Processes as Hazards

The five concepts described below are important to understanding natural hazards. They provide a conceptual framework for the detailed discussions of specific hazards in subsequent chapters of this book.

1. ***Hazards can be understood through scientific investigation and analysis.***

   Natural hazards such as earthquakes, volcanic eruptions, landslides, and floods are processes that can be identified and studied using the **scientific method**. Most hazards can be better understood from the past history of similar events, patterns in their occurrence, and types of precursor events.

2. ***An understanding of hazardous processes is vital to evaluating risk.***

   Hazardous processes are amenable to *risk analysis*, which considers both the probability that a damaging event will occur and the consequences of that event.

3. ***Hazards are commonly linked to each other and to the environment in which they occur.***

   Hazardous processes are linked in many ways. For example, earthquakes can produce landslides and tsunamis, and hurricanes often cause flooding and coastal erosion. Hazards are also associated with particular environments on Earth.

4. ***Population growth and socio-economic changes increase the risk from natural hazards.***

   The human and economic costs of natural disasters are increasing because of the growth in population, property development in hazardous areas, and land-use practices. Events that caused limited disasters in the twentieth century are causing catastrophes in the twenty-first century.

5. ***Damage and loss of life from natural disasters can be reduced.***

   Minimizing the adverse effects of hazardous events requires an integrated approach that includes scientific understanding, land-use planning and regulation, engineering, and proactive disaster preparedness.

Hazards can be understood through scientific investigation and analysis.

Science and Natural Hazards

Science is founded on investigations and experiments, the results of which are subject to verification. The scientific method involves a series of steps, the first of which is to formulate a question. With respect to a hazardous event, a geologist might ask: Why did a landslide that destroyed several homes occur? To answer this question, the geologist will spend time examining the failed slope. She may notice that a great deal of water is flowing from the toe of the landslide. If she also knows that a water pipe is buried in the slope, she may refine the question to: Did the water in the slope cause the landslide? This question is the basis for an hypothesis that may be stated as follows: The landslide occurred because a buried water pipe broke, causing a large amount of water to enter the slope and reduce the strength of the slope materials.

An **hypothesis** is a possible answer to a question and is an idea that can be tested and therefore accepted or rejected. In our example, we can test the hypothesis that a broken water pipe caused a landslide by excavating the slope to determine the source of the water. In science we test hypotheses in an attempt to disprove them. If we found there was no leaking water pipe in the slope on which the landslide occurred, we would reject the hypothesis and develop and test another hypothesis. Use of the scientific method has improved our understanding of many natural processes, including flooding, volcanic eruptions, earthquakes, tsunamis, hurricanes, and coastal erosion.

Scientists have identified where hazardous processes occur, their magnitude, and their frequency. They have also mapped the types and extents of different hazards. Coupled with knowledge of the frequency of past events, they use such maps to predict when and where floods, landslides, earthquakes, and other disasters will happen in the future. Scientists also search for types and patterns of precursor events. For example, *foreshocks* may precede a large earthquake, and a change in gas emissions may signal an imminent volcanic eruption.

Hazardous Processes Are Natural

Most hazardous Earth processes are natural and thus are not the direct result of human activity. Nothing that people do, for example, changes the behaviour of volcanoes. However, because these processes are natural, we face fundamental philosophical issues when making choices about how to minimize their adverse effects. We realize, for example, that flooding is a natural part of river dynamics and must ask ourselves if it is wiser to attempt to protect people and property with defensive structures or to restrict or prohibit development on floodplains.

Although we can, to a degree, control some hazards, many are completely beyond our control. For example, we have some success in preventing damage from forest fires by using controlled burns and advanced firefighting techniques, but we will never be able to prevent earthquakes. In fact, we may actually worsen the effects of some natural processes simply by labelling them as hazardous. Efforts to suppress wildfires, for example, have interfered with ecosystems in forests in which fire is a natural process and, in some cases, have increased the severity of subsequent fires. Rivers will always flood, but because we choose to live and work on floodplains, we have labelled floods as hazards, which has led to efforts to control them. Unfortunately, as we will discuss later, some flood-control measures intensify the effects of flooding, thereby increasing the risk of the event that we are trying to prevent (Chapter 9). The best approach to hazard reduction may be to identify hazardous processes and the areas where they occur. Every effort should be made to avoid putting people and property in harm's way, especially for hazards that we cannot control, such as earthquakes.

Prediction, Forecast, and Warning

A **prediction** of a hazardous event such as an earthquake involves specifying the date and size of the event. In contrast, a **forecast** is less precise and has uncertainty. A meteorologist may forecast a 40 percent chance of rain tomorrow, but she is not predicting the weather. Learning how to forecast disasters so that we can minimize loss of life and property is an important endeavour. In some cases, we have enough information to accurately forecast events. However, when information is insufficient to make accurate forecasts, the best we can do is identify areas where disasters can be expected in the future based on past history. If we know both the probability and the possible consequences of an event at a particular location, we can quantify the risk of the event, even if we cannot accurately predict when it will occur.

Damage inflicted by a natural disaster can be reduced if the event can be forecast and a warning issued. Attempting to do this involves the following elements:

- Identifying the location of a hazard
- Determining the probability that an event of a given magnitude will occur
- Identifying any precursor events
- Forecasting the event
- Issuing a warning

Location We can identify areas at risk from different hazardous processes. Major zones of earthquakes and volcanic eruptions have been identified by mapping (1) where earthquakes have occurred historically, (2) areas of young volcanic rocks, and (3) locations of active and recently active volcanoes. We can use past volcanic eruptions to identify areas that are likely to be affected by future ones. Volcanic hazard maps have been prepared for most Cascade volcanoes in the Pacific Northwest of the United States and for volcanoes in Japan, Italy, Colombia, and elsewhere. Detailed mapping of soils, rocks, groundwater conditions, surface drainage, and evidence for ground instability can pinpoint slopes that are likely to fail. We can also predict where flooding is likely to occur by mapping the extent of recent floods.

Probability of Occurrence Determining the probability of a particular event at a specific site is an essential part of hazard analysis. We have long discharge records for many rivers, and those enable us to develop probability models that can reasonably predict the number of floods of a given size that will occur within a particular period. Likewise, the probability of droughts can be determined from the history of past rainfall in the region, and the probability of earthquakes of specific magnitudes can be estimated from historic earthquake records. However, these probabilities are subject to the same elements of chance as throwing a particular number on a die or drawing an inside straight in poker. For example, although a flood may occur only once every 10 years, on average, it is possible to have two or more or no floods of this magnitude in that time, just as it is possible to throw a six twice in a row with a die. Probabilities of rare events within a specific region—for example, volcanic eruptions, tsunamis, and meteorite impacts—are much more difficult to estimate and are subject to large uncertainties.

Precursor Events Many disasters are preceded by *precursor events*. For example, the surface of the ground may creep for weeks, months, or years before a catastrophic landslide, and the rate of creep may increase just before final failure. Volcanoes sometimes swell or bulge before an eruption, accompanied by an increase in earthquake activity in the area. Foreshocks or unusual uplift of the land may precede an earthquake.

Identification of precursor events helps scientists predict when and where a disaster will happen. Documentation of landslide creep or swelling of a volcano may lead authorities to issue a warning and evacuate people from a hazardous area, as happened before the catastrophic eruption of Mount St. Helens in Washington State in 1980, for example.

Forecasting With some natural processes, it is possible to forecast accurately when a possible damaging event will occur. Government agencies can generally accurately forecast when large rivers will reach a particular flood stage. We can also forecast when and where hurricanes will strike land by tracking them at sea. Arrival times of tsunamis can be precisely predicted if a warning system detects the waves.

Warning Once a hazardous event has been predicted or a forecast made, the public must be warned. The flow of information leading to a public **warning** of a possible disaster, such as a large earthquake or flood, should move along a

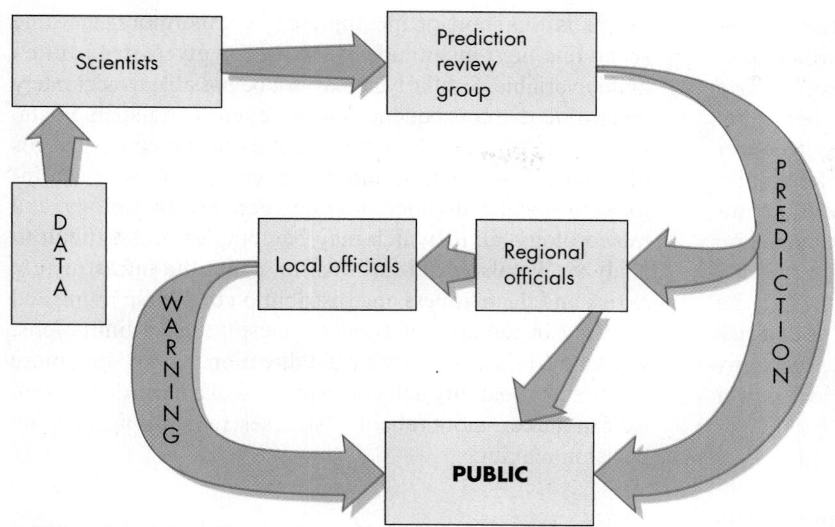

◀ **FIGURE 1.10 HAZARD PREDICTION OR WARNING** Possible path for issuing a prediction or warning of a natural disaster.

predefined path (Figure 1.10). However, the public does not always welcome such warnings, especially when the predicted event does not occur. In 1982, geologists issued an advisory that a volcanic eruption was likely near Mammoth Lakes, California. The advisory caused loss of tourist business and apprehension on the part of residents. The eruption did not occur and the advisory was eventually lifted. In July 1986, a series of earthquakes occurred over a four-day period near Bishop, California, in the eastern Sierra Nevada Mountains, beginning with a magnitude 3 event and culminating in a damaging magnitude 6.1 earthquake. Scientists concluded that there was a high probability that a larger earthquake would occur in the area in the near future and issued a warning. Local business owners, who feared the loss of summer tourism, felt that the warning was irresponsible; in fact, the predicted quake never occurred.

Incidents of this kind have led some people to conclude that scientific predictions are worthless and that advisory warnings should not be issued. Part of the problem is poor communication among scientists, the news media, and the public. Newspaper, television, and radio reports may fail to explain the evidence or the probabilistic nature of disaster forecasting and prediction, leading the public to expect black-and-white statements about what will happen. Although scientists are not yet able to accurately predict volcanic eruptions or earthquakes, they have a responsibility to publicize their informed judgments. An informed public is better able to act responsibly than is an uninformed public, even if the subject makes people uncomfortable. Ships' captains, who depend on weather advisories and warnings of changing conditions, do not suggest they would be better off not knowing about an impending storm, even though the storm might not materialize or might take an unexpected course. Just as weather warnings have proved very useful for planning, official warnings of earthquakes, volcanic eruptions, landslides, and floods are also useful to people when they decide where to live, work, or travel.

Let's consider again the prediction of a volcanic eruption in the Mammoth Lakes area. The location and depth of earthquakes suggested to scientists that magma was moving toward the surface. In light of the chance that the volcano could erupt, and considering the possible loss of life if it did, it would have been irresponsible for scientists not to issue the advisory. Although the predicted eruption did not occur, the advisory led to the development of evacuation routes and to consideration of disaster preparedness. This planning might eventually prove useful, because the most recent eruption in the Mammoth Lakes area occurred only 600 years ago and it is likely that one will occur in the future.

Forecasts and warnings are useful only if they provide people adequate time to respond in an appropriate manner. A minimum of several hours of warning is required in most instances, and much more time is needed if urban areas must be evacuated. Warnings of many hours to days are possible for hurricanes, many volcanic eruptions, large floods, and some tsunamis, but earthquakes and landslides commonly occur without any warning at all.

An understanding of hazardous processes is vital to evaluating risk.

Hazardous processes are amenable to risk analysis, which considers both the probability that a damaging event will occur and the consequences of that event. For example, if we were to estimate that, in any given year, Vancouver has a 1 percent chance of a moderate earthquake, and if we know the consequences of that earthquake in terms of loss of life and damage, we can then calculate the risk to society.

Determining *acceptable risk* is more complicated, because individuals, social groups, and countries have different attitudes about what level of risk is tolerable. Many people are willing to accept a level of risk that governments will not. A person may knowingly live on a floodplain, with the understanding that flooding of their property is a possibility. Governments, however, have a larger responsibility to society that requires a more cautious approach and a longer-term perspective. From their point of view, the risk associated with floodplain development, like a diamond, is forever, ensuring that losses ultimately will be incurred. Furthermore, governments are commonly left "holding the bag" after a disaster—they are essentially the insurer of last resort because individuals rarely have the financial ability to cover their losses. Governments also must allocate public monies to reducing risk from natural hazards, and those expenditures must compete with a plethora of public demands on resources.

Acceptable risk also depends on the situation. Driving an automobile is fairly risky, but most of us accept that risk as part of living in a modern world. In contrast, for many people, the acceptable risk represented by a nuclear power plant is very low because they consider the possibility of radiation poisoning unacceptable. Nuclear power plants are controversial because many people perceive them as high-risk facilities. Even though the probability of an accident owing to a geologic hazard, such as an earthquake, might be very low, the consequences could be devastating; accordingly, the risk is relatively high.

Institutions such as governments, banks, and insurers commonly view acceptable risk from a collective, or societal, perspective rather than a personal one. This perspective may be dominated by economic considerations. For example, an insurer may base policy premiums on the level of economic risk it faces from flooding or other natural hazards.

At the individual level, people have some control over the level of risk they are willing to accept. For the most part, you can choose where you live. If you choose to live in Vancouver, you may experience a damaging earthquake. If you move to Winnipeg, you should realize that you will live on the floodplain of the Red River, which has a long history of damaging floods. So why do people live in hazardous areas? Perhaps the allure of mountains and the ocean drew you to Vancouver, or you were offered an excellent job in Winnipeg. Whatever the case, individuals must weigh the pros and cons of living in a particular area and decide whether the risk is acceptable. This assessment should consider such factors as the frequency of damaging events, the potential damage the events could cause, and the extent of the geographic area at risk. The assessment should compare these factors to the potential benefits of living in the high-risk area. In this way, acceptable risk, which differs from person to person, can be determined. In fact, a trade-off generally exists between risk acceptance and the cost of protection against hazards. Society may be willing to accept some risk in allowing people to live on a floodplain or in using the floodplain for business activity because of the economic benefits of doing so. Commonly, protective dykes are built to provide some protection, but at a cost, of course. An economic cost–benefit analysis can be an essential part of the decision-making process for determining the most appropriate level of protection against floods and other hazards.

A frequent problem of risk analysis is that the data required to determine the probability or consequences of an event are either inadequate or lacking. Assigning probabilities to geologic events such as earthquakes and volcanic eruptions can be difficult, because the known record of past events is too short or incomplete.[17] Furthermore, the time separating high-magnitude events in any given area is often highly variable. Similarly, it may not be possible to accurately determine the consequences of an event from sparse data. For example, if we are concerned about the consequences of a release of radiation into the environment, we need information about the local biology, geology, hydrology, and meteorology, all of which may be complex and difficult to analyze. We also need information about the infrastructure at risk and the numbers and distribution of people living and working in the area of concern. Despite these limitations, risk analysis is a step in the right direction. As we learn more about the probability and consequences of a hazardous event, we can make a more reliable risk assessment for appropriate decision making.

Hazards are commonly linked to each other and to the environment in which they occur.

Many hazardous natural processes are directly or indirectly linked. For example, intense precipitation and surges accompanying hurricanes cause flooding, coastal erosion, and landslides. Volcanic eruptions on land cause volcanic debris flows (*lahars*) and floods, and some volcanic eruptions on islands can trigger tsunamis.

Natural hazards also are affected by Earth materials. Slopes developed on shale or loose glacial sediments, for example, are prone to landslides. In contrast, massive granite slopes are generally stable, although jointed granite may fail along fractures within the rock.

Population growth and socio-economic changes increase the risk from natural hazards.

Over much of human history, our numbers were small and nomadic, and losses from hazardous processes were not very significant (Table 1.3). As people began to cultivate crops and domesticate animals, populations increased and became more fixed, often in hazardous areas. Concentration of people and resources in fixed settlements increased losses from periodic earthquakes, floods, and other natural disasters. The rate of population growth increased nearly tenfold during the Early Industrial period (A.D.1600 to 1800). Since the Industrial Revolution, with modern sanitation and medicine, growth rates have increased another 10 times. The human population reached 6 billion in 2000,[18] and by 2014 it will be 7.3 billion—an increase of 1.3 billion people in only 13 years! Most of the increase in population has been, and will continue to be, in developing nations. India will have the largest population of all countries by 2050—about 18 percent of the world total; China will have about 15 percent of the world total.

Today billions of people live in areas vulnerable to damage by hazardous Earth processes (Figure 1.11). In addition, much of our economic productivity and wealth are located

TABLE 1.3 How We Became 7 Billion

40 000–9000 B.C.: HUNTERS AND GATHERERS

Population density about 1 person per 100 km^2 of habitable area*; total population probably less than a few million; average annual growth rate less than 0.0001% (doubling time about 700 000 years)

9000 B.C.–A.D. 1600: PREINDUSTRIAL AGRICULTURAL

Population density about 1 person per 1 km^2 of habitable area; total population several hundred million; average annual growth rate about 0.03% (doubling time about 2300 years)

A.D. 1600–1800: EARLY INDUSTRIAL

Population density about 7 persons per 1 km^2 of habitable area; total population by 1800 about 1 billion; annual growth rate about 0.1% (doubling time about 700 years)

A.D. 1800–2000: MODERN

Population density about 40 persons per 1 km^2 of habitable area; total population at the start of 2014 about 7.3 billion; annual growth rate in 2000 about 1.4% (doubling time about 50 years)

*Habitable area is assumed to be about 150 000 000 km^2.

Source: Data from Botkin, D. B., and E. A. Keller. 2000. Environmental Science, *3rd ed. New York: John Wiley and Sons. Seven billion data from U.S. Census Bureau, International Data Base.*

in hazard zones. Because more and more people are living in hazardous areas, the need for planning to minimize losses from natural disasters is increasing.

This rapid increase in population has been *exponential*—the population grows each year, not by the addition of a constant number of people, but rather by the addition of a constant percentage of the current population (Figure 1.12). The exponential growth in population can be expressed by the following equation:

$$N_t = N_0 e^{rt}$$

where N_t is the population at time t, N_0 is the starting population, r is the growth rate, and e is the base of the natural logarithm (2.71828).

This equation does not directly take into account changes in life expectancy, but it clearly shows that future population numbers are dependent on the rate of growth in population. At the current annual rate of growth, the population will increase from its current level of 7.3 billion to about 9.2 billion by 2050. In contrast, if the annual rate of increase were only half its present level, the population level in 2050 would be about the same as today.

An emerging issue related to natural hazards is the link between disasters and technological dependence. The ice storm in southern Quebec and Ontario in 1998 was a disaster mainly because electric power to millions of people was lost when transmission lines failed. The ice storm would not have been a large disaster 100 years earlier when people did not use electricity to keep themselves warm in winter. Related to the issue of overdependence on complex and fallible technological systems is the interrelation of these different systems. During the Quebec–Ontario ice storm, the loss of hydroelectric power required the use of generators to produce electricity. However, there were not enough generators to filter water at the same rate as is normally done, and the population came close to suffering a shortage of drinking water.

Inequities in health, education, and wealth between developed and developing countries aggravate these problems. Population growth in developing countries is far outstripping that in North America, Europe, Japan, and other wealthy jurisdictions. Most people in developing countries lack resources to protect themselves from hazardous events. Thus, when a disaster happens in a densely populated area in a developing country, the consequences are likely to be catastrophic. The same event in a fully developed country tends to kill far fewer people, although the economic cost is generally much greater.

Damage and loss of life from natural disasters can be reduced.

We deal with natural hazards primarily in reactive ways: following a disaster, we engage in search and rescue, firefighting, and the provision of emergency food, water, and shelter. These activities reduce loss of life and property and must, of course, be continued. However, a higher level of hazard reduction requires a *proactive* approach, in which we anticipate and prepare for disasters. Land-use planning that limits or prohibits construction in hazardous areas; building codes that demand hazard-resistant construction; and hazard modification or control, such as flood control channels, are some of the proactive measures that can be taken before disastrous events occur to lower our vulnerability and exposure to them.[12]

Reactive Response: Recovery from Disasters

The effects of a disaster on a population may be either direct or indirect. *Direct effects* include deaths, injuries, displacement of people, and damage to property and other infrastructure. *Indirect effects* are post-disaster impacts, including crop failure, starvation, emotional distress, loss of employment, reduction in tax revenues because of property loss, and higher taxes to finance the recovery. Many more people experience indirect effects than direct effects.[19,20] In our highly interconnected and interdependent world, a catastrophic natural disaster can have nearly global impacts. An example is the temporary loss of oil-refining capacity in Louisiana after Hurricane Katrina. The effect of this event was an immediate rise in gasoline prices throughout North America and Europe—an economic impact affecting hundreds of millions of people. A more serious interdependency can be seen in the potential impacts of global warming. An increase in average atmospheric temperature of more than 2°C may lead to serious failures in some cereal crops in many countries, which would lead to food shortages that would threaten global food supplies.[21] The same warming might cause sea levels to rise, flooding low-lying coastal areas where hundreds of millions

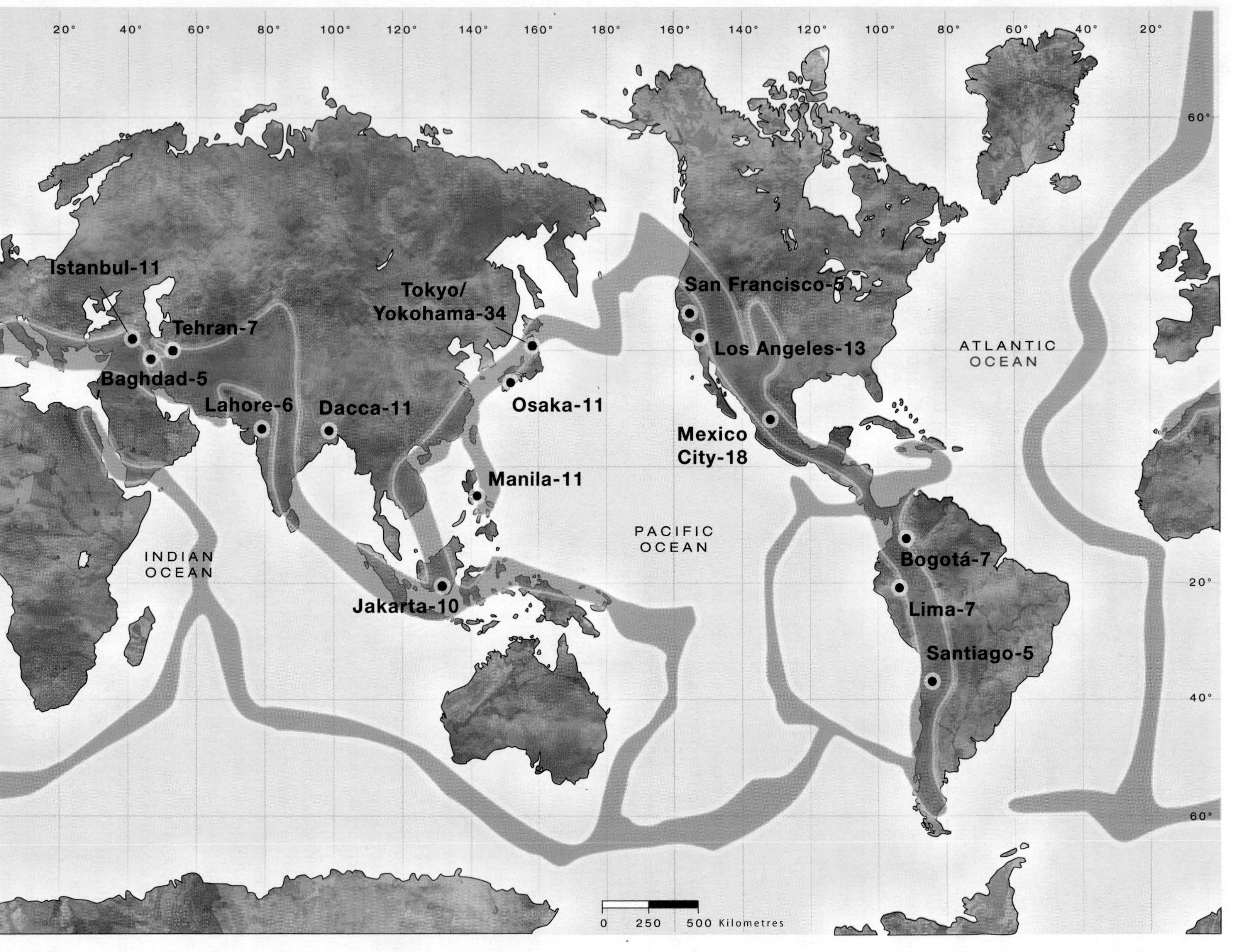

▲ **FIGURE 1.11 CONCENTRATION OF PEOPLE AND WEALTH IN HAZARDOUS AREAS** Many of the world's largest cities and much of our economic activity are concentrated in areas vulnerable to large earthquakes. In this figure, numbers are populations in millions and pink zones are areas prone to damaging earthquakes.

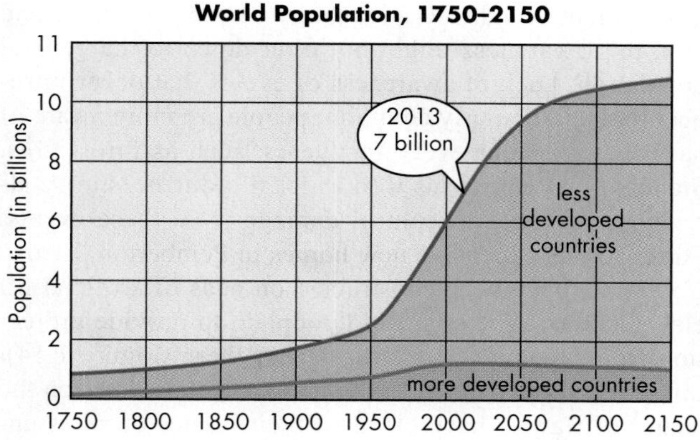

▲ **FIGURE 1.12 POPULATION GROWTH** The growth in global population has been exponential, but the rate of increase is now declining. The world's population at the beginning of 2014 was about 7.3 billion. *(Modified from U.S. Department of State)*

of people live. The result would be massive human migrations, creating refugees that would likely overwhelm the coping ability of many countries and lead to strife and war.

The stages of recovery following a disaster are emergency work, restoration of services and communication, and reconstruction (Figure 1.13). We can see the stages of recovery in the activities that followed the 1994 Northridge earthquake in the Los Angeles area. Emergency restoration began almost immediately with the repair of roads and utilities. Continuing restoration used funds from federal programs, insurance companies, and other sources that arrived in the first few weeks and months after the earthquake. Activity soon shifted from the restoration phase to the first phase of reconstruction, which lasted until about 2000. The effects of the earthquake on highway overpasses and bridges, buildings, and other structures were carefully evaluated, and new structures were built to a higher seismic standard. Large earthquakes are certain to occur again in the Los Angeles area; therefore, efforts to reduce the damage they cause must continue.

Now that Los Angeles is past the final phase of the reconstruction period, the lessons from two past disasters should be remembered: the Anchorage, Alaska, earthquake in 1964 and the flash flood that devastated Rapid City, South Dakota, in 1972. Restoration following the Anchorage earthquake began almost immediately with a tremendous influx of money from U.S. federal disaster programs, insurance companies, and other sources. Reconstruction was rapid and proceeded without much thought, as everyone competed for the available funds. Apartments and other buildings were hurriedly constructed in areas that had suffered ground rupture. Building sites were prepared by simply filling in cracks and re-grading the surface. By ignoring the potential benefits of careful land-use planning, Anchorage has made itself vulnerable to the same type of earthquake damage that it experienced in 1964. In contrast, in Rapid City, the restoration did not peak until approximately 10 weeks after the flood, and the community took time to carefully think through alternatives. As a result, Rapid City today uses the floodplain as a greenbelt, an entirely different use than before the 1972 flood. The change has reduced the flood risk substantially.[15,20]

The pace of recovery depends on several factors, the most important of which are the magnitude of the disaster

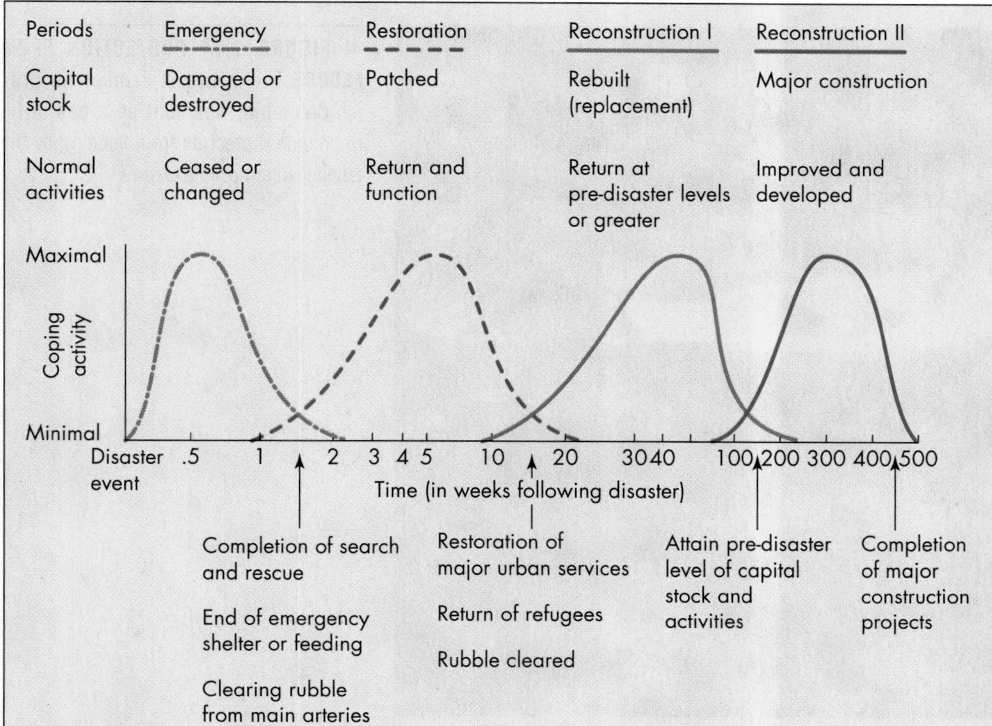

◄ **FIGURE 1.13 RECOVERY FROM DISASTER** Generalized model of recovery following a disaster. The first two weeks after a disaster are the period of emergency, during which normal activities cease or are changed. In the restoration phase, which typically lasts several months, normal activities return, although perhaps not at pre-disaster levels. Finally, during reconstruction, the capital stock is replaced, major new construction is completed, and normal activities return. *(From Kates, R. W., and D. Pijawka. 1977. "From rubble to monument: The pace of reconstruction." In* Disaster and Reconstruction, *eds. J. E. Haas, R. W. Kates, and M. J. Bowden, pp. 1–23. Cambridge, MA: MIT Press)*

and the social and economic impacts. Recovery is more rapid following smaller disasters such as the Northridge earthquake, than after catastrophes such as Hurricane Katrina and the 2004 Indian Ocean tsunami. Recovery also proceeds more rapidly in resilient communities and in those that have prepared in advance for natural disasters. Lack of preparation leads to a protracted recovery, often with severe economic consequences that last years, sometimes even decades. The complete failure of the U.S. and Louisiana governments to respond in a timely fashion to the destruction wrought by Hurricane Katrina resulted in a depopulation of New Orleans, from which the city still has not completely recovered nine years later. Recovery normally proceeds more rapidly in wealthy countries such as Canada, the United States, and Japan, than in countries that have few resources to deal with natural disasters, such as Haiti (2010 earthquake), Pakistan (2005 earthquake), and Indonesia (2004 tsunami). The global community has a responsibility to provide both immediate and long-term assistance to countries that do not have the ability to deal with catastrophes themselves.

Proactive Response: Avoiding and Adjusting to Hazards

The decisions we make, individually and collectively, in preparing for natural disasters depend in part on our perception of risk. Much research has been done in recent years to try to understand how people perceive different natural hazards. This understanding is important because the success of risk-reduction programs depends on the attitudes of the people who are likely to be affected. Although governments might be adequately aware of hazard and risk, this awareness might not filter down to the general population. Lack of awareness of events that occur infrequently is particularly common; people are more aware of hazards that occur every few years, such as forest fires. Standard procedures, as well as local ordinances, may already be in place to control damage from these events. For example, expensive new homes in Pemberton, British Columbia, have been constructed on pads of artificial fill elevated above the adjacent floodplain to provide protection from frequent floods that occur there (Figure 1.14). Similarly, some landowners in tsunami-prone areas on the island of Hawaii have elevated their homes on piles anchored in the ground. In the latter case, tsunamis have been sufficiently frequent in the past 100 years that the owners view the extra cost as a good investment.

One of the most environmentally sound and cost-effective adjustments to hazards involves **land-use planning**. People can avoid building on floodplains, in areas prone to active landslides, or in places where tsunamis or coastal erosion are likely to occur. In many Canadian and U.S. cities, floodplains have been delineated and zoned for a particular land use. Legal requirements for engineering geology studies at building sites may greatly reduce potential damage from landslides. Damage from tsunamis and coastal erosion can be minimized by requiring adequate setbacks of buildings from the shoreline or sea cliff. Although it might be possible to control physical processes in some instances, land-use planning is often preferable to a technological fix that may or may not work.

Insurance is another option for dealing with natural hazards. Flood and earthquake insurance is available in many

◄ **FIGURE 1.14 PROTECTION FROM FLOODS** This house in Pemberton, British Columbia, was built on a pad of fill to provide protection from flooding by the Lillooet River. *(John J. Clague)*

areas. However, huge insured losses stemming from recent hurricanes, earthquakes, and other disasters are forcing insurance companies to increase their premiums or deductibles in many hazard-prone areas or simply to discontinue some types of insurance. To reduce their exposure to disasters and catastrophes, insurance companies may themselves purchase *reinsurance* from one or more other insurance companies, which are referred to as "reinsurers."

Evacuation is a response to the hurricane hazard in states along the Gulf of Mexico and the eastern seaboard of the United States. Sufficient time is generally available for people to evacuate coastal areas, provided they heed warnings. However, if people do not react quickly or if the population in the affected area is large, evacuation routes may become clogged, as happened in Texas in September 2005 during Hurricane Rita.

Disaster preparedness is an option that individuals, families, cities, states, and entire nations can use to reduce risk. Of particular importance are public education and emergency preparedness training.

Attempts at *artificial control* of landslides, floods, lava flows, and other hazardous phenomena have met with mixed success. Seawalls may protect property to some extent, but they tend to narrow or even eliminate the beach. Retaining walls and other structures may protect slopes from landslides if well designed. They are necessary where potentially unstable slopes are excavated or where buildings border steep slopes. Common methods of flood control are channelization and construction of dams and levees. Unfortunately, flood-control projects tend to provide residents with a false sense of security; no method can completely protect floodplain residents from extreme floods.

The option to simply accept the risk and bear the loss in the event of a disaster is all too often chosen. Many people are optimistic about their chances of making it through any disaster and therefore will take little action on their own. They also believe that governments will step in with relief following a disaster, which is commonly the case. The do-nothing response is particularly true in areas where hazards, such as volcanic eruptions and earthquakes, are rare. The danger of such an approach was clearly shown in coastal New Jersey and New York when superstorm Sandy came ashore in October 2012. Preliminary estimates of direct damage are in the range of U.S.$75 billion, and only part of these losses is insured or will be assumed by governments.[5]

1.6 Many Hazards Provide a Natural Service Function

Ironically, the same natural events that injure people and destroy property also provide important benefits, which we will refer to as *natural service functions*. The following examples illustrate this point. Wildfires maintain forest health and provide openings that benefit many plants and animals. Cyclones are important sources of precipitation and thus water in large parts of the United States and Canada. Floods add new sediment to floodplains, creating the fertile soils that support agriculture. They cause erosion but also deliver sediment to beaches and flush pollutants from coastal estuaries. Some volcanic eruptions create new land, as in the case of the Hawaiian Islands, which are completely volcanic in origin. Nutrient-rich volcanic ash enriches soils, making them more productive for crops and wild plants. Earthquakes contribute to mountain building and thus are responsible for many of the scenic landscapes of the world. Some faults on which earthquakes occur serve as paths for groundwater, creating springs that are important sources of water. They can also be traps for oil and gas, without which the Industrial Revolution would never have happened (Figure 1.15).

1.7 Climate Change and Natural Hazards

Global and regional climate change may alter the incidence of some hazardous natural processes—notably storms, coastal erosion, landslides, drought, and fires (see Chapter 14). How might climate change affect the magnitude and frequency of these events? With global warming, sea levels will rise as the ocean's warmer surface waters expand and glaciers melt. Rising seas will induce or accelerate coastal erosion in some areas. Climate change may shift food production regions or force a change in the types of crops grown in specific areas. Deserts and semi-arid areas may expand, and warmer northern latitudes could become more productive. Permafrost is likely to degrade, causing problems for people who live at high latitudes. Some of these changes could force shifts in populations, which might bring about social and political upheaval.

Global warming will feed more energy from warmer ocean water into the atmosphere, which may increase the severity of thunderstorms, tornadoes, and hurricanes.[22] This trend may already be underway—2005 set a new record for direct economic losses from weather-related disasters, which cost at least $200 billion worldwide. This figure represents more than a 100 percent increase over the previous record of $100 billion set in 1998. Given concern about the effects of climate change, the scientific community is working hard to determine whether hurricanes and other severe storms, drought, heat waves, and other hazardous processes will become more common or severe in the near future.

Our ability to adjust to climate change will be determined, in large part, by the rate at which it happens. If climate changes slowly, we should be able to adjust our agricultural practices and settlement patterns without major economic and social disruption. If, however, the change occurs rapidly, we might not have the capacity to easily adapt.

◀ **FIGURE 1.15 HYDROCARBON TRAP** Faulting in the Santa Barbara Channel off the coast of California has created a linear trap against which oil has accumulated. Oil drilling platforms at the surface are located above the trap. *(Nik Wheeler/Corbis)*

Summary

Natural hazards are responsible for significant damage and loss of life worldwide each year. Natural processes that cause disasters are driven by energy derived from three sources: (1) Earth's internal heat, which produces slow convection in the mantle and is ultimately responsible for volcanic eruptions and earthquakes; (2) solar energy, which warms Earth's atmosphere and surface and is responsible for violent storms, floods, coastal erosion, and wildfires; and (3) the gravitational attraction of Earth, which is responsible for landslides, snow avalanches, and meteorite impacts.

Central to an understanding of natural hazards is awareness that disasters result from natural processes that have been operating for billions of years. These natural processes become hazards only when they threaten human life or property.

Hazardous events are repetitive, and study of their history provides information required for risk reduction. A better understanding of natural hazards and the risks they pose to humans can be obtained by integrating information on historic and prehistoric events and human exposure and vulnerability to hazardous events.

Geologic conditions and materials govern the type, location, and intensity of some natural events. The geologic cycle creates, maintains, and destroys Earth materials via physical, chemical, and biological processes. This cycle comprises a number of self-regulating sub-cycles, including the tectonic cycle, rock cycle, hydrologic cycle, and various biogeochemical cycles. The tectonic cycle describes large-scale geologic processes that deform Earth's crust, producing landforms, such as ocean basins, continents, and mountains. The rock cycle is a worldwide material recycling process driven by Earth's internal heat, which melts and metamorphoses crustal rocks. Weathering and erosion of surface rocks produce sediments and, ultimately, sedimentary rocks, which are added to the crust, offsetting materials returned to the mantle by plate tectonic processes. The hydrologic cycle is driven by solar energy and operates by way of evaporation, precipitation, surface runoff, and subsurface flow. Biogeochemical cycles involve transfers of chemical elements through a series of storage compartments or reservoirs, such as air or vegetation.

Five fundamental concepts establish a philosophical framework for studying natural hazards:

1. Hazards can be understood through scientific investigation and analysis.
2. An understanding of hazardous processes is vital to evaluating risk.
3. Hazards are commonly linked to each other and to the environment in which they occur.
4. Population growth and socio-economic changes increase the risk from natural disasters.
5. Damage and loss of life from natural disasters can be reduced.

Key Terms

biogeochemical cycle (p. 12)
catastrophe (p. 7)
disaster (p. 7)
forecast (p. 14)
geologic cycle (p. 11)
hazard (p. 7)
hydrologic cycle (p. 12)
hypothesis (p. 13)
land-use planning (p. 20)
mitigation (p. 7)
prediction (p. 14)
residence time (p. 12)
risk (p. 7)
rock cycle (p. 11)
scientific method (p. 13)
tectonic cycle (p. 11)
warning (p. 14)

Did You Learn?

1. Name the forces that drive Earth's internal and external processes.
2. What is the distinction between a natural hazard and a disaster, and between a disaster and a catastrophe?
3. What is the difference between hazard and risk?
4. Summarize the kinds of information that must be assembled to conduct a risk assessment.
5. What are the five fundamental concepts for understanding natural processes as hazards?
6. Explain the scientific method as it applies to natural hazards.
7. Explain the magnitude–frequency concept.
8. Define *risk* and *acceptable risk*. How do they differ?
9. Explain why population growth increases risk.
10. What are the stages of disaster recovery? How do they differ?
11. Describe four common adjustments to natural hazards.
12. Provide examples of related natural hazards.
13. Provide examples of natural service functions of natural hazards.

Critical Thinking Questions

1. How would you use the scientific method to test the hypothesis that sand on a beach comes from nearby mountains?
2. The argument has been made that we must curb human population growth in order to feed everyone. Even if we could feed 10 billion to 15 billion people, would we still want a smaller population? Why or why not?
3. The processes we call natural hazards have been occurring on Earth for billions of years and will happen for billions more. How then can we reduce loss of life and property damage from natural disasters?

MasteringGeology

MasteringGeology **www.masteringgeology.com**. Looking for additional review and test prep materials? Visit the Study Area in MasteringGeology to enhance your understanding of this chapter's content by accessing a variety of resources, including **Self-Study Quizzes, Geoscience Animations, GEODe Tutorials, RSS feeds, flashcards,** web links and an optional **Pearson eText.**

CHAPTER 2

▶ **VANCOUVER, BRITISH COLUMBIA** Vancouver, Canada's third largest city, lies at the western edge of the North American plate where the oceanic Juan de Fuca plate dives to the northeast beneath the continent. The fault separating these two plates produces giant earthquakes similar to those that happened in Indonesia in 2004 and Japan in 2011. When the "big one" hits, how will Vancouver and Victoria fare? (© *Gunter Marx/Alamy)*

Internal Structure of Earth and Plate Tectonics

Learning Objectives

The surface of Earth would be relatively smooth, with monotonous topography, were it not for internal forces that produce earthquakes, volcanoes, mountains, continents, and ocean basins. In this chapter, we focus on the interior of Earth. Your goals in reading this chapter should be to

- Understand the structure of Earth and its internal processes
- Know the ideas behind, and the evidence for, the theory of plate tectonics
- Understand the mechanisms that cause plates to move
- Understand the relationship between plate tectonics and natural hazards

Two Cities in Harm's Way

Southwestern British Columbia lies at the western edge of a large tectonic plate (the North American plate). Its two largest metropolitan areas—Vancouver and Victoria—are located 70 km and 40 km, respectively above the boundary between the North American plate and the oceanic Juan de Fuca plate, which is moving downward beneath the continent (Figure 2.1). The zone between the two plates, which is referred to as the Cascadia subduction zone, extends more than 1000 km southward from Vancouver Island to Northern California. The *mega-fault* that marks this plate boundary gradually descends eastward from near the ocean floor about 220 km west of Victoria to 70 km beneath Vancouver. Rarely—about once every 500 years on average—the fault ruptures and the western edge of North America moves up to several tens of metres westward relative to the ocean floor. This sudden slip produces a *mega-quake* of magnitude 8.5 to 9 that severely shakes the ground surface over an area of tens of thousands of square kilometres in the Pacific Northwest. The most recent of these giant earthquake happened in A.D. 1700, prior to European settlement of British Columbia. It spawned a huge tsunami that travelled across the Pacific Ocean to Japan, where damaging waves up to several metres high struck the east coast of Honshu. The tsunami was devastating to Native American coastal communities on the Pacific coasts of Vancouver Island, Washington, and Oregon.

Geologists in the United States and Canada have found traces of past giant earthquakes along the west coast of North America. The evidence allows them to forecast the effects of the next mega-quake, which in turn helps governments and the public to prepare for it. The main effects of such a mega-quake are strong ground shaking, liquefaction, landslides, and inundation of some low-lying coastal areas by the sea.

Earthquakes on the west coast of Canada are intimately linked to the plate tectonic environment of the region. Knowledge of plate tectonics is thus critically important to understanding this deadly natural hazard. In this chapter we summarize the plate tectonic paradigm, the processes that drive plate movements, and Earth's structure.

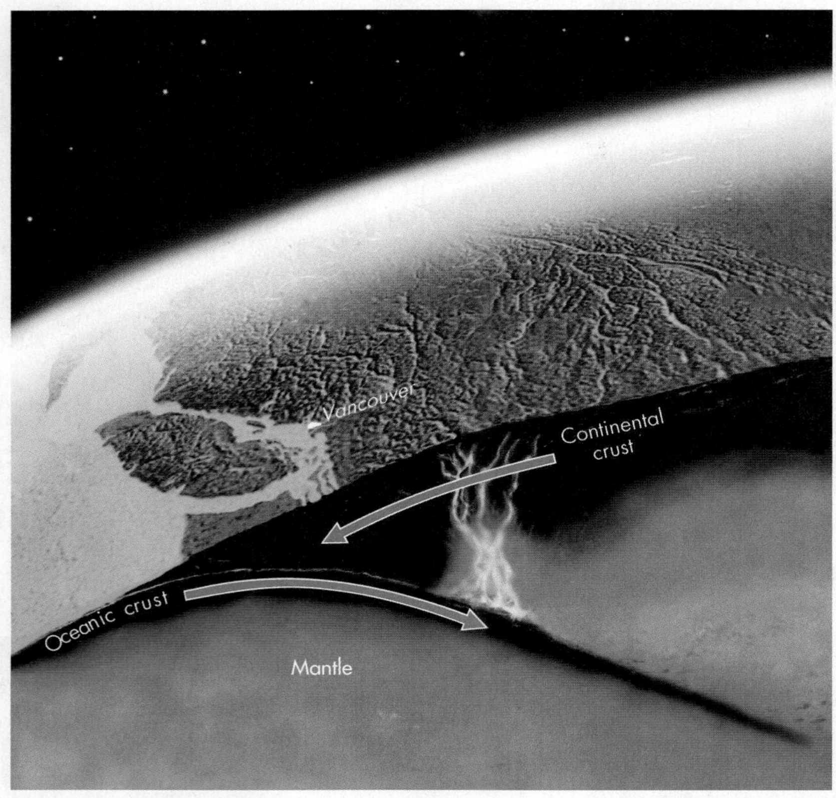

◄ **FIGURE 2.1 CASCADIA SUBDUCTION ZONE** Artist's rendering of subduction of the oceanic Juan de Fuca plate beneath the west coast of North America. As the two plates converge at a rate of about 45 mm/yr, the denser Juan de Fuca plate slides down into Earth's mantle. Collision of the two plates is responsible for earthquakes and the active volcanoes of the Cascade volcanic chain. *(B. Groulx and T. Poulton/Geoscape Calgary/ Poulton, T.; Neumar, T.; Osborn, G.; Edwards, D.; Wozniak, P. Geological Survey of Canada, Miscellaneous Report 72, 2002, poster. http://geoscape.nrcan.gc.ca/calgary/pdf/geoscape_calgary_view_e.pdf. © Department of Natural Resources Canada. All rights reserved.)*

2.1 Internal Structure of Earth

You might be familiar with the situation comedy *3rd Rock from the Sun*, the title of which refers to the position of planet Earth in the solar system. However, far from being a barren rock, Earth is a dynamic planet that is constantly changing. It has a rigid outer shell, a solid centre, and a thick layer of superheated liquid that slowly moves around the solid core. The processes that operate within Earth affect the surface of the planet, and they are responsible for the largest features on Earth's surface: continents and ocean basins. The configuration of the continents and ocean basins partly controls ocean currents. Heat carried by seawater around the globe affects climate, weather, and the distribution of plant and animal life on the planet. Earth's internal processes are also responsible for regional landforms, including chains of mountains and active volcanoes, and large areas of elevated topography such as the Tibetan Plateau and the Altiplano of South America. Mountains and plateaus significantly affect the circulation of air masses in the atmosphere—and therefore climate—thereby influencing all life on Earth. Thus, our understanding of the internal processes of Earth is of much more than academic interest. These processes are at the heart of the diverse environments that make our planet habitable.

Earth Is Layered and Dynamic

Earth has a radius of about 6300 km and a circumference of about 40 000 km (Figure 2.2a). The internal structure of Earth is shown in Figure 2.2b and can be considered in two fundamental ways:

- by composition, state (solid or liquid), and density (heavy or light)
- by strength (weak or strong)

We begin our exploration of the interior of our planet with a discussion of its structure from the first point of view (composition, state, and density). The major layers of Earth[1] are:

- A solid **inner core** with a thickness of more than 1300 km that is roughly the size of the moon, but with a temperature about as high as the surface of the sun.[2] Scientists think that the inner core is primarily metallic, composed mostly of iron (about 90 percent by weight), with minor amounts of sulphur, oxygen, and nickel.
- A liquid **outer core** with a thickness of just over 2000 km that has a composition similar to that of the inner core. The average density of the inner and outer core is approximately 10.7 g/cm^3. By comparison, the density of water is 1 g/cm^3 and the average density of Earth is about 5.5 g/cm^3.
- The **mantle**, which is nearly 3000 km thick and surrounds the outer core. It is composed mostly of solid iron- and magnesium-rich silicate rocks. The average density of the mantle is approximately 4.5 g/cm^3, which is less than half that of the underlying core.
- The **crust**, which ranges from a few kilometres thick to about 40 km thick and is the outer rock layer of Earth.

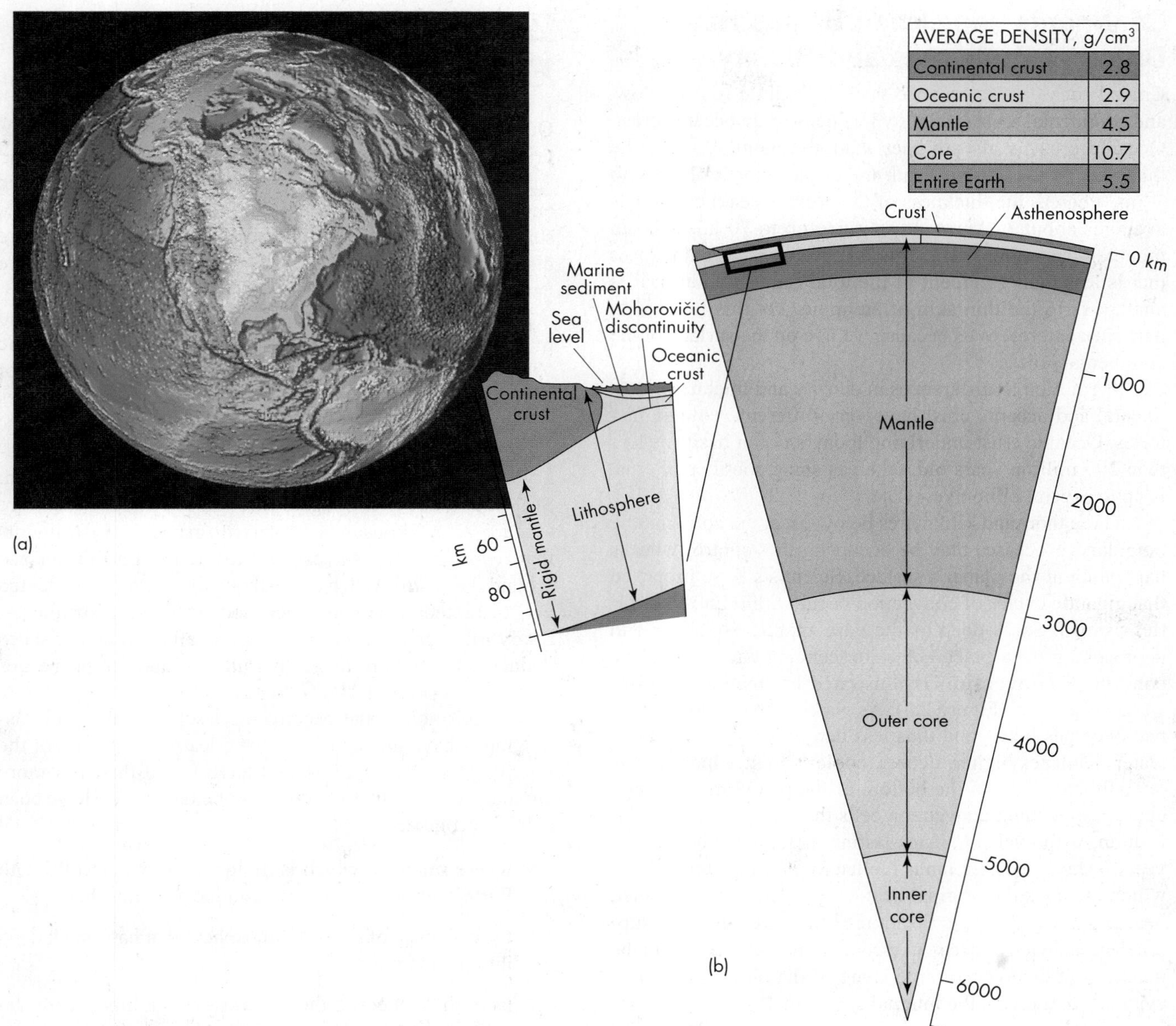

▲ **FIGURE 2.2 EARTH AND ITS INTERIOR** (a) A relief map of Earth as viewed from space. Land elevation increases as colour changes from green to yellow to red. The depths of the ocean floor increase as colour changes from lighter to darker shades of blue. (b) An idealized diagram showing the structure of Earth from its centre to its surface. Notice that the lithosphere consists of the crust and part of the mantle, and that the asthenosphere is located entirely within the mantle. Densities and thicknesses of the different layers have been estimated from the patterns and velocities of earthquake waves within Earth, from rocks formed within the lithosphere that have reached Earth's surface by tectonic processes, and from meteorites, thought to be pieces of old Earth-like planets. *((a) National Geophysical Data Center, National Oceanic and Atmospheric Administration; (b) From Levin, H. L. 1986.* Contemporary Physical Geology, 2nd ed. *Philadelphia: Saunders).*

The boundary between the mantle and the crust is referred to as the Mohorovičić discontinuity, or **Moho**. It separates the lighter rocks of the crust, which have an average density of approximately 2.8 g/cm^3, from the denser rocks of the mantle below.

The second way of considering the internal structure of Earth is by the strength of the rocks that form it. From a strength perspective, the cool, strong outermost layer of Earth is termed the **lithosphere** (*lithos* means "rock"). It consists of the crust and the rigid part of the mantle. The lithosphere has an average thickness of about 100 km, but ranges from a few kilometres thick beneath the crests of mid-ocean ridges to about 120 km thick beneath ocean basins and 20 to 400 km thick beneath the continents. The lithosphere is much stronger and more rigid than the material underlying it—the **asthenosphere** (*asthenos* means "without strength"). The asthenosphere constitutes all but the uppermost part of the mantle and is a slow-moving body of relatively weak, hot rock that behaves like a soft plastic material.

Continents and Ocean Basins Have Different Properties and History

Crustal rocks are less dense than the mantle rocks below, and continental crust is slightly less dense than oceanic crust. Oceanic crust is also thinner than continental crust—the thickness of the crust beneath the ocean floor is about 6 to 7 km, whereas the thickness of the crust beneath continents averages about 35 km and can reach up to 70 km beneath mountainous regions. The crust, on average, has a thickness that is less than 1 percent of the total radius of Earth; it is analogous to the thin skin of an apple. Yet this layer is of particular interest to us because we live on the surface of the continental crust.

In addition to differences in density and thickness, continental and oceanic crust have very different geologic histories. Oceanic crust underlying today's ocean basins is less than 200 million years old, whereas some continental crust is up to several billion years old.

Three thousand kilometres below us, at the core–mantle boundary, processes may be occurring that control what is happening at the planet's surface. Scientists have proposed that gigantic cycles of convection occur within Earth's mantle, rising from as deep as the core–mantle boundary and approaching the surface before descending back again. The concept of **convection** is illustrated by heating a pan of water on a stove (Figure 2.3). As water at the bottom of the pan becomes hotter and thus less dense, it rises. The rising water displaces higher, denser, cooler water, which moves laterally and sinks to the bottom of the pan. Earth's mantle appears to contain convection cells that operate in a similar fashion, with cycles perhaps lasting as long as 500 million years.[1] Mantle convection is fuelled by Earth's internal heat, which includes the original heat of formation of the planet, heat generated by crystallization of the core, and heat supplied by radioactive decay of elements such as uranium in the mantle. Let us now examine some of the observations and evidence that reveal the internal structure of Earth.

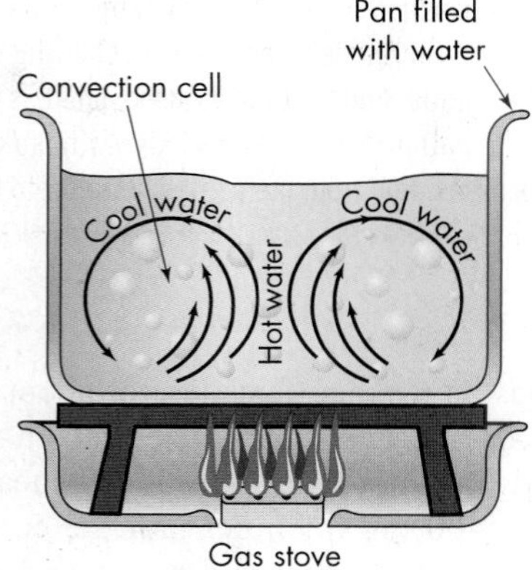

▲ **FIGURE 2.3 CONVECTION** Idealized diagram showing the concept of convection. As the pan of water is heated, less dense hot water rises from the bottom and displaces denser cooler water at the top, which then sinks down to the bottom. This process of mass transport is called convection, and each loop of rising and falling water is a convection cell.

2.2 How We Know about Earth's Internal Structure

Our knowledge of the structure of Earth's interior comes primarily from **seismology**, the study of earthquakes and earthquake waves.[3] When a large earthquake occurs, seismic energy is released and moves through Earth and along its surface as *seismic waves*. The properties of these waves are discussed in detail in Chapter 3, which deals with earthquake hazards.

Some waves travel through both solid and liquid materials, whereas others travel only through solids. The velocity of seismic waves differs according to the properties of the materials through which they move. When seismic waves encounter a boundary, such as the core–mantle boundary, some are *reflected* back, whereas others cross the boundary and are *refracted*—that is, they change direction. Still others fail to propagate through the liquid outer core (Figure 2.4). Thousands of **seismographs**—instruments that record seismic waves—are located at Earth's surface around the world. When an earthquake occurs, the reflected and refracted waves are recorded by these instruments. Study of seismographic records has allowed scientists to deduce the structure of Earth's interior and the properties of its layers (Figure 2.4).

As seismology has become more sophisticated and seismographs have improved, we have learned more about the internal structure of Earth and have found that it is more complex than originally thought. For example, we have been able to recognize:

- where **magma**, which is molten rock material beneath Earth's surface, is generated in the asthenosphere
- the existence of slabs of lithosphere that have sunk deep into the mantle
- large differences in the thickness of the lithosphere, reflecting differences in its age and history

2.3 Plate Tectonics

The term **tectonic** refers to the large-scale geologic processes that deform Earth's lithosphere and produce ocean basins, continents, and mountains. Tectonic processes are driven by forces deep within Earth.

Movement of Lithospheric Plates

What Is Plate Tectonics? The lithosphere is broken into large pieces called **lithospheric plates** that move relative to one another (Figure 2.5a).[4] A lithospheric plate may include a continent and part of an ocean basin, or it may be restricted to an ocean basin. Processes involved in the creation, movement, and destruction of these plates are collectively known as **plate tectonics**.

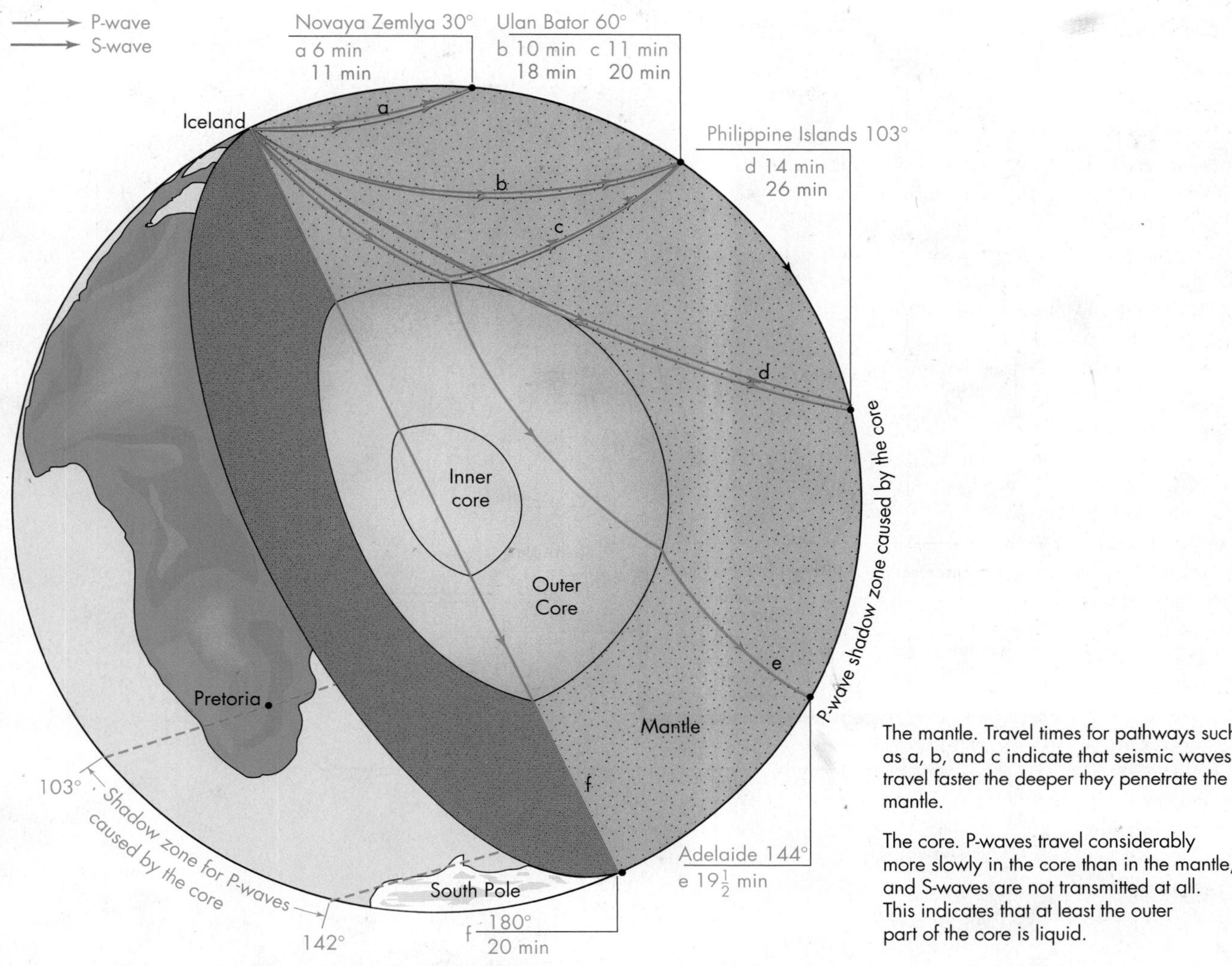

▲ **FIGURE 2.4 STRUCTURE OF EARTH INFERRED FROM SEISMIC WAVES** Scientists have inferred the physical properties and composition of Earth by studying the travel times and paths of seismic waves.

Plate Boundaries Are Delineated by Earthquakes and Active Volcanoes The boundaries between lithospheric plates are geologically active areas. Most earthquakes and many volcanoes are associated with plate boundaries (Figure 2.5b). Over geologic time, plates have formed and been destroyed, cycling materials from the interior of Earth to the surface and back into the mantle again. The continuous cycling of Earth materials by these processes is called the **tectonic cycle**.

Seafloor Spreading Is the Mechanism for Plate Tectonics As they move over the asthenosphere, lithospheric plates carry the continents embedded within them.[5] The idea that continents move is not new; it was first suggested by the German scientist Alfred Wegener in 1915. He presented evidence for **continental drift**, including the congruity of the shape of continents, particularly those bordering the Atlantic Ocean, and the similarity of fossils found in South America and Africa. Wegener's hypothesis was not taken seriously because scientists of the time could not envision a mechanism capable of moving continents around Earth. That mechanism was identified in the 1960s, when **seafloor spreading** was discovered. New crust is continuously added to the edges of lithospheric plates at **mid-ocean ridges**, or **spreading centres** (Figure 2.6). An example is the Mid-Atlantic Ridge, which runs the length of the Atlantic Ocean and passes through Iceland (Figure 2.7). Oceanic lithosphere is added at spreading centres but destroyed where one plate sinks beneath another at sites termed **subduction zones** (Figures 2.1 and 2.8). Continents do not move *through* oceanic crust; rather they are *carried along with it* as the plates move. Also, because the rate of production of new lithosphere at spreading centres is balanced by consumption of lithosphere at subduction zones, the Earth remains constant in size, neither growing nor shrinking.

Sinking Plates Generate Earthquakes The concept of a lithospheric plate sinking into the solid, but plastic upper mantle at a subduction zone is shown in diagrammatic form in Figure 2.8. Magma is generated as the descending plate

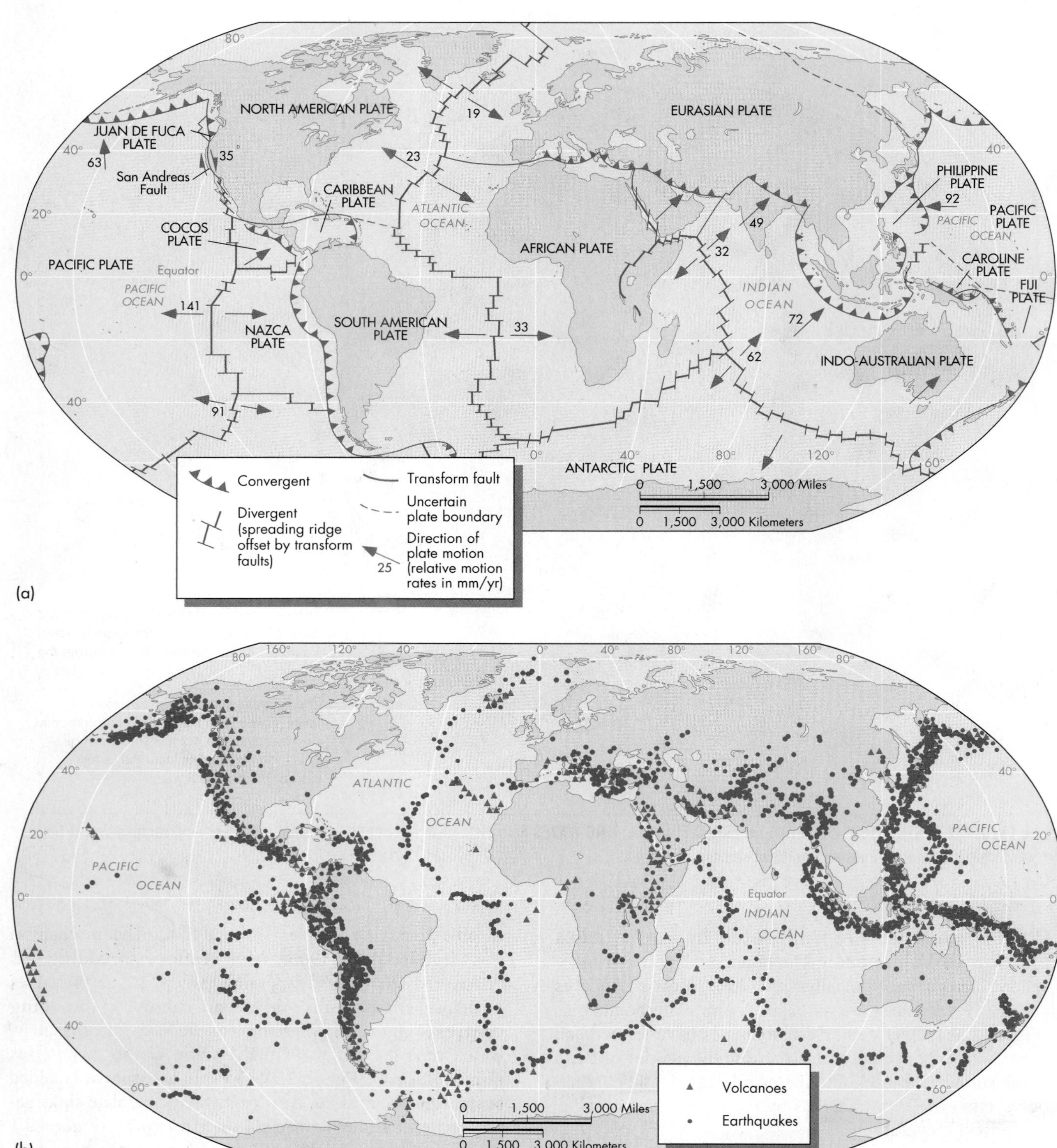

▲ **FIGURE 2.5 EARTH'S TECTONIC PLATES** (a) A map showing the major tectonic plates, plate boundaries, and directions of plate movement. (b) Map showing the locations of volcanoes and earthquakes. Note the correspondence between this map and the plate boundaries in (a). *(Based on Christopherson, R. W. 1994.* Geosystems, 2nd ed. *New York: Macmillan; Press, F., R. Siever, J. Grotzinger, and T. H. Jordan. 2003.* Understanding Earth, 4th ed. *New York: W.H. Freeman)*

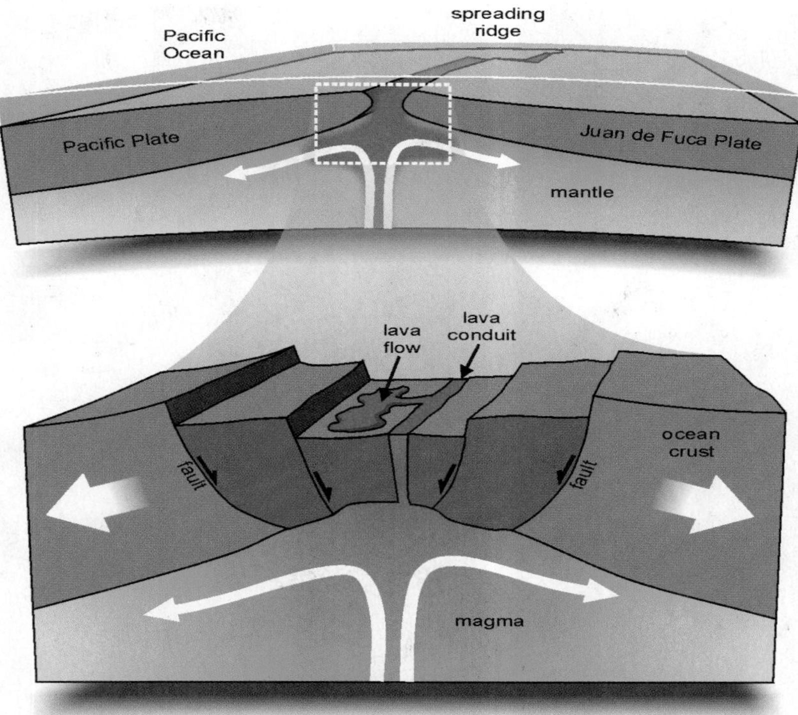

▲ **FIGURE 2.6 SPREADING CENTRE** New oceanic crust is created by upwelling of basaltic magma along fractures beneath mid-ocean ridges. The newly formed crust moves away from the ridge, forming the trailing edges of two oceanic plates. In this example from the northeast Pacific Ocean west of Vancouver Island, British Columbia, new crust formed at Juan de Fuca Ridge forms the trailing edges of the Pacific plate and the Juan de Fuca plate. *(Reprinted with permission from Tricouni Press)*

heats and releases fluids that move upward into overlying upper asthenosphere and lithosphere. The magma rises to the surface and is erupted at volcanoes, such as those that ring the Pacific Ocean basin above subduction zones (see Chapter 5). The path of the descending plate, or *slab* as it sometimes is called, is clearly marked by earthquakes. As the oceanic plate moves downward, earthquakes are produced both between it and the overriding plate, and within the subducting plate. The earthquakes occur because the sinking lithospheric plate is cooler and stronger than the surrounding asthenosphere; this difference causes rocks to break and seismic energy to be released.[6]

The path of a descending plate at a subduction zone is delineated by a dipping zone of earthquakes within the slab, referred to as a **Wadati-Benioff zone** (Figure 2.9) after the two seismologists who documented the pattern. The very existence of Wadati-Benioff zones is strong evidence that subduction of rigid 'breakable' lithosphere is occurring.[6]

Plate Tectonics Is a Unifying Theory The plate tectonic theory is to geology what Darwin's origin of species is to biology—a unifying concept that explains a large variety of phenomena. Geologists are still seeking the exact mechanism that drives plate tectonics, but they think it lies in convection within Earth's mantle. As rocks deep in Earth are heated, they become less dense and rise. Magma is injected into near-surface rocks at spreading centres. As the newly formed rocks move away from spreading centres, they cool and eventually become dense enough to sink back into the mantle at subduction zones (Figure 2.8).

Types of Plate Boundaries

There are three basic types of plate boundaries: divergent, convergent, and transform (Figure 2.10).[6] These boundaries are not narrow single cracks as shown on maps and diagrams, but rather broad zones of intense deformation ranging from a few kilometres to hundreds of kilometres across and extending through the crust.

Divergent boundaries occur where two plates move away from one another and new lithosphere is created (Figure 2.10 bottom). Typically, this process occurs at mid-ocean ridges and the process is called *seafloor spreading*. Mid-ocean ridges form where hot material from the mantle rises up to form a broad ridge, commonly with a central *rift valley* or *rift* where the plates are breaking and moving apart (Figure 2.8). Many of the cracks in

▲ **FIGURE 2.7 MID-ATLANTIC RIDGE** Image of the Atlantic Ocean basin showing details of the seafloor. Notice that the width of the Mid-Atlantic Ridge is about one-half the width of the ocean basin. *(Heinrich C. Berann/National Geographic Image collection)*

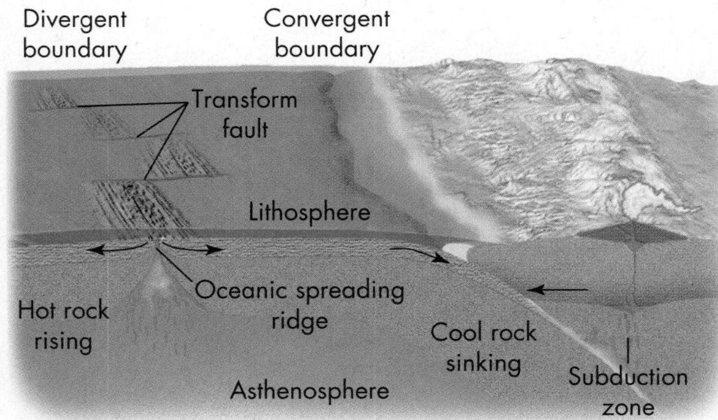

◄ **FIGURE 2.8 MODEL OF PLATE TECTONICS** New oceanic lithosphere is produced at a spreading ridge (divergent plate boundary). It returns to the mantle at a convergent plate boundary (subduction zone). *(Based on Lutgens, F., and E. Tarbuck. 1992.* Essentials of Geology. *New York: Macmillan)*

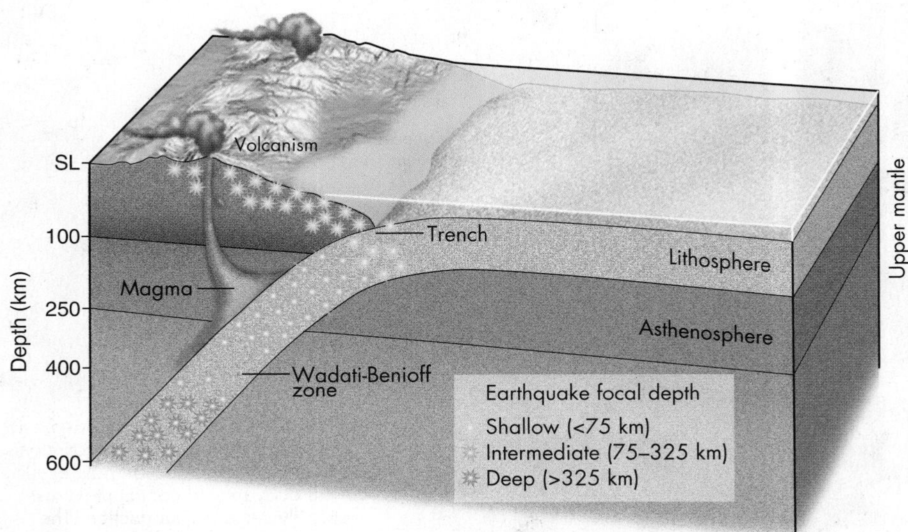

◄ **FIGURE 2.9 WADATI-BENIOFF ZONE** Idealized diagram of a subduction zone showing a zone of earthquakes that delineate the descending lithospheric plate.

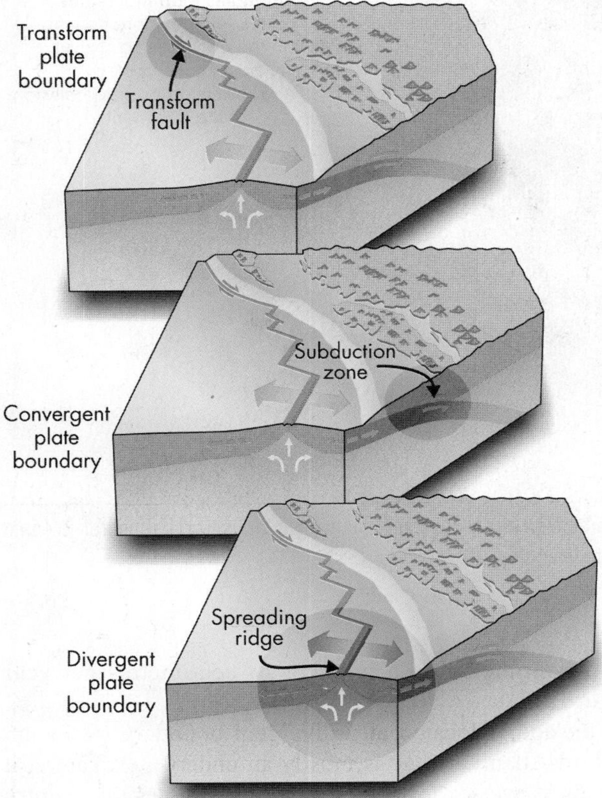

◄ **FIGURE 2.10 PLATE BOUNDARIES** Schematic diagram showing three types of plate boundaries: transform plate boundaries, along which adjacent plates move horizontally past one another; convergent plate boundaries, where one plate moves under another; and divergent plate boundaries, where two plates spread apart at a ridge and new oceanic crust is created. *(Reprinted with permission from Tricouni Press)*

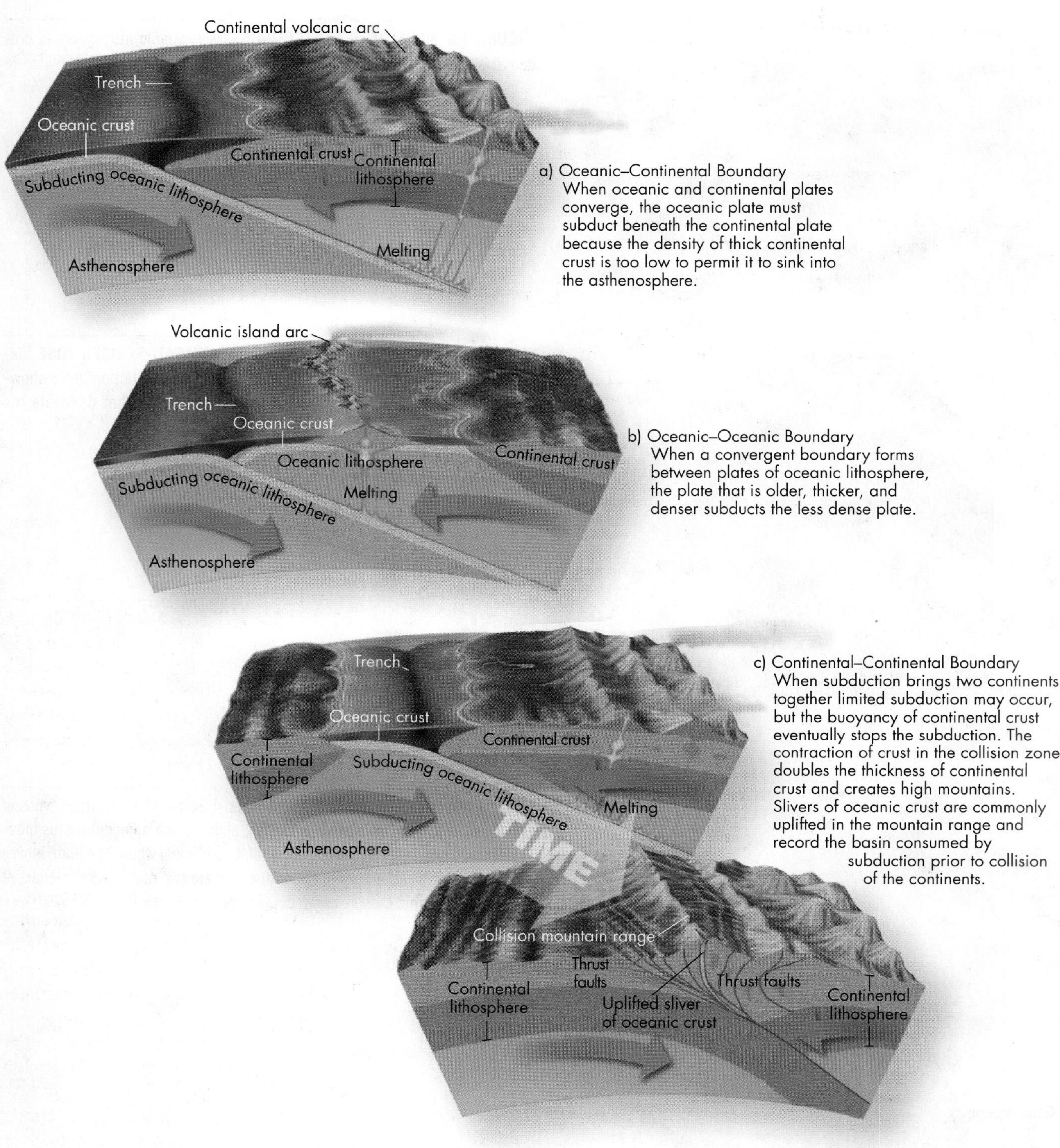

▲ **FIGURE 2.11 CONVERGENT PLATE BOUNDARIES** Idealized diagram illustrating characteristics of convergent plate boundaries: (a) continental–oceanic plate boundary; (b) oceanic–oceanic plate boundary; (c) continental–continental plate boundary.

the underwater rift zone are injected with magma from below. New lithosphere forms as the magma solidifies and the resulting basalt is slowly rafted, in a conveyor-belt fashion, away from the ridge crest. The tectonic plates on each side of the ridge move apart at a rate ranging from a few centimetres to a few tens of centimetres per year (Figure 2.5a).

Iceland is located at a divergent boundary. Although the Mid-Atlantic Ridge is mostly an underwater feature, it extends above sea level in Iceland and crosses the country in a northeasterly direction. A rift valley, characterized by frequent volcanic eruptions, marks the boundary between the North America plate, on which the western part of Iceland is located, and the Eurasian plate, on which the eastern part of the country sits. The two plates are moving apart at a rate of about 19 mm/yr.

Convergent boundaries occur where two plates collide head-on (Figure 2.10 middle and Figure 2.11). An oceanic–continental collision occurs where one of the converging plates is oceanic and the other is continental. The higher density oceanic plate is drawn down, or subducts, into the mantle beneath the leading edge of the lower density continental plate (Figure 2.11a). The collision of a continental with an ocean plate results in *compression*, with shortening and elevation of the edge of the continent; an analogy is pushing the edge of a table cloth or rug to produce folds. Shortening can cause folding of crustal rocks, as in the table cloth example, and faulting, both of which can thicken the lithosphere. This process of deformation produces mountain chains, such as the Andes in South America.

The south coast of British Columbia is situated at the edge of the North American plate at the north end of the Cascadia subduction zone (Figure 2.12). The subduction zone extends along the west coast of North America for more than 1000 km from northern California to central Vancouver Island. The subducting Juan de Fuca plate increases in temperature as it moves down beneath North America. At depths of 100 km to 120 km, it reaches temperatures in excess of 700°C and releases water, carbon dioxide, and other gases that rise into the lower part of the continental lithosphere. The superheated gases cause lower lithospheric rocks to melt, and the magma moves slowly up through the crust along fractures. Some of the magma reaches the surface, where it erupts and builds volcanoes. A chain of active volcanoes formed by repeated eruptions marks the inboard margin of the Cascadia subduction zone (Figure 2.13). Well-known volcanoes in this chain include Mount Baker, Mount Rainier, Mount St. Helens, Mount Hood, and Mount Lassen. Other important chains of active volcanoes include the Andes in South America, the Aleutian volcanoes in southwest Alaska, and the volcanoes of Indonesia, Japan, and the Caribbean.

An oceanic-to-oceanic collision occurs where both of the converging plates are oceanic. One plate subducts beneath the other, giving rise to a subduction zone backed by an arc-shaped chain of volcanoes known as an *island arc* (Figure 2.11b). An example of this type of plate boundary is the subduction zone separating the Pacific and Philippine plates in the western Pacific Ocean (Figure 2.5a).

A *submarine trench* is a deep, long, narrow depression on the ocean floor. It is a common feature found at subduction zones. Submarine trenches may be thousands of kilometres long and several kilometres deep and are sites of some of the deepest ocean waters on Earth. For example, the Marianas Trench at the edge of the Philippine plate has a depth of 11 km. Other major trenches include the Aleutian Trench south of Alaska and the Peru–Chile Trench west of South America.

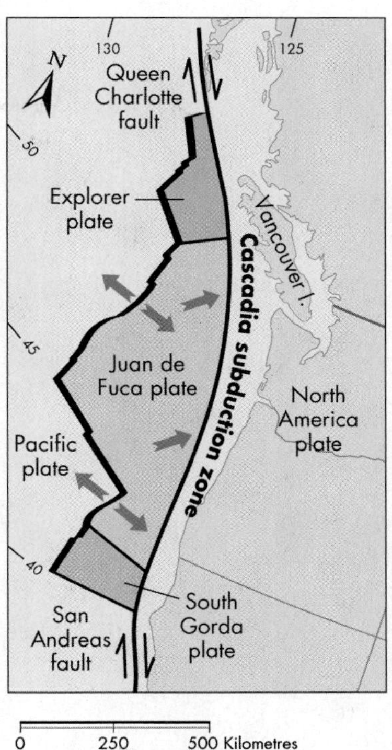

▲ **FIGURE 2.12 CASCADIA SUBDUCTION ZONE** Map of the Cascadia subduction zone, where the oceanic Juan de Fuca plate descends beneath the continental North American plate off the southwest coast of British Columbia. *(B. Groulx and T. Poulton/Geoscape Calgary/Poulton, T.; Neumar, T.; Osborn, G.; Edwards, D.; Wozniak, P. Geological Survey of Canada, Miscellaneous Report 72, 2002, poster. http://geoscape.nrcan.gc.ca/calgary/pdf/geoscape_calgary_view_e.pdf. © Department of Natural Resources Canada. All rights reserved.)*

If both colliding plates are continental, it is difficult for one to sink beneath the other because they have the same density. In such a situation, the plates meet along a continent-to-continent collision boundary delineated by high, faulted, and crumpled mountains (Figure 2.11c). This collision boundary is known as a *suture zone*. Continent-to-continent collisions are responsible for some of the highest mountain ranges on Earth, including the Himalayas in central Asia (Figure 2.14). Over the past 13 million years, collision of the Indian subcontinent with Asia to the north has elevated the Tibetan Plateau by 5 km.

Many older mountain belts formed in a similar way. For example, the Appalachian Mountains of eastern Canada and the United States formed during an ancient continent-to-continent plate collision 250 to 350 million years ago.

Transform boundaries occur where the edges of two plates slide horizontally past one another (Figure 2.10 top). The fault along which the movement takes place is known as a **transform fault**. If you examine Figures 2.5a, you will see that a spreading zone is not a single continuous ridge, but rather a series of ridges offset from one another along transform faults. Although transform faults are most common on the ocean floor, some occur within

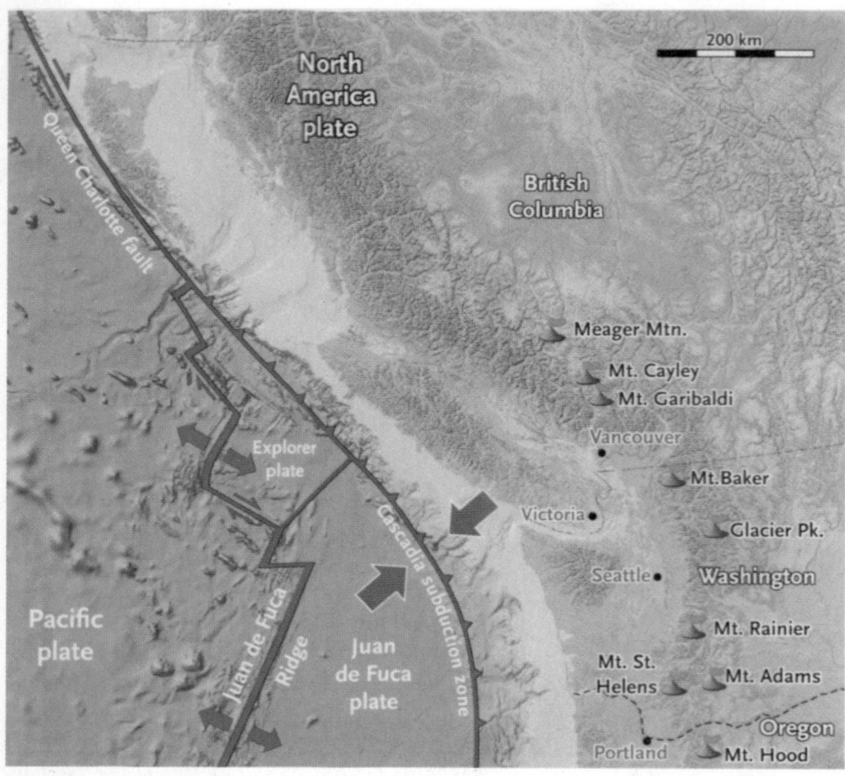

◄ **FIGURE 2.13 CASCADE VOLCANOES** A chain of active volcanoes lies at the back side of the Cascadia subduction zone. The volcanoes lie above the zone where fluids driven from the descending plate cause the overlying mantle and lowermost crust to partially melt. Some of the molten material rises and erupts at the surface. The heavy red arrows indicate directions where the Pacific and Juan de Fuca plates are moving away from the Juan de Fuca spreading ridge. The two heavy green arrows show the direction of convergence of the Juan de Fuca plate with the North American plate. The thin red arrows indicate directions of relative motion where plates are sliding past one another along transform faults. *(Reprinted with permission of Tricouni Press)*

continents. A well-known continental transform fault is the San Andreas fault in California, where the Pacific plate is sliding horizontally past the North American plate to the east (Figures 2.12 and 2.15). Other notable continental transform faults are the North Anatolian fault in Turkey and the Alpine fault in New Zealand. The Queen Charlotte fault, located on the ocean floor west of Haida Gwaii, British Columbia, is a transform fault that, like the San Andreas fault, separates the Pacific and North American plates (Figure 2.12).

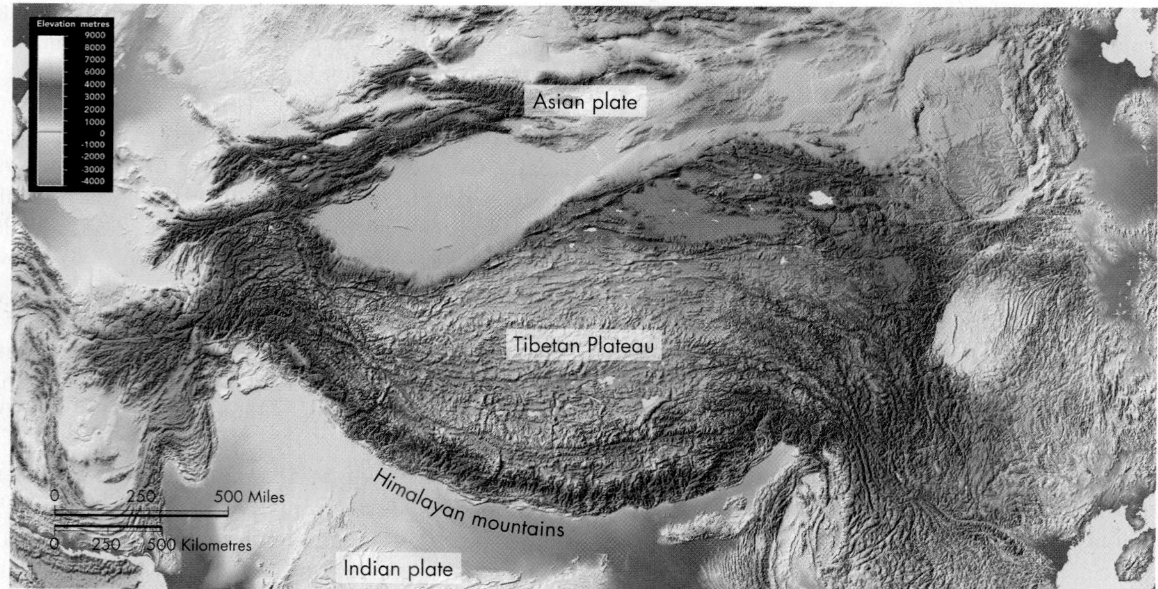

▲ **FIGURE 2.14 CONTINENT-TO-CONTINENT COLLISIONS** False-colour topographic image showing the Himalayan range and Tibetan Plateau. Collision of the Indian and Eurasian plates created the Himalayan range and uplifted the Tibetan Plateau to an average elevation of 5000 m above sea level. Numerous peaks in the Himalaya have elevations of more than 8000 m.

▲ **FIGURE 2.15 THE SAN ANDREAS FAULT** The surface trace of the San Andreas fault on the Carrizo Plain in southern California. This transform fault separates the Pacific and North American plates and is the source of many large earthquakes in California. *(R. E. Wallace/National Earthquake Information Center)*

In some places on Earth, three plates border one another. These sites are known as **triple junctions**. Two examples are the meeting point of the Juan de Fuca, North American, and Pacific plates off the northwest corner of Vancouver Island, and the junction of the spreading ridges associated with the Pacific, Cocos, and Nazca plates west of South America.

Rates of Plate Motion

In general, plates move a few centimetres per year, about as fast as your fingernails grow (Figure 2.5a). The Pacific plate moves past the North American plate along the San Andreas fault at a rate of about 3.5 cm per year, so that features such as rock units and streams are gradually displaced over time where they cross the fault. The Juan de Fuca plate is moving beneath the North America plate along the south coast of British Columbia at a rate of about 4 cm per year. During the past 5 million years, about 200 km of oceanic crust has subducted beneath North America. Although the central portions of plates move at a steady slow rate, movement may not be steady at plate boundaries, where collision, subduction, or both occur. In those places, the plates may be stuck against one another; movement is analogous to sliding one rough wood board over another. The boards may be stuck until the splinters that separate them break off, causing the boards to move quickly over each other. In extreme cases, the sudden slip along a subduction zone fault can amount to tens of metres, as happened for example during the Japanese (Tohoku) earthquake in March 2011.

Hot Spots

A number of places on Earth, called **hot spots**, are anchored in the slowly moving deep mantle, perhaps near the core–mantle boundary. Partly molten materials are hot and

A CLOSER LOOK 2.1

Paleomagnetism and Seafloor Spreading

When Alfred Wegener proposed the idea of continental drift in 1915, he could not conceive of a mechanism that could move continents. The mechanism was discovered a generation later. Mid-ocean ridges were identified in the 1950s, and in 1962 geologist Harry H. Hess published a paper suggesting that continents moved with respect to one another by the process of seafloor spreading, centered at mid-ocean ridges. He argued that new oceanic lithosphere is produced at spreading ridges and that lithospheric plates move away from the ridges, carrying along the embedded continents at their tops. These ideas produced a new paradigm that fundamentally changed our ideas about how Earth works.[3,6,7]

Proof of seafloor spreading came from three sources: (1) identification and mapping of mid-ocean ridges; (2) dating of volcanic rocks on the ocean floor; and (3) paleomagnetic mapping of ocean basins. Here we introduce and discuss Earth's magnetic field in some detail in order to show how scientists validated seafloor spreading and plate tectonics.

Earth has had a magnetic field for at least the past 3 billion years[2] (Figure 2.16). The planet's magnetic field can be described as a magnetic dipole with lines of magnetic force extending from the South Pole to the North Pole. A dipole magnetic field is one that has equal and opposite charges at its two ends. Earth's magnetic field has weaker non-dipole components, but these can be ignored for the purposes of this discussion. Convection in the iron-rich, fluid, hot outer core, along with the rotation of Earth, cause the outer core to rotate, which produces a flow of electric current. The flow of current within the core is responsible for Earth's magnetic field.[2,3] The magnetic field must be continuously generated or it would disappear in about 20 000 years, because the temperature of the core is too high to sustain permanent magnetization.[2]

The magnetic field of Earth is sufficiently strong that it permanently magnetizes some surface rocks. For example, magma that erupts at mid-ocean ridges becomes magnetized at the time it cools through a critical temperature. At that critical temperature, known as the *Curie point*, iron-bearing minerals such as magnetite and hematite within the volcanic rock become oriented parallel to the magnetic field.[3]

The term **paleomagnetism** refers to the study of the magnetism of rocks at the time their magnetic signature was acquired. Paleomagnetism is used to determine the magnetic history of Earth.

Earth's Magnetic Field Periodically Reverses Before the plate tectonic theory was formulated, geologists working on land had discovered that some volcanic rocks are magnetized in a direction opposite that of Earth's present field, suggesting that the polarity of the magnetic field was reversed at the time the rocks formed and cooled below the Curie point (Figure 2.16b). The magnetism of dated volcanic rocks of different age was then determined to document periods when Earth's magnetic field was reversed and periods when it was normal. You can verify the current "normal" magnetic field with a compass—the needle points to the north magnetic pole. During a period of reversed polarity, however, the needle would point to the south! The cause of **magnetic reversals** is not well understood, but is related to changes in the convective movement of the liquid material in the outer core and to processes occurring in the inner core. Reversals in Earth's magnetic field appear to be random, occurring, on average, every few hundred thousand years. The actual reversals may happen over a few thousand years, which is very fast in geologic time.

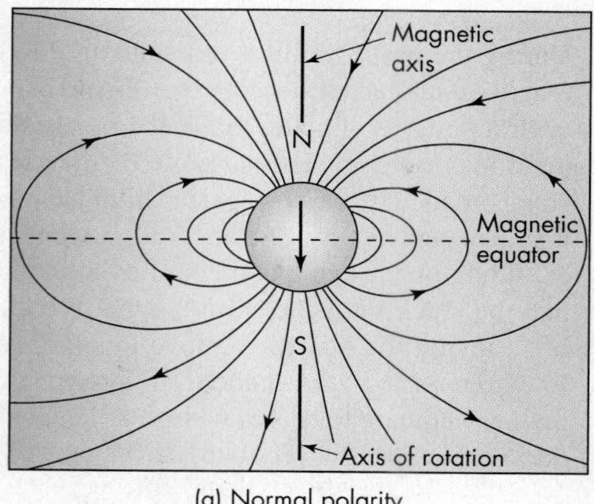

(a) Normal polarity

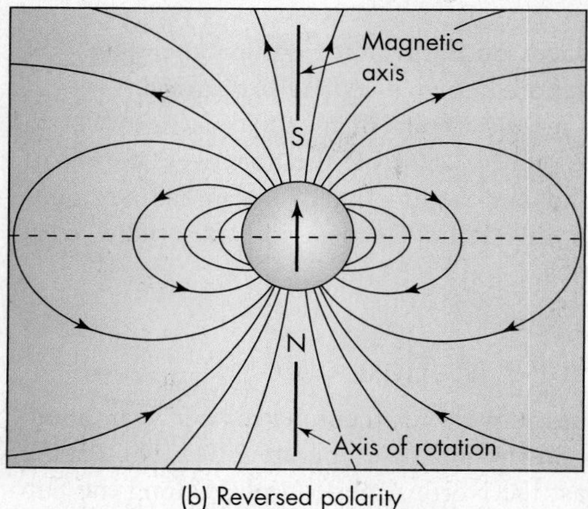

(b) Reversed polarity

▲ **FIGURE 2.16 EARTH'S MAGNETIC FIELD** Idealized diagram showing the magnetic field of Earth with (a) normal polarity and (b) reversed polarity. *(From Kennett, J. 1982.* Marine Geology. *Engelwood Cliffs, NJ: Prentice Hall. Reprinted and Electronically reproduced by permission of Pearson Education, Inc., Upper Saddle River, New Jersey.)*

What Produces Magnetic Stripes on the Seafloor? To further explore Earth's magnetic field, scientists completed magnetic surveys of the ocean floor by using ships to tow magnetometers, instruments that measure magnetic properties of rocks. The paleomagnetic record of the ocean floor is easy to read because basalt (see Chapter 5), which is produced at spreading centres, forms the floors of Earth's ocean basins. Basalt contains sufficient iron-bearing minerals to produce a good magnetic record.

The record the scientists found was unexpected. The rocks on the ocean floor are magnetically striped parallel to mid-ocean ridges[8] (Figure 2.17). The stripes represent alternating normal and reversed bands of basalt. Although not evenly spaced, the patterns are symmetrical about the axis of each mid-ocean ridge.

By measuring the widths of the alternating bands of normal and reversed magnetism and relating them to the ages of the basalts, scientists reasoned that the new crust was being created at mid-ocean ridges and gradually moving at right angles away from the ridge crests. Figure 2.17 illustrates their line of reasoning. The pattern shown is for the past 3.3 million years, which includes several periods of normal and reversed magnetism encoded into the basalts forming the ocean floor. Black stripes represent normally magnetized rocks, and blue stripes are rocks with reversed magnetization. The basic idea illustrated in Figure 2.17 is that rising magma at the mid-ocean ridge is extruded into the crust or erupted onto the seafloor, and the cooling rocks become normally magnetized.[8] When the field is reversed, the cooling rocks preserve a reversed magnetic signature, represented in Figure 2.17 by a blue stripe. Note that the patterns of magnetic anomalies in rocks on the two sides of the ridge are mirror images of one another. The only way of producing such a pattern is through the process of seafloor spreading. Thus, the pattern of magnetic reversals found in rocks on the ocean floor is strong evidence that spreading is happening.

Why Is the Seafloor No Older than 200 Million Years? Mapping of ocean-floor magnetic anomalies, when combined with age-dating of magnetic reversals in rocks on land, provides a database that leads to exciting inferences about the age of the oceans.[9]

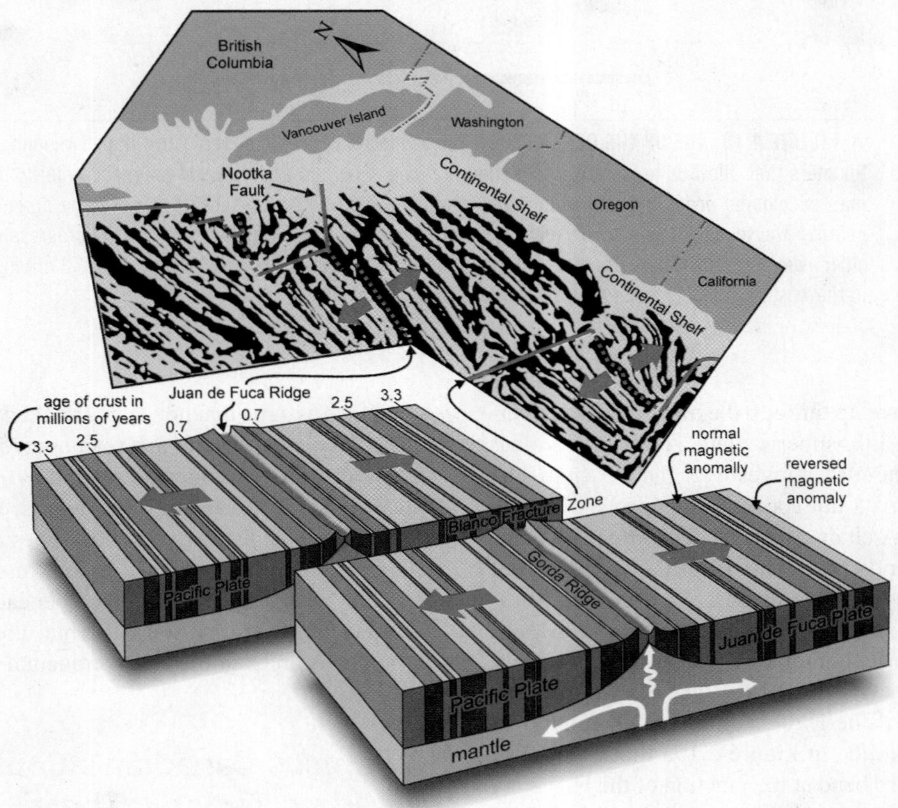

▲ **FIGURE 2.17 MAGNETIC STRIPING ON THE SEAFLOOR** (Top) Map showing the results of a magnetic survey of the seafloor in the eastern North Pacific Ocean, published in 1961. The black stripes record intervals of normal magnetic polarity, like today, and the intervening blue stripes represent intervals when Earth's magnetic field was reversed. (Bottom) An idealized diagram showing the formation of magnetic stripes by the conveyor-like movement of new crust away from the Juan de Fuca spreading ridge. The small red arrows are directions of motion along transform faults, which connect segments of the spreading ridge. *(Reprinted with permission of Tricouni Press)*

(continued)

Figure 2.18 depicts the age of the ocean floor inferred from this database. The pattern, which shows that the youngest seafloor basalts occur along active mid-ocean ridges, is consistent with the theory of seafloor spreading. As distance from the ridges increases, the age of the ocean floor also increases, to a maximum of about 180 million years. Thus, the present ocean floors of the world are no older than 200 million years. In contrast, some rocks on continents are much older, in fact as much as 4 billion years old, more than 20 times older than the oldest ocean floor! We conclude that the thick continental crust, by virtue of its buoyancy, is more long-lived than the crust of the ocean basins.

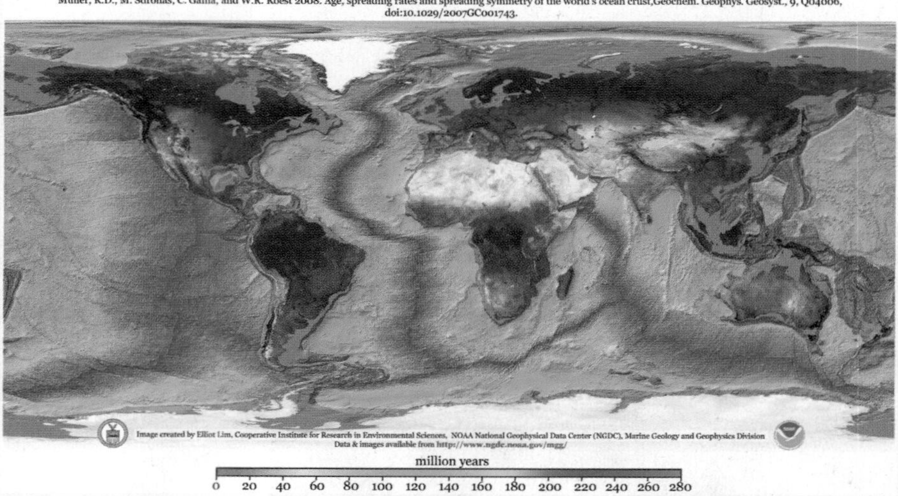

▲ **FIGURE 2.18 AGE OF THE OCEAN FLOOR** The seafloor has been mapped using ship-borne magnetometers that allow us to identify and delineate zones of normal and reversed magnetic polarity. This map is coloured according to absolute age in million years before present (Ma), based on magnetic polarity. The youngest ocean floor (red) is located along mid-ocean ridges; rocks become progressively older away from the ridges. The oldest ocean floor rocks are about 180 million years old and are found in the western Pacific Ocean. *(National Geophysical Data Center)*

buoyant enough to move up through the mantle into an overlying, slowly moving lithospheric plate. A chain of volcanoes is produced as the plate moves over the hot spot. The best-known example of a hot spot is Hawaii in the Pacific Ocean (Figure 2.19). A chain of volcanoes—the Hawaiian-Emperor chain—extends northwest from the big island of Hawaii. The age of volcanoes in this chain gradually increases in that direction. The youngest volcano is Mount Loihi, which is an active submarine volcano, presumably directly over a hot spot (Figure 2.19b). The oldest— Kauai, at the northwest end of the Hawaiian chain—is about 4 to 6 million years old. Notice in Figure 2.19a that the line of volcanoes makes a sharp bend at the junction of the Hawaiian and Emperor chains. The age of the volcanic rocks at the bend is about 43 million years. The bend is interpreted to record a time when the direction of movement of the Pacific plate changed.[10] With the exception of the Hawaiian Islands and some coral atolls (ring-like coral islands such as Midway Island), the Hawaiian-Emperor chain consists of submarine volcanoes known as *seamounts*. Seamounts are volcanic islands that were eroded by waves, currents, and submarine landslides and subsequently sank beneath the ocean surface.

If we assume that hot spots are fixed deep in the mantle, a chain of volcanic islands and seamounts such as the Hawaiian-Emperor chain provides additional evidence to support the plate tectonic theory. In other words, the ages of the volcanic islands and submarine volcanoes could systematically change as they do only if the plate is moving over a hot spot.

Not all hot spots are located beneath ocean basins. The Yellowstone National Park thermal area in Wyoming and Montana is an example of a continental hot spot.

A Famous Canadian's Contribution to the Plate Tectonic Theory

Canadian geophysicist J. Tuzo Wilson (Figure 2.20) was pivotal in advancing the plate tectonic theory. He was intrigued by Wegener's notion of a mobile Earth and was influenced by Harry Hess's ideas about magnetism on the seafloor. Wilson knew Hess in the late 1930s when Wilson was studying for his doctorate at Princeton University, where Hess was a dynamic young lecturer.

In 1963, Wilson developed a concept crucial to the plate tectonic theory. He suggested that the Hawaiian

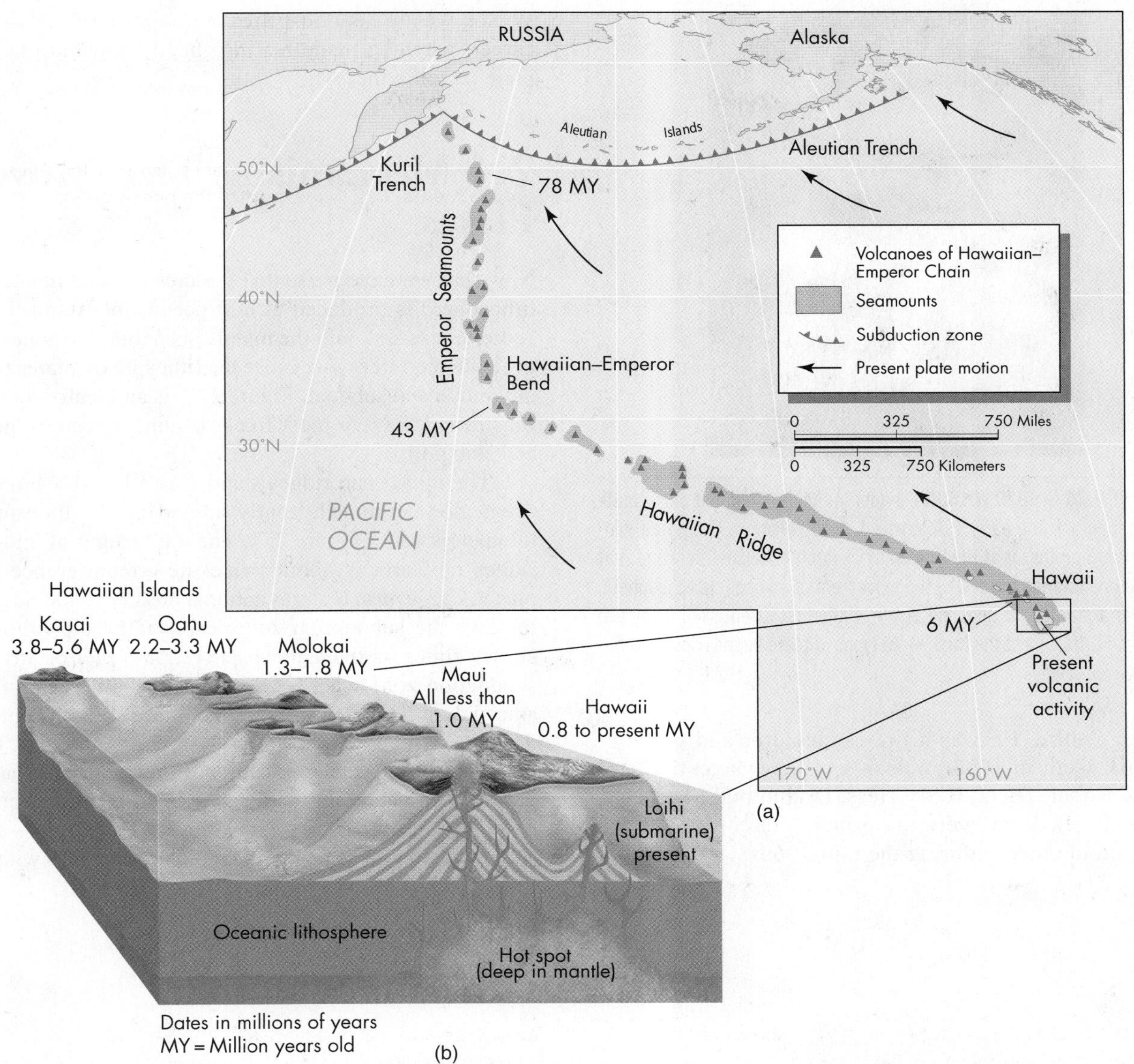

▲ **FIGURE 2.19 HAWAIIAN HOT SPOT** (a) Map showing the Hawaiian-Emperor chain of volcanic islands and seamounts. The Hawaiian Islands are the only volcanoes of the chain that reach above the ocean surface. The other volcanoes are seamounts. (b) Sketch map showing the Hawaiian Islands, which range in age from present-day to almost 6 million years. Notice that most of the mass of the volcanoes is below the ocean surface. *((a) From The Geological Society of America; (b) From Thurman, H. V. 1995.* Essentials of Oceanography 5th ed. *Upper Saddle River, NJ: Prentice Hall. Reprinted and Electronically reproduced by permission of Pearson Education, Inc., Upper Saddle River, New Jersey.)*

Island chain was formed when a plate moved over a stationary hot spot in the mantle. This hypothesis eliminated an apparent contradiction to the plate tectonic theory—the occurrence of active volcanoes located many thousands of kilometres from the nearest plate boundary. Hundreds of subsequent studies have proven Wilson right. However, in the early 1960s, his idea was considered so radical that his hot spot manuscript was rejected by all the major international scientific journals. Ultimately it was published in 1963 in a relatively obscure journal, the *Canadian Journal of Physics*, but it nevertheless became a milestone in plate tectonics.[11]

Wilson made another important contribution to the development of the plate tectonic theory two years later. He proposed that there must be a third type of plate boundary besides spreading ridges and subduction zones—specifically, faults that join segments of spreading ridges and slip horizontally.[12] A well-known example of such a "transform" fault boundary is the San Andreas fault. Unlike ridges and trenches, transform faults offset the crust horizontally, without creating or destroying crust.

Wilson was a professor of geophysics at the University of Toronto from 1946 until 1974, when he retired from teaching and became the director of the Ontario

▲ **FIGURE 2.20 J. TUZO WILSON** A leader in the development of the plate tectonic theory in the early 1960s was J. Tuzo Wilson, a Canadian geophysicist and professor at the University of Toronto. Wilson theorized that the Hawaiian Islands formed by the movement of an oceanic lithospheric plate over a mantle hot spot. He also was the first scientist to recognize transform faults, one of the three main types of plate boundaries. *(Courtesy Ontario Science Centre)*

Science Centre. He was a tireless lecturer and traveller until his death in 1993. Wilson and other scientists, including Robert Dietz, Harry Hess, Drummond Matthews, and Frederick Vine, were the principal architects of the plate tectonic theory during the mid-1960s. Interestingly, Wilson was in his mid-fifties, at the peak of his scientific career, when he made his insightful contributions to the plate tectonic theory.

2.4 Mechanisms That Move Plates

Now that we have presented the concept that new oceanic lithosphere is produced at mid-ocean ridges and that old, cooler plates sink into the mantle at subduction zones, let us evaluate the forces that cause the lithospheric plates to actually move and subduct. Figure 2.21 is an idealized diagram illustrating the two most likely driving forces: ridge push and slab pull.

The mid-ocean ridges stand 1 to 3 km above the deep ocean floor as linear, gently arched uplifts thousands of kilometres wide (Figure 2.7). The total length of mid-ocean ridges on Earth is about twice the circumference of the planet. *Ridge push* is a gravitational push from the ridge crest towards the subduction zone; essentially, the lithosphere slowly slides on the asthenosphere. *Slab pull* occurs at the subduction zone where the weight of the descending oceanic slab pulls the entire plate behind it. Which of the two processes, ridge push or slab pull, is more important? Calculations of the expected gravitational effects suggest that ridge push is less important than slab pull. In addition, scientists have observed that plates with large subducting slabs at their edges tend to move much more rapidly than plates with small

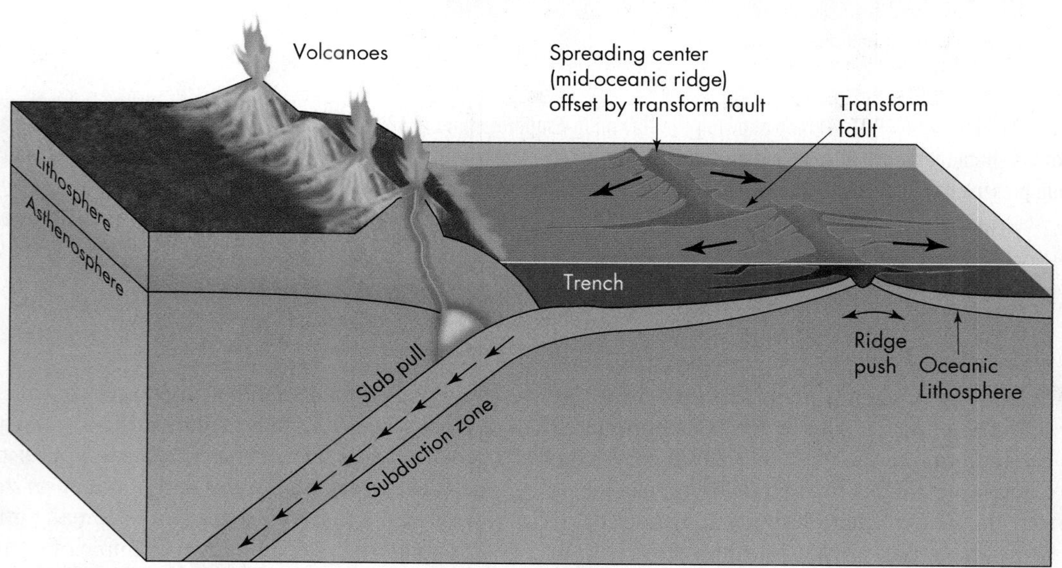

▲ **FIGURE 2.21 PUSH AND PULL IN MOVING PLATES** Diagram showing the mechanisms of ridge push and slab pull in facilitating the movement of lithospheric plates. Both are gravity-driven processes. The heavy lithosphere moves down the slope of the mid-ocean ridge and is pulled down through the lighter, hotter mantle at the subduction zone. *(Cox, A., and R. B. Hart. 1986.* Plate Tectonics. *Boston, MA: Blackwell Scientific Publications)*

ones. Thus, slab pull may be much more important than ridge push in moving plates.

2.5 Plate Tectonics and Hazards

The importance of the tectonic cycle to natural hazards cannot be overstated. All of us are affected by plate tectonics. As plates slowly move a few centimetres each year, so do the continents and ocean basins. Earth's resources (oil, gas, and minerals) are intimately associated with the tectonic history of the planet, and the linkage of hazardous events to plate tectonics is obvious. Most earthquakes and volcanoes that threaten people are near or at plate boundaries; most tsunamis are generated by subduction-zone earthquakes; and landslides are concentrated in mountains produced by plate collisions. Beyond this, several important conclusions can be drawn:

- The dominant hazard in Iceland, which lies on the Mid-Atlantic Ridge, is volcanic eruptions. There, active volcanoes underlie thick glacier ice. Subglacial lakes, which form as a result of heating from below, occasionally burst, producing large floods that threaten people and property along Iceland's coast. In addition, eruptions can inject large amounts of volcanic ash into the atmosphere, disrupting global air traffic, as happened during the eruption of Eyjafjallajokull in Iceland in April 2010 (see Chapter 5). That eruption cost airlines nearly 1 billion dollars over a six-day period.
- The earthquake hazard is appreciable anywhere that a transform fault crosses land; notable examples are California (San Andreas fault), the South Island of New Zealand (Alpine fault), and northern Turkey (North Anatolian fault). These areas are commonly hilly or mountainous and thus subject to landslides.
- Convergent plate boundaries, where one plate dives beneath another or where two plates collide, are particularly prone to earthquakes. Dangerous active volcanoes are common along subduction zones. Examples include volcanoes of the Aleutian Arc in Alaska, the Andes in South America, the mountainous spine of Japan, and the Cordillera of northwest North America. Earthquakes are also common where two continental plates collide. The high mountains that mark the suture zones of these collisions are prone to landslides and floods.

REVISITING THE FUNDAMENTAL CONCEPTS

Internal Structure of Earth and Plate Tectonics

1. **Hazards can be understood through scientific investigation and analysis.**
2. **An understanding of hazardous processes is vital to evaluating risk.**
3. **Hazards are commonly linked to each other and to the environment in which they occur.**
4. **Population growth and socio-economic changes increase the risk from natural hazards.**
5. **Damage and loss of life from natural disasters can be reduced.**

1. The plate tectonic theory revolutionized earth science in the 1960s and 1970s and continues to provide a paradigm for scientific advances. Along with a new understanding of Earth's internal structure provided by seismology and geophysics, the plate tectonic paradigm has catalyzed scientific research on hazards—notably earthquakes, tsunamis, and volcanoes. This vastly improved understanding has helped shed light on the risks that these hazards pose to people.
2. Risk assessment research has matured in recent decades. Although current understanding of risk is largely grounded in the social sciences, geologists and geophysicists have made important contributions to risk assessment through research on hazardous processes. This research has benefited from new ideas stimulated by the "plate tectonic revolution." For example, because scientists know the locations and types of plate boundaries and the rates of plate movements, they are able to better forecast potentially damaging earthquakes, which is essential for reducing earthquake risk.
3. Convection in Earth's mantle and associated lithospheric plate movements explain where and why earthquakes and volcanic eruptions occur. The plate tectonic theory also explains the spatial differences in the types of earthquakes and volcanoes on Earth. For example, the volcanoes that are most explosive and thus most dangerous are located along the back sides of subduction zones. The largest earthquakes on Earth also occur at subduction zones and are responsible for most disastrous tsunamis.
4. The risk posed by earthquakes, volcanic eruptions, and tsunamis is

(continued)

growing because human populations in areas near plate boundaries are increasing. Many megacities are located at or near active plate boundaries and are particularly vulnerable to these natural hazards. Even with appropriate mitigation and better education, the risk to people and property in these areas is increasing.

5. Loss of life and property damage from earthquakes, volcanic eruptions, and tsunamis can be minimized once these hazards are fully understood. The plate tectonic theory provides a framework for scientific research on these hazardous phenomena. This research eventually may lead to a better understanding of earthquakes and volcanic eruptions and their effects.

Summary

Our understanding of Earth's interior is based largely on the science of seismology. Scientists have identified and characterized the major layers of our planet, including the inner core, outer core, mantle, and crust. The uppermost layer of Earth, or lithosphere, is relatively strong and rigid compared to the asthenosphere beneath it. The lithosphere is broken into large pieces called plates that move relative to one another, typically at rates of a few centimetres per year. As these plates move, they carry along the continents embedded within them. Plate tectonic processes have produced continents, ocean basins, mountain ranges, and large plateaus. Ocean basins are created by seafloor spreading and destroyed by subduction; both processes result from convection in the mantle.

There are three types of plate boundaries: divergent (mid-ocean ridges or spreading centres), convergent (subduction and suture zones), and transform faults. At some locations, three plates meet at triple junctions.

Evidence in support of seafloor spreading includes paleomagnetic data, the configurations of hot spots and linear chains of volcanoes associated with them, and reconstructions of past continents.

The driving forces in plate tectonics are ridge push and slab pull. Scientists believe that the process of slab pull is more important than ridge push in moving tectonic plates from spreading centres to subduction zones.

Plate tectonics determines where and how frequent volcanic eruptions, earthquakes, tsunamis, and landslides happen on Earth.

Key Terms

asthenosphere (p. 27)
continental drift (p. 29)
convection (p. 28)
convergent boundary (p. 35)
crust (p. 26)
divergent boundary (p. 31)
hot spot (p. 37)
inner core (p. 26)
lithosphere (p. 27)
lithospheric plates (p. 28)
magma (p. 28)
magnetic reversal (p. 38)
mantle (p. 26)
mid-ocean ridge (p. 29)
Moho (p. 27)
outer core (p. 26)
paleomagnetism (p. 38)
plate tectonics (p. 28)
seafloor spreading (p. 29)
seismograph (p. 28)
seismology (p. 28)
spreading centre (p. 29)
subduction zone (p. 29)
tectonics (p. 28)
tectonic cycle (p. 29)
transform boundary (p. 35)
transform fault (p. 35)
triple junction (p. 37)
Wadati-Benioff zone (p. 31)

Did You Learn?

1. How do the inner core and the outer core differ?
2. Summarize the differences between the lithosphere and the asthenosphere.
3. What is the difference between the lithosphere and the crust?
4. Name the three major types of plate boundaries.
5. Explain why Earth has a magnetic field.
6. Explain why normal and reversed bands of magnetism on the two sides of mid-ocean ridges are symmetrical.
7. What are hot spots?
8. Name J. Tuzo Wilson's two main contributions to the theory of plate tectonics.
9. Explain the difference between ridge push and slab pull.
10. Describe how the plate tectonic theory explains the distribution of earthquakes and volcanoes on Earth.

Critical Thinking Questions

1. If convection did not occur within Earth's mantle, what would the surface of Earth look like? *Hint*: Consider the distribution of continents and oceans and the height of continents under such a scenario.
2. In 1915, Alfred Wegener postulated that the continents drift. He based this hypothesis partly on the obvious fit of the east coast of South America and the west coast of Africa. His hypothesis was ignored or ridiculed by contemporary scientists. Why do you think scientists were unwilling to favourably consider Wegener's hypothesis?

MasteringGeology

MasteringGeology **www.masteringgeology.com**. Looking for additional review and test prep materials? Visit the Study Area in MasteringGeology to enhance your understanding of this chapter's content by accessing a variety of resources, including **Self-Study Quizzes, Geoscience Animations, GEODe Tutorials, RSS feeds, flashcards,** web links and an optional **Pearson eText.**

CHAPTER 3

▶ **DEATH IN IRAN** These houses in Bam in southwest Iran were constructed of masonry and were unable to withstand the violent shaking of a magnitude 6.6 earthquake in 2003. *(Majid/Stringer/Getty Images)*

Earthquakes

Learning Objectives

Earthquakes are a deadly and destructive natural phenomenon. We cannot yet predict earthquakes: They commonly occur without warning, leaving no time to evacuate or to take precautions to limit property damage. Your goals in reading this chapter should be to

- Know what an earthquake is and how seismologists determine its magnitude
- Understand earthquake processes, such as faulting, tectonic creep, and the formation and movement of seismic waves
- Know which regions are most at risk from earthquakes and why
- Understand the effects of earthquakes, such as shaking, ground rupture, tsunamis, and liquefaction
- Identify how earthquakes are linked to other natural hazards
- Know the important natural service functions of earthquakes
- Understand how people can minimize earthquake risk and take measures to protect themselves

The Toll of Earthquakes

Earthquakes are one of the greatest natural hazards: During the twentieth century alone, more than 2 million people died in earthquakes and from the fires, tsunamis, and landslides triggered by earthquakes (Figure 3.1, Table 3.1). The twenty-first century promises to be even more devastating: Large earthquakes in Japan in 2011, Chile and Haiti in 2010, southwest China in 2008, northern Pakistan in 2005, and Iran in 2003 killed over 400 000 people. This number does not include the approximately 230 000 people in 11 countries who lost their lives to the tsunami caused by the giant earthquake off the west coast of Sumatra in December 2004.

The most disastrous earthquake in modern times occurred in China in July 1976, when an entire city was destroyed and more than 255 000 people were killed in less than six minutes. More recently, an earthquake in southwest China in May 2008 killed about 87 500 people; some villages were buried in landslide debris. These events, however, pale in comparison to an earthquake in north-central China in 1556 that killed an estimated 800 000 people and stands as one of the worst natural disasters in recorded history. Some famous cities of antiquity, such as Corinth; the Bronze Age cities of Troy, Mycenae, and Knossos; and Alexandria were partially or totally destroyed by earthquakes. The great city of Harappa in the Indus Valley was destroyed sometime after 2000 B.C., ending its dominance in the region. And the ancient cities of Megiddo and Jericho, lying along one of Earth's great fault systems that extends from the Red Sea along the Dead Sea rift valley, were also destroyed by earthquakes.

The consequences of an earthquake depend on its magnitude, depth, direction of fault rupture, and distance from populated areas; the nature of the local earth materials; and engineering and construction practices. Differences in some of these factors explain why, in January 2010, a magnitude (**M**) 7.0 earthquake killed about 240 000 people in Haiti,[1] whereas earthquakes of about the same size in California in 1989[2] (**M** 6.9) and 1994[3] (**M** 6.7) each claimed fewer than 70 lives, and one in western Washington[4] in 2001 (**M** 6.9) had no fatalities.

▲ **FIGURE 3.1 DEVASTATING EARTHQUAKE** Survivors search for relatives and friends among the rubble of the city of Beichuan in southwest China. The town was destroyed by a powerful earthquake in May 2008. The earthquake and the landslides it triggered killed about 87 500 people. *(© Chien-Min Chung/Corbis)*

What can explain the large differences in the number of casualties and damage from these four earthquakes? First, they occurred at different depths within Earth. The Haiti earthquake and the two California earthquakes took place at shallow depths in the *crust*, whereas the Washington earthquake was relatively deep. In deep earthquakes some of the energy dissipates by the time the seismic waves reach the surface, resulting in less damage.

Second, and more importantly, building and zoning regulations determined the damage these earthquakes caused. In California and Washington, most of the buildings that were damaged were constructed more than 25 years ago, before building codes included strict earthquake guidelines aimed at minimizing damage from ground shaking. Unfortunately, many developing nations

TABLE 3.1 Earthquakes Causing More Than 10 000 Deaths, 1900–2013

Date	Location	Deaths	Magnitude
April 4, 1905	Kangra, India	19 000	7.5
August 17, 1906	Valparaiso, Chile	20 000	8.2
October 21, 1907	Central Asia	12 000	8.0
December 28, 1908	Messina, Italy	72 000 to 110 000	7.2
January 13, 1915	Avezzano, Italy	33 000	7.0
February 13, 1918	Guangdong, China	10 000	7.4
December 16, 1920	Haiyuan, China	200 000	7.8
September 1, 1923	Kanto, Japan	143 000	7.9
May 22, 1927	Gulang, China	41 000	7.6
August 10, 1931	Xinjiang, China	10 000	8.0
January 15, 1934	Bihar, India	10 700	8.1
May 30, 1935	Quetta, Pakistan	30 000	7.6
January 25, 1939	Chillan, Chile	28 000	7.8
December 26, 1939	Erzincan, Turkey	33 000	7.8
October 5, 1948	Ashgabat, Turkmenistan	110 000	7.3
July 10, 1949	Khait, Tajikistan	12 000	7.5
February 29, 1960	Agadir, Morocco	12 000 to 15 000	5.7
September 1, 1962	Qazvin, Iran	12 000	7.1
August 31, 1968	Dasht-e Bayaz, Iran	7000 to 12 000	7.3
January 4, 1970	Tonghai, China	10 000	7.5
May 31, 1970	Chimbote, Peru	70 000	7.9
May 10, 1974	China	20 000	6.8
February 4, 1976	Guatemala	23 000	7.5
July 27, 1976	Tangshan, China	255 000	7.5
September 16, 1978	Iran	15 000	7.8
September 19, 1985	Michoacan, Mexico	10 000	8.0
December 7, 1988	Spitak, Armenia	25 000	6.8
June 20, 1990	Western Iran	40 000 to 50 000	7.4
September 29, 1993	Latur-Killari, India	10 000	6.2
August 17, 1999	Turkey	17 000	7.6

(continued)

TABLE 3.1 (continued)

Date	Location	Deaths	Magnitude
January 26, 2001	Gujarat, India	20 000	7.6
December 26, 2003	Southeast Iran	31 000	6.6
December 26, 2004	South Asia	230 000	9.1
October 5, 2005	Pakistan	86 000	7.6
May 12, 2008	Southwest China	87 500	7.9
January 12, 2010	Haiti	223 000	7.0
March 11, 2011	Japan	>15 870[1]	9.0

[1] Nearly all the casualties were the result of a tsunami triggered by the earthquake.

Source: From the U.S. Geological Survey. 2010. "Earthquakes with 1000 or more deaths since 1900." http://earthquake.usgs.gov/earthquakes/world/world_deaths.php. Accessed July 28, 2013.

do not have rigorous building codes or, if they do, the codes are commonly ignored or circumvented. Following the Haiti earthquake, studies revealed that builders used poor materials and ignored building codes.

Finally, in Port-au-Prince, Haiti, as in many cities, it was not just *how* structures were built that affected the number of casualties but also *where* they were built. Homes and other buildings located on the crests and slopes of hills above the city were subjected to particularly high levels of ground shaking during the earthquake because damaging seismic waves were amplified at these locations.

Although we cannot control the geologic environment or depth of earthquakes, we can do much to reduce the damage and loss of life they cause. The deaths in Haiti are especially tragic because much of the devastation could have been prevented if buildings had been properly constructed and the protection of people had been a priority. A lesson provided by the California and Washington earthquakes is that we can save lives by designing and building structures to withstand earthquake shaking.

3.1 Introduction to Earthquakes

Earthquakes result from the rupture of rocks along a **fault**, which is a fracture in Earth's crust. Rocks on opposite sides of a fault move suddenly, and energy is released in the form of **seismic waves**. The **epicentre** of the earthquake is the point on the surface of Earth directly above the fault rupture (Figure 3.2). The **focus**, or *hypocentre*, is the location of the initial rupture along the fault, directly below the epicentre. At the instant of the rupture, seismic waves radiate outward in all directions from the focus.

Millions of earthquakes occur worldwide every year, but only a small percentage are felt more than a few kilometres from their source, and even fewer cause damage (Table 3.2).

Let's begin our discussion of earthquakes by examining how they are measured and compared. **Seismologists**—scientists who study earthquakes—use the term **magnitude** to express the amount of energy an earthquake releases. Earthquake **intensity** provides a measure of the effects of the earthquake on people and structures. Earthquake ground motions are recorded by an instrument called a **seismograph**.

Earthquake Magnitude

The size, or magnitude, of an earthquake is expressed as a decimal number (e.g., 6.8). In 1931, the Japanese seismologist Kiyoo Wadati devised the first quantitative magnitude scale; in 1935 it was further developed by Charles F. Richter at the California Institute of Technology. The scale became known in the popular press as the **Richter scale**. It quantified the magnitude of local (California) earthquakes as the logarithm to the base 10 of the maximum signal wave amplitude recorded on a then-standard seismogram at a distance of 100 km from the epicentre.

Although some news reports still refer to the Richter scale, the term is no longer used by seismologists. Other, better magnitude scales have been developed since Richter's time. For example, a *body-wave scale* (M_b) is based on the strength of a type of earthquake wave that travels through Earth (see P wave, later in this chapter). The body-wave scale is used to measure the magnitude of deep earthquakes. Another scale, the *surface-wave scale* (M_s), is based on earthquake waves that travel along Earth's surface. Today, the most commonly used measure of earthquake size is **moment magnitude** (M_w), which is determined from the area that ruptured along a fault plane during the quake, the amount of movement or slippage along the fault, and the rigidity of the rocks near the focus. An increase from one whole number to the next higher one represents a 10-fold increase in the amount of shaking and about a 32-fold increase in the amount of energy released (Figure 3.3, Table 3.3). For example, the amount of ground motion from an **M** 7 earthquake is 10 times that of an **M** 6 earthquake, but the amount of energy released is 32 times as much. If we compare an **M** 5 with an **M** 7 earthquake, the differences are much greater. The energy released is 32 × 32, or about 1000 times greater. About 33 000 (32 × 32 × 32) shocks of **M** 5 are required to release as much energy as a single earthquake of **M** 8!

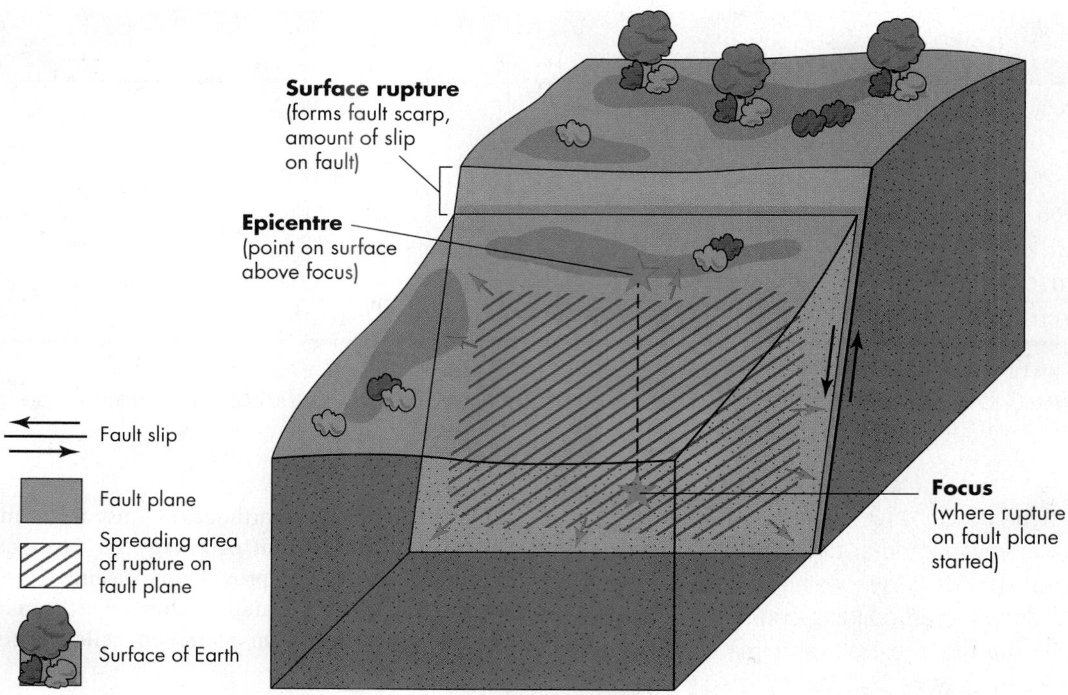

▲ **FIGURE 3.2 BASIC EARTHQUAKE NOMENCLATURE** This block diagram shows a fault plane (light tan surface), amount of displacement, rupture area (closely spaced diagonal lines), focus (lower red star), and epicentre (upper red star). The rupture starts at the focus and propagates up, down, and laterally along the fault plane (red arrows). During a great earthquake, slip may be 10 m to 30 m along a fault length of several hundred kilometres, and the rupture area may be more than 10 000 km^2.

Except for very large earthquakes, the Richter magnitude is approximately equal to the moment magnitude. Therefore, we will refer to the size of an earthquake simply as its magnitude, represented by the symbol **M**, without specifying Richter or moment magnitude.

Earthquakes are given descriptive adjectives to further characterize their magnitude (Table 3.3). Most damaging earthquakes are major (**M** 7–7.9) or strong (**M** 6–6.9). Major earthquakes can cause widespread and serious damage. Strong earthquakes can also cause considerable damage, but over a smaller area than a major quake. Fortunately, great (**M** 8) and giant (**M** 9) earthquakes are uncommon; the worldwide average is one **M** 8 quake per year and about three **M** 9 quakes per century (Table 3.4). By contrast, more than 1 million *very minor earthquakes* (**M** less than 3) occur each year. Most of these quakes are too small or too distant to be felt by people.

Earthquake Intensity

Intensity is another measure of earthquake size. The intensity of an earthquake at any location depends on magnitude, distance from the epicentre, and the nature of the ground at the site. The **Modified Mercalli Intensity Scale** measures the degree to which an earthquake affects people, property, and the ground. The 12 categories on this scale are assigned Roman numerals (Table 3.5). Each category contains a description of how people perceive the shaking from an earthquake and the extent of damage to buildings and other structures. For example, the 1971 Sylmar earthquake in the San Fernando Valley, California, had a single magnitude (6.6), but its Mercalli Intensity ranged from I to VII. Similarly, the 1949 Queen Charlotte Island earthquake, Canada's largest quake, had a magnitude of 8.1, with Modified Mercalli Intensities up to IX.

Earthquake intensities are commonly shown on maps. Conventional Modified Mercalli Intensity maps take days or even weeks to complete. They are based on newspaper articles, reports from damage assessment teams, and questionnaires sent to residents near the epicentre. The procedure, however, is changing. Intensity maps for earthquakes in parts of Canada and the United States are now created using the Internet. People access web pages of the Geological Survey of Canada and U.S. Geological Survey (USGS) to electronically submit forms detailing their experience in an earthquake. Based on this information, the two agencies rapidly generate earthquake intensity maps. The online *community internet intensity maps* produced by the USGS are updated every few minutes.

A major challenge during a destructive earthquake is to quickly determine where the damage is most severe. This information is now available in parts of California, the Pacific Northwest, and Utah, where there are dense networks of seismograph stations. The stations transmit direct measurements of ground motion during the earthquake.

TABLE 3.2 Major Earthquakes in Canada and the United States

Year	Location	Damage (U.S. $millions)	Number of Deaths
1811–1812	New Madrid, Missouri	Unknown	Unknown
1886	Charleston, South Carolina	23	60
1906	San Francisco, California	524	3000[1]
1925	Santa Barbara, California	8	13
1929	Sea floor off Newfoundland	Unknown	28
1933	Long Beach, California	40	115
1940	Imperial Valley, California	6	9
1946	Vancouver Island, British Columbia	Several million	2
1949	Haida Gwaii, British Columbia	Sparsely populated area	0
1952	Kern County, California	60	12
1959	Hebgen Lake, Montana (damage to timber and roads)	11	28
1964	Prince William Sound, Alaska (includes tsunami damage near Anchorage and on the Pacific Coast of Canada and the United States)	500	128
1965	Puget Sound, Washington	13	7
1971	Sylmar (San Fernando), California	553	65
1983	Coalinga, California	31	0
1983	Central Idaho	15	2
1987	Whittier, California	358	8
1989	Loma Prieta (San Francisco), California	6000	63
1992	Landers, California	271	1
1994	Northridge, California	20 000	60
2001	Nisqually, Washington	2000	0
2002	South-Central Alaska	Sparsely populated area	0
2012	Haida Gwaii, British Columbia	Sparsely populated area	0

[1]Deaths from the earthquake and subsequent firestorm.

Source: U.S. Geological Survey Earthquake Hazards Program. http://earthquakes.usgs.gov. Accessed July 28, 2013.

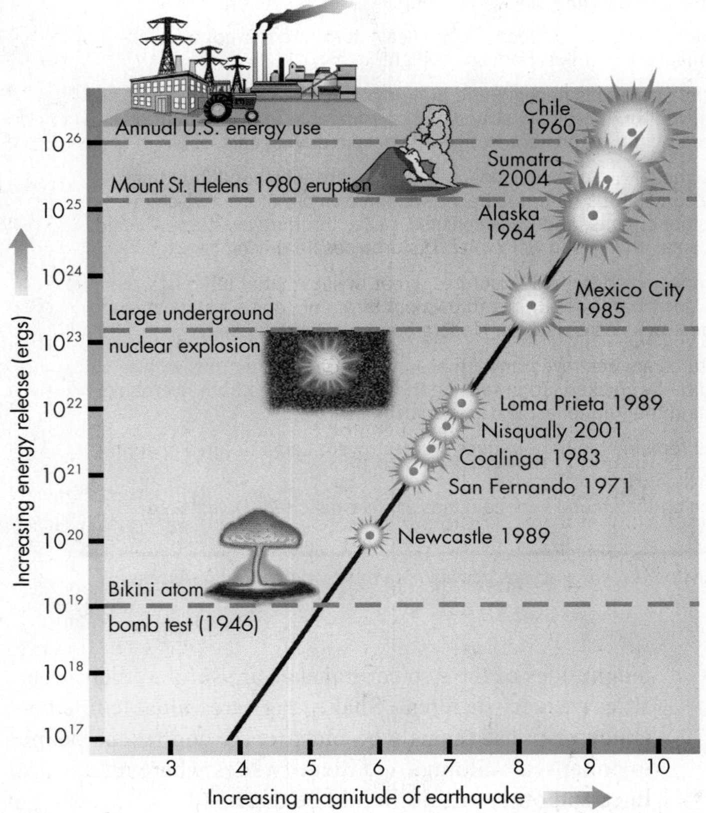

◀ **FIGURE 3.3 EARTHQUAKE ENERGY** This diagram shows the relation between earthquake magnitude and released energy, and comparisons with other energy sources. The energy released by the 1960 earthquake in Chile, the most powerful historic seismic event, was greater than the entire annual consumption of commercial energy in the United States. *(Clague, J., C. Yorath, R. Franklin, and B. Turner. 2006.* At Risk: Earthquakes and Tsunamis on the West Coast. *Vancouver, BC: Tricouni Press. Reprinted with permission.)*

TABLE 3.3 Relationships among Earthquake Magnitude, Displacement, and Energy

Magnitude Change	Ground Motion Change (Displacement[1])	Energy Change
1	10 times	About 32 times
0.5	3.2 times	About 5.5 times
0.3	2 times	About 3 times
0.1	1.3 times	About 1.4 times

[1]Displacement, vertical or horizontal, that is recorded on a standard seismograph.

Source: U.S. Geological Survey. 2009. "Earthquakes, facts and lists." http://neic.usgs.gov/neis/eqlists/eqstats.html. Accessed July 28, 2013.

TABLE 3.4 Magnitude and Frequency of Earthquakes Worldwide

Descriptor	Magnitude	Average Annual Number of Events
Great	8 and higher	1
Major	7–7.9	15
Strong	6–6.9	134
Moderate	5–5.9	1319
Light	4–4.9	13 000 (estimated)
Minor	3–3.9	130 000 (estimated)
Very minor	2–2.9	1 300 000 (estimated) (approximately 150 per hour)

Source: From the U.S. Geological Survey. 2009. "Earthquakes, facts and lists." http://neic.usgs.gov/neis/eqlists/eqstats.html. Accessed July 28, 2013.

TABLE 3.5 Modified Mercalli Intensity Scale (abridged)

Intensity	Effects
I	Felt by very few people.
II	Felt by only a few people at rest, especially on upper floors of buildings. Delicate suspended objects may swing.
III	Felt noticeably indoors, especially on upper floors of buildings, but many people do not recognize the shaking as an earthquake. Stationary cars may rock slightly. Vibration feels like the passing of a truck.
IV	During the day, felt indoors by many, outdoors by few. At night, some people awakened. Dishes, windows, doors disturbed; walls make cracking sound. Stationary cars rock noticeably. Sensation is that of a heavy truck striking a building.
V	Felt by nearly everyone; many people awakened. Some dishes and windows broken; a few instances of cracked plaster; unstable objects overturned. Disturbances of trees, poles, and other tall objects are sometimes noticed. Pendulum clocks may stop.
VI	Felt by all; many people frightened and run outdoors. Some heavy furniture moved; a few instances of fallen plaster or damaged chimneys. Damage is slight.
VII	Almost everybody runs outdoors. Damage is negligible in buildings of good design and construction; slight to moderate in well-built ordinary structures; considerable in poorly built or badly designed structures; some chimneys broken. Noticed by people driving cars.
VIII	Damage slight in specially designed structures; considerable in ordinary substantial buildings, with partial collapse; great in poorly built structures. Panel walls thrown out of frames; chimneys, factory stacks, columns, monuments, and walls collapse; heavy furniture overturned. Sand and mud ejected in small amounts; changes in well water. Disturbs people driving cars.
IX	Damage considerable even in specially designed structures; great in substantial buildings, with partial collapse. Well-designed frame structures thrown out of plumb; some buildings are shifted off foundations. Ground cracks conspicuous. Underground pipes are broken.
X	Some well-built wooden structures are destroyed; most masonry and frame structures with foundations destroyed; ground badly cracked. Train rails bent. Many landslides from riverbanks and steep slopes. Some sand and mud liquefies. Water is splashed over banks.
XI	Few, if any, masonry structures remain standing; bridges are destroyed. Large fissures open in ground. Landslides are common.
XII	Damage is total. Waves are seen on the ground surface. Lines of sight distorted. Objects are thrown into the air.

Source: U.S. Geological Survey Earthquake Hazards Program. http://earthquake.usgs.gov/learning/topics/mercalli.php. Accessed July 28, 2013.

This information, known as *instrumental intensity*, is used to immediately produce a *shake map*, which shows both perceived shaking and potential damage. Figure 3.4 shows shake maps for the 1994 **M** 6.7 Northridge and 2001 **M** 6.8 Nisqually earthquakes. Note that, although the magnitudes of the two earthquakes are similar, their intensities are very different. Shake maps are valuable to emergency response teams who must locate and rescue people in collapsed buildings and identify sites where natural gas lines and other utilities might be damaged.

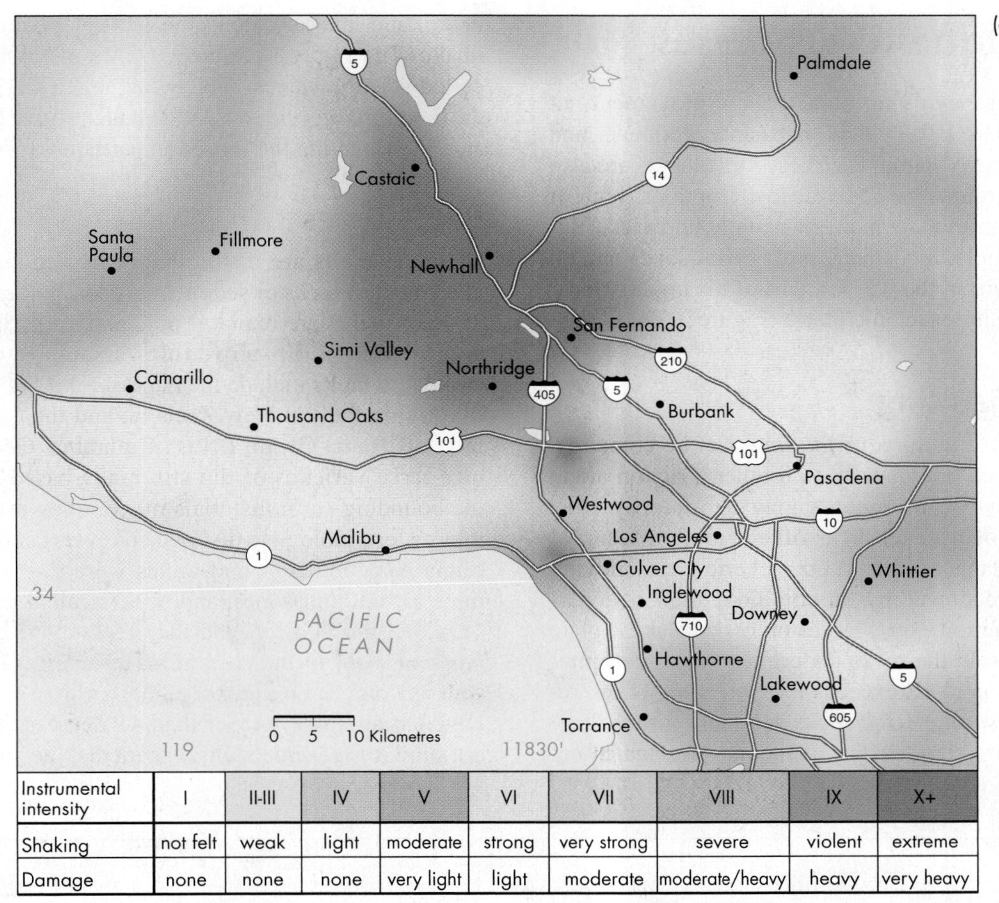

Instrumental intensity	I	II-III	IV	V	VI	VII	VIII	IX	X+
Shaking	not felt	weak	light	moderate	strong	very strong	severe	violent	extreme
Damage	none	none	none	very light	light	moderate	moderate/heavy	heavy	very heavy

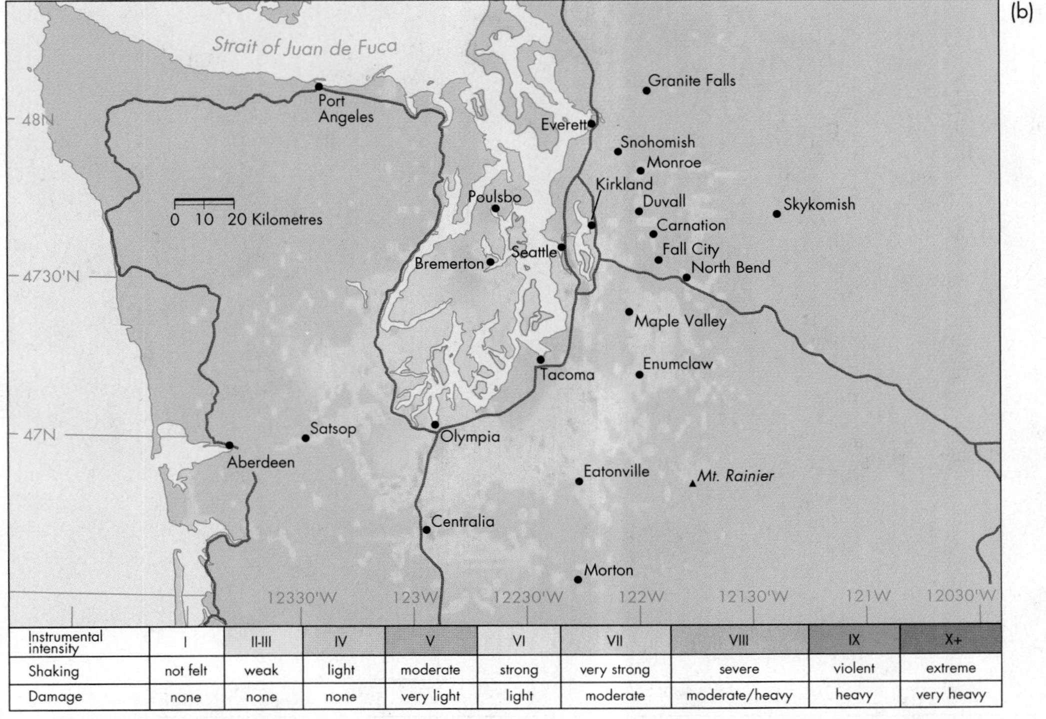

Instrumental intensity	I	II-III	IV	V	VI	VII	VIII	IX	X+
Shaking	not felt	weak	light	moderate	strong	very strong	severe	violent	extreme
Damage	none	none	none	very light	light	moderate	moderate/heavy	heavy	very heavy

▲ **FIGURE 3.4 SHAKE MAPS** Instrumental intensity maps of (a) the 1994 Northridge, California, earthquake (**M** 6.7) and (b) the 2001 Nisqually, Washington, earthquake (**M** 6.8). *((a) U.S. Geological Survey; courtesy of David Wald. (b) Pacific Northwest Seismograph Network, University of Washington)*

3.2 Earthquake Processes

Earth is a dynamic, evolving planet. Slow movements of lithospheric plates have formed ocean basins, continents, and mountain ranges (see Chapter 2). These movements are also responsible for earthquakes and volcanic eruptions, both of which are most common at and near plate boundaries (Figure 3.5). For example, in Canada, most earthquakes and all active volcanoes are in the western part of the country, near the boundaries of three tectonic plates (Figure 3.6).

Process of Faulting

The process of fault rupture, or *faulting*, can be compared to sliding two rough boards past each other. Friction along the boundary between the boards may temporarily slow their motion, but rough edges break off and motion occurs at places along the plane. Similarly, lithospheric plates moving past each other are slowed by friction along their boundaries. This braking action exerts forces on rocks near the plate boundary. As a result, the rocks undergo *strain* or deformation. When stress on rocks exceeds their breaking point, referred to as their *strength*, the rocks suddenly move along a fault. The rupture starts at the focus and propagates up, down, and laterally along the fault plane during the earthquake. The rupture produces waves of vibrational energy, called seismic waves, which can shake the ground. Faults are therefore *seismic sources*, and identifying them is the first step in evaluating the risk of an earthquake in a given area.

Fault Types

Geologic faults are distinguished by the direction of displacement of rocks or sediment bordering them (Figure 3.7). Displacements are mainly horizontal on **strike-slip faults** and vertical on **dip-slip faults**. Examples of well-known strike-slip faults include the San Andreas fault in California, the Alpine fault in New Zealand, and the Queen Charlotte fault off Haida Gwaii, British Columbia. Geologists recognize three varieties of dip-slip faults based on which way the bounding earth materials move. They use centuries-old mining terminology to distinguish reverse and normal faults. Many early underground mines were dug at an incline to mineralized fault zones, and miners called the block below their feet the *footwall* and the block above their heads the *hanging wall*. In the case of a **reverse fault**, the hanging wall has moved up relative to the footwall along a plane inclined at an angle steeper than 45 degrees. **Thrust faults** are similar to reverse faults, except that the angle of the fault

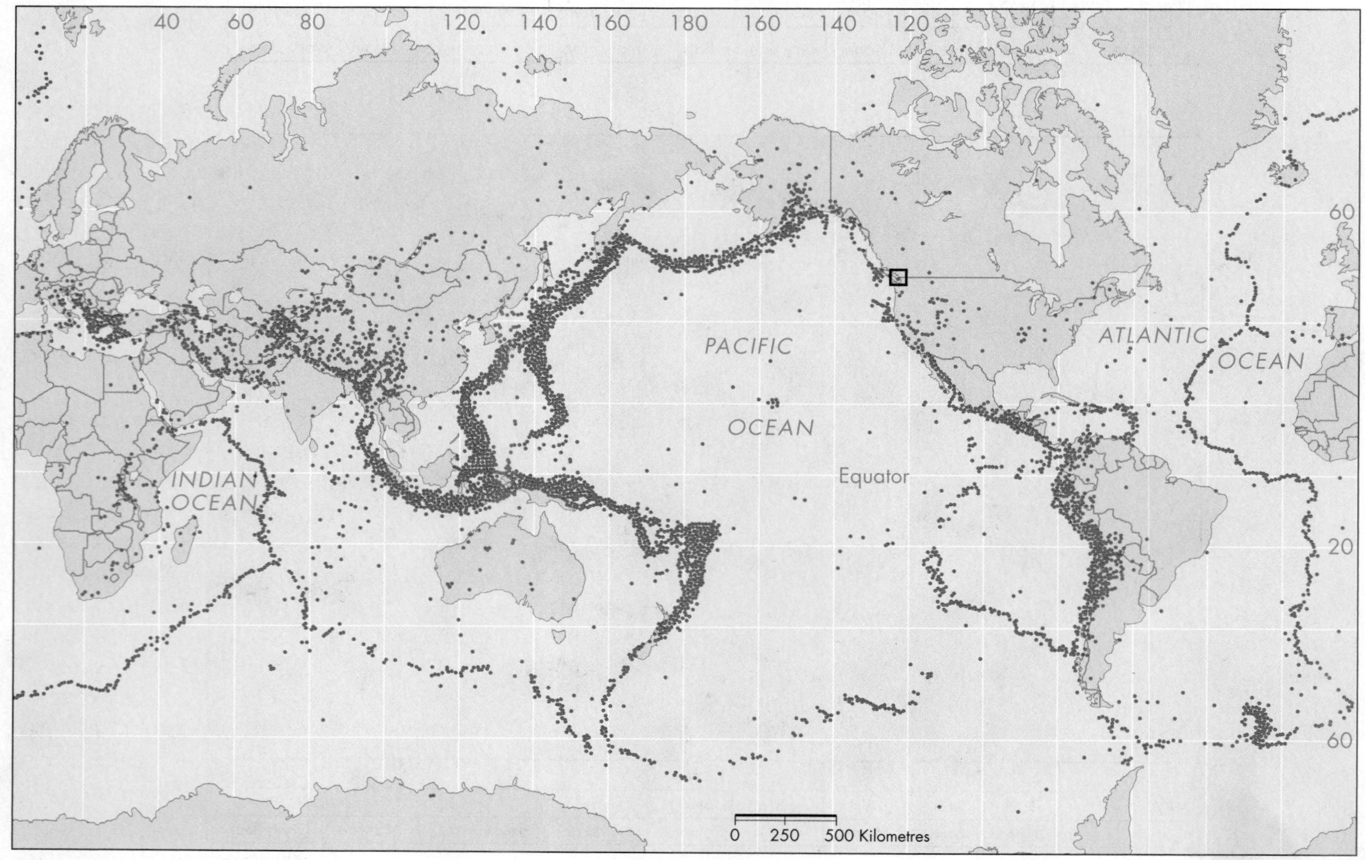

▲ **FIGURE 3.5 GLOBAL EARTHQUAKE DISTRIBUTION** A map of global seismicity (1963–1988, M 5+), showing epicentres of plate-boundary earthquakes (heavy concentration of dots within red zones) and intraplate earthquakes (isolated dots). Locations of plate boundaries are shown in Figure 2.5. Black square shows location of Figure 3.6. *(U.S. Geological Survey National Earthquake Information Center)*

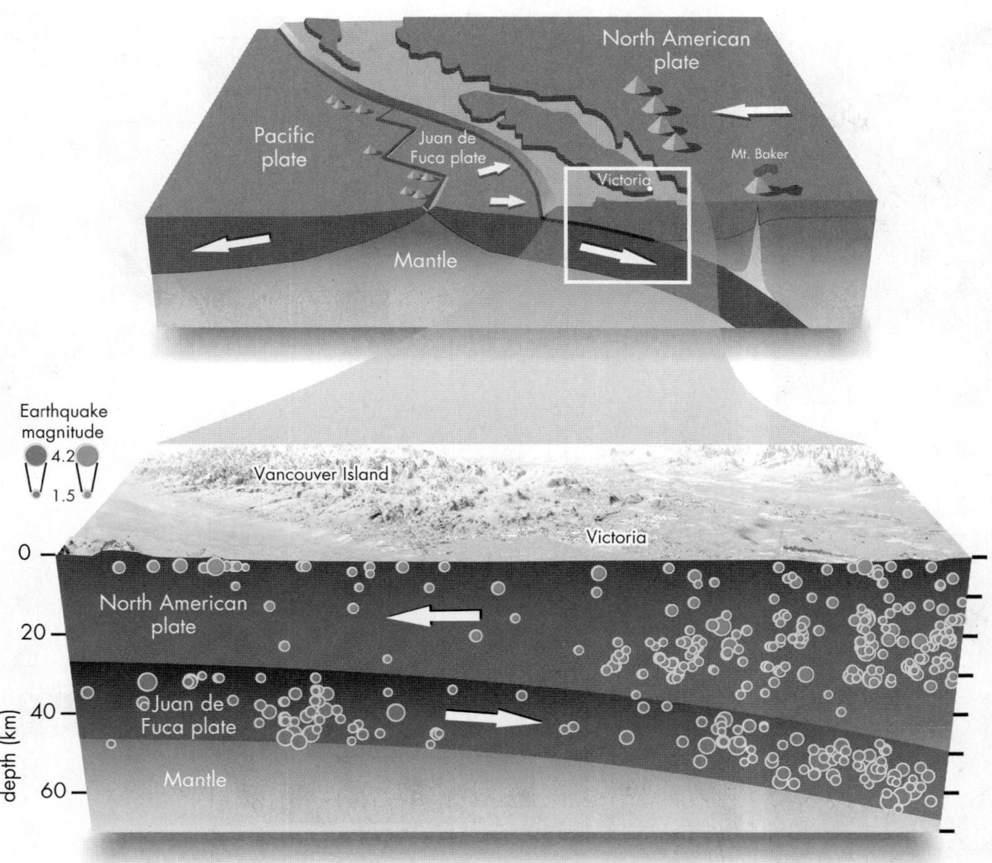

◄ **FIGURE 3.6 TECTONIC PLATES AND EARTHQUAKES, SOUTH-COASTAL BRITISH COLUMBIA** Section of Earth's lithosphere through southern Vancouver Island and the northern tip of Olympic Peninsula showing sources of earthquakes within the North American plate (blue dots) and the Juan de Fuca plate (red dots). White arrows show relative directions of plate motion. *(Based on Clague, J., C. Yorath, R. Franklin, and B. Turner. 2006.* At Risk: Earthquakes and Tsunamis on the West Coast. *Vancouver, BC: Tricouni Press)*

plane is 45 degrees or less. A **normal fault** is a dip-slip fault in which the hanging wall has moved downward relative to the footwall.

Until recently, geologists thought that most active faults extend to the ground surface. However, we now know that some faults are buried, or *blind*, and do not reach the surface. This discovery has made it more difficult to evaluate earthquake risk in some areas.

Fault Activity

Most geologists consider a particular fault to be *active* if it has moved during the *Holocene Epoch*—that is, the past 11 600 years—and *potentially active* if it shows evidence of movement during the *Pleistocene Epoch*, the past 2.6 million years. Faults that have not moved during the past 2.6 million years are generally classified as *inactive*. However, it can be difficult to determine when a fault was last active, especially if it has had no earthquakes within recorded history. In many cases, geologists must determine the *paleoseismicity* of the fault—that is, the prehistoric record of earthquakes. They do so by identifying faulted rock or sediment and determining the age of the most recent displacement.

Tectonic Creep and Slow Earthquakes

Some active faults exhibit **tectonic creep**, or *fault creep*, which is gradual movement along a fault without accompanying felt earthquakes. This process can slowly damage roads, sidewalks, building foundations, and other structures. Tectonic creep has damaged culverts under the University of California football stadium in Berkeley, California. Movement of 2.2 cm was measured beneath the stadium in only 11 years, and periodic repairs were necessary as the cracks developed.[5] More rapid creep has been recorded on the Calaveras fault near Hollister, California. A winery on that fault is slowly being pulled apart at a rate of about 1 cm per year.[6]

Slow earthquakes are similar to other earthquakes in that they are produced by fault rupture. The main difference is that the rupture, rather than being instantaneous, can last from days to months. The moment magnitude of slow earthquakes can be in the range of 6 to 7 when a large area of the fault ruptures, although the amount of slip is generally small, in the range of a centimetre or so. Slow earthquakes have been recognized through analysis of continuous geodetic measurements, similar to the handheld GPS devices that are used to identify your location, although with far superior accuracy. These instruments can differentiate movements of the ground surface in the millimetre range and have been used to identify small displacements from slow earthquakes.

Seismic Waves

Some seismic waves generated by fault rupture travel within the body of Earth and others travel along the surface.

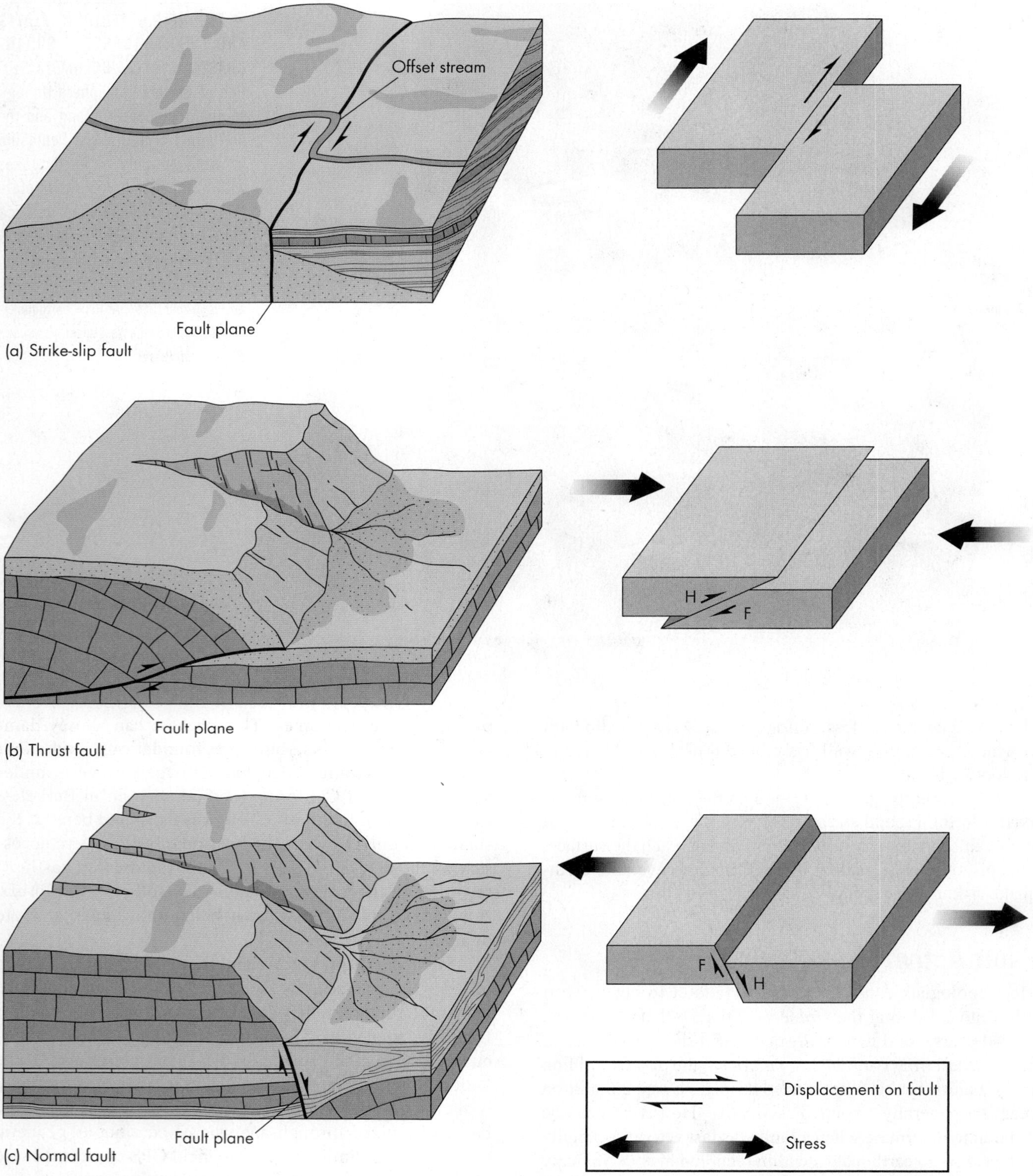

▲ **FIGURE 3.7 TYPES OF GEOLOGIC FAULTS** These schematic diagrams show the three common types of faults and their effects on the landscape. (a) Strike-slip fault with horizontal displacement along the fault plane. (b) Thrust fault in which the hanging wall (H) has moved up and over the footwall (F). (c) Normal fault in which the hanging wall (H) has dropped down. The coloured diagrams on the left show the landscape after movement along the fault (thick black line). The grey diagrams on the right show directions of stress (thick arrows) that produce the displacements (thin half arrows).

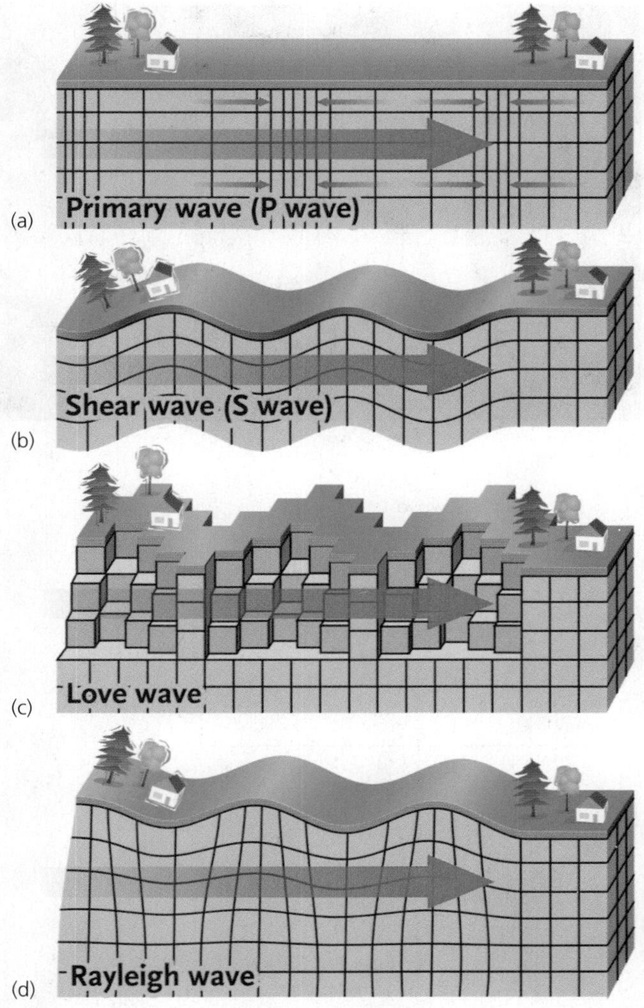

▲ **FIGURE 3.8 SEISMIC WAVES** Waves produced by earthquakes are of four types. Compressional or primary (P) waves are push-pull; shear or secondary (S) waves are like snapping a rope; Love waves whip the ground from side to side; and Rayleigh waves roll the ground like waves at sea. The large arrows indicate the direction of travel of the waves. *(Reprinted with permission of Tricouni Press)*

Body waves include P waves and S waves. **P waves,** also called *compressional* or *primary* waves, are the faster of the two types of body waves (Figure 3.8a). They can travel through any type of material, although their velocity in solids is much higher than in liquids. The average velocity of P waves in Earth's crust is 6 km/s, in contrast to 1.5 km/s in water. P waves propagate by repeated compression and expansion of the medium through which they are moving—essentially a push-pull effect. Interestingly, when P waves reach Earth's surface, they may produce sound that can be heard by people and some animals.[7] However, the sound that most people hear during an earthquake is the noise of objects vibrating, not the P waves themselves.

S waves, also called *shear* or *secondary* waves, travel only through solid materials (Figure 3.8b). They have an average velocity through Earth's crust of 3 km/s, half the velocity of P waves. S waves produce a back-and-forth motion at right angles to the direction the waves are moving. This movement is similar to the whipping motion of a long jump rope held by two people on a playground. When liquids are subjected to sideways shear, they are unable to spring back, explaining why S waves cannot travel through them.[7]

Complex **surface waves** form when P and S waves reach Earth's surface and move along it (Figures 3.8c and 3.8d). They travel more slowly than either P or S waves and cause much of the damage near the epicentre. Surface waves produce a complex horizontal and vertical movement or a rolling motion and, consequently, can damage walls and foundations of buildings, bridges, and roads. People near the epicentre of a strong earthquake have reported seeing these waves rippling across the land surface. One type of surface wave, called a *Love wave,* causes horizontal shaking that is especially damaging to foundations. Another type is the slower-moving *Rayleigh wave,* which travels with an elliptical motion, like rolling ocean waves; surface materials move vertically as the wave moves forward.

3.3 Earthquake Shaking

As mentioned at the beginning of this chapter, a variety of factors determine the severity of shaking people experience during an earthquake and what damage the earthquake causes. The main factors are (1) earthquake magnitude; (2) distance to the epicentre and focal depth; (3) the direction of rupture; and (4) local soil and rock conditions. Damage and loss of life from an earthquake also critically depend on engineering and construction practices and socio-economic conditions in the affected area.

Distance to the Epicentre and Focal Depth

Earthquake shaking generally decreases with distance from the epicentre. The epicentre is located using the P and S waves detected by seismographs (Figures 3.9a and 3.9b). The analog or digital record of these waves is called a *seismogram* (Figure 3.9c). Seismic waves appear on a seismogram as an oscillating line that resembles an electrocardiograph of a person's heartbeat (Figure 3.9c). Because P waves travel faster than S waves, they appear first on a seismogram. Seismologists use the difference between the arrival times of the first P and S waves to determine the distance of the epicentre from the seismograph. For example, in Figure 3.9c the first S waves arrived 50 seconds after the first P waves. Knowing the velocities of P and S waves, the seismologist concludes that the epicentre is about 420 km away. Seismographs around the world record different arrival times for P and S waves from the same earthquake; stations farthest from the epicentre record the greatest difference between P and S wave arrival times (Figure 3.9d).

From this information, it is apparent that the epicentre of an earthquake can be located by using P and S arrival times at seismographs at different locations (Figure 3.10). The distance

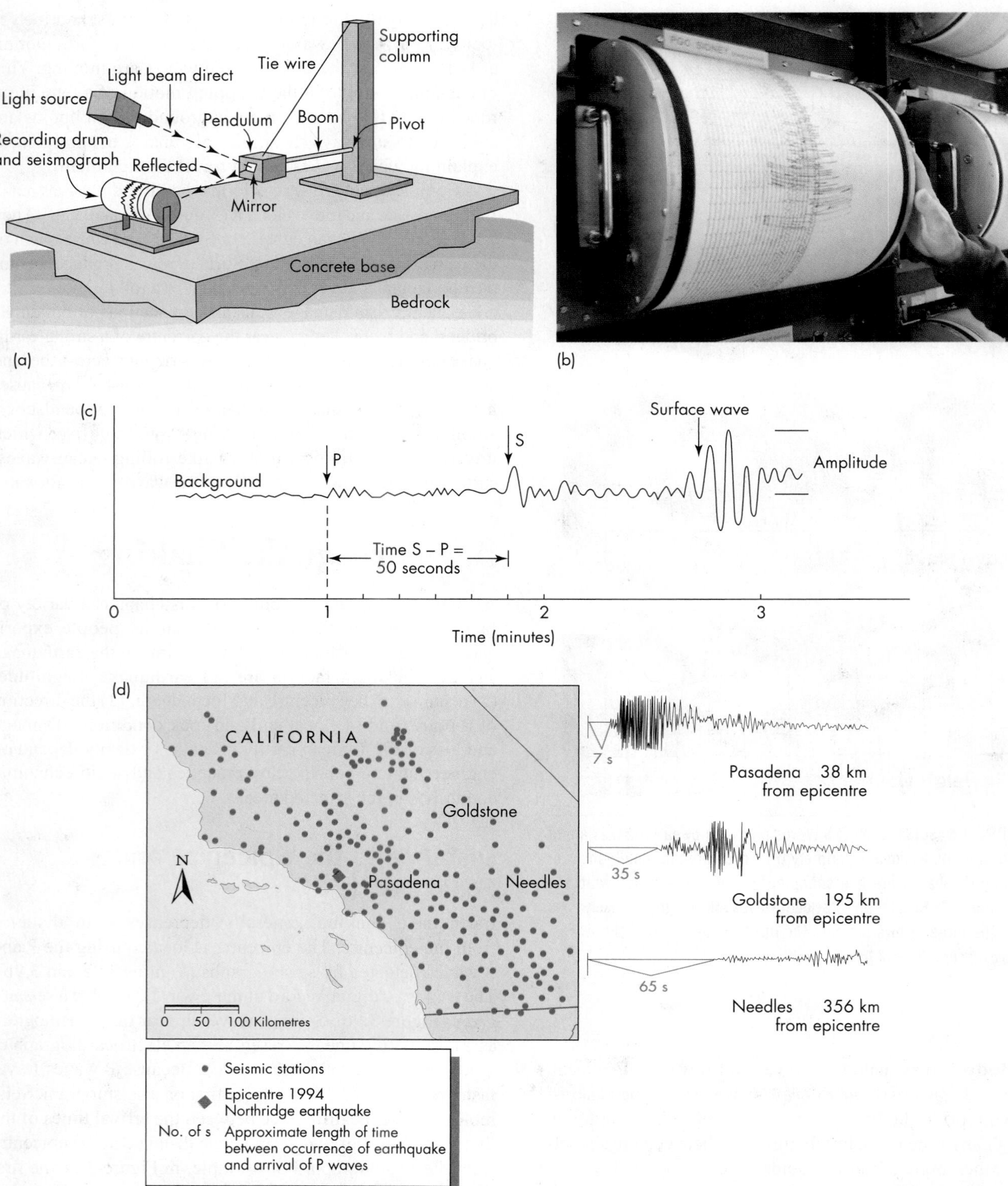

▲ **FIGURE 3.9 SEISMOGRAPHS AND SEISMOGRAMS** (a) A sketch showing how a seismograph works. (b) The recording drum of a seismograph at the Pacific Geoscience Centre near Victoria, British Columbia, showing the seismogram of the Washington State earthquake of February 28, 2001 (**M** 6.8). (c) A seismogram of an earthquake, showing successive arrivals of P, S, and surface waves. The difference of 50 seconds in the first arrivals of P and S waves tells us that the epicentre is about 420 km from the seismograph. (d) Three seismograms of the 1994 Los Angeles (Northridge) earthquake, recorded at different distances from the epicentre. The seismic waves take longer to reach the seismograph and their amplitudes decrease with increasing distance from the epicentre. The greater the amplitude of the waves on the seismogram, the stronger the ground shaking. *(Ian McKain/AP Images); (Based on Southern California Earthquake Center)*

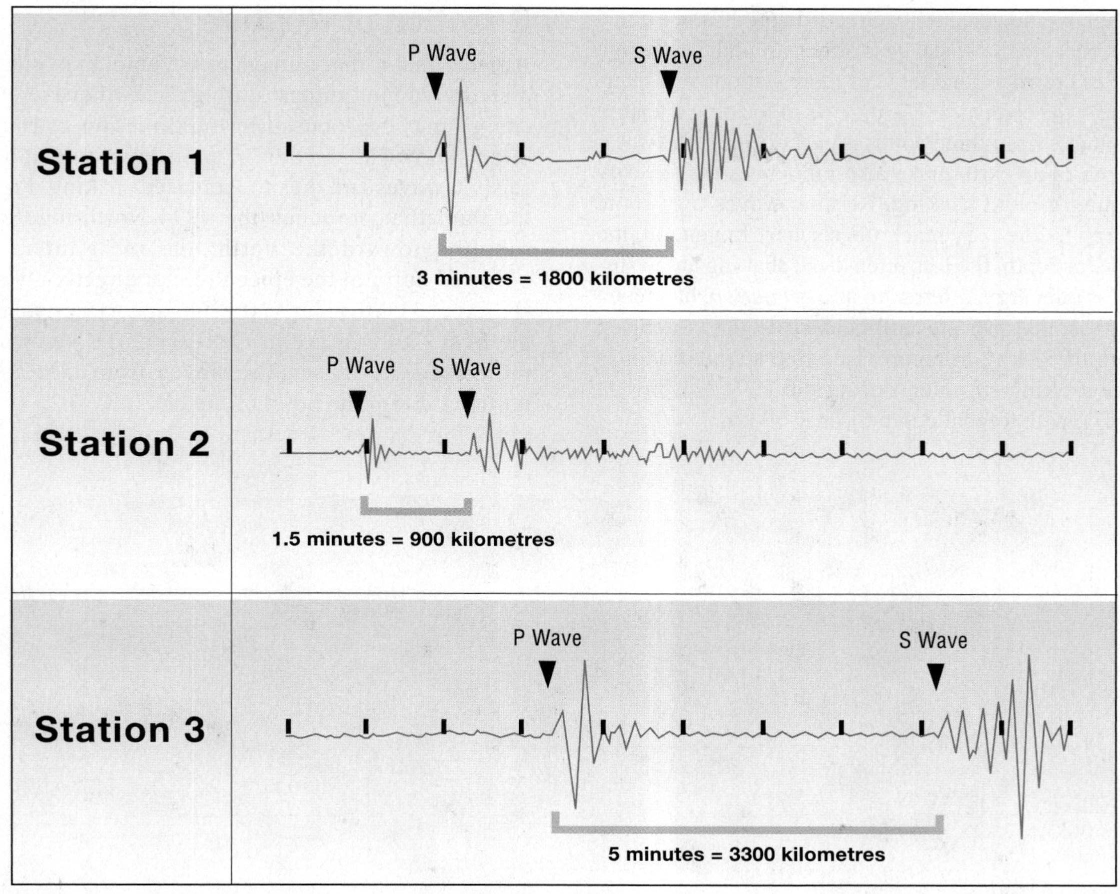

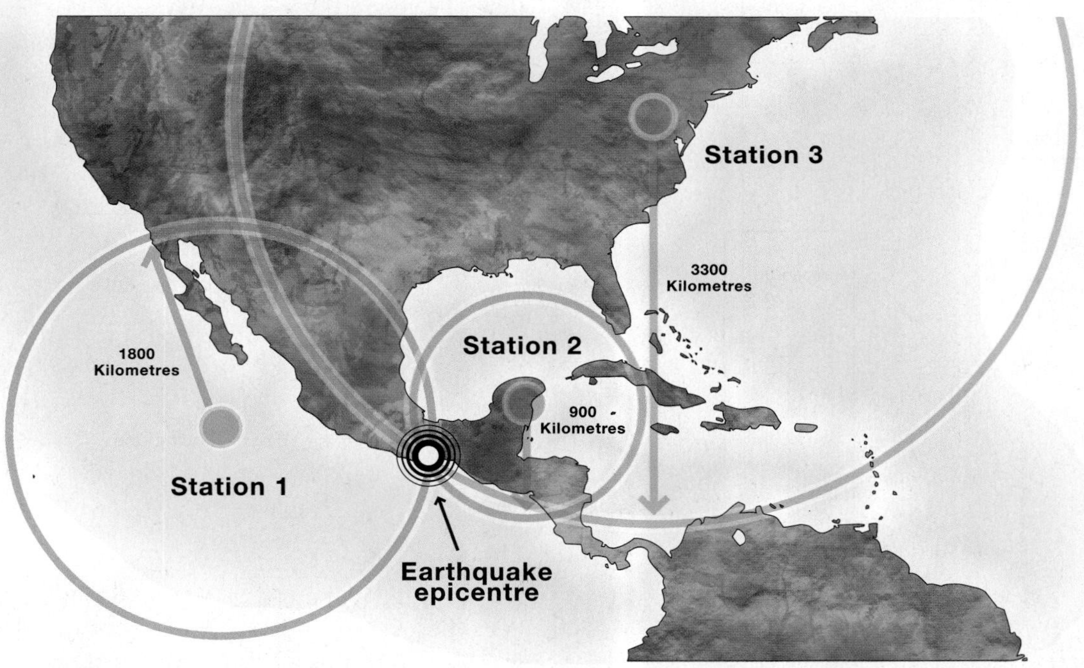

▲ **FIGURE 3.10 LOCATING AN EARTHQUAKE** The epicentre of an earthquake can be determined from the arrival times of P and S waves at three or more widely separated sites. P and S waves travel at different velocities; thus the time between their first arrivals at a site provides a measure of the distance from the recording seismograph to the epicentre. A circle with a radius equal to the distance from the epicentre is drawn around each seismograph station. The intersection of the three circles is the epicentre. *(Adapted from* How Are Earthquakes Located? *IRIS Education & Outreach Series, No. 6. http//www.iris.edu)*

to the epicentre is calculated for each seismograph, and a circle with a radius equal to that distance is drawn around the seismic station. Circles drawn around three or more stations will intersect at a point—the epicentre. The process of locating a feature by using distances from three points is called *triangulation*.

The depth of an earthquake also influences the severity of the resulting ground shaking. Seismic waves lose some of their energy before they reach the surface. In general, the greater the focal depth, the less intense the shaking at the surface. This loss of energy, referred to as *attenuation*, happened in the 2001 Washington State earthquake (**M** 6.8), which had a focal depth of 52 km. In comparison, the attenuation was less, and the shaking stronger, during the 1994 Northridge quake (**M** 6.7), with a focal depth of only 19 km.

Direction of Rupture

Another factor that influences the amount of shaking is the direction of fault rupture during the earthquake. Earthquake energy may be focused in the direction of rupture. This effect, known as *directivity*, contributes to amplification of seismic waves and thus to increased shaking. For example, the fault that produced the 1994 Northridge earthquake ruptured toward the north; the most intense shaking occurred north of the epicentre, not directly over the focus (Figure 3.11). Similarly, the damage to Christchurch from the **M** 6.3 earthquake in February 2011 was worsened by the directivity of seismic waves from the fault rupture northward into the heart of the city.

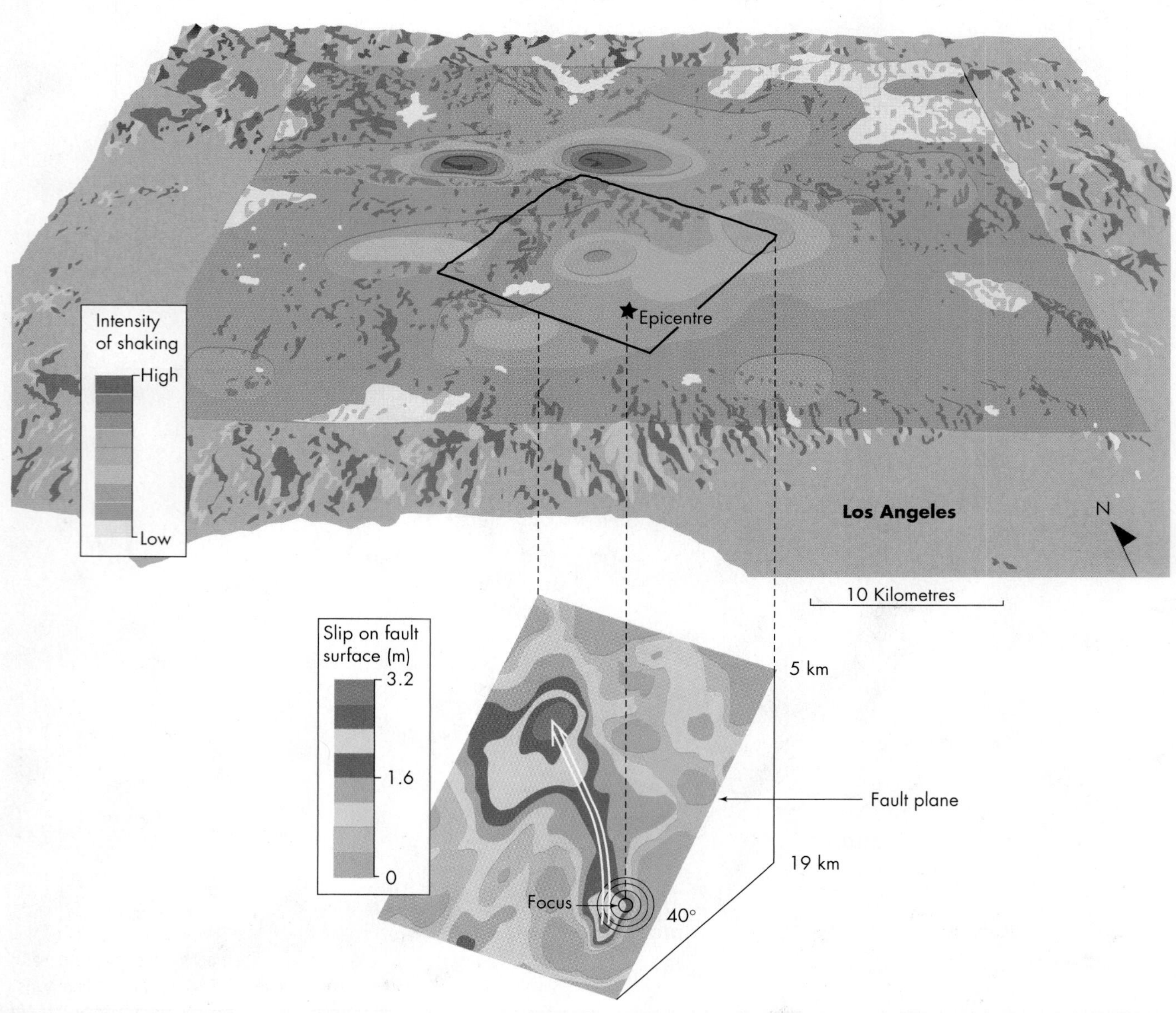

▲ **FIGURE 3.11 DIRECTIVITY OF FAULT RUPTURE** View to the north of the Los Angeles basin, showing the epicentre and intensity pattern of the 1994 Northridge earthquake. The red area 10 km north of the epicentre experienced the most intense shaking. The lower part of the figure shows the focus and a section of the fault plane that ruptured during the earthquake. Colours on the fault plane indicate the amount of slip during the quake. Slip was greatest in the reddish-purple zone northwest of the focus. The rupture started at the focus and progressed to the north in the direction of the white arrow. *(U.S. Geological Survey. 1996.* USGS Response to an Urban Earthquake, Northridge '94. *U.S. Geological Survey Open File Report 96–263)*

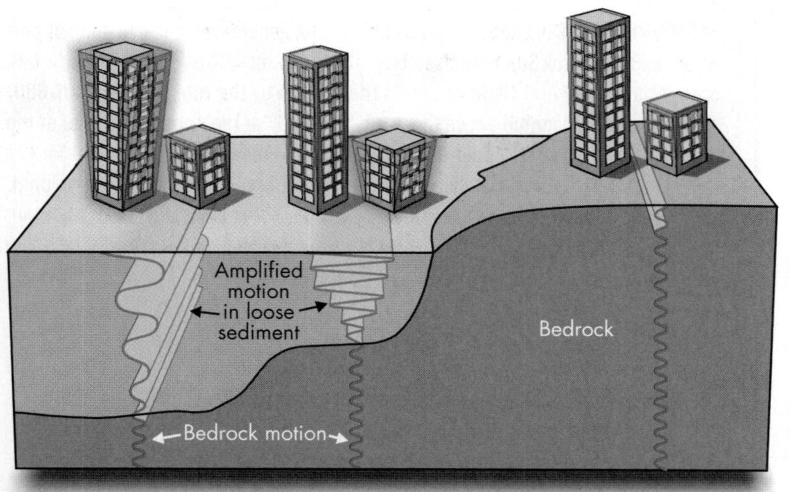

◄ FIGURE 3.12 AMPLIFICATION OF GROUND SHAKING This schematic diagram shows the relationship between near-surface geologic materials and ground motion amplification during an earthquake. Amplification is highest in water-saturated sediment. *(Reprinted with permission of Tricouni Press)*

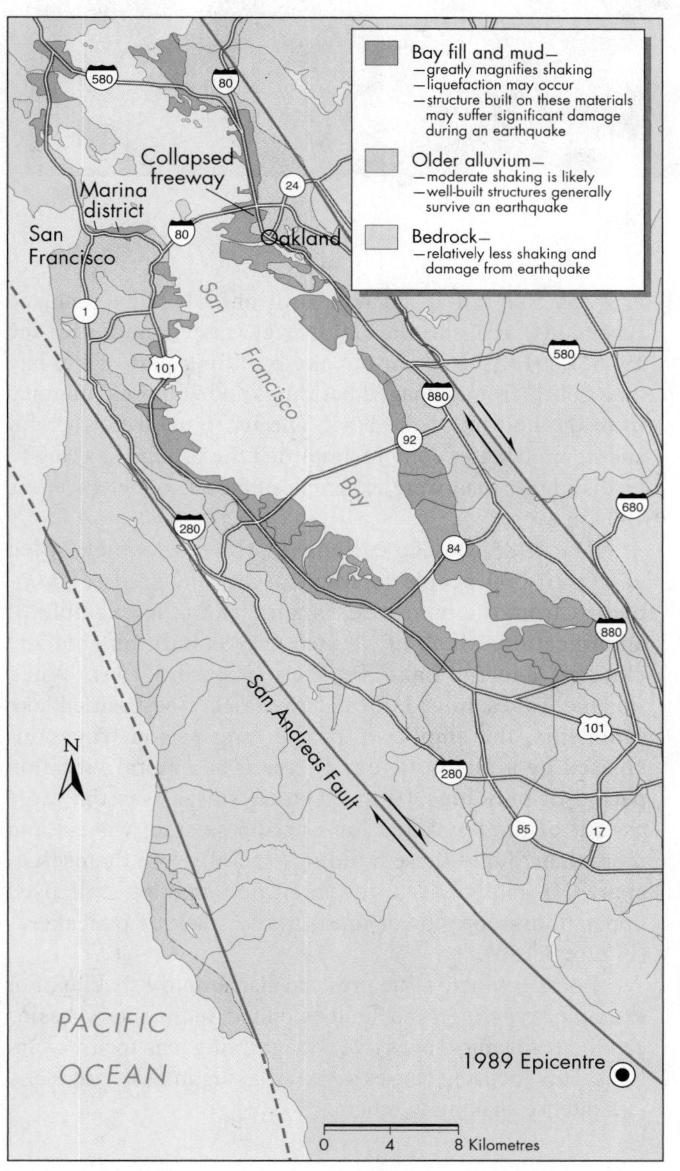

◄ FIGURE 3.13 LOMA PRIETA EARTHQUAKE A simplified geologic map of the San Francisco Bay area, showing the locations of the San Andreas fault zone and the epicentre of the 1989 Loma Prieta earthquake. Fault zones are shown as solid and dashed red lines; arrows indicate displacement directions. The most severe shaking occurred along the San Francisco Bay shoreline and where the bay had been filled in to create new land (bay fill shown in dark orange). Shaking caused the collapse of the Nimitz Freeway in Oakland and damaged buildings in the Marina district of San Francisco (northwest part of map). *(Modified after T. Hall from U.S. Geological Survey)*

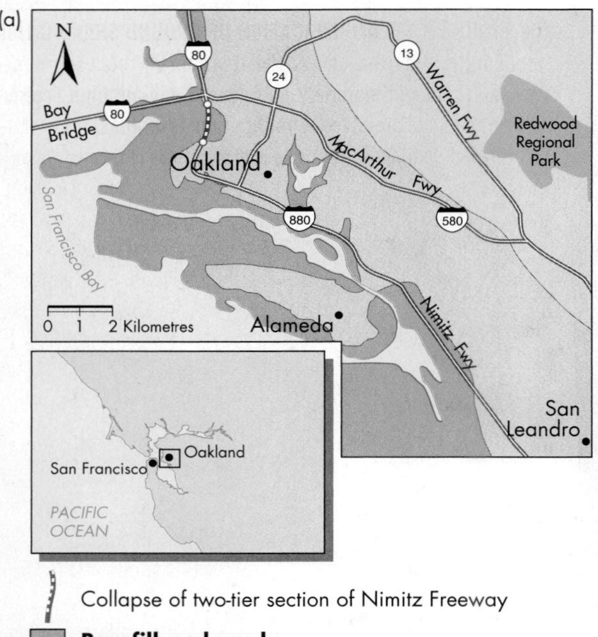

◄ **FIGURE 3.14 COLLAPSE OF A FREEWAY** (a) A generalized geologic map of part of the east shore of San Francisco Bay, showing areas of bay fill and mud (dark orange) and alluvium (light orange). The section of the Nimitz Freeway (I-880) that collapsed is located above the word "Oakland" in the upper left corner of the map. (b) The collapsed upper deck of the Nimitz Freeway. The road level for the lower deck was approximately at the level of the break in the concrete column. *((a) Based on Hough, S. E., P. A. Friberg, R. Busby, E. F. Field, K. H. Jacob, et al. 1990. "Sediment-induced amplification and the collapse of the Nimitz Freeway."* Nature *344:853–855. (b) Dennis Laduzinski)*

Local Soil and Rock Conditions

Local geology strongly influences the amount of ground motion during an earthquake. The dense granitic and metamorphic rocks of the Canadian Shield transmit earthquake energy very efficiently, so even moderate earthquakes can cause damage over large areas. In contrast, seismic energy, and thus ground shaking, attenuate rapidly away from the epicentres of most earthquakes in the extremely heterogeneous, folded, and faulted crust of western North America. Seismic waves also move much more slowly through unconsolidated sediments than through bedrock. They slow even further if the unconsolidated material has a high water content. For example, seismic waves typically slow as they move from bedrock to stream, or *alluvial* sands and gravels, and then slow again as they move through marine silts and clays (Figure 3.12). As P and S waves slow, some of their forward-directed energy is transferred to surface waves. This effect, known as **amplification**, increases the amount of ground motion in an earthquake.

For example, ground shaking during the 1989 Loma Prieta, California, earthquake (**M** 7.1) was particularly strong along the shores of San Francisco Bay (Figure 3.13). Amplified ground motions collapsed the upper deck of the Nimitz Freeway in Oakland, killing 41 people (Figure 3.14), and caused extensive damage in the Marina district of San Francisco. The portion of the Nimitz Freeway that collapsed was built on soft San Francisco Bay muds, and the Marina district was created after the 1906 San Francisco earthquake by filling part of the bay with debris from damaged buildings and with mud pumped from the bottom of the bay.[8] The loose nature of the fill and its high water content amplified the ground shaking to levels higher than were experienced closer to the epicentre, 100 km away.

The 1985 Mexico earthquake (**M** 8.0), which killed nearly 10 000 people in Mexico City hundreds of kilometres from the epicentre, is another tragic example of amplification. Much of Mexico City is built on silts and clays of ancient Lake Texcoco (Figure 3.15a). When seismic waves moved through the thick, fine-grained lake sediments, the amplitude of the long-period waves increased by a factor of four or five. The natural vibration period of buildings 10 to 20 storeys high was the same as that of the amplified long-period seismic waves, and more than 500 of these buildings literally tore themselves apart.[9] In many cases, the amplified shaking collapsed upper floors onto lower ones, like a stack of pancakes[10] (Figure 3.15b).

Local geologic structures can also amplify shaking. For example, *synclines* and fault-bounded sedimentary basins can focus seismic waves like a magnifying lens focuses sunlight. This focusing causes severe shaking in some areas and less intense shaking in others.

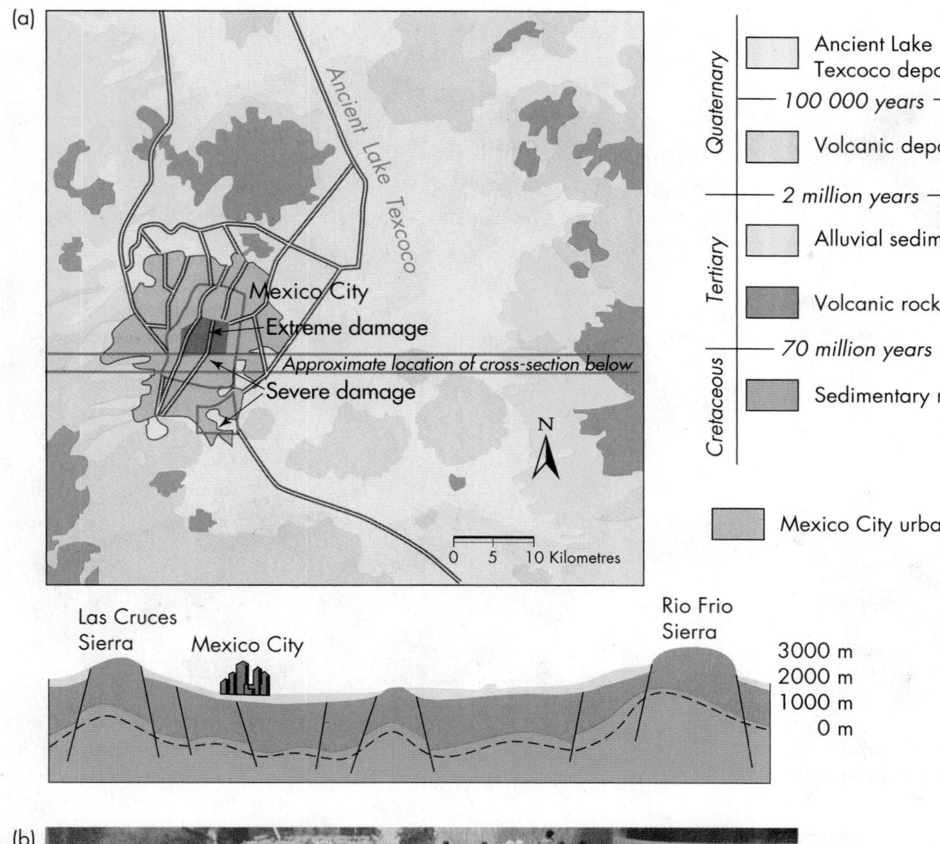

◀ **FIGURE 3.15 EARTHQUAKE DAMAGE IN MEXICO CITY** (a) A generalized geologic map and cross-section of Mexico City, showing the distribution of ancient lake deposits (yellow), where damage during the 1985 earthquake was greatest. Damage was extreme in the solid red area and severe in the zone delineated by the red line. (b) One of many buildings that collapsed during the 1985 earthquake. *(T. C. Hanks and Darrell Herd, U.S. Geological Survey)*

3.4 The Earthquake Cycle

Observations of the famous 1906 San Francisco earthquake (**M** 7.8) led to the recognition of what has since been termed the **earthquake cycle**. The idea behind the earthquake cycle is that stored elastic strain drops abruptly after an earthquake and then slowly accumulates until the next quake.

Strain is deformation resulting from stress. *Elastic strain* can be thought of as temporary deformation. When the stress is released, the elastically deformed material returns to its original shape. If the stress continues to increase, the deformed material will eventually rupture, making the deformation permanent. For example, a stretched rubber band or a bent archery bow will break with a continued increase in stress. When the rubber band or bow breaks, the broken ends snap back, releasing the pent-up energy. A similar effect, referred to as *elastic rebound*, occurs after an earthquake (Figures 3.16 and 3.17).

Seismologists speculate that some earthquake cycles have three or four stages.[11] The first stage is a long period of inactivity along a fault segment. In the second stage, accumulated elastic strain produces small earthquakes. A third stage, characterized by *foreshocks*, may occur hours or days before a large earthquake. Foreshocks are small- to moderate-size earthquakes that precede the main quake. However, some large earthquakes occur without foreshocks, and this

Road

Time 1

No strain and no displacement

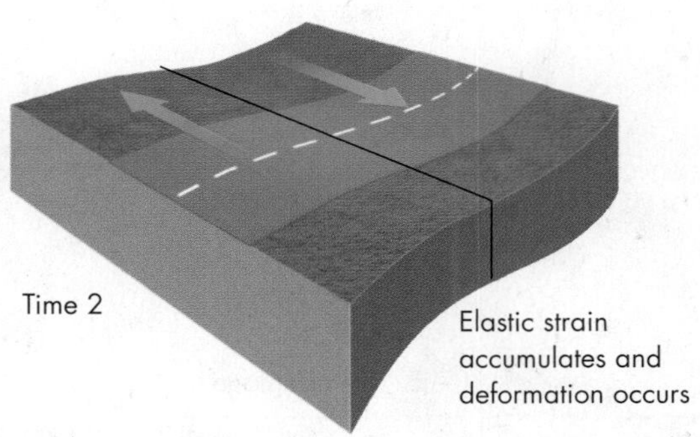

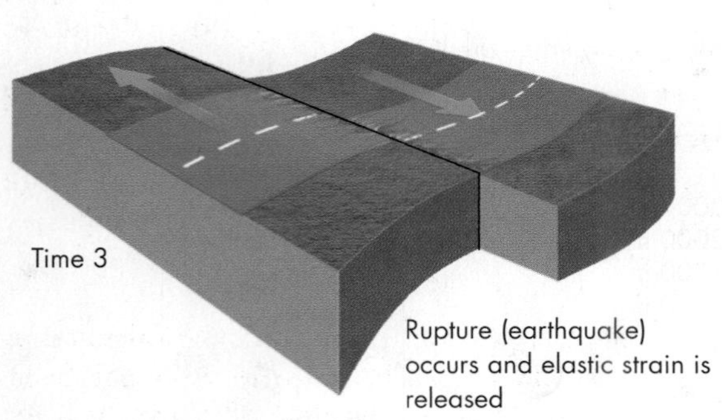

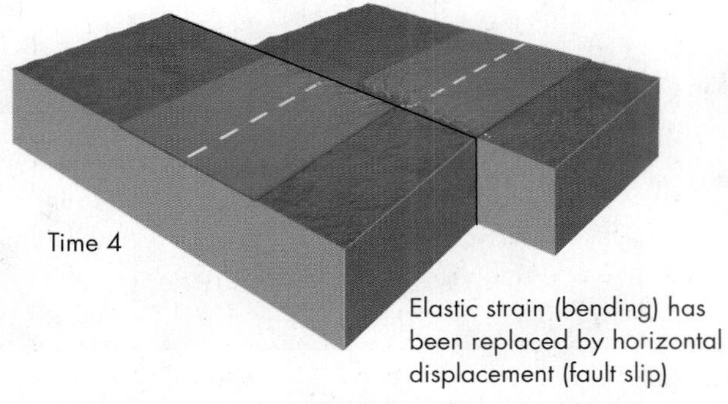

▲ **FIGURE 3.16 THE EARTHQUAKE CYCLE** An idealized diagram illustrating the earthquake cycle along a strike-slip fault. Elastic strain builds up as stress is applied from the movement of the two crustal blocks bounding the fault. Arrows at times 2 and 3 show the direction of stress. At time 3 rupture occurs and the two crustal blocks abruptly move relative to one another. The offset of the road at time 4, immediately after the earthquake, indicates the amount of displacement. The entire cycle, from time 1 to time 4, may last from hundreds to thousands of years.

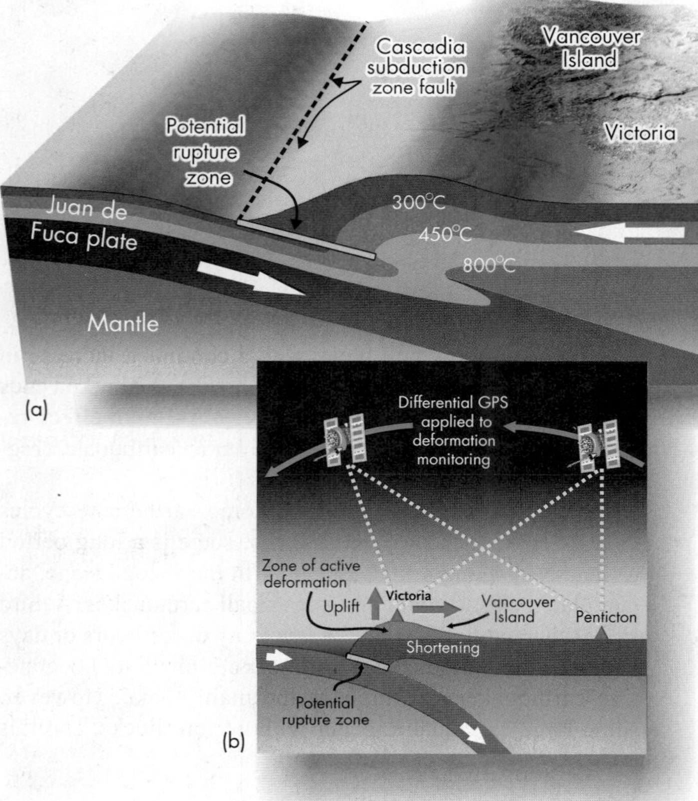

◄ **FIGURE 3.17 ACCUMULATION OF ELASTIC STRAIN ALONG A SUBDUCTION ZONE** The Juan de Fuca plate is presently locked to the continental North American plate off the west coast of North America. (a) The locked part of the fault zone ranges in temperature from about 150°C to 350°C. This zone is where the next great Cascadia earthquake will be initiated. (b) Vancouver Island is being flexed upward because of compression and elastic shortening of the North American plate above the locked interface. This deformation has been detected from satellite measurements of small changes in the relative position of points on Earth's surface. Sometime within the next several hundred years, the plates will suddenly unlock, triggering a great earthquake. *(Geoscape Victoria: Waiting for the 'Big One' Natural Resources Canada. http://geoscape.nrcan.gc.ca/victoria/bigone_e.php. © Department of Natural Resources Canada. All rights reserved.)*

third stage is absent. The fourth stage is the *mainshock*—the major earthquake—and its aftershocks.[11] An *aftershock* is a smaller earthquake that occurs a few minutes to a year or more after the mainshock and has an epicentre in the same general area as the mainshock. The earthquake cycle is hypothetical, and periods between major earthquakes are variable, but the four stages described above have been recognized along many active faults.

3.5 Geographic Regions at Risk from Earthquakes

Earthquakes are not randomly distributed. Most occur in well-defined zones along the boundaries of Earth's lithospheric plates (see Chapter 2 and Figure 3.5). Many of these earthquake-prone areas also are densely populated. In Chapter 1, we discussed the concept of "risk," which can be expressed as the product of hazard (in this case, the probability that a damaging earthquake of a specified magnitude will occur) and the vulnerable population and infrastructure. Risk is much greater in densely populated, earthquake-prone regions than in sparsely populated ones. In this context, countries situated along the "Pacific Ring of Fire," which borders the Pacific Ocean, are at high risk of earthquakes: southwest Canada, the western United States, Mexico, most countries in Central America, the Andean countries of South America, Japan, Taiwan, Philippines, Indonesia, and New Zealand. Areas surrounding the Himalayan mountain ranges, including southern China, northern India, and northern Pakistan, are also at high risk, as are many Middle Eastern countries (e.g., Israel, Turkey, Iran, Iraq, and Afghanistan) and parts of southern Europe (Italy, Greece) (Table 3.1). Earthquakes are most common in North America in the Pacific coastal areas of British Columbia, Washington, Oregon, California, and Alaska. Large cities with high levels of seismic risk include Anchorage, Vancouver, Seattle, Portland, San Francisco, and Los Angeles. Other seismically active areas in North America include southwest and eastern Yukon Territory, the California–Nevada border area, northern Utah, southern Ontario and Quebec, South Carolina, the central Mississippi River valley, some of the Arctic islands, and the U.S. territories of Puerto Rico and the Virgin Islands (Figure 3.18). The fact that some of these areas are at, or very close to, plate boundaries is not surprising. What may be more surprising are the high-risk zones that are not close to present-day plate boundaries but have experienced historical large earthquakes.

Plate-Boundary Earthquakes

Plate-boundary earthquakes, which occur on faults separating lithospheric plates, are of three types. The first type is strike-slip earthquakes, which occur on transform faults where lithospheric plates slide horizontally past one another. Examples are earthquakes on the San Andreas and Queen Charlotte faults, which separate the North American and Pacific plates. The second type is thrust earthquakes on faults separating converging plates, for example those at the Cascadia subduction zone off the coasts of British Columbia, Washington, and Oregon. The third type is normal dip-slip earthquakes on faults associated with divergent plate boundaries, such as the Mid-Atlantic Ridge.

Strike-Slip Earthquakes California straddles two lithospheric plates: the Pacific plate west of the San Andreas fault zone and the North American plate to the east. The motion of these plates results in frequent, damaging earthquakes. The 1989 Loma Prieta earthquake on the San Andreas fault system south of San Francisco caused 63 deaths, 3757 injuries, and an estimated U.S.$5.6 billion in property damage.[2,12] Deaths and injuries would have been much greater if many people had not stayed home to avoid the crowds and congestion of the third game of the World Series in Oakland. The Loma Prieta earthquake was not a great earthquake—the so-called "big one" that Californians fear. If a great earthquake (**M** 8 and higher) were to occur today in a densely populated part of California, damage would exceed U.S.$100 billion and several thousand people would be killed.[13]

Thrust Earthquakes **Subduction earthquakes** occur on the large thrust faults that bound subducting and overriding lithospheric plates, for example the Juan de Fuca and North American plates along the west coast of North America.[14] Strain energy accumulates in rocks adjacent to the plate-boundary fault until it is suddenly released during a subduction earthquake (Figure 3.17). Subduction earthquakes are the largest on Earth—some, such as the giant Sumatra earthquake in 2004, are larger than **M** 9. These giant earthquakes displace the seafloor upward and laterally over areas of many tens of thousands of square kilometres and, in so doing, trigger large tsunamis. The last subduction earthquake in the Pacific Northwest occurred in January 1700, nearly 100 years before the first Europeans visited the area.[15] Geological evidence from tidal marshes in northern California, Oregon, Washington, and British Columbia, together with written records of tsunami damage in Japan, indicate that the quake had a magnitude larger than 9.0 (Figure 3.19).[16,17] The **M** 9.2 earthquake in southern Alaska in 1964 and the **M** 9.1 quake off Sumatra in 2004 are recent examples of giant subduction events. They are, respectively, the second- and third-largest seismic events ever recorded, with the largest being an **M** 9.5 subduction earthquake off Chile in 1960. The 2010 **M** 8.9 Chile earthquake, although smaller than the 1960 quake, was also a subduction event generated at the fault that separates the down-going Nazca plate to the west from the overriding South American plate to the east. The fault rupture was more than 500 km long, parallel to the coast. The earthquake and its tsunami killed about 500 people.[18]

The second-largest historic earthquake in Canada (**M** 7.8) occurred on October 27, 2012, on the Queen Charlotte fault west of Haida Gwaii, British Columbia. The Queen Charlotte fault is the transform fault marking the boundary between the North American and Pacific plates; it

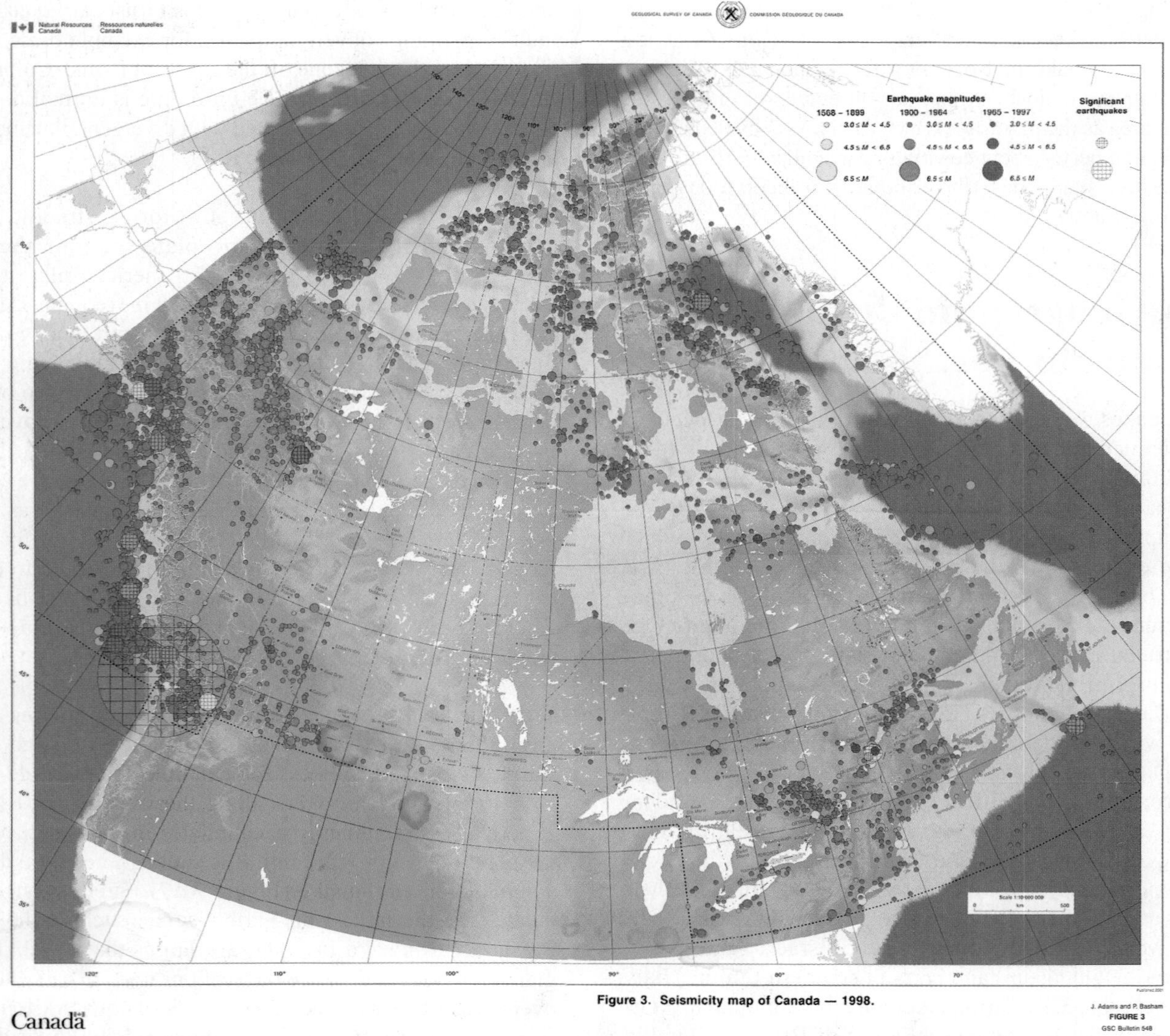

▲ **FIGURE 3.18 EARTHQUAKES IN CANADA** A map showing epicentres of historic earthquakes in Canada (**M** ≥ 2.5). *(Reproduced or adapted with the permission of Natural Resources Canada, courtesy of the Geological Survey of Canada (Bulletin 548))*

extends from the north end of Vancouver Island to southeast Alaska. Historic earthquakes on this fault have been mainly strike-slip events, which is consistent with the fact that the two plates are sliding horizontally past each other. This earthquake, however, was a thrust event—the Pacific plate moved upward against the North American plate to the east. An explanation for the seemingly anomalous slip is that the section of the fault on which the earthquake occurred is oriented obliquely to the direction of convergence of the two plates, resulting in localized strong compression of the two plates. The earthquake released the elastic strain that had been accumulating in this compressional plate tectonic environment for at least a century. Fortunately, it occurred in a remote area, and although it produced strong ground shaking near the source and a small tsunami on the west coast of southern Moresby Island, there was little damage and there were no injuries.

Normal Fault Earthquakes Earthquakes are common at divergent plate boundaries. Most divergent plate boundaries are located in oceans, far from land, and the earthquakes are typically no larger than **M** 6. As a result, they pose no risk to people. There are, however, two areas where divergent plate boundaries occur on land: Iceland on the axis of the Mid-Atlantic Ridge, and the Ethiopian and East African rifts, which extend southward from Ethiopia through Tanzania and Malawi. Earthquakes occur in both areas, although few are large enough to be damaging.

Intraplate Earthquakes

Intraplate earthquakes, which occur within lithospheric plates rather than at their boundaries, can be large and extremely damaging. They can occur on strike-slip, reverse, or

◄ **FIGURE 3.19 EARTHQUAKE STRATIGRAPHY** Exposure of sediments in the bank of the Niawiakum River in southwest Washington State. Evidence of five large earthquakes can be seen in this tidal channel. Fossil tidal marsh soils (dark horizons, indicated by arrows) represent old marsh surfaces, each of which subsided 1 to 2 m during an earthquake. After the earthquakes, the marsh surfaces were buried by tidal mud (light-coloured layers). The dates on the left side of the photograph are approximate ages of the earthquakes. The uppermost buried soil records the last subduction earthquake in January 1700. During the next subduction earthquake, the modern marsh (top surface) will subside and become covered with sea water. Over time, tidal mud will bury the marsh, providing the substrate for a new marsh. The underlying sediments will record this story. Divisions on shovel handle (circled) are 10 cm. *(Brian Atwater/U.S. Geological Survey)*

normal faults; in many cases, the rupture involves a combination of strike-slip and dip-slip movements.

Because large intraplate quakes are infrequent, people generally are unprepared for them and buildings might not be able to withstand their strong shaking. At least two **M** 7.5+ intraplate earthquakes occurred in the winter of 1811–1812 in the central Mississippi Valley, an area with no prior history of earthquakes of this magnitude. They destroyed the town of New Madrid, Missouri, and were felt in nearly every city of eastern North America from New Orleans to Quebec City, an area of more than 2.5 million km^2.[18] The ground motions rang church bells in Boston, more than 1600 km away! Forests were flattened; fractures opened in the ground, forcing people to cut down trees to cross the fractures; and some land sank several metres, causing local flooding. Journals and newspapers reported that local uplift of the land reversed the flow of the Mississippi River for a short time.[19]

The two earthquakes occurred within the New Madrid seismic zone, which is the seismically active part of an ancient downwarped area of thin lithosphere known as the Mississippi Embayment. Two hypotheses have been proposed to explain the thin lithosphere in this area. One is that the thinning took place about 600 million years ago, when a divergent plate boundary developed in what is now the southeast United States.[18] The other hypothesis is that the lithosphere thinned when the North American plate passed over a hot spot in the mantle about 95 million years ago. In any case, geologic and geodetic observations in the Mississippi Embayment indicate that the seismic activity is very recent, perhaps less than 10 000 years old.

The time between large earthquakes, or the *recurrence interval*, in the Mew Madrid seismic zone is probably several hundred years.[19,20] Taking into account likely material amplification, faults in this seismic zone appear to be capable of producing earthquake intensities comparable to those associated with large earthquakes in California. Thus, the interior of the North American plate is far from "stable." In recognition of this, the U.S. Federal Emergency Management Agency (FEMA) and vulnerable states and municipalities have adopted a new building code designed to mitigate major earthquake hazards in the region.

Moderate to large intraplate earthquakes also occur in southern Ontario and Quebec.[21,22] Most of these quakes are associated with the ancient rifted edge of the North American continent, which lies along the St. Lawrence River valley from Montreal to Sept Iles, Quebec; related rift structures also occur along the Ottawa and Saguenay river valleys. Strain accumulates on these structures because of ocean-floor spreading on the Mid-Atlantic Ridge at the east margin of the North American plate. The seismicity involves thrust earthquakes at depths of 5 to 30 km and occurs in three clusters: western Quebec, which experienced moderate earthquakes in 1732 (**M** 5.8), 1935 (**M** 6.2), and 1944 (**M** 5.8); Charlevoix, northeast of Quebec City (1663, 1791, 1860, 1870, and 1925; **M** 6.0–7.0); and the lower St. Lawrence region near Baie-Comeau, Quebec, a diffuse cluster of mostly small earthquakes.[21,22] The most recent damaging earthquake in central Canada occurred in the Saguenay region of Quebec in 1988 (**M** 5.9). Two more recent earthquakes in central Canada are the **M** 5.0 Val-de-Bois quake on June 23, 2010, and the **M** 4.6 Shawville quake on May 17, 2013. These two earthquakes were centred, respectively, about 60 km north and 60 km west-northwest of Ottawa and were felt as far away as Toronto and Montreal.

The largest earthquake in the Canadian Arctic occurred in Baffin Bay off the northeast corner of Baffin Island on November 20, 1933. It had a magnitude of 7.3 or 7.4 and thus was the largest historic earthquake on the eastern continental margin of North America and also the largest known earthquake north of the Arctic Circle.[23] It spite of its size, the earthquake caused no damage because of its offshore location and the sparse population in the area.

3.6 Effects of Earthquakes and Links with Other Natural Hazards

Ground shaking is not the only cause of earthquake death and damage. Other primary and secondary effects of earthquakes are also damaging and provide excellent examples of how natural hazards are linked. Primary effects of earthquakes include ground shaking and surface rupture. Secondary effects include liquefaction, land-level change, landslides, fire, tsunamis, and disease. We discuss each of these effects below, with the exception of tsunamis, which are covered in Chapter 4.

▲ **FIGURE 3.20 FAULT SCARP** This scarp was produced by ground rupture during the 1992 Landers earthquake (**M** 7.3) in California. The rupture extended over a distance of 70 km in the Mojave Desert. *(Edward A. Keller)*

Shaking and Ground Rupture

Large earthquakes produce strong ground shaking and, in some instances, surface rupture (see Survivor Story). The intensity of seismic shaking is commonly expressed as a ratio of *ground acceleration* to the acceleration from gravity (g). For example, a large earthquake might produce ground accelerations of 0.6 g, which is 60 percent of the acceleration from gravity (980 cm/s^2).

Strong ground shaking can be especially damaging to a building if the horizontal component of motion is strong or if the frequency of the shaking matches the natural vibrational frequency of the building, a phenomenon called **resonance**. In general, high-frequency, short-period seismic waves damage low buildings, whereas low-frequency, long-period waves damage tall buildings or long linear structures such as bridges.

Most ground cracks that appear during earthquakes are caused by liquefaction and mass movement (discussed below), but the ground can rupture along the fault that generated the earthquake. Ground rupture commonly produces a low cliff called a *fault scarp* that extends for hundreds of metres to kilometres along the fault (Figure 3.20).

The 1906 San Francisco earthquake (**M** 7.8) produced up to 6.5 m of horizontal ground displacement along the San Andreas fault north of San Francisco. The maximum intensity of the quake was XI on the Modified Mercalli Intensity Scale.[7] At this intensity, ground motion can uproot and snap large trees, throw people to the ground, and collapse large buildings, bridges, dams, tunnels, pipelines, and other rigid structures.[24]

The great Alaska earthquake (**M** 9.2) in 1964 damaged roads, railroads, airports, and buildings throughout the southern part of the state. The 1989 Loma Prieta (**M** 6.9) and 1994 Northridge (**M** 6.7) earthquakes were much smaller than the Alaska earthquake, with more localized strong shaking, but they were far more costly. In inflation-adjusted dollars, the Loma Prieta quake was five times as costly and the Northridge quake about 20 times as costly as the Alaska quake. The Northridge earthquake was one of the worst disasters in U.S. history, with losses estimated at U.S.$20 billion.[9] It caused so much damage because there was so much to be damaged. The quake occurred in a densely populated area with trillions of dollars of property at risk.

Large earthquakes in Canada have caused little damage by California standards. The **M** 8.1 Haida Gwaii (Queen Charlotte Islands) earthquake in 1949 was larger than the 1906 San Francisco earthquake, but no more than 2000 people lived in the area of strong ground shaking at the time, so damage was minor and there was no loss of life. Similarly, the epicentre of the **M** 7.8 Haida Gwaii earthquake in 2012 was in a remote area and caused almost no damage. The epicentre of the 1946 Vancouver Island earthquake (**M** 7.3) was far enough away from towns and other infrastructure that the damage from that event was also limited. Today, however, the same area supports a much larger population and far more economic wealth than in 1946; thus if the same event were to occur today, damage would likely be in the billions of dollars.

Liquefaction

Intense seismic shaking can cause water-saturated loose sediment to change from a solid to a liquid. This process, called **liquefaction**, takes place at shallow depths when pore water pressures become so elevated that the water suspends sediment particles, allowing the deposit to flow (Figure 3.22). Once the pressures decrease, the liquefied sediment compacts and regains its solid form. Liquefaction is generally accompanied by lateral shifts of a layer of solid sediment on one or more layers of fluidized sediment. Watery sand and silt may also flow upward along fractures in overlying solid materials and erupt onto the surface as *sand blows*, also called *sand volcanoes* (Figure 3.23). Liquefaction can cause the land surface to settle irregularly, with potential damage to foundations of buildings, water and sewer lines, and other buried utilities. Liquefaction of poorly compacted sediment has caused multi-storey apartment buildings to tip over, highway bridges to collapse, and dams to fail (Figure 3.24).[25]

SURVIVOR STORY

Shaky Honeymoon

In late May 2009, John and Jane Moon were enjoying the last day of their honeymoon at a resort on the island of Roatan, located off the Caribbean coast of Honduras (Figure 3.21). They were asleep in their house, which was built on stilts like most structures in the resort.

At 2:45 A.M. local time, the Moons' house started shaking violently as the seismic waves of a large (**M** 7.3 earthquake) moved across Roatan. The quake was centred only 64 km to the northeast and 10 km below the seafloor. It occurred on a strike-slip fault in the Caymen Trench, which forms the boundary between the North American and Caribbean plates.

As Jane put it, "It was pitch black but you could feel the whole house shaking violently from side to side; it lasted like what felt like an eternity, but according to the U.S. Geological Survey it was only 30 seconds." She adds, "After the shaking stopped, there was an almighty noise, like a rumble, or deep resonating sound. The noise afterwards was incredible...it chilled you to the bone."

A tsunami watch was put into effect for Honduras, Guatemala, and Belize, after the Pacific Tsunami Warning Center in Honolulu issued a warning of the possibility of a local tsunami. This fact was not lost on the Moons.

"When the shaking stopped we staggered through to the back of the house and into the kitchen. My fear at this point was that a tsunami could slam into the building at any second; it was quite terrifying. I kept thinking, why us? And isn't this typical on the last day of our honeymoon!"

Jane, an earth scientist who completed her M.Sc. degree under the supervision of one of the authors (Clague), knew that if a tsunami were to arrive, it would be in minutes rather than hours. She and John decided to stay put. "It was scary, not knowing if this was the right decision, but the house was 3 m above the beach and had miraculously survived the earthquake," she said. "I couldn't even face opening a door to look out at the sea for fear of what we might see."

▲ **FIGURE 3.21 NEWLYWEDS JANE AND JOHN MOON AT ROATAN, HONDURAS** Their beach house can be seen in the distance, just above the head of the dolphin. *(Jane Moon)*

"We waited for over two hours in the house by which time I guessed that the tsunami risk had passed, or perhaps the tsunami had already happened, but under our building without us knowing. We went back to bed to wait until light to assess the damage more fully. We had quite a number of aftershocks, decreasing in magnitude."

The powerful earthquake was felt in Guatemala, El Salvador, Belize, and parts of Nicaragua, Costa Rica, Colombia, Cuba, and Jamaica. It caused at least seven fatalities and injured 40 people. More than 200 buildings collapsed or were damaged across northern Honduras, Guatemala, and Belize.

—John J. Clague

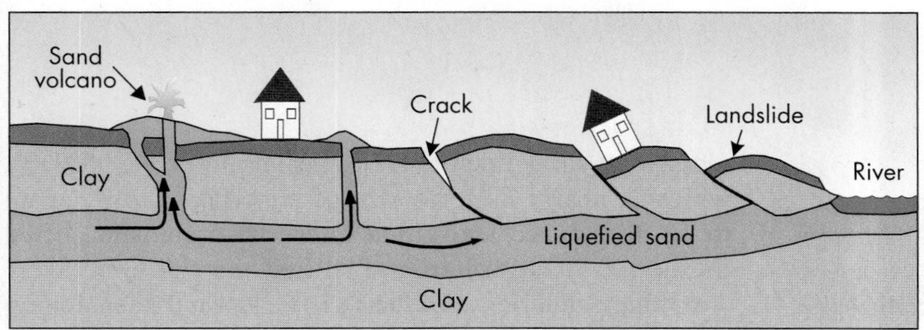

◄ **FIGURE 3.22 LIQUEFACTION** Liquefaction of water-saturated silt or sand during an earthquake may cause the ground to fracture and subside, damaging or destroying buildings and other human works. *(Reprinted with permission of Tricouni Press)*

▲ **FIGURE 3.23 SAND BLOWS IN CHRISTCHURCH** The Christchurch, New Zealand, earthquake of February 22, 2011, caused widespread damaging liquefaction in parts of the city. This photograph, taken soon after the earthquake, shows a car enveloped in liquefied sand. *(© NZPA/Corbis)*

▲ **FIGURE 3.24 LIQUEFACTION DAMAGE** Damage to apartment buildings in Niigata, Japan, caused by liquefaction during a strong earthquake on June 16, 1964. About one-third of the city subsided as much as 2 m because of liquefaction. *(Courtesy NOAA National Geophysical Data Center)*

◀ **FIGURE 3.25 COSEISMIC UPLIFT ALONG THE SEATTLE FAULT** (a) A LIDAR (Light Detection and Ranging) image of part of Bainbridge Island in Puget Sound, Washington State. The image accurately depicts the ground surface with vegetation removed. The dashed red line shows the trace of the Seattle fault. (b) Slip along the fault about 1000 years ago raised an intertidal rock platform 7 m above sea level. *(Reprinted with permission of Tricouni Press)*

Land-Level Changes

Some earthquakes raise or lower the land over large areas. Coseismic land-level changes can cause substantial damage to structures built along shorelines and streams.

The most recent great earthquake at the Cascadia subduction zone in 1700 caused the Pacific coasts of Vancouver Island, Washington, and Oregon to subside up to 2 m, inundating coastal marshes and forests with sea water.[16,26] The 1964 Alaska earthquake had similar effects, causing measurable surface deformation over an area of 300 000 km^2, which is nearly half the size of Alberta.[27] Uplift of up to 10 m exposed and killed intertidal organisms, lifted docks out of the water, and shifted the shoreline away from fish canneries. In other areas, the land subsided up to 2.4 m, partially flooding several coastal communities. Finally, a large earthquake near Seattle, Washington, about 1000 years ago raised the land surface up to 7 m (Figure 3.25).[28]

◄ **FIGURE 3.26 EARTHQUAKE-TRIGGERED LANDSLIDE** This large landslide was one of thousands triggered by the 2002 **M** 7.9 Denali earthquake in Alaska. The landslide deposit covers part of Black Rapids Glacier. *(U.S. Geological Survey/U.S. Department of the Interior)*

Landslides

Earthquakes and landslides are two of the most closely linked natural hazards. Ground motions produced by a large earthquake can cause rock or sediment to fail and move downslope. Hundreds, even thousands, of landslides have been triggered by single earthquakes in mountainous areas (Figure 3.26). Some earthquake-triggered landslides can cause great loss of life. A landslide caused by a strong earthquake in El Salvador in January 2001 buried the community of Las Colinas, killing hundreds of people. An earthquake in Peru in 1970 triggered a huge landslide that buried the cities of Yungay and Ranrahirca, killing about 8000 people.[29]

Landslides also were responsible for much of the loss of life in the 2008 Wenchuan earthquake in southwest China. The massive earthquake affected a mountainous area inhabited by 15 million people.[30,31] After two minutes of strong shaking, nearly 90 000 people had lost their lives, 375 000 were injured, and millions were homeless. Parts of towns were destroyed by landslides triggered by the earthquake;[32] many rivers were blocked and "quake lakes" formed behind the landslide barriers (Figure 3.27).[33] It was feared that the landslide dams would eventually fail, producing floods that would threaten millions of people living downstream. Entire villages had to be evacuated because of the possibility of this flooding.

◄ **FIGURE 3.27 LANDSLIDE-DAMMED LAKE** The 2008 Wenchuan earthquake in China produced many landslides that blocked stream courses, impounding lakes. In this case, a channel was excavated in the landslide dam with the help of explosives to lower the level of the lake and reduce the possibility of an outburst flood. *(Color China/AP Images)*

◄ **FIGURE 3.28 EARTHQUAKE AND FIRE** Fires associated with the 1995 Kobe, Japan, earthquake caused extensive damage to part of the city. *(Corbis Images)*

Fires

Fire is another major hazard linked to earthquakes. Ground shaking and surface rupture can sever electrical power and gas lines, starting fires. These fires may be difficult to put out because firefighting equipment may be damaged; streets, roads, and bridges blocked; and water mains broken. Appliances such as gas water heaters may topple when shaken, producing gas leaks that ignite. Earthquakes in both Japan and the United States have been accompanied by devastating fires (Figure 3.28). The San Francisco earthquake of 1906 has been referred to as the San Francisco Fire, because 80 percent of the damage was caused by a firestorm that ravaged the city for several days after the quake.[9] The 1989 Loma Prieta earthquake caused large fires in the Marina district of San Francisco.

Disease

Some large earthquakes cause outbreaks of disease. A loss of sanitation and housing, contaminated water supplies, and disruption of public health services all contribute to the spread of disease. Earthquakes rupture sewer and water lines, causing water to become polluted by disease-causing organisms.

A tragic example of disease linked to an earthquake is the cholera epidemic that struck Haiti after the catastrophic earthquake of January 2010. The epidemic began in mid-October 2010 in the Artibonite Valley, about 100 km north of Port-au-Prince. A sewage spill from a Nepali UN peacekeeping base may have contaminated a river from which some of the infected people had drunk water. Within ten weeks, cholera had spread to all of Haiti's ten provinces, and within five months, nearly 5000 Haitians had succumbed to the disease and thousands more had been hospitalized.[34] As of the end of November 2012, almost 8000 deaths had been reported and over 5 percent of the population of the country had been sickened by the disease.

3.7 Natural Service Functions of Earthquakes

Earthquakes are so destructive that it's hard to imagine they could have any benefits. However, like many natural hazards, earthquakes do provide natural service functions. They contribute to the development of groundwater and energy resources, the formation of valuable mineral resources, and the development of landforms.

Groundwater and Energy Resources

Geologic faults, on which earthquakes occur, influence the flow of water, oil, and gas. Fault zones provide paths for the downward flow of surface water. They also channel groundwater to surface discharge points, called *springs*.

In other settings, faults create natural subsurface barriers that slow or redirect the flow of water, oil, or natural gas. Faulting commonly pulverizes rock to form an impervious clay barrier. It also can place impervious rock against rock containing water, oil, or natural gas. Such subsurface barriers are responsible for oases in arid areas of southern California and for many underground accumulations of oil and gas in Alberta, Texas, Oklahoma, and elsewhere.

Mineral Resources

Faulting can contribute to the accumulation of economically valuable minerals. Some valuable minerals are preferentially deposited along faults in what are called *veins*. Veins associated with major fault zones may contain enough gold, silver, or platinum to be mined.

Landform Development

Earthquakes occurring episodically over hundreds of thousands to millions of years form scenic landscapes. The uplift of rocks along faults can produce hills, mountain ranges, and coastal and stream terraces. Weak rocks along fault zones may be eroded by streams to form valleys. For example, many of the valleys in British Columbia, including the Fraser Canyon, Okanagan Valley, and Rocky Mountain Trench, are located along fault zones.

A spectacular example of the role that earthquakes have played in creating scenic landscapes is the Rocky Mountains of British Columbia and Alberta. The mountains were elevated by countless earthquakes along a series of thrust faults during the Eocene Epoch (55–34 million years ago). The beauty of the Rocky Mountain national parks and their appeal to tourists are based in geology.

3.8 Human Interaction with Earthquakes

Several human activities are known to trigger small to moderate earthquakes or increase their frequency. These quakes provide important lessons that help us better understand seismic hazards and risk.

People can cause earthquakes by:[35]

- building a dam and flooding a valley
- injecting liquid waste deep into the ground
- pumping oil or gas from the ground
- hydraulic fracturing ("fracking")
- detonating underground nuclear explosions

Reservoirs

Flooding a valley behind a dam may induce earthquakes. The weight of the water loads the underlying crust and can create or extend fractures. These effects may produce new faults or reactivate existing ones. Several hundred small earthquakes occurred in the decade following completion of Hoover Dam and the filling of Lake Mead in Arizona and Nevada. One of these quakes had a magnitude of 5 and two were about **M** 4.[35] Reservoir filling in India and China has triggered earthquakes as large as **M** 6.0. Two of the quakes killed hundreds of people and severely damaged the dams impounding the reservoirs.[24]

Deep Waste Disposal

In the early 1960s, an unplanned experiment by the U.S. Army provided the first direct evidence that injecting fluids into the crust can cause earthquakes. Several hundred earthquakes occurred in the vicinity of Denver, Colorado, between April 1962 and November 1965. One of the quakes, an **M** 4.3 event, was large enough to knock bottles off shelves in stores. A geologist traced the source of the earthquakes to the Rocky Mountain Arsenal, a chemical warfare plant at the northeast side of Denver. Liquid waste from the plant was being injected under pressure down a deep disposal well to a depth of 3600 m. The injected liquid increased fluid pressures and caused slippage of numerous fractures in the host rocks. The geologist demonstrated a high correlation between the rate of waste injection and the times and locations of the earthquakes. When the waste disposal stopped, so did the earthquakes[36] (Figure 3.29). Deep injection of fluids in wells has also triggered earthquakes in Ohio and Texas.[37,38]

Recognizing that injection of fluids into the crust could trigger earthquakes was an important development because it drew attention to the relation between fluid pressure and earthquakes. Subsequent studies of subduction zones and active fold belts showed that high fluid pressures are present in many areas where earthquakes occur. One hypothesis to explain this association is that fluid pressure rises until rocks break, triggering an earthquake and discharging fluid upward. After the event, fluid pressure begins to build up again, beginning a new cycle.

Pumping of Oil or Gas

Over the past three decades, many small earthquakes have occurred in areas of oil and gas production in western Alberta and northeast British Columbia.[39,40] All of these earthquakes, with one possible exception, had magnitudes less than 4, and most of them had shallow foci. Earthquakes of this size can rattle buildings, but are not likely to cause damage. One of the better documented examples is a series of earthquakes associated with the Strachan gas field in the Rocky Mountain foothills southwest of Rocky Mountain House, Alberta.[41] A strong correlation exists between rates of gas production in this field and earthquake frequency, but seismic activity began several years after production began (Figure 3.30). Production-triggered seismicity also is suspected around Turner Valley and Brazeau River in western Alberta. Injection of fluids into gas-bearing rocks to enhance gas recovery has been implicated in a swarm of small earthquakes in the Fort St. John area in northeast British Columbia.

Hydraulic Fracturing

Induced *hydraulic fracturing*, commonly called "fracking," is a technique used to enhance hydrocarbon production by releasing natural gas or petroleum from a subsurface rock layer.[42] Fracking involves injecting a highly pressurized fluid into hydrocarbon-bearing rocks to create or propagate fractures. The fluid contains grains of sand or ceramics that prevent the fractures from closing when the injection is stopped and the pressure of the injected fluid is reduced. Hydrocarbons within the host rock seep into the induced fractures and, from there, into a horizontal well bore. The first use of hydraulic fracturing was in 1947, but the technique only became economical in 1998 when it was used to extract natural gas from shale in Texas.

Hydraulic fracturing has caused small earthquakes in some areas in the United States and Canada. According to the U.S. Geological Survey, few fracking operations have induced earthquakes that are large enough to be of concern to the public. However, although the magnitudes of these earthquakes have been small, there is no guarantee that larger quakes will not occur. The British Columbia Oil and Gas Commission commissioned a study of 38 small (M 2.2–3.8)

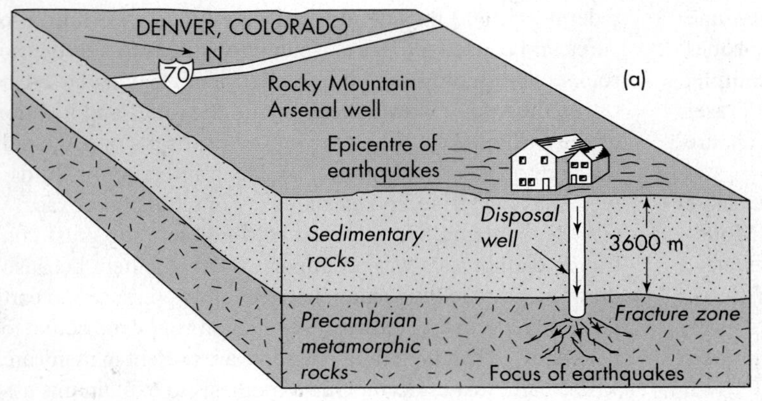

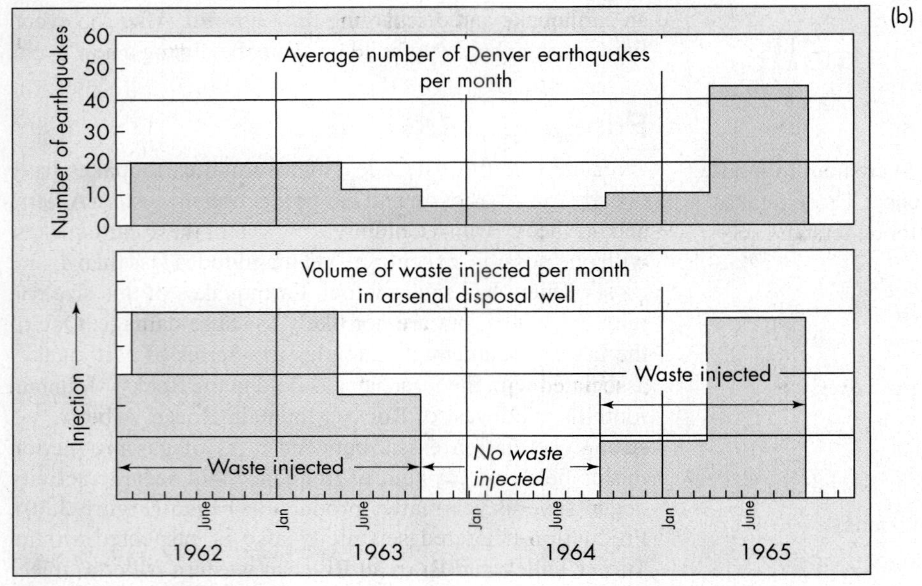

◄ **FIGURE 3.29 HUMAN-CAUSED EARTHQUAKES** (a) A generalized block diagram showing the deep disposal well at Rocky Mountain Arsenal in Colorado. (b) A graph showing the relation between earthquake frequency and the amount of injected liquid waste. Earthquakes were most numerous when waste fluids were injected into the ground. *(After Evans, D. M. 1966. "Man-made earthquakes in Denver."* Geotimes *10:11–18. Reprinted by permission of the American Geosciences Institute.)*

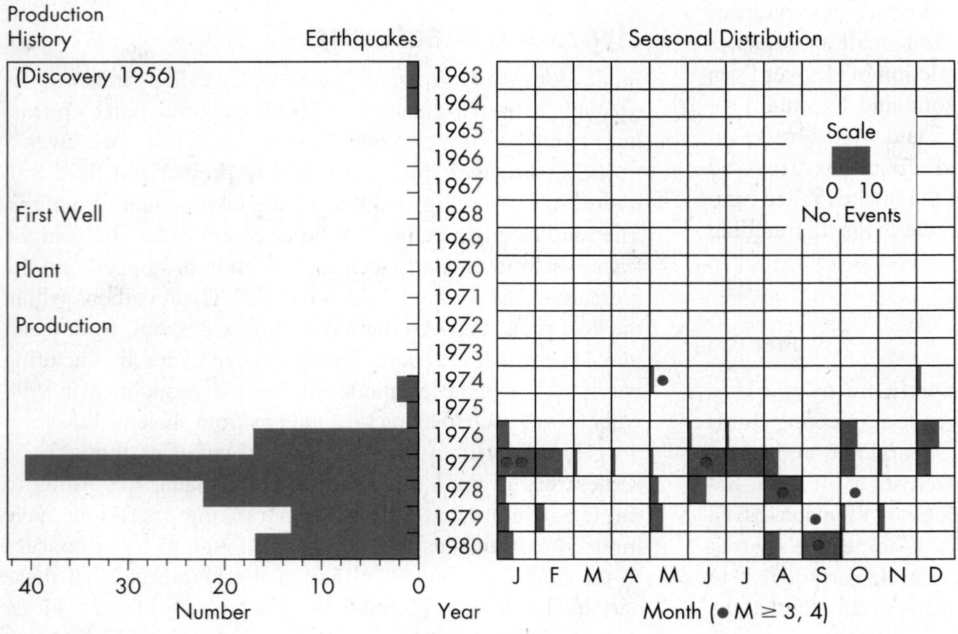

◄ **FIGURE 3.30 SEISMICITY INDUCED BY HYDROCARBON EXTRACTION** Frequency of earthquakes in the vicinity of Rocky Mountain House, Alberta, in relation to significant events in the history of hydrocarbon production from the Strachan gas field. *(Wetmiller, R. J. 1986. "Earthquakes near Rocky Mountain House, Alberta, and their relationship to gas production facilities."* Canadian Journal of Earth Sciences *23:172–181. Reprinted by permission of NRC Research Press.)*

earthquakes that occurred between 2009 and 2011 at a location in northeast British Columbia and concluded that they were caused by fluid injection during hydraulic fracturing near pre-existing faults.[43]

Nuclear Explosions

Underground testing of nuclear weapons in Nevada in the 1950s triggered numerous earthquakes up to **M** 6.3.[35] Analysis of aftershock sequences suggests that the explosions released some natural strain within Earth. This idea has led some scientists to suggest that nuclear explosions can be used to prevent large earthquakes by releasing strain before it reaches a critical threshold.

3.9 Minimizing the Earthquake Hazard

An important reason that earthquakes cause so much damage and loss of life is that they generally strike without warning. Much research is being done to anticipate earthquakes, but the best we can do at present is forecast the likelihood that an earthquake will happen in a particular area or on a particular segment of a fault. Earthquake forecasts assist planners and officials responsible for seismic safety. They do not, however, help residents of a seismically active area prepare for a specific earthquake.

Earthquake Hazard Reduction Programs

The Geological Survey of Canada and the U.S. Geological Survey, in conjunction with university and other scientists, are developing programs to reduce earthquake hazards in Canada and the United States. The programs have five major goals:[44]

1. Operate national seismograph networks. The Canadian National Seismograph Network is a Canada-wide system of seismographs located to detect and locate all earthquakes larger than **M** 3.5; the detection threshold is **M** 2.0 in the populated southern part of the country and in seismically active areas. Digital data from the instruments are automatically acquired and archived in real time, and seismologists are alerted to the occurrence of earthquakes by email.
2. Develop an understanding of earthquake sources by obtaining information on the physical properties and mechanical behaviour of faults and developing quantitative models of the earthquake process.
3. Determine earthquake potential by conducting detailed studies of seismically active areas. Scientists identify active faults, document past earthquakes from geologic evidence, and determine average slip rates of active faults. This information is used to make probabilistic forecasts of earthquakes.
4. Predict effects of earthquakes on buildings and other structures from estimates of shaking intensity and amounts of surface rupture. This information is needed to evaluate losses from earthquakes of different magnitudes (Case Study 3.1).
5. Communicate research results in order to educate individuals, communities, and governments, thereby reducing loss of life and property from earthquakes.

Estimation of Seismic Risk

Hazard maps are used to portray earthquake risk. The simplest maps show the locations of epicentres of historic earthquakes of different magnitudes (Figure 3.18). Other more complex maps display probabilities of earthquakes of different sizes, or the amount of shaking likely to occur. For example, the seismic ground acceleration map of Canada (Figure 3.32) shows zones of different ground accelerations, expressed as a 10 percent probability of exceedance in 50 years.[47] It is based largely on the instrumented records of earthquakes in Canada (Figure 3.18). The map is used to establish engineering standards for public buildings in earthquake-prone areas, notably the south coast of British Columbia and the St. Lawrence Valley in southern Ontario and Quebec.

Another common type of hazard map depicts areas where secondary earthquake effects, notably liquefaction, landslides, and tsunamis, are most likely to occur. Maps of this type have been prepared for many populated areas in the western United States (Figure 3.33).

Following a damaging earthquake in the Los Angeles area in 1971, the State of California passed legislation requiring that a geological site investigation be completed in "special study zones," encompassing the rupture zones of potentially and recently active faults.[9,48] The purpose of the legislation is to determine whether an active fault passes through a proposed building site.

The State of California also classifies faults in order to evaluate seismic risk. Classification is based on the largest earthquake the fault can produce and on the fault's long-term *slip rate*. The slip rate is the average displacement rate on the fault measured over thousands of years and numerous earthquakes. Classifying faults provides more information than simply determining whether a fault has been active in the past 11 600 years, but the long-term slip rates of most major faults in North America are unknown or very poorly known.[49]

Long-term slip rates can be estimated by **paleoseismologists**—geologists who examine the recent geologic record of faulting to determine times and sizes of prehistoric earthquakes. Paleoseismologists excavate trenches across faults at key locations, identify offsets of sediment layers, and date the offsets by the *radiocarbon dating* method. The complete history of movement on a fault cannot be determined from evidence in a single trench; rather, data must be collected and synthesized from many trenches. Geologists in Canada, the United States, Japan, and New Zealand have been world leaders in paleoseismological research.

3.1 CASE STUDY

The Denali Earthquake: Estimating Potential Ground Rupture Pays Off

On November 3, 2002, a magnitude 7.9 earthquake occurred on the Denali fault in a remote area of south-central Alaska south of Fairbanks. The earthquake ruptured a 340-km length of the fault, producing horizontal displacements of up to 8 m. The quake triggered thousands of landslides and was accompanied by intense ground shaking, but it caused little structural damage and no deaths—primarily because few people live in the affected area.[45]

The Denali earthquake demonstrated the value of geologic studies for seismic hazard assessments. Geologists studied the fault in the early 1970s as part of planning for the Trans-Alaska (Alyeska) pipeline. They determined that the fault zone at what would eventually be the pipeline crossing is several hundred metres wide and could experience 6 m of horizontal displacement in an **M** 8 earthquake. These estimates were used in the engineering design of the pipeline. Where it crossed the fault zone, the pipeline was elevated on long horizontal steel beams with Teflon shoes that allow the pipe to slide horizontally up to 6 m. These slider beams were installed along a zigzag path for added flexibility and to facilitate horizontal movement during an earthquake (Figure 3.31). The 2002 earthquake occurred, as expected, within the mapped fault zone and caused the pipeline to shift horizontally about 4.3 m. Because of the engineering design, the pipeline suffered little damage and there was no spillage of oil.

Although the U.S.$3 million cost for the design and construction of the pipeline across the fault might have seemed excessive in 1970, in retrospect it was money well spent. The pipeline carries tens of millions of dollars' worth of oil each day and supplies approximately 17 percent of the domestic oil supply of the United States. A break in the pipeline during the 2002 earthquake would have had a costly impact on U.S. oil supply. Repair of the pipeline and cleanup of the oil spill alone would have cost several million dollars.[46]

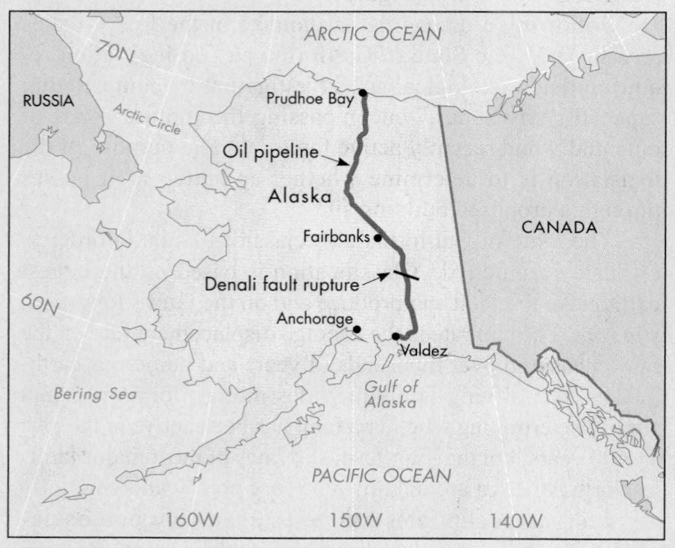

▲ **FIGURE 3.31 TRANS-ALASKA OIL PIPELINE SURVIVES A LARGE EARTHQUAKE** The Alyeska pipeline was designed to withstand several metres of horizontal displacement where it crosses the Denali fault. The design was put to the test in November 2002 when it shifted horizontally 4.3 m during the **M** 7.9 Denali earthquake. Slider beams, Teflon shoes, and built-in bends of the pipeline accommodated the earthquake rupture and demonstrated the importance of geologic investigations for seismic hazard assessments. *(Alyeska Pipeline Service Company)*

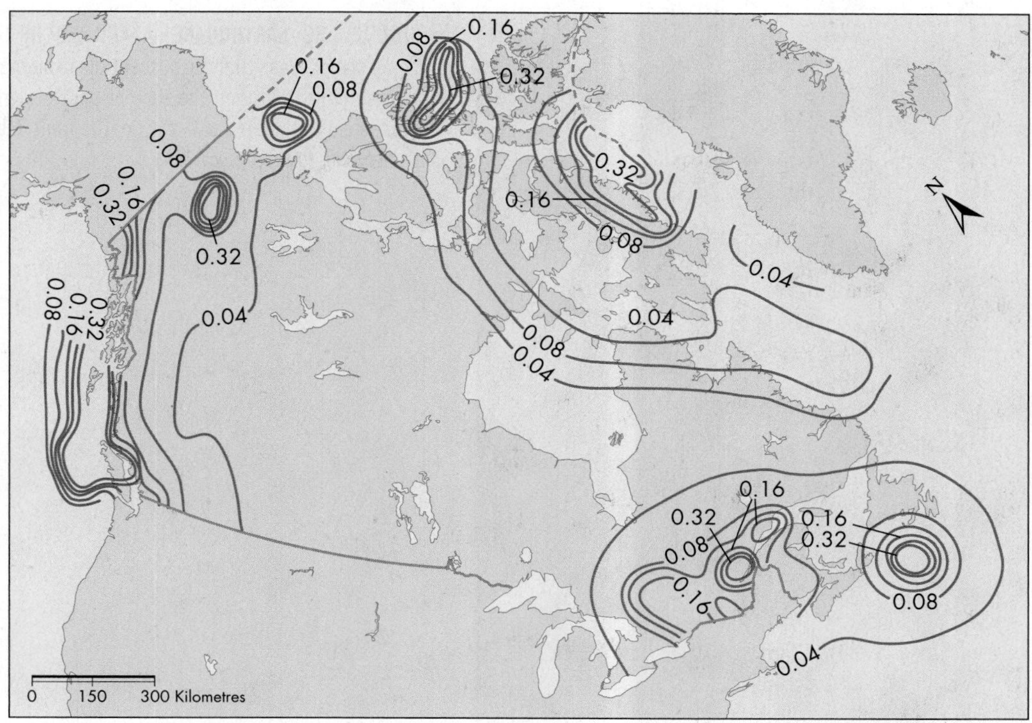

▲ **FIGURE 3.32 SEISMIC GROUND ACCELERATION MAP** This map shows peak horizontal ground accelerations that have a 10 percent chance of occurring in the next 50 years in Canada. Ground acceleration is a measure of the intensity of earthquake shaking and is used by engineers to design safe buildings and other structures. *(Basham, P. W., D. H. Weichert, F. M. Anglin, and M. J. Berry. 1985. "New probabilistic strong ground motion maps of Canada."* Bulletin of the Seismological Society of America *75:563–595. © Seismological Society of America)*

Forecasts and Prediction

Two approaches have been used to anticipate future earthquakes. One is called forecasting and the other, prediction. A **forecast** states that an earthquake of a specified magnitude has a certain probability of occurring in an area within a specified number of years. A **prediction** specifies that an earthquake of a given magnitude will occur in a defined region within a restricted period of time (hours, days, weeks). Predicting earthquakes was once thought to be as easy as "one-two-three." First, deploy instruments to detect precursors; second, detect and recognize the precursors that tell you when and how big an earthquake will be; and third, after reviewing the data, publicly predict the earthquake. Unfortunately, earthquake prediction is much more complex than first thought.[50]

Japanese seismologists made the first attempts at earthquake prediction by using the frequency of very small *microearthquakes* (magnitude less than 2), precise repeated geodetic surveys to detect tilting of the land surface, and repeat measurements of the local magnetic field of Earth. They found that earthquakes in the areas they studied were nearly always accompanied by swarms of microearthquakes several months before the major shocks. Furthermore, they found that tilting of the ground correlated strongly with earthquake activity.[35]

In 1975, Chinese scientists predicted a major earthquake (**M** 7.0) that saved hundreds of thousands of lives. The prediction was based primarily on a series of progressively larger foreshocks that began four days before the mainshock. This prediction appears to have been a lucky coincidence, because Chinese scientists previously had issued many unsuccessful predictions and they subsequently failed to predict major quakes. Nonetheless, the result was beneficial.[49] The earthquake destroyed or damaged about 90 percent of the buildings in the city of Haicheng. Most of the 1 million people living in the city were saved because they had been evacuated the day before the quake struck. Unfortunately, most large earthquakes are not preceded by foreshocks.

People have proposed phenomena ranging from lunar tides to unusual animal behaviour as earthquake precursors. Odd animal behaviour includes unusual barking of dogs, chickens that refuse to lay eggs, horses or cows that run in circles, rats perched on power lines, and snakes that emerge from the ground in the winter and freeze. To date, no scientific studies have shown a convincing correlation between unusual animal behaviour or lunar tides and earthquakes.

A significant dilemma in making a prediction is that it must be reliable; there is no room for error. If the prediction does not come to fruition, people will lose confidence in the predictor, a classic example of the "cry wolf" phenomenon. On the other hand, false assurances by experts that a damaging earthquake is not imminent, in spite of signs that it might happen, can be tragic (Case Study 3.2). Any prediction of the location, time, and magnitude of an earthquake should be accompanied by an estimate of the probability that it

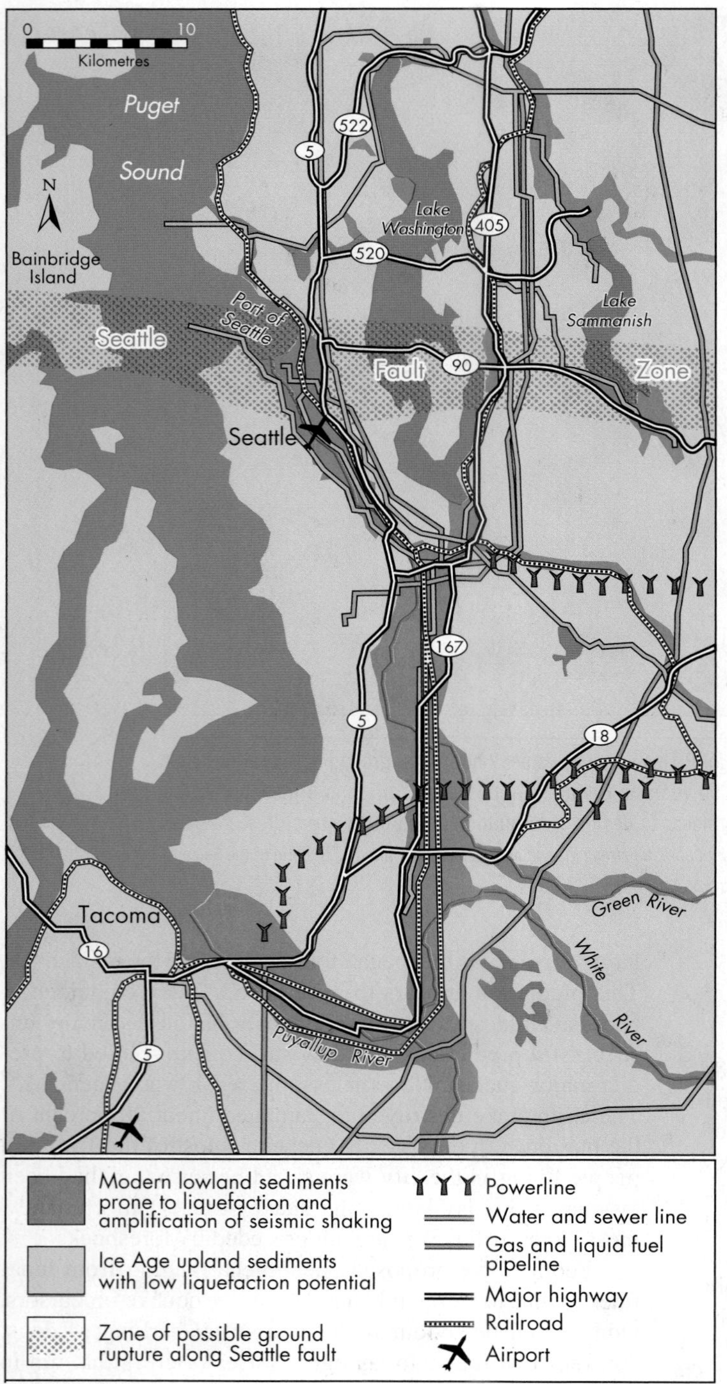

◄ **FIGURE 3.33 EARTHQUAKE HAZARD MAP OF THE SEATTLE AREA** The map shows areas that are susceptible to liquefaction and ground motion amplification, and the zone of possible ground rupture and displacement associated with the Seattle fault. Major infrastructure is also shown. *(U.S. Geological Survey)*

will happen. Such predictions, if they are possible, will most likely be based on **precursors**, measurable changes that precede an earthquake. The following are some examples:

- The pattern and frequency of earthquakes, such as foreshocks that preceded the Haicheng earthquake, discussed above
- Land-level change
- Seismic gaps along faults
- Physical and chemical changes in Earth's crust

Land-Level Change *Uplift* or *subsidence* may precede a large earthquake and provide clues to forecast it. For example, several centimetres of uplift occurred in the decade before the 1964 Niigata earthquake (**M** 7.5) in Japan.[51] Similar amounts of uplift occurred over the five years before the 1983 Sea of Japan earthquake (**M** 7.7).[51,52]

Uplifts of 1 m to 2 m also preceded large earthquakes in Japan in 1793, 1802, 1872, and 1927. On the morning of the 1802 earthquake, the sea withdrew about 300 m because of an uplift of about 1 m. The earthquake struck four hours later, destroying many houses and uplifting the land another metre.[51]

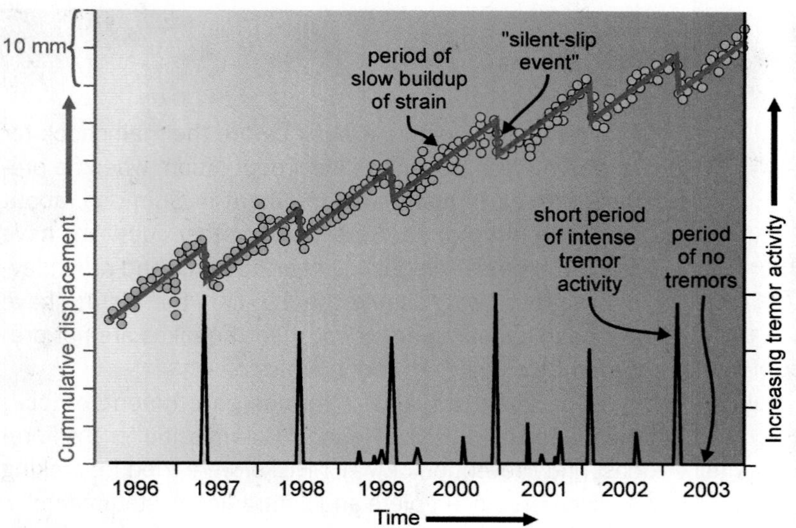

◄ **FIGURE 3.34 EPISODIC TREMOR AND SLIP EVENTS** A diagram showing slow-slip events along the Cascadia subduction zone and tremor activity recorded in the Victoria area, British Columbia, between 1996 and 2003. The blue circles represent day-to-day changes in the position of a precisely located GPS instrument near Victoria with respect to another GPS station near the inland city of Penticton, British Columbia; the latter is assumed to be stationary and fixed to the North American plate. The red line segments show the progressive movements between the slip events, marked by reversals every 13 to 16 months. Since 1996, Victoria has moved about 40 millimetres closer to Penticton. The bottom graph shows the number of hours of tremor activity per 10-day period for Vancouver Island. Tremor and slip events last from 10 to 20 days. *(Reprinted with permission from Tricouni Press)*

Canadian geophysicists identified a possible precursory sign of great plate-boundary earthquakes at the Cascadia subduction zone in the Pacific Northwest. By using data from continuously monitored Global Positioning System (GPS) stations, they recognized small amounts of movement of the ground over periods of one to two weeks caused by slip above the megathrust fault at depths of 20 to 40 km (Figure 3.34).[53,54] Simultaneously, they identified non-earthquake tremors at about the same depth and same time as the slip events. Although it might seem as if these slip events would relieve strain across the fault, the opposite is true: strain increases up-dip, along the locked part of the plate-boundary fault during each episode of slip. In effect, the silent-slip events ratchet up the strain on the locked portion of the fault. The slip events occur at intervals of 13 to 15 months, last from 10 to 20 days, and migrate laterally along the length of the subduction zone. Geophysicists have argued that a future silent-slip event will probably trigger a giant plate-boundary earthquake, and thus its associated tremors indicate times of increased probability of such an event. The trick will be to recognize the specific slip event that tips the balance. If that were possible, the tremors might give people a few days to prepare for "the big one."

Seismic Gaps Portions of long faults that mark plate boundaries, such as the San Andreas fault in California, the Anatolian fault in Turkey, and the Queen Charlotte fault off the west coast of the Queen Charlotte Islands, have had no earthquakes for long periods, even though other parts of the faults have recently been active. These seismically quiet zones are termed *seismic gaps*. It is reasonable to assume that all parts of these faults are active and will ultimately slip. Therefore, zones that have been quiet in the recent past may be more likely to experience an earthquake than those that have produced quakes recently. Some seismic gaps, notably those along the Anatolian fault, have produced devastating earthquakes in the past few decades, but many have not. Moreover, some parts of seismically active faults have produced more than one large earthquake within short intervals.

Since 1965, the seismic-gap concept has been used successfully in forecasts of at least 10 large, plate-boundary earthquakes. One of these quakes took place in Alaska, three in Mexico, one in South America, and three in Japan. Seismic gaps along the San Andreas fault in California include one near Fort Tejon that last ruptured in 1857 and one along the Coachella Valley, a segment that has not produced a great earthquake for several hundred years. Both gaps may produce large earthquakes in the next few decades.[11,55]

Physical and Chemical Phenomena Local changes in gravity and the magnetic and electrical properties of near-surface crustal materials have preceded large earthquakes. For example, changes in *electrical resistivity* have been reported before earthquakes in the United States, Eastern Europe, and China.[51] Changes have also been observed in groundwater levels, temperature, and chemistry.[56] For example, significant increases in the radon content of well water were reported in the month or so before the 1995 Kobe, Japan, earthquake.[56] Many of these physical and chemical changes may occur when rocks expand and fracture in the hours or days preceding an earthquake.

Status of Earthquake Prediction and Forecasting

Optimistic scientists believe that we will someday be able to make not only reliable long- and intermediate-range forecasts but also short-range predictions (seconds to weeks) of the locations and magnitudes of damaging earthquakes. Although progress on earthquake prediction has not matched expectations, intermediate- to long-range forecasting, including hazard evaluation and probabilistic analysis of areas along active faults, has progressed faster than expected (see the Professional Profile).[60] For example, the 1983 Borah Peak earthquake (**M** 7) on the Lost River fault in central Idaho has been lauded as a success story in intermediate-range earthquake forecasting. Previous investigations demonstrated that the fault was active and that a large earthquake was likely to happen in the future.[61]

3.2 CASE STUDY

The 2009 L'Aquila Earthquake

The disastrous L'Aquila earthquake (**M** 6.3) occurred in the Apennines in central Italy on April 6, 2009. The epicentre was near the medieval city of L'Aquila, which, together with surrounding villages, suffered most of the damage (Figure 3.35). Almost 300 people died, making the earthquake the deadliest in Italy since 1980. The quake destroyed or damaged thousands of buildings in L'Aquila. Poor building standards and construction materials may have contributed to the large number of victims. According to rescuers, some concrete elements of the fallen buildings "seemed to have been made poorly, possibly with sand."[57] An official of Italy's Civil Protection Agency stated that "in California, an earthquake like this one would not have killed a single person."[58] The mainshock was followed by dozens of significant aftershocks. The strongest, on April 7, had a magnitude of 5.3 and caused further damage. At the time of the writing of this book, more than three years after the earthquake, much of the centre of L'Aquila is cordoned off and off-limits to the public. Reconstruction is proceeding slowly.

The L'Aquila earthquake was more than a disaster. It ignited a controversy with global fallout. A month before the earthquake, Italian laboratory technician Giampaolo Giuliani appeared on Italian television and predicted a major quake in the region, based on increased levels of radon in well waters.[59] He was accused of being alarmist by Italy's director of civil defence and was forced to remove his findings from the internet. Giuliani was also reported to police a week before the mainshock for spreading fear among the local population, when he predicted an earthquake was imminent in Sulmona, about 50 km from L'Aquila. Scientists in other countries have studied the relation between earthquakes and radon levels in groundwater since the 1970s—the results have not been encouraging, as most earthquakes are not preceded by changes in radon emissions.

Six scientists and a former government official, members of the Italian National Commission for the Forecast and Prevention of Major Risks, were tried for making "inaccurate, incomplete and contradictory statements" a week before the earthquake, when residents in L'Aquila and surrounding communities were becoming increasingly apprehensive in light of frequent foreshocks.[59] The trial, which lasted from September 2011 to October 2012, culminated with a verdict of guilty of involuntary manslaughter. The seven members of commission who were convicted and sentenced to six years in prison had different backgrounds. Three of the seven were seismologists, one was a volcanologist with experience in civil protection, two were earthquake engineers, and one was a civil servant with a background in fluid mechanics.

Much has been written about the decision and the international reaction to it, with many learned groups around the world decrying the verdict and the court sentence. This reaction seems to stem from the assumption that the scientists were being condemned because they failed to predict the earthquake, which geologists know is not possible. The story, however, is more nuanced—the decision was based on the scientists' failure to communicate the risk, not predict the earthquake.

◄ **FIGURE 3.35 EARTHQUAKE DAMAGE IN L'AQUILA** Thousands of buildings, many of them medieval treasures, were damaged or destroyed by the L'Aquila earthquake in April 2009. This photograph shows typical earthquake damage in the heart of the city. *(Newscom)*

The seven members of the commission met a week before the earthquake to consider the implications of a swarm of small tremors that were causing anxiety among citizens of L'Aquila and surrounding communities. The commission was asked to consider the possibility that the earthquake swarm might be foreshocks and that an earthquake large enough to damage or destroy buildings was imminent.

The three seismologists were clearly the relevant experts. Based on past experience and the scientific literature, they were likely aware that the probability of a larger damaging shock was low and thus concluded that a major earthquake was unlikely. At the commission meeting in L'Aquila, one of the seismologists stated, "It is unlikely that an earthquake like the one in 1703 [the penultimate damaging earthquake in the region] could occur in the short term, but the possibility cannot be totally excluded." Another said that "in the seismically active area of L'Aquila, it is not possible to affirm that earthquakes will not occur." These comments seem reasonable in view of our current knowledge of earthquakes.

Unfortunately, these experts did not make their opinions known to the public. And worse, one hour before the meeting convened, the civil servant, who had no background in seismology, stated in an interview with a local television station that the situation in L'Aquila was "certainly normal" and posed "no danger." He added that "the scientific community continues to assure me that, to the contrary, it's a favorable situation because of the continuous discharge of energy." This statement was misleading and scientifically incorrect. Tremors release little stress and do not reduce the probability of a major earthquake.

The court case centred on these statements. The citizens of L'Aquila were reassured that they were not in danger. A L'Aquila resident said that the civil servant's statement "was deadly for a lot of people here," largely because they were inclined to sleep outside if there was a significant possibility that a damaging earthquake might occur. Instead, people stayed inside and in many cases were killed or injured when their homes collapsed.

Five members of the panel, including the three seismologists, were travelling to the commission meeting and were unaware that the civil servant had made the statements he did in the TV interview. And they were travelling back to Rome when he answered press questions at a second press conference after the commission meeting. They had not been invited to the press conference. Thus, as Dr. Max Wyss, director of the World Agency of Planetary Monitoring and Earthquake Risk Reduction in Geneva, Switzerland, so eloquently put it, "it appears that the three seismologists and the two earthquake engineers were convicted for saying nothing even though they were never given the opportunity to say anything."

Many scientists worldwide have argued that the conviction of scientists and engineers was wrong because they were asked to provide an expert opinion on a question with no easy answer. However, the issue was one of risk communication and not earthquake prediction. In this regard, the court should have recognized the different roles played by the commission members. Arguably, the "obligation to avoid death, injury and damage, or at least to minimize them," as alleged by the prosecutor, lay on the shoulders of the civil servant, not the scientists and engineers who did not participate in the press conferences.

The earthquake killed two people and caused approximately U.S.$15 million in damage. Movement during the earthquake created a scarp up to several metres high along the 36 km length of the rupture zone. Field studies after the earthquake found that the new fault scarp coincided with the previously mapped fault trace, validating the usefulness of careful mapping of scarps produced by prehistoric earthquakes. Where the ground has broken before, it might break again!

Government officials and news media that publicize and thus give credence to earthquake predictions that have not been scientifically reviewed do a disservice to the public. In 1990, a pseudoscientific prediction of an earthquake in New Madrid, Missouri, was acted on by some government and business leaders and was publicized by local, state, and national media, even after the prediction was dismissed by the National Earthquake Prediction Evaluation Council.[62] Schools and businesses closed, public events were cancelled, people evacuated their homes, and more than 30 television and radio vans converged on the predicted epicentre.[63] The earthquake didn't happen. Widely publicizing earthquake predictions that have not been independently vetted by seismologists is the geologic equivalent of yelling "fire" in a crowded movie theatre.

The difficulty of predicting, or even forecasting, earthquakes is shown by the experience of the U.S. Geological Survey at Parkfield, California. The town of Parkfield, located on the San Andreas fault, sustained large (about **M** 6) earthquakes at about 20-year intervals between 1857 and 1966, gaining the dubious title of California's earthquake capital. On the basis of this history, the U.S. Geological Survey in 1984 forecast that a large earthquake would strike the Parkfield area sometime between 1987 and 1993. In 1985, it initiated a long-term earthquake monitoring project, termed

PROFESSIONAL PROFILE

Gail Atkinson, Seismologist

Dr. Gail M. Atkinson (Figure 3.36), Canada Research Chair in geophysics and professor in the Department of Earth Sciences at the University of Western Ontario, is an internationally recognized engineering seismologist. She is known for her work in engineering ground motion studies, earthquake source and attenuation processes, and seismic hazard analysis. Her numerous papers have been widely used in engineering applications and have changed earthquake engineering practice. "There are many populated areas that are at considerable risk of earthquakes," Atkinson says. "We need to continue to learn how to assess earthquake hazard."

Atkinson has been involved in seismic hazard analyses and ground motion studies for major engineering projects, including nuclear power plants, dams, tailings dams, offshore structures, liquid natural gas facilities, pipelines, waste-disposal facilities, and buildings in Canada, the United States, and overseas. She is an active member of the Canadian National Council of Earthquake Engineering (CANCEE), the body responsible for developing seismic design regulations for the National Building Code of Canada. Her work is highly regarded in the United States, where she served as president of the Seismological Society of America from 2001 to 2003 and as chair of its government relations committee. She participates actively in the U.S. National Earthquake Hazards Reduction Program (NEHRP), having been awarded over U.S.$1 million in NEHRP research grants over the past decade.

Atkinson is also president of the Polaris Consortium, a $10 million multi-institutional project involving 90 portable geophysical observatories that collect and analyze data on Earth's structure, resources, and earthquake hazards. Polaris stands for Portable Observatories for Lithosphere Analysis and Research Investigating Seismicity. For the past seven years, Atkinson and a group of colleagues have been studying earthquakes in Canada and around the world to determine how to deal with these disasters. The team set up three seismographic arrays in Ontario, British Columbia, and the Northwest Territories. Each array has 30 seismographs, and every seismograph is linked to a satellite, which sends data directly to Atkinson and other members of the team via the internet. This system allows the team to see earthquakes across Canada and the world in real time on their computers.

▲ **FIGURE 3.36 GAIL ATKINSON, SEISMOLOGIST** Dr. Gail Atkinson, a professor in the Department of Earth Sciences at the University of Western Ontario, is one of Canada's leading engineering seismologists. She and her students study how earthquake shaking might affect critical infrastructure such as nuclear power plants, dams, and pipelines. *(Gail Atkinson)*

Atkinson hopes these seismographs will enable her team to know what actually happens during an earthquake. She says that the key to understanding earthquakes is to understand their waves. "There is no way to predict where or when an earthquake will happen, but the ability to know the difference between P and S waves will change our warning systems," Atkinson says.

Although both P and S waves leave the earthquake source at the same time, the P waves move faster. As the waves move farther from the source, the P and S waves get farther apart.

"Though the time between the P and S waves may only be a matter of seconds, our hope is that with the ability to determine which is a P wave and which is an S wave, a better warning system can be put in place," Atkinson says. This warning system could include automatic shutoffs for gas lines and nuclear power plants that would help decrease damage. The warnings could give people just enough time to get to a safer location.

—John J. Clague

the Parkfield Experiment, to better understand what happens on the San Andreas fault before, during, and after a quake. The experiment involved more than 100 researchers from the U.S. Geological Survey, universities, and government laboratories. The scientists installed a dense network of instruments to capture the anticipated earthquake and reveal the quake process in unprecedented detail.[64] Then they waited and waited. The time of the forecast proved to be incorrect, although finally, on September 28, 2004, a magnitude 6 quake struck close to the town. During the time of the experiment, much larger, unanticipated earthquakes occurred on the San Andreas fault at Loma Prieta (1989) and Landers (1992). These quakes may have temporarily reduced the stress on the Parkfield portion of the fault.

Parkfield is the site of another important earthquake research project—the San Andreas Fault Observatory at Depth, or SAFOD. In 2004, building on more than 15 years of experience gained from the Parkfield earthquake experiment, the U.S. Geological Survey, with the support of the National Science Foundation, drilled a hole into the San Andreas fault zone to a depth of more than 3 km.[65] Several other holes were drilled off the main hole in the summer of 2007. The SAFOD project employed advanced directional-drilling technology developed by the petroleum industry to angle the main hole eastward at depth through the entire fault zone until relatively undisturbed rock was reached.

The objectives of the SAFOD project were to obtain samples of rock within the fault zone near the initiation point of previous **M** 6 earthquakes, and to install instruments to monitor small earthquakes, fluid pressure, temperature, and stress. From the samples and measurements, scientists are learning the composition of the fault-zone materials and are determining the stresses that initiate earthquakes and control their propagation. They are also testing hypotheses on the roles that high-pressure fluids and chemical reactions play in controlling fault strength and earthquake recurrence. By observing quakes "up close," SAFOD marks a major advance in predicting earthquakes.

Earthquake Warning Systems

Technically, it is possible to develop a system that would provide about 30 seconds of advance warning to Seattle, Vancouver, and Portland of the arrival of damaging earthquake waves from a great earthquake at the Cascadia subduction zone. Such a system would be based on the principle that radio waves travel much faster than seismic waves. The Japanese have a functioning system that provides earthquake warnings for their high-speed "bullet" trains; derailment of one of these trains by an earthquake could kill hundreds of people. The system functioned as designed during the 2011 Tohoku earthquake.

The warning times that these systems provide range from as little as 15 seconds to as long as 1 minute, depending on the location of the earthquake epicentre. This time could be sufficient for many people to shut off gas valves, shut down machinery and computers, and take cover.[66] An earthquake warning system, however, is not a prediction tool, because it only warns that an earthquake has already occurred.

Some people believe that the damage to scientific credibility caused by false alarms would be far greater than the benefits of a brief warning of an approaching earthquake. In the case of the Japanese high-speed rail system, about 5 percent of the warnings are false alarms. Others have expressed concern for liability issues resulting from warning system failures, and damage and suffering stemming from actions taken based on false early warnings.

3.10 Perception of and Adjustment to the Earthquake Hazard

Perception of the Earthquake Hazard

Severe ground shaking can be traumatic. The large number of people, especially children, who suffered mental distress following the 1971 Sylmar and 1994 Northridge earthquakes in California attests to the emotional and psychological effects of earthquakes. These events caused a number of families to move away from Los Angeles.

Typically, one community's experience with a large earthquake does not stimulate other communities to improve their preparedness. Intense shaking during the Northridge earthquake disabled part of the local seismograph network. The malfunction delayed emergency response because the location of the epicentre was not immediately known. The intense shaking also caused many poorly designed freeway bridges and buildings to collapse. A year later, the residents of Kobe, Japan, experienced nearly identical problems during a similar earthquake. Communication problems and damage to critical infrastructure prevented the Japanese government from quickly and effectively responding to the disaster. Japan is arguably one of the most earthquake-prepared countries in the world, yet significant relief did not reach Kobe until about 10 hours after the earthquake!

Two large (**M** 7.6 and **M** 7.2) earthquakes in Turkey in 1999 provided another example of the problems earthquakes pose to modern society. The first earthquake, on August 17, levelled thousands of concrete buildings. About 250 000 people were left homeless and more than 17 000 died. Many modern buildings collapsed during the intense shaking of the earthquake, whereas older buildings were left standing. As in the case of the 2001 Bhuj earthquake in India, poor construction practices contributed to the collapse of the newer buildings. Allegations were made that some Turkish contractors bulldozed collapsed buildings soon after the earthquake to remove evidence of shoddy construction. If those allegations are true, the contractors tied up equipment that could have been used to rescue people trapped in other collapsed buildings.

The lessons learned from California, Japan, and Turkey are bitter ones. Our vulnerability to catastrophic loss from large earthquakes is large and growing. Unreinforced concrete and masonry buildings not designed to withstand strong ground motion may collapse during a large earthquake. As a rule, reinforced concrete buildings constructed to meet modern seismic provisions of building codes should remain standing.

Community Adjustments to the Earthquake Hazard

It is not possible or desirable to prevent people from living in earthquake-prone areas; thus countries, provinces, territories, states, municipalities, and individuals must take steps to minimize seismic risk. The steps include careful location of critical facilities, structural protection, education, and ready availability of insurance and emergency relief. Individuals must also take steps to protect themselves.

Location of Critical Facilities Important public and community buildings and other structures, commonly referred to as *critical facilities*, must be located as safely as possible. Critical facilities include hospitals, schools, power plants, communication systems, and police and fire stations.

Selecting safe locations requires site-specific investigation of earthquake hazards, such as the potential for liquefaction and landslides. Planning commonly involves *microzonation*, which is the identification of areas subject to different earthquake hazards. Microzonation is necessary because the ground response to seismic shaking within even a small area can differ greatly. In urban areas, where individual properties may be worth millions of dollars, detailed maps of ground response to seismic shaking help engineers and architects design buildings and other structures that can better withstand earthquakes. Microzonation requires a significant investment of time and money, but it provides important information on how specific sites will respond if a strong earthquake does occur. Although we will never be able to completely eliminate death and injury from earthquakes, we can reduce the number of casualties by safely locating and building critical facilities.

Structural Protection The statement "earthquakes don't kill people, buildings kill people" succinctly captures the importance of building design and construction in reducing damage and injury from earthquakes. Buildings, bridges, pipelines, and other structures must be constructed to withstand at least moderate shaking in seismically active regions. Appropriate construction practices require that governments adopt and enforce building codes with seismic provisions similar to those in the National Building Code of Canada, the International Building Code, and the International Residential Code. When applying these provisions, architects and engineers try to balance earthquake risk reduction against the additional costs of earthquake-resistant design.[49]

Of equal importance to seismic provisions in building codes is the inspection and strengthening of existing structures to withstand the ground motions of large earthquakes. Strengthening existing structures, referred to as *retrofitting*, is costly, but it is necessary in buildings that might collapse during an earthquake. For example, a recent study showed that more than half of the hospitals in Los Angeles County could collapse in a large earthquake.[49] The cost of retrofitting these buildings is likely to be more than U.S.$8 billion, but the alternative—doing nothing—is unacceptable. Retrofitting is especially important in earthquake-prone developing countries and in communities in Canada and the United States that have only recently included modern seismic provisions in their building codes.

Improvements in building codes and retrofitting have been successful in reducing earthquake fatalities in the United States. Their effect can be appreciated by comparing two earthquakes of approximately equal magnitude, the 1988 Armenia (**M** 6.8) and 1994 Northridge (**M** 6.7) earthquakes. The loss of life in Armenia was staggering—at least 25 000 people were killed. In stark contrast, only 60 people died in the Northridge earthquake. Some towns near the epicentre of the earthquake in Armenia were almost totally levelled. Most buildings were constructed of unreinforced concrete and collapsed during the shaking, crushing or trapping their occupants. This is not to say that the Northridge quake was not a disaster. It certainly was—it left 25 000 people homeless, dropped several freeway overpasses, injured at least 8000 people, and caused about U.S.$20 billion in damage. However, because most buildings in Los Angeles are constructed with wood frames or reinforced concrete, thousands of deaths were avoided.

Education As with all other natural hazards, education is an important element of earthquake preparedness. An educational program might include distribution of pamphlets and videos to the public; instruction on earthquakes in primary and secondary schools; workshops and training sessions for engineers, architects, geologists, and community planners; and provision of information on the internet. Education can also take place through earthquake and tsunami drills in schools and through earthquake disaster exercises for government officials and emergency responders.

Availability of Insurance and Emergency Relief Insurance and emergency relief are vital to help a community, province, territory, state, or country recover from an earthquake disaster. Losses from a major earthquake can be huge—tens of billions of dollars. After the 1906 San Francisco earthquake, only 6 of the 65 companies that insured property in the city were able to pay their liabilities in full.[9] Potential insured losses from a large earthquake in a densely urbanized area, such as Vancouver, Seattle, San Francisco, Los Angeles, or San Diego, are enormous and probably beyond the capability of the insurance industry to handle. Insurance claims paid after the 1994 Northridge earthquake were U.S.$15.3 billion, but this amount is dwarfed by the more than CAD$200 billion in losses from the 1995 Kobe quake.[9] In the early 1990s, a large international reinsurance company completed a study of damage likely to be caused by an **M** 6.5 earthquake with a source 10 km beneath Vancouver. The bottom line of the report is that losses could total tens of billions of dollars.[67] Yet an **M** 6.5 earthquake is by no means the largest that could occur in Vancouver. An earthquake of this magnitude, for example, happened beneath Seattle in 1965.

The State of California has partially addressed the problem of catastrophic earthquake losses by making state-subsidized earthquake insurance available through the nonprofit California Earthquake Authority. Even with state subsidies, however, barely 25 percent of California residents have earthquake insurance, in part because of the high deductible.[9] The U.S. government has no federally subsidized program of earthquake insurance. Likewise, Canada does not offer any form of earthquake insurance, in spite of the high seismic risk in some parts of the country. However, the Canadian government would likely step in with relief funds in the event of a catastrophic earthquake.

Personal Adjustments Before, During, and After an Earthquake

Close to 180 million people in six Canadian provinces, three territories, and 39 American states live in seismically active areas. In terms of damage alone, billions of dollars could be saved if our buildings and their contents were better secured to withstand shaking from earthquakes. Earthquake preparedness is the responsibility of individuals, communities, and governments. An informed and prepared populace will result in greatly reduced injury and death when a large quake occurs.

Before the Shaking Starts Most earthquake casualties result from partial building collapse and falling objects, such as chimneys and light fixtures.[68] Consequently, it is important to know the safe spots in each room of a house or apartment. Safe spots include inside walls, supported archways, and the undersides of sturdy tables and desks. Danger spots include windows, fireplaces, and areas adjacent to heavy hanging objects and tall unsecured furniture. Do not rush outside during an earthquake, as debris from the outside of the building and power lines may fall to the ground.

Families should discuss and prepare an emergency plan.[69] Start by discussing what to do at home, at school, or at work if an earthquake happens. Prepare a list of what needs to be done ahead of time. People who live alone should develop a plan for themselves with links to neighbours and friends. Select an appropriate out-of-the-area contact who can be notified about the family's status after the earthquake. This contact, ideally a family member or close family friend, can pass on news to other family members if individuals are separated. Learn first aid and CPR, and keep a list of emergency telephone numbers. Ask for the emergency plan at your children's school. Know the safe places to be—and where not to be—in your home during an earthquake.

You can take simple, inexpensive measures to prepare your home for a quake:[69,70]

- Learn how to shut off your gas, water, and electricity.
- Make sure your house is bolted to its foundation.
- Make sure chimneys are strong and well braced.
- Keep breakable and heavy objects on bottom shelves of cabinets.
- Secure heavy furniture that could topple, such as bookcases, cabinets, and wall units.
- Strap water heaters to walls.
- Secure appliances that could move enough to rupture gas or electricity lines.
- Do not place heavy pictures and other items over beds.
- Locate beds and chairs away from chimneys and windows.
- Put secure latches on cabinet doors to prevent dishes and glassware from spilling out.
- Put anti-skid pads under televisions, computers, and small appliances or secure them with hook-and-loop-type fasteners or other such products.
- Keep flammable and other hazardous liquids, such as paints, in the garage or in a shed away from the house.
- Check chimneys, roofs, walls, and foundations for structural damage.
- Put plywood in the attic on joists around each chimney to prevent masonry from coming through the ceiling.
- Maintain a supply of emergency food, water, and other supplies in a secure, easily accessible area. Your food and water should be sufficient to last for three to five days. Choose foods that require no refrigeration, cooking, or preparation and are compact and lightweight.
- Store an emergency kit in your car, tool shed, or garage in case you have to evacuate your home and cannot go back in.

During the Shaking The best advice during an earthquake is not to panic. Books, dishes, glass, furniture, and other objects may come crashing down. If you are indoors, stay there. Move away from windows, glass partitions, mirrors, fireplaces, bookcases, tall furniture, and light fixtures. Get under a desk or table, or place yourself in an archway or inside corner of a room. Once there, protect your head and face. Avoid doorways, as doors may slam shut and cause injury. Do not use an elevator or run from the building. If you are in an elevator, get out as soon as you can. If you are outdoors, move to an open area away from buildings, windows, trees, and power lines. If driving, pull over to the side of the road away from power lines, bridges, overpasses, and buildings. Stay in your vehicle until the shaking stops. If you are in a crowded public place, try to take cover where you will not get trampled. Do not run for exits; sidewalks next to tall buildings are particularly dangerous because of falling glass and other materials. Keep away from windows, skylights, and display shelves laden with heavy objects.

When the Shaking Stops Try to remain calm, even though you may feel dizzy and sick.[68] Take several deep breaths, look around, and then leave the building, carefully watching for fallen and falling objects. Check yourself and others nearby for injuries. Administer first aid, but do not move seriously injured individuals unless their lives are in immediate danger. Place a HELP sign in your window if you need assistance. Put on sturdy shoes and protective clothing to prevent injury from debris and hunt for hazards. Look for fires, gas and water leaks, arcing electrical wires, and broken sewage lines. If you suspect damage, turn off the utility at the source. However, do not turn off the gas if there is no damage. If you smell gas, extinguish all fires, do not use matches, and do not operate electrical switches. Open the windows, shut off the gas valve, leave the building, and, if possible, report the leak to authorities. Check your home for damage, including the roof, chimney, and foundation.

The next step is to check and secure your food and water supplies. If tap water is available, fill a bathtub or other containers. Emergency water may also be obtained from water heaters, melted ice cubes, and toilet tanks. Purify water by boiling it for several minutes. If you have lost your power, strain the water through paper towels or several layers of clean cloth. Wear proper hand and eye protection to clean up any spilled hazardous materials. Do not use barbecues, camp stoves, or unvented heaters indoors. Do not flush toilets if the sewage line is damaged. Do not use the telephone unless there is a severe injury or fire; chances are that phones will not work anyway. Emergency phone numbers are found on the inside cover of most telephone books. Do not use your vehicle unless it is urgent, as it is important to keep the streets clear for emergency vehicles. Do not leave your car if downed power lines are across it. Avoid waterfront areas in case of a tsunami. Turn on your portable radio or television for instructions and news reports. Rely on emergency authorities for guidance. Be prepared for aftershocks.

For more information on earthquake preparedness, you can contact the

- Geological Survey of Canada
- U.S. Federal Emergency Management Agency (FEMA)
- United States Geological Survey
- Provincial or state emergency management agencies

REVISITING THE FUNDAMENTAL CONCEPTS

Earthquakes

❶ Hazards can be understood through scientific investigation and analysis.

❷ An understanding of hazardous processes is vital to evaluating risk.

❸ Hazards are commonly linked to each other and to the environment in which they occur.

❹ Population growth and socio-economic changes increase the risk from natural hazards.

❺ Damage and loss of life from natural disasters can be reduced.

1. Scientists' understanding of earthquakes has come a long way in recent decades. They have identified faults that are likely to produce earthquakes in the future and determined the likelihood of large earthquakes within a region. Scientists are not yet able to predict the exact time, place, or magnitude of an earthquake, but are searching for precursory signs that an earthquake is imminent.
2. Risk assessment is a mature field of research in seismology. We have a good sense of the probability and maximum size of an earthquake within a region over periods ranging from a few decades to a century or more. We also can determine the consequences of such an earthquake. These two key components are required for risk assessment.
3. Earthquakes are linked to several other natural hazards. For example, subduction earthquakes are responsible for deadly tsunamis, and large crustal earthquakes can trigger hundreds, or even thousands, of landslides. Health risks may also increase following a large destructive earthquake due to the spread of disease from pathogens in contaminated water.
4. The vulnerability of society to earthquakes is growing because human populations in earthquake-prone areas are rapidly increasing. Also, in many countries, buildings are not being constructed to withstand seismic shaking. Deaths and damage from even a moderate-size earthquake in an urban environment differ considerably. The loss of life from a magnitude 6.5 earthquake in an urban area in British Columbia would probably be small, but property damage would be billions of dollars. In contrast, an earthquake of a similar size in a city in Iran, Turkey, India, China, or other areas might kill tens of thousands of people. Because of population growth, earthquakes that used to produce disasters are now producing catastrophes.
5. Injury and loss of life from earthquakes can be minimized in a variety of ways, the most important of which is to design and build structures that will not collapse during a quake. State-of-the-art scientific information must be incorporated in building codes, and architects, engineers, and contractors must ensure that buildings are constructed to the highest possible standard. Also important are public education and public support of science that provides a better understanding of earthquakes and seismic hazards. We are far from being able to predict earthquakes, but we generally know where earthquakes occur and how likely they are to happen.

Summary

Seismologists use a variety of magnitude (**M**) scales to measure the amount of energy released during an earthquake. In the case of the commonly used moment magnitude scale, an increase from one whole number to the next larger number represents a 10-fold increase in the amount of ground shaking and about a 32-fold increase in the amount of energy released. The Geological Survey of Canada and the U.S. Geological Survey calculate preliminary magnitudes for large earthquakes and later revise these values after further analysis.

Scientists determine the intensity of an earthquake from its effects on people and structures. Earthquake intensity depends on the severity and duration of shaking, distance to the focus, and the local geological environment. The Modified Mercalli Intensity Scale has 12 intensity categories based on people's experiences and on property damage. Instrumental intensity is determined by using a dense network of seismographs. Information on intensity helps emergency responders focus relief efforts on areas that have experienced the most intense shaking. Other measurements, such as ground acceleration, are needed to design structures that can withstand shaking during future quakes.

Earthquakes occur when rocks or sediments are displaced rapidly along faults. Displacements are mainly horizontal along strike-slip faults and mainly vertical along dip-slip faults. Thrust faults are low-angle, dip-slip faults along which material above the fault plane (hanging wall) is displaced up and over material below it (footwall). Fault rupture may extend to the surface, creating a scarp, or it may terminate at depth as a blind fault.

A fault is considered active if it has moved at some time in the past 11 600 years and potentially active if it has moved in the past 2.6 million years. Some faults exhibit tectonic creep, which is slow continuous movement that is not accompanied by felt earthquakes.

Strain accumulates in rocks along a fault as the sides pull in different directions between earthquakes. Eventually, the accumulated strain exceeds the strength of the rocks and they rupture. Waves of energy, called seismic waves, radiate outward in all directions from the rupture.

Seismic waves compress (P waves) or shear (S waves) rock and sediment; some travel across the ground as surface waves. P waves travel the fastest, but S and surface waves cause most shaking damage. The severity of ground shaking is affected by the type and thickness of the materials through which seismic waves travel, the direction of fault rupture, and focal depth.

Buildings most likely to be damaged by earthquakes are (1) those that are constructed on artificially filled land or water-saturated, granular sediments that amplify shaking or liquefy, (2) those that are not designed to withstand significant horizontal acceleration of the ground, and (3) those that have natural vibrational frequencies that match the frequencies of the seismic waves.

Seismologists recognize an earthquake cycle comprising three or four phases. The cycle begins with a period of seismic inactivity, during which elastic strain builds up in the rocks bordering a fault. The second phase is marked by small or moderate earthquakes that happen when the accumulated strain locally exceeds the strength of the rocks. The third phase, which occurs in only some earthquake cycles, is characterized by foreshocks. The final stage is brief: the fault ruptures, producing elastic rebound and seismic waves.

Plate-boundary earthquakes occur on faults separating two tectonic plates and are of three types: strike-slip earthquakes on transform faults such as the San Andreas fault; large thrust earthquakes at subduction zones such as the Cascadia subduction zone; and normal, dip-slip earthquakes at spreading centres such as the Mid-Atlantic Ridge. Intraplate earthquakes occur locally within the North American plate, for example, in eastern Yukon Territory, western and mid-continent United States, southern Ontario and Quebec, parts of eastern and Arctic Canada, and South Carolina. The largest historic intraplate earthquakes in North America happened in the central Mississippi Valley in 1811 and 1812.

Large earthquakes produce violent ground motion that may damage or destroy buildings, bridges, dams, tunnels, pipelines, and other rigid structures. Secondary effects of earthquakes include liquefaction, regional subsidence and uplift of the land, landslides, fires, tsunamis, and disease. Large earthquakes release accumulated strain on faults and temporarily reduce the probability of another large quake in the same area. Some faults channel groundwater flow to springs, trap gas and oil, and expose or contribute to the formation of valuable mineral deposits.

People have increased earthquake activity by flooding valleys behind dams, by raising fluid pressures along faults through disposal of liquid waste in deep wells, by injecting fluids into the ground to enhance recovery of hydrocarbons, by extracting gas and oil from the crust, and by detonating underground nuclear explosions.

Regional earthquake hazards and risk can be determined through detailed geologic mapping of fault zones, excavation of trenches across faults to determine earthquake history and frequency, and analysis of sediments sensitive to shaking. Earthquake risk can be reduced by updating and enforcing the seismic provisions of building codes and by retrofitting existing vulnerable structures.

Earthquake prediction is a long-term goal of seismologists, but is decades from being achieved. Scientists have successfully made long- and intermediate-term forecasts of earthquakes by using probabilistic methods, but they have been unable to make consistent, accurate, short-term predictions. A problem in predicting earthquakes is that they are variable in time and space, with clusters of events separated by longer periods with low activity.

Warning systems and earthquake prevention are not reliable alternatives to earthquake preparedness. Communities in seismically active areas must develop emergency plans that allow them to effectively respond to a catastrophic earthquake. Effective emergency plans include earthquake education, disaster-response protocols, and availability of earthquake insurance. Individuals who live in or visit seismically active areas must learn how to react if a large earthquake occurs.

Key Terms

amplification (p. 62)
body wave (p. 57)
dip-slip fault (p. 54)
earthquake (p. 49)
earthquake cycle (p. 63)
epicentre (p. 49)
fault (p. 49)
focus (p. 49)
forecast (p. 77)
intensity (p. 49)
intraplate earthquake (p. 66)
liquefaction (p. 68)
magnitude (p. 49)
Modified Mercalli Intensity Scale (p. 50)
moment magnitude (p. 49)
normal fault (p. 55)
P wave (p. 57)
paleoseismologist (p. 75)
plate-boundary earthquake (p. 65)
precursor (p. 78)
prediction (p. 77)
resonance (p. 68)
reverse fault (p. 54)
Richter scale (p. 49)
S wave (p. 57)
seismic wave (p. 49)
seismograph (p. 49)
seismologist (p. 49)
slow earthquake (p. 55)
strike-slip fault (p. 54)
subduction earthquake (p. 65)
surface wave (p. 57)
tectonic creep (p. 55)
thrust fault (p. 54)

Did You Learn?

1. What is an earthquake?
2. Explain the difference between the epicentre and focus of an earthquake.
3. What is moment magnitude? How is it related to the amount of ground shaking and energy released by an earthquake?
4. What is instrumental intensity? What is its relation to a shake map?
5. Explain how faulting occurs.
6. What are the differences in the rates of travel of P, S, and surface waves?
7. Explain how seismologists locate earthquakes.
8. Summarize the four stages of the earthquake cycle.
9. What are foreshocks and aftershocks?
10. Where are earthquakes most likely to occur in the world? Where in North America?
11. Explain why earthquakes at transform, convergent, and divergent plate boundaries are different.
12. List the major primary and secondary effects of earthquakes.
13. Why are the largest earthquakes not always the most damaging?
14. What is liquefaction and in what type of earth materials does it occur?
15. Explain how earthquakes are linked to other natural hazards.
16. Provide examples of the possible benefits of earthquakes.
17. Explain how humans can cause earthquakes.
18. Provide examples of phenomena that may be earthquake precursors.
19. What is the difference between an earthquake prediction and a forecast?
20. How can people minimize their risk from earthquakes?

Critical Thinking Questions

1. You live in an area where a large earthquake might happen. The community is debating the merits of developing an earthquake warning system. Some people worry that false alarms will be common; others argue that the cost of the system is far greater than the benefits it will provide. What are your views on these points? Do you think the public should pay for an earthquake warning system, assuming such a system is feasible? What are potential implications of not developing a warning system if a large earthquake results in damage that could otherwise have been partially avoided?
2. You are considering buying a house in Victoria, British Columbia. You know that large earthquakes can occur in the area. What questions would you ask before purchasing the home? Consider the effects of earthquakes, the type of earth material underlying the property, and the age of the house. What might you do to protect yourself both financially and physically if you decide to buy the house?
3. You are working in a developing country where most buildings are unreinforced and built of bricks. The last damaging earthquake in the area happened 200 years ago and killed thousands of people. How would you describe the earthquake risk to your family members who live there with you? What steps could you take to reduce the risk?

MasteringGeology

MasteringGeology **www.masteringgeology.com**. Looking for additional review and test prep materials? Visit the Study Area in MasteringGeology to enhance your understanding of this chapter's content by accessing a variety of resources, including **Self-Study Quizzes, Geoscience Animations, GEODe Tutorials, RSS feeds, flashcards,** web links and an optional **Pearson eText.**

CHAPTER 4

▶ **TSUNAMI DEVASTATION** Aftermath of the March 11, 2011, tsunami in Otsuchi, Japan. The ferry in the centre of the photo was carried inland by the tsunami and dropped on a house. *(Yomiuri Reuters)*

Tsunamis

Learning Objectives

In this chapter we focus on one of Earth's most destructive natural hazards—**tsunamis**. These destructive waves are commonly called "tidal waves," but they are not tidal; the name is misleading. Tsunamis are common in some coastal regions and very rare in others. For years, scientists attempted to get public officials to expand tsunami warning systems outside the Pacific Ocean basin, but it took the catastrophe of the Indian Ocean tsunami in 2004 for governments and communities to take the tsunami hazard seriously. However, as often occurs after such catastrophes, translating increased awareness of a hazard into improved warning, preparedness, and mitigation is proceeding at a slow pace. This problem came into sharp focus on March 11, 2011, when a giant earthquake occurred beneath the seafloor off northern Honshu, Japan. The earthquake triggered a tsunami that reached elevations of up to 40 m above sea level along the northern Honshu coast and claimed more than 16 000 lives. It also caused meltdowns at three nuclear reactors at the Fukushima-Daiichi nuclear power plant. This chapter explains tsunamis and assesses the hazard they pose to people. Your goals in reading this chapter should be to

- Know what a tsunami is
- Understand the process of tsunami formation and propagation
- Understand the effects of tsunamis and the hazards they pose to coastal regions
- Know what geographic regions are at risk from tsunamis
- Recognize the links between tsunamis and other natural hazards
- Know what national, regional, and local governments, and individuals can do to reduce the tsunami risk

Giant Earthquake and Tsunami in Japan

The March 2011 earthquake off the Pacific coast of Honshu in Japan, also known as the 2011 Tōhoku earthquake, had a magnitude of 9.0.[1,2] It is fourth largest earthquake in the world since modern record keeping began in the late nineteenth century. The earthquake triggered a powerful tsunami that reached heights of up to 40 m above sea level in Iwate Prefecture and travelled up to 10 km inland in the Sendai area (Figures 4.1 and 4.2).[3]

The Japanese National Police Agency reported that the tsunami claimed nearly 16 000 lives, although this is likely an underestimate as 2800 people were still missing at the time this book was written.[1,2] The economic toll was huge: more than 380 000 buildings were destroyed or seriously damaged and another 690 000 buildings were partially damaged. The World Bank estimated the direct damage from the earthquake and tsunami to be U.S.$235 billion, making it the most expensive natural disaster in history. Shortly after the catastrophe, Japanese Prime Minister Naoto Kan said, "In the 65 years after the end of World War II, this is the toughest and the most difficult crisis for Japan."[1]

The tsunami also damaged three nuclear reactors in the Fukushima-Daiichi nuclear power plant, which led to their meltdown and the evacuation of thousands of residents.[2] A total of eleven nuclear reactors at four plants were automatically shut down following the earthquake as a precautionary measure. Unfortunately, Japanese officials had not considered the possibility of a tsunami as large as that of March 2011. Cooling is needed to remove heat from reactor cores after an emergency shut down and to maintain temperatures in spent fuel pools. Diesel generators powered the backup cooling system at the Fukushima-Daiichi plant, but the tsunami breached protective seawalls and destroyed them. The loss of the backup cooling system led to overheating of the reactor cores, leading to three explosions that released radioactivity. The three reactors suffered meltdowns and continued to leak coolant water long afterward. Radiation levels were up to 1000 times normal inside the plant and eight times normal outside the plant.[1] Many radioactive hot spots were found outside the evacuation zone, including in Tokyo. Radioactive matter also contaminated food supplies in several places.

◄ **FIGURE 4.1 TSUNAMI STRIKES JAPAN** A surging mass of muddy, debris-laden water races across the coastal plain at Natori, near Sendai, Japan, on March 11, 2011. The tsunami destroyed nearly everything in its path and killed more than 16 000 people in coastal communities on Honshu. *(Reuters/Kyodo)*

◄ **FIGURE 4.2 TSUNAMI INUNDATION AND RUN-UP** Google Earth image of the Sendai plain, Japan, showing the area of inundation (blue colours) and run-up (red and orange colours) during the tsunami in March 2011. Narrow bays focused the waves and experienced the largest run-ups. *(Mori, N., T. Takahashi, T. Yasuda, and H. Yanagisawa. 2011. "Survey of 2011 Tohoku earthquake tsunami inundation and run-up."* Geophysical Research Letters. *Reprinted with permission of John Wiley & Sons.)*

The source of the earthquake was the giant *thrust fault* that marks the surface of the Pacific plate where it moves beneath northern Honshu (Figure 4.3). *Subduction earthquakes*, which result from the slip of two plates at a subduction zone, are fairly common in this region, but the 2011 event was unusual because of its size. Few subduction earthquakes in Japan have magnitudes greater than 8.5, yet the 2011 earthquake released nearly twice the energy of a magnitude 8.5 quake. Portions of northeastern Japan moved 2.4 m closer to North America during the earthquake. Remarkably, the seafloor east of the epicentre, about 200 km east of the Japanese coast, moved 50 m toward the east and rose about 7 m.

The earthquake shifted Earth's rotational axis between 10 and 25 cm.[1] The speed of Earth's rotation increased, shortening the day by 1.8 microseconds due to the redistribution of the planet's mass. Dr. Richard Gross, a leading NASA researcher, explained that even such a small difference is important to NASA because it affects the way that spacecraft being sent to Mars are navigated.[1] If NASA does not take such a change into account, the chance the mission might fail will be greater.

The upward movement of the seafloor off northern Honshu caused the tsunami that brought destruction to people living along the adjacent coast. The tsunami also propagated throughout the Pacific Ocean, reaching the Pacific coasts of both North and South America, from Alaska to Chile. Chile, which is 17 000 km from the source of the earthquake, was struck by waves up to 2 m high. Warnings were issued and evacuations carried out in many countries bordering the Pacific. The West Coast and Alaska Tsunami Warning System in the United States issued a tsunami warning for most of the coast of California, the entire coast of Oregon, and the coast of western Alaska. A lower-level tsunami advisory was issued for other coastal areas of Alaska, and the outer coasts of Washington and British Columbia. In California and Oregon, the tsunami surge was up to 2.4 m high and

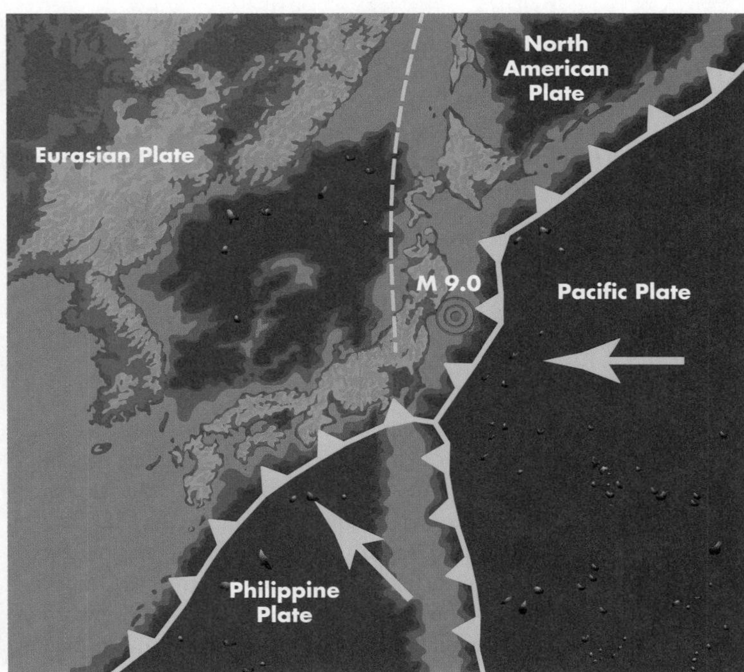

◄ **FIGURE 4.3 GEOLOGY OF A CATASTROPHE** Geologic setting of the giant Tōhoku earthquake. The earthquake occurred at the subduction zone off the northern coast of Honshu. The earthquake triggered a tsunami that killed over 16 000 people. The red circle locates the epicentre of the earthquake. *(Base map generated by GeoMapApp, http://www.geomapapp.org/)*

caused over $10 million in damage to docks and harbours.[1] The surge on the west coast of Vancouver Island was up to 1 m high, but caused no major damage. Damage to public infrastructure on the Hawaiian Islands was estimated to be $3 million, but private property owners suffered greater losses—the Four Seasons Resort Hualalai alone experienced tens of millions of dollars in damage.[1]

As in the case of the 2004 Indian Ocean earthquake and tsunami, the damage caused by the Japanese tsunami was far more deadly and destructive than the earthquake itself. Entire towns were destroyed, including Minamisanriku, where 9500 people died. Perhaps surprisingly, given that Japan has a long experience with tsunamis, a Japanese government study found that only 58 percent of people in coastal areas in Iwate, Miyagi, and Fukushima prefectures heeded tsunami warnings and evacuated to higher ground.[1] Of those who attempted to evacuate after hearing the warning, only 5 percent were caught in the tsunami; of those who did not heed the warning, about half were swept away.

The Tōhoku earthquake and tsunami taught us several lessons:

- Japan is arguably the most tsunami-prepared nation in the world. However, it was woefully unprepared for the tsunami that happened on March 11, 2011. Although Japanese geologists had documented past tsunamis of the same size as the 2011 tsunami, that knowledge had not reached public officials, emergency responders, and the public. Rather, people were under the false impression that only much smaller earthquakes and tsunamis, similar to those that Japan has experienced in recent times, were possible. As a consequence, more than half the fatalities were individuals who did not heed the tsunami warning to move to higher ground, or who were unable to do so. On the other hand, the loss of life could have been far greater had not so many people responded appropriately when the warning was issued. This tsunami was just as large as the one that struck Sumatra in 2004, yet the loss of life was only about 10 percent of that in Sumatra. This fact highlights the differences in the tsunami vulnerability of developed nations such as Japan and developing ones such as Indonesia.
- Earthquake and tsunami education is necessary for people who live on or visit coastlines that are vulnerable to tsunamis. Warning systems, without education, are of limited value.
- Tsunamis can have unanticipated serious secondary effects. The destruction of the reactors at the Fukushima-Daiichi nuclear power facility was not considered a possibility prior to the earthquake. Yet it could have been prevented if the back-up diesel generators had been placed on higher ground or if the protective tsunami walls had been built to withstand a tsunami the size of that in 2011. Similarly, many of the secondary economic losses, such as those caused by the slow-down in the Japanese auto industry, had not been anticipated, and yet could have been mitigated with appropriate planning.
- Recent scientific research documenting historically unprecedented tsunamis had not yet found its way into the decision-making process. We need to more effectively and rapidly move relevant scientific findings into the public arena. A high-tech tsunami warning system alone is not enough.

- Finally, the first defence against injury and loss of life from a tsunami is a simple one. When people living in a low-lying coastal area experience a large earthquake, they must understand that a tsunami could be on the way and immediately move to higher ground.

4.1 Introduction to Tsunamis

In winter, tourists flock to the west coast of Vancouver Island, British Columbia, to witness the full fury of a Pacific storm. Waves, metres high, rush ashore in turbulent fury, driven by storm-force winds blowing off the ocean. Yet powerful though they are, these waves pale in comparison to the waves of a large tsunami.

A tsunami is a series of waves caused by the displacement of a large volume of water, typically in an ocean or a large lake.[4] Tsunami waves do not resemble normal sea waves because their wavelength is much longer. Their wavelengths in the open ocean are about 200 km compared to tens of metres in the case of wind-driven waves, and their amplitudes are less than 1 m. At the shore, however, tsunami wave heights may reach tens of metres in large events. Several types of events can trigger a tsunami, including a large earthquake, landslide, explosive volcanic eruption, or an impact in the ocean of an asteroid or comet. Earthquakes or landslides trigger most disastrous tsunamis. Earthquake, landslide, and volcanic triggers are discussed below; asteroid and comet impacts are considered in Chapter 15.

Recent examples of damaging tsunamis include the following:[4,5]

- The 1755 (~**M** 8–9) Lisbon, Portugal, earthquake produced a tsunami that, along with the severe ground shaking and resulting fire, killed an estimated 20 000 people. The tsunami crossed the Atlantic Ocean, and waves up to 7 m high struck the West Indies.
- The violent eruption in 1883 of Krakatoa in the Sunda Strait between Java and Sumatra produced a tsunami over 35 m high that destroyed 165 villages and killed more than 36 000 people.
- The 1946 (**M** 8.1) earthquake near the Aleutian Islands in Alaska produced a tsunami that killed about 160 people on the Hawaiian Islands.
- The 1960 (**M** 9.5) Chile earthquake triggered a deadly tsunami that killed 61 people in Hawaii after travelling 15 hours across the Pacific Ocean.
- The 1964 (**M** 9.2) Alaska earthquake was responsible for several tsunamis that killed about 130 people in Alaska and California.
- The 1993 (**M** 7.8) earthquake in the Sea of Japan caused a tsunami that killed 120 people on Okushiri Island, Japan.
- The 1998 (**M** 7.1) Papua New Guinea earthquake triggered a submarine landslide that produced a local tsunami that killed more than 2100 people.
- The 2004 (**M** 9.3) Sumatra earthquake generated a tsunami that killed about 230 000 people (see Case Study 4.1).
- The 2010 (**M** 8.8) Chilean earthquake caused a tsunami that killed about 150 people in coastal communities between Valparaiso and Concepción.
- The 2011 (**M** 9.0) Japanese earthquake generated a tsunami that killed at least 16 000 people along the northeast coast of Honshu.

Earthquake-Triggered Tsunamis

An earthquake can cause a tsunami by displacing the seafloor or the floor of a large lake, or by triggering a large landslide. Displacement of seafloor is probably the most common of these mechanisms and occurs when a block of Earth's crust moves rapidly up or down during an earthquake. In general, it takes an earthquake of magnitude 7.5 or larger to generate a damaging tsunami. The upward or downward movement of the seafloor displaces the overlying water and initiates a four-stage process that culminates with the tsunami rushing ashore, commonly far from the source (Figure 4.4):

- Displacement of the seafloor during an earthquake sets in motion oscillatory waves that transmit energy outward and upward from the source. These waves intercept the ocean surface and spread outward, much like ripples in a pond after it has been struck by a pebble.
- In the deep ocean, the waves move rapidly and are spaced far apart. Their velocity is equal to the square root of the product of the acceleration of gravity and the water depth. The acceleration of gravity is approximately 10 m/sec^2. If we multiply the average ocean depth of 4000 m by 10 m/sec^2 and then take the square root of that number, we arrive at a velocity of 200 m/sec, which is equivalent to a velocity of 720 km/h or about the maximum air speed of a jet passenger aircraft! In the deep ocean, the spacing between the crests of waves may be more than 200 km and the height of the waves is generally less than 1 m. You would not notice a passing tsunami in a boat in the deep ocean.
- As the tsunami nears land, both the water depth and the velocity of the tsunami decrease. Near land, the forward speed of a tsunami may be about 45 km/h—too fast to outrun, but not nearly as fast as in the open ocean. The decrease in velocity is accompanied by a decrease in the spacing between wave crests and an increase in wave height.
- As the first tsunami wave approaches the shore, it transforms into a turbulent, surging mass of water, which then rapidly moves inland. It may be several metres to several tens of metres high and destroy everything in its path. During some tsunamis, the trough of the wave arrives first, causing the sea to recede and exposing the seafloor. A popular misconception is that a tsunami consists of a single immense wave that curls over and

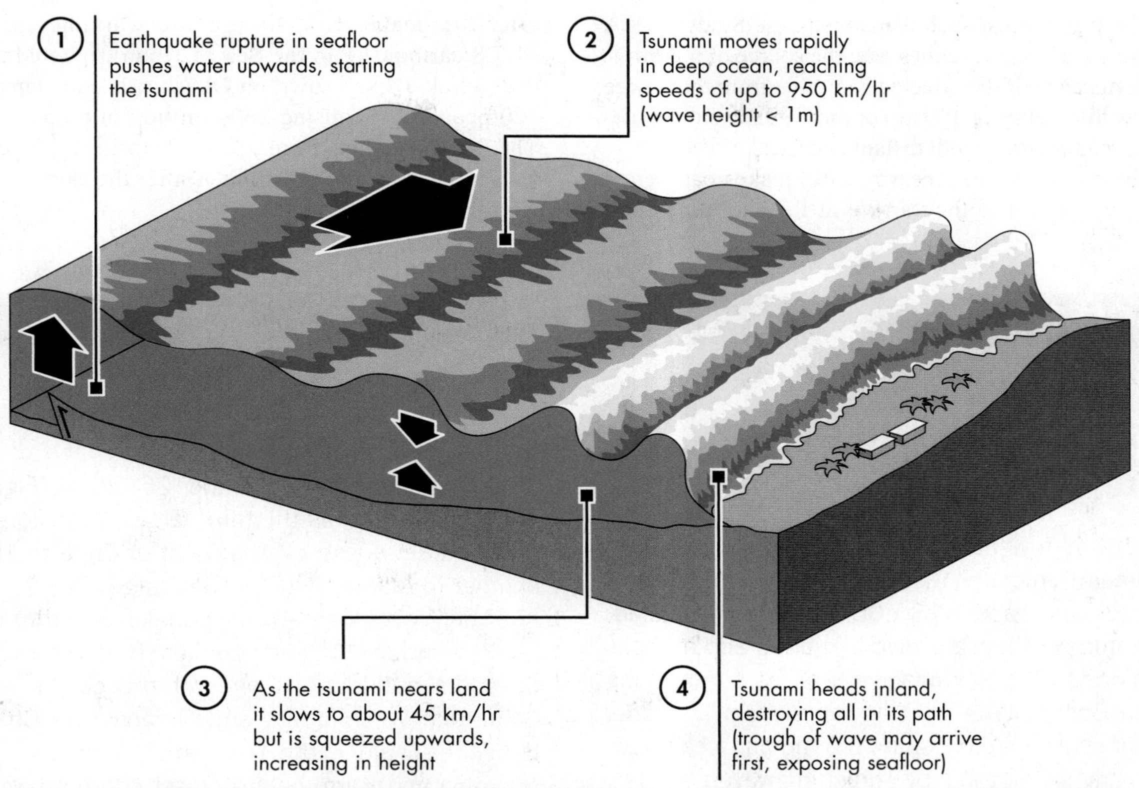

▲ **FIGURE 4.4 FORMATION AND DEVELOPMENT OF A TSUNAMI** Idealized diagram showing the process of how a tsunami is produced by an earthquake and travels away from the source toward a coastline. *(Adapted from the United Kingdom Hydrographic Office)*

crashes on the shore. Instead, the waves are typically turbulent, onrushing surges of debris-laden water (Figure 4.5). In some instances, however, the level of the sea or lake simply rises very rapidly, with little turbulence, as water moves inland. When one wave overtakes another, a steep wall of water, or *bore*, can be created. A tsunami also is not a single wave, but rather a series of waves separated by minutes to more than an hour.[6] Most of the deaths in Crescent City, California, during the great Pacific tsunami of 1964 occurred when residents who had returned to their homes after the first wave had withdrawn were swept away by a second, larger wave. The **run-up** of a tsunami is the maximum horizontal and vertical distances that the largest wave reaches as it travels inland. Once a wave has reached its farthest extent inland, the water returns back to the open ocean in a strong and often turbulent flow. A tsunami can also generate other types of waves, referred to as *edge waves*, that travel along the shore. The interaction between edge waves and later incoming tsunami waves can be complex. This interaction may produce wave amplification, causing the second or third wave of a tsunami to be larger than the first (see Survivor Story). Commonly, a series of tsunami waves will strike a coast over a period of several hours.[6]

Earthquakes can produce distant or local tsunamis, and in some cases both. A **distant tsunami**, or *tele-tsunami*, travels thousands of kilometres across the open ocean and strikes remote shorelines with little loss of energy. Great subduction

▲ **FIGURE 4.5 TSUNAMI STRIKES THAILAND** A huge wave surges into the tourist resort of Phuket on the morning of December 26, 2004. The tsunami killed nearly 10 000 people in Thailand, including almost 1000 foreign tourists and Thai citizens in Phuket. *(AP Images)*

earthquakes trigger most such tsunamis (Case Study 4.1). A **local tsunami** affects shorelines near the source of the earthquake. The distance of the affected shoreline from the source can be a few kilometres or 100 km or more. Great subduction earthquakes can produce both distant and local effects.

Tsunamis generated by nearby earthquakes can be especially deadly because they arrive at the shoreline soon after the quake, with little or no warning. In 1993, an **M** 7.8 earthquake in the Sea of Japan triggered a tsunami that struck a small town on Okushiri Island, Japan, killing 120 people and causing $600 million in property damage. The waves ranged from 15 to 30 m high.[7] The tsunami arrived only two to five minutes after the earthquake, so no warning could be issued.

4.1 CASE STUDY

Catastrophe in the Indian Ocean

Prior to 2004, few people knew what the Japanese word "tsunami" meant. That changed in late December that year when about 230 000 people were killed, many hundreds of thousands were injured, and millions were displaced in over a dozen countries surrounding the Indian Ocean. With no warning system in place, residents of coastal areas around the Indian Ocean were picked up and swept away by onrushing waves.

The source of the tsunami was an **M** 9.1 quake that struck just off the Indonesian island of Sumatra on Sunday morning, December 26, 2004 (Figure 4.6). The earthquake was the third largest in historic time—only an earthquake off the coast of Chile in 1960 and another in Alaska in 1964 were larger.[8]

The December 26 earthquake was a subduction quake, similar to earthquakes that happen at the Cascadia subduction zone off the coasts of British Columbia, Washington, and Oregon (see Chapter 3). It occurred along the fault that separates the Indo-Australia and Burma plates, west and northwest of the

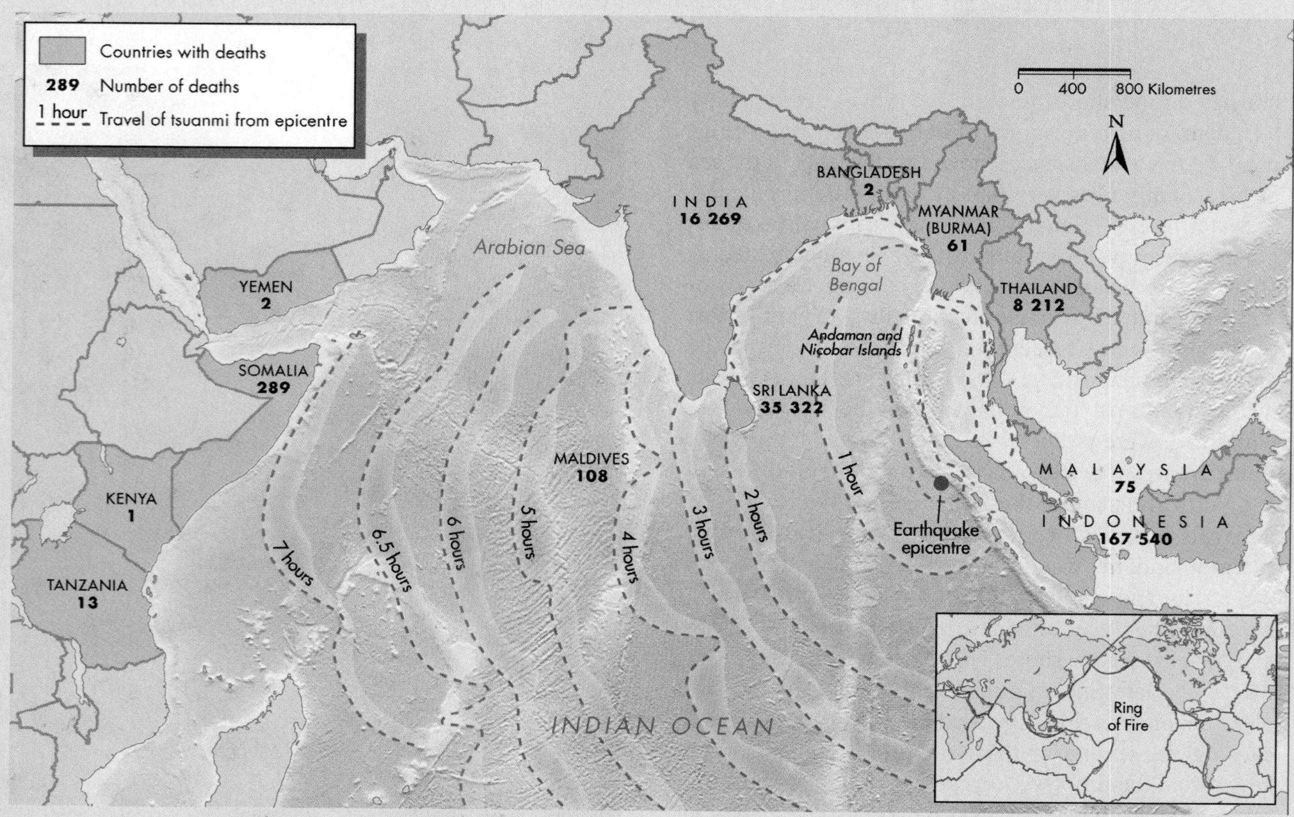

▲ **FIGURE 4.6 DEADLIEST TSUNAMI IN HISTORY** The Indian Ocean tsunami of December 26, 2004, was by far the deadliest tsunami in history. It formed off the northwest coast of the island of Sumatra and spread death and destruction across the Indian Ocean to the east coast of Africa. Dashed lines are the approximate positions of the lead wave or trough of the wave train at different times after the earthquake. *(Data from Casualties summarized in Telford, J., and J. Cosgrave. 2006.* Joint Evaluation of the International Response to the Indian Ocean Tsunami: Synthesis Report. *London: Tsunami Evaluation Coalition; Tsunami travel time data from NOAA)*

island of Sumatra. There, the Indo-Australia plate slowly moves eastward and downward beneath the Burma plate along the Sunda Trench west of Thailand and Indonesia.

The Indo-Australia and Burma plates were locked before the earthquake of December 26, 2004. Strain had been accumulating along the subduction zone for more than 150 years as a result of convergence of the two plates, and the accumulated strain was released by the earthquake.

The fault separating the two plates ruptured over a distance of more than 1200 km. Measurements and computer models indicate that the seafloor slipped as much as 5 m vertically and up to 15 m horizontally along the fault. Parts of the Andaman and Nicobar islands were elevated by these movements, whereas land along the western coast of Sumatra subsided up to 2 m, moving parts of the coastline below sea level.

The movement on this fault displaced the entire mass of overlying water and produced a series of waves that moved rapidly away from the seafloor. The waves reached nearby Indonesian islands within minutes and other countries hours later (Figure 4.6). Countries

(a)

(b)

◄ **FIGURE 4.7 BANDA ACEH BEFORE AND AFTER THE 2004 TSUNAMI** QuickBird satellite images of Banda Aceh, a provincial capital on Sumatra (a) on June 23, 2004, before the tsunami, and (b) on December 28, 2004, two days after the tsunami. All the buildings in this area were destroyed, including part of the bridge at the lower right. *(Digital Globe)*

(continued)

4.1 CASE STUDY *(Continued)*

bordering the Indian Ocean did not have a tsunami warning system like the one in the Pacific, and people were, for the most part, caught by surprise.

Scientists at the Pacific Tsunami Warning Center in Hawaii identified the earthquake and recognized that it might trigger a tsunami, but were not immediately able to determine the size of the quake.[9] Once they recognized that a tsunami was likely, they contacted Indonesian colleagues and had the U.S. State Department relay their concerns to nations surrounding the Indian Ocean. By then, however, it was too late; even if their warnings had reached authorities in time, there was no system in place in the affected countries to notify coastal residents. Had an effective warning system existed, tens of thousands of lives could have been saved.

Deaths from the Indian Ocean tsunami probably exceeded 230 000, but the exact number will never be known. More than three-quarters of these deaths were in Indonesia, which suffered both the most intense earthquake ground shaking and the largest tsunami. Other countries with catastrophic loss of life included Sri Lanka, India, and Thailand. All houses, businesses, and other buildings in some areas were completely destroyed (Figures 4.7 and 4.8). Tourist areas of Thailand were also hit hard—several thousand visitors in tourist resorts at and around Phuket were killed.

People reacted differently to the approaching waves. Some seemed mesmerized by them, whereas others recognized the danger and ran in panic. About 100 tourists and employees at a hotel in Phuket were saved when 10-year-old Tilly Smith sounded the warning. Tilly was on vacation with her family and recognized the signs of a tsunami from a lesson at her school in Oxshott, England, only two weeks before. She had learned that the sea sometimes recedes before the arrival of the first tsunami wave. Tilly observed an unprecedented withdrawal of the sea from the shore near her hotel and told her mother. When her mother didn't react, she started screaming that they were in danger and should get off the beach. Tilly finally convinced her family, as well as others, to return to the hotel. Shortly thereafter, the beach and hotel were hit by powerful waves. Her mother later admitted that she didn't know what a tsunami was. Her daughter's school lesson had saved her life and the lives of others.[10]

On the Nicobar Islands near the epicentre (Figure 4.6), Abdul Razzak, a port official, was awakened by the earthquake. He remembered from a *National Geographic* television program that tsunamis were often produced by undersea earthquakes. Razzak sent two co-workers on a motorcycle to warn villages, and he ran to nearby areas yelling to people to go to the hills. About 1500 people obeyed the warnings and evacuated to the hillsides where they watched in horror as the waves rolled in and destroyed their villages.[11]

On the Andaman Islands to the north, about 840 people in five Aboriginal tribes escaped injury. They had knowledge passed down from their ancestors about the relation between strong earthquakes and tsunamis. In one

(a)

(b)

▲ **FIGURE 4.8 DAMAGE AND DEBRIS FROM THE 2004 INDIAN OCEAN TSUNAMI** (a) In some areas, the tsunami completely destroyed all but the most sturdy buildings, such as this mosque in Aceh Province, Indonesia. (b) In other areas, such as this part of the Indonesian resort town of Pangandaran, the tsunami piled up huge amounts of debris, making it difficult to locate victims. *((a) Spencer Platt/Getty Images; (b) Dimas Ardian/Getty Images)*

instance, a Jarawa tribal elder led his people to the safety of a hilltop following the earthquake tremors, having been taught as a child to follow this procedure. Members of the Onge tribe also fled to the hills because their ancestors had taught them that if the level of the stream in their village suddenly dropped, it meant that the sea was pulling back and was "preparing to strike like a fist." In contrast, at least 48 recent settlers from the Indian mainland were killed on the Andaman Islands by the tsunami.[12] Long-time residents had a cultural memory of the natural hazard, whereas more recent immigrants did not.

In Khao Lak, Thailand, it was elephants, not people, who sounded the warning and saved lives.[13] Elephants started trumpeting about the time the earthquake occurred over 600 km to the west. Nearly an hour later, the elephants that were not taking tourists for rides broke loose from their strong chains and ran inland. Elephants carrying tourists ignored their handlers and climbed a hill behind the resort beach where about 4000 people were soon to be killed by the tsunami. When handlers saw the tsunami in the distance, they got other elephants to lift tourists onto their backs with their trunks and proceed inland. The elephants did this even though they were only accustomed to people mounting them from a wooden platform. The elephants managed to take the tourists to safety before the tsunami arrived.

SURVIVOR STORY

Swept Away by a Tsunami

Christine Lang (Figure 4.9), a Grade 2 teacher and accomplished swimmer, left rainy Vancouver for a Christmas vacation in December 2004. What she thought would be the trip of a lifetime became a catastrophe in the Thai tourist resort on Phi Phi Island. Christine was one of more than 2000 tourists enjoying the Christmas break on Phi Phi Island when a tsunami, powered by a 9.1 magnitude earthquake off the coast of Sumatra, swept ashore.

On the morning of December 26, Christine and a friend were strolling down the main street of the tourist resort when something strange happened. An older Thai woman raced from a jewellery store, screaming hysterically while running inland away from the beach. Almost immediately, clerks sprinted from the currency exchange service. Confused, Christine stopped in her tracks. "What's going on?" she asked her friend.

Other shopkeepers filled the street, abandoning their businesses. As they darted past her, Christine realized they were not only running from their shops, but also from the sea. She looked toward the beach, but a row of buildings blocked her view.

Christine was beset by questions. "What's happening? Is there a fire? Has a boat crashed into the pier? Is this a terrorist attack? What are these people running from?"

Screams filled the air. More and more people joined the mob. A German man wrenched Christine's arm and screamed, "Run!" She and her friend began to race inland with the crowd. They passed the Phi Phi Hotel, the only high building on the island, and arrived at an abandoned intersection. At that moment, a monstrous wall of water, two-storeys high, barrelled toward them. Christine tried to escape by running toward the Phi Phi Hotel but was caught in the turbulent water.

She braced herself, teeth and fists clenched, yelling, "No! No! No!" The tsunami pulled her in every direction. "I felt like a rag doll in a washing machine," she said.

She was slammed against a building. Although underwater, she could hear the screech of metal and the crack of wood timbers. She struggled to reach the surface, but was pinned beneath debris. Aching for air, she realized she was drowning. Thoughts raced through Christine's mind: "I don't want to die. Not now. There's so much more I want to do. I can't believe I'm going to die on my Christmas vacation! I'm really drowning. How can this be? I need air. I need air."

A second rush of water blasted into Christine's death trap, freeing her arms and legs. Her body was swept forward like a torpedo. She slipped into unconsciousness, but awoke underwater and spotted a circle of light coming from above. Instinct and adrenalin kicked in, and she swam furiously toward the halo in the distance. Her head broke the surface as she gulped for air. Oxygen filled her lungs. "I can breathe," Christine screamed, "I'm not going to die here!"

▲ **FIGURE 4.9 TSUNAMI SURVIVOR** Christine Lang was engulfed in the great Indian Ocean tsunami when it struck Phi Phi Island, Thailand, on December 26, 2004. She survived the terrifying ordeal. *(Courtesy of Christine Lang)*

A CLOSER LOOK 4.1

Geologic Evidence for Tsunamis

Canadian and U.S. geologists have assembled a large body of evidence that the Pacific coasts of Vancouver Island, Washington, Oregon, and northern California occasionally experience tsunamis generated by huge earthquakes at the Cascadia subduction zone. The most recent of the Cascadia tsunamis, which dates to January 1700, might have reached elevations of a few tens of metres at places along the Pacific coast.

How did geologists document these tsunamis? When a large tsunami runs ashore, it carries sand and gravel eroded from the seafloor. These coarse sediments are carried inland and deposited as a sheet of sand or gravel on the land surface. The resulting coarse deposit can be very different in character from the underlying earth materials (Figure 4.10). Geologists have learned that the best places to look for tsunami deposits are tidal marshes and low-lying coastal lakes—low-energy environments in which the dominant sediment is silt, clay, and peat. When a tsunami runs across a tidal marsh or enters a lake, it leaves a layer of sand or gravel, in some cases with shells or other marine fossils. The fossils are evidence that the sediments have come from the sea. When the tsunami is over, deposition of fine sediment resumes and the coarse sediment layer is preserved

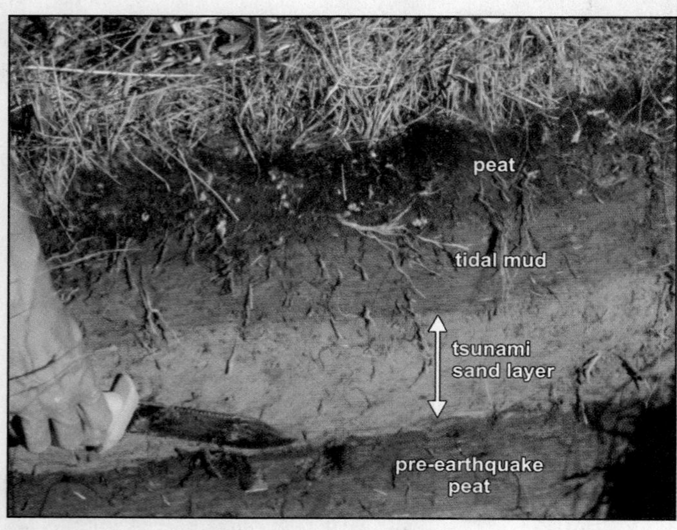

◄ **FIGURE 4.10 TSUNAMI SAND** This layer of sand in a tidal marsh near Tofino on the west coast of Vancouver Island was deposited by a tsunami generated by a great earthquake at Cascadia subduction zone in 1700. The sand layer becomes thinner and finer with increasing distance inland. It overlies peat and is overlain by tidal mud, both of which are the normal sediments that accumulate in the marsh. *(Tricouni Press tricouni@telus.net/Vancouver, City on the Edge)*

Landslide-Triggered Tsunamis

Some landslides that happen on the seafloor or in a lake can produce tsunamis. Large landslides that fall from mountain slopes into a large body of water can also generate tsunamis. In many cases, earthquakes trigger these landslides. An example is the deadly tsunami in July 1998 on the north shore of Papua New Guinea. The tsunami occurred on the heels of an **M** 7.1 earthquake, which was centred beneath the seafloor north of the island. The earthquake was felt at Sissano Lagoon, 50 km away, and shortly thereafter a tsunami arrived with waves up to 15 m high. Coastal villages were swept away, leaving 12 000 people homeless and 2100 dead.[16,17] A submarine landslide triggered by the earthquake caused the tsunami. The event highlighted the devastation that can result from a tsunami produced by a nearby earthquake and submarine landslide.[5,17] The earthquake alone probably would not have generated a large tsunami.

A similar event occurred on the east coast of Canada in November 1929. An **M** 7.2 earthquake at the southern edge of the Grand Banks, 250 km south of Newfoundland, triggered a huge submarine slump that, in turn, set off a tsunami.[18,19] The tsunami travelled across the Atlantic Ocean, registering on tide gauges as far away as South Carolina and Portugal. It damaged more than 40 coastal communities and claimed 27 lives in Newfoundland (Figure 4.12). Burin Peninsula was hardest hit. Three waves thundered up narrow channels and into bays over a half-hour period on the evening of November 18. The tsunami lifted small boats and schooners, snapped anchor chains, and tossed the craft onshore or engulfed them. Houses floated from their foundations; some were splintered, whereas others were swept back and forth by the flooding and ebbing waters. Damage was made worse by the fact that the tsunami arrived near the peak of a very high tide. Maximum wave heights in

within the mud or peat sequence. In such cases, the stage is set for archiving the next tsunami that comes ashore.

Some tidal marshes and low-lying lakes in Washington and Oregon contain several layers of tsunami sand and thus provide a record of prehistoric great earthquakes and tsunamis.[14,15] Each tsunami sand layer in a tidal marsh lies directly on a former marsh surface that subsided during the great earthquake that happened immediately before the tsunami came ashore (Figure 4.11). Scientists have used the spatial distribution of anomalous sand layers in coastal areas to infer the size of the tsunamis that deposited them. They have also dated sequences of tsunami sand layers using the *radiocarbon dating* method and learned that very large tsunamis generated by giant earthquakes at the Cascadia subduction zone occur, on average, about every 500 years. This information is important for assessing the risk that coastal communities on the Pacific coast face.

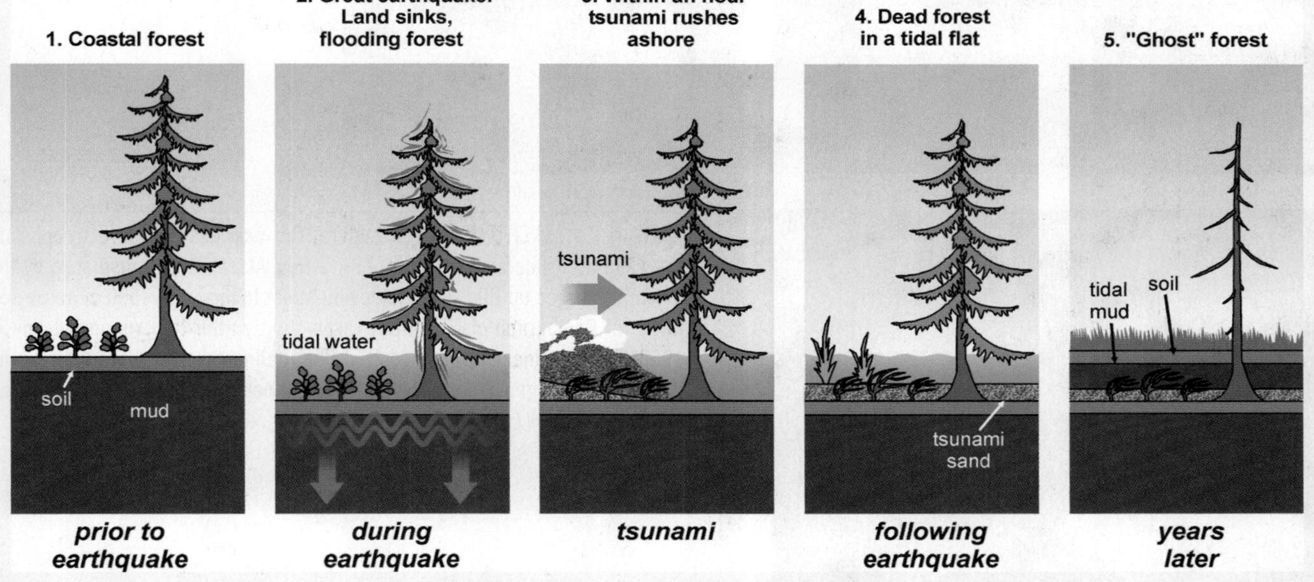

▲ **FIGURE 4.11 GEOLOGIC EVIDENCE FOR SUBDUCTION ZONE EARTHQUAKES AND TSUNAMIS** Schematic diagram of tidal marsh sediments showing the geologic signature of a great Cascadia earthquake and its attendant tsunami. A tidal marsh subsides up to 2 m during a great earthquake, just before it is overrun by the tsunami triggered by the quake. The tsunami leaves a layer of sand on the subsided marsh surface. After the earthquake, the inundated surface becomes a platform on which intertidal silt and clay are deposited. The definitive signature of the earthquake is a marsh peat (the former vegetated marsh surface), abruptly overlain by a sand sheet (the tsunami deposit) that, in turn, is sharply overlain by tidal mud. *(Reprinted with permission from Tricouni Press)*

communities that suffered the greatest damage ranged from 3 to 7.5 m.

The most spectacular landslide-generated tsunami of the twentieth century occurred on July 10, 1958, at Lituya Bay, Alaska.[20] A large earthquake triggered a rockslide at the head of the bay. The rockslide plunged into the bay and displaced seawater that ran up the opposite valley wall to an elevation of 525 m, destroying the forest in its path (Figure 4.13). A 30 m wave surged 11 km to the mouth of the bay, where it swept away three fishing boats anchored inside a low, forested spit. One of the three boats, the *Badger* (Figure 4.13), was swept over the spit and into the ocean beyond. Remarkably, the crew of the *Badger* survived to tell their harrowing story.

Geologists have suggested that even larger tsunamis than the 1958 Lituya Bay event have been produced by collapses of the flanks of volcanoes on Hawaii in the Pacific Ocean and the Canary Islands in the Atlantic Ocean.[21,22] Massive hummocky landslide deposits have been mapped on the seafloor adjacent to the Hawaiian Islands; some of them extend dozens of kilometres from the shore. The deposits were emplaced during the collapse of the flanks of Mauna Loa and Kohala, two of the large volcanoes on Hawaii. Geologists infer that these collapses triggered large tsunamis. In support of this idea, they note possible tsunami deposits far above the present shore on the slopes of Hawaii and Lanai. The deposits contain fragments of coralline limestone that otherwise occur only at and below sea level. The inference is that the fragments were deposited by waves up to several hundred metres high. The Hawaiian Islands would be devastated if an event of this type were to occur today. It is unlikely, however, that distant parts of the Pacific Rim would be affected in the same way. A tsunami generated by a landslide, even a very large one, loses energy over distances of thousands of kilometres, and wave run-ups on the west coast of North America would not be catastrophic.

◄ **FIGURE 4.12 TSUNAMI STRIKES NEWFOUNDLAND** Coastal communities on Burin Peninsula, Newfoundland, bore the brunt of a tsunami triggered by a submarine landslide on the Grand Banks in 1929. This photograph shows buildings in Lord's Cove that were tossed and smashed by the tsunami. *(Photo by Harris M. Mosdell, from the W. M. Chisholm collection, provided by A. Ruffman, GeoMarine Associates, Halifax, Nova Scotia)*

(a)

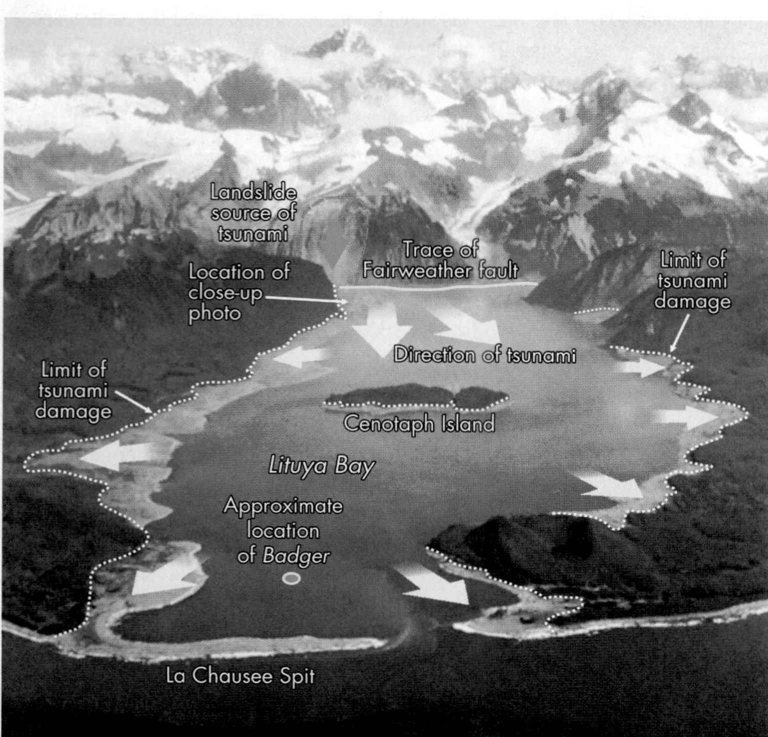

(b)

◄ **FIGURE 4.13 LITUYA BAY TSUNAMI** (a) The rocky headland directly opposite the 1958 landslide at the head of Lituya Bay, Alaska. Water displaced by the landslide surged up this slope, removing forest to more than 500 m above sea level. (b) Photograph of Lituya Bay taken shortly after the tsunami of July 7, 1958. The prominent trimline (dotted line) delineates the upper limit of the tsunami. The surging waters obliterated forest below this line. *(Reprinted with permission from Tricouni Press)*

Volcanic-Triggered Tsunamis

Tsunamis caused by volcanic eruptions are much less common than earthquake- or landslide-triggered tsunamis. However, the second most deadly tsunami in history was caused by a huge eruption of Krakatoa (also spelled Krakatau), an active volcano in the Sunda Strait between the Indonesian islands of Java and Sumatra. The massive eruption on August 26 and 27, 1883, was among the largest in recorded history.[23,24] It produced about 21 km^3 of fragmented rock and ash, and destroyed two-thirds of the island of Krakatoa. The cataclysmic explosion, which was heard up to 5000 km away, triggered a tsunami that destroyed 165 villages along the shores of Sunda Strait, damaged 132 others, and killed over 36 000 people.

Smaller tsunamis can be triggered by large volcanic mudflows that enter the sea during explosive eruptions. These tsunamis rapidly attenuate, or decrease, in size with distance from the volcano and thus are less hazardous than tsunamis caused by large earthquakes. Several active volcanoes in the Aleutian Arc in southern Alaska can produce small tsunamis by this mechanism.

4.2 Regions at Risk

Although most ocean shores and some lakeshores could experience a tsunami, some coasts are at much more risk than others due to their location with respect to earthquakes, landslides, and volcanoes. Coasts in proximity to a major subduction zone or directly across the ocean basin from a subduction zone capable of generating **M** 9 earthquakes are at greatest risk. About 85 percent of recorded tsunamis have been in the Pacific Ocean because of their association with large earthquakes at subduction zones that surround much of the Pacific.[25] Areas at greatest risk from tsunamis in the Pacific Basin are Japan, Kamchatka, Hawaii, islands in the southern and west Pacific, Chile, Peru, Mexico, and the northeast Pacific coast from Alaska to northern California (Figure 4.14). Other regions judged to have a high risk include parts of the Mediterranean and the eastern Indian Ocean.

Tsunamis generated by underwater landslides unrelated to earthquakes and by the flank collapse of volcanoes are less common, but they still represent a hazard to the east and west coasts of the United States and Canada, and to Hawaii and Alaska.

The Pacific coast of North America has been struck by one large tsunami during historic time, and geological research has provided evidence for many others over the past few thousand years. The Alaska earthquake of March 27, 1964, the second largest of the twentieth century (**M** 9.2), was responsible for tsunamis that killed 130 people, some as far away as California. The main tsunami swept southward across the Pacific Ocean at a velocity of about 830 km/h, reaching Antarctica in only 16 hours. It caused extensive damage on Vancouver Island, B.C., and claimed lives as far south as Crescent City in northern California. The town of Port Alberni, at the head of Alberni Inlet on Vancouver Island, was particularly hard hit.[26] Three main waves struck Port Alberni over a three-hour period on the morning of March 28. The second and most destructive wave surged 1 km inland, forcing the police to use boats to rescue guests from the upper floor of a local hotel. About 260 homes in Port Alberni were damaged by this tsunami, 60 extensively.

The risk that residents of the west coast of Canada and the United States face from tsunamis has been investigated by

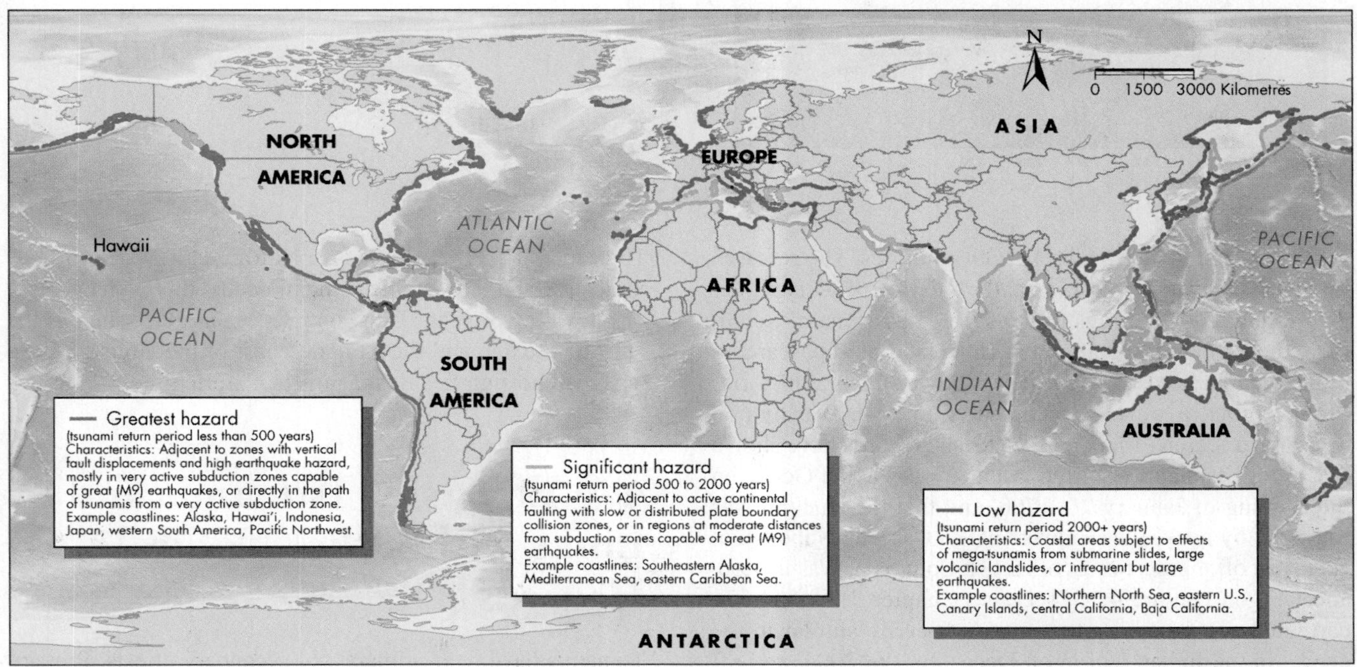

▲ **FIGURE 4.14 GLOBAL TSUNAMI HAZARD** Map showing the relative risk of the world's coastlines to a tsunami at least 5 m high. The map is generalized because tsunami run-up differs considerably over short distances depending on the form of the seafloor directly offshore and the topography and vegetation landward of the beach. *(Risk Management Solutions. 2006.* 2004 Indian Ocean Tsunami Report. *Newark, CA: Risk Management Solutions, Inc. Reprinted with permission. All rights reserved.)*

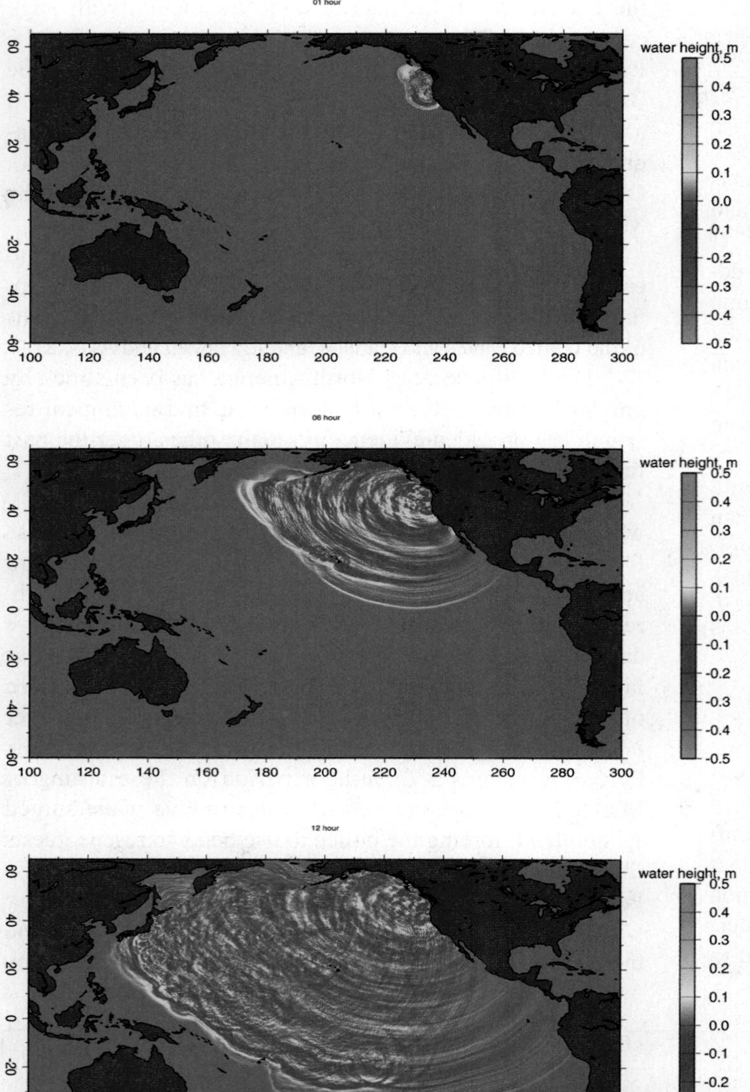

◀ **FIGURE 4.15 NORTH PACIFIC TSUNAMI OF JANUARY 1700** This output from a computer model of the tsunami triggered by the giant earthquake at the Cascadia subduction zone on January 26, 1700, is based on arrival times and run-up at several sites in Japan. The panels show wave height and location, from top to bottom, 1 hour, 6 hours, and 12 hours after the earthquake. *(Reprinted with permission of Kenji Satake)*

geologists and geophysicists in both countries. The research involves excavation and radiocarbon dating of tsunami deposits, dating of the rings of trees killed by tsunamis, computer modelling of tsunami propagation and run-up, searches of historical records in Japan, and study of the oral histories of Aboriginal peoples. Geological studies in Washington State and a search of Japanese historical records have shown that a huge tsunami occurred in the North Pacific Ocean on the evening of January 26, 1700, and that the tsunami was triggered by an **M** 9 earthquake at the Cascadia subduction zone, just off the west coast of British Columbia, Washington, and Oregon (Figure 4.15; see also Chapter 3).[27,28,29] The earthquake and ensuing tsunami were eerily similar to the events in the Indian Ocean on December 26, 2004, and off the east coast of Honshu, Japan, on March 11, 2011.

Some places, such as Japan and Hawaii, experience frequent tsunamis. Honshu, the largest of the Japanese islands, is struck by a 10 m tsunami, on average, once every 10 years.[30] Historical records in Hawaii dating back to 1813 indicate that a measurable tsunami occurs there, on average, once every two years.[31] In contrast, other areas, such as the Gulf of Mexico and the Atlantic coast of the United States, have had no significant tsunamis in historic time.

4.3 Effects of Tsunamis and Links with Other Natural Hazards

Tsunamis have both primary and secondary effects. Primary effects are related to the impact of the onrushing water and its entrained debris and to the resulting flooding and erosion. The energy of the fast-moving, turbulent water is sufficient to tear up beaches, most coastal vegetation, and houses and other buildings. These effects diminish with distance from the coast. Much of the damage to both the landscape and human structures results from the return of water to the ocean. What is often left behind is bare or debris-covered ground (Figure 4.8). The majority of tsunami deaths are from drowning. Death and injury also result from physical impacts, either by floating debris or by being washed into stationary objects such as buildings or trees.

Secondary effects of tsunamis are those that occur in the hours, days, and weeks following the event. Immediately after a tsunami, fires may start in developed areas from ruptured natural gas lines or from ignition of flammable chemicals released from damaged tanks. Such fires increased the damage caused by the 2011 tsunami in Japan. Also, as mentioned in the introductory section of this chapter, the 2011 Japanese tsunami destroyed backup diesel generators required to remove heat from reactors at the Fukushima-Daiichi nuclear power plant, which led to overheating of the reactor cores and their meltdown. Other secondary effects are pollution of freshwater supplies by contaminated seawater, damaged wastewater treatment systems, and rotting animal carcasses and plants. Outbreaks of disease may occur when people who have survived the tsunami come into contact with polluted water and soil. In the case of the 2004 Indian Ocean tsunami, public health officials were initially concerned that there would be outbreaks of waterborne illnesses such as malaria and cholera, but there were few cases of these diseases due to quick action by relief agencies and saltwater inundation of mosquito breeding grounds. Loss of shelter exposes people to insects, extreme weather, and other environmental hazards.[32] The trauma of a tsunami can produce long-lasting mental health problems in survivors. An interesting secondary effect of the 2011 Japanese tsunami was the huge amount of debris it produced. Much of the debris was washed out to sea and carried across the Pacific Ocean by currents. Some of the debris has washed ashore along the outer coasts of British Columbia, Washington, and Oregon, requiring continuing cleanup of beaches in these areas.

There are several links between tsunamis and other natural hazards. As mentioned previously, tsunamis are closely linked to offshore earthquakes, some landslides, explosive eruptions of island volcanoes, and asteroid and comet impacts. Coastal communities near the epicentre of a tsunami-producing earthquake experience casualties and property damage from both ground shaking and the tsunami itself. Powerful tsunami waves can dramatically change a coastline through erosion and sediment deposition, as illustrated by the 2004 Indian Ocean tsunami. A combination of tsunami erosion and coseismic subsidence drastically altered the shoreline near Banda Aceh (Figure 4.7).

4.4 Minimizing the Tsunami Hazard

Like many natural hazards, tsunamis cannot be prevented. However, the damage they cause can be greatly reduced through a variety of actions, including:[33]

- Detection and warning
- Structural control
- Construction of tsunami inundation maps
- Land use
- Probability analysis
- Education
- Tsunami readiness

Detection and Warning

The first warning of a possible tsunami comes when an earthquake of **M** 7.5 or larger happens in an offshore area. Not all such earthquakes produce tsunamis, and few that are smaller than **M** 7.5 do. We are able to detect tsunamis in the open ocean and accurately estimate their arrival time to within a few minutes. The travel times of tsunamis produced by earthquakes off Japan, Kamchatka, and Alaska are sufficiently long that low-lying coastal areas of British Columbia, Washington, and Oregon can be evacuated following alerts.

Three types of warning systems exist for tsunamis in the Pacific Ocean: a Pacific-wide system (the Pacific Tsunami Warning Center) located in Hawaii (Figure 4.16b); regional systems, including the West Coast and Alaska Tsunami Warning System, located in Alaska; and local systems in Chile and Japan.[33] All three types of systems use a network of seismographs to provide real-time estimates of earthquake magnitude and location before issuing a tsunami warning. The Pacific Tsunami Warning Center uses more than 100 coastal tidal gauges and sensors connected to floating buoys to verify that a tsunami was indeed produced following an earthquake. The bottom sensors, known as *tsunameters*, detect small changes in the pressure exerted by the increased volume of water as a tsunami passes over them. They transmit the pressure measurements to buoys at the ocean surface, which relay the data to the National Oceanic and Atmospheric Administration (NOAA) Geostationary Operational Environmental Satellite (GOES). The satellite in turn relays the information back to Earth, where it is sent through NOAA communication systems to warning centres (Figure 4.16a). After the Indian Ocean tsunami, similar systems were established in the Indian and Atlantic oceans, including warning sensors for Puerto Rico and the east coasts of the United States and Canada. By international agreement, information from the United States' warning system is shared with warning centres in 23 other countries. In Canada, tsunami information is disseminated through the British Columbia Provincial Emergency Program.

When the source of the tsunami is less than about 100 km away, there generally is insufficient time to warn and safely evacuate people, and no warning is possible when a local landslide triggers a tsunami. People close to the source, however, will probably feel the earthquake and can immediately move to higher ground. Certainly, if one observes the water receding, it is a sign to run inland or to higher ground if possible. Some coastal communities in British Columbia, Hawaii, Alaska, Washington, Oregon, California, Japan, New Zealand, and elsewhere also have warning sirens to alert people that a tsunami may soon arrive.

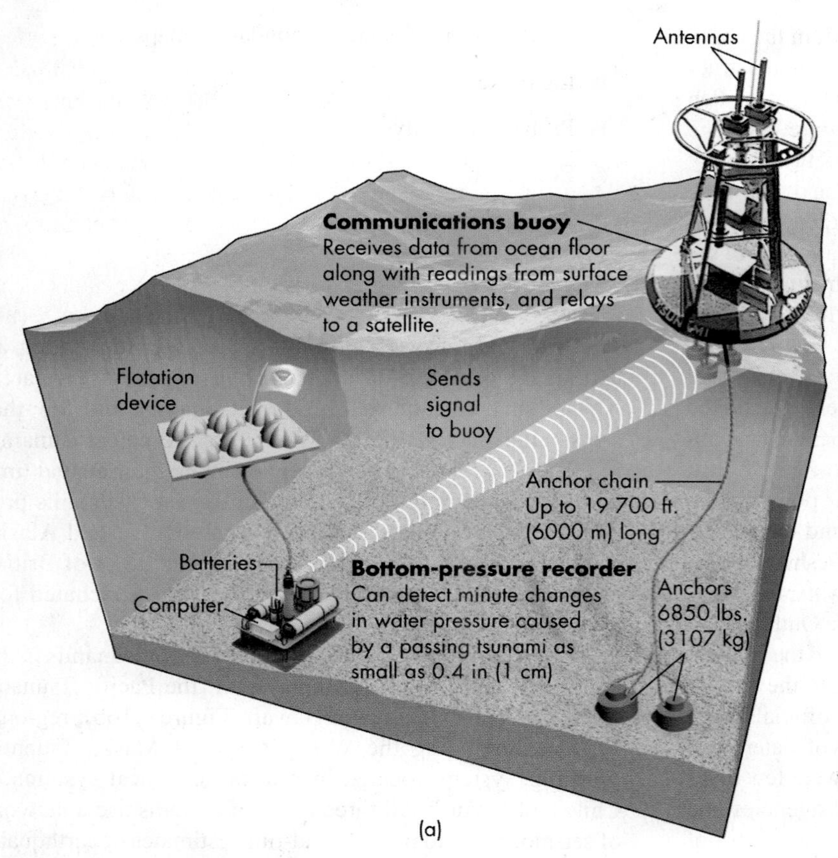

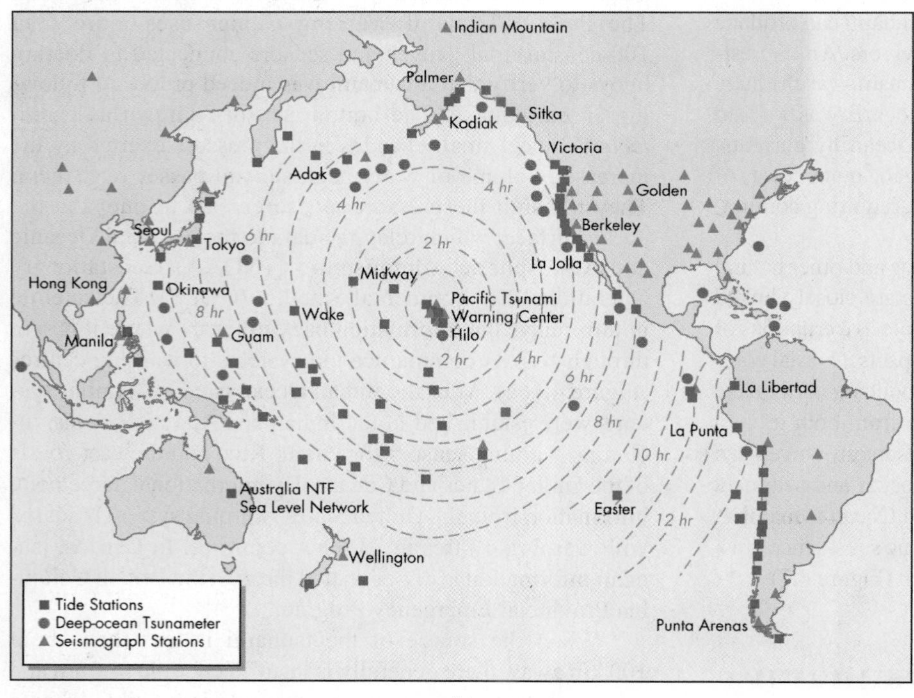

◄ **FIGURE 4.16 TSUNAMI WARNING SYSTEM IN THE PACIFIC** (a) A bottom sensor detects a tsunami and a tethered buoy transmits the information to a tsunami warning centre. (b) The Pacific Tsunami Warning Center in Hawaii acquires information from three sources: a network of seismographs, more than 100 tide gauges, and 30 DART ocean-bottom pressure sensors linked to surface buoys. The dashed lines show the time it would take a tsunami to reach Hawaii from locations in the Pacific Ocean. *(Modified after NOAA National Weather Service)*

Structural Control

Tsunamis that are even a metre or two high have such power that houses and small buildings are unable to withstand their impact.[4] However, larger structures, such as high-rise hotels and critical facilities, can be engineered to greatly reduce or minimize the impact of a tsunami. For example, the cities of Hilo and Honolulu in Hawaii have special requirements for construction of buildings in the tsunami run-up zone. Elevation of buildings and other types of flood-proofing (for example, installing seals for basement windows and bolting

▲ **FIGURE 4.17 HOUSE ON STILTS** This house in Hilo, Hawaii, is elevated on posts to provide protection against tsunamis. A small tsunami would pass beneath the living level of the home, leaving it undamaged. *(John J. Clague)*

houses to their foundations) provide protection where water depths are 1 m or less and currents are not strong. Structures can be elevated to higher levels to provide greater protection, but the cost may be prohibitive. Some houses on the coast of the Hawaiian Islands have been built on pilings, their floors elevated 2 to 3 m above ground level to allow water to move freely beneath them (Figure 4.17). However, the current building codes and guidelines in most jurisdictions exposed to tsunamis do not adequately address the effects of tsunamis on buildings and other structures.[34]

Dykes and walls can be constructed to prevent waves from reaching threatened residential and commercial areas. However, these barriers are expensive and should be built to the highest possible elevation that can be reached by a tsunami.[26] In some cases, offshore barriers can deflect tsunami waves or lessen their energy before they reach the shore. Again, these are expensive structures and may provide only limited protection. They are economically feasible only where large populations are at risk and where the threatened shoreline is at the head of a bay or inlet. Japan has constructed offshore breakwaters and onshore concrete walls to protect many of its coastal towns and cities from tsunamis.

Tsunami Inundation Maps

It is a fairly straightforward exercise to produce a run-up or inundation map after a damaging tsunami. Such a map was prepared for the island of Oahu, Hawaii, following the 1946 Aleutian tsunami (Figure 4.18). The map illustrates the tremendous variability of run-up in a typical tsunami—from 0.6 m in Honolulu to 11 m at Kaena Point and Makapuu Point at the northwest and southeast corners of the island.

Risk to coastal communities can be assessed by determining the frequency and size of past tsunamis from historical records and geological data. Maximum tsunami heights along a reach of the coast are estimated using computer-generated models and the distribution of historic and prehistoric tsunami deposits. Maps can then be produced showing areas likely to be inundated by tsunamis of different sizes. Local communities can use the maps to guide or restrict development in tsunami-prone areas and to educate people living in these areas about the risk they face. Many American and Canadian communities, for example Ucluelet and Port Alberni on Vancouver Island, have such maps.

Numerical models provide reliable forecasts of tsunamis propagating through the ocean and striking coastal communities. NOAA's Center for Tsunami Research provides software products specifically designed to support the Pacific Tsunami Warning Center's forecasting operations. It also conducts tsunami inundation modelling to assist coastal communities in their efforts to assess and mitigate tsunami risk. Computer models provide estimates of tsunami arrival times, currents, and forces on structures.

Land Use

Following the 2004 Indian Ocean tsunami, scientists discovered that tropical vegetation affected tsunami damage. The destruction along shorelines that experienced the highest waves was near total (Figure 4.19). Damage in areas where the waves were smaller, however, was more variable—some coastal villages were destroyed, whereas others suffered much less damage. Many villages that were spared were protected from the energy of the tsunami by either coastal mangroves or several rows of plantation trees.[35] Land-use and land-cover studies have documented the advantages of locating villages inland from a protective buffer of coastal vegetation. Yet, coastal mangroves are commonly not valued and are removed and replaced by homes, tourist hotels, and other buildings. These structures are generally located close to the beach and thus are particularly vulnerable to tsunamis. Although it might not be practical to move tourist areas inland, protective vegetation could be retained or planted between the development and the ocean to provide protection from at least small tsunamis.

Probability Analysis

The risk of a particular event can be defined as the product of the probability of the event and the consequences, should it occur. Thus determining the likelihood or probability of a tsunami is an important component of risk analysis. A hazard analysis may rely on evidence from past tsunamis rather than attempt to calculate the probability of a future event. The aim of this approach is simply to derive a tsunami inundation map for use in identifying evacuation routes. In contrast, probability analysis provides information not only on the likelihood of a tsunami, but its location, the extent of the run-up, and the possible severity of damage. The approach taken in a probabilistic analysis of tsunami hazard involves:

- Identification and specification of potential earthquake sources and their associated uncertainties

▲ **FIGURE 4.18 TSUNAMI RUN-UP ON OAHU** Map of Oahu, Hawaii, showing vertical run-up of the 1946 tsunami that originated from an earthquake in the Aleutian Islands, Alaska. Values are in feet. *(Modified after Walker, D. 1994.* Tsunami Facts. *SOEST Technical Report 94-03. School of Ocean and Earth Science and Technology. Reprinted with permission.)*

- Specification of factors that will attenuate or reduce tsunami waves as they travel from the source area
- Statistical analysis of past tsunamis, their sources, attenuation, and other factors, similar to an earthquake hazard analysis

The probabilistic approach to tsunami hazard assessment is still being refined and improved. One difficulty is that tsunamis at a particular location are generally rare events. If past events are too rare to develop a robust frequency–magnitude relation, a statistical technique known as *Monte Carlo* simulation can be used. The objective of the simulation is to determine tsunami return periods and probabilities for both distant and local sources. The technique selects a random sample of earthquakes of different magnitudes and determines the tsunami that would be propagated by each of these quakes. A mathematical model is then constructed based on the simulated events to estimate tsunami amplitude or run-up along a particular coast. The analysis must be done for each of the potential seismic sources for that particular coastline.[36]

Education

Even the most reliable tsunami warning system is unlikely to be effective if people do not respond in orderly and intelligent ways. Because tsunamis are infrequent, people's recollections, as with any rare natural phenomenon, fade with time, leading some communities into a false sense of security. Education is therefore essential if communities are to become more resilient (see Professional Profile).

A public education program must provide tsunami information at regular intervals, perhaps annually, and must include instructions on how to get information during a warning, where to go, and what to take. Educational initiatives should be included in school curricula to ensure that future generations understand the hazards and potential impacts of tsunamis. Education about tsunamis should not be limited only to those living on or near the coast but should be extended to all communities because people from inland regions often travel to tsunami-prone areas.

A range of educational initiatives can be undertaken in coastal communities. Activity sheets containing graphics,

◄ **FIGURE 4.19 TSUNAMI DAMAGE TO TREES** IKONOS satellite images of a low-lying coastal area (a) before and (b) after the 2004 Indian Ocean tsunami. Note the near-total destruction of vegetation by the tsunami. *(IKONOS Satellite images courtesy of the Centre for Remote Imaging, Sensing and Processing (CRISP) and GeoEye. Copyright 2007. All rights reserved)*

pictures, data, questions, and other relevant information can be used in schools to educate students about tsunami hazards. Evacuation routes can be publicized and marked by signs. Citizens should be consulted about land use in the tsunami inundation zone before decisions are made about siting or relocating critical facilities such as hospitals and police stations, schools and other high-occupancy buildings, and petroleum-storage-tank farms. Tsunami information can be printed in newspapers and telephone books, along with phone numbers of local emergency service offices. Citizens should also be regularly informed about local warning systems.

It is important to educate coastal residents and visitors that there is a difference between a **tsunami watch**, which is a notification that an earthquake that could cause a tsunami has occurred, and a **tsunami warning** indicating that a tsunami has been detected and is moving across the ocean toward their area. For a distant tsunami, several hours might pass between the time a warning is issued and when the waves arrive. In a local tsunami, there may be very little lead-time, so attention must be given to nature's warning signs, such as earthquake shaking or recession of water from a shoreline. People must also be educated that a tsunami is a train of waves and that the second and third waves may be larger than the first. Finally, people must be told that that the water returning to the sea once a wave has reached its inland limit can be just as dangerous as the incoming wave.

PROFESSIONAL PROFILE

Jose Borrero, Tsunami Scientist

Tsunami scientist Jose Borrero has witnessed firsthand the combined powers of land and sea unleashed (Figure 4.20). As a researcher in the University of Southern California Department of Civil Engineering, he has travelled to areas hit by some of the most massive natural disasters in recent decades.

Soon after a 10 m tsunami killed 2100 people in New Guinea in 1998, Borrero and his colleagues were there, documenting the extent of the wave damage.

"It looked like a hurricane came through. We saw dead bodies on the beach, ghost towns that looked completely bombed out—it was shocking," Borrero says.

The researchers travelled up and down the coast looking for flattened trees, measuring waterlines on house walls, and asking local people what they remembered about the time and size of the waves. Back home, they used the wave damage data, seismometer measurements, and computer models to reconstruct the events leading up to the tsunami. Their work revealed that an underwater landslide triggered by an earthquake had created the powerful tsunami.

After the 2004 Indonesian earthquake and tsunami, Borrero and a team of *National Geographic* filmmakers were among the first outsiders to reach one of the hardest hit areas—the city of Banda Aceh, Indonesia. What Borrero saw horrified him. "It was the worst of the worst of what I saw in New Guinea, except that instead of being confined to just one town, it went on for 200 miles." In some areas, 10 m waves had pushed boats onto balconies and stripped the trees off the side of a mountain 25 m up.

Borrero and colleagues used the data to develop a report about the risk Sumatra faces from future earthquakes and tsunamis along the next segment of the earthquake fault. "If you know what's possible, you can make a plan. We can give that information directly to cities and towns so they can make evacuation routes and public awareness programs," he says.

Tsunami education, Borrero says, is critical for coastal areas located directly on top of earthquake faults that produce tsunamis. "Tsunami warning systems only work for areas that are more than 2 hours away from the area of ground shaking. The only way to alert people is through education. If you feel an earthquake, and you're near the coast, don't sit and wait—head for high ground."

Growing up surfing in California, Borrero always wanted a career where he could work directly with the ocean. "There's this primal fear people have of giant walls of water. Many people have told me they have this recurring nightmare where they're trapped in a tsunami and can't escape. I've never had dreams like that; maybe understanding it keeps me from worrying too much."

—Kathleen Wong

▲ **FIGURE 4.20 DR. JOSE BORRERO** A research professor in the Department of Civil Engineering Tsunami Research Center at the University of Southern California, Dr. Borrero was part of an international damage assessment team that studied the effects of the 2004 Indonesian tsunami. Dr. Borrero is shown here with boats that were destroyed by the tsunami in Banda Aceh, Indonesia. *(© USC Tsunami Research Group)*

Tsunami Readiness

In order for a community to be prepared, or **tsunami-ready**, it must:

- Have an emergency operation centre with 24-hour capability
- Have ways to receive tsunami warnings from the Pacific Tsunami Warning Center, Canadian Meteorological Service, National Weather Service, Coast Guard, or responsible provincial and state agencies
- Have ways to alert the public
- Develop a tsunami preparedness plan that includes emergency drills
- Promote community awareness of tsunami hazards through educational programs

The educational component is of particular importance. Most people do not even know when a tsunami watch or warning has been issued. If they do, they may not be aware of the appropriate response. For example, in 2005, there was an earthquake in the Pacific Ocean far from the city of Santa Barbara, which has been certified a tsunami-ready community. As it turned out, a tsunami did not occur, but a tsunami watch was instituted for the California coast. Nothing was said about the size of the possible tsunami, and so some people on hearing the notice drove to the top of a nearby mountain pass more than a kilometre above sea level. The pass was great for a view, but people certainly did not need to drive that far or that high to evacuate the potential danger zone. The media reported that some people perched on a sea cliff at night, while others climbed palm trees to see if the waves were coming. This experience suggests that coastal communities, even "tsunami-ready" ones, are not adequately prepared for tsunamis.

4.5 Perception and Personal Adjustment to Tsunami Hazard

The above discussion suggests that many people do not know the signs of an approaching tsunami or what to do if a watch or warning is issued. If a warning is issued, a person can take the following actions:

- If you are at the beach and experience a strong earthquake, leave the beach and low-lying coastal area immediately.
- If the trough of a tsunami wave arrives first, the ocean will recede from the shoreline. This recession is nature's warning sign that a large wave is on the way and you should run from the beach.
- Most tsunamis consist of a series of waves, and there can be up to an hour between waves. It is therefore important to stay out of dangerous areas until authorities issue a notice that all is clear.
- If you hear a tsunami siren in a coastal community, move away from the beach to higher ground (at least 20 m above sea level) and listen to a radio for emergency information.
- If you are aware that a tsunami warning has been issued, do not go down to the beach to watch the tsunami. If you can see the wave, you may be too close to escape. Remember that these waves move fast and can be deadly. A 2 m tall person is very small compared to a 15 m tsunami wave.

REVISITING THE FUNDAMENTAL CONCEPTS

Tsunamis

1. **Hazards can be understood through scientific investigation and analysis.**
2. **An understanding of hazardous processes is vital to evaluating risk.**
3. **Hazards are commonly linked to each other and to the environment in which they occur.**
4. **Population growth and socio-economic changes increase the risk from natural hazards.**
5. **Damage and loss of life from natural disasters can be reduced.**

1. Although we cannot yet predict earthquakes that cause tsunamis, a tsunami can be detected and tracked once it has been generated. Seismologists are able to infer from the first seismic waves recorded by seismometers whether or not an earthquake has spawned a tsunami. If a tsunami has been produced, it can be tracked and its magnitude determined from data transmitted by tidal stations and ocean-bottom sensors. Such systems are now operating in the Pacific and other oceans of the world. The tsunami detection and warning system in the Pacific Ocean works well, whereas others are not yet totally reliable.
2. The risk from a tsunami can be defined as the product of the probability of an event of a particular size and the consequences of that event. Our understanding of earthquakes has progressed to the point where we can estimate the probability of a great earthquake likely to generate a tsunami at a given subduction zone. We also are able to determine what areas along a coastline might be inundated and consequences in terms

(continued)

of property damage and possible loss of life. Such risk assessments have been done for some tsunami-prone areas. As tsunami science progresses, better risk assessments will become possible.

3. Significant links exist between tsunamis and some other natural hazards. For example, a tsunami may be linked to a damaging earthquake and it may cause coastal erosion. A tsunami may also have secondary effects such as fires and disease that result, respectively, from fires generated from spilled flammable materials and contamination of water.
4. The human population in coastal areas vulnerable to tsunamis is increasing rapidly. Coastal cities are rapidly expanding and the tourist industry is bringing people to shorelines at risk from tsunamis. In addition, many tsunami-prone coastal areas have lost their natural vegetation, which formerly served as a barrier between the coast and inland development. Thus, even though the incidence of tsunamis has not increased, their impact on people has. As a result, what used to be disasters may now become catastrophes.
5. Damage and loss of life can be reduced by informing people of the signs of an approaching tsunami. Coastal communities can become tsunami-ready through education and the use of sirens, radio and television warnings, and cell phone text messages. In some areas, buffers of coastal vegetation can be planted or preserved to reduce tsunami damage.

Summary

The devastating tsunamis in the Indian Ocean in 2004 and in Japan in 2011 were international wake-up calls that we are not yet prepared for tsunamis and that we need effective tsunami warning systems in the world's major ocean basins. These systems must be designed to reach both coastal residents and visitors, and they must be coupled with an effective education program so that people are more aware of the hazard.

A tsunami is produced by the sudden vertical displacement of water in the sea or a lake. Processes that can produce tsunamis include large earthquakes, landslides, explosive volcanic eruptions, and the impact of extraterrestrial objects. The largest and most damaging tsunamis over the past few millennia have been caused by giant earthquakes associated with Earth's major subduction zones.

Distant tsunamis can travel thousands of kilometres across the ocean to strike a remote shoreline. In contrast, a local tsunami travels a much shorter distance to a nearby coast and can strike with little warning.

Tsunami effects are both primary and secondary. The primary effects are related to the direct impact, inundation, and erosion by the tsunami as it moves inland from the shore. Virtually nothing at the shore can survive a large tsunami. In 2004 in Indonesia, the force of the waves flattened concrete buildings. Secondary effects of tsunamis include possible water pollution, fires in urban areas, and disease.

Tsunamis are linked to the earthquakes that cause them, and tsunami inundation may directly follow the ground shaking and land-level change produced by quakes. Both earthquakes and tsunamis may cause fires. A tsunami can also change the coastline through erosion and deposition of sediment. Following a large earthquake and tsunami, parts of the coast may scarcely resemble what they looked like prior to the event.

Many strategies are available to minimize tsunami hazard. These include detection and warning, structural control, construction of tsunami run-up maps, land-use practices, probability analysis, education, and tsunami readiness. We are able to detect distant tsunamis in the open ocean and accurately estimate their arrival time to within a few minutes. It is more difficult to provide adequate warning of a local tsunami because it arrives at the shore soon after an earthquake. In this case, ground shaking or a sudden withdrawal of water from the coast may signal that a tsunami will soon arrive. In addition, warning sirens can alert people in a community to move inland or to nearby higher ground.

Without adequate education, tsunami watches and warnings are ineffective because many people do not know the appropriate action to take to save themselves and others. Through education, people will learn the natural warning signs of an approaching tsunami. Further, they will understand that tsunamis are a series of waves and the second or third wave may be the largest. The water returning to the ocean following tsunami inundation can cause as much damage as the incoming water.

Most communities along coasts with significant tsunami hazards are not adequately prepared for this underestimated natural hazard. Adequate preparation includes improved perception of the hazard, preparation and implementation of a tsunami preparedness plan, and promotion of community awareness and education.

Key Terms

distant tsunami (p. 95)
local tsunami (p. 96)
run-up (p. 95)
tsunami (p. 90)
tsunami-ready (p. 111)
tsunami warning (p. 109)
tsunami watch (p. 109)

Did You Learn?

1. Describe how a tsunami forms and propagates.
2. Explain the difference between a distant tsunami and a local tsunami.
3. What are the major primary and secondary tsunami effects?
4. Explain the relation between plate tectonics and tsunamis.
5. What geographic regions are most at risk from tsunamis?
6. How are tsunamis detected in the open ocean?
7. Name several natural hazards that are linked to tsunamis.
8. Explain the difference between a tsunami watch and a tsunami warning.
9. Describe the methods used to minimize the tsunami hazard.
10. What is meant by the phrase "tsunami-ready"?

Critical Thinking Questions

1. You are in charge of developing an education program aimed at raising a community's understanding of tsunamis. What sort of program would you develop and what would it be based on?
2. Can a tsunami warning system be effective if a tsunami reaches a populated coastal area less than 30 minutes after it first forms? Such a scenario is possible, for example, in the case of some communities on Vancouver Island if a great earthquake were to occur at the Cascadia subduction zone.
3. What do you think should be the role of the media in helping make people more aware of the tsunami hazard? How should scientists be involved in increasing awareness?
4. You live in a coastal area that experiences large but infrequent tsunamis. You are working with the planning department of your community to improve its tsunami readiness. What issues do you think are most important for achieving this goal and how would you convince the community that it is necessary or in its best interest to be tsunami-ready?

MasteringGeology

MasteringGeology **www.masteringgeology.com**. Looking for additional review and test prep materials? Visit the Study Area in MasteringGeology to enhance your understanding of this chapter's content by accessing a variety of resources, including **Self-Study Quizzes, Geoscience Animations, GEODe Tutorials, RSS feeds, flashcards,** web links and an optional **Pearson eText.**

CHAPTER 5

▶ **MOUNT UNZEN** The light-coloured mountain at the top is Mount Unzen, one of Japan's 19 active volcanoes. The volcano erupted violently on June 3, 1991, after two centuries of inactivity. A searing pyroclastic flow travelled 4.5 km from the crater, killing 43 scientists and journalists, and prompting local authorities to evacuate 12 000 people from their homes. Between 1991 and 1994, more than 10 000 small pyroclastic flows and lahars damaged parts of Fukae Town and Shimabara City in the foreground. This 1993 photograph shows a channel that was constructed in an attempt to funnel lahars to the sea. *(© Michael S. Yamashita/Corbis)*

Volcanoes and Volcanic Eruptions

Learning Objectives

There are about 1500 active volcanoes on Earth, almost 400 of which have erupted in the past century. Volcanoes occur on all seven continents as well as in the ocean. While you are reading this paragraph, at least 20 volcanoes are erupting on our planet. Your goals in reading this chapter should be to

- Know the different types of volcanoes
- Understand the relation between volcanoes and plate tectonics, and know where volcanoes occur on Earth
- Understand the effects of volcanic eruptions
- Know how volcanoes are linked to other natural hazards
- Recognize the benefits of volcanic eruptions
- Know the premonitory signs of volcanic eruptions
- Know the adjustments that people can make to avoid death and reduce damage from volcanic eruptions

Mount Unzen, 1991

One of the most destructive of Japan's 19 active volcanoes is Mount Unzen in the southwest corner of the country. More than 200 years ago, Mount Unzen erupted and killed an estimated 15 000 people. The volcano then lay dormant until June 3, 1991, when another violent eruption forced the evacuation of 12 000 people. By the end of 1993, Mount Unzen had produced about 0.2 km^3 of lava and more than 10 000 superheated flows of hot gas, ash, and large rock fragments, more such flows than any other volcano in recent time (Figure 5.1).[1,2] The 1991 eruption also produced damaging flows of volcanic debris and water, which geologists refer to as lahars. A specially designed channel was constructed to contain the lahars, but the flows overran the channel, burying many homes in mud (Figure 5.2).

The Mount Unzen story emphasizes three of this book's fundamental principles: the 1991 eruption had historical precursors; two different hazardous processes—pyroclastic flows and lahars—occurred in tandem; and evacuation of thousands of people greatly reduced the loss of life.

5.1 Introduction to Volcanoes

Volcanic activity, or volcanism, is directly related to plate tectonics, and most active volcanoes are located near plate boundaries (see Chapter 2).[3] Mid-ocean ridges and subduction zones are sites where molten rock, or **magma**, reaches the surface and erupts as **lava**. Approximately two-thirds of all active volcanoes on land are located along the *Ring of Fire*, which surrounds the Pacific Ocean (Figure 5.3). The volcanoes lie above the subduction zones bordering the Pacific, Nazca, Cocos, Philippine, and Juan de Fuca plates.

How Magma Forms

It is a common misconception that lithospheric plates move around on an ocean of molten rock. This is not true. With the exception of the outer core deep within the interior of Earth, the planet is composed of solid rock. So where does the magma come from that is erupted at the surface in volcanoes?

▲ **FIGURE 5.1 PYROCLASTIC FLOW FROM MOUNT UNZEN** The grey cloud moving rapidly downslope toward the village in the foreground is an incandescent pyroclastic flow consisting of hot gases, volcanic ash, and large rocks erupted from the volcano in June 1991. The firefighters in the truck and at the lower right are running for their lives. Fortunately, the pyroclastic flow stopped before reaching the village. *(AP Images/SUB-YOMIURI)*

◀ **FIGURE 5.2 DAMAGE FROM LAHARS** During and after the 1991 eruption of Mount Unzen, a series of lahars swept down the Mitzunashi River valley into inhabited areas on the coastal plain below the volcano. One of the lahars overtopped a channel that had been constructed to contain them, engulfing many houses and other buildings. *(© Michael S. Yamashita/Corbis)*

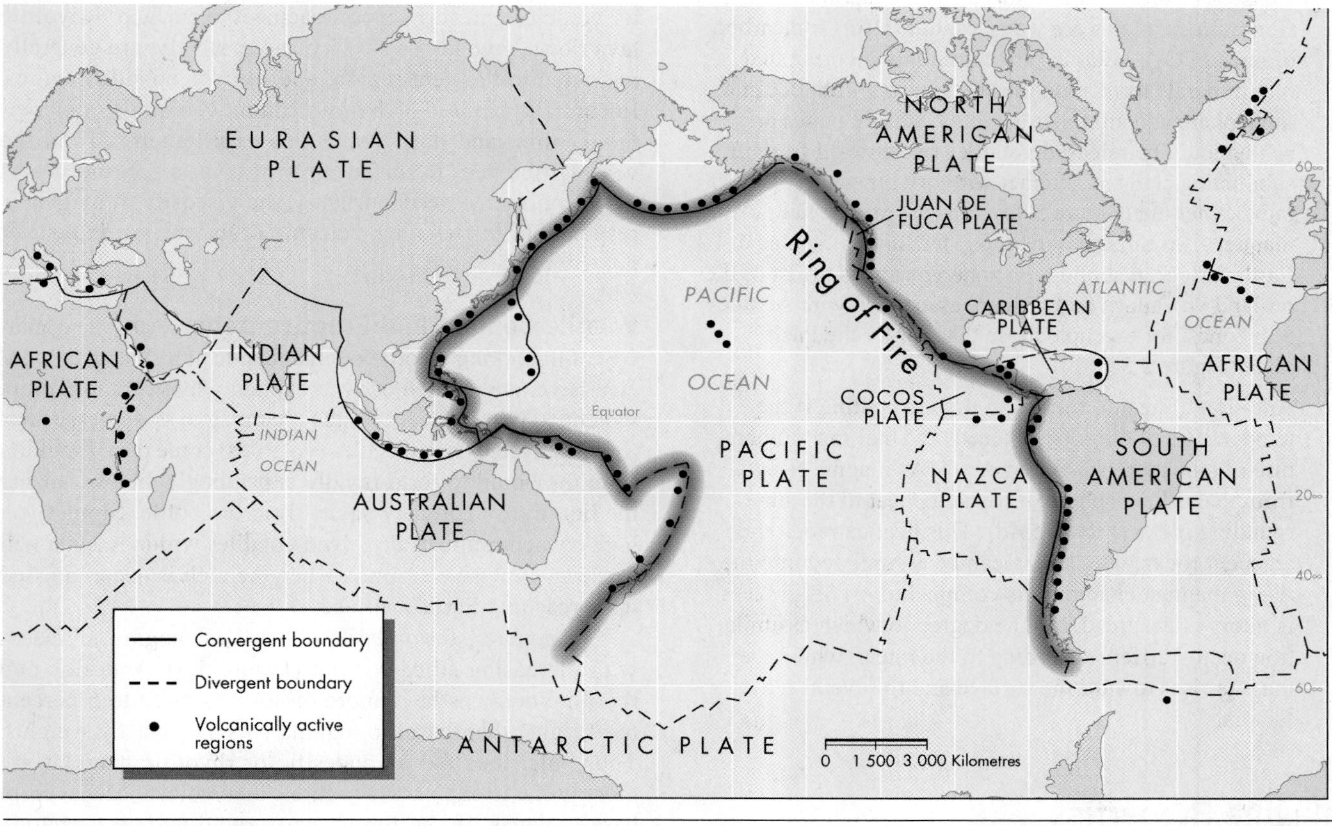

▲ **FIGURE 5.3 THE RING OF FIRE** The orange band in this drawing is the Ring of Fire, a belt of active volcanoes and frequent earthquakes bordering the Pacific plate. The Ring of Fire includes most of the world's onshore active volcanoes. *(Modified from Costa, J. E., and V. R. Baker. 1981.* Surficial Geology. *Fairfax, VA: Techbooks. Reprinted with permission of John Wiley & Sons.)*

Most magmas come from the asthenosphere, where rock is close to its melting temperature (see Chapter 2). Figure 5.4 illustrates the three main ways in which silicate rocks can melt: (1) decompression; (2) addition of volatiles (e.g., dissolved gas); and (3) addition of heat. These processes are closely linked to plate tectonic processes.

1. **Decompression melting** occurs when the pressure exerted on hot rock within the asthenosphere is reduced. The melting temperature for mantle silicate rocks at Earth's surface is approximately 1200°C. Within most tectonic settings, this temperature is exceeded within the upper 100 km of Earth. However,

like temperature, pressure also increases with depth. It is this great pressure generated by the weight of overlying rock that keeps Earth's mantle in a solid state far above the surface melting temperature of the rock. If the pressure is reduced without a change in temperature, the rock might melt. Decompression melting happens at divergent plate boundaries, continental rifts, and hot spots. At divergent plate boundaries and continental rifts, the lithosphere is stretched and thinned, which causes the mantle to well up toward the surface where pressures are lower, which causes decompression melting (Figure 5.4a). At hot spots, plumes of hot rock deep in the asthenosphere well up to shallow depths, generating magma (Figure 5.4b).

2. Addition of volatiles lowers the melting temperature of rocks by helping to break chemical bonds within silicate minerals. Rocks that are close to their melting temperature will melt in the presence of added volatiles. **Volatiles** are chemical compounds that exist in a gaseous state at Earth's surface and evaporate easily. Common examples are water vapour (H_2O) and carbon dioxide (CO_2). Volatiles are commonly incorporated into minerals formed on the seafloor or within oceanic lithosphere and are released when oceanic plates are subducted. The released volatiles rise upward from the subducting slab and interact with dry lithosphere to induce melting (Figure 5.4c). Magma produced in this manner rises through the lithosphere and is erupted at Earth surface at subduction zone volcanoes. Most of the on-land volcanoes on Earth are associated with subduction zones, for example those that border the Pacific plate (Figure 5.3).

3. Addition of heat to rocks can induce melting if the temperature of the rocks exceeds the melting temperature of silicate rocks at that depth. As magma rises from the asthenosphere, it transfers heat to the surrounding rocks (Figure 5.4d). This heat can melt the adjacent rocks, which then can be assimilated into the rising magma, changing its composition. This process is likely widespread, but the degree to which assimilation occurs differs according to the rate at which the magma rises toward the surface and to several other factors.

Magma Properties

Composition Magma is composed of melted silicate minerals and dissolved gases. The main elements in magma are oxygen (O), silicon (Si), aluminum (Al), iron (Fe), magnesium (Mg), calcium (Ca), sodium (Na), and potassium (K). The two most abundant elements are Si and O; when combined, they are referred to as silica (SiO_2). The names that geologists apply to volcanic rocks depend on the amount of silica present in the rock. *Basalt*, the most common volcanic rock on Earth, contains between 45 percent and 52 percent SiO_2 by weight. *Andesite* contains more silica than basalt (52–63 percent) and is not as common. *Dacite* (63–68 percent silica) and *rhyolite* (>68 percent silica) are even less common. Magma also contains small but significant amounts of dissolved gases, mostly water vapour and carbon dioxide. Most active volcanic areas have a variety of interesting surface features connected to their underground "plumbing system"—fractures and chambers through which magma, volcanic gases, and hot waters flow.

Viscosity A fluid's resistance to flow is termed *viscosity*. Silica-rich magma, like molasses or refrigerated honey, does not flow easily and has a high viscosity. In contrast, magma with a relatively low silica content flows more easily, like warm honey, and has a low viscosity. Viscosity is affected by temperature as well as composition (Figure 5.5). Basaltic lava, which has a relatively low silica content, has a low viscosity when it erupts onto the surface; however, as it flows away from the vent and cools, its viscosity increases, causing the flow to move more slowly and change in form.

The variability in magma viscosity strongly influences both the mobility of the magma in the subsurface and its velocity and form if its reaches the surface. Rhyolitic lava flows have high viscosity, move slowly, are generally restricted to the vent region, and form steep-sided domes. In contrast, basaltic lava flows can move rapidly, are commonly thin, and may travel tens of kilometres from the vent. Differences in surface flow of lavas affect the shapes of volcanoes. More importantly, the viscosity of magma is responsible for whether volcanic eruptions are violent or not (Figure 5.5).

Volatile Content and Eruptive Behaviour The analogy of uncorking a bottle of champagne or a shaken carbonated beverage is often used to explain why volcanoes erupt violently. When uncorked, the champagne is depressurized and dissolved CO_2 molecules (volatiles) come out of solution from the liquid to form rapidly expanding bubbles, causing the liquid to vigorously squirt from the bottle. Similarly, a high concentration of dissolved volatiles within magma will cause an explosive eruption when the melt is decompressed upon reaching Earth's surface.

In general, the volatile content of magma increases with increasing silica content (Figure 5.5). Andesitic and rhyolitic magmas have more dissolved gas (2 to 5 percent by weight) than basaltic magma (<1 percent by weight). Thus volcanoes fed by andesitic or rhyolitic magmas are more prone to explosive eruptions than volcanoes that erupt basaltic lavas. In the former case, rapid bubble formation and degassing of the more viscous magma during an eruption breaks apart the molten material into small fragments that are violently ejected from the volcano and rain down as *pyroclastic debris*, or **tephra**. These terms encompass all types of fragmental volcanic particles that are explosively ejected from a volcano. The particles range from fine dust to large angular *blocks* and smooth-surfaced *bombs* greater than 64 mm in size. Gravel-size tephra (2–64 mm) is referred to as *lapilli*. An accumulation of tephra is referred to as a **pyroclastic deposit** that can be cemented or fused to form a *pyroclastic rock*.

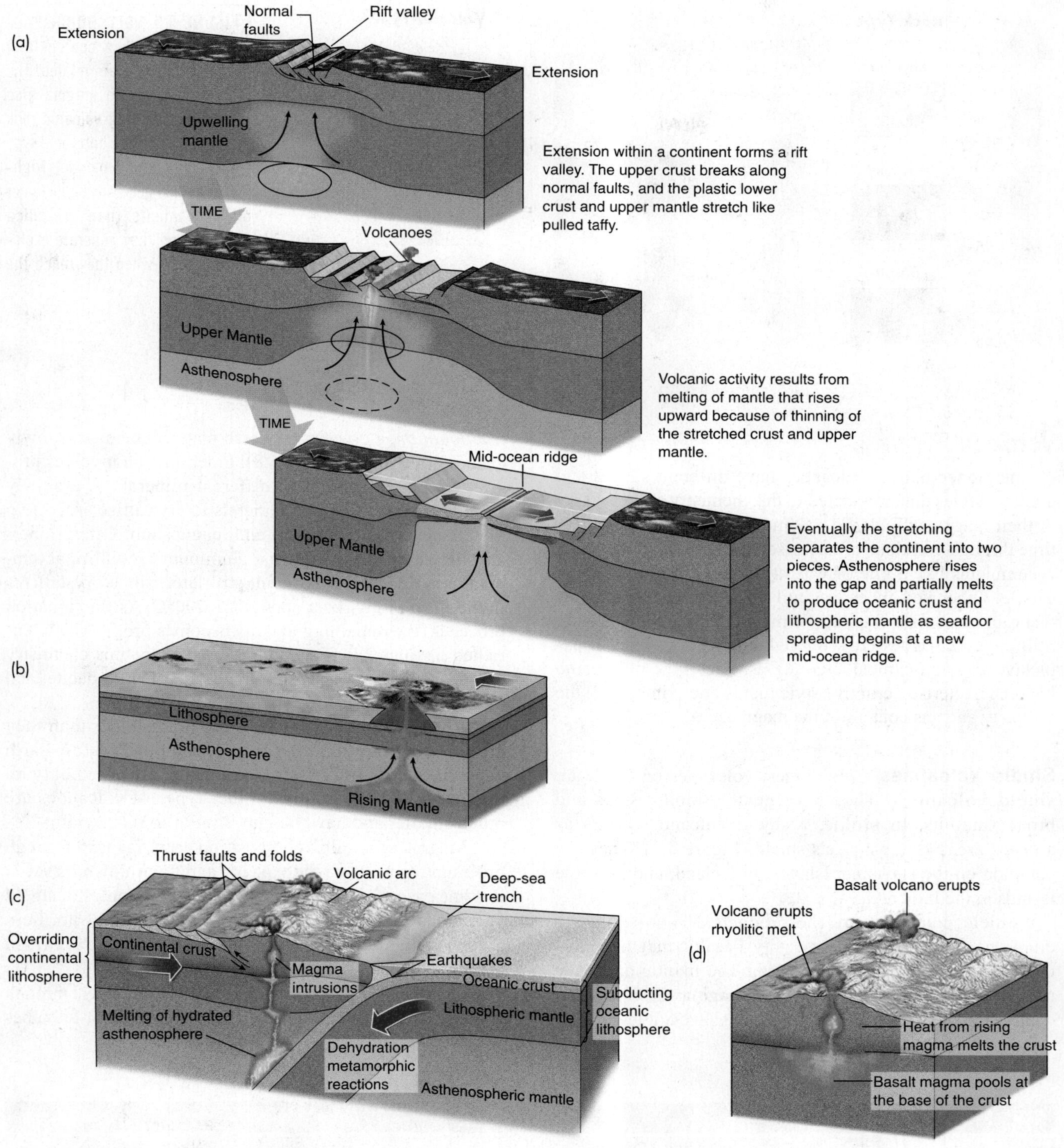

▲ **FIGURE 5.4 MAGMA FORMATION** Magma is generated primarily within the asthenosphere. (a) Decompression melting occurs when the overburden pressure on rocks in the asthenosphere is lowered due to thinning of the overlying lithospheric plate. (b) Decompression melting also occurs when superheated rocks well up from deep within the asthenosphere at a hot spot. (c) Addition of volatiles to the asthenosphere from a down-going oceanic plate causes melting at a subduction zone. (d) Heat within rising magma causes the adjacent lithosphere to melt.

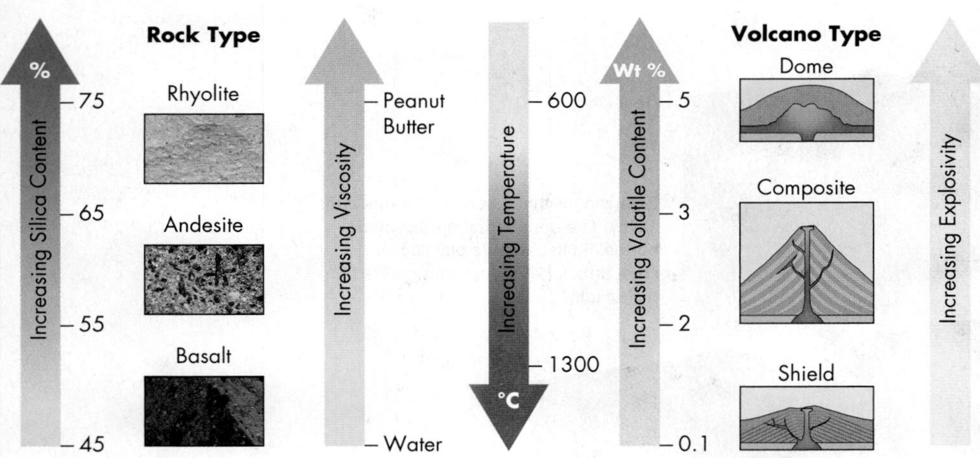

◄ **FIGURE 5.5 VOLCANO CHARACTERISTICS** The silica content and viscosity of magma are related. The silica content of the magma also defines the resulting volcanic rock type, as well as the shape and type of the associated volcano. High-silica magmas generally have higher volatile contents than low-silica magmas, and they generate explosive eruptions when they reach the surface.

Volcano Types

As mentioned above, volcanoes have different shapes and eruptive styles that are related to the chemistry and viscosity of their magmas (Table 5.1). Some volcanoes build up over time through eruptions of low-viscosity basaltic magma that accumulates like a pile of wax at the bottom of a dripping candle. Other volcanoes are built of cinder-like fragments that rain out from a fountain of magma erupted into the air. Still other volcanoes grow through alternating cycles of explosive and non-explosive eruptions. Each type of volcano has a characteristic eruptive style that is due primarily to the viscosity and gas content of the magma.

Shield Volcanoes

The largest volcanoes on Earth are **shield volcanoes**. They have gently sloping sides and broad summits. In profile, a shield volcano appears as a broad arc, like a warrior's shield (Figure 5.6). They are common on the Hawaiian Islands, in Iceland, and on some islands in the Indian Ocean.

Shield volcanoes are the products of non-explosive eruptions of very hot, low-viscosity, basaltic magmas (Figure 5.7). These magmas come from the mantle and have not mixed with silica- and gas-rich magmas derived from continental crust. Basaltic magmas also have not undergone *fractionation*, a process by which magma chemistry slowly changes over time (Figure 5.8). Fractionation involves progressive crystallization of different minerals in a slowly cooling magma. The first minerals to crystallize are rich in the elements iron, calcium, and magnesium. Later, minerals with abundant sodium and aluminum crystallize at temperatures of 700–900°C, and, still later, silica crystallizes in the form of quartz at about 600–700°C. As fractionation proceeds, the remaining magma becomes progressively enriched in silica, alumina, and gases. The magma chemistry evolves from that of basalt, to andesite, then to dacite, and finally to rhyolite.

Although shield volcanoes are much wider than they are tall, they are among the tallest mountains on Earth when measured from their bases which are commonly on the ocean floor. In contrast, other types of volcanoes are much smaller and have heights similar to their widths.

Magma of basaltic composition erupts as lava through openings, or *vents*, in the volcano and flows down its sides, in some cases for dozens of kilometres. Examples of shield volcanoes in Canada include Mount Edziza in northern British Columbia and the Ilgachuz Range in central British Columbia.[4]

Lava can flow many kilometres away from a vent through underground *lava tubes* (Figure 5.9). The walls of the tubes

TABLE 5.1 Types of Volcanoes

Volcano Type	Shape	Silica Content of Magma	Viscosity	Rock Type	Eruption Type	Example
Shield volcano	Gentle arch or dome with gentle slopes	Low	Low	Basalt	Lava flows, some explosive activity	Mauna Loa, Hawaii, Figure 5.6
Composite volcano or stratovolcano	Cone-shaped with steep sides	Intermediate	Intermediate	Andesite	Combination of lava flows and explosive activity	Volcan Osorno, Chile, Figure 5.10
Volcanic dome	Dome shaped	High	High	Rhyolite	Highly explosive	Mono Craters, California, Figure 5.15
Cinder cone	Steep cone, commonly with summit crater	Low	Low	Basalt	Explosive activity	Mount Edziza, British Columbia, Figure 5.16

◄ **FIGURE 5.6 SHIELD VOLCANO** Mauna Loa, a classic shield volcano formed from countless eruptions of basalt. Note that the profile of the volcano is very gently curved, like that of a warrior's shield. The volcano's diameter is more than 80 km. *(Duane DeVecchio)*

◄ **FIGURE 5.7 LAVA FOUNTAIN** Fiery fountain of very hot, low-viscosity, basaltic lava erupting from Kilauea volcano on the Big Island of Hawaii. *(U.S. Geological Survey)*

insulate the magma and keep it hot and fluid. Sometimes, lava completely drains from the tubes, leaving long, sinuous caverns. Lava tubes form natural conduits for the movement of groundwater and may cause engineering problems when they are encountered during construction projects.

Composite Volcanoes Mount Fuji, Mount Rainier, and Volcan Osorno (Figure 5.10) are examples of **stratovolcanoes**, also called *composite volcanoes*. The term stratovolcano is derived from the interlayered lavas and pyroclastic deposits that characterize these conical volcanoes. Stratovolcanoes erupt less frequently than active shield volcanoes, but the eruptions involve andesitic or dacitic magmas and are explosive (Figure 5.11). Lavas are relatively silica-rich and viscous, and thus they rarely flow more than a few kilometres from vents.

Many active volcanoes in Alaska, Washington, and Oregon are stratovolcanoes. British Columbia also has stratovolcanoes, including Mount Meager and Mount Garibaldi, but they are dormant or inactive, and their original conical forms have been destroyed by erosion.

Don't let the beauty of stratovolcanoes fool you—their eruptions can be deadly. Stratovolcanoes have been responsible for most of the death and destruction caused by volcanoes throughout history. The 1980 eruption of Mount St. Helens in Washington State, which killed 57 people, was accompanied by a huge landslide and a gigantic lateral blast, similar in form to the blast of a shotgun.[5] Ash and pyroclastic surges and flows from Mount Vesuvius destroyed the Roman towns of Pompeii and Herculaneum in A.D. 79, killing 10 000 to 25 000 people (Figures 5.12 and 5.13). Today, 3 million people live within sight of this active volcano. We should consider Vesuvius and other active stratovolcanoes armed and dangerous (Case Study 5.1).

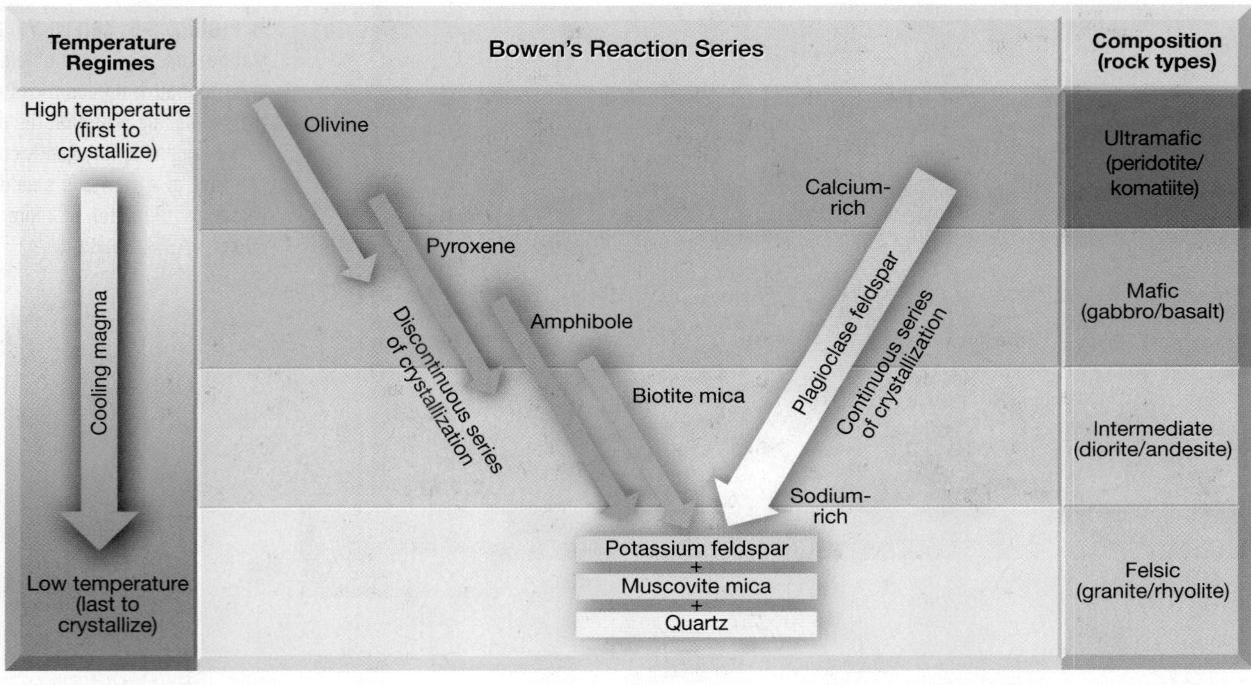

▲ **FIGURE 5.8 FRACTIONATION** Sequence of crystallization of minerals in magma as temperature decreases. Olivine, pyroxene, and calcium-rich plagioclase crystallize at the highest temperatures. Quartz, potassium feldspar, and muscovite crystallize later, at lower temperatures. *(Tarbuck, E. J., F. Lutgens, and D. Tasa. 2008.* An Introduction to Physical Geology, *9th ed. Pearson Education Inc., Upper Saddle River, NJ. Reprinted and Electronically reproduced by permission of Pearson Education, Inc., Upper Saddle River, New Jersey.)*

(a)

(b)

▲ **FIGURE 5.9 LAVA TUBES** Magma can move many kilometres below ground in tunnels called lava tubes. The tunnels form below cooled crusts of basaltic lava. Both active and drained tubes can be hazards if their roofs collapse. (a) A hole in an active lava tube on Kilauea volcano, Hawaii. (b) An older, drained tube near the summit of Kilauea—the Thurston (Nakuku) lava tube. *((a) Jeffrey B. Judd/U.S. Geological Survey; (b) Volcano Hazards Team)*

Volcanic Domes Lassen Peak and Mono Craters in California are examples of **volcanic domes**, steep-sided mounds of lava that form around vents from the eruption of highly viscous, silica-rich magmas (Figure 5.15).

Cinder Cones **Cinder cones**, also called *scoria cones*, are relatively small volcanoes made up of nut- to fist-sized pieces of red or black basalt (Figure 5.16).[6] The cones are round to oval in surface form and commonly have a crater at

◀ **FIGURE 5.10 STRATOVOLCANO** Volcan Osorno in Chile is a beautiful, snow-clad composite volcano that last erupted in 1869. *(© Paul Street/Alamy)*

▲ **FIGURE 5.11 EXPLOSIVE ERUPTION** Mount St. Helens erupts violently on May 18, 1980, sending huge quantities of ash and gases high into the atmosphere. Explosive eruptions like this are characteristic of stratovolcanoes. *(© Lightroom Photos/Alamy)*

▲ **FIGURE 5.12 DESTRUCTION OF POMPEII** Painting showing the Roman town of Pompeii during the eruption of Mount Vesuvius (top of the painting) in A.D. 79. The town was buried in suffocating hot ash and overrun by pyroclastic flows. *(C.G. image © BBC)*

their top. Tephra from extinct cinder cones is the "lava rock" used widely in commercial landscaping. Cinder cones are found on the flanks of larger volcanoes, along some normal faults, and along cracks or fissures.

Eruptions in 1943 in the Itzicuaro Valley of central Mexico, about 320 km west of Mexico City, offered a rare opportunity to observe the birth and rapid growth of a volcano at a location where none had existed before. An astounding event occurred on February 20, 1943, following several weeks of earthquakes and thunderous sounds coming from beneath the ground surface. As Dionisio Pulido was preparing his cornfield for planting, he noticed that a hole he had been trying to fill for years had reopened in the ground at the base of a knoll. As Señor Pulido watched, the surrounding ground swelled, rising more than 2 m, and sulphurous gases and ash began billowing from the hole. By that night, the hole was ejecting glowing-red rock fragments high into the air. After only five days, the cinder cone had grown to more than 100 m high, and blocks and ash continued to erupt from the vent.

▲ **FIGURE 5.13 PLASTER CASTS OF VOLCANO VICTIMS** Archeologists have excavated the remains of more than 1500 people who asphyxiated and burned during the A.D. 79 eruption of Mount Vesuvius. Over one-third of the victims at Pompeii died from roof collapses, falling roof slates, and bombs thrown out of the volcano. About two-thirds of the victims were killed by pyroclastic flows, probably from a combination of debris impact and suffocation through ash inhalation. This photograph shows two adults and a child in their death positions. Pompeii was covered with up to 3 m of ash and pumice, and Herculaneum was buried beneath nearly 20 m of volcanic debris. *(Paco Gómez Garcia/AGE Fotostock)*

In June 1943, basaltic lava began to flow from a fissure at the base of the now 400-m-high cone. The lava flow soon overran the nearby village of San Juan Parangaricutiro, leaving little but the church steeple exposed (Figure 5.17). No one was killed in these eruptions, and within a decade Paricutín became a dormant volcano. Nevertheless, during its nine years of eruption, more than 1 billion m^3 of ash and 700 million m^3 of lava erupted from Señor Pulido's cornfield. Crops failed and livestock became sick and died. Although several villages were relocated to other areas, some residents have moved back to the vicinity of Paricutín. Locating property boundaries proved to be difficult because everything was covered by ash and lava, resulting in land ownership disputes.[6]

Maars The violent interaction of magma and groundwater produces **maars**. The conversion of water into steam drives a violent explosion that forms a crater, similar to one made by a meteorite impact.

Ice-Contact Volcanoes Many volcanoes and lava flows erupt beneath or against glaciers. Several active volcanoes lie at the base of ice caps in Iceland. These subglacial volcanoes periodically erupt, melting large quantities of ice and thus producing huge outburst floods, termed *jökulhlaups*. The most recent such eruption and flood happened in Iceland in 2010. Ice-contact volcanoes are also common in some areas formerly covered by Pleistocene ice sheets. Many excellent examples are found in British Columbia, which was repeatedly covered by the *Cordilleran ice sheet*.[7] These ice-contact volcanoes have odd shapes that were produced by rapid chilling of lavas and pyroclastic material when they came in contact with water and ice. Rapid cooling of the lavas produced forms called *pillows*, which broke up as they rolled down the slopes of the submerged volcanoes, forming other types of volcanic deposits called *pillow breccias*. Where the eruptions did not melt through the overlying ice sheet, subglacial mounds were produced. In other cases, the eruptions melted through the ice, producing lava flows that covered pillow breccias. Such steep-sided, flat-topped volcanoes have been termed *tuyas,* after Tuya Butte in northern British Columbia, where they were first recognized.[7] Similar mountains in Iceland are called table mountains because of their flat tops.

An interesting ice-contact lava flow of late Pleistocene age is found in Garibaldi Provincial Park, 65 km north of Vancouver, British Columbia (Figure 5.18). Mount Garibaldi is a deeply dissected, dormant volcano that has a long history of eruptions extending back more than 2 million years. At the end of the Pleistocene, about 12 000 years ago, a lava flow issued from a satellite cone on the flank of Mount Garibaldi. The flow came into contact with a decaying valley glacier in Cheakamus Valley at what is termed The Barrier.[8] There, the chilled front of the lava flow froze and formed a steep rock face several hundred metres high. Since the glacier in Cheakamus Valley disappeared about 11 000 years ago, the fractured face of The Barrier has repeatedly collapsed, producing large landslides in the valley below; the last of these landslides occurred in the winter of 1855–1856.

Volcanic Features

A volcano is a complex system, much more than a mountain that simply expels lava and pyroclastic debris from its top. Volcanoes or volcanic areas commonly include craters, calderas, volcanic vents, geysers, and hot springs.

Craters, Calderas, and Vents **Craters** are depressions at the tops of volcanoes that form by explosion or collapse of the summit area. They are usually hundreds of metres to a couple of kilometres in diameter. Circular to oval depressions up to a few dozen kilometres in diameter that form during explosive ejection of magma and subsequent collapse of a volcano are termed **calderas** (Figures 5.19 and 5.20). Caldera-forming eruptions are the largest and most deadly type of volcanic eruption, but fortunately they are rare; eruptions that produce calderas several kilometres in diameter occur, on average, once every 200 to 1000 years. The violent eruptions that form large *resurgent calderas* are discussed below.

Volcanic vents are openings through which lava and pyroclastic debris erupt. Some vents are roughly circular; others are elongated cracks called *fissures*. Extensive fissure eruptions have produced huge accumulations of nearly horizontal lava flows called *flood basalts*. Thick flood basalts underlie about 150 000 km^2 of the interior plateaus of southern British Columbia, Washington, Oregon, and Idaho, an area nearly three times the size of Switzerland. The lavas range in age from about 17 million to 6 million years old.

5.1 CASE STUDY

Planning for a Disaster in Naples

Naples, Italy's third largest city, is at risk from an eruption of Mount Vesuvius (Figure 5.14). The stratovolcano is a distinctive "humpbacked" mountain, consisting of a large cone (*Gran Cono*) partially encircled by the steep rim of a summit caldera produced by the collapse of an earlier and originally much higher structure called Monte Somma. Vesuvius erupted explosively in A.D. 79, killing 10 000 to 25 000 residents of Pompeii and Heraculaneum. It has erupted more than 40 times since then, most recently in 1944,[9] but none of these eruptions rivalled the A.D. 79 event in size. Over the past few centuries, quiescent periods have ranged from 18 months to 7½ years; the current lull in activity has been the longest in nearly 500 years. Vesuvius is today regarded as one of the most dangerous volcanoes in the world because of its tendency toward frequent, explosive eruptions and because several million people live within sight of it.

Emergency planning for a future eruption of Vesuvius assumes that the worst-case scenario is an eruption of similar size and type to one in 1631, which generated about 0.1 km^3 of magma.[10] In this scenario, pyroclastic flows could reach about 7 km from the crater, and much of the surrounding area would suffer from ash fall. Because of prevailing winds, towns to the south and east of the volcano are most at risk from ash fall. Ash accumulations exceeding 100 kg/m^2—at which point roofs might collapse—may extend as far as Avellino to the east or Salerno to the southeast. Planners assume that the ash fall hazard to the northwest, in the direction of Naples, would barely extend past the slopes of the volcano.[11] However, the specific areas that would actually be affected by ash fall would depend on the wind pattern during the eruption.

The area around the volcano is now densely populated. The emergency plan assumes 14 to 20 days' advance warning of an eruption and envisions the evacuation of 600 000 people living in the *zona rossa* ("red zone"), which is at greatest risk from pyroclastic flows.[10] The evacuation—by trains, ferries, cars, and buses—is planned to take about seven days, and the evacuees would be sent to other parts of the country. The dilemma that would face regional officials implementing the plan is when to start the massive evacuation, because if it were left too late many people could be killed, whereas an early evacuation would be unnecessary and expensive if the volcano did not erupt. In 1984, 40 000 people were evacuated from the Campi Flegrei area, another volcanic complex near Naples, but no eruption occurred.[10]

Ongoing efforts are being made to reduce the population living in the red zone by demolishing illegally constructed buildings, establishing a national park around the upper flanks of the volcano to prevent further construction, and offering financial incentives to people to move away. The goal over the next 20 years is to reduce to two or three days the time needed to evacuate the area.[12]

The volcano is closely monitored by the Osservatorio Vesuvio in Naples with extensive networks of instruments that measure earthquakes, changes in the local gravitational field, and small ground movements.[13] In addition, the chemistry of gases emitted from *fumaroles* on Vesuvius is regularly determined. All of this monitoring is being done to track magma rising underneath the volcano.

◄ **FIGURE 5.14 NAPLES AND ITS SLEEPING GIANT** Mount Vesuvius looms over the city of Naples. The volcano has erupted more than 40 times in the past 2000 years and poses a hazard to the more than 3 million people who live around it. *(Tom Pfeiffer/www.canodiscovery.com)*

◄ **FIGURE 5.15 VOLCANIC DOME** Mono Craters in California are volcanic domes consisting of rhyolitic and dacitic lavas. The volcanoes range in age from about 600 to 40 000 years old. *(Long Valley Observatory, U.S. Geological Survey)*

◄ **FIGURE 5.16 CINDER CONE** Eve Cone, which is part of the Mount Edziza shield volcano in northwest British Columbia, has a small crater at its summit. *(Natural Resources Canada 2010, courtesy of the Geological Survey of Canada/C. J. Hickson)*

◄ **FIGURE 5.17 RAPID GROWTH OF A VOLCANO** In this 1943 photograph, the Paricutín cinder cone in central Mexico is erupting a cloud of volcanic ash and gases. A lava flow has nearly buried the village of San Juan Parangaricutiro. *(Tad Nichols)*

▲ **FIGURE 5.18 ICE-CONTACT LAVA FLOW IN BRITISH COLUMBIA** The Barrier is a steep, unstable face several hundred metres high that formed when the Clinker Peak lava flow came into contact with a late Pleistocene glacier in Cheakamus Valley, north of Vancouver, British Columbia. The Table, in the background, is a pillar of flat-lying lava flows that erupted into the base of the ice sheet that covered British Columbia at the end of the Pleistocene, about 15 000 years ago. *(Austin Post/U.S. Geological Survey)*

◄ **FIGURE 5.19 CALDERA, CRATER LAKE, OREGON** This water-filled caldera formed during a violent eruption of Mount Mazama volcano about 7700 years ago. *(QT Luong/terragalleria.com)*

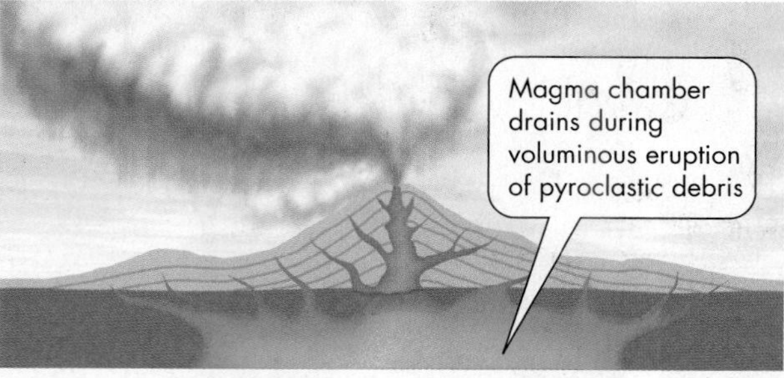

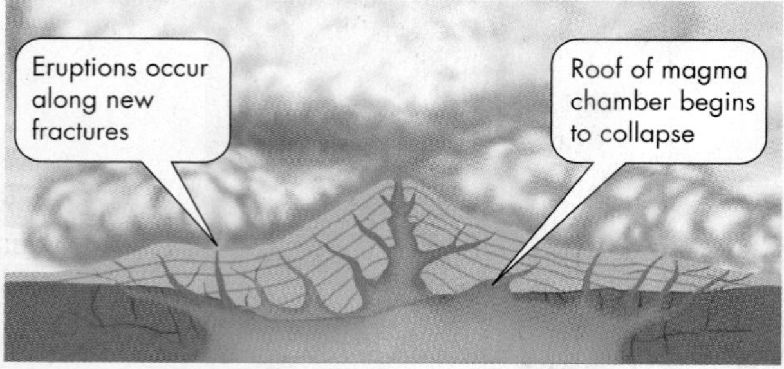

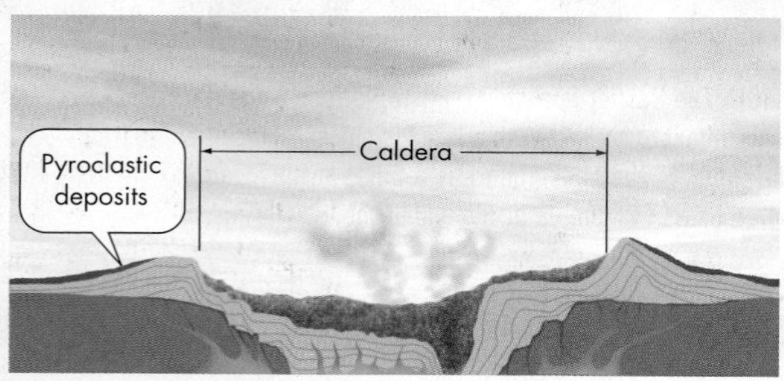

▲ **FIGURE 5.20 CALDERAS FORM BY EXPLOSION AND COLLAPSE** Calderas generally form by collapse of a magma chamber below a composite volcano during and shortly after an explosive eruption. The largest calderas on Earth, located in Yellowstone National Park (Wyoming), California, and New Zealand, are tens of kilometres across. *(Modified from Smith, G. A., and A. Pun 2006.* How Does Earth Work? Physical Geology and the Process of Science. *Upper Saddle River, NJ: Pearson Prentice Hall. Reprinted and Electronically reproduced by permission of Pearson Education, Inc., Upper Saddle River, New Jersey.)*

Hot Springs and Geysers Groundwater becomes heated when it comes into contact with hot rock. The hot water can discharge at the surface as a *hot spring*, or *thermal spring*. Less commonly, groundwater boils in an underground chamber to produce periodic, steam-driven releases of steam and hot water called *geysers*. World-famous geyser fields are found in Iceland, New Zealand, and Yellowstone National Park in Wyoming (Figure 5.21).

Resurgent Calderas and Super Eruptions Calderas many dozens of kilometres across are produced by very rare but extremely violent eruptions referred to as **super eruptions**. Such eruptions, and the **supervolcanoes** from which they originate, are substantially larger than any in historic times—they produce more than 1000 km^3 of ash and fragmented rock.[14] This volume is approximately 1000 times the amount of ash produced by the 1980 eruption of Mount St. Helens and is enough to cover all of Prince Edward Island to a depth of about 175 m. A super eruption occurs when a large volume of magma rises to shallow depths in the crust over a mantle hot spot. Pressure builds in the large and growing magma pool until the crust is unable to contain the pressure. Less commonly, super eruptions occur at convergent plate boundaries. At least 10 such eruptions have occurred in the past million years, three of which were in North America. Ash deposits from such an eruption can be 100 m thick near the crater's rim and a metre or so thick 100 km away.[15]

Toba caldera on the Indonesian island of Sumatra is the site of the largest super eruption of the past 2 million years. The caldera is 30 km wide and 100 km long; it formed 75 000 years ago when about 2800 km^3 of pyroclastic material were blown high into the atmosphere. Sulphur aerosols produced by this cataclysmic eruption lowered sea-surface temperatures by 3–3.5°C for about six years and may have caused nearly complete deforestation of southeast Asia. Genetic evidence suggests that the human species nearly became extinct due to the Toba eruption—the total number of people on Earth fell to between 2000 and 20 000 immediately after the eruption, probably because of famine.[16] Researchers have proposed that the eruption created population bottlenecks in human species existing at the time and accelerated differentiation of the isolated populations. Eventually, all but two species became extinct—Neanderthals (*Homo neanderthalensis*) and modern humans (*Homo sapiens*).[16]

The most recent super eruptions in North America occurred about 600 000 years ago at Yellowstone National Park in Wyoming and about 700 000 years ago at Long Valley, California. The latter event produced the Long Valley caldera and covered a large area with ash (Figure 5.22). Measurable uplift of the land and a multitude of earthquakes up to **M** 6 in the early 1980s suggested that magma was moving upward beneath Long Valley, prompting the U.S. Geological Survey to issue a potential volcanic hazard warning that was subsequently lifted.

A caldera-producing eruption can develop quickly, in a few days to a few months. Subsequent, intermittent, lesser volcanic activity can linger for 1 million years. The Yellowstone event created hot springs and geysers, including Old Faithful, and sporadic eruptions occurred in the Long Valley caldera until very recently. Both sites are still capable of eruptions because magma is present at shallow depths beneath the caldera floors. Both are considered resurgent calderas because their floors have slowly domed upward since the explosive eruptions that formed them.

A concept related to supervolcanoes is *large igneous provinces*, which are extensive areas of basalt produced by fissure eruptions that are historically unprecedented.[17] These huge outpourings cover several thousand square kilometres

◄ **FIGURE 5.21 OLD FAITHFUL** An eruption of Old Faithful Geyser in Yellowstone National Park, Wyoming. Although the geyser's name implies predictability, eruption intervals vary from day to day and year to year and often change after earthquakes. *(John J. Clague)*

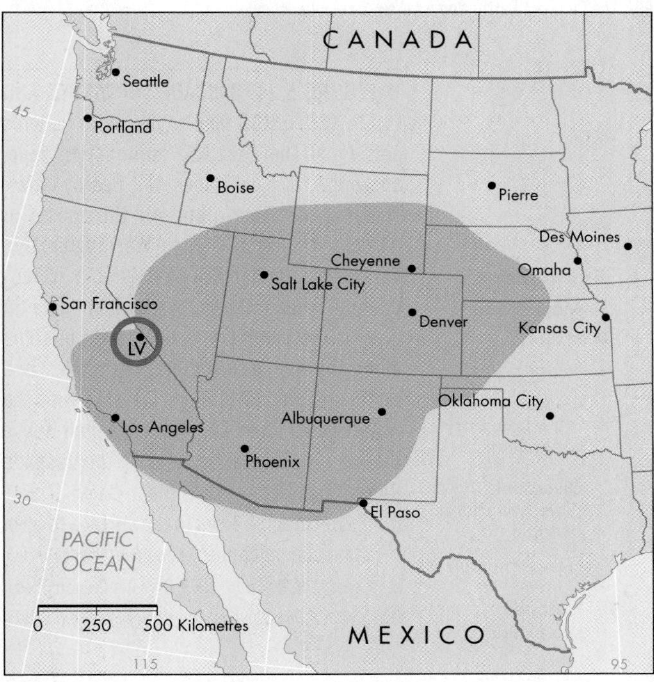

▲ **FIGURE 5.22 WIDESPREAD ASH FALL HAZARD** The area in orange was covered by ash from the Long Valley caldera eruption approximately 700 000 years ago. The red circle around Long Valley (LV) has a radius of 120 km and encloses the area that would receive more than 1 m of ash if a similar eruption were to occur again. *(From Miller, C. D., D. R. Mullineaux, D. R. Crandell, and R. A. Bailey. 1982.* Potential Hazards from Future Volcanic Eruptions in the Long Valley–Mono Lake Area, East-central California and Southwest Nevada—A Preliminary Assessment. *U.S. Geological Survey Circular 877)*

and have volumes in the order of millions of cubic kilometres. In most cases, the lavas of large igneous provinces were laid down over periods of several million years. These eruptions released massive amounts of carbon dioxide and other gases and thus may have altered climate.

Many volcanologists think that Iceland may be a large igneous province that is currently forming. The last major lava outpouring in Iceland occurred in 1783–1784 from the Laki fissure which is about 40 km long. This eruption generated approximately 14 km^3 of basaltic lava.[17]

Volcanoes and Plate Tectonics

Volcanism is directly related to plate tectonics. More specifically, the tectonic setting determines the type of volcano that will be present (Figure 5.23):

1. ***Subduction zones***

   Stratovolcanoes occur at subduction zones and thus are the most common type of volcano found around the Pacific Rim. For example, volcanoes in southwest British Columbia, Washington, Oregon, and California are associated with the Cascadia subduction zone (Figure 5.24). More than 80 percent of terrestrial volcanic eruptions have come from volcanoes above subduction zones.[18] The dominant volcanic rock in this setting is andesite, which is produced by the mixing of basaltic magmas with continental crust, and by fractionation.

2. ***Mid-ocean ridges***

   In some areas, plates move away from one another instead of colliding. Commonly, spreading occurs in the oceans along mid-ocean ridges, where basaltic magma derived directly from the asthenosphere, part of Earth's upper mantle (Chapter 2), rises to the ocean floor to create new crust. This magma mixes very little with other materials; therefore the lavas are made almost entirely of low-viscosity basalt. Where spreading ridges occur on land, such as in Iceland, shield volcanoes are formed. A spreading ridge is slowly extending southward beneath East Africa, producing a rift zone marked by many active volcanoes. The rift zone will eventually become a seaway with an axial spreading ridge.

3. ***Hot spots beneath the oceans***

   Volcanoes are also found where hot mantle material wells up beneath a plate at a stationary point rather than at the

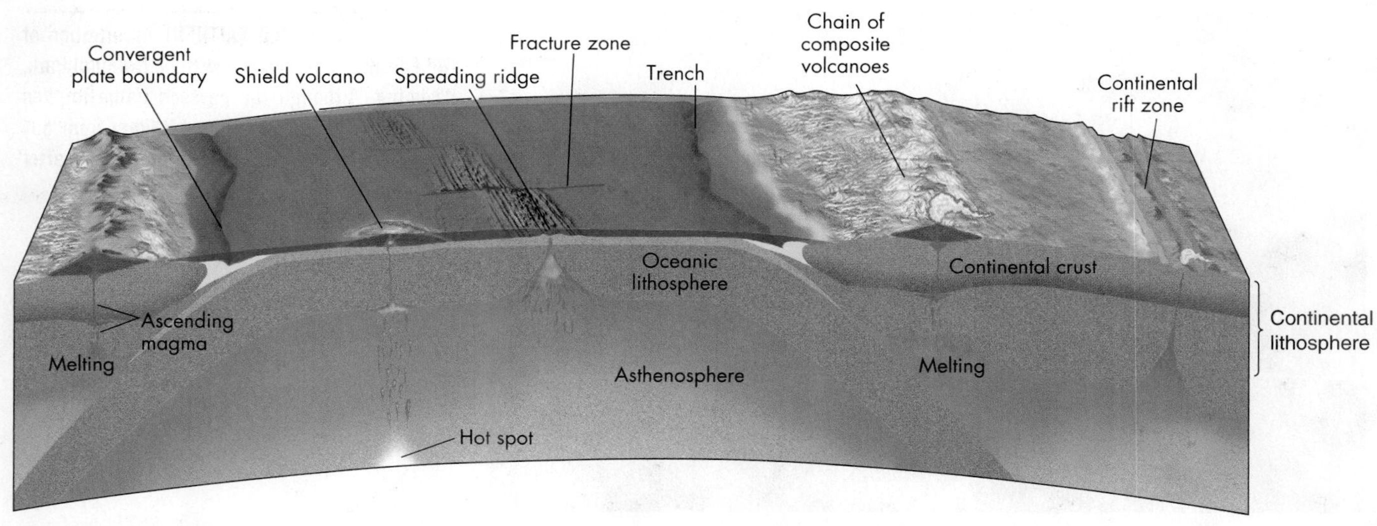

▲ **FIGURE 5.23 VOLCANIC ACTIVITY AND PLATE TECTONICS** An idealized diagram showing plate tectonic processes and their relation to volcanic activity. Numbers refer to explanations in text. *(Modified from Skinner, B. J., and S. C. Porter. 1992.* The Dynamic Earth, *2nd ed. New York: John Wiley)*

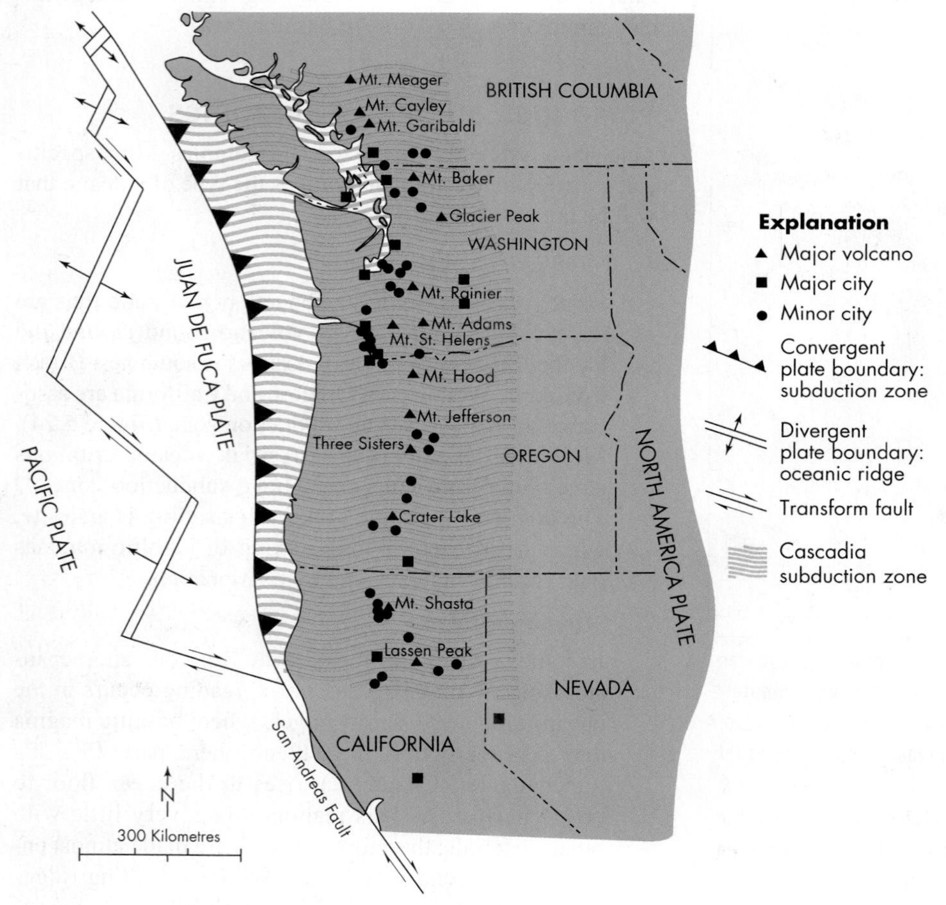

◄ **FIGURE 5.24 CASCADE VOLCANOES AND PLATE TECTONICS** Map of the plate tectonic setting of the Cascadia subduction zone, showing major volcanoes and nearby cities. The Juan de Fuca plate, off the coasts of northern California, Oregon, Washington, and Vancouver Island in British Columbia, is subducting beneath the North American plate (in green). The subduction zone starts offshore along the line with the black triangles and extends eastward through the area with the wavy orange lines. Crustal rocks melt above the Juan de Fuca plate beneath the Cascade Range and the southernmost Coast Mountains. The magma rises to the surface to form the Cascade volcanoes. *(Modified from Crandell, D. R., and H. H. Waldron. 1969.* Disaster Preparedness. *Washington, DC: Office of Emergency Preparedness)*

boundary between two plates. The upwelling mantle material, focused on a single spot, creates a volcano. However, the plate continues to slowly move and a series of volcanoes form, becoming progressively older in the direction of plate movement. These chains of volcanoes are termed *hot spot* volcanoes. The most famous example is the Hawaiian Islands, located within the Pacific plate (see Chapter 2).[19]

The Hawaiian Islands have been built up from the seafloor through submarine eruptions similar to those at mid-ocean ridges. The source of the magma appears to be a hot spot that has been stationary for many millions of years. Over time, plate motion has produced a chain of volcanic islands running southeast to northwest. The island of Hawaii is currently near the hot spot and is experiencing active

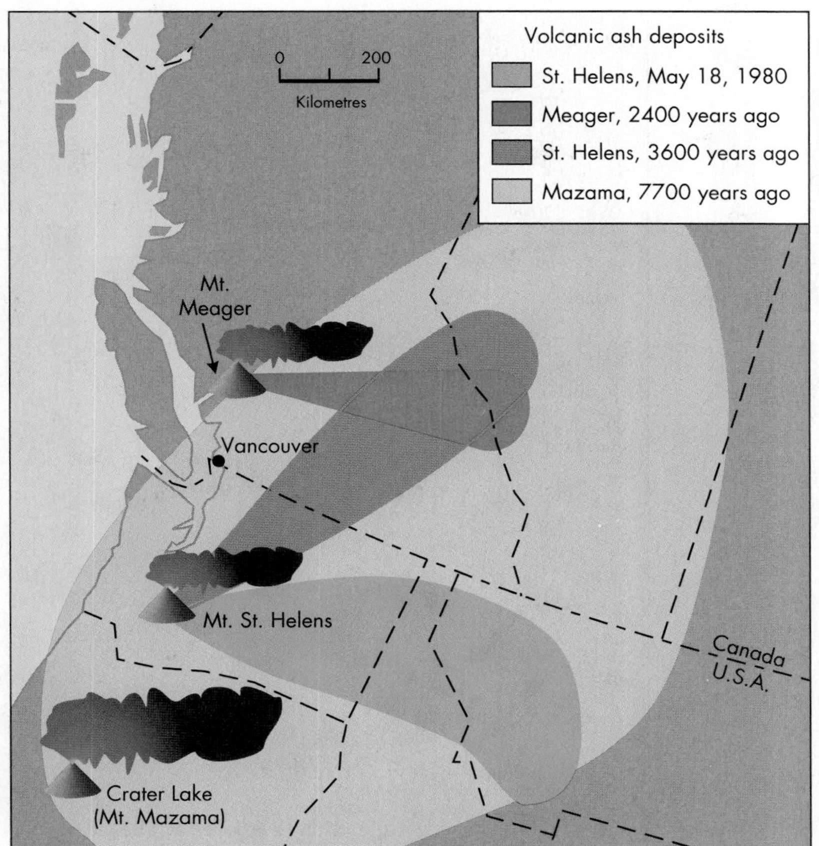

◄ **FIGURE 5.25 VOLCANIC ASH FALLS** This map shows the distribution of ash deposited during the cataclysmic eruption of Mount Mazama 7700 years ago. The "ash plumes" of three younger eruptions of Mount St. Helens and Mount Meager are plotted for comparison. *(Reprinted with permission from Tricouni Press)*

volcanism and growth. Islands to the northwest, such as Molokai and Oahu, appear to have moved off the hot spot because their volcanoes are no longer active. The oldest volcanoes of the Hawaiian chain lie west of Kauai.

4. ***Hot spots beneath continents***

   Caldera-forming eruptions occur in this tectonic setting. They may be extremely explosive and violent, and they are associated with dacitic and rhyolitic magmas. As mentioned earlier, the largest of these eruptions in North America occurred more than a half million years ago at Yellowstone National Park and Long Valley. A smaller, caldera-forming eruption happened about 7700 years ago at what is now Crater Lake, Oregon (Figure 5.19). Crater Lake lies within the caldera produced by the explosion of Mount Mazama. Ash from this eruption has been found throughout western North America and as far north as central British Columbia, Alberta, and Saskatchewan (Figure 5.25).[20]

5.2 Geographic Regions with Active Volcanoes

As we have established, most volcanoes on Earth's land surface occur along the Ring of Fire, which borders the Pacific Ocean (Figure 5.3). We also have seen, however, that active volcanoes occur in other settings: hot spots (Hawaii, Long Valley, Yellowstone), mid-ocean ridges (Iceland), and continental rift zones (East Africa). The highest risk of volcanic activity in Canada and the United States is in the mountainous regions of the Pacific Northwest, northwest and central British Columbia, the Aleutian Islands, and Yellowstone Park (Figure 5.26). More than 90 percent of North America is free of risk from local volcanic activity, but the effects of a large caldera explosion in the western United States would likely be felt far from the source in the form of ash fall.

Young volcanoes in Canada are restricted to British Columbia and southern Yukon Territory (Figure 5.27). Several eroded stratovolcanoes in southwest British Columbia form the northern part of the Cascade volcanic chain, which owes its origin to subduction of the oceanic Juan de Fuca plate beneath North America. They include Mount Garibaldi, Mount Cayley, and Mount Meager; only Mount Meager has had an eruption in postglacial time (the past 10 000 years). Similar volcanoes are also present in the Alaska–Yukon boundary area and north of the Aleutian Trench, where the Pacific plate is subducting beneath North America.

A large number of geologically youthful volcanoes occur within a broad north-trending belt in northwest British Columbia and southernmost Yukon. Most of these volcanoes are small and the product of a single eruption of basaltic magma, but Mount Edziza and Hoodoo Mountain have formed from numerous eruptions over a long period. A volcano near the boundary between Alaska and Yukon Territory (Mount Churchill) was the source of two large explosive eruptions about 1200 and 1900 years ago.[21] The

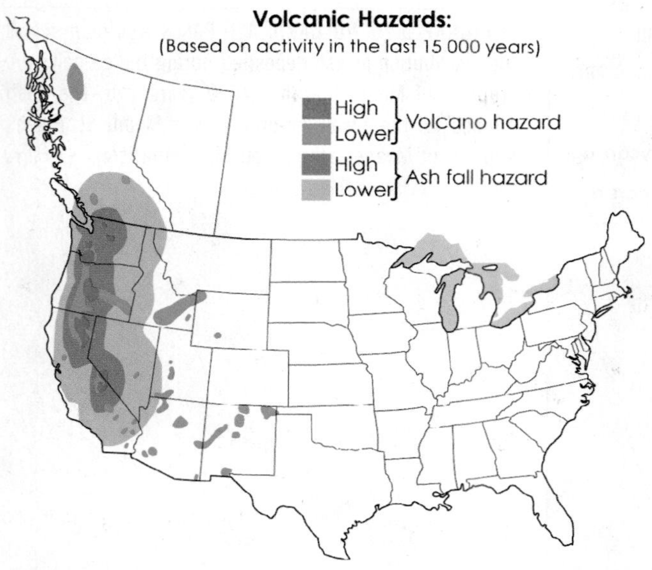

▲ **FIGURE 5.26 CANADIAN AND U.S. VOLCANIC HAZARD** Volcanic hazard for Canada and the contiguous United States based on activity during the past 15 000 years. The red colours show high and lower local volcanic hazard. The grey area is at risk of receiving 5 cm or more of ash fall from large explosive eruptions. *(Adapted from U.S. Geological Survey)*

▲ **FIGURE 5.27 VOLCANOES AND THEIR TECTONIC ENVIRONMENTS IN WESTERN CANADA** Recent volcanic activity in western Canada occurs in three tectonic settings: (1) along convergent plate boundaries, where one plate subducts beneath another (for example, Mount Garibaldi and Mount Meager); (2) in regions where the North American plate is rifting (for example, Mount Edziza and the Iskut River area); and (3) at hot spots where upwelling magma breaks through the crust (Nazko). The green lines delineate transform faults. Zones of different types of volcanism are circled. *(Reproduced or adapted with the permission of Natural Resources Canada 2013, courtesy of the Geological Survey of Canada (Bulletin 548))*

younger eruption left a visible layer of volcanic ash over an area of 340 000 km^2 in the southern Yukon and westernmost Northwest Territories. Researchers have recently found traces of the ash as far away as Greenland and Ireland.[22] The youngest eruption in Canada happened in the late nineteenth century in a remote area near the British Columbia–Alaska boundary. A scoria cone and a lava flow were produced by this eruption.[23] A similar scoria cone (Tseax River cone) and lava flow, north of Terrace, British Columbia, are products of another eruption, in the eighteenth century. The lava flow from Tseax River cone is about 23 km long and fills a 10 km length of the flat-floored Nass River valley.[24] Scientists believe that volcanism in northwest British Columbia is the result of crustal rifting caused by right-lateral faulting along the Pacific–North America plate boundary to the west (the Queen Charlotte and Fairweather faults).

A third group of volcanoes extends in a westerly direction from Nazko in central British Columbia to the Pacific Coast near Bella Coola. These volcanoes increase in age to the west—Nazko volcano is only 8000 years old, whereas the dykes that fed former volcanoes on the coast are up to 14 million years old. The simplest explanation for this age trend is that the volcanoes formed over a hot spot in the mantle and moved to the west, in a conveyor-belt fashion, off the hot spot.[25] Small, mainly basaltic volcanoes, some of which are postglacial in age, occur in Wells Gray Provincial Park in eastern British Columbia. Many of these volcanoes erupted beneath ice sheets that at times covered the area.

The volcano that poses the greatest risk to Canada is, ironically, in the United States. Glacier-cloaked Mount Baker in northern Washington is a great stratovolcano formed by numerous eruptions over the past 50 000 years[26] (Figure 5.28). The relatively un-dissected conical shape of Mount Baker reflects its status as an active volcano. It most recently erupted in 1843, and venting of gases and hot fluids from the summit crater in the late 1970s provided a timely reminder that the volcano is active. The hazards of greatest concern at Mount Baker are ash fall, landslides, lahars, and the filling of river valleys with sediment.

A major eruption of Mount Baker could spread ash over Vancouver, Abbotsford, Chilliwack, and other cities and towns in southern British Columbia. The ash would at least temporarily paralyze air and ground traffic.

The rocks around Mount Baker's summit crater have been extensively altered to soft clay minerals by circulating

▲ **FIGURE 5.28 MOUNT BAKER** This active volcano looms above the Vancouver skyline. A major eruption would melt the glaciers on the summit and flanks of the volcano, causing floods and lahars in the valleys below. *(C. J. Hickson [left], John J. Clague [right])*

hot, acidic groundwater. The altered rocks could fail if the volcano became inflated with magma at the onset of an eruption. Landslides could also occur on the steep flanks of the volcano at other times, without an accompanying eruption. Such landslides could range in size from small debris avalanches and debris flows to massive collapses of the entire summit or a side of the volcano.

During an eruption of Mount Baker, hot volcanic debris would mix with water melted from snow and ice on the summit and flanks of the mountain to form large lahars that would rush down nearby river valleys. In the worst-case scenario, a large lahar might reach the Bellingham area, as apparently happened about 7000 years ago. Lahars might also enter Baker Lake, a reservoir near the base of Mount Baker, and displace enough water to overtop the dam or cause it to fail.

Large landslides or lahars would increase sediment loads to rivers, including the Skagit and Nooksack rivers, and cause them to build up their beds. Settlements and other infrastructure on the valley floors might be buried or flooded as the level of the riverbeds rose.

5.3 Volcanic Hazards

Worldwide, 50 to 60 volcanoes erupt each year, including two or three in the United States, mainly in Alaska.[1] Many eruptions are in sparsely populated areas, causing little, if any, loss of life or economic damage. In contrast, eruptions near densely populated areas can be catastrophic[27] (Table 5.2). Approximately 500 million people live close to volcanoes, and in the past 100 years nearly 100 000 people have been killed by volcanic eruptions—28 500 in the 1980s alone.[1,27] Densely populated countries with many active volcanoes, such as Japan, Mexico, the Philippines, and Indonesia, are particularly vulnerable.[2] Several active or potentially active volcanoes in western North America are near cities with populations of more than 100 000—Vancouver northwest of Mount Baker, Seattle and Tacoma northwest of Mount Rainier, and Portland west of Mount Hood.

Volcanic hazards can be subdivided into primary (direct) effects of eruptions and secondary (indirect) effects. Lava flows, ash fall, volcanic bombs, pyroclastic flows, pyroclastic surges, lateral blasts, and poisonous gases are *primary effects* (Figure 5.29). *Secondary effects* include lahars, debris avalanches, other landslides, groundwater and surface water contamination, floods, fires, and tsunamis. Large eruptions can also cool Earth's atmosphere for a year or so.[15]

The size of a volcanic eruption can be quantified by using a scale called the Volcanic Explosivity Index (VEI), which takes into account the volume of material erupted, the height of the eruption cloud, the duration of the main eruptive phase, and other parameters to assign a number on a linear scale from 0 to 9. The May 1980 eruption of Mount St. Helens, which destroyed about 630 km^2 of land, expelled 1.4 km^3 of magma, and produced an eruption cloud 24 km high, has a VEI value of 5. In contrast, the last large eruption from the Yellowstone caldera, which occurred about 600 000 years ago and expelled more than 1000 km^3 of magma, has a VEI value of 8. In general, the larger the size of a volcanic eruption, the greater the hazard.

Lava Flows

Lava flows are one of the most common products of volcanic activity. They happen when magma reaches the surface and flows out of a central crater or when it erupts from a fissure or vent on the flank of a volcano. The three major types of lava take their names from the volcanic rocks they form: basalt, which is by far the most abundant of the three; andesite; and rhyolite.

Lava flows can be fluid and move rapidly, or they can be viscous and move slowly. Basaltic lavas, which have low viscosity and high eruptive temperatures, have velocities of up to a few kilometres per hour on steep slopes. They harden with a smooth, commonly ropy surface texture, called

TABLE 5.2 Selected Historic Volcanic Eruptions

Volcano or City	Year	Effect
Vesuvius, Italy	A.D. 79	Destroyed Pompeii and killed 16 000 people; city was buried by volcanic ash and rediscovered in 1595
Skaptar Jökull, Iceland	1783	Killed 10 000 people (many died from famine) and most of the island's livestock; also killed crops as far away as Scotland
Tambora, Indonesia	1815	Global cooling; produced "year without a summer"
Krakatoa, Indonesia	1883	Tremendous explosion; 36 000 deaths from tsunamis
Mount Pelée, Martinique	1902	Pyroclastic flow killed 30 000 people in a matter of minutes
La Soufrière, St. Vincent	1902	Killed 2000 people
Mount Lamington, Papua New Guinea	1951	Killed 6000 people
Villarica, Chile	1963–1964	Forced 30 000 people from their homes
Mount Helgafell, Heimaey Island, Iceland	1973	Forced 5200 people to evacuate their homes
Mount St. Helens, Washington	1980	Debris avalanche, lateral blast, and lahars killed 57 people and destroyed more than 100 homes
Nevado del Ruiz, Colombia	1985	Eruption generated lahars that killed at least 23 000 people
Mount Unzen, Japan	1991	Pyroclastic flows and lahars killed 43 people and destroyed hundreds of homes; 12 000 people evacuated
Mount Pinatubo, Philippines	1991	Tremendous explosions, pyroclastic flows, and lahars combined with a typhoon killed more than 300 people; 60 000 people evacuated
Montserrat, Caribbean	1995	Explosive eruptions, pyroclastic flows; south side of island evacuated, including capital city of Plymouth; several hundred homes destroyed
Chaitén, Chile	2008	Explosive eruptions, pyroclastic flows; 5000 people evacuated; aviation in South America disrupted for weeks
Eyjafjallajökull, Iceland	2010	Large eruption of ash; disrupted air travel in the United Kingdom and northern Europe for several weeks.

Source: Data are partially derived from Ollier, C. 1969. Volcanoes. *Cambridge, MA: MIT Press.*

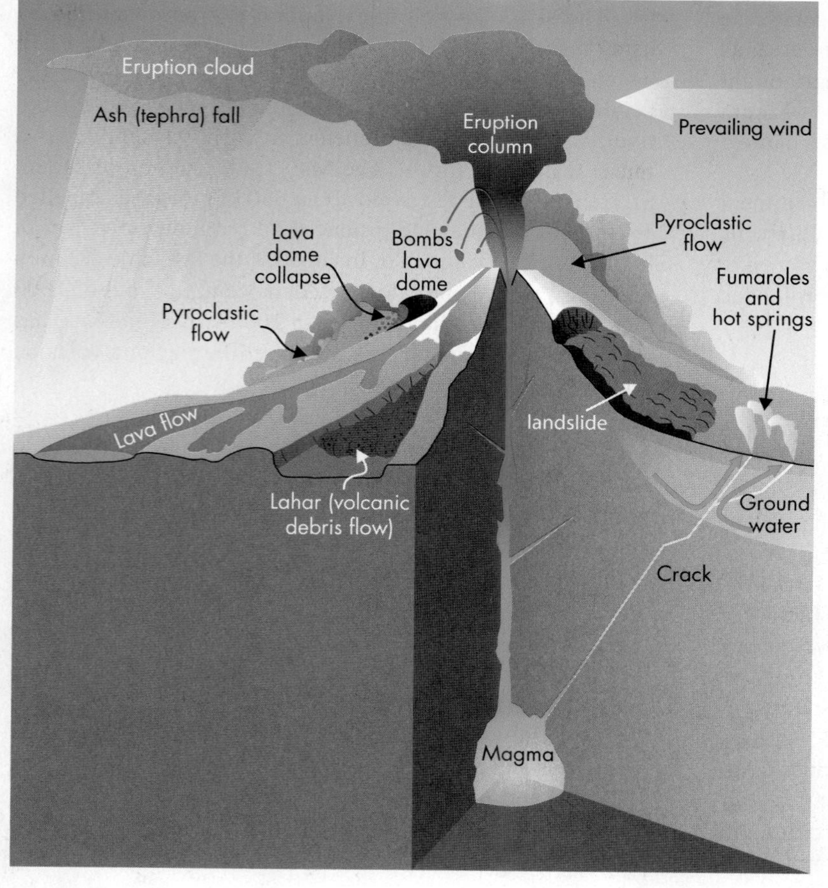

◄ **FIGURE 5.29 VOLCANIC HAZARDS** Hazards associated with explosive stratovolcanoes such as Mount Meager, Mount St. Helens, and Mount Rainier. *(Reprinted with permission of Tricouni Press)*

▲ **FIGURE 5.30 PAHOEHOE LAVA FLOW** The ropy surface texture of a pahoehoe lava flow engulfing a home at Kalapana, Hawaii, in 1990. This flow and others produced by eruptions of Kilauea volcano destroyed more than 100 structures, including the Kilauea National Park Service Visitors Center. *(© Paul Richards/Bettmann/Corbis)*

pahoehoe (pronounced pa-hoy-hoy) (Figure 5.30). Cooler, more viscous, basaltic lavas move at rates of as little as a few metres per day and harden with a rough, blocky surface texture, termed **aa** (pronounced ah-ah) (Figure 5.31). Most lava flows travel slowly enough that people and animals are able to move out of their way.[28] Property and environmental damage, however, can be extensive. For example, lava flows can dam rivers, kill resident fish, and form a barrier to migrating fish, as happened along the Nass River in central British Columbia during an eruption of Tseax volcano a few hundred years ago.[24,29] Native oral traditions suggest that this eruption may have killed as many as 2000 people, perhaps due to suffocation from carbon dioxide or to fires.

▲ **FIGURE 5.31 AA LAVA FLOW** This aa lava flow is moving over an older flow of pahoehoe lava. Aa has a blocky surface texture that develops on cooler, slower-moving basaltic lava than pahoehoe. *(© J. D. Griggs/Corbis)*

If so, this eruption would rank as Canada's worst natural disaster.

Lava flows from the flank of Kilauea volcano in Hawaii, which began in 1983 and continue sporadically to this day, have caused major property damage. This series of eruptions is the longest and largest in Kilauea's recent history.[3] By 1992, more than 50 structures in the village of Kalapana had been destroyed by lava flows, including the Kilauea National Park Visitors Center. The village of Kalapana has virtually disappeared, and it will be many decades before much of the land is productive again.

Pyroclastic Flows and Surges

The most lethal eruptive phenomena are **pyroclastic flows**, which are avalanches of hot gas, ash, and volcanic rock fragments that cascade down the slopes of a volcano during an explosive eruption (Figure 5.1). Pyroclastic flows are also known as ash flows or *nuées ardentes* (French for "glowing clouds"). Some pyroclastic flows form when the towering column of ash rising above an erupting volcano collapses. Others form when a less energetic eruption boils over the edge of a crater or caldera, or when a lava flow or dome on a steep slope collapses. Large pyroclastic flows can move at speeds of up to 150 km/h and run dozens of kilometres from their source.[11] They destroy buildings and other objects in their paths by impact, burial, or incineration. People caught in a pyroclastic flow have no chance of survival.

Pyroclastic flows have killed more people than any other volcanic phenomenon in the past 2000 years.[30] The most deadly event in recent history occurred in 1902 on the Caribbean island of Martinique. On the morning of May 8, a pyroclastic flow roared down Mount Pelée and through the town of St. Pierre, killing an estimated 30 000 people[31] (Figure 5.32). The only two survivors were severely burned.

Pyroclastic surges are dense clouds of hot gas and rock debris produced by explosive interaction of water and magma. They are more violent and travel much faster than pyroclastic flows—pyroclastic surges have been clocked at more than 360 km/h.

Lateral Blasts

Gas, ash, and rock fragments can be blown horizontally from the side of a volcano at the start of an eruption. These **lateral blasts** eject debris at tremendous speeds and can be very destructive. A lateral blast during the initial stage of the May 18, 1980, eruption of Mount St. Helens flattened forest up to 25 km from the vent.[5]

Ash Falls

A tremendous quantity of finely broken volcanic rock and gas is blown high into the atmosphere during many volcanic explosions (Figure 5.11). The particles are carried downwind and rain out to produce **ash fall**. Volcanic ash can carpet hundreds or even thousands of square kilometres of land around a volcano, creating several hazards:

▲ **FIGURE 5.32 PYROCLASTIC FLOW DESTROYS ST. PIERRE** The town of St. Pierre on the Caribbean island of Martinique was obliterated by an incandescent pyroclastic flow in May 1902. Only two of the town's 30 000 residents survived. *(© Israel C. Russell/National Geographic Society/Corbis)*

- Vegetation, including crops and trees, may be destroyed. The long-term impact of ash on forestry and agriculture, however, can be beneficial because ash enriches the soil and increases soil moisture.
- Surface water may be contaminated. Very fine ash particles clog the gills of fish and kill other aquatic life. Ash can also temporarily increase the acidity of the water. Studies of steelhead and salmon in streams west of Mount St. Helens showed that fish populations suffered severely after the 1980 eruptions. In addition to the increased acidity of surface waters, harmful or lethal secondary effects include the loss of fish-spawning habitat and riparian vegetation.
- Buildings may be damaged as ash piles up on roofs (Figure 5.33). As little as 1 cm of ash adds an extra 2.3 tonnes of weight on an average house with a 140 m^2 roof.
- Respiratory illnesses such as asthma and bronchitis are aggravated by contact with volcanic ash and associated aerosols.[28] Coarser particles can lodge in the nose and eyes. **Silicosis** has been attributed to long-term exposure to volcanic ash.
- Ash can damage mechanical and electrical equipment. It is abrasive and, at great distances from the volcano, is fine enough to work its way into the moving parts of machinery. Electrical power can be disrupted because transformers are poor conductors of heat and will overheat and explode when covered by only a few millimetres of ash.
- Ash can affect aircraft flying at high altitudes. The effect of ash on jet engines was discovered in 1982 when a Boeing 747 jet on a flight over Indonesia encountered ash erupted from Galunggung volcano. All four engines failed, but the plane was able to make a successful emergency landing in Jakarta. In 1989 a KLM 747 jet on its way to Japan flew through a cloud of volcanic ash erupted from Redoubt Volcano in Alaska. Power to all four of its jet engines was lost and the plane began a silent, 4270 m fall toward the Talkeetna Mountains below.[32] The 231 passengers on board endured a tense five minutes before the captain was able to restart the engines. When engine power was regained, the aircraft was only 1220 m above the highest mountain peaks. Fortunately, the pilot was able to make an emergency landing in Anchorage, Alaska. Repairs to the aircraft cost an estimated U.S.$80 million.[11] Ash abrades the outer parts of jet engines and can melt inside the engines at the high temperatures at which they operate. The ash also abrades the exterior of the aircraft, frosting cockpit windows and landing lights. Due to such concerns, air flights to, from, and within northern Europe were disrupted in April and May of 2010 when Iceland's Eyjafjallajökull erupted (Case Study 5.2).

◄ **FIGURE 5.33 TEPHRA ON ROOFS** Fall-out of ash and cinders can increase the load on roofs, causing structural collapse. Shown here are numerous houses buried or partly buried in cinders in Heimaey, Iceland, following an eruption in 1973. *(Kai Honkanen/Alamy)*

5.2 CASE STUDY

Icelandic Eruption Paralyzes Air Travel in Europe

Eyjafjallajökull is one of Iceland's smaller ice caps, located on the southern tip of the island. The ice cap covers the caldera of a volcano about 1700 m high that, before 2010, last erupted in 1921. The 2010 eruptions, although relatively small in terms of explosive volcanic eruptions, severely disrupted air travel in western and northern Europe. The first eruption phase ejected andesitic tephra several hundred metres into the air. Fine ash from this phase of the eruption rose no more than 4 km into the atmosphere and had no effect on air travel.

On April 14, 2010, however, the eruption entered a highly explosive phase and an estimated 250 million m^3 of ash was lofted about 9 km into the atmosphere (Figure 5.34). Over the course of the next week, ash drifted over large areas of northern Europe and about 20 countries closed their airspace, affecting more than 100 000 travellers. A high proportion of flights within, to, and from Europe were cancelled, creating the greatest air travel disruption since World War II. Air travel continued to be disrupted on a smaller scale into May 2010. The airline industry worldwide lost nearly CAD$2 billion during the disruption.

The 2010 Eyjafjallajökull eruption was not as large as the 1980 Mount St. Helens or 1991 Pinatubo eruptions, yet its impact on aviation was far greater. Several factors made this eruption so disruptive to air travel:

1. The volcano was directly under the jet stream.
2. The direction of the jet stream was to the southeast and unusually stable throughout the time of the second phase of the eruption.
3. The eruption happened underneath 200 m of glacier ice. Water generated by the melt of the ice flowed back into the erupting volcano, which significantly increased the explosive power of the eruption and created a cloud of highly abrasive glass-rich ash.
4. The volcano's explosive power was sufficient to inject ash directly into the jet stream.

(a)

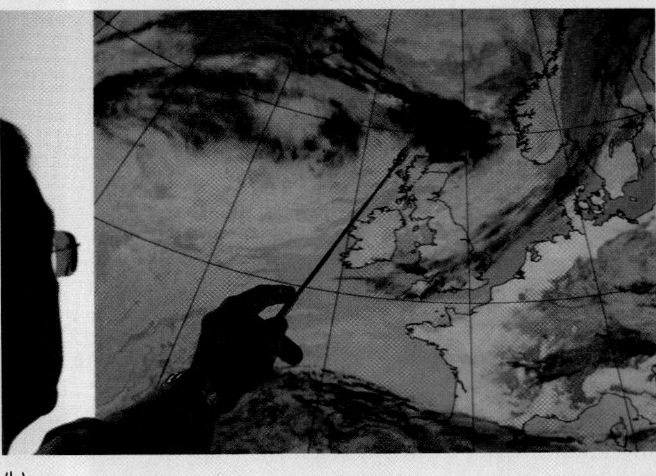

(b)

▲ **FIGURE 5.34 ICELANDIC VOLCANIC ERUPTION** (a) The eruption of Eyjafjallajökull in April 2010 blanketed the Icelandic countryside with volcanic ash. (b) A meteorologist points to an enhanced colour satellite image highlighting a volcanic ash plume moving from Iceland toward the United Kingdom on April 15, 2010. All flights in and out of Britain's airports were grounded. *((a) Arctic Images/Corbis; (b) Matt Cardy/Getty Images)*

Poisonous Gases

A number of gases, including water vapour (H_2O), *carbon dioxide* (CO_2), *carbon monoxide* (CO), *sulphur dioxide* (SO_2), *hydrogen sulphide* (H_2S), *chlorine* (Cl), and *hydrofluoric acid* (HF), are emitted during volcanic eruptions. In some cases, these gases reach toxic concentrations. A notable example is the disaster at Lake Nyos in the Cameroons in 1986. Lake Nyos is a deep crater lake on a dormant volcano. On an August night in 1986, with little warning other than a loud rumbling, the lake released a misty cloud of dense carbon dioxide gas. Nearly odourless, the gas cloud flowed from the volcano into valleys below, displacing the air. It spread

▲ **FIGURE 5.35 POISONOUS GAS FROM DORMANT VOLCANO** (a) In 1986 Lake Nyos released a large volume of carbon dioxide. (b) The gas asphyxiated 1742 people and about 3000 cattle. (c) Gas being released from the bottom waters of Lake Nyos with a degassing fountain in 2001. *((a) © T. Oban/Corbis/Sygma; (b) © Peter Turnley/Corbis; (c) University of Savoie)*

silently through five villages, suffocating 1742 people, an estimated 3000 cattle, and numerous other animals[33] (Figures 5.35a and 5.35b).

Carbon dioxide continues to accumulate at the base of Lake Nyos and another release could occur in the future.[33] Although the area around the lake was closed after the 1986 disaster to all but scientists studying the hazard, thousands of people have returned to farm the land. Scientists have installed an alarm system at the lake that will sound if carbon dioxide levels become dangerously high. A pipe extends from the lake bottom to a degassing fountain on the surface of the lake (Figure 5.35c), allowing carbon dioxide gas to escape slowly into the atmosphere. The fountain is releasing a little more carbon dioxide gas than is naturally seeping into the lake; thus the hazard is being slowly reduced. Additional pipes with degassing fountains will be necessary to completely eliminate the hazard.

Sulphur dioxide can react with water in the atmosphere to produce sulphuric acid and acid rain downwind of an eruption. Sulphuric acid damages crops, and the increased acidity of water collected in cisterns can leach heavy metals into drinking water. Sulphur dioxide vented from the Laki volcano in Iceland in 1783 killed people and livestock and damaged crops. Many of the survivors died from starvation.

Fluorine killed and disfigured livestock after the 1845 and 1970 eruptions of Hekla volcano in Iceland.[34] The 1783 Laki eruption contaminated pastures in Europe with fluorine and caused the death of grazing cattle in as little as two days.[35]

Volcanoes can also produce a type of smog known as *vog* (volcanic material, *v*, and fog, *og*). Eruptions of Kilauea have emitted sulphur dioxide, steam, and other volcanic gases, which at times react with water vapour to produce vog and acid rain. The southeast part of the island of Hawaii can be blanketed with a thick, blue acidic haze that far exceeds air quality standards for sulphur dioxide. Public health warnings have been issued because small, acidic aerosol particulates and sulphur dioxide can induce asthma attacks and cause other respiratory problems. Residents and visitors have reported breathing difficulties, headaches, sore throats,

◄ **FIGURE 5.36 LAHAR DISASTER IN COLOMBIA** More than 23 000 people in Armero, Colombia, were killed on November 13, 1985, when a lahar from erupting Nevado del Ruiz volcano swept through the town. This photograph, taken about one month after the disaster, shows the lahar track and, to the left and right, remnants of Armero. *(U.S. Geological Survey; photo by R. J. Janda, December 9, 1985)*

watery eyes, and flu-like symptoms when exposed to vog. In addition, acid rain has made the water in some shallow wells and household rainwater-collection systems undrinkable. Acidic rainwater extracts lead from metal roofing and water pipes, and may have caused elevated lead levels in the blood of some residents.[36]

Edifice or Sector Collapse

The flank of a volcano may collapse during an eruption or even at times when there is no eruptive activity. Flank collapse can increase the size and strength of an eruption. A massive landslide during the early moments of the May 18, 1980, eruption of Mount St. Helens catastrophically depressurized the volcano and caused a very different eruption than had been predicted.[5] As magma moves up into the throat of a volcano from the magma chamber below, the volcano inflates, much like blowing up a balloon. The volcano's slopes become over-steepened and unstable during the magma's ascent. Ground shaking associated with steam venting, magma ascent, or an earthquake can trigger a collapse, which, if large enough, is called a **sector collapse**.

Debris Flows and Other Mass Movements

The most serious secondary effects of volcanic activity are debris flows, also known by their Indonesian name, **lahars**. Lahars are produced when large amounts of loose volcanic ash and other pyroclastic material become saturated with water and rapidly move downslope (Figure 5.36). Even relatively small eruptions of hot volcanic material may quickly melt large volumes of snow and ice on a volcano. The rapid melting produces a flood of meltwater that mixes with pyroclastic material eroded from the slope of the volcano to create a lahar with the consistency of wet cement. Lahars can travel many kilometres down valleys from the flanks of the volcano where they form.[28]

One of the world's worst historical volcanic disasters was a lahar. Nevado del Ruiz in Colombia erupted on November 13, 1985. Despite the fact that the volcano was monitored and warnings were issued, a series of lahars killed more than 23 000 of the 30 000 residents of the town of Armero (Figure 5.36).[37] A similar lahar in 1845 killed the same town's entire population of 1400 people.

Lahars can occur days, weeks, or years after an eruption. Explosive eruptions may remove all vegetation from areas around a volcano, making them vulnerable to lahars and other landslides. In addition, the flanks of the volcano may be covered with loose tephra, which is easily mobilized into debris flows by heavy rains or rapid snowmelt.

Gigantic lahars have moved down the flanks of volcanoes in the Pacific Northwest in both historic and prehistoric times. Two of these lahars occurred on Mount Rainier (Figure 5.37). Approximately 5600 years ago, the "Osceola mudflow" moved 1.9 km^3 of sediment up to 80 km from the summit.[38] Deposits of the younger, 500-year-old "Electron mudflow" reached about 56 km from the volcano into now-populated suburbs of Seattle.[38] Observers of a lahar the size of the Electron mudflow might see a wall of mud the height of a house moving toward them at close to 30 km/h. With the flow moving at this speed, the observers would need a car headed in the right direction to escape being buried alive.[28] Hundreds of thousands of people now live on the deposits of the Osceola and Electron mudflows (Figure 5.38), and there is no guarantee that a similar flow will not occur again.

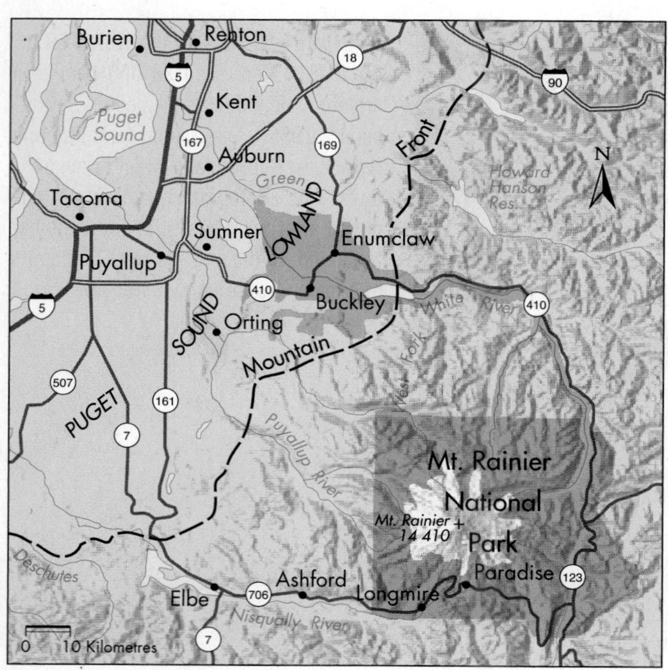

◄ **FIGURE 5.37 LAHARS FROM MOUNT RAINIER** A map of Mount Rainier and vicinity showing the extent of the 5600-year-old Osceola mudflow in the White River valley (coloured dark orange) and the 500-year-old Electron mudflow (coloured pale orange) in the Puyallup River valley. *(From Crandell, D. R., and D. R. Mullineaux. 1969.* Volcanic Hazards at Mount Rainier, Washington. *U.S. Geological Survey Bulletin 1238)*

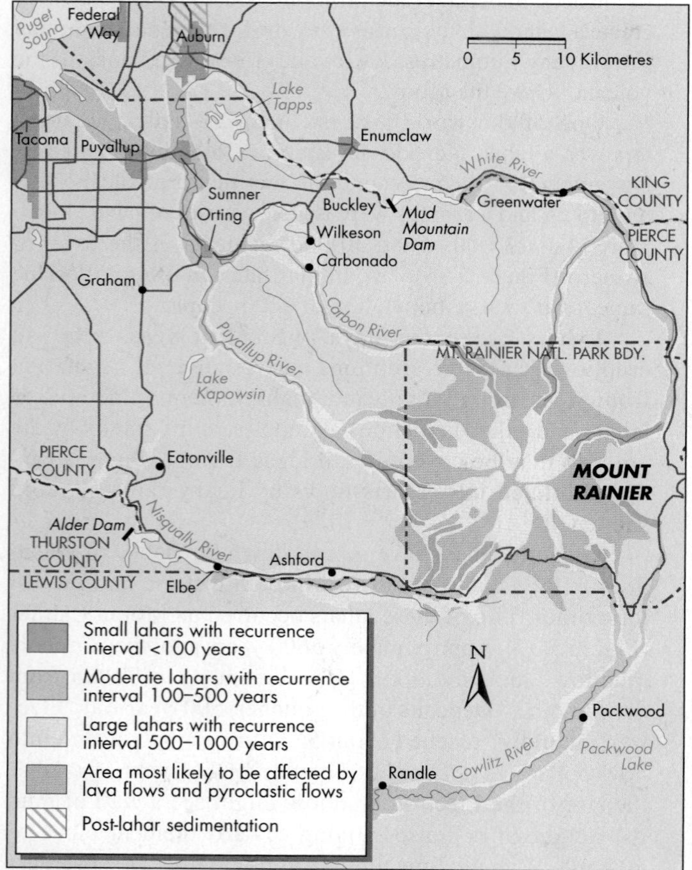

◄ **FIGURE 5.38 LAHAR HAZARD POTENTIAL** A map of Mount Rainier and vicinity showing areas at risk from lahars, lava flows, and pyroclastic flows. The Seattle and Tacoma suburbs of Puyallup, Sumner, Orting, and Auburn are at risk. *(Hoblitt, R. P., J. S. Walder, C. L. Driedger, K. M. Scott, P. T. Pringle, and J. W. Vallance. 1998.* Volcanic Hazards from Mount Rainier, Washington, Revised 1998. *U.S. Geological Survey Open-File Report 98-428)*

A CLOSER LOOK 5.1

Mount St. Helens 1980–2008: From Lateral Blasts to Lava Flows

The May 18, 1980, eruption of Mount St. Helens in Washington State (Figure 5.39) exemplifies the many types of things that happen during an explosive volcanic eruption. And the story is not yet over!

Mount St. Helens awoke in March 1980, after 120 years of dormancy, with earthquakes and small explosions created by the boiling of groundwater as it came in contact with hot rock. By May 1, a prominent bulge had developed on the north flank of the mountain (Figure 5.40a). The bulge grew at a rate of about 1.5 m per day until, at 8:32 A.M. on May 18, an **M** 5.1 earthquake triggered a huge debris avalanche (Figure 5.40b), causing 2.3 km^3 of rock, the entire area of the bulge, to break away and shoot down the north flank of the mountain, displacing water in nearby Spirit Lake (see Professional Profile). The avalanche then struck and overrode a ridge 8 km to the north and made an abrupt turn before moving 18 km down the Toutle River valley.

During the flank collapse, Mount St. Helens erupted with a lateral blast from the area of the former bulge (Figures 5.40b and 5.40c). The blast moved at speeds of more than 480 km/h, faster than a bullet train, and levelled timber over an area of about 600 km^2 (Figure 5.41a).[39] Shortly thereafter, pyroclastic flows began to sweep down the north slope of the volcano.

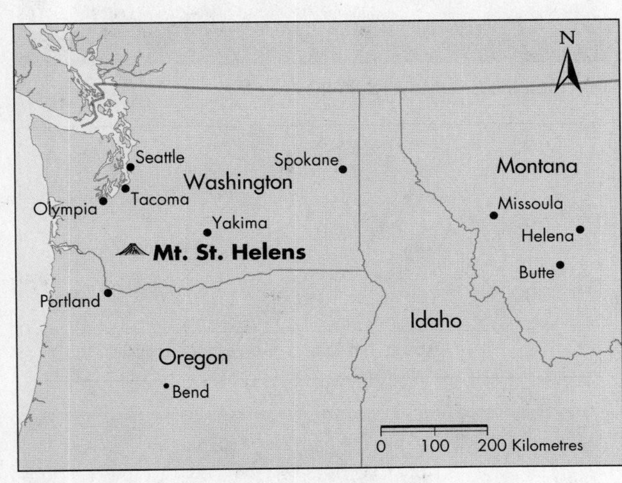

(a)

(b)

◄ **FIGURE 5.39 MOUNT ST. HELENS BEFORE AND AFTER** (a) A map showing the location of Mount St. Helens in Washington State. (b) The volcano before (in photo held aloft) and after the May 18, 1980, eruption. Much of the north side of the volcano slid away during the eruption, and the summit was lowered by about 400 m. The lateral blast shown in Figure 5.40 originated in the amphitheatre-like area at the top centre of the photograph. *((b) Jim Richardson/National Geographic Image Collection)*

(continued)

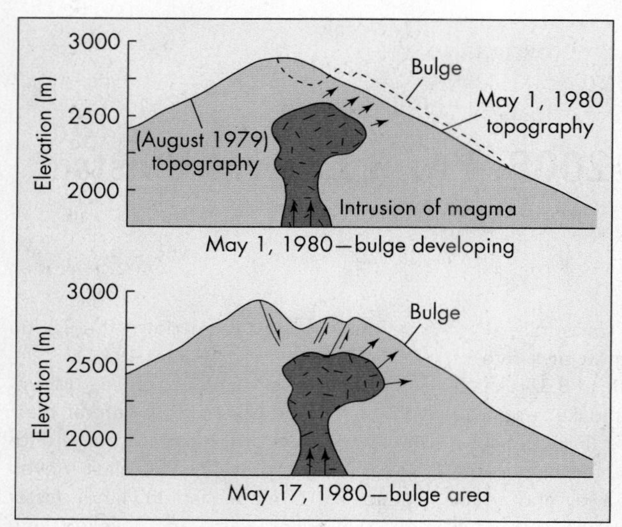

(a) Before eruption May 1 to 17, 1980

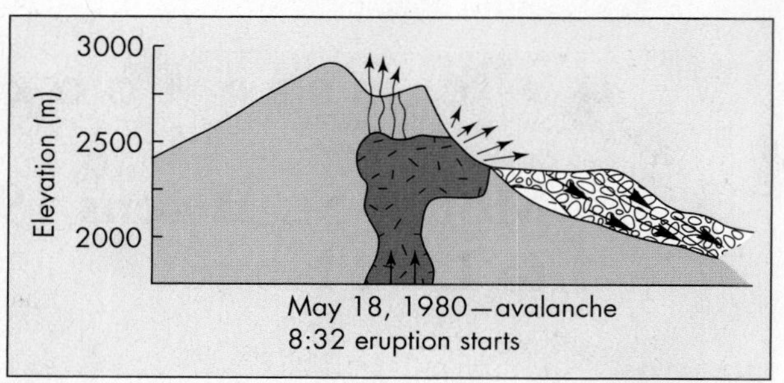

(b) Eruption starts May 18,1980

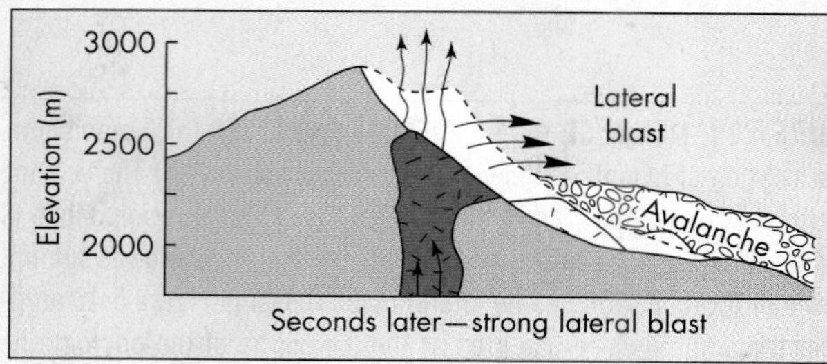

(c) Seconds after eruption starts

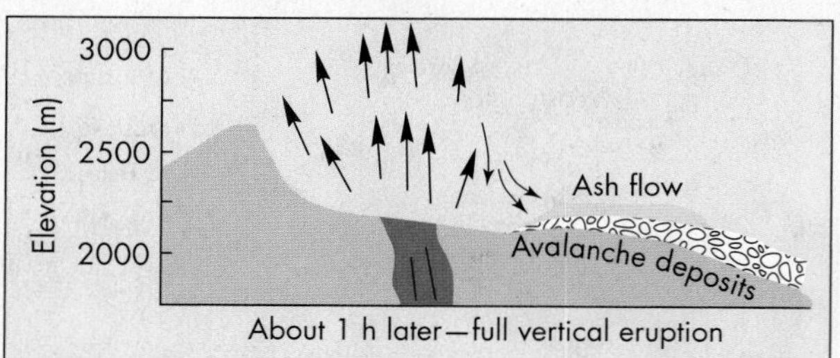

(d) About an hour after eruption starts

▲ **FIGURE 5.40 MOUNT ST. HELENS ERUPTS** Diagrams and photographs showing the sequence of events at Mount St. Helens on the morning of May 18, 1980. Photographs in (b) and (c) were taken less than 10 seconds apart. The bottom photograph shows an ash column rising high into the atmosphere. *(Drawings inspired by a lecture by James Moore, U.S. Geological Survey; (b) and (c) © 1980 by Keith Ronnholm, Geophysics program, University of Washington, Seattle; (d) Jack Smith/AP Images)*

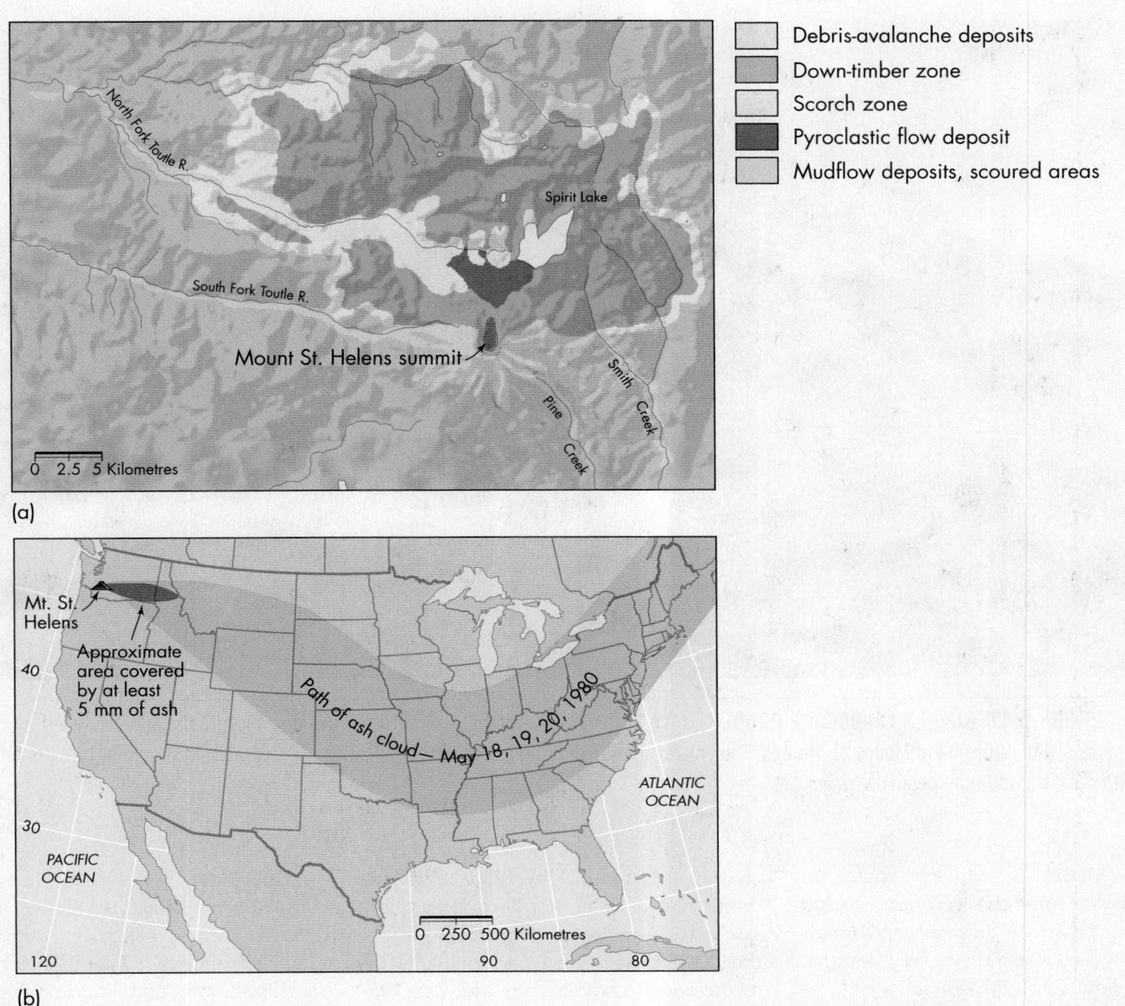

▲ **FIGURE 5.41 ERUPTION DEPOSITS AND ASH CLOUD** (a) A map of the area around Mount St. Helens, showing the extents of debris avalanche deposits, tree blow-down, pyroclastic flow deposits, and lahars associated with the May 18, 1980, eruption. (b) The path of the ash cloud (orange) from the 1980 eruption. The area covered by at least 5 mm of ash is shown in red. *(Data are from U.S. Geological Survey publications)*

Within an hour of the lateral blast, a large column of ash and gases had reached an altitude of approximately 19 km, extending more than 4 km into the stratosphere (Figure 5.40d). The eruption continued for more than nine hours; in that time about 1 km^3 of volcanic ash was ejected from the volcano. The ash fell over a large area of Washington, northern Idaho, and western and central Montana; traces of ash were recorded in southwest Canada. The ash cloud drifted eastward to New England (Figure 5.41b), and in less than three weeks had circled Earth.

The entire northern slope of the volcano, which is the upper part of the watershed of the North Fork of the Toutle River, was devastated. Forested slopes were transformed into a grey barren landscape of volcanic ash, rocks, blocks of melting glacier ice, narrow gullies, and hot steaming pits[39] (Figure 5.42).

The first of several lahars occurred minutes after the start of the eruption. The lahars raced down the valleys of the North and South forks of the Toutle River at speeds of 29 km/h to 55 km/h, threatening the lives of people camped along the river.[39] On the morning of May 18, 1980, two young people on a fishing trip were sleeping about 36 km downstream from Spirit Lake in the Toutle Valley. They were awakened by a loud rumbling noise from the river, which was covered by felled trees. They attempted to run to their car, but water from the rising river poured over the road, preventing their escape. A mass of mud then crashed through the forest toward the car, and the couple climbed onto its roof to escape the mud. They were safe only momentarily, as the mud pushed the vehicle over the bank and into the river. Leaping off the roof, they fell into the river, which was by now a rolling mass of mud, logs, collapsed train trestles, and other debris. The two were carried downstream approximately 1.5 km, disappearing several times beneath the flow, before another family of campers spotted and rescued them.

(continued)

▲ **FIGURE 5.42 BARREN LANDSCAPE PRODUCED BY ERUPTION** A photograph showing the desolate landscape produced by the May 18, 1980, eruption of Mount St. Helens. The entire valley, from the centre left to the lower right corner of the photograph, is filled with debris avalanche deposits. *(John J. Clague)*

When Mount St. Helens could be viewed again after the eruption, its top was gone. What was originally a symmetrical snow-clad volcano was now a huge, steep-walled amphitheatre facing northward (Figures 5.39b and 5.42), and what was originally Spirit Lake was now filled with sediment and trees. The horizontal blast, debris avalanche, pyroclastic flows, and lahars had devastated an area larger than Toronto. The eruption killed 57 people, and associated flooding destroyed more than 100 homes. Enough timber to build 50 000 houses was flattened by the blast. The total damage from the eruption was estimated to exceed U.S.$1 billion.

Nineteen small eruptions occurred during the first six years following the cataclysmic eruption in May 1980. Lava flows during these small eruptions built a lava dome to a height of over 250 m above the 1980 crater floor. Earthquake activity increased several times between 1989 and 2004, accompanied by small explosions, ash flows, and lahars.

On September 23, 2004, following a period of quiescence, Mount St. Helens returned to life when magma began moving up toward the crater floor.[40] The eruption continued until mid-2005 and was observed continuously by staff of the U.S. Geological Survey Cascades Volcano Observatory. The most significant eruption during this period occurred on March 8, 2005, when a plume of steam and ash rose 11 000 m into the atmosphere.

On January 16, 2008, a plume of steam flowed from a fracture on top of the lava dome, accompanied by an increase in seismic activity. Scientists suspended activities in the crater and on the flanks of the volcano, but the risk of a major eruption was deemed low. Soon thereafter, the outflow of steam and seismic activity ceased.

Today the mountain is monitored with a network of automated seismographs and GPS satellite receivers to detect earthquakes and ground deformation, aircraft to collect gas samples, acoustic instruments to detect lahars, and video cameras to continuously record events.[40] The volcano remains active.

5.4 Links between Volcanoes and Other Natural Hazards

Volcanoes are intimately linked to several other natural hazards. We have already stressed the relation between volcanoes and plate tectonics, specifically the relation between plate tectonic setting and types of eruptive activity. Volcanoes are also linked to fire, earthquakes, landslides, tsunamis, floods, and climate change.

Volcanic eruptions are not a major cause of fire, but it is not difficult to imagine the link between these phenomena. Plants and human-built structures commonly catch fire when they are struck by lava flows, pyroclastic flows, and superheated bombs. In January 2002, Africa's most destructive volcanic eruption in 25 years sent residents of the

PROFESSIONAL PROFILE

Catherine Hickson, Volcano Scientist

Why does someone become a volcanologist? For some, it's an interest sparked by a grade-school project with baking soda and vinegar; for others, it is a special episode on the Discovery Channel. For a rare few, it is witnessing an eruption of historic size. Dr. Catherine Hickson (Figure 5.43) got her start in volcanology in just such a way. "It was really a matter of luck rather than 'good' timing that found me close to Mount St. Helens that fateful day," says Hickson. She was 14 km east of Mount St. Helens when it erupted on May 18, 1980. "We were forced to flee, but I was captivated by the colossal events that unfolded in minutes. The landscape around us was changed from a pristine wilderness to a grey moonscape."

Of particular interest to Hickson was the "directed blast," more properly called a pyroclastic surge. The surge that devastated more than 360 km^2 of terrain around the mountain was driven by both magma and steam. Hickson explains: "As magma entered the volcano, it heated the surrounding rocks and began melting the capping glaciers. The mountain stewed in this acidified brew for weeks and then finally gave way." That fateful morning, the energy of the magma and the pressurized water was unleashed, taking 57 lives.

"Volcanoes and eruptions are a fascinating field of scientific study, but for me, the impact of eruptions on humans is equally important," says Hickson. Through her experiences at Mount St. Helens, she was given initial insight into the brutal reality of what happens when people are faced with a volcanic threat. "The eruption changed me. Since then I have tried to devote time and energy to making our scientific knowledge of natural hazards understandable to the public and policymakers. How do we make emergency and land-use planners listen to what geologists are saying and, most importantly, take action?" she asks.

▲ **FIGURE 5.43 CATHERINE HICKSON, VOLCANO SCIENTIST** Catherine Hickson was camping just east of Mount St. Helens when it erupted on May 18, 1980. The eruption was a life-defining moment for her—she subsequently trained for a career in volcanology and currently works to reduce the risk that people face from volcanic eruptions. *(C. J. Hickson)*

Hickson has made countless presentations to professionals and the public over the years. In 2002, she created a multinational project in South America with the help of funding from the Canadian International Development Agency and the governments of Argentina, Bolivia, Chile, Colombia, Ecuador, Peru, and Venezuela. This successful project built capacity in Andean countries to reduce the risk of hazardous natural processes, including volcanic eruptions.

"What we really need to do," Hickson notes, "is to think about the long-term threat of natural hazards on not only our current built environment but also the environment our children and grandchildren will be living in—let's not leave them a legacy of destroyed cities because we were careless in our development practices."

—*John J. Clague*

Democratic Republic of Congo fleeing raging fires ignited by lava flows. One flow sparked an explosion that killed 60 people.

Earthquakes commonly precede or accompany volcanic eruptions as magma rises through Earth's crust to the surface. Weeks of earthquakes preceded the first eruption at Paricutín volcano west of Mexico City. Some earthquakes preceding an eruption may be large enough to cause damage.

Landslides are possibly the most common secondary effect of volcanic activity. As discussed above, sector collapses and lahars can do great damage and take many lives. In some cases, they can trigger tsunamis, and the waves can reach great heights close to the point of failure. Like other landslide-triggered tsunamis, however, they attenuate rapidly as they move away from the source and do not cause catastrophic damage thousands of kilometres away.

Volcanic eruptions beneath glaciers commonly cause devastating jökulhlaups. Iceland is at greatest risk from these eruption-triggered floods. Recurrent floods from beneath the Vatnajökull ice cap have damaged highways, bridges, and farmland.

Lastly, volcanic eruptions can affect climate. A cloud of ash and sulphur dioxide remained in the atmosphere for more than a year after the 1991 Mount Pinatubo eruption (Figure 5.44). The ash particles and aerosol droplets scattered incoming sunlight and slightly cooled the atmosphere during the year following the eruption.[41]

(a)

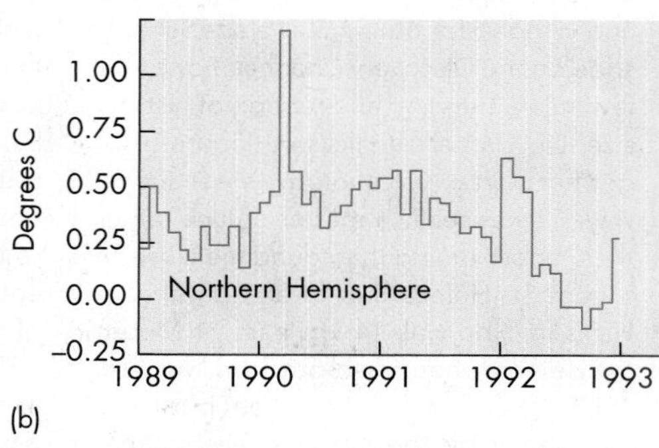

(b)

▲ **FIGURE 5.44 VOLCANIC ERUPTION TEMPORARILY COOLS CLIMATE** (a) An eruption of Mount Pinatubo in 1991 ejected huge amounts of volcanic ash and sulphur dioxide up to about 30 km into the atmosphere. Ash particles remained in the upper atmosphere and circled Earth for more than a year. (b) The sulphur dioxide and, to a lesser extent, ash temporarily lowered Earth's average surface temperature *((a) D. Harlow/U.S. Geological Survey, Denver, CO; (b) http://www.cru.uea.ac.uk/. Courtesy of Climatic Research Unit. Reprinted with permission.)*

5.5 Natural Service Functions of Volcanoes

Although active volcanoes pose a serious threat to those who live near them, they provide important natural service functions. Perhaps their greatest gift to us occurred billions of years ago when gases and water vapour released from volcanoes began to form the atmospheric and hydrologic systems that allowed life to appear and evolve. Volcanoes also provide us with fertile soils, a source of power, recreational opportunities, and new land.

Volcanic Soils

Volcanic eruptions provide lavas and pyroclastic deposits, which, when weathered, are an excellent growth medium for plants. The nutrients produced by weathering of volcanic materials allow such crops as coffee, maize, pineapple, sugar cane, and grapes to thrive. However, these fertile soils encourage people to live in hazardous areas. So although volcanic soils provide an important resource, nearby volcanic activity can make it difficult to safely use that resource.

Geothermal Power

The internal heat associated with volcanoes can be used to generate power for nearby urban areas. Geothermal electrical power is being generated in Hawaii, California, Nevada, Italy, New Zealand, and other areas, and much of the city of Reykjavík, Iceland, is heated with geothermal energy. An important benefit of geothermal energy is that it can be a renewable resource. However, the steam driving the system cannot be removed faster than it is restored naturally or it will become depleted, as has happened at Rotorua in New Zealand. As well, geothermal energy cannot be exploited without some adverse environmental impacts. Collector wells, surface pipes, and electrical transmission lines can be unsightly and occupy large tracts of land (Figure 5.45), and the water produced by the conversion of steam to power may be acidic or contain high concentrations of heavy metals.

Recreation

Heat associated with volcanoes provides recreational opportunities. Many health spas and hot springs are located in volcanic areas. Volcanoes also provide opportunities for hiking, snow sports, and education. About 3 million tourists visit Yellowstone National Park each year to see its geysers and hot springs, and Kilauea volcano attracts more than 1 million visitors annually. Volcanoes in Wells Grey and Garibaldi provincial parks in British Columbia are also popular tourist attractions.

Creation of New Land

In our discussion of the benefits of volcanoes, we would be remiss not to mention that eruptions are responsible for creating some of the land we inhabit. Iceland, Hawaii, and many other oceanic islands would not exist without volcanoes!

◄ **FIGURE 5.45 GEOTHERMAL POWER** The Wairakei geothermal field on the North Island of New Zealand was one of the first in the world to be developed. The pipes seen in this photograph transmit steam produced from superheated waters to turbines that generate electricity. *(N. Minton/Shutterstock)*

5.6 Minimizing the Volcanic Hazard

Volcanic eruptions are beyond our control—there is nothing we can do to affect their timing and severity. We can, however, take actions to minimize losses of life and property from volcanic eruptions.

Forecasting

Forecasting volcanic eruptions is an important part of efforts to reduce volcanic risk. An eruption forecast is a statement of the probability that a volcano will erupt in a particular way and within a defined time. It is analogous to a weather forecast and is not as precise as a prediction.[3]

Scientists gather information about precursor phenomena to improve their ability to forecast volcanic eruptions. One problem is that most forecasting techniques require experience with actual eruptions before precursor phenomena can be fully understood. Earth scientists have a good track record of predicting eruptions on the Hawaiian Islands because they have had so much experience with eruptions there.

Eruption forecasts rely on information gained by

- Monitoring seismic activity
- Monitoring thermal, magnetic, and hydrologic conditions at the volcano
- Monitoring the land surface to detect tilting or swelling of the volcano
- Monitoring volcanic gas emissions
- Studying the geologic history of the volcano[27,42]

Monitoring Seismic Activity Experience with volcanoes, such as Mount St. Helens and those on the Big Island of Hawaii, suggests that earthquakes are an early warning of an impending volcanic eruption. In the case of Mount St. Helens, earthquake activity started in mid-March, two months before the cataclysmic eruption on May 18, 1980. At Kilauea, earthquakes have been used to monitor the movement of magma as it approaches the surface. Magma bodies have also been mapped using geophysical techniques.

Small steam explosions and earthquakes began several months before the 1991 Mount Pinatubo eruption.[3] Mount Pinatubo, however, was an eroded ridge that did not have the classic shape of a stratovolcano, and it had not erupted in 500 years. Most people living near the mountain were unaware that it was a volcano, and they were not particularly concerned about the earthquakes. After the initial steam explosions, scientists began monitoring earthquakes on the volcano and studying past volcanic activity. Earthquakes increased in number and magnitude before the catastrophic eruption, and their foci migrated from deep beneath the volcano to shallow depths beneath the summit.[3]

Geophysicists have proposed a generalized model of seismic activity that could help predict future eruptions of stratovolcanoes (Figure 5.46).[42] Several weeks before a volcano comes to life, increasing pressures exerted by magma high within the crust create numerous fractures in the plugged throat of the volcano. The fracturing causes small earthquakes that gradually increase in frequency over a period of one to several weeks. The fracturing and seismic activity accelerate a few days prior to the eruption. Based on the accelerated seismic activity, a seismologist monitoring the earthquakes can issue an alert that an eruption might be imminent. Unfortunately, the short warning time may be insufficient to accomplish a large-scale evacuation of a major city such as Naples, which lies in the shadow of Mount

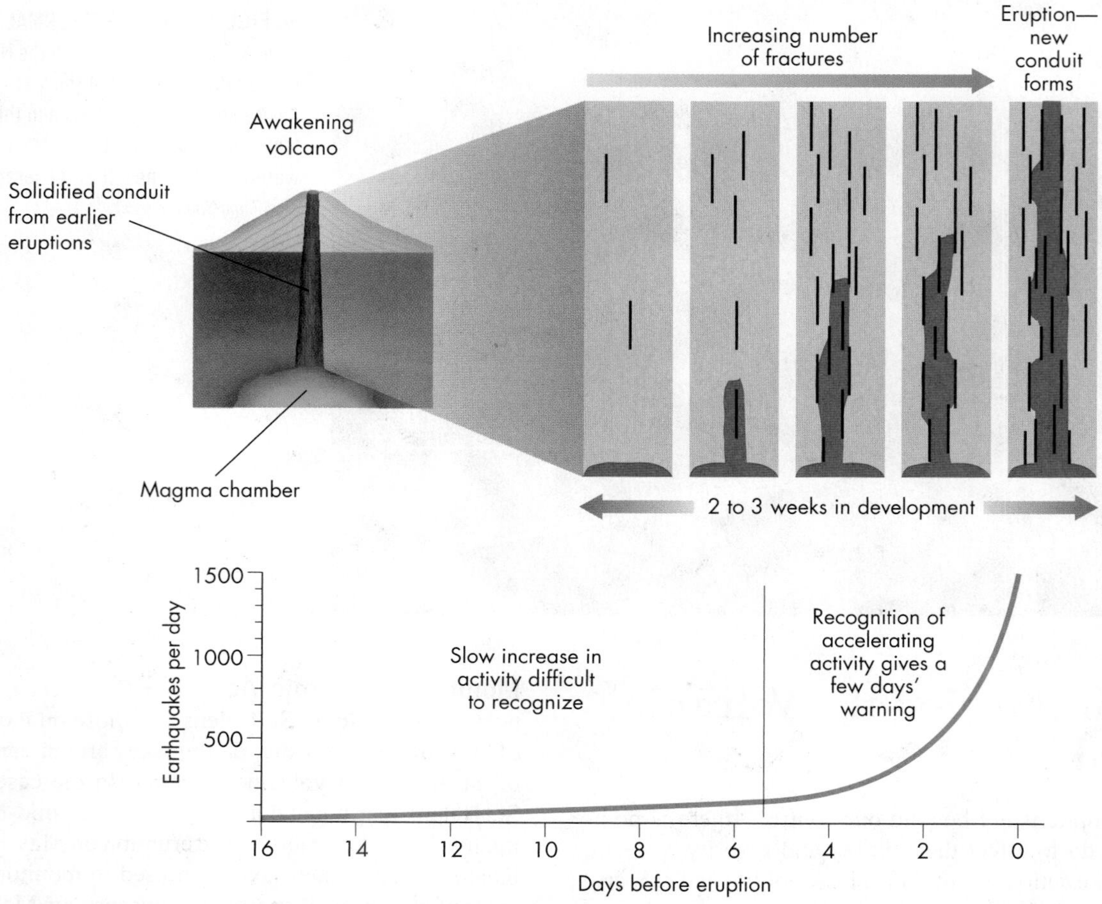

▲ **FIGURE 5.46 A VOLCANO REAWAKENS** Increased seismic activity is a good indicator of an approaching volcanic eruption. As a dormant stratovolcano reawakens, magma (red colour) fractures rock as it rises into the throat of the volcano. The fracturing and accompanying seismic activity increase slowly at first, but accelerate a few days prior to the eruption. *(Adapted from Kilburn, C. R. J., and P. R. Sammonds. 2005. "Maximum warning times for imminent volcanic eruptions."* Geophysical Research Letters *32:L24313 Reprinted with permission of John Wiley & Sons.)*

Vesuvius. Thus to forecast eruptions, seismic activity must be used in concert with other eruption precursors discussed below. Unlike earthquakes, volcanic eruptions always provide warning signs.[43]

Thermal, Magnetic, and Hydrologic Monitoring

Before an eruption, a large volume of magma moves into a holding reservoir beneath the volcano and changes local magnetic, thermal, hydrologic, and geochemical conditions. A rise in heat flow may be detected by satellite sensors or infrared aerial photography. Increased heat may also melt snow or glacier ice, which can be detected by remote sensing. This method was used with some success at Mount St. Helens before the main eruption on May 18, 1980.

When older volcanic rocks are heated by new magma, their magnetic properties may change. These changes in rock magnetization can be detected by ground or aerial magnetic surveys.[27,44]

Land Surface Monitoring Some volcanic eruptions can be forecast by monitoring small changes in the volcano's surface. Hawaiian volcanoes, especially Kilauea, have supplied most of the data. The summit of Kilauea tilts and swells before an eruption and subsides during the eruption (Figure 5.47). Tilting of the summit, in conjunction with an earthquake swarm, was used to predict an eruption near the farming community of Kapoho on the flank of the volcano, 45 km from the summit. As a result, the inhabitants were evacuated before lava overran and destroyed most of the village.[45] Because swelling and earthquakes have preceded past eruptions, scientists are able to reliably predict the activity of Hawaiian volcanoes. Monitoring of ground movements, such as tilting, swelling, and opening of cracks, or of changes in the water level of lakes on or near a volcano may indicate magma movements in advance of an eruption.[27] Today, satellite-based radar and GPS satellite receivers can be used to monitor changes in volcanoes, including surface deformation, without sending people into hazardous areas.[46]

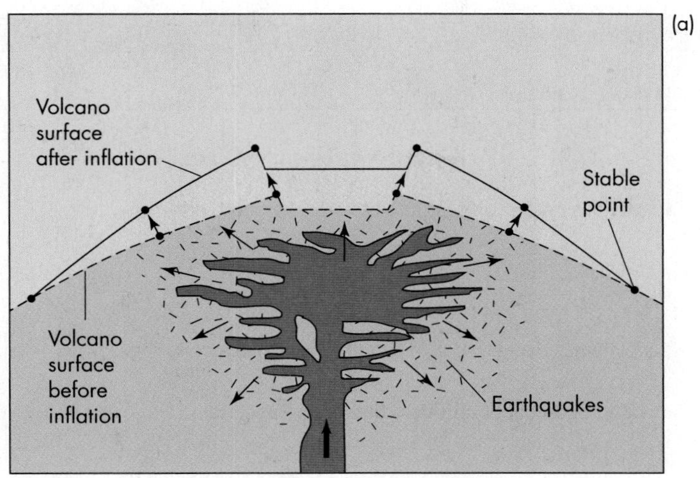

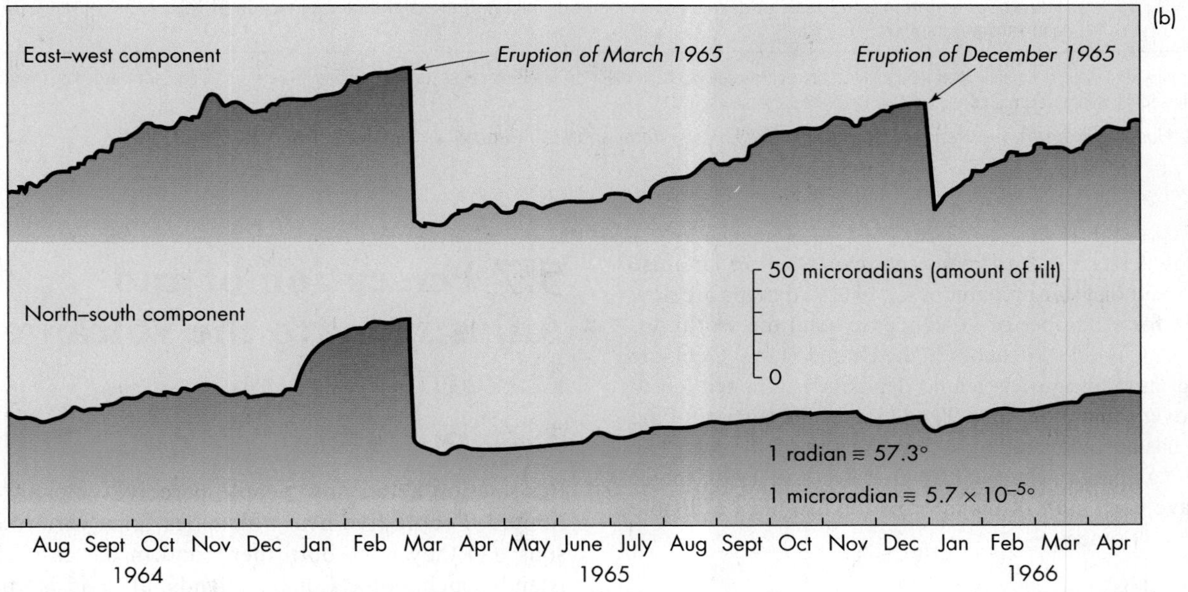

▲ **FIGURE 5.47 INFLATION AND TILTING BEFORE ERUPTION** (a) An idealized diagram of Kilauea, showing inflation and surface tilting as magma moves up into the throat of the volcano. The red area is the underground magma chamber that fills before an eruption. (b) Graphs showing tilting of the surface of Kilauea in two directions (east-west and north-south) from 1964 to 1966. Note the slow increase in tilting before eruptions and the rapid lowering of the surface during eruptions. *((a) Wright, T., and T. C. Pierson. 1992.* Living with Volcanoes. *U.S. Geological Survey Circular 1073; (b) Fiske, R. S., and R. Y. Koyanagi. 1968.* The December 1965 Eruption of Kilauea Volcano, Hawaii. *U.S. Geological Survey Professional Paper 607)*

Monitoring Volcanic Gas Emissions Changes in the relative amounts of carbon dioxide and sulphur dioxide or changes in gas emission rates may indicate movement of magma toward the surface, making them a useful tool for predicting an eruption. Gas emissions are commonly measured in conjunction with monitoring of deformation and seismicity to increase the reliability of predictions. The most precise compositional data still require field sampling under potentially dangerous conditions, although remote measurement of gas composition and flux using sensors has become much more reliable over the past two decades. Measurements of volcanic gases proved useful in studying eruptions at Mount St. Helens and Mount Pinatubo. The volume of sulphur dioxide emitted from Mount Pinatubo increased by more than a million times two weeks before the eruption in 1991.

Geologic History An understanding of the recent geologic history of a volcano is helpful in predicting its future eruptive behaviour. The primary tools for reconstructing a volcano's history are geologic mapping and age dating of lava flows and pyroclastic deposits. Geologic mapping

TABLE 5.3 U.S. Geological Survey Volcanic Alert Levels and Aviation Colour Codes[a]

Ground Alert Level	Volcanic Condition	Aviation Colour Code
NORMAL	(1) Typical background, noneruptive state.	GREEN
	Or	
	(2) If downgraded from higher alert level, activity has ceased and returned to a background, noneruptive state.	
ADVISORY	(1) Elevated unrest above known background level.	YELLOW
	Or	
	(2) If downgraded from a higher alert level, activity has decreased significantly with close monitoring for possible renewed increase.	
WATCH	(1) Heightened or escalating unrest with increased potential of eruption, time frame uncertain.	ORANGE
	Or	
	(2) Eruption underway with limited hazards; for aviation—no or minor volcanic ash emissions.	
WARNING	Hazardous eruption imminent, underway, or suspected; for aviation—significant emission of volcanic ash into atmosphere.	RED

[a]Note: For most eruptions, the ground alert level and aviation colour code will be at the same levels; however, for some eruptions there will be a greater hazard on the ground or to aviation and different levels will be assigned for the two environments.

Source: Modified from http://volcanoes.usgs.gov/activity/alertsystem/ and U.S. Geological Survey Fact Sheet 2006-3139.

underpins derived hazard maps that can assist in land-use planning and disaster preparedness.[27] Hazard maps are now available for a number of volcanoes around the world, including the Cascade volcanoes in the United States. Geologic mapping and dating of volcanic deposits at Kilauea led to the discovery that more than 90 percent of the surface of the volcano has been covered by lava in the past 1500 years. The town of Kalapana, destroyed by lava flows in 1990, might never have been built if this information had been available before development.

Volcanic Alert or Warning

At what point should the public be warned that a volcanic eruption might occur? This question has partially been answered by volcanologists and policymakers. The U.S. Geological Survey recently established an alert notification system for volcanic activity.[47] This system has two components: ground-based volcano alert levels and aviation-based colour code levels (Table 5.3). Each component has four levels; for most eruptions, the volcano alert level and aviation code level will be the same.[47] For some eruptions, however, the hazard to those on the ground and those in the air will differ and different alerts or codes will be issued. For example, if a lava flow from the Kilauea volcano on Hawaii threatens a community, the U.S. Geological Survey would issue a Warning because a hazardous eruption is underway, but only an Orange aviation code because there are minor volcanic ash emissions.

Although the system is a good start, the hard questions remain: When should evacuation begin? When is it safe for people to return? These are questions that public officials will have to answer when volcanic watches or warnings are issued.

5.7 Perception of and Adjustment to the Volcanic Hazard

Perception of Volcanic Risk

Information about how people perceive volcanic risk is limited. People live near volcanoes for a variety of reasons: (1) they were born there, and in the case of some islands, such as the Canary Islands, all land is volcanic; (2) the land is fertile and good for farming; (3) people are optimistic and believe an eruption is unlikely; (4) people are fatalistic or risk takers; (5) they are unaware of any risk; and (6) they cannot choose where they live (for example, their choice may be limited by economics). A study of risk perception in Hawaii found that a person's age and length of residence are significant factors in their knowledge of volcanic hazards and possible adjustments to them.[48] One reason the evacuation of 60 000 people before the 1991 eruption of Mount Pinatubo was successful was that the government had educated people about the dangers of violent eruptions and lahars. A video depicting these events was widely shown before the eruption and helped convince local officials and residents that they faced a real and immediate threat.[3]

Scientific understanding of volcanoes is constantly improving. However, good science is not sufficient (Survivor Story). Probably the greatest risk reduction will come from an increased understanding of human and societal issues that arise during an emerging **volcanic crisis**, when scientists predict that an eruption is likely in the near future. In such a crisis, improved communication among scientists, emergency

SURVIVOR STORY

A Close Call with Mount St. Helens

Don Hamilton (Figure 5.48) was lucky to be nowhere near a telephone on the evening of May 17, 1980.

The next day, at 8:32 A.M., Mount St. Helens would erupt, spewing enormous amounts of rock and ash into the atmosphere and covering the surrounding terrain with up to hundreds of metres of debris.

But that evening, Hamilton was sitting on the porch of a lodge at Spirit Lake, no more than 16 km from the peak of the mountain, visiting the lodge's resident, a local eccentric named Harry Truman. By that time, earthquakes and seismic observations had geologists clamouring about an imminent eruption, and Truman, then 84 years old, had become a minor celebrity for his stalwart refusal to evacuate.

Hamilton was working for the now-defunct *Oregon Journal*, and that night he might have stayed with Truman if he had had a telephone. But Hamilton needed some way to file his story, so he left for Portland.

As it turned out, leaving was the best decision of his life. When the volcano erupted the next morning, Hamilton was safely in bed in Portland. Truman and his lodge were buried in a massive landslide.

In the weeks leading up to the eruption, Hamilton had witnessed several earthquakes while visiting the old man. "I was up there a few times in his lodge, up on his front porch, when some pretty good quakes hit," Hamilton says. "I could see the road rippling. It was the weirdest thing I'd ever seen."

By this time, the media had already caught up with Truman, who had appeared on the "Today Show" and in *The New York Times*. But for all his bravado, Hamilton notes, there were cracks in Truman's resolve. "I saw some genuine fear in his face. He looked drawn. His eyes were bugging out a little bit. The earthquakes made him stop and pay attention."

But in spite of all the warnings, most people weren't prepared for the sheer magnitude of the eruption. "The geologists were certainly warning us that this was a very dangerous and unstable situation," he says. "But I don't think anybody really anticipated the enormity of what happened. There wasn't a lot of documented history of this kind of thing."

▲ **FIGURE 5.48 OREGON REPORTER ESCAPES DEATH ON MOUNT ST. HELENS** A fateful decision to leave Harry Truman's lodge on the slopes of Mount St. Helens on May 17, 1980, saved Don Hamilton's life. Don recounted his story 25 years later when Mount St. Helens erupted again in March 2005. *(Don Hamilton)*

The morning of the eruption, Hamilton and his brother, a photographer for the *Oregon Journal*, chartered a flight to Mount St. Helens and were lucky enough to have a pilot who was willing to loosely interpret the Federal Aviation Administration (FAA) restrictions on the air space around the volcano.

Hamilton describes the view from the plane as utterly extraordinary. The peak of Mount St. Helens had been blown off in the initial blast. A massive column of volcanic ash rose straight into the clouds above.

As they got closer they saw massive lahars as well as cars, trucks, and bridges that had been destroyed. "The road that I'd been going up for two weeks was completely washed out," Hamilton says.

For residents of Portland and many other areas near the volcano, ash from the eruption became a regular part of life that summer. "People were wearing surgical masks. Ash was piled up on the side of the road. It just stayed there all summer," Hamilton says. "For that whole summer you just lived with it."

For Hamilton, the timing couldn't have been worse. "It was the only summer I had a convertible," he says. "I kept the top up most of the time that summer."

—Chris Wilson

managers, educators, media, and private citizens is particularly important. The goal is to prevent a volcanic crisis from becoming a disaster or catastrophe.[46]

Adjustments to Volcanic Hazards

Apart from evacuation, the primary human adjustments to volcanic eruptions are attempts to deflect lava flows from populated areas. Several methods, such as bombing, hydraulic chilling, and wall construction, have been employed to deflect lava flows, with mixed success.

Bombing has proved most effective against flows in which fluid lava is confined to a relatively narrow channel bounded by solidified lava. The purpose of the bombing is to block the channel and cause the lava to pile up and break through upstream, where it will take a less damaging route. Successive bombing at higher and higher points on a flow may be necessary to control the threat.

Poor weather conditions, abundant smoke from burning vegetation, and falling ash reduce the effectiveness of bombing.[49] Bombing is ineffective in the case of large unconfined flows.

The world's most ambitious hydraulic chilling effort was initiated in January 1973 on the Icelandic island of Heimaey. Basaltic lava flows from Mount Helgafell nearly closed the harbour of Vestmannaeyjar, the island's main town and Iceland's main fishing port. The situation prompted immediate action.

Three favourable conditions existed: (1) the slow movement of the lava flows provided time to initiate a control effort; (2) transport by sea and roads allowed for the delivery of pipes, pumps, and heavy equipment; and (3) water was readily available. Initially, the edges and surface of the flow were cooled with water discharged from fire hoses (Figure 5.49). Then, bulldozers were moved up on the slowly advancing flow to make a path for a large water pipe. The plastic pipe did not melt because water was flowing in it. Small holes in the pipe accelerated cooling of hot spots along parts of the flow. Watering had little effect the first day, but the flow then began to slow down and stop.

These actions had a positive effect. They restricted lava movement, reduced property damage, and kept the harbour open. When the eruption stopped five months later, the harbour was still usable.[50] In fact, the shape of the harbour was actually improved because the front of the flow provided additional protection from the sea.

(a)

(b)

(c)

◄ **FIGURE 5.49 FIGHTING LAVA FLOWS ON THE ISLAND OF HEIMAEY, ICELAND** (a) A lava fountain as seen from the harbour of Vestmannaeyjar. (b) An aerial view of Vestmannaeyjar. White steam appears above the advancing black lava flow. The steam comes from water being applied to cool and slow the front of the flow. An arcing stream of water from a water cannon is visible in the lower right corner of the photograph. (c) An aerial view showing the front of the blocky lava flow moving into the harbour. *((a) Solarfilma ehf; (b) James R. Andrews; (c) James R. Andrews)*

REVISITING THE FUNDAMENTAL CONCEPTS

Volcanoes and Volcanic Eruptions

❶ **Hazards can be understood through scientific investigation and analysis.**

❷ **An understanding of hazardous processes is vital to evaluating risk.**

❸ **Hazards are commonly linked to each other and to the environment in which they occur.**

❹ **Population growth and socio-economic changes increase the risk from natural hazards.**

❺ **Damage and loss of life from natural disasters can be reduced.**

1. As with earthquakes, the long-term frequency of eruptions of a dormant or active volcano can be estimated from geologic evidence. Although a prediction of when the next eruption will occur is still not possible, significant advances have been made in the past three decades in understanding eruptive behaviour. The 1981 eruption of Mount St. Helens and the 1991 eruption of Mount Pinatubo taught us a lot about the events leading up to an eruption. Together with recent technological advances (advanced seismometers, tiltmeters, satellite remote sensing, and volatile monitoring), we are able to anticipate some eruptions and constrain the time that they will occur to within a few weeks.
2. The risk from a volcanic eruption is the product of its probability and its consequences. The rocks from which a volcano is constructed yield information about the frequency, type, and magnitude of previous eruptions, and consequently provide insight into future events. However, the likelihood of an eruption in the immediate future is based on monitoring possible precursory events such as earthquakes and changes in the form of the volcano. Consequence scenarios can be developed based on the type and size of a potential eruption, the population at risk, and the location, type, and value of infrastructure that might be affected.
3. Volcanic eruptions are linked to landslides, lahars, floods, and in some cases tsunamis. Volcanic eruptions on high mountains melt ice and snow, and thus generate volcanic mud flows (lahars). Lava flows may block streams, causing upstream flooding and producing temporary dams that, if overtopped, may lead to downvalley flooding. Large landslides may occur on stratovolcanoes during or between eruptions.
4. The human population close to active volcanoes is constantly growing; accordingly volcanic risk is increasing. In A.D. 79, when Mount Vesuvius erupted, perhaps half the residents of Pompeii and Herculaneum died (approximately 2000 people). Today, almost 3 million people live within sight of Mount Vesuvius. The economic implications of a future large eruption are obvious, and although an evacuation plan exists for the region, an unexpected eruption similar to that in A.D. 79 could result in catastrophic loss of life.
5. The locations of active volcanoes are known and, with advances in prediction and evacuation plans, loss of life in future eruptions can be minimized. We have been more successful in recent years in encouraging people to leave once a volcanic warning has been issued. Buildings on and near volcanoes can be designed to withstand the weight of volcanic ash, and the paths of lahars are readily identifiable, allowing people living along them to be evacuated in the lead-up to an eruption.

Summary

The viscosity of magma is related to its temperature and silica content and is important in determining the eruptive style of a volcano. Shield volcanoes are the largest volcanoes on Earth. They are common at mid-ocean ridges, such as in Iceland, and over mid-plate hot spots, such as on the Hawaiian Islands. Shield volcanoes are characterized by non-explosive eruptions of basaltic lava. Stratovolcanoes occur at subduction zones, particularly around the Pacific Rim. Many of the volcanoes in Japan, the Aleutian Islands of Alaska, the Cascade Mountains of Canada and the United States, Central America, and the Andes of South America are of this type. These volcanoes are characterized by explosive eruptions and comprise mainly lava flows and pyroclastic deposits of andesitic composition. Volcanic domes are smaller, highly explosive volcanoes that occur inland of subduction zones and consist largely of rhyolite and dacite.

Features of volcanoes include vents, craters, and calderas. Other features related to volcanic activity are hot springs and geysers. Calderas form from infrequent large and violent eruptions. They may present a volcanic hazard for a million years or more after they form. Recent uplift and earthquakes at the Long Valley caldera in eastern California and continuing thermal activity in Yellowstone National Park are reminders of this potential hazard.

Volcanic activity is directly related to plate tectonics. Most volcanoes are located at plate boundaries where magma is produced by the spreading or subduction of lithospheric plates. Two-thirds of the volcanoes on land are associated with subduction of lithospheric plates along the Ring of Fire surrounding most of the Pacific Ocean. Specific geographic regions of North America at risk from volcanic eruptions include parts of British Columbia and Alaska, Long Valley in California, the Yellowstone area, northern California, Oregon, and Washington.

The primary effects of volcanic activity include lava flows, lateral blasts, pyroclastic surges and flows, ash fall and bombs, and emission of poisonous gases. Secondary effects include debris flows (lahars), which are generated when melting snow and ice or precipitation mix with ash and other pyroclastic material. Lahars can travel far from their source and are extremely destructive. All these phenomena have occurred historically in Alaska and Washington and will occur there in the future.

Volcanoes are associated with other natural hazards, including fire, earthquakes, landslides, tsunamis, and climate change. However, they also provide benefits, including fertile soils, a source of power, recreational opportunities, and new land.

Efforts to reduce volcanic risk must be centred on human and societal issues, including communication and education. The objective of these efforts is to prevent a volcanic crisis from becoming a disaster or catastrophe. Seismic, thermal, magnetic, hydrologic, and land surface monitoring of active volcanoes, combined with an improved understanding of their recent geologic history, may eventually result in reliable forecasting of volcanic activity.

Perception of volcanic risk is a complex social and economic issue. Some people have little choice but to live near a volcano, and others ignore or downplay the risk. The primary human adjustment to an impending eruption is evacuation. Bombing, hydraulic chilling, and construction of barriers have been used in attempts to control lava flows, but these methods have had mixed success and require further evaluation. Community-based education plays an important role in informing people about the hazards of volcanoes.

Key Terms

aa (p. 135)
ash fall (p. 135)
caldera (p. 125)
cinder cone (p. 122)
crater (p. 125)
decompression melting (p. 117)
lahar (p. 139)
lateral blast (p. 135)
lava (p. 114)
lava flow (p. 133)
maar (p. 124)
magma (p. 114)
pahoehoe (p. 135)
pyroclastic deposit (p. 118)
pyroclastic flow (p. 135)
pyroclastic surge (p. 135)
sector collapse (p. 139)
shield volcano (p. 120)
silicosis (p. 136)
stratovolcano (or composite volcano) (p. 121)
super eruption (p. 128)
supervolcano (p. 128)
tephra (p. 118)
volatiles (p. 118)
volcanic crisis (p. 150)
volcanic dome (p. 122)
volcanic vent (p. 127)

Did You Learn?

1. What is magma? What is it composed of?
2. Explain how magma chemistry, gas content, and viscosity contribute to volcano explosivity.
3. What determines the viscosity of magma?
4. List the major types of volcanoes and the characteristics of the magma associated with each.
5. Describe the eruption styles of different types of volcanoes.
6. Explain the relation between plate tectonics and volcanoes.
7. What is a geyser and how does it operate?
8. Explain why caldera eruptions are so dangerous.
9. List the primary and secondary effects of volcanic eruptions.
10. Differentiate among ash falls, lateral blasts, pyroclastic flows, and pyroclastic surges.
11. Name the main gases emitted during a volcanic eruption.
12. Explain how volcanic eruptions produce lahars.
13. What kinds of information help geologists and geophysicists forecast volcanic eruptions?
14. Describe the methods that have been used to control lava flows.

Critical Thinking Questions

1. While looking through some boxes in your grandparents' house, you find a sample of volcanic rock that your grandfather collected. He cannot remember where he collected it. You take it to school, and your geology instructor tells you that it is a sample of andesite. What can you tell your grandparents about the type of volcano the rock probably came from, its geologic environment, and the type of volcanic activity that likely produced it?
2. People's perception of volcanic risk and what they will do in the event of an eruption depend on both their proximity to the volcano and their knowledge of volcanic processes and how to respond. With this knowledge in mind, develop a public relations program that could alert people to a potential volcanic eruption. Keep in mind that, in past tragedies, people have ignored hazard maps that show they live in dangerous areas.

3. You are going to take group of secondary school students to Hawaii to see Kilauea volcano. Some of the students have seen documentaries of the 1980 eruption of Mount St. Helens and are concerned for their safety. Others are fearless and want to collect a sample of molten lava. What do you tell the fearful students? What do you tell the fearless ones?

MasteringGeology

MasteringGeology **www.masteringgeology.com**. Looking for additional review and test prep materials? Visit the Study Area in MasteringGeology to enhance your understanding of this chapter's content by accessing a variety of resources, including **Self-Study Quizzes, Geoscience Animations, GEODe Tutorials, RSS feeds, flashcards,** web links and an optional **Pearson eText.**

CHAPTER 6

▶ **DEADLY LANDSLIDE** At least 76 people in the mining community of Frank perished on April 29, 1903, when a huge mass of rock broke away from Turtle Mountain and cascaded down into the valley. The scarp and the debris sheet left by the landslide are visible in this photograph. *(Natural Resources Canada 2013, courtesy of the Geological Survey of Canada)*

Landslides

Learning Objectives

Landslides, the downslope movement of Earth materials, constitute a serious natural hazard in many parts of the world. Most landslides are small and slow, but some are large and fast. Both may cause significant loss of life and damage to property, particularly in urban areas. Your goals in reading this chapter should be to

- Understand slope processes and the different types of landslides
- Know the forces that act on slopes and how they affect slope stability
- Know which geographic regions in North America are at risk from landslides
- Know the effects of landslides and their links with other hazardous natural processes
- Understand how people can affect the landslide hazard
- Be familiar with adjustments people can make to avoid death and damage caused by landslides

The Frank Slide

The local Blackfoot and Kutenai peoples called Turtle Mountain the "Mountain That Moves," probably because of the small rock falls that routinely rain down from the North Peak, especially during the spring thaw.[1] In the months leading up to the tragic landslide on April 29, 1903, miners working a coal seam at the base of the mountain reported strange rumblings. Coal slipped from the mine walls, and timbers used to prop up the walls splintered into pieces. Yet, for the 20 miners who went underground on the night of April 28, and for the hundreds of residents of Frank, Alberta, who were comfortable in their beds, there was no reason to think that tragedy was imminent. Then, at 4:10 on the morning of April 29, in total darkness, a slab of rock 900 m wide, 650 m wide, and 150 m thick broke away from the east face of Turtle Mountain and crashed into the valley below.[2] The fragmented rock streamed across the valley floor and covered an area of about 3 km^2, in places to a depth of 30 m (Figure 6.1). Most of the town, including its main business and residential areas, was unscathed, but seven miners' cottages, a dairy farm, a ranch, a shoe store, a livery stable, and a construction camp, along with the surface infrastructure of the coal mine, were destroyed.[3] The landslide dammed the Crowsnest River, forming a small lake, and buried approximately 2 km of the Canadian Pacific Railway and 3 km of the Frank and Grassy Mountain Railway.

Frank was home to approximately 600 people in 1903; at least 76 people in the path of the slide were killed, including three miners. Seventeen men trapped in the Frank mine escaped by tunneling 6 m through coal and 3 m of landslide debris to the surface.[2] Several people in the direct path of the landslide survived, including three young girls—Fernie Watkins was found alive among the debris; Marion Leitch, 15 months old at the time, was thrown from her house to safety on a pile of hay; and Gladys Ennis, 27 months old, was found choking in a pile of mud by her mother. Marion's story inspired the tragedy's most enduring myth—that the landslide had destroyed the entire town, with the exception of one little girl, dubbed "Frankie Slide," who, according to various accounts, was found alive on a rock, on a bale of hale, in a crib, in an attic, in a pile of debris, or in her dead mother's arms![2]

▲ **FIGURE 6.1 FRANK SLIDE** The 1903 Frank slide is Canada's best-known landslide and a classic example of a rock avalanche. It travelled more than 3 km in 100 seconds (average velocity of about 30 m/s) and buried part of the town of Frank, Alberta, killing at least 76 people. Coal mining at the base of the mountain likely triggered the landslide. *(By Steven Fick and Mary Vincent; March/April 2003. Reprinted with permission. Photo by David Barbour/GeoSolutions Consulting Inc.)*

After the tragedy, Frank was evacuated, but people soon returned and both the mine and the railway were back in operation within a month. Frank continued to grow until the Alberta government forced people to abandon the southern part of the town in 1911, following a report on the mountain's stability by a royal commission. Mining ceased in 1918 and the town, which today has a population of about 300, became a bedroom community for other working coal mines in the region.

The Frank slide became an immediate sensation in 1903, capturing media attention and becoming a regional tourist attraction for several years. The Alberta Government designated the landslide a Provincial Historical Site in 1977, and the Frank Slide Interpretive Centre was opened in 1985. The interpretive centre attracts about 50 000 visitors annually.

The Frank slide is Canada's best-known landslide and has been the subject of much study by geologists. It is an excellent example of a **rock avalanche**, a type of landslide involving sudden failure of a large mass of rock that rapidly fragments and travels as a streaming mass at high speeds. Where unimpeded by topography, as at Frank, rock avalanche debris spreads out as a thin sheet far from its source.

There is concern that another large landslide could occur at Turtle Mountain. A potentially unstable rock mass has been identified on the South Peak of the mountain, and many homes are located directly below it. In response to this threat, the Alberta Geological Survey began to monitor Turtle Mountain in 2005 to further characterize its structure and stability. It installed a network of over 40 state-of-the-art sensors on the mountain as an early warning system.[4] The continuous data stream from this network provides valuable insights into the mechanics of the slowly moving rock mass.

Contributing Factors

Factors responsible for landslides include causes and triggers. A *cause* is an internal or external factor that, over time, reduces the stability of a slope and brings it to the point of failure. A *trigger*, on the other hand, is an event that sets off the landslide; it can be considered "the final straw."

Although the Frank slide is one of the largest historic landslides in Canada, its cause and trigger are still debated. Several factors played important roles.[5]

1. *Geology* The mountain is a deeply eroded *anticline* of Paleozoic limestone thrust over weaker Mesozoic coal-bearing shale. Water infiltrated and slowly dissolved the carbonates. Over time, this process decreased the stability of the slope.
2. *Glaciation* During Pleistocene glaciations, valley glaciers eroded shale on the lower slope of the mountain. Glacial erosion steepened the overall slope of the mountain, reducing its strength.
3. *Mining* Mining of coal at the base of Turtle Mountain may have reduced the stability of the limestone higher on the mountain. Seams of coal were mined in large openings, or *stopes*, excavated deep within the mountain.
4. *Weather* It is possible that weather was the "final straw" that triggered the slide. The region experienced heavy snowfalls in March 1903, followed by unusually warm temperatures in late April. On the night of the landslide, temperatures dropped precipitously, and it is likely that water from melted snow entered fissures at the top of the mountain and froze, exerting pressures on the unstable rock mass.

Lessons Learned

Several lessons can be learned from the Frank landslide.

1. *Large landslides cannot be prevented.* Landslides of the size of the Frank slide are not preventable. They are also difficult or impossible to predict. With detailed continuous monitoring, it might be possible to detect the characteristic slow creep-like movement of a slide before a catastrophic failure, but it is not possible or economically feasible to monitor every slope where there is a possible threat to people or property. Furthermore, most slopes that are slowly moving never fail catastrophically.
2. *Geology is important.* Geology is an underlying cause of most or all landslides. The type of rock or sediment, the presence of joints and faults, and the degree of fracturing of rock are all important in localizing landslides. Of course, other factors, such as climate, seismicity, and topography, may also be important.
3. *Human activity can trigger landslides.* Landslides can be triggered by human activities that increase the stress on a slope or reduce material strength. Although not an underlying cause of the Frank slide, coal mining might have reduced the strength of the unstable rock mass on Turtle Mountain and triggered failure. Other human activities that can trigger landslides are undercutting of toes of slopes, loading of upper parts of slopes, or the introduction of water into slopes by, for example, irrigation or landscaping.

6.1 Introduction to Landslides

Landslide and **mass wasting** are terms used to describe the downslope movement of rock or sediment due to gravity. Landslides can be slow or fast and can involve small or large volumes of sediment or rock. Although many scientists limit use of the term "landslide" to mean rapid movements of soil or rock, we use the term in its broadest sense to also include slowly moving (or creeping) bodies of coherent rock or soil. In this sense, "landslide" and "mass wasting" are synonymous.

Types of Landslides

Earth materials may fail and move downslope in many ways. Scientists classify landslides to reflect these differences (Table 6.1). Four important variables underpin most landslide classifications: (1) the mechanism of movement (fall, topple, slide, flow, or complex movement); (2) the type of material (rock, consolidated sediment, or organic soil); (3) the amount of water present; and (4) the rate of movement.[6] In general, movement is considered rapid if it can be discerned with the naked eye; otherwise, it is classified as slow. Movement rates range from a few millimetres per year in the case of slow creep to many dozens of metres per second for some rockslides and rock avalanches.[7]

A **fall** involves bounding of rock or blocks of sediment from the face of a cliff (Figures 6.2 and 6.3). A **slide** is the downslope movement of a coherent block of rock or sediment along a discrete failure plane (Figures 6.2 and 6.4). A **slump** is a particular type of slide in which the failure plane is curved upward (Figures 6.2 and 6.5). A **flow** is the slow to rapid downslope movement of sediment in which particles move semi-independently of one another, commonly with the aid of water (Figures 6.2 and 6.6). **Debris flows** are mixtures of mud, debris, and water. They range in consistency from thick mud soups to wet cement, such as you might see flowing out of a cement truck. Debris flows typically move rapidly—some travel down established stream valleys and may leave their channels when they flow across fan surfaces; others take long narrow tracks or *chutes* on steep hillsides. Some debris flows transport as little as a few hundred cubic metres of material; at the other end of the spectrum, cubic kilometres of material can be transported by debris flows following the failure of the flank of a volcano (see Chapter 5).

TABLE 6.1 Common Types of Landslides

Mechanism	Type of Landslide	Characteristics
Fall	Rock fall	Individual rocks bound downward or fall through the air.
Slide	Slump	Coherent blocks of rock or sediment slide on an upward-curved surface; also called a rotational landslide.
	Debris slide or avalanche	Sediment or soil slides on an inclined surface; also called an earth slide.
	Rockslide	Large blocks of bedrock slide on an inclined surface, typically bedding planes, foliation surfaces, or joints.
	Rock avalanche	A type of rockslide in which the fragmented rock mass flows at very high velocities, commonly for long distances.
Flow	Creep	Very slow downslope movement of rock and soil. Sackung is deep-seated creep of large masses of fragmented rock along poorly defined slip surfaces.
	Earthflow	A flow of wet, deformed soil and weathered rock.
	Debris flow	A cement-like mixture of rock, sand, mud, plant debris, and water travels rapidly down a stream channel or ravine; includes mudflows and lahars.
Complex		A combination of two or more types of landslides.

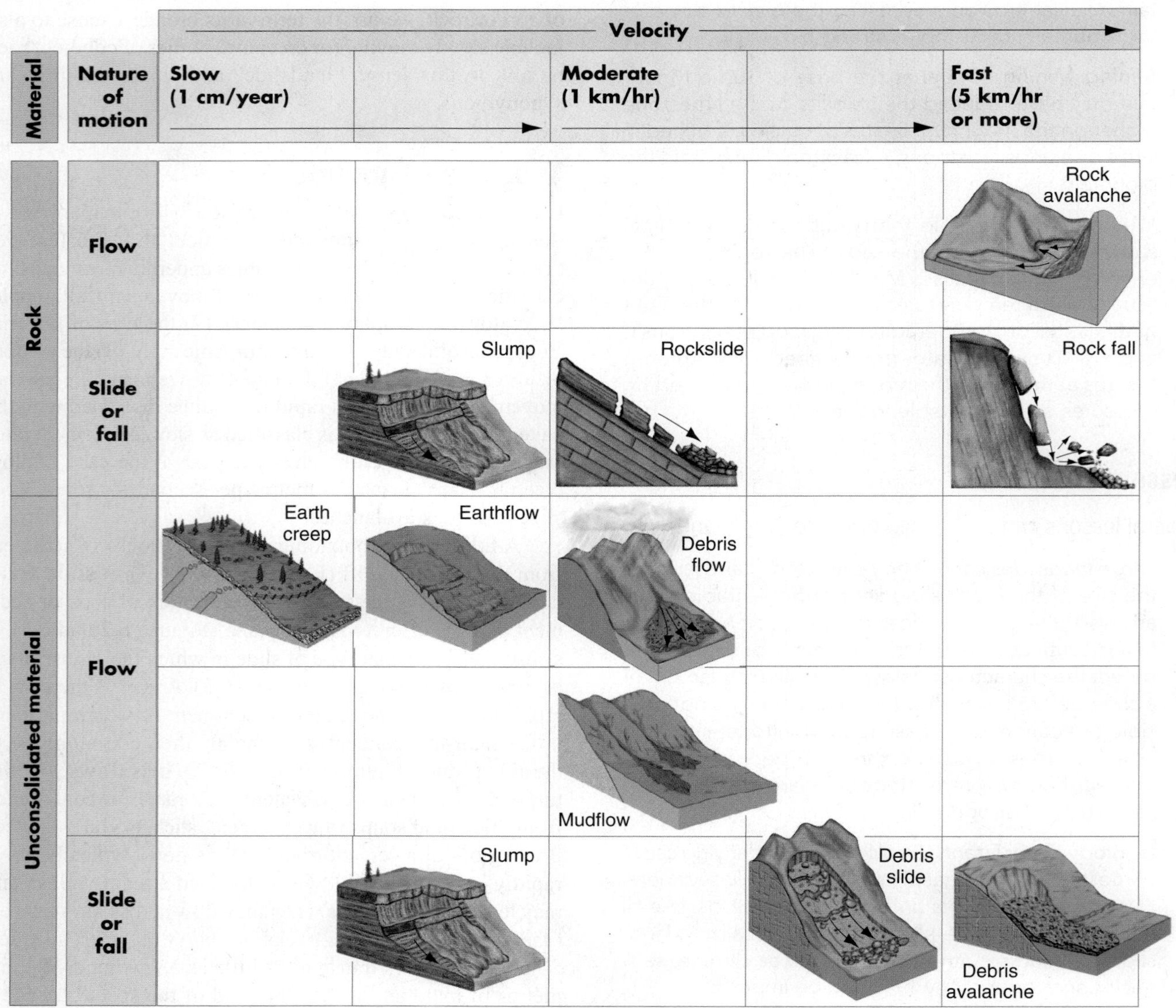

▲ **FIGURE 6.2 TYPES OF LANDSLIDES** Landslides are classified according to (1) type of movement (fall, slide, topple, flow), (2) type of material that fails (i.e., rock or unconsolidated sediment), (3) amount of water or air involved in the movement, and (4) velocity.

▲ **FIGURE 6.3 ROCK FALL** Rock fall on the Trans-Canada Highway near Yale, British Columbia. A rock fall is a fast-moving mass of fragmented rock that bounds down a steep slope. *(Duncan Wyllie)*

▲ **FIGURE 6.4 LARGE ROCKSLIDE BURIES HIGHWAY EAST OF VANCOUVER** A photograph of the Hope Slide, which buried Highway 3 in southern British Columbia to a depth of up to 80 m on January 9, 1965. At 48 million m^3, the Hope Slide is one of the two largest historic landslides in Canada. *(Vancouver Sun)*

◄ **FIGURE 6.5 SLUMP** Slumps are landslides with curved failure surfaces. In this example near Quesnel, British Columbia, a large coherent block of silt rotated backward as it slid downward and outward. *(John J. Clague)*

Very slow flow of rock or sediment, at rates ranging from millimetres to dozens of centimetres per year, is termed **creep**. A special type of creep involves movement of large masses of rock, up to many billions of cubic metres, along ill-defined, deep failure planes. The German word **sackung** (loosely translated as "slope sag") has been applied to these huge, slow-moving landslides. A slow creep-like movement in which a rock mass pivots about a point is termed a **topple**. Topples are common in rocks with joints or bedding planes that dip steeply into the slope.

Many landslides are complex combinations of sliding and flow. An example is the 1965 Hope Slide, one of the two largest historic landslides in Canada (Figure 6.4). Late on the night of January 9, 1965, a huge mass of rock, enough to fill more than 20 indoor stadiums, slid away from the flank of Johnston Peak, 17 km southeast of Hope, British Columbia.[8] The rock fragmented as it raced down the slope toward B.C. Highway 3 in the valley below. On reaching the valley floor, the debris obliterated Outram Lake and drove a wave of muddy sediment 150 m up the opposite valley wall. The muddy debris fell back down the slope and flowed both up and down the valley. The landslide buried more than 3 km of Highway 3 beneath as much as 80 m of debris and killed four motorists.

High-velocity flows of fragmented rock are referred to as rock avalanches (Figure 6.7). Rock avalanches can

◄ **FIGURE 6.6 EARTHFLOW** This landslide near the community of Pavilion, British Columbia, is a slow-moving earthflow involving weathered sedimentary rocks. The hummocky surface and natural levees at the margins of the landslide are indicative of highly viscous, flow-like motion. *(John J. Clague)*

◄ **FIGURE 6.7 ROCK AVALANCHE** A photo-draped digital elevation model (DEM) of a rock avalanche on the flank of Mount Munday in the Coast Mountains of British Columbia. The landslide began as a rockslide but rapidly evolved into a high-velocity flow that streamed across the glacier in the foreground. The distance from the top to the bottom of the photo is 4 km. *(Evans, S. G. 2006. "Single-event landslides resulting from massive rock slope failure; characterizing their frequency and impact on society." In S. G. Evans, G. Scarascia-Mugnozza, A. L. Strom, and R. L. Hermanns (eds.),* Landslides from Massive Rock Slope Failure, *pp. 53–73. NATO Science Series IV, Vol. 49, Springer, Dordrecht)*

achieve speeds of more than 300 km/hr, travel long distances, and override large obstacles in their paths (see A Closer Look 6.1). They also can be deadly. One of the worst landslide disasters in history occurred on May 31, 1970, in the Cordillera Blanca of Peru. A large earthquake released an avalanche of rock and glacier ice from the steep north side of Nevados Huarascán, the sixth highest peak in the Western Hemisphere (ca. 6750 m asl). The mass of rock and ice fragmented as it fell from the mountain and rapidly transformed into a huge debris flow (50–100 million m^3) that streamed at high speed 16 km westward to the valley of Rio Santa, and from there another 15 km northward to near Caraz[9] (Figure 6.8). Along its path, the debris overrode a hill and completely destroyed the city of Yungay, claiming about 8000 lives. Eight years earlier, on January 10, 1962, an ice avalanche from the same source area on Nevados Huascarán flowed down the same valley and destroyed the city of Ranrahirca, killing about 4000 people[9] (Figure 6.8).

Most **subaqueous** (underwater) **landslides** are also complex events. A slump or slide on the submerged slope of a delta or at the edge of the continental shelf can change into a debris flow or a *turbidity current* that travels great distances from the point of failure. A famous example of a

◄ **FIGURE 6.8 LANDSLIDE DISASTERS IN PERU** Avalanches of ice and rock from the north face of Nevados Huascarán in Peru in 1962 and 1970 claimed more than 12 000 lives in the towns of Yungay and Ranrahirca. The extents of the two landslides are shown in orange and yellow in this graphic. *(Data from the World Glacier Monitoring Service, Zurich, Switzerland, and based on a figure by UNEP's DEWA/GRID-Europe, Geneva, Switzerland. Reprinted with permission.)*

complex subaqueous landslide is the Grand Banks landslide off Newfoundland in 1929. A large earthquake triggered a huge (200 km^3) initial slump or slide on the seafloor at the edge of the *continental shelf*. The failed mass rapidly transformed into a turbid flow of mud, sand, and water (a turbidity current) that travelled hundreds of kilometres along the sea floor at speeds of 60 to 100 km/h, breaking 12 submarine telegraph cables.[10] Some complex landslides may form when water-saturated sediments flow from the lower part of a slope and undermine the upper part, causing *slump blocks* to form.

Forces on Slopes

To understand landslides, we must examine the forces that determine the stability of a slope. Slope stability can be evaluated by determining the relation between **driving forces** that move rock or sediment down a slope and **resisting forces** that oppose such movement (Figure 6.11). The largest downslope driving force is the weight of the slope material and the water it contains. That weight can include anything placed on the slope, such as fill material and buildings. The resisting force is the **shear strength** of the slope material—that is, its resistance to failure by sliding or flow along potential slip planes. Potential slip planes are surfaces of weakness in the slope material, such as bedding planes in sedimentary rocks, foliation in metamorphic rocks, and fractures in all types of rock.

Slope stability is evaluated by computing a **factor of safety** (FS), defined as the ratio of the resisting forces to the driving forces. If the factor of safety is greater than 1, the resisting forces exceed the driving forces and the slope is considered stable. If the factor of safety is less than 1, the driving forces exceed the resisting forces and a slope failure can be expected. Driving and resisting forces are not static; as local conditions change, these forces may change, increasing or decreasing the factor of safety.

Driving and resisting forces on slopes are determined by interrelations of the following variables:

- Type of material
- Slope angle and topography
- Climate
- Vegetation
- Water
- Time

The Role of Material Type Material composing a slope can affect both the type and the frequency of landslides. Important material characteristics include mineral composition, degree of cementation or consolidation, and the presence of planes of weakness. Planes of weakness may be sedimentary bedding planes, metamorphic foliation, joints, or zones along which rock or soil has moved before, such as an old landslide slip surface or a fault. These planes can be especially hazardous if they are inclined more than about 15° and intersect or are parallel to the slope of a hill or mountain.

A CLOSER LOOK 6.1

Estimating the Velocity of Landslides from Their Run-Up and Superelevation

Simple equations can be used to estimate the velocity of a rock avalanche or debris flow where it runs up a hill or valley wall or where it superelevates while rounding a bend. *Superelevation* refers to the tendency of some flows to rise higher on the outside of a bend in their paths than on the inside. We apply two widely used equations to estimate the velocity of the 1959 Pandemonium Creek landslide in Tweedsmuir Provincial Park, British Columbia (Figures 6.9 and 6.10).[11]

The Pandemonium Creek landslide travelled a distance of 9 km from a steep cirque headwall (location A in Figure 6.9) to Knot Lakes (location K). The rock mass fell from the cirque headwall and disintegrated on the steep crevassed surface of the glacier below. Below the glacier, the debris became constricted between two prominent *lateral moraines* (Figure 6.9). The combination of a steeper slope and constriction by the moraines caused the debris to rapidly accelerate along this part of the path. The landslide then crossed Pandemonium Creek and struck the north side of the valley at an angle of about 60° to the direction of travel. The leading edge of the debris ascended a 28° slope to a point 335 m above the valley floor (location B in Figure 6.9). It then turned about 90° and "bob-sledded" down the valley (C) to an alluvial fan, where part of the debris turned another 90° and flowed into Knot Lakes. The velocity of the landslide as it ran up the valley wall to location B can be estimated from the velocity–potential energy equation[12, 13]

$$v^2 = 2gh$$

where v is the velocity, h is the maximum vertical run-up (in this case 335 m), and g is gravitational acceleration (9.80 m/s^2). This equation, which neglects frictional and other energy losses, yields a velocity of 290 km/hr.

Velocities of the debris along the "bobsled run"—the seven bends along the valley of Pandemonium Creek below the run-up zone (Figure 6.10)—were estimated from superelevation data using the equation

$$v^2 = r_c g \tan \theta \cos \alpha$$

where v is velocity, r_c is the mean radius of curvature of the bend, θ is the transverse slope of the top surface of the debris as it rounds the bend, and α is the path slope angle.[12] The velocity estimates range from about 75 to 135 km/hr.

◀ **FIGURE 6.9 HIGH-VELOCITY LANDSLIDE** The 1959 Pandemonium Creek landslide in Tweedsmuir Provincial Park, British Columbia, travelled 9 km from its source at location A to Knot Lakes (K). The velocity of the landslide can be estimated from the run-up of the debris at location B and superelevation of debris at bends in the flow path above and below location C. *(Photo BC5145-163. © Province of British Columbia. All rights reserved. Reprinted with permission of the Province of British Columbia. www.ipp.gov.bc.ca)*

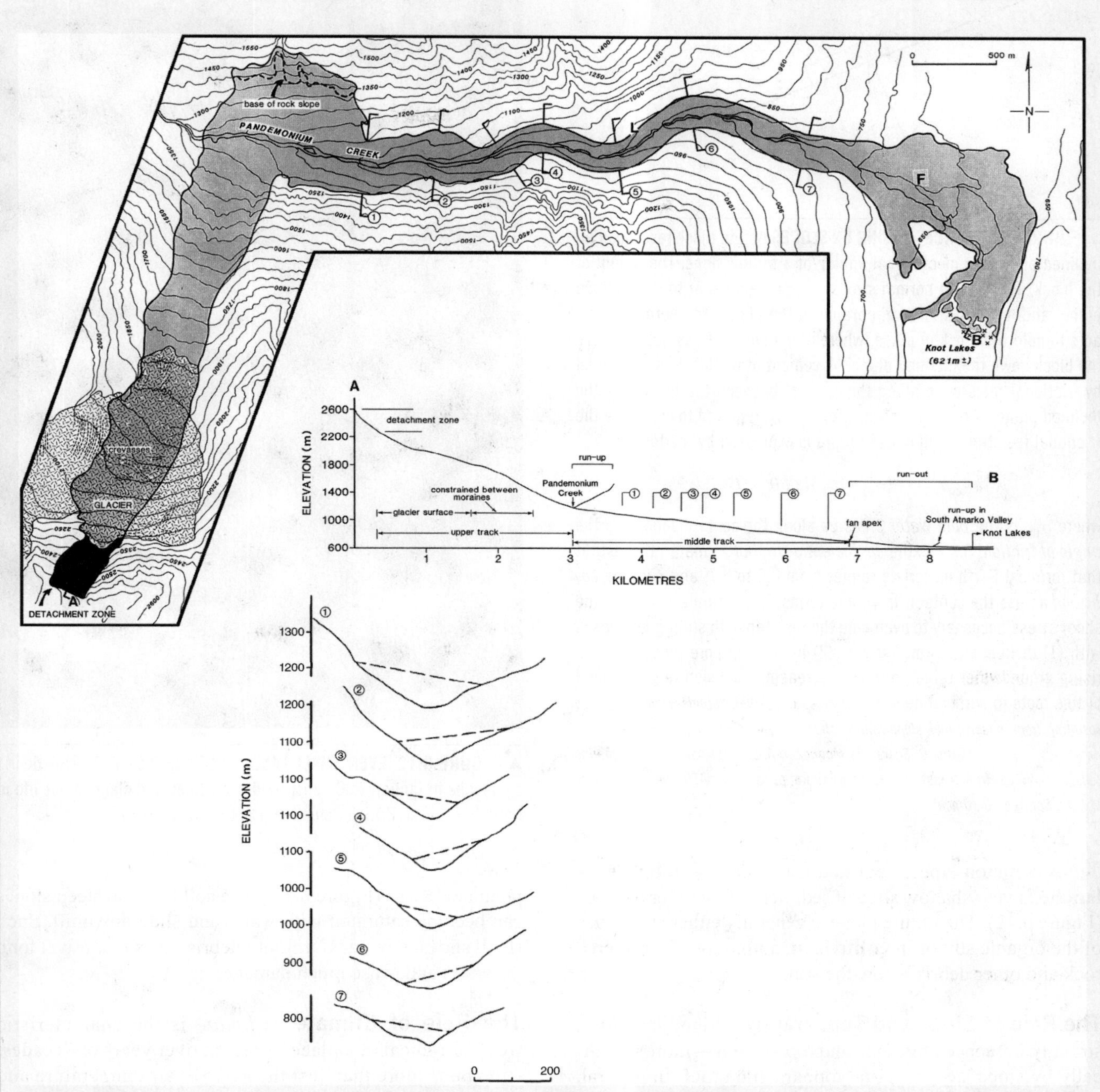

▲ **FIGURE 6.10 MAP, LONGITUDINAL PROFILE, AND CROSS-SECTIONS OF THE PANDEMONIUM CREEK LANDSLIDE** The map shows the path of the Pandemonium Creek landslide (compare with Figure 6.9). F = Pandemonium Creek fan. The longitudinal profile is drawn without vertical exaggeration. Cross-valley profiles 1–7 illustrate how the debris superelevated along bends in the path of the landslide along Pandemonium Creek. *(Courtesy of NRC Research Press)*

In the case of slides, the shape of the slip surface is strongly controlled by the type of material that fails. Slides have two basic patterns of movement, *rotational* and *translational*. Rotational slides, or slumps (Figure 6.5), have curved slip surfaces, whereas translational slides have planar slip surfaces.

Rotational sliding tends to produce small topographic benches that tilt upslope (Figures 6.2 and 6.5). Slumps are most common in unconsolidated sediment and in mudstone, shale, or other weak rock types. The inclined slip planes of translational slides include fractures in all rock types, *bedding planes* in sedimentary rocks, weak clay layers, and *foliation planes* in metamorphic rocks. The material that moves along these planes can be large blocks of bedrock or sediment.

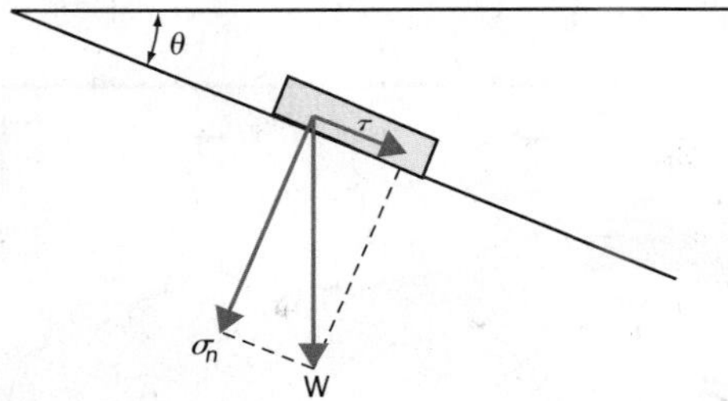

▲ **FIGURE 6.11 FORCES ACTING ON SLOPES** A rigid block resting on an inclined plane is a mechanical analog of a failing slope. The weight of the block, *W*, exerts a normal stress, σ_n, perpendicular to the inclined plane, and a shear stress, τ, parallel to the plane. The normal stress acts to hold the block in place, whereas the shear stress acts to move the block down the inclined plane. Movement of the block is restrained by frictional resistance along the contact between the block and the inclined plane. The critical shear stress, τ_{crit}, required to overcome the frictional resistance and cause failure is expressed by the formula

$$\tau_{crit} = (\sigma_n - \sigma_{pw})\tan\theta + C$$

where σ_{pw} is the *pore water pressure* along the contact zone; θ is the *angle of friction; tan* θ is the *coefficient of friction*, a material constant that for most Earth materials ranges from 0.6 to 0.7; and *C* is the *cohesion* across the contact. In simple terms, this formula says that the shear stress necessary to overcome the resistance to sliding decreases with (1) decreasing normal stress, (2) increasing pore pressure (e.g., rising groundwater table), and (3) decreasing cohesion (e.g., removal of tree roots in surficial deposits. *(Evans, S. G. 2006. "Single-event landslides resulting from massive rock slope failure; characterizing their frequency and impact on society." In S. G. Evans, G. Scarascia-Mugnozza, A. L. Strom, and R. L. Hermanns (eds.),* Landslides from Massive Rock Slope Failure, *pp. 53–73. NATO Science Series IV, Vol. 49, Springer, Dordrecht)*

▲ **FIGURE 6.12 EVEN SMALL LANDSLIDES CAN BE FATAL** This debris avalanche in January 2005 destroyed two homes and claimed one life in North Vancouver, British Columbia. *(Ian Lindsay/Vancouver Sun)*

A common type of translational slide is a **debris avalanche**, a very shallow slide of sediment or soil over bedrock (Figure 6.12). The failure plane is generally either at the base of the organic soil or in **colluvium**, a mixture of weathered rock and other debris below the soil.

The Role of Slope and Topography Slope stability is strongly influenced by *slope* and *topography*—more specifically by slope steepness and topographic relief. In general, the steeper the slope, the greater the driving forces that promote failure. For example, a study of landslides that occurred during two rainy seasons in the San Francisco Bay area of California established that 75 to 85 percent of them occurred on steep slopes in urban areas.[14]

Topographic *relief* refers to the height of a hill or mountain above the land below. Areas of high relief are hilly or mountainous, have dozens to thousands of metres of relief, and are generally prone to landslides. In Canada and the United States, the mountains of the western part of the continent, the Appalachian Mountains, walls of river valleys, and coastal bluffs have the greatest frequency of landslides. All types of landslides occur on steep slopes within these areas, and even gentle slopes developed on some rocks may creep imperceptibly. Even small slides can be lethal if they occur in populated areas (Figure 6.12). The soil layer on steep slopes can become saturated with water and slide downhill. Such small slides may transform into debris flows that travel long distances and cause much damage.

The Role of Climate *Climate* is the characteristic weather typical of a place or region over years or decades. Climate is more than just the average air temperature and amount of precipitation. It also includes the type of precipitation and its seasonal patterns, such as winter rain and snow along the Pacific coast, winter blizzards in the Arctic and Great Plains, summer thunderstorms in the southwest United States, heavy snows along the shores of the Great Lakes, and hurricane activity on the Mexican, Gulf, and Atlantic U.S. coasts. Temperate climates characterize much of the Atlantic and Pacific coasts, whereas most of the continental interior has strong seasonal contrasts, with cold winters and hot summers. Climate influences the amount and timing of water that infiltrates or erodes a hillslope and the type and abundance of hillslope vegetation.

In arid and semi-arid climates, vegetation tends to be sparse, soils are thin, and bare rock is exposed in many areas.

(a)

(b)

◄ **FIGURE 6.13 WATER ERODING THE TOE OF A SLOPE CAUSES INSTABILITY** (a) Stream-bank erosion shown in the lower left corner of this photograph caused slope failure and damage to a road in the San Gabriel Mountains, California. (b) Wave erosion caused this landslide in Cove Beach, Oregon. Further movement threatens the homes above. *((a) Edward A. Keller; (b) Gary Braasch/Stone/Getty Images)*

Free-face and **talus** slopes are common in these areas, and landslides include rock falls, debris flows, and shallow soil slips.

In subhumid and humid climates, abundant vegetation and thick soil cover most slopes. Landslides in these areas include deep complex landslides, soil creep, rockslides, slumps, and debris flows. We will now discuss the roles of vegetation and water on slopes in more detail.

The Role of Vegetation The type of vegetation in an area is a function of climate, soil type, topography, and fire history, each of which also independently influences what happens on slopes. Vegetation is a significant factor in slope stability for three reasons:

1. Vegetation provides a protective cover that reduces the impact of falling rain. It allows rainwater to infiltrate into the slope while retarding surface erosion.
2. Plant roots add strength and cohesion to slope materials. They act like steel rebar reinforcements in concrete and increase the resistance of a slope to landsliding.[15]
3. Vegetation adds weight to a slope, which can increase the likelihood that the slope will fail.

The Role of Water Water is nearly always involved in landslides, so its role is particularly important.[16] Water affects slope stability in three ways:

1. Many landslides, such as shallow *soil slips* and debris flows, happen during rainstorms when slope materials become saturated.
2. Other landslides, such as slumps, develop months or even years following deep infiltration of water into a slope.
3. Erosion of the toe of a slope by a stream reduces the mass of resisting material and thus decreases the slope's stability.

Stream or wave erosion reduces the factor of safety by removing material from the base of a slope (Figure 6.13). This problem is particularly critical if the base of the slope is an old, inactive landslide that is likely to move again (Figure 6.14). It is important to recognize old landslides when road cuts and other excavations are planned so that potential problems can be isolated and corrected before construction.

Water also contributes to the *liquefaction* of fine granular sediments. When disturbed, water-saturated silts and sands can lose their strength and flow as a liquid. The shaking of thick clays beneath Anchorage, Alaska, during a great earthquake in 1964 caused layers of sand at depth to liquefy, triggering large, destructive landslides.[17] Spontaneous liquefaction of *Leda clay* in the St. Lawrence Lowland in southern Quebec and Ontario triggered large landslides that destroyed many homes and killed more than 70 people in the twentieth century (Figure 6.15). Leda clay was deposited in a large marine embayment that extended up the St. Lawrence Valley at the end of the Pleistocene. At that time, low-lying areas up to about 200 m above present sea level were inundated by the sea and covered by silt and clay deposited by streams flowing from the wasting ice sheet that covered most of central and eastern Canada. Landslides in Leda clay occur on river valley slopes when

▲ **FIGURE 6.14 REACTIVATION OF A SLIDE** (a) Part of this beach (end of thin arrow) in Santa Barbara, California, was buried during a reactivation of an older landslide. (b) Close-up of the head of the landslide where it destroyed two homes. The thick black arrow in (a) points to the location of this picture. *(Don Weaver)*

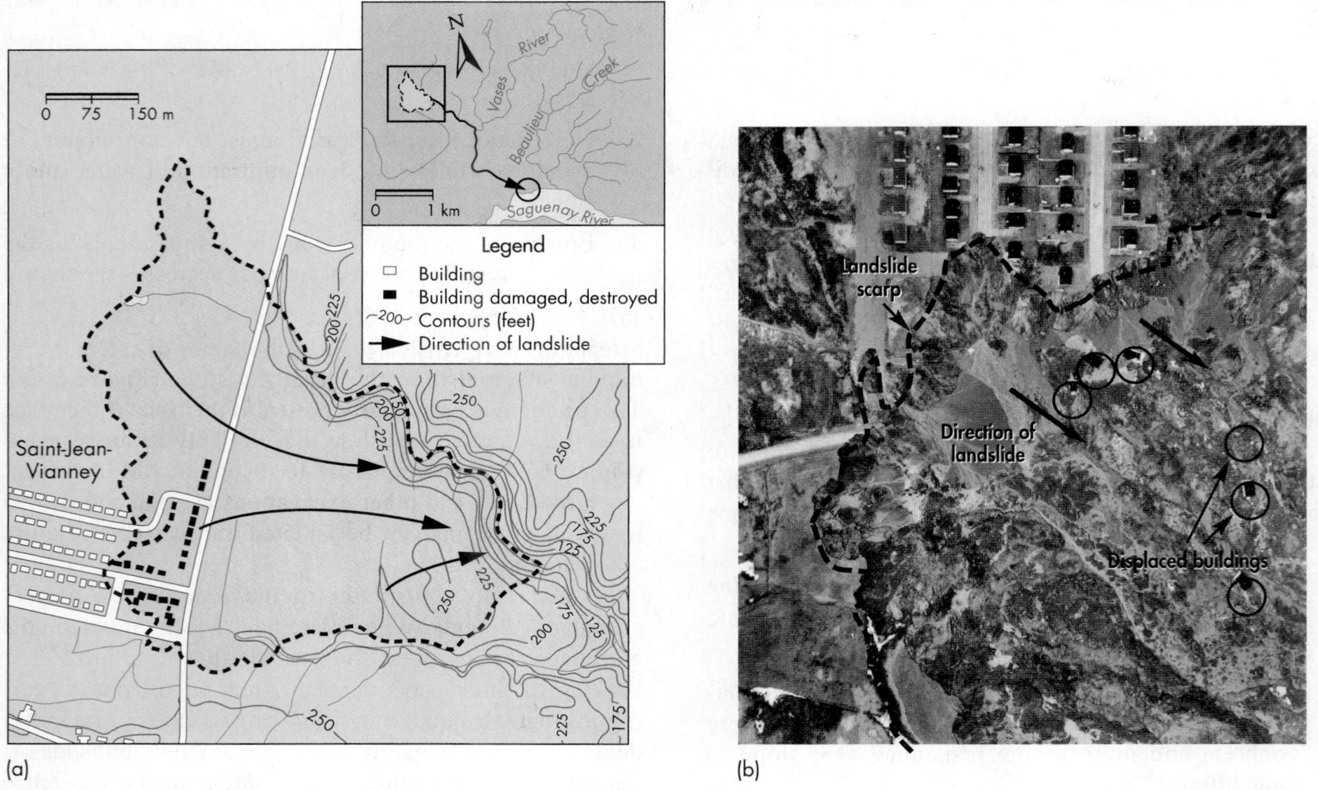

▲ **FIGURE 6.15 QUICK CLAY FAILURE** (a) A map and (b) oblique aerial view of the Saint-Jean-Vianney landslide, which happened in southern Quebec on May 4, 1971. Approximately 40 houses were engulfed by the landslide and 31 lives were lost. Some displaced houses can be seen in the debris below the scarp. *((a) Reproduced or adapted with the permission of Natural Resources Canada 2013, courtesy of the Geological Survey of Canada (Bulletin 548); (b) © Gouvernement du Québec)*

initially solid material transforms into liquid mud (Figure 6.16).[18] Liquefaction starts in a small area and spreads to a much larger area. The *quick clay* failures in Quebec and Ontario are especially interesting because liquefaction can occur without earthquake shaking. The sensitivity of Leda clay to liquefaction stems from the structure of the sediment. In the natural state, the constituent mineral grains are loosely packed and the forces binding the grains are

◀ **FIGURE 6.16 LIQUEFACTION** Mounds of sand and silt resulting from liquefaction during an earthquake in Christchurch, New Zealand, in February 2011. Firm, water-saturated sediments at depth liquefied and flowed to the surface along cracks in the ground. Some sensitive silty sediments on slopes can liquefy and trigger "quick clay failures," with or without earthquake ground motions. *(Reuters)*

weak; an analogy is a "house of cards." When the sediment is disturbed, the mineral binding forces drop to zero and the house of cards collapses.

Researchers have identified very large landslides in Leda clay west of Ottawa, Ontario.[19] One of the landslides, in the valley of Quyon River, which is a tributary of the Ottawa River, covers an area of about 31 km^2 and has an approximate volume of 600 million m^3. The Quyon River valley landslide and many others west of Ottawa occurred about 1000 years ago and may have been triggered by a moderate or large earthquake at that time.

The freezing of water in fractures in rock can destabilize slopes and trigger rock falls. Water increases about 9 percent in volume when it freezes and exerts large forces along fractures in rock. Monitoring of steep slopes along the Trans-Canada Highway in British Columbia has shown that rock fall is most common at times when temperatures fluctuate frequently above and below the freezing point[20] (Figure 6.17).

Water is also implicated in shallow landslides (thaw flow slides) in areas of *permafrost* (Figure 6.18). **Thaw flow slides** typically occur during warm spells in summer when the *active layer* (seasonal thawed layer) is thickest. The skin of water-saturated sediment slides away from the frozen ground below and flows downhill.[21]

The Role of Time Forces acting on slopes change with time. For example, both driving and resisting forces change seasonally as the water table fluctuates. Much of the weathering of rocks, which slowly reduces their cohesion and strength, is caused by the chemical action of water in contact with soil and rock near Earth's surface. Soil water is commonly acidic because it reacts with carbon dioxide in the atmosphere and soil to produce weak carbonic acid. Slope failure may occur without an obvious trigger when the resisting forces finally decrease below the driving forces. The failure may be preceded, over days or months, by an increase in the rate of creep.

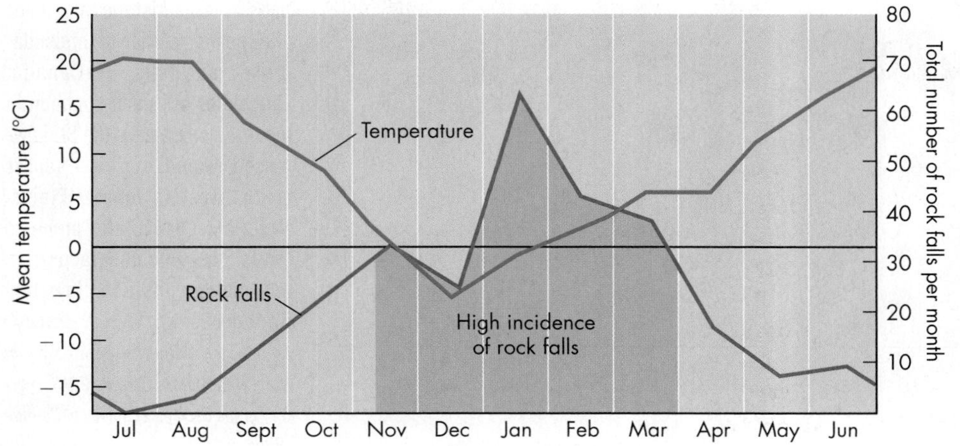

◀ **FIGURE 6.17 EFFECTS OF WEATHER ON ROCK-FALL ACTIVITY** A plot of rock-fall frequency and weather in the Fraser Canyon, southwest British Columbia, 1933–1970. Rock fall is most frequent at times when temperatures fluctuate above and below freezing. *(Based on Peckover, F. L., and J. W. G. Kerr. 1977. "Treatment and maintenance of rock slopes on transportation routes."* Canadian Geotechnical Journal *14:487–507)*

◀ **FIGURE 6.18 THAW FLOW SLIDE** Flow of thawed water-rich sediment on a gentle slope in permafrost terrain in the Mackenzie Mountains, Northwest Territories. The flow came to rest on the floodplain of Dekale Creek, where it has formed a fan (lower left). *(Natural Resources Canada 2010, courtesy of the Geological Survey of Canada)*

6.2 Geographic Regions at Risk from Landslides

As you might imagine, landslides occur wherever there are significant slopes. Mountainous areas have a higher risk for landslides than most areas of low relief. The latter generalization is supported by a map of major landslide areas in Canada (Figure 6.19). Areas where landslides are most common are the western Cordillera of British Columbia, Yukon, and Alberta, and the Appalachians of Quebec and New Brunswick. The landslide hazard in other parts of the country is more localized, notably in the St. Lawrence Valley where many slopes are developed in Leda clay, and along the valleys of the large rivers that cross the Prairies. Earth materials that are particularly prone to landsliding

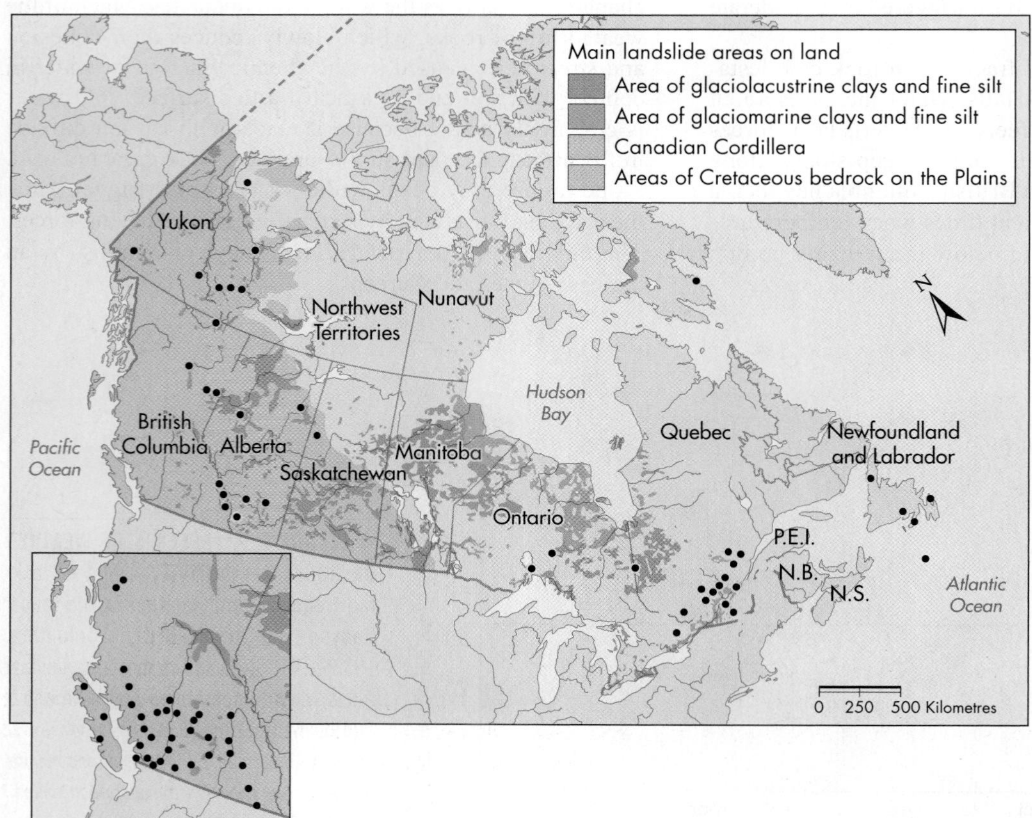

◀ **FIGURE 6.19 LANDSLIDES IN CANADA** This map shows locations of large historic landslides and the distribution of landslide-prone materials in Canada. Landslides are most common in mountainous areas, the St. Lawrence Lowland, and river valleys on the western Interior Plains. Materials particularly prone to landsliding include lacustrine and marine silts and clays and Cretaceous shales. *(Reproduced or adapted with the permission of Natural Resources Canada 2013, courtesy of the Geological Survey of Canada (Bulletin 548))*

include lacustrine and marine silts and clays, Cretaceous shales, and Quaternary volcanic rocks. Landslide-prone areas in the United States are similar: the mountainous areas of the west coast, the Rocky Mountains, the Alaska Range, and the Appalachians. The U.S. Plains are relatively free of landslides.

Three factors are expected to increase landslide incidence and risk in some areas of the world in this century:

1. Urbanization and development will increasingly expand into landslide-prone areas.
2. Tree cutting will continue in landslide-prone areas.
3. Changing global climate patterns will result in increases in precipitation in some regions.[22]

6.3 Effects of Landslides and Links with Other Natural Hazards

Effects of Landslides

Landslides and related phenomena cause substantial damage and loss of life (Figure 6.20). In North America, 30 people, on average, are killed each year by landslides. The total cost of landslide damage exceeds U.S.$1 billion per year and, in some years, is more than U.S.$3 billion.[23]

The direct, damaging effects of landslides include being struck by or buried in debris (see the Survivor Story). Landslides can damage homes, roads, and utilities that have been constructed at the top, base, or side of a hill. They regularly block roads and railroads, delaying travel for days or more. One massive landslide, the 1965 Hope Slide, blocked a major highway in southern British Columbia for weeks, and another in April 1903 buried part of the town of Frank, Alberta, killing more than 76 people (see story at start of this chapter). Landslides may even block shipping lanes. In 1980 a debris flow from the Mount St. Helens volcanic eruption filled the channel of the Columbia River with more than 34 million m^3 of sediment. The sediment stopped cargo ships from reaching Portland, Oregon, until dredging was completed.

Indirect damaging effects of landslides include flooding upstream of landslide dams, downstream floods caused by the rapid breaching of landslide dams (Case Study 6.1), and blockage of salmon migration routes. Partial blockage of the Fraser River in southern British Columbia by rock fall during construction of the Canadian National Railway in 1914 prevented salmon from reaching their spawning grounds. The runs still have not recovered to their pre-1914 levels, resulting in large economic losses to the Fraser River salmon fishery.[24]

Links between Landslides and Other Natural Hazards

Landslides are linked to just about every other natural hazard you can think of. Earthquakes, volcanoes, storms, and fires can cause landslides; these relationships are discussed in other chapters in this book. As mentioned above, landslides can be responsible for flooding if they form a debris dam across a river.

A large landslide can also trigger a tsunami or cause widespread flooding if it displaces water out of a lake or bay. In 1963 more than 240 million m^3 of rock slid into the Vaiont reservoir in northeast Italy. The landslide displaced water over the dam, flooding the valley below and killing more than 1900 people.[25] Most of the deaths were in the town of Longarone, which was flooded with 70 m of water.

◄ **FIGURE 6.20 KILLER LANDSLIDE** On February 17, 2006, a landslide engulfed the farming village of Guinsaugon on the Philippine island of Leyte. The muddy debris killed more than 2000 people, nearly the entire population of the town. *(U.S. Navy photo by Photographer's Mate 1st Class Michael D. Kennedy)*

SURVIVOR STORY

Landslide

Danny Ogg thought he was "a goner" when the first boulder hit his truck.

He was driving on Interstate 70 west of Denver when a giant granite boulder crashed across his windshield, leaving him with only one thought: "I'm going to be buried alive."

The boulder was one of about 20 or 30 that rocketed down the steep slope onto the interstate highway, which winds through the Rocky Mountains, early one spring morning in 2004 (Figure 6.21). The largest rock weighed 30 t.

▲ **FIGURE 6.21 BOULDERS ON U.S. INTERSTATE 70** Granite blocks, some as large as 4 m^3, line the shoulder of an interstate highway near Glenwood Springs, Colorado. Heavy rain and subsequent freezing conditions dislodged rock from a free face more than 100 m above the highway. One car-sized boulder bounced across the highway and landed in the nearby Colorado River. A highway crew had to drill or blast apart the largest blocks and push them over to the shoulder so that traffic could get through. *(U.S. Geological Survey)*

Ogg, a 50-year-old driver from Tennessee, says he had no indication that a landslide was underway until the first rock hit his tractor-trailer. After that, it was "bang bang bang, crunch crunch," as he puts it, as his truck collided with the falling debris and eventually hit the median. The vehicle's fuel tank was torn open in the process, spilling hundreds of gallons of diesel.

His sense of time was skewed in the terrifying moments of the landslide. "After I hit the first boulder, it took two, three minutes, I guess. It was fast but it wasn't," he remarks.

Ogg suffered an injured back and some torn cartilage and ligaments, but he was lucky, considering the much more serious injuries he could have incurred.

"I thought I was a goner," he says. "I don't know how long I was in there. It addled me. I was silly as a goose, as we say in Tennessee."

His sense of location was also askew: "You know you ain't dead yet, but you don't know where you are," he continues. "I didn't know if I was hanging off the mountain or what."

Ogg notes that some of the boulders left potholes in the road a foot or two deep, and a *Rocky Mountain News* article reported that some of them bounced as high as 6 m after impact.

Although landslides are difficult to predict and avoid, Ogg says he will be more vigilant now when driving in the mountains. "When you go around the mountain now, you look up."

—Chris Wilson

A spectacular landslide-triggered tsunami occurred in Lituya Bay, Alaska, on July 7, 1958 (Figure 4.13). A strong earthquake triggered a rockslide on a steep slope high above the head of the bay. The rockslide plunged into the bay and displaced a huge mass of seawater that raced up the opposite valley wall to an elevation of 525 m, completely destroying the forest in its path.[26] In four minutes, a 30 m high wave surged 11 km to the mouth of the bay, where it swept away three fishing boats anchored just inside a low forested spit. Smaller, though still damaging, landslide-triggered tsunamis have occurred in *fiords* in Norway, Alaska, and British Columbia.

Earthquakes can trigger subaqueous landslides that in turn generate devastating tsunamis. In 1988, a submarine landslide triggered by an **M** 7.1 earthquake caused a tsunami that killed more than 2200 people in several communities on the north coast of Papua New Guinea (see Chapter 4).

The most destructive tsunami in Canadian history was triggered by a submarine landslide. On November 18, 1929, an **M** 7.2 earthquake occurred about 20 km beneath the seafloor at the southern edge of the Grand Banks, 250 km south of Newfoundland (see Chapter 4). The shaking triggered a huge submarine slump, which, in turn, set off a

6.1 CASE STUDY

Mount Meager Landslides and Consequent Flooding

Mount Meager is a volcanic massif in the southern Coast Mountains of British Columbia, 150 km north of Vancouver. The Mount Meager massif has been deeply eroded and is partly covered by snow and glaciers. The steep slopes of the massif are unstable and subject to frequent large landslides. Three landslides larger than 200 million m^3 have occurred in the past 10 000 years (8700, 4400, and 2400 years ago), and at least six landslides larger than 1 million m^3 have happened in the past 100 years.[27] The high frequency of landslides at Mount Meager led one geologist to label Meager Creek, which flows along the southeast side of the massif, "the most dangerous valley in Canada."

Mount Meager is remote, but timber is harvested in valleys bordering the massif and the area is a popular destination for outdoor enthusiasts. In addition, a pumice mine operates seasonally on the east side of the mountain and several "run-of-the-river" hydroelectric projects are in the planning stage. The nearest community is Pemberton, which is located along the Lillooet River 65 km southeast of Mount Meager. Given the considerable distance of Pemberton from the volcano, you might assume that any landslide at Mount Meager would have no impact on the community. However, geologists working in the area over the past decade discovered that this is not the case. They learned that when the huge landslides occurred at Mount Meager 8700, 4400, and 2400 years ago, they transformed into volcanic debris flows (*lahars*) that flowed down the Lillooet River valley all the way to Pemberton. A landslide of the same size in the future would likely devastate the community, as well as Mount Currie, a town to the east. However, with only three events of this size in 10 000 years, the probability of another in the near future is very low.

These huge landslides, however, are not the only source of concern; smaller landslides at Mount Meager also pose a risk to Pemberton and Mount Currie, as illustrated by events in early August 2010. On August 6 of that year, about 48 million m^3 of volcanic rock slid away from the south flank of Mount Meager into the valley of Capricorn Creek (Figure 6.22).[28] The rock mass impacted the volcano's weathered and saturated flank and transformed into a debris flow that travelled at high speed down Capricorn Creek to Meager Creek, about 7 km away. The debris then ran across Meager Creek and climbed high on the south wall of the valley. There, the debris mass bifurcated: one lobe ran nearly 4 km upstream and a second travelled 3 km downstream into the Lillooet River valley (Figure 6.23). Landslide debris blocked the Lillooet River for a couple of hours. A larger mass of debris came to rest at the mouth of Capricorn Creek, where it dammed Meager Creek, creating a lake 1.5 km long. There was immediate concern that this

▲ **FIGURE 6.22 SOURCE OF 2010 MOUNT MEAGER LANDSLIDE** This photograph shows the scar of the August 2010 landslide on the south flank of Mount Meager. The landslide occurred in weak, altered volcanic rocks. *(John J. Clague)*

▲ **FIGURE 6.23 MOUNT MEAGER LANDSLIDE** Photograph showing the path and deposit of the 2010 Mount Meager landslide. The source of the landslide was weak volcanic rocks at the head of Capricorn Creek. About 48 million m^3 of debris ran down Capricorn and Meager creeks into Lillooet River valley (foreground). The landslide briefly stemmed the flow of both Meager Creek and the Lillooet River. *(John J. Clague).*

(continued)

6.1 CASE STUDY *(Continued)*

lake would overtop the debris dam and drain catastrophically, producing an outburst flood that would endanger Pemberton. Acting on this concern, the British Columbia government ordered a night-time evacuation of 1500 residents in the valley. Fortunately, the dam breached slowly enough that the flood wave remained within the Lillooet River channel in Pemberton. No lives were lost in the landslide, but despite its relatively remote location, direct costs of the event were about $10 million.[28] A landslide from the same source area in 1998 also dammed Meager Creek and formed a lake of comparable size to that in 2010. Given the history of landslides at Mount Meager, it seems likely such events will occur in the future and outburst floods remain a threat to people living in the valley.

Perhaps a more serious problem is the secondary effect of the large amounts of sediment that are delivered to Meager Creek and the Lillooet River by the frequent landslides at Mount Meager. Loose landslide debris is eroded and carried downvalley by the Lillooet River past Pemberton and Mount Currie, ultimately reaching Lillooet Lake. Repeat surveys of the Lillooet River channel done over the past decade show that the bed of the river along its dyked reach is rising due to deposition of sediment. The height between the river bed and the top of the dykes that protect Pemberton from flooding is gradually decreasing due to sediment accumulation in the channel, in effect increasing the probability of a destructive flood during a high rainfall period. Pemberton and Mount Currie were most recently flooded by the Lillooet River in October 2003.

Solutions to the Landslide and Flood Problem

Landslides at Mount Meager are large and cannot be prevented, and the communities in the Lillooet River valley will not relocate. The danger landslides pose to the communities, however, can be assessed and potentially minimized. The following are possible strategies for dealing with the problems identified above:

1. *Monitor the mountain.* It makes sense to monitor Mount Meager for signs of instability and incipient landslides. Airborne and satellite remote sensing systems are capable of detecting small movements of the ground that might herald a landslide.
2. *Install an effective warning system.* Sensors to detect landslides in real time could be installed at critical locations on the Mount Meager massif. The shortcomings of such a system are the considerable cost of installing and maintaining it and the requirement that the system be fully integrated into the emergency operations of Pemberton and Mount Currie. These communities do not have the resources to install and operate an early warning system.
3. *Dredge the Lillooet River channel.* The long-term risk posed by the build-up of sediment in the channel of the Lillooet River could be reduced by removing the sediment, thereby increasing the amount of water the channel can convey before the dykes are overtopped and the valley is flooded. This strategy, like the previous one, is beyond the financial means of the affected communities and would not provide a long-term solution to the problem. Also, it would be difficult and costly to dispose of the sand dredged from the channel.
4. *Provide additional flood protection to communities.* The dykes that protect Pemberton from flooding could be raised to reduce the risk of flooding. They also could be extended downriver to Mount Currie, which is currently completely unprotected. This engineered solution would be expensive and does not address the long-term problem of the build-up of sediment in the river channel. Another engineered solution is to raise the levels of houses that are vulnerable to flooding. Some homeowners have done just that—they have placed their houses on pads of gravel elevated 1 m or more above the level of the floodplain. Pemberton is also redirecting community growth away from the floodplain and onto higher ground.

Lessons Learned

Several lessons can be learned from the situation in the Lillooet River valley. First, some landslides can pose a threat to communities situated tens of kilometres from the source. Such is the case of very large rockslides and rock avalanches that transform into highly mobile debris flows. Second, landslides can block streams and impound lakes that might subsequently overflow and rapidly drain, creating dangerous downstream floods. Third, frequent landslides generate large amounts of sediment that, when conveyed by rivers away from the source, may increase the risk of flooding.

tsunami that propagated across the Atlantic Ocean, registering on tide gauges as far away as South Carolina and Portugal. The tsunami damaged more than 40 coastal communities on Burin Peninsula in Newfoundland and claimed 27 lives.[29]

6.4 Natural Service Functions of Landslides

Although most hazardous natural processes provide important natural service functions, the potential benefits of landslides are few. Benefits include the creation of new habitats in forests and aquatic ecosystems.

Landslides, like fire, are a major source of ecological disturbance in forests. For some old-growth forests, this disturbance can be beneficial by increasing both plant and animal diversity.[30] Landslide-dammed lakes provide new habitat for fish and other aquatic organisms, and large landslides provide open spaces in forests favoured by some animals and plants.

Landslides can produce sediments that contain valuable minerals. Weathering frees mineral grains from rocks, and landslides transport these minerals downslope. Heavier minerals, particularly gold and diamonds, can be concentrated at the base of the slope and in adjacent streams. Gold and diamonds have been mined from landslide deposits, although they are much more common in fluvial sediments.

6.5 Human Interaction with Landslides

Mass wasting is a natural process, and landslides can happen without any human involvement. However, expansion of urban areas and transportation networks and exploitation of natural resources have increased the number and frequency of landslides in many areas. For example, grading of land surfaces for housing developments can initiate landslides on previously stable hillsides.

The effect of human activities on the magnitude and frequency of landslides ranges from nearly insignificant to very significant. In instances where human activities, such as road construction and deforestation, increase the number and severity of landslides, we need to learn how our practices cause slope failures and how we can minimize their occurrence. Below are descriptions of some human activities that cause landslides.

Timber Harvesting

The possible cause-and-effect relation between timber harvesting and erosion is a major environmental and economic issue around the world. Two controversial practices are *clearcutting*, which involves harvesting all trees from large tracts of land, and construction of logging roads to remove cut timber from the forest. Landslides, especially debris avalanches, debris flows, and earthflows, are responsible for much of the erosion in these areas.

One 20-year study in Oregon found that shallow slides are the dominant form of erosion. It also found that timber-harvesting activities, such as clearcutting and road building, did not significantly increase landslide-related erosion on geologically stable land. In contrast, logging on unstable slopes or weak rocks increased landslide erosion by several times compared to slopes that had not been logged.[31]

Road construction in timber harvesting areas can pose an especially serious problem because roads interrupt surface drainage, alter subsurface movement of water, and can adversely change the distribution of materials on a slope by cut-and-fill or grading operations.[31] Based in part on past errors, geologists, engineers, and foresters have developed improved management practices to minimize the adverse effects of logging. New forest-management practices include harvesting smaller cut blocks, selective logging of cut blocks, helicopter logging to minimize the number and length of access roads, and controlled surface drainage along roads. In spite of these advances, landslides continue to be associated with timber harvesting.

Urbanization

Human activities are most likely to cause destructive landslides in urbanized areas with high densities of people, roads, and buildings. Examples from Rio de Janeiro, Brazil, and Los Angeles, California, illustrate this point.

Rio de Janeiro, with a population of more than 6 million people, may have more problems with slope stability than any other city of its size.[32] Several factors contribute to the serious landslide problem in Rio: (1) the granite peaks that spectacularly frame the city (Figure 6.24a) have steep slopes of fractured rock that are covered with thin soil, (2) the area is periodically inundated by torrential rains, (3) cut-and-fill construction has seriously destabilized many slopes, and (4) vegetation has been removed from the slopes.

The landslide problem started early in the city's history, when many of the slopes were logged for lumber and fuel and to clear land for agriculture. Landslides associated with heavy rainfall followed this deforestation. More recently, a lack of building sites on flat ground has led to increased urban development on steep slopes. Many slopes have been undercut and their vegetation and soil removed. In addition, fill placed on slopes to expand the size of building sites has increased the load on already unstable land. Given that Rio de Janeiro also experiences tremendous rainstorms, it becomes apparent that the city has a serious problem.

In February 1988, an intense rainstorm dumped more than 12 cm of rain on Rio de Janeiro in four hours. The storm caused flooding and debris flows that killed at least 90 people and left 3000 people homeless. Most of the deaths were caused by debris flows in hill-hugging shantytowns where housing is precarious and control of storm water runoff nonexistent. However, more affluent mountainside areas were not spared from destruction. In one area, a landslide destroyed a nursing

◄ **FIGURE 6.24 LANDSLIDES IN RIO DE JANEIRO** (a) Panoramic view of Rio de Janeiro, Brazil, showing the steep "sugarloaf" mountains and hills (right centre of the image). Steep slopes, fractured rock, shallow soils, and intense rainfall contribute to the landslide problem in Rio, as do such human activities as urbanization, logging, and agriculture. Nearly all the bare rock slopes were at one time vegetated; that vegetation has been removed by landslides and other erosional processes. (b) Aerial view of the Morro dos Prazeres slum in Rio de Janeiro, where a landslide destroyed several homes on April 8, 2010 *(Getty Images)*

(a)

(b)

home and killed 25 patients and staff. Restoration costs for the entire city exceeded $100 million. A similar event occurred in Rio de Janeiro in April 2010, when heavy rains caused numerous landslides that destroyed 60 homes and killed more than 200 people (Figure 6.24b). If future disasters are to be avoided, the city and the Brazilian government must take extensive and decisive measures to control storm runoff, increase slope stability, and limit development on dangerous slopes.

Southern California has also experienced a large number of landslides associated with hillside development, and Los Angeles has the dubious honour of illustrating the economic importance of studying urban geology.[33] It took natural processes millions of years to produce the valleys, ridges, and hills of the Los Angeles basin. In a little more than a century, people have radically altered them. More than 40 years ago, F. B. Leighton, a geological consultant in southern California, wrote: "With modern engineering and grading practices and appropriate financial incentive, no hillside appears too rugged for future development."[33] Almost overnight, we can convert steep hills into flat lots and roads. Unfortunately, the grading process in which benches, referred to as pads, are cut into slopes for home sites has been responsible for many landslides in southern California.

Grading codes that minimize landslide risk have been in effect in the Los Angeles area since 1963. The codes were adopted in the aftermath of destructive and deadly landslides in the 1950s and 1960s. Since these codes have been in effect and detailed engineering geology studies have been required before development, the number of hillside homes damaged by landslides and floods has been greatly reduced. Although initial building costs are greater because of the strict codes, the costs are more than offset by the reduction of losses in subsequent years.

People may destabilize formerly stable slopes by removing rock or sediment, watering lawns and gardens, installing septic systems, or adding fill and buildings (Figure 6.25). As a rule, any project that steepens or saturates a slope, increases its height, or places an extra load on it may cause a landslide.[34]

Landslides have been a problem in some urban areas of Canada. In Vancouver, damaging debris avalanches and debris flows have occurred on slopes underlain by glacial

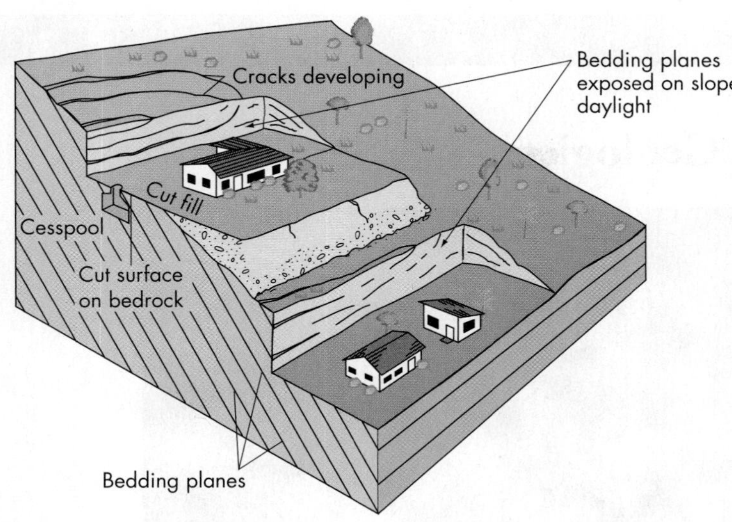

◄ **FIGURE 6.25 URBANIZATION AND LANDSLIDE POTENTIAL** This diagram shows how building on slopes can increase the chance of landslides. The diagonal lines at the left side of the block diagram are bedding planes in sedimentary rock. Excavation into the hillside behind the houses has removed support. Fill (yellow) used to extend the flat pad for building adds weight to the slope. Cracks shown in the upper part of the diagram are an early sign that a landslide is likely to occur. Wastewater from septic fields adds water to the hillslope. *(Reprinted, with permission, from Leighton, F. B. 1966. "Landslides and urban development." In* Engineering Geology in Southern California. *Los Angeles Section of the Association of Engineering Geologists)*

sediments, fill, and soils.[35] An example is a small debris avalanche that destroyed two homes and claimed one life in a North Vancouver neighbourhood on January 19, 2005. Fill in the backyard of a home at the top of a steep slope failed during heavy rain and swept down the slope and into the houses below (Figure 6.12). The landslide entrained soil and trees as it moved downslope, adding to the destruction. Another similar landslide had occurred in the same area and under similar circumstances in 1979. Landslides have also damaged and destroyed homes built too close to the edges of valley walls in Edmonton.

6.6 Minimizing Landslide Hazard and Risk

To minimize landslide risk, it is necessary to identify areas where landslides are likely to occur, employ engineered structures to prevent them, warn people of impending failures, and control active slides (see Professional Profile). As discussed below, the preferred and least expensive option is to avoid developing sites where landslides are occurring or are likely to occur.

Identification of Potential Landslides

Recognizing areas with a high potential for landslides is the first step in minimizing the hazard. Surface features indicative of unstable or potentially unstable slopes include:

- Crescent-shaped cracks or terraces on a hillside
- A scalloped or recessed crest of a valley wall
- A tongue-shaped area of bare soil or rock on a hillside
- Large boulders or piles of talus at the base of a cliff
- An area of tilted, or **jack-strawed**, trees
- Trees that are convex at their base but straight higher up
- Exposed bedrock with layering that is parallel to the slope
- Tongue-shaped masses of sediment at the base of a slope or at the mouth of a valley
- A hummocky, or irregular and undulating, land surface at the base of a slope

Earth scientists search for these indicators in the field and on aerial photographs. They then assess the hazard and produce several kinds of maps (Figure 6.26).

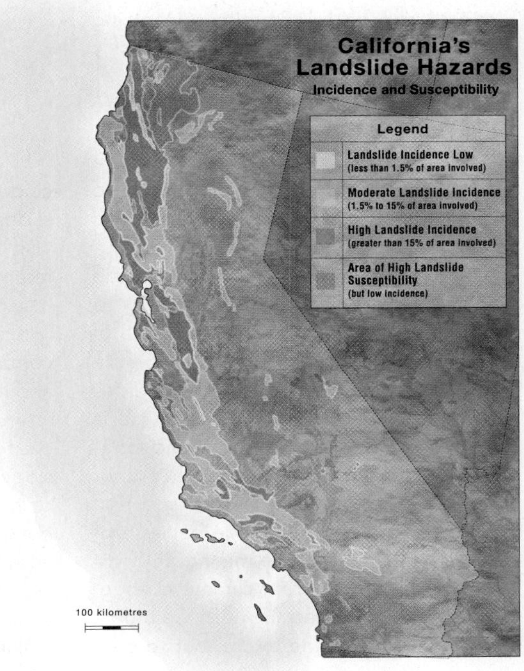

▲ **FIGURE 6.26 LANDSLIDE HAZARD MAP** A generalized map showing areas of different landslide incidence in California. *(Modified from the National Atlas and USGS Open-File Report 97-289)*

PROFESSIONAL PROFILE

Matthias Jakob, Engineering Geologist

Matthias Jakob (Figure 6.27) has a passion for falling, bouncing, flowing, and creeping rock and soil. Born near the German Alps and now a senior geoscientist with a geotechnical consulting firm in Vancouver, British Columbia, Jakob is fascinated with the processes that form mountains and even more so with the processes that wear them down.

Jakob has studied landslides around the world, from North America to Italy, Germany, Bolivia, Ecuador, Argentina, Chile, and Venezuela. He has gained first-hand experience with a large variety of landslides in a range of climates and geologic settings.

"Every time you work in a new region, you have to retrain yourself, understand the special local conditions, and adapt your methods and interpretations accordingly. A good landslide scientist is the one who constantly questions his or her hypotheses," Jakob insists. "Geological idiosyncrasies at any given site can be overwhelming at times and may lie outside one's expertise. Yet, time after time, ignorance of geological subtleties has led to catastrophic loss. A multidisciplinary approach, sometimes involving highly specialized scientists, is commonly needed to solve the geo-puzzle."

▲ **FIGURE 6.27 GEOLOGIST EXAMINES RECENT LANDSLIDE** Matthias Jakob, an engineering geologist based in Vancouver, British Columbia. *(Courtesy of Matthias Jakob)*

"In our business, you quickly realize that landslides are sometimes ignored. Landslides at any given location are rare events and human memory is short. Too often, I hear statements like 'I have lived here all my life and nothing has ever happened.' Such an assertion may not be true and, even if it is, it may be foolhardy to assume that an inhabited area is free of hazards."

Deciphering the history of landslides is a crucial part of Jakob's work. He is particularly interested in short- and long-term changes in the source area of landslides, which may alter hazard and risk. The source area can change because of human activity, such as mining and forestry, or indirectly from climate change. "We think that coastal British Columbia will get wetter in the future, but to conclude that more rain will cause more landslides is overly simplistic," Jakob says. "Different types of landslide respond differently to changes in rainfall. An earthflow may move after one year of above-normal rainfall, whereas debris flows respond to a combination of higher antecedent rainfall and higher rainfall intensities."

Jakob is a firm believer that climate is changing and that the changes will affect slope stability. He plans to focus much of his future research on the interplay between slope stability and hydrology. He notes that "increasingly warmer winters are responsible for severe insect damage to forests in British Columbia's interior, making the region more vulnerable to wildfires. Widespread burning commonly increases rates of natural erosion and may increase the size and frequency of debris flows."

Jakob states that "time is of the essence in this line of work." Steep slopes and mountain valleys in western Canada are under increasing development pressure. New development, in tandem with climate change, will increase landslide risk and pose a challenge for landslide specialists, particularly in British Columbia, where the number of lawsuits dealing with geotechnical practice exceeds those of all other Canadian provinces combined.

—John J. Clague

One type of map is the direct result of the landslide inventory described above. It may show areas that have experienced slope failure, or it may be a more detailed map discriminating landslide deposits of different age. Information concerning past landslides may be combined with land-use considerations to develop a *slope stability map* for use by engineering geologists or a *landslide hazard map* with recommended land uses for planners. Preparing a *landslide risk map* is more complicated because it involves evaluating the probability that a landslide will occur and an assessment of potential losses.[36] In this context, it is important to reiterate the difference between landslide *hazard* and *risk*. Hazard reflects the expected type, size, and likelihood of landslides in a particular area. Risk is the vulnerability of the area to landslides and takes into account population and economic infrastructure. There is no risk if landslides occur in a natural area without people or development. In contrast, all other things being equal, risk increases as population and development grow.

Prevention of Landslides

Prevention of large, natural landslides is difficult or impossible, but common sense and good engineering practices help to minimize the hazard. For example, loading the top of slopes, cutting into sensitive slopes, or increasing the flow of water on slopes should be avoided or done with caution.[34] Common engineering techniques for landslide prevention include surface and subsurface drainage, removal of unstable slope materials, and construction of retaining walls or other supporting structures.[7]

Drainage Control Surface and subsurface drainage control is generally effective in stabilizing a slope. The objective is to keep water from running across or infiltrating the slope. Surface runoff may be diverted around the slope by surface drains. Infiltration may also be controlled by covering the slope with an impermeable layer, such as cement, asphalt, or plastic. Groundwater may be removed by installing subsurface drains. The simplest drains, which are commonly used in fractured bedrock, are horizontal or inclined drill holes or tunnels (Figure 6.28a). More complex drains, which are sometimes used in soft rocks or unconsolidated sediments, consist of pipes with holes along their length. The pipes are surrounded with permeable gravel or crushed rock and positioned underground to intercept and divert water away from a potentially unstable slope.[7]

Two examples from British Columbia illustrate the importance of drainage in stabilizing landslides. The Trans-Canada Highway crosses the Drynoch landslide, a large slow-moving earthflow in the Thompson River valley in southern B.C. For many years, slow movement of the earthflow displaced and damaged the highway, necessitating regular and costly repairs. The British Columbia Ministry of Highways and Transportation installed a system of surface and subsurface drains to reduce the amount of water infiltrating the slope. Soon thereafter, movement rates decreased substantially and, today, little maintenance of the highway is required where it crosses the landslide.

(a)

(b)

(c)

▲ **FIGURE 6.28 WAYS OF INCREASING SLOPE STABILITY** (a) Holes drilled in fractured rock to drain groundwater and thereby reduce water pressures. (b) Rock bolts anchoring fractured rock on a steep cut face, reducing the possibility of rock fall. (c) Metal mesh draped over a rock face to intercept small rock fall. *(John J. Clague)*

The Downie slide is a huge (1.4 km^3) sackung in the Columbia River valley north of Revelstoke, British Columbia. The toe of the landslide was inundated when Revelstoke Dam was completed in 1983. After dam construction, BC Hydro and Power Authority (BC Hydro) conducted an exhaustive study of the landslide to determine its three-dimensional

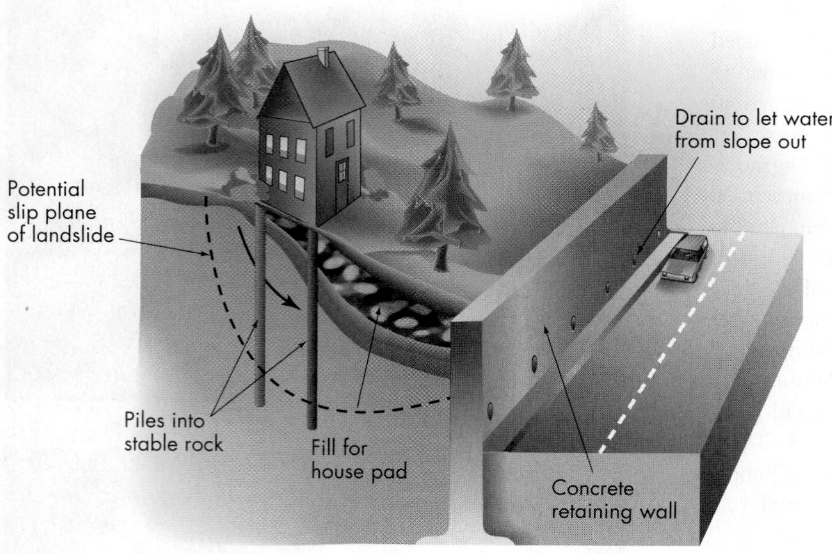

◀ **FIGURE 6.29 HOW TO SUPPORT A SLOPE** The types of slope support shown in this illustration include a deeply anchored concrete retaining wall along a road, concrete piles sunk into stable rock, and subsurface drains that reduce water pressure in the slope.

geometry, its movement rates, and the effects of inundation of its toe beneath 100 m of water. BC Hydro staff decided that the best approach to ensure against catastrophic slope failure would be to drain water away from rocks at the base of the landslide. Work on the Downie slide was carried out between 1977 and 1982. A drainage system consisting of hundreds of metres of tunnels and thousands of metres of drill holes was established to drain the water away. Since 1982, when the work was completed, the landslide mass has practically stopped moving and is considered to be more stable than it was before the dam was built.

Grading Carefully planned grading can improve slope stability. In a cut-and-fill operation, material from the upper part of a slope is removed and placed near the base. The overall gradient of the slope is thus reduced, and placement of material at the toe of the slope increases the resisting force. However, this method is not practical on a very high, steep slope. Instead, the slope may be cut into a series of benches or steps, each of which contains surface drains to divert runoff. The benches reduce the overall slope and are good collection sites for falling rock and small slides.[7]

Slope Supports One of the most common stabilization strategies is to support slopes with retaining walls constructed of concrete, brick, or stone-filled wire baskets called *gabions*. The walls can be supported with *piles*, which are long concrete, steel, or wooden beams driven into the ground (Figure 6.29). To function effectively, the walls must be anchored well below the base of the slope, be backfilled with permeable gravel or crushed rock (Figure 6.30), and contain drainage holes to reduce water pressure in the slope (Figure 6.29).

◀ **FIGURE 6.30 RETAINING WALL** This retaining wall of concrete cribbing was installed and then back-filled to help stabilize the road cut. *(Edward A. Keller)*

Potentially unstable rock slopes can also be secured with rock bolts, which are steel or iron pins up to several metres long inserted into holes drilled into the slope and anchored with facing plates (Figure 6.28b). Heavy metal screens and rock fences can also be placed over steep cut slopes to catch small rock fragments before they reach roads or railways (Figure 6.28c). Screens provide no protection against rockslides, but they can be effective for small rock falls.

Preventing landslides can be expensive but well worth the effort. The benefit-to-cost ratio for landslide prevention ranges from 10 to 2000. That is, for every dollar spent on landslide prevention, between $10 and $2000 are saved.[37]

The cost of *not* preventing a landslide is illustrated by the massive Thistle landslide, which occurred southeast of Salt Lake City in April 1983. The landslide moved down a mountain slope and across a canyon to form a natural dam about 60 m high. The dam created a lake that flooded the community of Thistle, the Denver–Rio Grande Railroad switchyard and tracks, and two major U.S. highways (Figure 6.31).[38] The total direct and indirect costs of the landslide and associated flooding exceeded U.S.$400 million.

The Thistle slide involved reactivation of a portion of an older, larger landslide. It had been known for many years that the older landslide mass moved at times of high precipitation. It thus came as no surprise that the 1983 landslide was triggered by extremely high amounts of precipitation during a strong El Niño year. A review of the evidence suggests the Thistle slide could have been prevented by installing a network of subsurface and surface drains at a cost of U.S.$300 000 to $500 000, a small amount compared to the cost of the damage.[38] Because the benefit-to-cost ratio in landslide prevention is so favourable, it seems prudent to evaluate active and potentially active landslides in areas where considerable damage may be expected and possibly prevented.

◄ **FIGURE 6.31 LANDSLIDE BLOCKS A VALLEY** The costliest landslide in U.S. history, the 1983 slide at Thistle, Utah, was a reactivation of an older slide. The landslide blocked the Spanish Fork River and impounded a lake that inundated the community of Thistle, the Denver–Rio Grande Railroad, and two major American highways. Nearly the entire landslide deposit moved again in 1999. *(Michael Collier)*

Landslide Warning Systems

Landslide warning systems do not prevent landslides; rather they provide time to evacuate people and their possessions, stop trains, and reroute traffic. Hazardous areas can be visually inspected for surface changes, and small rock falls on roads and other areas can be noted for quick removal. Human monitoring has the advantage of reliability and flexibility, but it becomes difficult during adverse weather and in hazardous areas.[39] Rock fall along transportation routes in British Columbia has been monitored for decades by the B.C. Ministry of Transportation and Highways, CP Railway, and CN Railway.

Other warning methods include electrical systems, tiltmeters, and geophones that pick up vibrations from moving rocks. Many Canadian and U.S. railways have rock fences on slopes above their tracks that are linked to signal systems. When a large rock hits the fence, a signal is sent and trains can be stopped before they are in danger. This and other mitigation measures have significantly reduced the number of railroad accidents, injuries, and fatalities in North America.[39] BC Hydro conducts regular surveys of slopes along its hydroelectric reservoirs to measure surface displacements and to ensure that movements in areas of known landslides, such as the Downie slide, are not accelerating.

A pilot landslide warning system was developed in western Washington State by the U.S. Geological Survey and Pierce County. Acoustic flow monitors, which can detect the passage of a debris flow, were installed in valleys draining Mount Rainier.[40] A large debris flow from the mountain, similar to one that occurred about 500 years ago, would devastate more than a dozen communities and kill thousands or possibly tens of thousands of people.

Shallow wells can be drilled into slopes and monitored to signal when slopes contain a dangerous amount of water. In some regions, a rain-gauge network is useful for warning when a precipitation threshold has been exceeded and debris avalanches or debris flows are likely to occur. Surface movements can be monitored through periodic surveys of benchmarks on landslides. It is now possible to detect changes in positions of surface points, with an accuracy of a few millimetres or less, by using satellite global positioning technology.

6.7 Perception of and Adjustment to Landslide Hazards

Perception of Landslide Hazards

The common reaction of homeowners, when asked about landslides, is "It could happen somewhere else, but not here." Just as flood hazard mapping does not prevent development in flood-prone areas, landslide hazard maps do not prevent many people from moving into hazardous areas. Prospective hillside occupants who are initially unaware of the hazard may not be swayed by technical information. The infrequency of large landslides reduces awareness of the hazard, especially where evidence of past events is not readily visible. Unfortunately, it often takes a disaster to bring the problem to the attention of many people. In the meantime, people in many parts of North America and other areas continue to build homes in areas subject to future landslides.

Adjustments to the Landslide Hazard

Some people will continue to build and live on dangerous slopes, requiring that we make adjustments to minimize death and damage from landslides. Adjustments include locating critical facilities outside landslide-prone areas and implementing remedial corrective measures.

Siting of Critical Facilities As in the case of earthquakes (Chapter 3), safely siting critical public facilities, such as hospitals, schools, electrical-generating facilities, and police stations, is crucial. We must ensure that these buildings are not located on or directly below unstable hillsides.

Remedial Corrective Measures The best way to stop a slow-moving landslide is to attack the process that caused it. In most cases, the culprit is high water pressure within or below the slide. The pressure can be reduced by installing an effective drainage system. A reduction in water pressure increases the resisting force of the slope material, thereby stabilizing the slope.[16]

Personal Adjustments: What You Can Do to Minimize Your Landslide Risk

Consider the following advice if purchasing property on a slope:

- Seek an evaluation of the property by a professional geologist or engineering geologist.
- Avoid homes at the mouth of a valley or canyon, even a small one, because such sites may be in the path of debris flows.
- Consult local agencies, such as city or county engineering departments, that may be aware of landslides in your area.
- Look out for "little landslides" on the property—they usually get larger with time.
- If purchasing a home, look for cracks in walls; also look for retaining walls that lean or are cracked. Be wary of doors or windows that stick or floors that are uneven. Foundations should be checked for cracks or tilting. If cracks in the walls of a house or foundation can be followed across the ground beyond the structure, be especially concerned that the ground may be moving.

- Be wary of leaks in a swimming pool or septic tank, trees or fences tilted downslope, or utility wires that are taut or sagging.
- Avoid slopes with small springs; suspect ground that is wet or has unusually lush vegetation.
- Walk the property and, if possible, surrounding properties, looking for linear or curved cracks, even small ones, that might indicate instability.
- Look for the surface features that geologists use to identify potential landslides.
- Although it might be cost-effective to correct a potential landslide problem, the expense can still be considerable and much of the fix is below ground where you will never see the improvement. Overall, it is better not to purchase land that can move.

The presence of one or more of the above features does not prove that a landslide is present or that one will occur. However, further detailed investigation is warranted if any of these features are present.

REVISITING THE FUNDAMENTAL CONCEPTS

Landslides

❶ Hazards can be understood through scientific investigation and analysis.

❷ An understanding of hazardous processes is vital to evaluating risk.

❸ Hazards are commonly linked to each other and to the environment in which they occur.

❹ Population growth and socioeconomic changes increase the risk from natural hazards.

❺ Damage and loss of life from natural disasters can be reduced.

1. The potential for a slope to fail and a landslide to occur can be estimated by documenting the three-dimensional form of the slope, including its steepness; by determining its geology; and by knowing the amount, duration, and seasonal distribution of rainfall and snow that the slope receives. Weak planes in the rock mass, such as bedding planes, foliation, joints, and faults, are particularly important factors in landslides. Erosion or excavation of the toe of the slope and loading of the upper part of the slope may also destabilize it. Geologists look for evidence that a slope has failed in the past or is slowly creeping—for example ground cracks, tilted vegetation, and revegetated scarps. Any such evidence calls for caution when considering whether buildings, roads, or other structures should be built on the slope.
2. Today, landslide hazard assessments are conducted on and adjacent to sloping building sites before construction begins, at least in wealthy nations such as Canada and the United States. The assessments consider the likelihood and consequence of landslides, and therefore are elements of risk analysis.
3. Landslides are linked to severe weather, earthquakes, volcanoes, coastal and river erosion, and tsunamis. In most cases, landslides are the result, and not the cause, of other natural hazards. However, as the 1980 Mount St. Helens eruption (Chapter 5) and the 1929 Grand Banks tsunami (Chapter 4) showed, volcanic eruptions and tsunamis can be triggered by landslides.
4. As population increases, fewer suitable construction sites on level and gently sloping land remain. As a result, growth is accommodated by developing areas where slopes are steeper and potentially unstable. As the loss of life and homes in Rio de Janeiro over the past few decades indicates (Section 6.5), rapid growth of cities can lead to construction on steep, unstable slopes. Had people not been forced to live on such steep slopes in Rio, the loss of 200 lives from landslides in April 2010 could have been avoided.
5. Injury and loss of life from landslides can be minimized by identifying slopes that are unstable or potentially unstable and leaving them in a natural state, or by engineering them to increase their factor of safety. Evaluation of construction sites for potential landslides and remediation if landslides are possible before construction begins are standard practice in Canada. Surface runoff can be directed away from slopes, thereby minimizing the driving force that water provides to the mass wasting process. Slopes can be graded to decrease the slope angle, retaining walls can be constructed to support the slope, and debris basins can be built to catch debris flows before they reach developed areas.

Summary

Slopes are dynamic, evolving systems on which rock or surficial sediments may move downward at rates ranging from millimetres per year to several dozen metres per second. Slope failure involves the falling, sliding, toppling, or flow of materials; most landslides are characterized by combinations of sliding and flow.

The stability of a slope is determined by its geology, slope angle and height, climate, vegetation, water, and by time. The cause of a landslide can be determined by examining the relations between driving forces, which induce failure, and resisting forces, which oppose movement. The most important driving force is the weight of the slope materials, and the most important resisting force is the shear strength of the slope materials. The factor of safety of a slope is the ratio of resisting forces to driving forces. A ratio greater than one means that the slope is stable, whereas a ratio of one or less indicates that failure is likely. The type of rock, sediment, or soil forming a slope influences both the type and the frequency of landslides.

Water is an important factor in slope failure. Moving water in streams, lakes, or oceans may erode the base of slopes, increasing the driving forces. Excess groundwater increases both the weight of the slope material and the water pressure within the slope, which in turn decreases the resisting forces.

Landslides can occur anywhere there are slopes. In North America, they are most common in the mountainous areas of the west, the Appalachians, and the St. Lawrence Lowland. When landslides occur in populated areas, they can cause significant damage and loss of life. Landslides are linked to other hazards, especially floods, earthquakes, tsunamis, volcanic eruptions, and wildfires.

The effects of land use on slope stability range from insignificant to very significant. We should avoid developing areas where landslides occur independently of human activity or, alternatively, use appropriate protective measures. Where land use has increased the number and severity of landslides, we need to learn how to minimize their recurrence. Filling large water reservoirs alters groundwater conditions along their shores and may cause slope failure. Logging operations on weak, unstable slopes may also increase the incidence of landslides. Grading of slopes for development has created or increased landslide problems in many urban areas of the world.

To minimize landslide hazard and risk, it is necessary to identify unstable and potentially unstable slopes, adopt preventative measures, and correct problems when they arise. Hazardous and potentially hazardous slopes can be identified by mapping and monitoring. Grading codes, enacted in response to landslides triggered by cut-and-fill operations, have reduced landslide damage in urban areas. Prevention of large natural landslides is difficult, but careful engineering practices can minimize the risk in places where such slides are possible. Engineering techniques for landslide prevention include drainage control, proper grading, and construction of retaining walls and other support structures. Efforts to stop or slow existing landslides must attack the processes that led to failure—usually by initiating a drainage program that lowers water pressure in the slope. Even with all these measures, losses from landslides are expected to increase through the twenty-first century.

Most people perceive landslide risk as minimal, unless they have prior experience with landslides. Furthermore, hillside residents, like floodplain occupants, are not easily swayed by technical information. Nevertheless, the wise person will have a geologist inspect property on a slope before purchasing it.

Key Terms

colluvium (p. 166)
creep (p. 161)
debris avalanche (p. 166)
debris flow (p. 159)
driving forces (p. 163)
factor of safety (p. 163)
fall (p. 159)
flow (p. 159)
jack-strawed (p. 177)
landslide (p. 156)
mass wasting (p. 159)
resisting forces (p. 163)
rock avalanche (p. 158)
sackung (p. 161)
shear strength (p. 163)
slide (p. 159)
slump (p. 159)
subaqueous landslide (p. 162)
talus (p. 167)
thaw flow slide (p. 169)
topple (p. 161)

Did You Learn?

1. Define landslide.
2. Name and describe the three main mechanisms of landsliding.
3. Define factor of safety. Why is it important?
4. How do slumps (rotational slides) differ from soil avalanches and rockslides (translational slides)?
5. Explain how the angle of a slope affects its stability.
6. Describe why vegetation is important in slope stability.
7. What role does time play in slope stability?
8. Name the regions of Canada where landslides are most common.
9. How are landslides linked to other natural hazards?
10. Summarize the ways in which urbanization can increase or decrease the stability of slopes.
11. List surface features that indicate a slope might be unstable.
12. What can be done to prevent landslides?

Critical Thinking Questions

1. Engineering geologists are commonly asked to say whether a slope is safe to build on. In some cases, the possibility of a landslide is so large that the decision is easy. More commonly, however, there is a small possibility that the slope could fail in the months or years after construction. Because preventing development of the slope incurs an economic loss, what do you think is the tolerable risk threshold on which the decision should be based?
2. Your consulting company is hired by Parks Canada to determine landslide risk in Banff National Park, Alberta. Develop a plan of attack that outlines what must be done to achieve this objective.
3. Why do you think that few people are concerned about landslides? Assume you have been hired by a municipality to make its citizens more aware of the landslide risk on the steep slopes in the community. Outline a plan of action and defend it.
4. The snow-clad Cascade volcanoes of western North America experience numerous landslides, most of which are not triggered by volcanic eruptions. You have been hired to establish a warning system for subdivisions, businesses, and highways in valleys flanking Mount Baker volcano, near the Washington–British Columbia border. How would you design a warning system that will alert citizens to evacuate hazardous areas?

MasteringGeology

MasteringGeology **www.masteringgeology.com**. Looking for additional review and test prep materials? Visit the Study Area in MasteringGeology to enhance your understanding of this chapter's content by accessing a variety of resources, including **Self-Study Quizzes, Geoscience Animations, GEODe Tutorials, RSS feeds, flashcards,** web links and an optional **Pearson eText.**

CHAPTER 7

▶ **AVALANCHE!** A large snow avalanche races down a mountain slope in the Bow River Valley, Canadian Rockies. *(Clair Israelson)*

Snow Avalanches

Learning Objectives

The number of snow avalanche injuries and fatalities has been increasing because more people are participating in winter recreational activities, notably skiing and snowboarding. Some roads and rail lines in mountains are also at risk from snow avalanches. Your goals in reading this chapter should be to

- Understand the causes and triggers of snow avalanches
- Know the different types of avalanches
- Know the geographic regions where avalanches occur
- Recognize links between avalanches and other hazards, as well as the natural service functions of avalanches
- Understand how humans interact with avalanche hazards
- Know what can be done to minimize the risk from avalanches

The Chilkoot Disaster

The Klondike gold rush began in 1896 with the discovery of gold at Bonanza Creek, Yukon Territory. By 1898, thousands of prospectors were heading to the Klondike by almost every route imaginable. None of the routes, however, was shorter or cheaper than the Chilkoot Trail, which linked Skagway, Alaska, to the Yukon. As a result, this trail became the most popular way for prospectors to reach the Klondike.

The Chilkoot Trail extends from tidewater at Skagway, Alaska, into northwest British Columbia. The most difficult stretch of the trail is the approach to Chilkoot Pass, which, at 1067 m above sea level, marks the boundary between the United States and Canada. The approach to the pass, although not long, is steep, in places nearly 40°. The pass was crossed by thousands of prospectors during the spring and summer of 1898 (Figure 7.1).

Disaster struck some of these prospectors on April 3, 1898. Heavy snow had fallen throughout February and March of that year, but warm weather at the beginning of April melted the surface of the snowpack, producing icy conditions. Then, on April 2, the snow began to fall again. Guides knew conditions were perfect for avalanches and warned prospectors to stay away. The lure of gold was too strong, however, and some men continued up the pass in spite of the warnings. The first of a series of avalanches occurred at 2:00 A.M. on April 3, burying 20 prospectors. Another avalanche buried three more at 9:30 A.M. All were rescued, but fear of more avalanches convinced some 220 other people living and working on the trail to evacuate. Sometime between 11:00 A.M. and noon, a third massive avalanche thundered down the mountainside, covering an area of 4 ha, burying a group of evacuees under up to 9 m of snow.

About 1500 volunteers worked for four days to rescue the injured and recover bodies. The avalanche claimed 60 lives, making it one of the most widely reported events of the Klondike gold rush and one of the deadliest avalanches in North American history.

◄ **FIGURE 7.1 FORTUNE SEEKERS ASCENDING CHILKOOT PASS** Heavily laden prospectors trudge toward the summit of Chilkoot Pass during the Klondike gold rush. Sixty of the fortune seekers were killed by a large avalanche on April 3, 1898. *(E.A. Hegg/Library and Archives Canada)*

7.1 Introduction to Snow Avalanches

As you learned in Chapter 6, avalanches are rapid downslope movements of snow, ice, rock, or soil. In this chapter, we are concerned with **snow avalanches**, which are masses of snow, generally more than a few cubic metres in volume, that separate from the intact snowpack and slide or flow downslope. For convenience in this chapter, we use the word "avalanche" as a short form for "snow avalanche," while recognizing that ice, rock, and soil can also avalanche down steep slopes.

An avalanche may travel as a coherent block, deforming and fragmenting to some degree, or it may rapidly disaggregate into small particles that move independently of one another. Some avalanches are too small to bury a person; at the other end of the spectrum, some are larger than 1 million m^3 and are capable of destroying a village or many hectares of forest (Figure 7.2, Table 7.1). A 300 000 m^3 avalanche in Montroc, France, in 1999 crushed ski chalets under 100 000 t of snow 5 m deep, killing 12 people. During World War I, more than 60 000 soldiers died in avalanches during the mountain campaign in the Alps. Many of these avalanches were triggered by artillery fire.

Most avalanches occur in remote, uninhabited mountains during fall, winter, and spring, and consequently go unnoticed. Although generally not witnessed, avalanches are common—by one estimate, about 1.5 million avalanches large enough to bury a person occur annually in western Canada;[1] yet only about 100 avalanche accidents are reported in this region each year. If we assume that approximately 20 percent of all avalanche accidents are reported, it follows that only about 1 in 3000 potentially destructive avalanches injure people or damage property. This number may increase in the future as areas in which avalanches happen are further developed and used by winter recreationists.

▲ **FIGURE 7.2 LARGE AVALANCHE** A large avalanche moves down a chute in the San Juan Mountains, Colorado. It started when a slab of snow near the ridge crest suddenly failed. The avalanche descended 830 m to the valley floor and then climbed up the opposite slope, burying 245 m of U.S. Highway 550 beneath 1 m of snow. *(Mark Rawsthorne)*

Avalanches, like landslides, are driven by gravity. The two differ, however, in many important respects—for example, in the failure and transport mechanisms and in the

TABLE 7.1 Avalanche Size

Size	Run-Out	Potential Damage	Length and Volume
Sluff	Small snow slide that normally does not bury a person	Relatively harmless	Length < 50 m Volume < 100 m^3
Small	Avalanche stops on the slope	Can bury, injure, or kill a person	Length < 100 m Volume < 1000 m^3
Medium	Avalanche runs to the bottom of the slope	Can bury and destroy a car, damage a truck, destroy small buildings, or break trees	Length < 1000 m Volume < 10 000 m^3
Large	Avalanche runs out over areas significantly less steep than 30°; may reach the valley bottom and run up the lower part of the opposing slope	Can bury and destroy large trucks and trains, large buildings, and forested areas	Length > 1000 m Volume > 10 000 m^3

physical properties of the failed material. Our discussion of avalanches begins with a consideration of snow climatology, proceeds to avalanche initiation, and then to a discussion of avalanche motion.

Snow Climatology

Snowfall and snow accumulation depend on the season and on geographic factors such as latitude, altitude, and proximity to an ocean or large body of freshwater. Snow accumulates when temperatures are at or below freezing. The length of the snow season depends mainly on latitude and altitude. Snow may fall and remain on the ground for only a few weeks at low elevations in the mid-latitudes or for almost 12 months at high latitudes or in high mountains. In general, snowfall is rare between 35° N latitude and 40° S latitude, except at high elevations. Unusually heavy snowfalls occur around some large lakes, notably the Great Lakes. Parts of southern Ontario and upstate New York are located in "snow belts" bordering the Great Lakes, which are the source of much of the precipitation that falls there as snow.

The amount of snow on the ground depends on many factors, the most important of which are the slope of the land, elevation, amount of snowfall, and winds. Snow accumulates on slopes less than about 45°; it sloughs away on steeper slopes. Temperature decreases with altitude; thus high mountains, even near the equator, have permanent snow cover. Examples include Mount Kilimanjaro in Tanzania and Volcán Cayambe in Ecuador. The cold air masses of the Arctic and Antarctic hold little water vapour; thus these regions receive little snowfall. The snow that is present, however, does not melt at sea level in parts of these regions. Similarly, some mountains in Bolivia, Chile, and Argentina are high (4500–6900 m above sea level) and cold, but they lie east of the Atacama Desert, are hyper-arid, and consequently receive little snow. Winds redistribute snow, producing build-ups, or *slabs*, that are unstable. Thus, the amount of snow on the ground can differ considerably over short distances.

Avalanche Initiation

Some avalanches, termed **point-release avalanches**, begin with failure of a small amount of loose fluffy snow and grow as they move downslope. The sliding snow causes failures in the adjacent snowpack, producing a distinctive, downslope-widening trough, like an inverted V (Figure 7.3). Point-release avalanches commonly happen after heavy

▲ **FIGURE 7.3 POINT-RELEASE AVALANCHE** A point-release avalanche, as the name implies, results from initial failure of a small amount of snow. More snow becomes incorporated into the avalanche as it moves downslope, giving rise to the distinctive inverted-V shape seen in this example. *(B. Jamieson/Geological Survey of Canada)*

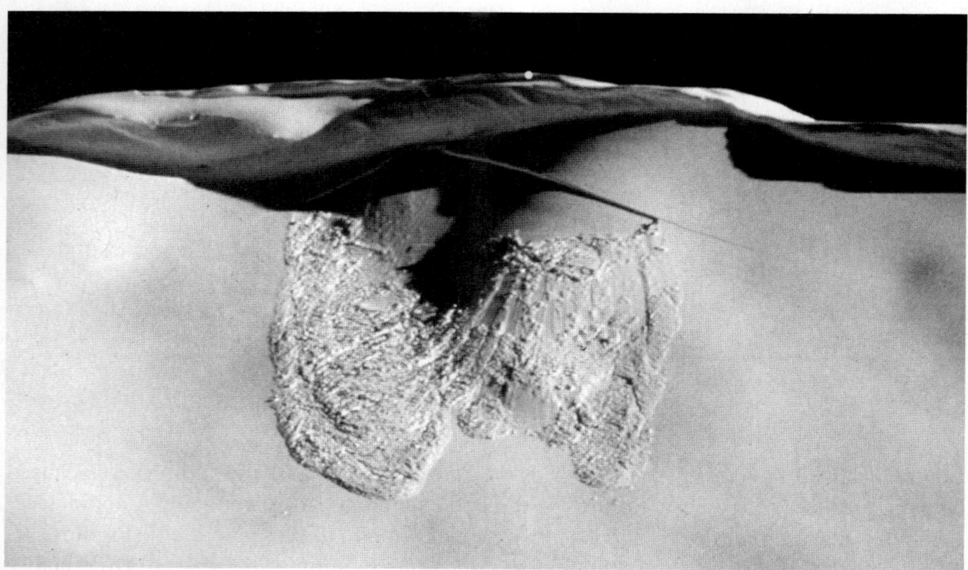

◄ **FIGURE 7.4 SLAB AVALANCHE** This avalanche occurred when a slab of snow slipped along a weak layer, shearing away from near-vertical crown and lateral scarps. The slab rapidly disintegrated as it moved downslope. *(B. Jamieson/Geological Survey of Canada)*

snowfall. The thick, loose snow is unstable, both because of its mass and because the snow crystals have little time to bond.

Layers of cohesive snow can fail as a slab, leaving steep crown, lateral, and toe scarps, and a smooth basal failure plane (Figure 7.4). A **slab avalanche** begins with fracturing of the snowpack along a weak layer at depth. Gravity causes the snowpack to creep downslope, with the top of the snowpack moving faster than the bottom. If a weak layer is present (Figure 7.5), it may deform, or shear, more than the layers above and below it. This weak layer may slowly deform over hours or days and be followed by sudden failure. The initial point failure propagates along the weak surface, causing a slab to break away along bounding crown and lateral fractures[2] (Figure 7.4). Slab avalanches are very dangerous; thus backcountry recreationists and avalanche forecasters look for weak layers when assessing local avalanche risk.

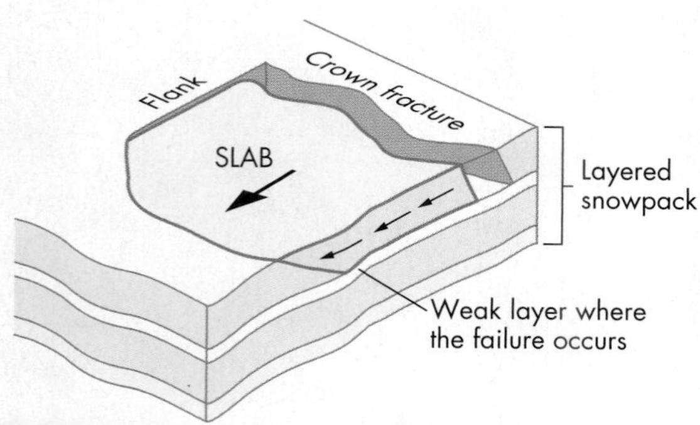

▲ **FIGURE 7.5 SLAB AVALANCHE WEAK LAYER** Slab avalanches begin when snow fails along a mechanically weak layer parallel to the surface. This diagram shows some of the terms used by avalanche scientists and safety personnel when referring to slab avalanches. *(Reproduced or adapted with the permission of Natural Resources Canada 2013, courtesy of the Geological Survey of Canada (Bulletin 548))*

Weak Layers

The structure of the snowpack determines its potential to avalanche.[3,4,5] Slab avalanches require a buried weak layer and an overlying stronger slab. The relationship between avalanche potential and properties of snow that can be easily measured—strength, grain size, grain type, and temperature—are complex and not fully understood. Additional complexity results from the fact that the thickness and characteristics of the snow cover vary in both time and space. A few generalizations can be made, however.

New snow that has not had time to bond to the layer below, especially if it is light and powdery, is susceptible to sliding. Compacted snow is less likely to move than light powdery snow. Snow that is above the level of boulders and plants on a slope has no natural objects to anchor it and is therefore more dangerous than thinner snow. Of course, snow that completely buries surface irregularities is just what skiers and snowboarders desire.

Weather determines the evolution of the snowpack. Important meteorological factors include heating by solar radiation, radiative cooling, temperature gradients in the snowpack, and snowfall amounts and type.[6] Most avalanches happen during or soon after a storm.

Weak layers in the snowpack can form in three main ways:

1. ***Wind***

   Blowing snow can build up on sheltered lee slopes, and wind can stabilize the snowpack on other slopes. A **wind slab** is a body of thick, poorly bonded snow deposited on a slope by wind. Snow can be deposited by winds rising over the crest of a slope or the top of a mountain. The leeward slope, where the snow accumulates, is said to be *top-loaded*. Wind blowing parallel to a ridge crest can also deposit snow, producing *cross-loaded* wind slabs. Wind may increase during a snowstorm, depositing a dense layer of broken and packed snow crystals on a delicate layer of unbroken crystals. The layer of unbroken crystals becomes a weak buried surface along which failure may occur later.

2. ***Formation of hoar at depth in the snowpack***

   Hoar consists of ice crystals that are deposited on and within the snowpack when the air is moist and cold. Layers of hoar generally have lower strength than the rest of the snowpack and thus cause many avalanches. Hoar forms when water vapour condenses directly into ice, without going through an intermediate liquid phase. Depth hoar forms from air occupying spaces between snow crystals. Its depth depends on the temperature profile in the snow, although most layers of depth hoar occur near the base of the snowpack. Cold air on a thin snowpack is a particularly favourable situation for developing depth hoar.

3. ***Formation of hoar at the surface***

   Surface hoar, also called *hoar frost*, consists of ice crystals that form at the surface of the snowpack on cold clear nights. The ice crystals, which may be feather-shaped and larger than snowflakes, change only very slowly once buried. The overlying and underlying snow layers commonly gain strength over days and weeks, leaving the buried surface hoar as a weak layer.

Avalanche Motion

During the first few seconds of a slab avalanche, the failed snow mass is a coherent slab comprising fractured blocks of snow. Within a few tens of metres, the slab disintegrates into smaller fragments and then individual grains of snow.[2] At velocities of about 35 km/h, dry avalanches generate a cloud of powdered snow that billows above the flowing mass (Figure 7.2). The powder cloud is much less dense (3–15 kg/m^3) than the flowing snow (50–150 kg/m^3). Wet avalanches, unlike dry avalanches, contain intergranular liquid water and are denser than dry avalanches (300–400 kg/m^3), but do not achieve the high velocities of some large dry avalanches.[1] Some dry avalanches have been clocked at 200 km/h, leaving almost no time for a person downslope to get out of the way. These avalanches may have sufficient momentum to climb opposing slopes and destroy the forest on them. They may also displace air, causing a damaging air blast that arrives seconds before the avalanche.

Avalanche Triggering

Most avalanches occur naturally during or soon after snowstorms. Others happen when normal daytime heating or an inflow of warm air raises the temperature of the upper part of the snowpack.

In most recreational accidents, a person triggers the avalanche.[7] Field studies indicate that when the snowpack is near the threshold of failure, slab avalanches can be triggered simply by the extra weight of a person traversing the slope. The person's weight increases the shearing force in the weak layer below, triggering failure.

Some avalanches are triggered intentionally, generally with explosives, as part of avalanche-control programs. Some skiers attempt to release unstable snow in a controlled manner, which is an inadvisable practice.

Terrain Factors

An avalanche path has three parts:

1. The **start zone**, where the snowpack fails
2. The **track**, along which the avalanche accelerates and achieves its highest velocity
3. The **run-out zone**, where the avalanche decelerates and snow is deposited

The three parts of the path are easy to identify where avalanches follow ravines, gullies, and *chutes* on forested slopes (Figure 7.6). They are much more difficult to distinguish on slopes above the treeline or where the snow is not channelled by topography.

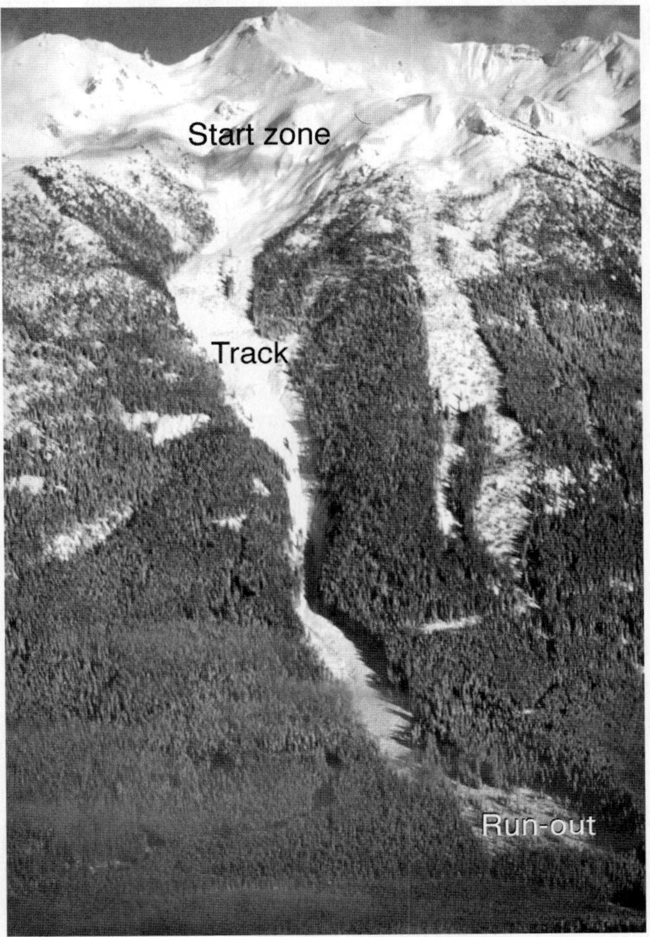

▲ **FIGURE 7.6 COMPONENTS OF AN AVALANCHE PATH** A large avalanche path with a start zone, a track extending through forest, and a run-out zone on the valley floor. *(B. Jamieson/Geological Survey of Canada)*

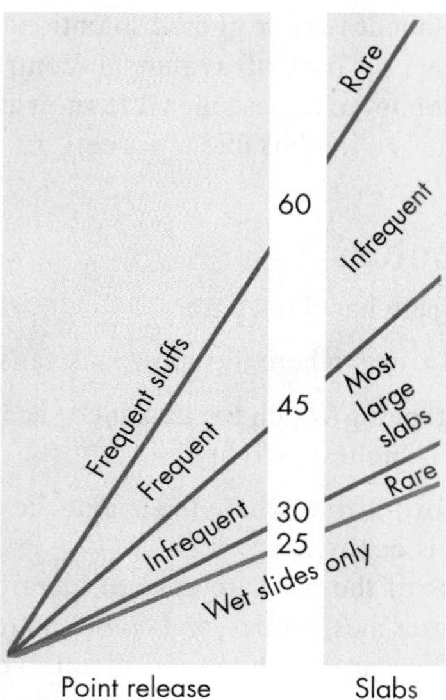

▲ **FIGURE 7.7 RELATION BETWEEN SLOPE ANGLE AND AVALANCHE TYPE AND FREQUENCY** Slopes less than 25° and steeper than 60° have a low avalanche risk. Little snow accumulates on steep slopes, and snow does not easily slide on gentle ones. Most large avalanches happen on slopes of 30° to 45°. Wet avalanches, however, are more common on slopes less than 30°. *(Reproduced or adapted with the permission of Natural Resources Canada 2013, courtesy of the Geological Survey of Canada (Bulletin 548))*

Slope Angle The most important terrain factor for avalanche formation is slope steepness. Numerous, small, loose avalanches, called **sluffs**, occur on slopes steeper than 60° (Figure 7.7). Most sluffs are too small to bury a person, but they can cause climbers to fall or they can trigger larger avalanches on slopes below.

Most large avalanches are released from slopes between 30° and 45°, although some happen on slopes of 45° to 60°.[7] Slopes between 30° and 45° are favoured by skiers, snowboarders, and snowmobilers, which is why recreationists commonly trigger avalanches.

Fewer than 5 percent of dry avalanches occur on slopes less than 30°. Wet snow slides, however, can happen on slopes that are less than 25°[1] (Figure 7.7), because the snow layer contains liquid water between the grains of recrystallized snow and therefore is heavier.

Orientation of Slope The location of a slope with respect to wind and incident sunlight affects avalanche occurrence. Slopes in the lee of wind can accumulate large amounts of snow in wind slabs and cornices, increasing the avalanche danger. Deposits of wind-blown snow on leeward slopes commonly consist of interleaved stronger and weaker layers, reflecting variable wind speeds during deposition. This snow stratigraphy creates ideal conditions for slab avalanches. Slopes facing the sun are commonly more prone to avalanches during sunny, warm weather, whereas shaded slopes are more likely to release avalanches in cold weather. In middle latitudes in the Northern Hemisphere, more avalanches happen on shady slopes with northern and northeastern aspects than on slopes with other orientations.

Other Terrain Factors Several other terrain factors play a role in triggering avalanches.[8,9] Convex slopes are statistically more dangerous than concave ones. The reasons for this difference lie partly in human behaviour—skiers and snowboarders favour convex slopes over concave ones—and partly in the fact that the *tensile strength* of snow is lower than the *compressive strength*. Avalanches are also more common on slopes with smooth surfaces, such as grass or smooth rock, than on treed or rough slopes. Rough surfaces, such as talus slopes, must accumulate enough snow to bury surface irregularities before dangerous avalanches can occur. Vegetation may anchor the snowpack, although in some situations buried vegetation can create weak zones within it. Avalanches rarely initiate within tree-covered areas, but they do enter and destroy forests (Figure 7.8).

▲ **FIGURE 7.8 DOWNWED TREES AVALANCHE PATH COLORADO** Avalanches generally initiate in non-forested areas, but they can run into and destroy mature forest. *(© Aurora Photos/Alamy)*

The avalanche hazard is commonly greater at high elevations because snowfall and winds are greater there and trees are fewer or absent. Gullies or ravines can have dangerous wind slabs at their heads or margins. These features also funnel avalanches, increasing their destructive force and making escape difficult.

7.2 Geographic Regions at Risk of Avalanches

Avalanches can occur almost anywhere with snow and a sufficiently steep slope for the snow to slide on. Generally, however, the snowpack must be at least 50 cm thick; areas with deeper snow will produce more and larger avalanches.[1]

Avalanche activity in Canada is greatest in the mountains of Alberta, British Columbia, and Yukon Territory (Figure 7.9). Avalanches are also common in parts of Newfoundland and Labrador, in the Gaspé region of Quebec, along the north shore of the St. Lawrence River, and along the east margin of the Arctic Islands. They are less common in central and eastern Canada and rare on the Prairies and in the Arctic. Most avalanches in the United States occur in the high mountains extending from the Sierra Nevada and Cascade ranges in the west to the Rocky Mountains in the east.

▲ **FIGURE 7.9 AVALANCHES IN CANADA** This map shows the regional occurrence of avalanches in Canada. It is necessarily generalized and does not take into account, for example, isolated steep areas and isolated areas of heavy snowfall or strong winds. AMSD is annual mean snow depth. *(Reproduced or adapted with the permission of Natural Resources Canada 2013, courtesy of the Geological Survey of Canada (Bulletin 548))*

7.3 Impacts of Avalanches and Links with Other Natural Hazards

Impacts of Avalanches

More than 600 people have died in avalanches in Canada since the earliest reported accidents in the mid-nineteenth century.[1] Until the early twentieth century, most avalanche fatalities were people killed while building railways or roads, or working at mine sites (Figure 7.10). The greatest loss of life occurred in the late 1800s and early 1900s. During construction of the Canadian Pacific Railway and in the early years of its operation, "white death" avalanches claimed approximately 250 railway workers in Rogers Pass in Glacier National Park, British Columbia.[10] The worst accident was in March 1910, when a large avalanche killed 62 members of a work crew near Mount Cheops. In response to this and previous accidents, the 8 km Connaught Tunnel was constructed beneath the mountain pass to bypass some of the most dangerous avalanche paths.

The number of industrial and transportation fatalities is now low, largely because measures are routinely taken to protect roads and rail lines from avalanches. In contrast, the number of recreational accidents has increased dramatically since the 1930s. On average, 12 people have died in avalanches in Canada each year over the past decade; most fatal accidents have involved recreationists who were buried in avalanches that they or a member of their group triggered.

A tragic example involving snowmobilers happened in December 2008 in the Flathead Valley, 20 km east of Fernie, British Columbia. A party of 11 men was caught in two avalanches in Harvey Pass, a popular local backcountry snowmobile destination. The first avalanche buried part of the group, and it was followed by a second avalanche that buried the remaining men as they tried to assist their friends. Two of the buried riders were able to free themselves from the snow within about 20 minutes, and they used their avalanche beacons to locate and rescue a third man after an additional 20 minutes of digging. The remaining eight men died. All the victims were residents of the small town of Sparwood, just north of Fernie. The disaster prompted a call for more awareness of avalanche safety in Canada. It also brought attention to the increase in recent years of fatal accidents involving snowmobile operators. The increase in accidents is due partly to the tremendous growth in this winter recreational activity and partly to the dangerous behaviour of some snowmobile operators, including the practice of "high-marking," in which the operator rides up a steep slope as far as possible, then turns the machine around without rolling it or losing power and returns to the base of the hill.

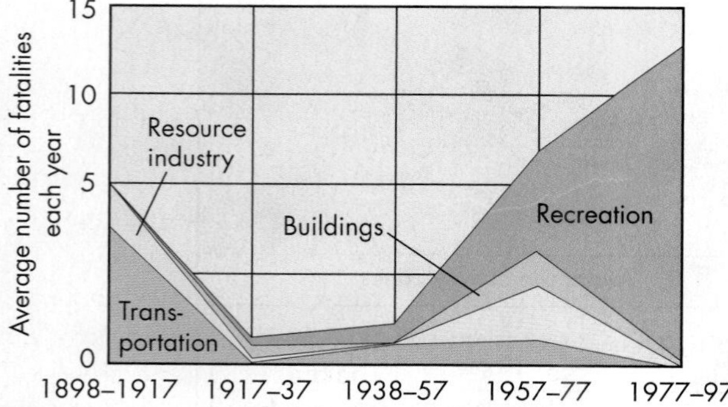

▲ **FIGURE 7.10 AVALANCHE DEATHS IN CANADA** The distribution of avalanche fatalities in Canada according to the activity of the victims at the time of the accident. Categories include recreation (skiing, snowboarding, and snowmobiling); transportation, including highway and railway construction; resource industry, including accidents at work camps; and residential and commercial buildings. Some accidents have been excluded because they do not fit any of these categories. In the past 50 years, the number of recreational accidents has increased dramatically, whereas the number of other accidents has decreased. *(Reproduced or adapted with the permission of Natural Resources Canada 2013, courtesy of the Geological Survey of Canada (Bulletin 548))*

Avalanches also cause traffic delays, with significant economic losses.[11] For example, the Trans-Canada Highway in Rogers Pass, British Columbia, is closed an average of about 100 hours each winter because of avalanches. A typical two-hour closure of the highway can result in monetary losses of $50 000 to $90 000, depending on the number of trucks that are delayed.[1] Thus, the average annual cost of avalanches at Rogers Pass solely because of traffic delays is several million dollars. This sum does not include the costs of preventing avalanches and clearing the highway during closures. Taking into account the many other highways in British Columbia affected by avalanches, the total annual cost of traffic delays in that province alone exceeds $5 million. This figure is conservative because it does not include railway delays or the effect of closures on timely deliveries and lost business.

Property damage from avalanches in Canada in most years is less than $500 000, although much damage probably is unreported. Damage from avalanches in the European Alps in some winters is much larger. Typical infrastructure damage includes residential and commercial buildings at ski resorts, ski lifts, and vehicles.

Avalanches damage forests by uprooting, breaking, and injuring trees (Figure 7.8). They also remove soil required for forest regeneration. Individual avalanches can destroy or damage up to $500 000 worth of timber.[1] Clearcut forestry practices can make the problem worse by creating new source areas; avalanches initiated in the cut-block may run into commercially valuable forests. In some cases, structures can be installed to support the snowpack and thus inhibit avalanches, but this solution is expensive and not always successful.

Links between Avalanches and Other Natural Hazards

Avalanches have few direct links to other natural hazards. Earthquakes in mountainous areas in winter, however, can trigger numerous avalanches, as happened during the **M** 7.9 Denali, Alaska, earthquake in November 2002.

◄ **FIGURE 7.11 ECOLOGICAL BENEFITS OF AVALANCHES** Avalanches are an important shaper of vegetation in high mountain valleys, such as this one in Kananaskis Valley in the Alberta Rocky Mountains. They renew vegetation on the valley walls seen in this photograph. *(John J. Clague)*

Some earthquake-triggered avalanches may block roads and rail lines, impeding rescue and relief efforts. Perhaps of greater importance is the indirect link between changing climate and avalanche frequency. Climate change may increase winter snowfall in some areas or increase the severity of winter storms, which can alter avalanche frequency. Some regions may experience more variable winter weather, with frequent thaws that also can change avalanche frequency.

7.4 Natural Service Functions of Avalanches

Avalanches provide the same ecological benefits as landslides (Chapter 6). They are a source of ecological disturbance that can increase local plant and animal diversity. In many mountain valleys, avalanches maintain broad strips of herbaceous and shrub vegetation between closed subalpine forests that, collectively, can cover more than half the valley walls (Figure 7.11). These open areas are important to many animals, including grizzly bears, marmots, pikas, and ground squirrels, and to a variety of subalpine plants.

7.5 Human Interaction with Avalanches

Avalanches are common in all but the driest mountain ranges on Earth. They are a hazard only when humans share their space. In long-settled mountains, such as the European Alps, people have adjusted to avalanches by avoiding them as much as possible, for example by establishing settlements outside areas prone to avalanches. More recently, driven by the boom in leisure activities and tourism, settlement has greatly expanded in the Alps, encroaching into avalanche-prone areas, often with disastrous results. In less settled mountains, such as those of western North America and southern New Zealand, recreational avalanche accidents were once uncommon. That situation has changed with the rapid growth in outdoor winter recreation in these areas (see Case Study 7.1).

7.6 Minimizing Avalanche Risk

Avalanche risk can be reduced by placing buildings, roads, and other infrastructure outside dangerous areas; by using engineered structures to slow or deflect avalanches; by reinforcing exposed structures; by triggering controlled avalanches with explosives; and with forecasting.

Risk can be mitigated by appropriately locating permanent structures. If a risk remains after structures have been built, other measures must be taken. Protection along vulnerable transportation corridors involves forecasting and a control program that includes temporary road and rail closures. Deflecting structures and reinforcement can provide additional protection for fixed structures.

Location of Infrastructure

Avalanche risk can be estimated by determining the distribution, frequency, and sizes of avalanches in a given area. Hazardous areas can be shown on maps (Figure 7.12) that provide planners information they require to locate buildings, roads, and other structures in areas where the hazard is

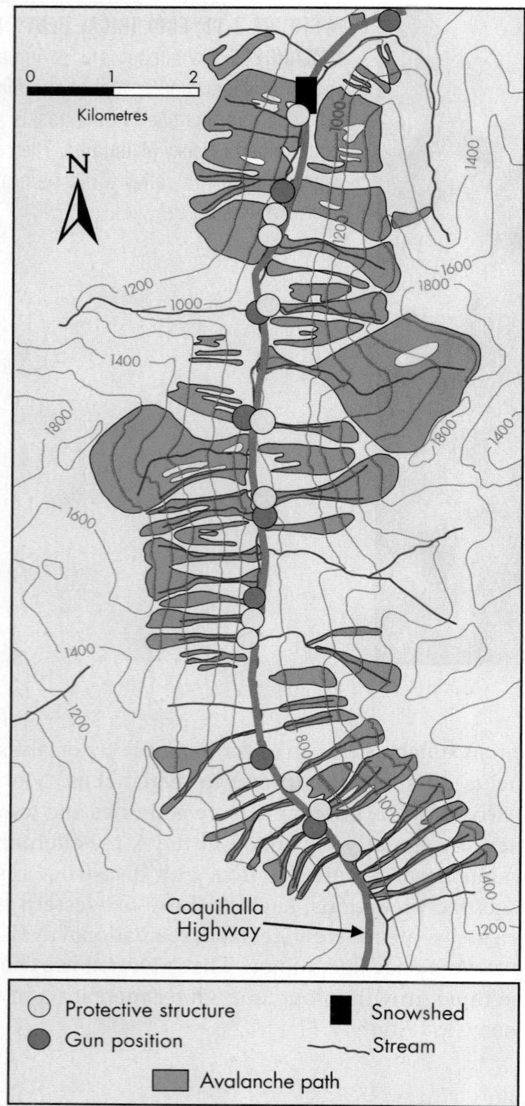

▲ **FIGURE 7.12 MAP OF AVALANCHE PATHS** Avalanche paths are shown in pink on this map of a section of Coquihalla highway between Hope and Merritt, British Columbia. Protective structures were built to reduce the chance of avalanches reaching the highway. The green lines are contours, which are lines of equal elevation; in this case, the vertical spacing between contours is 200 m. *("Geoscape Vancouver: Mountain corridors," www.geopanorama.rncan.gc.ca/vancouver/mountain_e.php © Natural Resources Canada. All rights reserved)*

low or where mitigation measures can be taken to reduce the risk. Avalanche frequency, like the frequency of floods and earthquakes, is commonly described in terms of a *recurrence interval*, the average time between avalanches of a particular size or extent.

In Switzerland and some places in the United States, new residential buildings are generally not allowed in a *red zone*, defined as the area where avalanches capable of impact pressures of 30 kPa (kilopascals) or more occur, on average, once every 300 years. Some jurisdictions have a *blue zone* that lies outside the red zone and comprises areas where avalanches with recurrence intervals of 30 to 300 years might produce impact forces of no more than 30 kPa. Legislation may require that buildings located in the blue zone have reinforced walls facing the avalanche path or that they be protected by deflection structures. No special engineering is required outside the blue zone.

In Canada, a single line is commonly drawn on a map to delineate areas at risk from avalanches. Areas inside this line are subject to avalanches with a specified recurrence interval, such as 300 years. Currently, however, Canada has no consistent policies for identifying and delineating avalanche hazard zones.

Structures in the Start Zone

Fences or nets are installed in some avalanche start zones to support the snowpack and prevent large avalanches (Figure 7.14). Such structures are considered unsightly by some people and are also expensive. They are only practical for protecting inhabited structures, busy roads, and critical infrastructure. Smaller, less expensive structures may allow re-growth of forest along avalanche paths. Europeans have considerable experience with start-zone structures, but such structures are not commonly used in North America.

Structures can be built that reduce the accumulation of wind-blown snow in the avalanche start zone. These structures are sophisticated versions of snow fences seen in places along highways. Snowdrifts accumulate along the fences, reducing the amount of snow that reaches the avalanche start zone. Other structures, resembling upward-sloping roofs, carry wind-blown snow below the normal start zone.

Structures in the Track and Run-Out Zone

A variety of structures are used to slow or deflect avalanches in the track and run-out zone. They include deflecting berms, splitting wedges, avalanche sheds, and mounds. *Berms* deflect avalanches away from buildings or parallel to roads (Figure 7.15). Soil or concrete *splitting wedges* are located on the exposed sides of buildings to force the avalanche around the structure. *Avalanche sheds*, or *galleries*, allow avalanches to run over roads or railways (Figure 7.16). *Mounds*, somewhat reminiscent of large "moguls" along ski runs, are constructed to slow avalanches and thus reduce their run-out. Unfortunately, mounds may become buried early in the winter, allowing subsequent avalanches to travel unimpeded over them.

Control through the Use of Explosives

Explosives are used to release accumulations of snow. Highways, railroads, and ski runs are temporarily closed during this procedure. Explosive charges are projected from compressed-air cannons, fired by military artillery, or dropped from helicopters on ridges. Approximately 30 000 charges are

7.1 CASE STUDY

Deadly Avalanche in Glacier National Park

A huge avalanche killed seven young backcountry skiers in Glacier National Park, British Columbia, on February 1, 2003 (Figure 7.13). The accident highlights the risks that even well-prepared people face when they venture into the backcountry during winter.

The skiers were a group of 14 Grade 10 students and three adults from a private school south of Calgary, Alberta. The three adults included two teacher-guides and a volunteer; all three had backcountry experience.

The group had spent the afternoon before the accident at an alpine cabin 20 minutes from the Trans-Canada Highway. They skied near the cabin and, supervised by Andrew Nicholson and Dale Roth, avalanche-certified teachers, dug avalanche pits, did snowpack testing, and performed compression tests on nearby slopes. They set out storm boards to collect the overnight snowfall and, in the morning, compared the samples with samples of old snow already on the ground. They then skied to the park visitor centre, where Nicholson talked with officials about snow conditions and was given a daily bulletin that included weather conditions, satellite imagery, and avalanche danger ratings. The report for February 1 stated that, below the treeline, where the group planned to stay, the avalanche danger was "Moderate—Natural avalanches are unlikely. Human triggers are possible." A snow layer deposited on December 6 suffered "compression test failures" according to the daily report. Another layer, deposited on January 20, "easily released" during tests. The deep November layer, comprising two laminated crusts of ice bounding a layer of unstable crystals, had worried avalanche forecasters all winter.

The group set out up a valley near Rogers Pass, one of the most avalanche-prone areas in western Canada. The Canadian army routinely fires howitzers in Rogers Pass in winter to trigger controlled avalanches and keep the highway and railway open. During the ascent, the group followed the standard practice for travelling in avalanche zones by maintaining 10 to 15 m spacing between pairs of skiers. About 15 minutes into the trip, the guides stopped and quizzed each student on avalanche safety protocol. However, their route offered no protection if an avalanche occurred on the high slopes above them, which included the 2600 m peak of Mount Cheops. Aware of the considerable avalanche danger, the guides conferred with the students, who wanted to continue to ski for Balu Pass. Nicholson and Roth made the final decision to proceed.

▲ **FIGURE 7.13 KILLER AVALANCHE** A view to the west up the valley of Connaught Creek in Glacier National Park, British Columbia. The site of the avalanche that killed seven young skiers from Alberta on February 1, 2003, is shown by the yellow arrow. *(Kip Wiley)*

Rich Marshall and his wife, Abby Watkins, professional guides from Golden, British Columbia, watched the group ski up the valley. They had stopped for tea in some trees at an elevation of about 1680 m and could see the students about 100 m below them. At 11:45 A.M., just as the group was skiing past Mount Cheops on the valley bottom, something, perhaps the weight of snow blown over the shoulder of the mountain, became too much for the January snow layer. It failed and avalanched down the mountain. This avalanche, by itself, would have just dusted the valley bottom. Unfortunately, it caused a slab about 0.5 km wide to fail along the weak November crust, and approximately 900 t of snow cascaded down the slope and onto the valley floor. "We see an avalanche of that magnitude at least once a year, but usually not in that pass," said Eric Dafoe, a public safety coordinator for Parks Canada, "and usually during a big storm period when no one is around."

Marshall was closing his thermos at 11:45 when he heard a sharp crack from across the valley. He saw the snowpack at an elevation of approximately 2400 m, not far below the summit of Mount Cheops, give way and begin to descend toward the students and their guides. "Avalanche! Avalanche! Avalanche!" he yelled.

The students had only seconds before the avalanche, moving with enough force to flatten 4 ha of

(continued)

7.1 CASE STUDY (*Continued*)

forest, struck them. First came a "wall of snow," one student later told wardens, and then "blackness." Marshall and Watkins were dusted by the avalanche and then sped toward the group. Investigators would later conclude that the couple saved at least five lives.

Each student carried a shovel and a probe and wore an *avalanche transceiver*, which emits a beeping signal that can be received by another beacon. The closer Watkins got to a buried student, the louder and more frequent were the beeps. She and Marshall moved quickly, but some of the group had been moved more than 200 m by the avalanche. Once free, some of the survivors tried to dig out their friends. Nicholson, who was carrying a satellite phone, called the Rogers Pass warden, and within 40 minutes, 10 rescuers were on the scene. The number grew quickly to 40 and included park staff, mountain guides, military personnel, and heli-ski guides. But seven of the students were too deeply buried. It would be 1 hour and 20 minutes from the time the avalanche hit before the last body was recovered.

In the days following the disaster, grieving parents questioned the decision to continue to Balu Pass. If the chances of an avalanche were "considerable" just above where the students were to ski, why take the risk? Strathcona-Tweedsmuir School head Tony Macoun insisted that the risks were weighed by the entire group and that every reasonable precaution had been taken. "They were as prepared as they could have been," said Ingrid Healy, an assistant head at the school, "as *anyone* could have been."

used each year for avalanche control in Canada, about half in ski areas.[1]

Explosions in start zones usually, but not always, release avalanches. They are no guarantee that the snowpack will not fail naturally or be triggered by a person soon afterward.

Forecasting

An **avalanche forecast** is an assessment of the probability and size of avalanches within a defined area under existing or future conditions.[12,13] Forecasts are based on four types of information:

1. ***Occurrences of avalanches***

   Observed occurrences of avalanches provide an indication of the likelihood that additional avalanches will occur under similar conditions.

2. ***Stability and strength tests***

   Several field tests have been developed to assess snowpack stability. All these tests are subjective,

▲ **FIGURE 7.14 AVALANCHE SUPPORT STRUCTURE** Support structures in an avalanche start zone above Davos, Switzerland. Such structures prevent large avalanches from starting, but they are expensive and are generally used only where people and property are at high risk. *(B. Jamieson/GeologicalSurvey of Canada)*

◄ **FIGURE 7.15 PROTECTION AGAINST AVALANCHES** Avalanche braking mounds and catch dam at Neskaupsta∂ur, Iceland. *(© Tómas Jóhannesson, Icelandic Meteorological Office)*

▲ **FIGURE 7.16 AVALANCHE SHED** An avalanche shed on the Coquihalla highway, west of Coquihalla Pass, British Columbia. The shed was constructed at a cost of CAD$12 million to carry avalanches over the highway. *(John J. Clague)*

and results will differ for different users. Nevertheless, they provide some indication of the likelihood that the snowpack may fail. The *shovel test* is a traditional field test that involves pulling the upper part of a column of undisturbed snow about 0.3 m^2 with a shovel. Fractures produced along weak snowpack layers by the force of the shovel identify potential failure planes for slab avalanches. The *compression test* involves the application of a vertical force, commonly with the back of a shovel blade, on a column of undisturbed snow similar to that used in shovel tests. Any weak layers will fracture along the exposed vertical face of the column (Figure 7.17). In the *rutschblock test*, a skier steps onto a column of undisturbed snow, 2 m across the slope and 1.5 m along the side. The skier pushes on the column with his or her skis and then jumps on it. A score of 1 to 7 is assigned to the snowpack based on the loading required to release a block from the upper part of the column.

3. ***Snowpack observations***

   Forecasting also involves observations of snowpack characteristics, including the thickness and properties of visible weak layers.

4. ***Weather***

   An important aspect of avalanche forecasting is tracking weather conditions, including temperature, precipitation, and wind.

▲ **FIGURE 7.17 SNOW COMPRESSION TEST** A field compression test is used to locate weak layers in the snowpack and provide an index of stability. Note the weak layer that has fractured to the right of the man's elbow (dark band). *(B. Jamieson/Geological Survey of Canada)*

Observations of recent avalanches provide the most direct evidence of avalanche risk. Weather observations provide the least direct evidence.

Observations of recent avalanches, weak snowpack layers, heavy snowfall, and warming are important to forecasters because they are avalanche indicators. If few or no avalanches have occurred, if field tests suggest the snowpack is stable, and if the weather is cold and clear, the danger of avalanches is low and a forecaster will probably allow a highway or ski area to remain open. However, forecasters must commonly make decisions based on contradictory indicators—some suggesting stability and others indicating instability. They consider all the data, weigh some observations more than others, and assess uncertainties and the risk associated with different forecasts. Their decisions are ultimately based on experience and training. Forecasts are constantly revised as new information becomes available.

Computer software has been developed to assist forecasters. Most algorithms compare current measurements of precipitation, temperature, wind, and other variables with average values of the same variables spanning many years. Avalanches would be considered unlikely if most past days with similar conditions were avalanche-free.

Another method of forecasting involves rule-based or expert systems that incorporate the knowledge of avalanche experts.[12] Rules are constructed about the effects of snow stratigraphy, precipitation, wind, temperature, field tests, and other factors to establish a measure of avalanche danger. For example, the avalanche danger might be increased by one unit if wind speed exceeds 25 km/h or decreased by one unit if the temperature drops below 0°C. No prior data are required to apply such rules, but the rules must be tested with current data and revised if necessary.

Modelling

Attempts to model snow avalanches date back to the early twentieth century, notably the pioneering work done in preparation for the 1924 Winter Olympics in Chamonix, France. Major advances in modelling the phenomenon were made by A. Voellmy in his book *Ober die Zerstorunskraft von Lawinen (On the Destructive Force of Avalanches)*, published in 1955. He used a simple empirical formula that treats an avalanche as a sliding block of snow moving with a drag force proportional to the square of the speed of its flow. Since the 1990s, many more sophisticated models have been developed, especially in Europe. Much of the European work was carried out through the SATSIE (Avalanche Studies and Model Validation in Europe) research project, funded by the European Commission.

7.7 Avalanche Safety

Winter travelling in the backcountry is never 100 percent safe. The practice of good avalanche safety is a process involving route selection, examination of the snowpack and weather conditions, and human factors.[13] People involved in winter recreation in the mountains need to ask themselves the following three questions:

1. Is the slope to be crossed prone to avalanches?
2. Is the snowpack unstable?
3. What are the consequences of being caught in an avalanche in this terrain?

The first question can be answered by using such terrain factors as slope angle and orientation, discussed earlier in the chapter. The second question can be answered by using public bulletins, observations of recent avalanches in the area, stability tests, and snowpack and weather observations. The Canadian Avalanche Association, the American Avalanche Association, and Parks Canada provide advice on avalanche dangers and appropriate human responses (Table 7.2; see Professional Profile). The third question requires an appreciation of snowpack and terrain factors. If an avalanche occurs, is it likely to be small or large? Is it likely to be a slab or point-release avalanche? Will the avalanche be in wet or dry snow? Could I be swept over a cliff or into trees or boulders?

The Canadian Avalanche Centre has adopted a simple rule-based support tool—the **Avaluator**—to help amateur recreationists, including backcountry skiers, snow boarders, and snowmobilers, make critical decisions both before and during backcountry trips. The Avaluator consists of a two-sided pocket card (Figure 7.18) and a companion booklet that focus on four key decision and travel skills—trip planning, identifying avalanche terrain, slope evaluation, and good travel habits. For trip planning, the Avaluator provides a chart that combines the avalanche forecast danger rating and a terrain rating (the Avalanche Terrain Exposure Scale, ATES). The colours in the chart offer a general assessment of the risk of the trip and provide guidance on the required level of preparedness for safely managing the expected avalanche hazard. ATES defines three terrain classes—simple, challenging, and complex, each with its specific set of criteria.[14] For example, the criteria associated with Class 3 (complex) terrain are "exposure to multiple overlapping avalanche paths or large expanses of steep, open terrain; multiple avalanche starting zones or terrain traps below; minimal options to reduce exposure; complicated glacier travel with extensive crevasse bands or icefalls." The Avaluator includes two separate checklists that allow the user to keep track of present avalanche conditions and assess the nature of the local terrain. The checklists consist of simple questions relating to what are termed "warning signs"—for example, "Are there signs of slab avalanches from today or yesterday?" and "Are there gullies, trees or cliffs that increase the consequences of being caught in an avalanche?" The user totals the number of yes answers on the two checklists and uses a chart similar to the trip planner to arrive at a recommendation—normal caution, extra caution, or not recommended.

Good Habits Minimize Risk

Avalanche danger reports should be considered and all warnings heeded. Never follow in the tracks of others without making your own evaluation; snow conditions are almost certain to have changed since the tracks were made. Observe the terrain and note obvious avalanche paths where vegetation is missing or damaged, where there are few surface anchors for snow, and below cornices or ice formations. Avoid travelling below others who might trigger an avalanche. Minimize traverses of steep slopes. Maintain separation—ideally one person should cross the slope and enter an area safe from avalanches before the next person crosses. Route selection should also consider what dangers lie above and below the traverse, ascent, or descent, and the consequences of an unexpected avalanche. Stop or camp only in safe locations. Wear warm gear to delay hypothermia if buried. Plan escape routes. Never travel alone. The party should be large enough to perform a rescue, although additional people increase the disturbance to the slope. If you find yourself in a potentially dangerous avalanche situation, you should seriously question the choice of the route, ask why your safety is being put in jeopardy, and consider alternatives to pressing on.

TABLE 7.2 Avalanche Danger Scale

Danger Level (Colour)	Avalanche Probability and Trigger	Recommended Action in Backcountry
Low (green)	Natural avalanches unlikely; human-triggered avalanches *unlikely*	Travel is generally safe; normal caution is advised.
Moderate (yellow)	Natural avalanches unlikely; human-triggered avalanches *possible*	Use caution in steeper terrain on certain slope aspects.
Considerable (orange)	Natural avalanches possible; human-triggered avalanches *probable*	Be cautious in steeper terrain.
High (red)	Natural and human-triggered avalanches *likely*	Travel in avalanche terrain is not recommended.
Extreme (red with black border)	Widespread natural or human-triggered avalanches *certain*	Travel in avalanche terrain should be avoided and travel should be confined to low-angle terrain well away from avalanche run-outs.

Source: Data from Canadian Avalanche Centre. 2008. http://www.avalanche.ca/cac/bulletins/danger-scale

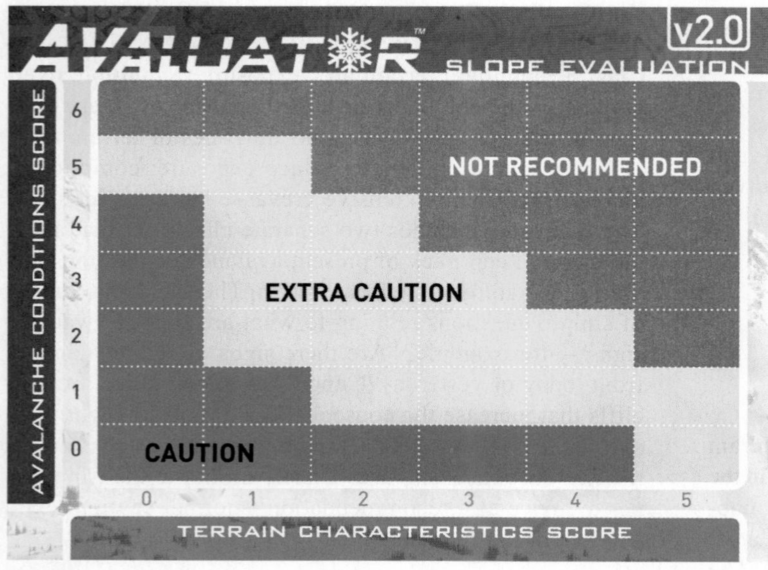

(a)

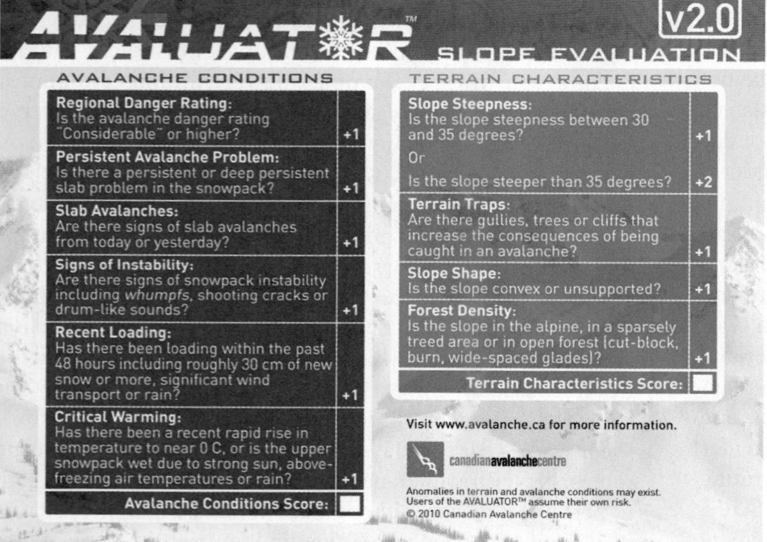

(b)

◄ **FIGURE 7.18 THE AVALUATOR** The decision aids of the Avaluator include (a) a trip planner, which provides a general assessment of the expected risk of a trip into avalanche terrain, and (b) a slope evaluation tool for assessing the seriousness of individual route options during a backcountry trip. *(From the Canadian Avalanche Centre. Reprinted with permission.)*

Too often, inappropriate human factors enter into decisions about risk. Some people focus on a peak, an attractive slope, or other destination and continue in spite of strong evidence of avalanche danger. Groups may spread out, leaving slower but knowledgeable people out of decisions about the route or avalanche danger. Careful decision-making may be abandoned at the end of the day, when people are tired and in a rush to reach their destinations. Many avalanches happen during clear weather when skiers, snowboarders, and snowmobilers are drawn to beautiful open slopes, high mountain passes, and peaks. Inexperienced winter recreationists can reduce their risk by travelling with more experienced people or by hiring a guide.

Canada has two organizations that support avalanche safety—the Canadian Avalanche Association and its sister organization, the Canadian Avalanche Centre. The Canadian Avalanche Association was established in 1981 to provide programs and services for professional avalanche risk management in Canada. The Canadian Avalanche Centre was established in 2004 to deliver a public avalanche warning program and coordinate public avalanche awareness and education.

Another group that plays an important role in winter safety in Canada is the Canadian Ski Patrol System. CSPS consists of 5000 alpine and Nordic skiers and snowmobilers who volunteer their time to help those in need in winter, mainly at organized ski areas. It offers avalanche search and rescue training in collaboration with the Canadian Avalanche Centre.

PROFESSIONAL PROFILE

Grant Statham, Parks Canada

Grant Statham (Figure 7.19) works for Parks Canada as an avalanche risk specialist. The Parks Canada avalanche risk management program spans seven national parks, employs 30 avalanche forecasters, and is responsible for avalanche control along highways, public avalanche warnings, and search and rescue. Statham's role is to coordinate programs between different parks to ensure they are consistent and to provide advice to Parks Canada's Executive Board. His day-to-day work involves the design and maintenance of avalanche warning systems, and he works closely with other agencies and professionals to develop training standards. In his spare time, Statham works as an independent mountain guide, teaches avalanche courses, and climbs.

When he was 17 years old, Statham moved to Banff and began climbing frozen waterfalls. He soon recognized that "avalanches are by far the biggest hazard that ice climbers face, and I developed an immediate interest and respect for them." At 18, he began working as a ski patroller, and soon thereafter began training as a mountain guide. "That is where my formal study of avalanches began, in the mountains and in the classroom," he notes.

And Statham has had first-hand experience with avalanches. He says, "You cannot be an experienced avalanche professional without having been around and involved in avalanches." In 1988, he just missed getting caught in a large avalanche that would have pushed him over a 300-m-high cliff on Mt. Diadem in Jasper National Park. Five years later, six people in the heli-ski group he was guiding were caught in a large avalanche in the Cariboo Mountains in British Columbia. Fortunately, no one was buried. And in 2010, he was hit by an avalanche in the Monashee Mountains in British Columbia. The avalanche carried two people into the trees, but again no one was buried.

▲ **FIGURE 7.19 GRANT STATHAM** Grant Statham works for Parks Canada as an avalanche risk specialist. He also is an independent mountain guide, teaches avalanche courses, and climbs. *(John J. Clague)*

In spite of the risk, Statham loves his job. "There is no other job like mine in the world," he says. "I get to use many years of field experience with avalanches to design systems and influence policies that enable people to get out and enjoy the mountains in winter. I am able to teach and share my experience in a broader way than I could have ever imagined, and I get to work outside in the environment I love so much. I enjoy keeping abreast of the latest research and trying to figure out ways to incorporate this research into practical systems that can help everyone." He adds, "I am proud to work for Parks Canada because of our long history of leadership in avalanche safety and the profound respect this agency has for avalanche risk."

What advice does Statham have for winter recreationists in avalanche country? "Understand avalanche terrain. People get hung up on the complexities of the snowpack, which is inherently difficult to predict. Experienced mountain guides understand that the secret to managing avalanche risk is to control your exposure to it. Terrain is the 'gas pedal' of the mountains—when the avalanche danger is low, push down harder on the pedal; when the danger is higher, ease off the pedal." He adds, "There are always places to go in the mountains—every day of the year—the secret is understanding where those places are and in choosing the right terrain to match the conditions."

—John J. Clague

7.8 Avalanche Rescue and Survival

The motion of the snow during an avalanche kills about 25 percent of avalanche victims.[1] If the victim survives this battering, he or she must be quickly located and extracted from the snow. Between 55 percent and 65 percent of victims buried in the open are killed.[6] Survival depends mainly on the length of time the person is buried and the burial depth.

Research carried out in Italy on a sample of 422 buried skiers indicates how low the chances of survival can be.[15] Survival dropped very rapidly from 92 percent within 15 minutes of burial, to 30 percent in 35 minutes, to almost 0 percent in 2 hours. Buried victims die of suffocation, hypothermia, or injuries. In most backcountry situations, few victims survive long enough for other members of the group to get help. Reports of live recoveries by searchers called to the accident site are rare. Therefore, the victim's best, and generally only, chance of survival depends on an effective search by other members of the group. It is vital that everyone surviving an avalanche conduct an immediate search, rather than waiting for help to arrive. Help can be called once victims are located and dug out. Even in a well-equipped country, such as France, it typically takes 45 minutes for a helicopter rescue team to arrive, by which time most of the victims are likely to have died. In some cases, avalanche victims are not located until summer when the snow melts.

The chance of survival also depends on depth of burial. Only 5 to 10 percent of avalanche victims survive burial in more than 1.5 m of snow, and few survive deeper burial.

Chances of a buried victim being found alive and rescued are increased when everyone in the group carries standard avalanche safety equipment and has been trained in how to use it. Safety equipment includes avalanche cords, transceivers, probes, and shovels.[16]

Avalanche Cords

An **avalanche cord** is the oldest avalanche safety device and was widely used before transceivers became available. The principle is simple—an approximately 10-m-long red cord, similar to a parachute cord, is attached to a person's belt. While skiing, snowboarding, or snowshoeing, the cord is dragged along behind. In the event of an avalanche, the light cord remains on top of the snow and, being red, is easily visible to rescue personnel. Typically the cord has markings every metre, which indicate the distance to the victim.

Avalanche Transceivers

An avalanche may cover a large area, and if victims are completely covered by snow, the chances they will be found alive are low. Survival becomes much more likely if the victims are wearing **avalanche transceivers**, also referred to as beacons, beepers, ARVA (*Appareil de recherche de victimes en avalanche*, in French), and LVS (*Lawinen-Verschütteten-Suchgerät*, Swiss German). These portable devices emit a beep via a radio signal. They are switched to receive mode to locate a buried victim up to 80 m away, commonly in 5 to 15 minutes. Analog receivers provide audible beeps that rescuers interpret to estimate the distance to the victim. Digital models give visual indications of the direction and distance to victims and require little practice to use.

Probes

Collapsible rods can be extended to probe for victims up to several metres deep in snow. Probing can be done with sectional rods, skis, ski poles, or branches. It is a time-consuming process if a thorough search is done without a beacon. Five probers are about 100 times slower than one searcher with a transceiver. In the United States, 86 percent of the 140 victims found by probing since 1950 were already dead.[6] And of victims buried more than 2 m, only 4 percent were found and rescued alive.

Shovels

As an avalanche decelerates, the snow becomes compressed into a compact mass. Shovels are essential to dig to the victim, because the deposit is too dense to dig with hands or skis. Shovels are also useful for digging snow pits to evaluate snowpack stability.

Avalanche Dogs

In the late 1930s, the Swiss Army started training search dogs in avalanche rescues. Since then, training techniques have been refined and many avalanche victims owe their lives to dogs trained in avalanche rescue. A well-trained **avalanche dog** works a snowfield rapidly, searching for human scent rising through the snow. It can perform a general search of one hectare in approximately 30 minutes, whereas it takes 20 people without transceivers four hours to search the same area.[17] However efficient, though, trained dogs are not infallible. Weather, snow conditions, the capabilities of a particular dog, and scent diffusion all determine the depth at which a victim can be located. An experienced dog will be able to locate victims buried between 2 and 4 m under most conditions.

Unfortunately, a trained dog generally does not arrive at the accident scene in time to recover people alive. There are reports of buried avalanche victims being found alive by dogs in the United States, but none in Canada. Dogs are most useful in locating hikers and skiers who have fallen due to injury or hypothermia and become covered by snowfall, or those who hole up in a snow cave after having become lost or exhausted.

Avalanche Survival

If you are ever caught in an avalanche, try to ski or board toward the side of the moving mass of snow. When you fall, jettison your equipment and attempt swimming motions. As the snow comes to rest, try to preserve an airspace in front of your mouth, and try to thrust an arm, leg, or object above the surface (assuming you are still conscious). Once the snow stops, enlarge the air space if you are able to move; however, limit movement to minimize oxygen consumption. Rapid breathing may cause the snow in your face to glaze, creating a seal that limits oxygen.

The low odds of surviving an avalanche emphasize the importance of the mitigation measures described above. Information on current weather and snow conditions, avalanche education, and a thoughtful decision-making process are essential to minimize personal avalanche risk.

REVISITING THE FUNDAMENTAL CONCEPTS

Snow Avalanches

1. **Hazards can be understood through scientific investigation and analysis.**
2. **An understanding of hazardous processes is vital to evaluating risk.**
3. **Hazards are commonly linked to each other and to the environment in which they occur.**
4. **Population growth and socio-economic changes increase the risk from natural hazards.**
5. **Damage and loss of life from natural disasters can be reduced.**

1. The avalanche hazard in a given area can be determined from the observed occurrences of avalanches, the thickness and properties of the snowpack, and weather conditions. Several field tests have been developed to assess snowpack stability. Of particular concern are weak layers within the snowpack.
2. Snow avalanches are amenable to risk analysis because the hazard can be evaluated, and the consequences, should an avalanche occur, can be determined. Weather conditions and the characteristics of the snowpack provide information on the likelihood of avalanches. Topography dictates where avalanches are likely to occur. The consequences of an avalanche to a skier or snowmobiler traversing snow-covered terrain, or to buildings or other structures in the path of potential avalanches, can be determined.
3. Snow avalanches are most closely linked to weather, in particular to times of heavy snowfall or snowfall following a period during which an icy crust forms on the snowpack. Strong winds can create wind slabs that are prone to failure. Avalanches can occur wherever snow accumulates on a slope steep enough for the snow to slide away. However, avalanches are most common in mountains in temperate maritime climates. They often follow well-defined tracks through forest and in ravines and swales in the landscape. Above the treeline, however, they can occur almost anywhere.
4. A century ago, most avalanche injuries and fatalities happened along roads and rail lines. Now, however, the number of industrial and transportation fatalities is low because measures have been taken to protect highways and rail lines from avalanches. In contrast, the number of recreational injuries and fatalities has increased because of the much greater use of mountainous terrain in winter by skiers, snowboarders, and snowmobilers. In effect, the exposure of people to avalanches has increased.
5. Buildings, roads, and other infrastructure often can be located outside high avalanche hazard areas. Where avalanche zones cannot be avoided, infrastructure can be protected with avalanche deflecting structures, snow fences, avalanche sheds, or structural reinforcement. Dangerous snow build-ups can also be released in a controlled environment with the use of explosives. Avalanche forecasts, such as those issued by the Canadian Avalanche Centre, can save lives by alerting winter recreationists of conditions conducive to avalanches.

Summary

A snow avalanche is the sudden failure and rapid movement of a mass of snow down a mountainside. The snow may move as a coherent block, or it may rapidly disaggregate into small particles that move independently of one another. Avalanches present obvious dangers to backcountry recreationists and to people working in resource industries in steep mountain areas. They also damage property and forest, and cause expensive rail and road traffic delays.

Avalanches are of two types. Point-release avalanches involve failures of small masses of snow; they grow in size as they travel downslope, producing a distinctive, downslope-widening trough, like an inverted V. Slab avalanches are failures of a mass of cohesive snow along a weak layer at depth. They leave steep crown, lateral, and toe scarps and a smooth planar basal failure plane.

Weather determines the evolution and stability of the snowpack. The most important weather-related factors are heating by solar radiation, radiative cooling, temperature gradients in the snowpack, and snowfall amounts and type.

Weak layers in the snowpack form in three main ways. First, wind contributes to the rapid build-up of snow on sheltered lee slopes. The resulting wind slabs consist of thick, poorly bonded snow that is susceptible to failure. Wind may also increase during a snowstorm, depositing a layer of broken and packed snow crystals on a delicate layer of unbroken snowflakes that fell earlier. The layer of broken crystals becomes a buried weak surface along which failure may later occur. Second, weak layers of crystals formed by the sublimation of water vapour can form within the snowpack (depth hoar). Hoar can form at different depths, depending on the temperature profile in the snowpack. Third, large ice crystals form on the snow surface on cold clear nights (surface hoar). The crystals may change only slowly once buried and may become a weak layer within the snowpack.

Most avalanches start naturally during or soon after snowstorms. Others happen when normal daytime heating or the inflow of a warm air mass heats the upper layer of the snowpack. In most recreational accidents, a person triggers the avalanche.

An avalanche path consists of three parts—the start zone, where the snowpack fails; the track, along which the avalanche accelerates and achieves its highest velocity; and the run-out zone, where the avalanche decelerates and snow is deposited. The three zones are easy to identify where avalanches follow gullies or chutes that have carried previous avalanches and where tracks are on forested slopes. The zones are more difficult to distinguish on slopes above the treeline and where the snow is not channelled.

Large avalanches generally occur on slopes between 30° and 45°. These slopes are also popular with skiers, snowboarders, and snowmobilers, which is why recreationists commonly trigger avalanches.

Avalanches in Canada are most common in the mountains of Alberta, British Columbia, and Yukon Territory. However, they can occur anywhere in Canada where slopes of sufficient steepness accumulate and release snow. Most avalanches in the United States occur in the high mountains extending from the Sierra Nevada and Cascade ranges in the west to the Rocky Mountains in the east.

Avalanche risk can be reduced by locating infrastructure outside known danger zones, by placing defensive structures within avalanche tracks, by reinforcing exposed structures, by setting off controlled avalanches with explosives, and with forecasting. Defensive structures include fences and nets in the start zone, and berms, splitting wedges, avalanche sheds, and mounds in the run-out zone.

Avalanche forecasting is an assessment of the likelihood and size of avalanches under existing or future conditions. Forecasts are based on observed occurrences of avalanches, stability and strength tests, observations of snowpack characteristics, and observations of weather conditions. Numerical and rule-based computer software has been developed to assist forecasters. The computational algorithms compare current measurements of precipitation, temperature, wind, and other meteorological variables with average values of the same variables spanning many years.

Avalanche safety involves route selection, examination of the snowpack, consideration of weather conditions, and human factors. People involved in winter recreation in the mountains need to ask and answer the following three questions: Is the slope to be crossed prone to avalanches? Is the snowpack unstable? What are the consequences of being caught in an avalanche on the terrain to be crossed?

The chance of surviving an avalanche depends on the length of time and depth that a person is buried. The probability that a buried victim will be found alive and rescued is increased when everyone in a group carries standard avalanche safety equipment and has been trained in how to use it. Safety equipment includes avalanche cords, transceivers, probes, and shovels.

Key Terms

avalanche cord (p. 204)
avalanche dog (p. 205)
avalanche forecast (p. 198)
avalanche transceiver (p. 204)
Avaluator (p. 201)
hoar (p. 191)
point-release avalanche (p. 189)
run-out zone (p. 191)
slab avalanche (p. 190)
sluff (p. 192)
snow avalanche (p. 188)
start zone (p. 191)
track (p. 191)
wind slab (p. 190)

Did You Learn?

1. Name the two types of avalanches and explain how they form.
2. Describe how weak layers form in a snowpack.
3. Explain the role that wind plays in avalanches.
4. Describe or sketch the three parts of an avalanche's path.
5. Identify the factors that affect snowpack stability.
6. How does depth hoar form and what effect does it have on the stability of the snowpack?
7. Name the two types of wind slabs and explain how they form.
8. What regions in North America are most at risk from avalanches?
9. What are the natural service functions of avalanches?
10. Identify the kinds of engineered structures that are used to reduce avalanche risk.
11. What is an avalanche shed?
12. Describe the procedures used to forecast avalanches.
13. Explain how a snow compression test is performed.
14. What is the Avaluator?
15. Name the safety and rescue equipment that every backcountry winter recreationist should carry.

Critical Thinking Questions

1. Do you support the idea that all skiers, snowboarders, and snowmobilers should be required to carry avalanche beacons in mountains outside ski areas in winter? If not, why not?
2. You have been hired by a ski resort to advise it on avalanche safety in an area that is slated for development. How would you identify areas that should be off-limits to skiing? How would you handle areas that are likely to be attractive to skiers but, under some circumstances, will have a high avalanche danger?
3. You are part of a ski party returning from a long, tiring day of skiing. Your party is properly dressed for severe winter conditions, has been trained in avalanche awareness, and is equipped with avalanche probes and transceivers. You are two hours from your vehicles and it has started to snow heavily, with strong gusty winds. To reach your cars, you must pass along a narrow section of valley with slopes rising at an angle of about 35° from the valley floor. You are aware that the avalanche danger has increased since the beginning of the day but are tired and anxious to reach your destination. What actions do you take?

MasteringGeology

MasteringGeology **www.masteringgeology.com**. Looking for additional review and test prep materials? Visit the Study Area in MasteringGeology to enhance your understanding of this chapter's content by accessing a variety of resources, including **Self-Study Quizzes, Geoscience Animations, GEODe Tutorials, RSS feeds, flashcards,** web links and an optional **Pearson eText.**

CHAPTER 8

▶ **FLOODING IN PIAZZA SAN MARCO, VENICE, ITALY** A high tide on November 1, 2004, flooded 80 percent of Venice in northern Italy. Flooding is becoming more common as Venice subsides and sea level rises. *(AP Images)*

Subsidence and Soil Expansion and Contraction

Learning Objectives

Subsidence, or sinking of the land, and soil expansion and contraction are important geologic processes that cause extensive damage in some areas. Your goals in reading this chapter should be to

- Understand subsidence and soil and expansion and contraction and what causes them
- Know the geographic regions at risk from subsidence and soil expansion and contraction
- Understand the hazards associated with subsidence
- Recognize links between subsidence and other hazards, as well as the natural service functions of karst
- Understand how people interact with subsidence and soil hazards
- Know what can be done to minimize the risk of subsidence and soil expansion and contraction

Venice Is Sinking

Italy's beautiful and famous city of Venice faces a serious geologic problem. The city is sinking, or subsiding, up to 2 mm per year.[1] Venice is built on 17 small islands connected by more than 400 bridges. The city's location and numerous canals are part of its attraction, but the presence of so much water and the ongoing slow subsidence of the city make Venice extremely prone to flooding.

The land on which Venice is built is only centimetres above sea level. Although subsidence has been occurring naturally for millions of years, pumping of groundwater from the 1930s to the 1960s significantly increased the rate at which the city is sinking.[2] The pumping and thus the human contribution to this natural hazard ended in the 1970s, but, unfortunately, natural subsidence is still occurring, and the problem is exacerbated by sea-level rise. The response has been to raise buildings and streets, but this solution is not viable in the long term.

To combat the flooding, engineers designed and built 78 hinged, steel floodgates across the three tidal inlets that connect the Venice lagoon to the Adriatic Sea.[3] Each of the floodgates is 28 m high and 20 m wide, and swings upward from the seafloor to block the tidal inlets during storms, thus preventing wind-driven surges of seawater from the Adriatic Sea from entering the lagoon. The total cost of the project was about $6 billion. However, the floodgates will not slow the subsidence and will not completely prevent flooding; thus their long-term value has been questioned.[4] As sea level continues to rise, the future of Venice is uncertain.

8.1 Introduction to Subsidence and Soil Expansion and Contraction

Subsidence is a slow or rapid, nearly vertical, downward movement of Earth's surface. The subsiding area can be circular, linear, bead-like, or irregular. Subsidence can also involve an imperceptible lowering of the land surface over a large area.

Subsidence is commonly associated with the dissolution of limestone, dolostone, marble, gypsum, or rock salt at depth. The resultant **karst** landscape is irregular in form and has closed depressions. Other natural causes of subsidence include thawing of frozen ground, compaction of recently deposited sediment, shrinkage of expansive soils, earthquakes, and deflation of magma chambers. Human-induced subsidence, discussed in section 8.6, can result from withdrawal of fluids from subsurface reservoirs, collapse of soil and rock over mines and other subsurface excavations, and draining of wetlands. Some **soils** expand when they become wet and contract when they dry out. Freezing and thawing of pore water also produce volume changes in soils.

Before we proceed, we need to define "soil." We use the word to refer to a mixture of unconsolidated mineral and organic material at Earth's surface that has been modified by physical, chemical, and biological weathering processes. Engineers use the word in a much different sense: any unconsolidated sediment, whether at Earth's surface or at depth. An engineering soil is not formed by weathering processes at the interface between the atmosphere and geosphere.

Soil expansion and contraction are not as dramatic as some subsidence—you will never see a headline stating "Dozens Die in Saskatoon as Soil Expands." However, these processes are one of the most widespread and costly of natural hazards.

Karst

Karst results from dissolution and subsidence of near-surface rocks. Four common sedimentary rocks (rock salt, gypsum, limestone, and dolostone) and one metamorphic rock (marble) can dissolve when water moves through them below the surface. Rock salt and gypsum dissolve when they come in contact with neutral waters, whereas limestone, dolostone, and marble dissolve readily only in acidic waters. Most karst occurs in limestone because that rock is much more common than rock salt, gypsum, dolostone, or marble.[5]

Water containing carbon dioxide is acidic. Acidification occurs in the atmosphere where rainfall interacts with carbon

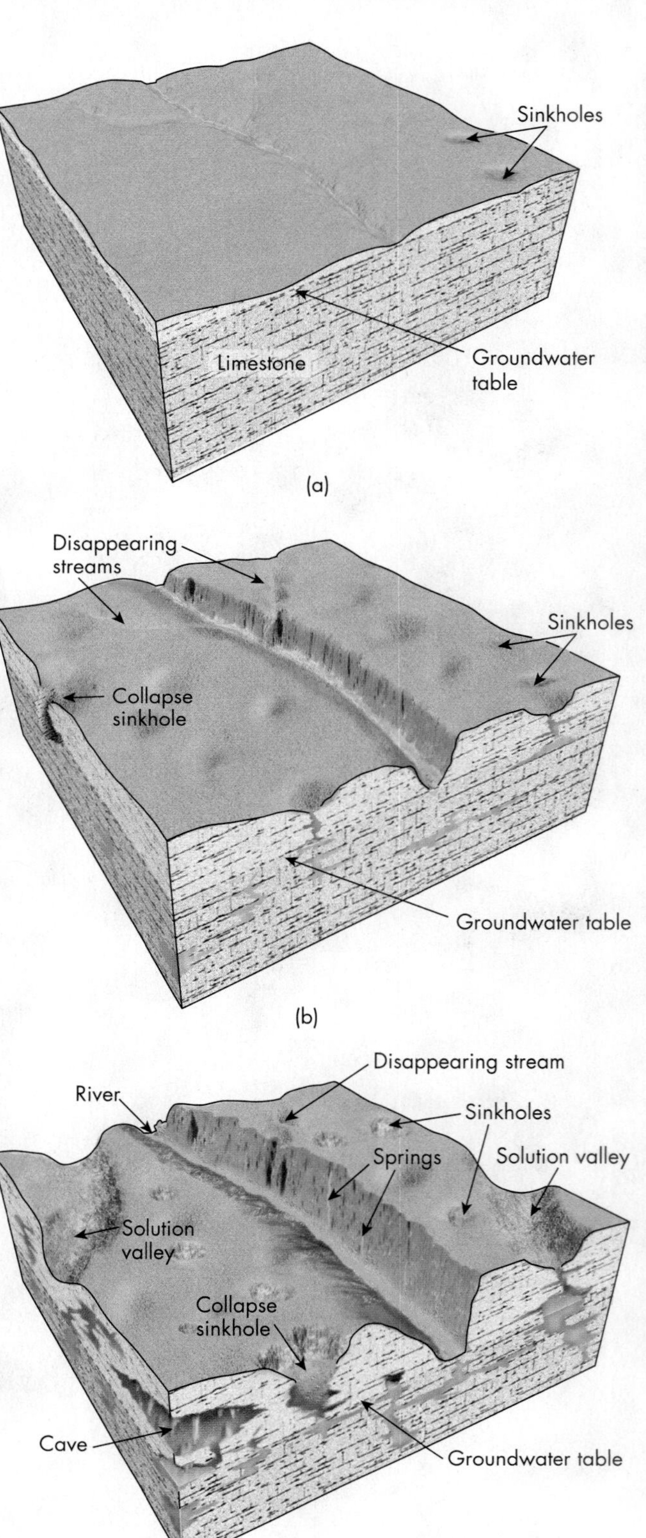

▶ **FIGURE 8.1 DEVELOPMENT OF KARST TOPOGRAPHY** (a) In the early stage of karst formation in a limestone terrain, water from the surface seeps through fractures and along layers in the soluble rock. Weak acid in the water slowly dissolves the rock. (b) The water table falls as a river incises into the rock. Caves start to form and then collapse to become sinkholes. Some surface streams disappear underground. (c) In later stages of karst development, a downward-eroding river continues to lower the water table. Large caverns and sinkholes develop and eventually merge to form solution valleys and plains without surface streams. In humid tropical climates, intense dissolution removes nearly all the rock, leaving behind pillars of limestone referred to as tower karst. *(Based on an illustration by D. Tasa in Tarbuck, E. J., and F. K. Lutgens. 2005.* Earth: An Introduction to Physical Geology, *8th ed. Upper Saddle River, NJ: Pearson Prentice Hall)*

◄ **FIGURE 8.2 SMALL SINKHOLE** Most sinkholes, such as this one in Boyle County, Kentucky, are small. *(Jim Currens/Kentucky Geological Survey)*

◄ **FIGURE 8.3 COLLAPSE SINKHOLE** This aerial view shows the large sinkhole that continues to grow as workers try unsuccessfully to retrieve sunken sports cars from the depression in Winter Park, Fla., May 11, 1981. *(AP Images)*

dioxide and in soil where carbon dioxide is produced by bacterial decomposition. The respiration of most soil bacteria is like respiration in humans—both involve consumption of oxygen and release of carbon dioxide.

Areas underlain by thin-bedded, fractured, or well-jointed limestone are especially vulnerable to dissolution. Surface waters easily infiltrate to depth in these rocks along fractures or cracks between sedimentary layers. Acidic percolating waters enlarge fractures by dissolving rock (Figure 8.1). Dissolution of limestone can result in an average lowering of the land surface of about 10 cm per century.[6]

The chemical reactions leading to limestone dissolution are the following:

$$\text{water} + \text{carbon dioxide} = \text{carbonic acid}$$
$$H_2O + CO_2 = H_2CO_3$$

$$\text{carbonic acid} + \text{limestone} = \text{calcium} + \text{bicarbonate}$$
$$H_2CO_3 + CaCO_3 = Ca^{++} + 2HCO_3^{-1}$$

Limestone dissolution produces empty spaces, or voids, of a range of sizes beneath the land surface. Over time, contiguous voids join to produce caves and caverns. Where these large openings are close to the surface, pits known as **sinkholes** may develop, either individually (Figures 8.2 and 8.3) or in large numbers. A surface pockmarked with a large number of sinkholes is termed a **karst plain**.

Sinkholes In karst areas, sinkholes range from one to several hundred metres in diameter and are of two basic types (Figure 8.4):

1. *Solution sinkholes.* These pits form by solution of buried bedrock. Acidic groundwater is concentrated

▲ **FIGURE 8.4 FORMATION OF SINKHOLES** (a) Dissolution of soluble bedrock takes place along vertical and horizontal fractures, leading to the formation of small caves. (b) Dissolution continues to enlarge fractures, and some caves become large caverns with roofs partly supported by groundwater. Slow subsidence and enhanced solution in surface depressions initiates a solution sinkhole. (c) A subsequent lowering of the water table leaves cave and cavern roofs unsupported, and a cavern roof collapses to form a collapse sinkhole.

in holes that develop along bedding planes, joints, and fractures.

2. *Collapse sinkholes***.** This type of sinkhole develops by the collapse of surface or near-surface rock or sediment into an underground cavern system.

Some sinkholes open into a subterranean passage through which water escapes during rainstorms. Others are filled with rubble that blocks the flow of water to the subsurface. Blocked sinkholes may contain small lakes, but most such lakes eventually drain when the water seeps through the debris.[7] Artificial ponds and lakes constructed over sinkholes may drain suddenly into subterranean cavities (see Survivor Story).

Cave Systems Many karst areas are characterized by beautiful rolling hills separated by lower areas underlain by extensive **cave systems** (Figure 8.1). As solution pits in limestone enlarge, a system of caves or larger caverns can form. Mammoth Cave in Kentucky, Carlsbad Caverns in New Mexico, and Castleguard Cave in the Alberta Rocky Mountains are three famous cave systems in North America. Caves develop at or near the groundwater table where water that is saturated with calcium and bicarbonate from the chemical weathering of limestone is replaced with water that is not saturated with these ions. Caves grow larger as groundwater moves through fractures in the rock or between sedimentary layers. Later, if the groundwater table falls, seeping water will deposit calcium carbonate on the sides, floor, and ceiling of the cave as *flowstone, stalagmites*, and *stalactites* (Figure 8.5).

Tower Karst Large, steep limestone pillars that rise above the surrounding landscape are known as *tower karst*. They

◄ **FIGURE 8.5 CAVE FORMATIONS** Carlsbad Caverns, New Mexico, contains stalactites, which hang from the ceiling, stalagmites, which grow up from the ground, and flowstone, which forms as water flows slowly down the walls or across an inclined surface. These beautiful features form as waters saturated with calcium and bicarbonate slowly precipitate calcium carbonate on the sides, floor, and ceiling of a cave or cavern. *(Bruce Roberts/ Science Source)*

SURVIVOR STORY

Sinkhole Drains Lake

The pristine waters of Scott Lake in central Florida attract birds, otters, alligators, and are teeming with large-mouth bass, carp, bluegill, and catfish. Well-heeled corporate executives have built multimillion-dollar mansions along the shores of the 115-hectare lake.

Within the space of a week in 2006, nearly the entire lake disappeared. The story began on June 13, when the area received over 20 cm of rain—but instead of rising, the level of the lake began dropping. Residents on the southeast shore of Scott Lake watched as the normally placid waters began to bubble and then form a whirlpool over a single point. Boathouses collapsed and docks were left standing high and dry (Figure 8.6). Dave Curry, who has lived at the lakeshore for more than 30 years, knew the cause immediately—a sinkhole.

▲ **FIGURE 8.6 DAVE CURRY VIEWS SCOTT LAKE** Property owner Dave Curry walks out on a boat dock that marked the level of Scott Lake in south-central Florida before it drained in June 2006. The lake drained when a sinkhole opened beneath it. *(Curry Controls Company)*

As in most of Florida, the sediment around Scott Lake is less than 10 m thick. Beneath the sediment cover is a layer of limestone, which supports the great Floridan aquifer. Several years of drought had lowered the water level in the aquifer, undermining support for the sediment above. The added weight of the heavy rains on June 13 caused the lake bottom to collapse into a hole more than 1.5 m in diameter and 16 m deep.

"This happens all the time in Florida," Curry says. He had watched the lake drain once before, in 1969, probably due to a previous sinkhole.

This time, Scott Lake lost more than 4 million m^3, over 95 percent of its volume. The event was catastrophic for wildlife. "I counted seven alligators down in that one hole. They couldn't get out; they ended up being washed into the Floridan aquifer along with the fish and the turtles," Curry says.

Those fish not sucked into the sinkhole were left gasping on the muddy lake bottom. Curry added, "There were huge dead fish lying all around the banks. But the smell didn't last too long because thousands of buzzards and all kinds of other bird life I'd never seen before came in and cleaned it up in a few days. Otters were catching some of the live fish in distress."

Houses nearest the sinkhole fared the worst. The foundations of two houses cracked, windows shattered, and fissures opened in lawns and swimming pools. The shoreline steepened when the lake drained, causing the houses to slip.

Since then, mud and other debris formed a natural plug over the sinkhole and Scott Lake has refilled. But local residents did not rely on Mother Nature to correct the problem. They sealed the hole with concrete.

—***Kathleen Wong***

are residual landforms of a highly eroded karst landscape and are common only in humid tropical regions, notably Cuba, Puerto Rico, and Southeast Asia.

Disappearing Streams Karst regions are underlain by a complex network of subterranean channels. Surface streams may disappear into caves and continue underground as *disappearing streams*.

Springs Natural discharges of groundwater at the surface are known as **springs**. Many springs in karst areas have large flows, especially during rainy periods. They are an important resource, but many are drying up because of over-pumping of groundwater. Springs are also vulnerable to contamination; pollutants at the surface can enter underground drainage systems relatively easily and migrate to springs.

Permafrost

At high latitudes and at high elevations in mountains, subsurface materials remain frozen throughout the year. Permanently frozen ground, referred to as **permafrost**, may be continuous across the landscape, whereas in slightly warmer climates, it may exist as discontinuous patches or thin layers.[8] To qualify as permafrost, sediment or rock must remain cemented with ice for at least two years. More than half of Canada is underlain by permafrost (Figure 8.7), and most of northern Russia and parts of Scandinavia are frozen.

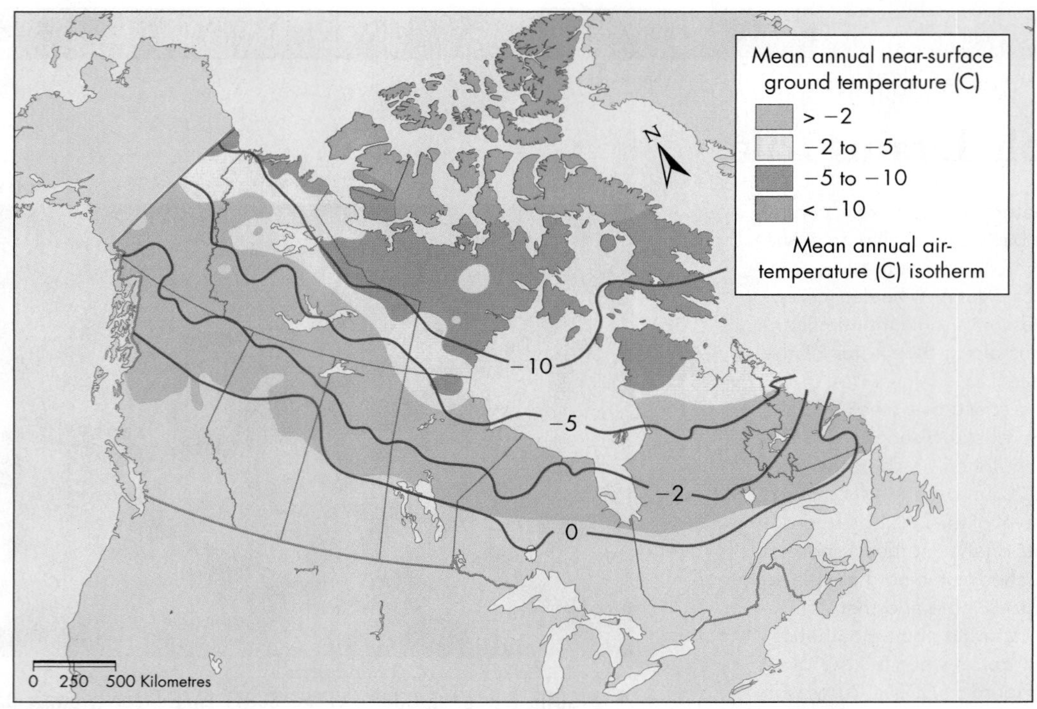

▲ **FIGURE 8.7 DISTRIBUTION OF PERMAFROST** Map showing spatial variations in mean annual air temperature and near-surface ground temperature in Canada's permafrost region. Permafrost is sporadic within the zone of air temperatures between 0°C and –2°C, and is extensive or continuous in areas with mean air temperatures less than –2°C. *(Smith, S. L., M. M. Burgess, and J. A. Heginbottom. 2001. "Permafrost in Canada: A challenge to northern development." In G. R. Brooks (ed.),* A Synthesis of Geological Hazards in Canada. Geological Survey of Canada Bulletin 548. *Reproduced with the permission of the Minister of Public Works and Government Services Canada, 2013.)*

Permafrost typically exists in any climate where the mean annual air temperature is below 0°C. Exceptions are found in northern Scandinavia and northeast Russia west of the Ural Mountains, where snow acts as an insulating blanket.

If the mean annual air temperature is only slightly below 0°C, permafrost will exist only in areas that are sheltered from solar radiation. In the Northern Hemisphere, those areas are slopes with a northerly *aspect* (the horizontal direction that a slope faces). Typically, permafrost remains discontinuous in a climate where the mean annual surface temperature is between −5 and 0°C. *Discontinuous permafrost* can be further subdivided into *extensive discontinuous permafrost*, where permafrost covers between 50 and 90 percent of the landscape and is found in areas with mean annual temperatures between −2° and −4°C, and *sporadic permafrost*, where the permafrost cover is less than 50 percent and mean annual temperatures are typically between 0° and −2°C. At mean annual surface temperatures below −5°C, aspect and other factors are insufficient to thaw permafrost; thus it is *continuous*.

Typically, below-ground temperatures in permafrost areas are less variable from season to season than the air temperature directly above the ground surface (Figure 8.8). Temperatures tend to increase with depth due to the upward movement of geothermal heat toward the surface. The thickness of the permafrost layer in any area depends on mean annual surface temperature, on geothermal heat flux, and on the type of material that is frozen. The maximum known thickness of permafrost is about 1500 m in the northern Lena and Yana River basins in Siberia.

Under normal conditions, frozen ground thaws to a depth of a few tens of centimetres to, at most, a few metres during summer. This *active layer* then refreezes in the fall. However, more extensive thawing of permafrost can occur when the ground is disturbed by human activity, wildfire, or when the climate warms. Thawing can produce subsidence, especially if the sediment contains a large amount of ice. Land subsidence of several metres or more is possible from thawing of permafrost.[9] An irregular surface produced by permafrost thaw is known as *thermokarst*. Climatic warming over the past four decades has thawed large areas of Arctic permafrost in Canada, forming unstable, water-filled depressions (Figure 8.9). In some areas, the *permafrost table* is dropping at rates of close to 20 cm per year.[10]

Some sediments, both in areas of permafrost and in areas that experience seasonal freezing but lack permafrost, expand when they freeze. **Frost-susceptible sediments** (engineers refer to them as "frost-susceptible soils") expand when they freeze due to the 9 percent increase in volume that occurs when water changes to ice. This volume increase causes an upward movement of sediment particles and the

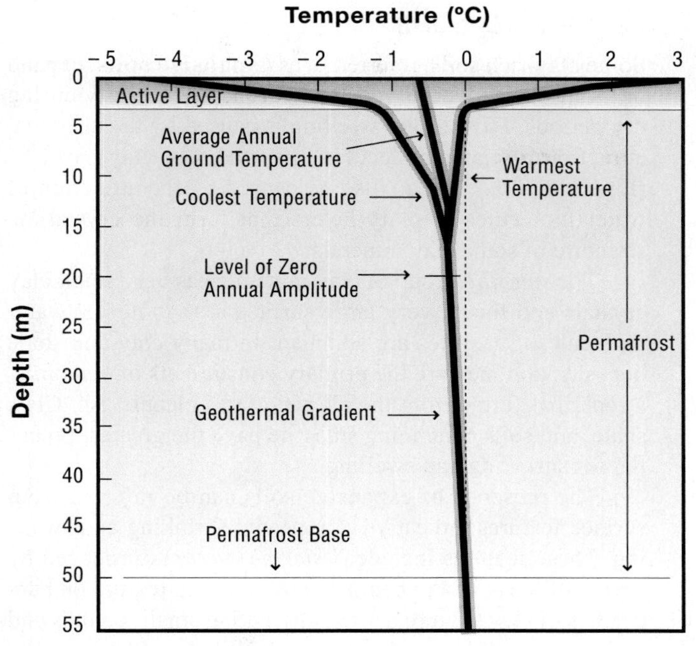

◄ **FIGURE 8.8 GROUND TEMPERATURE AS A FUNCTION OF DEPTH NEAR YELLOWKNIFE** The curves in this diagram show the coolest temperatures of the year (left), the warmest temperatures of the year (right), and the mean annual ground temperature (middle). The seasonal difference in temperature decreases with increasing depth to about 15 m, where there is no discernable change in temperature over the course of a year. Every summer, a portion of the ground just below the surface—the active layer—rises above 0°C and thaws, only to refreeze in the fall. *(Burgess, M. M., and S. L. Smith. 2000. "Shallow ground temperatures." In L. D. Dyke and G. R. Brooks (eds.),* The Physical Environment of the Mackenzie Valley, Northwest Territories: A Base Line for the Assessment of Environmental Change. Geological Survey of Canada Bulletin 547. *Reproduced with the permission of the Minister of Public Works and Government Services Canada, 2013.)*

land surface, a process known as **frost heaving**. This process is most common in silty sediments. Upward movement is generally much greater than 9 percent because additional water is drawn toward the zone of freezing, thereby increasing the amount of ice that forms. The ice can segregate into lenses within the sediment.

Piping

As groundwater percolates slowly through loose sediments, it can pick up particles of silt and sand in the subsurface and slowly carry them laterally to a point of water discharge on a slope (a spring). Over time, shallow subterranean tunnels and cavities may develop in the sediments and ultimately collapse to produce surface depressions and ravines.[11] This process is called **piping**. It is most common in silty and sandy sediments. Karst-like depressions formed by piping are present on inhabited benches underlain by glacial lake silts in Kamloops, British Columbia.[12] Occasionally, new depressions form, damaging structures that overlie the developing subterranean cavities.

Sediment Compaction

Peat and other organic sediments, and loose clay, silt, and sand are susceptible to subsidence. Subsidence occurs as the sediment compacts or loses water. Sediment compaction can occur naturally or as the result of human activities.

Fine Sediment Clay, silt, and sand commonly contain abundant *pore water*—water occupying spaces among the particles that constitute the sediment. Over time, the amount

◄ **FIGURE 8.9 THERMOKARST** Thermokarst ponds are common on this lowland bordering Hudson Bay in Manitoba. This scene shows a mosaic of frozen peat and shallow ponds that occupy depressions where permafrost has thawed. The ponds range in width from several metres to several hundred metres. *(Lynda Dredge/Geological Survey of Canada)*

of pore water decreases and the sediment compacts. Rapid deposition and compaction are especially common on river deltas. In a natural system, compaction of delta sediments is balanced by new sediment deposition, keeping the surface of the delta, called the *delta plain,* from sinking below sea or lake level. However, most deltas are favoured sites for cities, farms, and other infrastructure, all of which must be protected from flooding with *dykes*. But protective dykes prevent new sediment from accumulating on the delta plain; thus the land surface gradually subsides. Sedimentation on the Mississippi River delta plain, for example, was stopped by levees built on both sides of the river by the U.S. Army Corps of Engineers. These levees have protected communities, like New Orleans, from river flooding, but they have also kept new sediment from being added to the delta plain. Similarly, dykes protect 250 000 people and billions of dollars of infrastructure on the Fraser River delta south of Vancouver, British Columbia. Much of that delta plain is at or just below mean sea level and is slowly subsiding because it is no longer being replenished with new sediment. Construction of the Aswan dam upstream of the Nile River delta and the diversion of two-thirds of the river water into canals have stopped sediment from reaching much of the delta plain.[13]

Collapsible Sediments Some windblown silt deposits, referred to as *loess*, and some silty and sandy stream deposits in arid regions have a large amount of pore space and grains that are loosely bound or are water-soluble. These deposits may remain dry long after they form. Large amounts of infiltrating water, however, will weaken the bonds and dissolve minerals that hold the sediments together. The entire deposit may then collapse, lowering the land surface, in some cases, over a metre.[14] Sediments that are prone to this behaviour are referred to as **collapsible sediments** (engineers refer to them as "collapsible soils").[15]

Organic Sediments A variety of **organic sediments** accumulate in marshes, bogs, swamps, and other wetlands. These sediments contain large amounts of water and partially decayed leaves, stems, roots, and, in colder regions, moss. *Peat*, which contains more than 50 percent organic matter, compacts to a fraction of its original thickness when it loses water.

Bacterial decomposition of peat and other organic sediments converts organic carbon compounds to carbon dioxide, methane, and water. Water and wind erosion and burning of peat lands also destroy dry organic sediments. The decomposition, erosion, and burning of organic sediments cause the irreversible subsidence of drained wetlands. Subsidence of New Orleans has been aggravated by draining of wetlands and destruction of organic sediments.[16]

A dramatic example of wetland subsidence is the Florida Everglades. Droughts combined with land drainage, primarily for agriculture and urban development, caused more than half of the Everglades to subside between 0.3 m and 3 m during the twentieth century.[17]

Expansive Soils

Some clay-rich soils, referred to as **expansive soils**, expand or swell during wet periods and shrink significantly during dry periods. Most of the swelling is caused by the chemical attraction of water molecules to surfaces of clay particles (Figure 8.10a).[18] It can also be caused by the attraction of water molecules to platy layers that form the crystalline structure of some clay minerals.

The *smectite* group of clay minerals has very small clay crystals and thus a very large surface area to attract water molecules. Smectites are abundant in many clay and shale deposits, and they are the primary constituents of *bentonite*, a rock that forms from the alteration of volcanic ash. Clay, shale, and soils containing smectite have the greatest potential for shrinking and swelling.

The presence of expansive soils can be inferred from surface features indicative of repeated shrinking and swelling. These features include *desiccation cracks* produced by drying of the soil, a popcorn-like weathering texture on bare patches of clay (Figure 8.10c), alternating small mounds and depressions, tilting and cracking of blocks of concrete in sidewalks and foundations (Figure 8.10d), wavy bumps in asphalt pavement (Figure 8.11), and random tilting of utility poles and gravestones.

Structural damage to buildings located on expansive soil is caused by changes in the moisture content and therefore the volume of the soil. Factors that affect the moisture content of an expansive soil include climate, vegetation, topography, and drainage.[19] Regions with a pronounced wet season followed by a dry season are more likely to experience problems than regions where precipitation is more evenly distributed throughout the year.

Vegetation exerts a strong influence on the moisture content of a soil. Trees use local soil moisture for respiration and photosynthesis. Withdrawal of water from the ground may produce soil shrinkage (Figure 8.10b).

Earthquakes

Earthquakes can lower or raise the ground surface over large areas. As mentioned in Chapter 3, the 1964 Alaskan earthquake produced subsidence over tens of thousands of square kilometres, while other areas were elevated. Some coastal areas subsided up to 2 m, flooding some communities. The geologic record shows that similar great earthquakes at the Cascadia subduction zone in the Pacific Northwest have repeatedly lowered the outer coasts of Vancouver Island, Washington, and Oregon[20] (see Chapter 3).

Deflation of Magma Chambers

Subsidence can also result from volcanic activity. As magma moves upward underneath or into a volcano, the surface of the volcano may be forced upward. When the volcano erupts, the underground magma chamber is partly or completely emptied and the surface subsides. Ground-

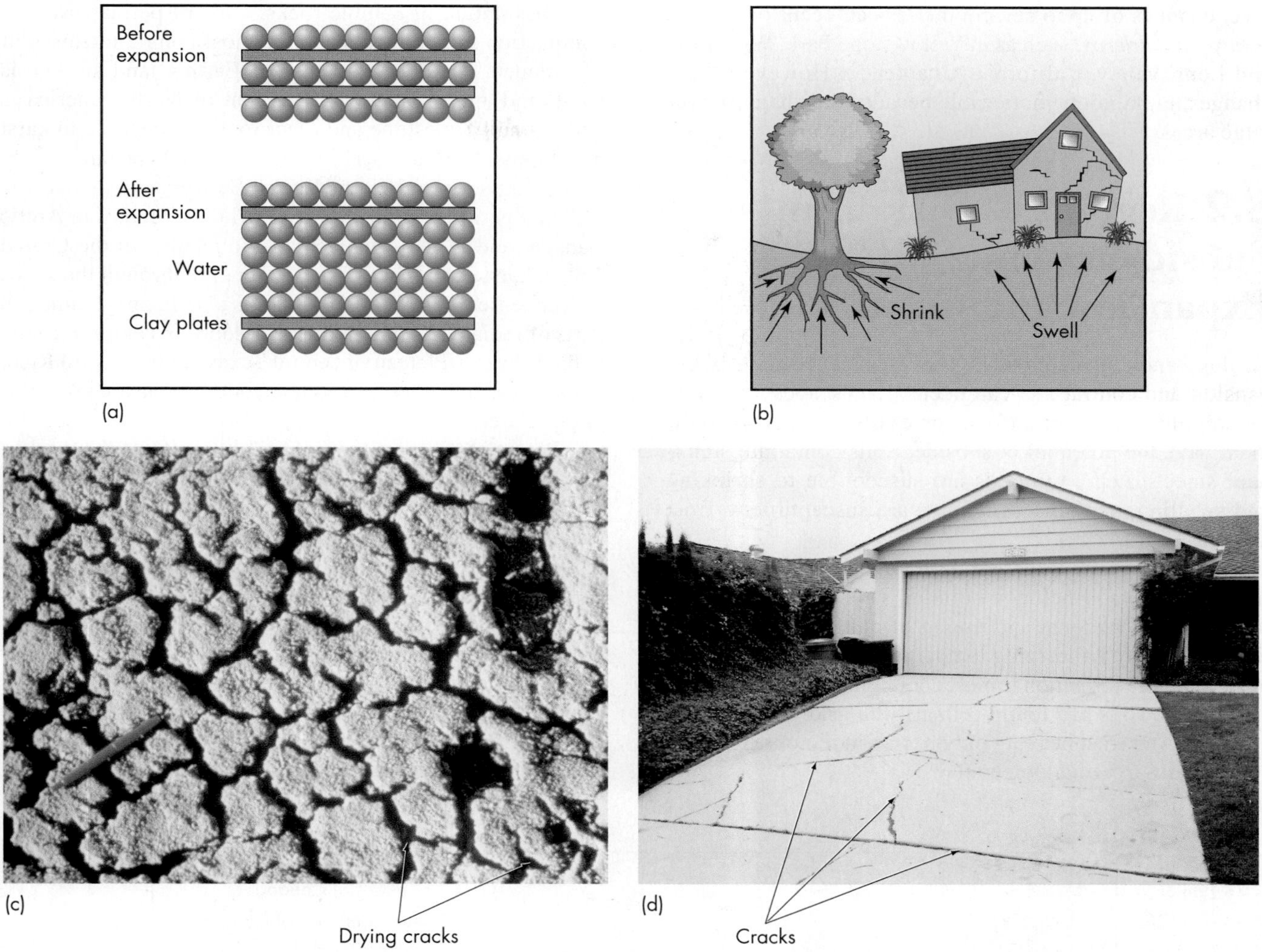

Before expansion

After expansion

Water

Clay plates

(a)

Shrink

Swell

(b)

(c)

Drying cracks

(d)

Cracks

▲ **FIGURE 8.10 EXPANSIVE SOILS** (a) Smectite is a group of clay minerals that expand as water molecules are added to the clay particles. (b) Effects of the shrinking and swelling of clay at a home site. (c) Drying of an expansive soil produces this popcorn-like surface texture and a network of polygonal desiccation cracks. (d) Shrinking and swelling of expansive soil cracked the concrete in this driveway. *((a&b) Based on Mathewson, C. C., and J. P. Castleberry, II.* Expansive Soils: Their Engineering Geology. *College Station, TX: Texas A&M University. (c) U.S. Geological Survey. (d) Edward A. Keller))*

◄ **FIGURE 8.11 DAMAGE FROM EXPANSIVE SOIL** Uneven shrinking and swelling of expansive clays in layers of steeply dipping bedrock produced the rolling surface of this road and sidewalk in Colorado. *(David C. Noe, Colorado Geological Survey)*

level changes of up to several metres can occur over giant *resurgent calderas*, such as in Yellowstone Park, Wyoming, and Long Valley, California (Chapter 5). However, these changes might not be noticeable because they happen over large areas.

8.2 Regions at Risk from Subsidence and Soil Expansion and Contraction

Geology and climate dictate where subsidence and soil expansion and contraction can occur. Landscapes underlain by soluble rocks, permafrost, or easily compacted sediment have the potential to subside. Soils containing abundant smectite clay minerals are susceptible to shrinking and swelling, and silty sediments are susceptible to frost heaving.

Geology alone does not determine if a sinkhole will form or if the soil below a building will expand. Climate, principally the amount and timing of rainfall and the duration and pattern of freezing temperatures, is also a determinant. Sinkhole formation is most common in humid climates. Collapsible soils are found primarily in arid and semiarid climates. And frost heaving only occurs in regions with seasonal below-freezing temperatures.

Dissolution of soluble rocks, thaw of permafrost, and compaction of sediments are the most common causes of subsidence. Nearly 10 percent of Earth's land surface is karst, and approximately 25 percent of North America is underlain by limestone and other rocks susceptible to karst development (Figure 8.12). Karst in Canada occurs mainly on Vancouver Island; in the Rocky, Selwyn, and Mackenzie Mountains; in the Hudson Bay lowlands; locally in Arctic Canada; and in the Appalachian Mountains. In the United States, karst occurs in a region extending through the states of Tennessee, Virginia, Maryland, and Pennsylvania; in parts of Indiana, Kentucky, and Missouri; in central Florida; in the Edwards Plateau of central Texas; and in Puerto Rico. Subsidence and other karst-related phenomena are problems in these areas.

Permafrost covers more than 20 percent of the world's land surface.[8] As the climate in the Arctic continues to warm, the area of permafrost is decreasing, and many towns and settlements are being threatened. Examples include Barrow in Alaska; Tuktoyaktuk and Pangnirtung in Canada; and Yakutsk, Norilsk, and Vorkuta in Russia.[8]

Subsidence caused by compaction of sediment is most pronounced on deltas, such as those of the Mississippi River in the United States, the Mackenzie and Fraser rivers in Canada, and the Nile River in Egypt.

Peat and other organic sediments underlie temperate- and cold-climate wetlands of Alaska, Canada, Siberia,

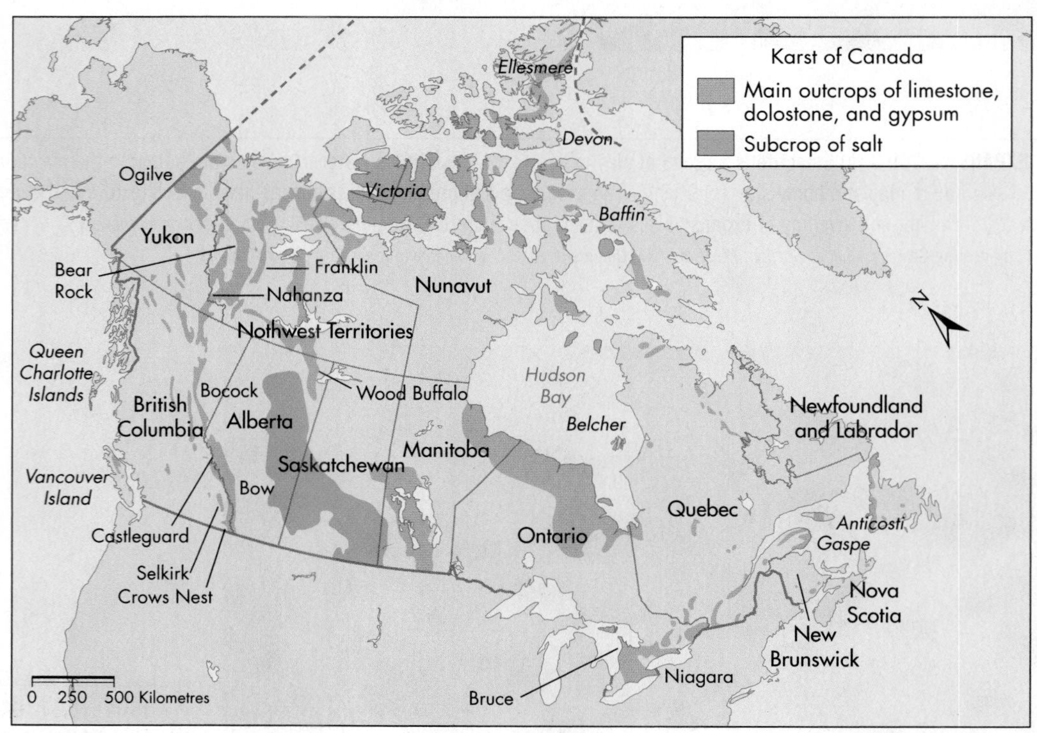

▲ **FIGURE 8.12 DISTRIBUTION OF KARST** Karst occurs in areas underlain by limestone, dolostone, and gypsum. This map shows the distribution of these rocks in Canada. Large areas of the conterminous United States, Alaska, and Puerto Rico also are underlain by these soluble rocks. *(© 2004,* Encyclopedia of Caves and Karst Science, *J. Gunn (ed.). Reproduced with permission)*

and northern Europe. These wetlands are variously called "bogs," "fens," "moors," and "muskeg." Coastal wetlands are particularly vulnerable to subsidence.

Expansive soils are a problem primarily in parts of southern Canada and the western United States. Earthquake-related subsidence is a risk on the west coast of North America. Deflation of a magma chamber can cause subsidence in any area of active volcanism.

8.3 Effects of Subsidence and Soil Expansion and Contraction

Subsidence and soil expansion and contraction cause much economic damage each year. Karst areas are subject to sinkhole collapse and groundwater pollution. Expansion and shrinkage of soils damage highways, buildings, pipelines, and other structures. Additional damage results from subsidence of delta plains, wetlands, and many areas underlain by thawing permafrost.

Sinkhole Formation

Sinkholes have caused considerable damage to highways, homes, sewage lines, and other structures. Natural and artificial fluctuations in the water table trigger most sinkhole collapse. Dissolution of limestone takes place mainly near the water table, and near-surface caverns enlarge when the water table is high. As long as a cavern is filled with water, the buoyancy of the water supports the weight of the overlying material. If, however, the water table drops, some of the buoyant support is lost and the cave roof may collapse. This situation was dramatically illustrated in Winter Park, Florida, on May 8, 1981, when a large collapse sinkhole began to develop. The sinkhole grew rapidly and within 24 hours had swallowed a house, part of a community swimming pool, half of a six-lane highway, parts of three businesses, and parking lots containing several vehicles (Figure 8.3).[21] Damage exceeded U.S.$2 million.

Groundwater Use and Contamination

Karst areas have abundant groundwater that is intensively used by people but is also easily polluted. Sinkholes and caves provide direct connections between surface water and groundwater (Figure 8.13). They make the groundwater vulnerable to pollution and to water-table lowering during droughts. One common source of pollution is waste carelessly discarded in sinkholes. Groundwater can also be contaminated where polluted water from surface streams flows into caves and fractures. Polluted surface water reaches the groundwater table without having been naturally filtered by sediment or rock.

Water-table fluctuations in karst areas affect people, plants, and wildlife. For example, groundwater is extracted from karst on the Edwards Plateau for use in towns and cities throughout central Texas. Frequent droughts in this region rapidly lower the water table and reduce or stop water flow from springs, threatening unique plants and animals that are found only in and around the springs.

Permafrost Thaw

Early settlers in Yukon Territory and Alaska built their homes directly on permafrost, only to find that heat from the floors of the buildings thawed the soil, causing irregular

◄ **FIGURE 8.13 WATER FROM A SUBTERRANEAN STREAM** Groundwater and surface water in karst areas are intimately connected. This waterfall discharges groundwater from Falling Spring northwest of West Union, Iowa. *(Scott Camazine/Science Source)*

◄ **FIGURE 8.14 THAWING PERMAFROST DESTROYS BUILDING** This apartment building in Cherskii in eastern Siberia partially collapsed because the permafrost on which it was built is thawing. Structural damage from thawing permafrost is becoming common in Canada, Russia, and Alaska. *(Professor V. E. Romanovsky, University of Alaska Fairbanks)*

ground settlement. By the middle of the twentieth century, most buildings in these areas were placed on columns, or piles, sunk into the permafrost. By elevating the floors of the buildings, the piles kept the structures' heat away from the ground. These protective strategies, however, assumed the climate would not change. That assumption has been proven wrong by developments over the past several decades.

Climate at high latitudes has warmed markedly in recent years, thawing some permafrost and causing roads to cave in, airport runways to fracture, and buildings to crack, tilt, or collapse (Case Study 8.1 and Figure 8.14). About 300 apartment buildings have been damaged in two Siberian cities alone.[8] The State of Alaska now spends around 4 percent of its annual budget repairing damage caused by seasonal and permanent thawing of permafrost.[22] The town of Tuktoyaktuk, in Canada's north, is suffering a two-pronged attack—it is being threatened not only by permafrost thaw, but also by coastal erosion. Waves generated in summer in the Beaufort Sea are attacking ice-rich sediments exposed along Tuktoyaktuk's shoreline. This erosion is forcing the community of 800 people to locate new construction away from the shore on the south side of town, and the entire town may ultimately have to be relocated.

Coastal Flooding and Loss of Wetlands

Flooding of low-lying coastal areas and destruction of wetlands are two major effects of ground subsidence. Subsidence of the Mississippi Delta during the past century has contributed to wetland loss and the sinking of New Orleans. Much of New Orleans is subsiding relative to sea level at an average rate of 8 mm per year[23] and is near or below the level of both the Gulf of Mexico and adjacent Lake Pontchartrain. Only a ring of levees and floodwalls surrounding the city keeps it from being flooded by the river, the lake, and the Gulf.

Outside the levees and floodwalls are wetlands that help protect the city and its suburbs from hurricanes. The progressive loss of wetlands through subsidence and draining contributed to the damage caused by Hurricane Katrina in 2005 (see Case Study 8.2). Without intervention, the remaining marshes will disappear by 2090 and New Orleans will be directly on the Gulf of Mexico.[16]

Soil Volume Changes

Soil expansion and frost heaving cause significant problems (Figure 8.16) and are responsible for several billion dollars of damage annually to highways, buildings, and other structures in Canada and the United States. In many years this cost exceeds that of all other natural hazards combined.

Every year more than 250 000 new houses are constructed on expansive soils in North America. About 60 percent of these houses will experience some minor damage, such as cracks in the foundation, walls, driveway, or walkway, and 10 percent will be seriously damaged, some beyond repair (Figures 8.10 and 8.11).[24] Underground water lines in expansive soils may rupture when there is a significant change in soil moisture. The resulting loss of water pressure makes the line vulnerable to contamination and requires customers to boil their water before consuming it.

Frost heaving has much the same effects as swelling of expansive soils. Foundations and pavement crack, and subsurface pipes may break. Frost heaving also moves larger particles in the soil upward at a faster rate than smaller particles. This upward movement, which can be as much as 5 cm per year, might simply be an annoyance—for example requiring removal of large stones in a garden or farm field. It can also be destructive—for example, fence posts and utility poles can be jacked out of the ground by frost. Even coffins have risen to the surface due to frost heaving.[25]

8.1 CASE STUDY

Permafrost Thaw in Canada's North

The ground in the Canadian Arctic is thawing beneath northerners' feet. Thawing of ice-rich permafrost has produced thermokarst terrain in many parts of the Arctic, changing drainage and vegetation, and causing lakes to expand or, conversely, to drain. Landsliding has raised suspended sediment concentrations in lakes and streams, adversely impacting aquatic habitats. Thawing of permafrost could also release significant amounts of carbon dioxide sequestered in wetlands.

Thawing permafrost has also damaged buildings, roads, and other infrastructure, leading to higher maintenance and mitigation costs. The problem in Inuvik, Northwest Territories, is one of the most serious in the country. Inuvik was established in 1953 as a replacement administrative centre for the hamlet of Aklavik, which was prone to flooding and had no room to expand (Figure 8.15). The Mackenzie Valley, in which Inuvik is located, experienced the largest increase in average air temperature in Canada over the past century. In the past 35 years, the ground temperature around Inuvik has risen to about –1°C from about –3°C,[26] and thawing permafrost is damaging some of the buildings in the town.

An example of the seriousness of the problem is the now-closed Arctic Tern female offender facility in Inuvik. The facility was built on ice-rich sediments that have thawed since it opened. The concrete floor of the building has settled irregularly up to 20 cm, opening gaps in the floor and walls.[27] An official of the Northwest Territories Association of Communities estimated that the cost of repairing this and other damaged buildings in the town, which has a population of only 3500 people, could reach $121 million.[27]

In November 2012, the mayor of Inuvik, Floyd Roland, spoke to media about the problem and the impact the changing climate is having on his town:

> We're in a zone that is permafrost rich, as we call it, and we've had to adapt our construction techniques up here to avoid causing any melting and

▲ **FIGURE 8.15 INUVIK, A TOWN BUILT ON PERMAFROST** Inuvik is a town of 3500 people in the Northwest Territories, located about 200 km north of the Arctic Circle. A distinctive feature of the town is "utilidors"—above-ground conduits carrying water and sewage. Utilidors are necessary because the ground underlying Inuvik is frozen. An increase in ground temperature from about –3°C to –1°C in the past 35 years has increased permafrost thaw, damaging some buildings. *(© Staffan Widstrand/Corbis)*

8.1 CASE STUDY *(Continued)*

slumping of permafrost areas—we've noticed an increase in the last number of years.

We've been dealing with permafrost conditions, both the immediate building of the community, the infrastructure as well remediation as things start to change, the seasons start to change, and you have to adapt for that when it comes to our runways, our roadways.

The permafrost was more stable—our seasons have definitely changed up here. When I grew up as a young boy here our winter season was coming from middle of September; for example the rivers would start to freeze.

And we'd have much colder winters where nowadays it's sometimes middle of October and we don't get the extreme cold that we normally did.[28]

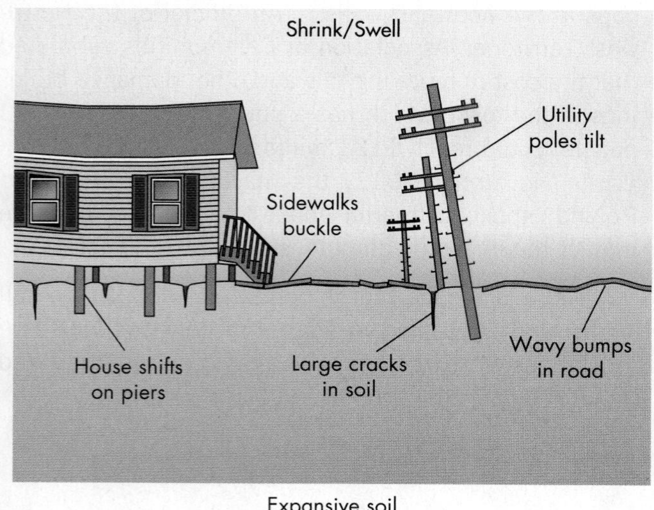

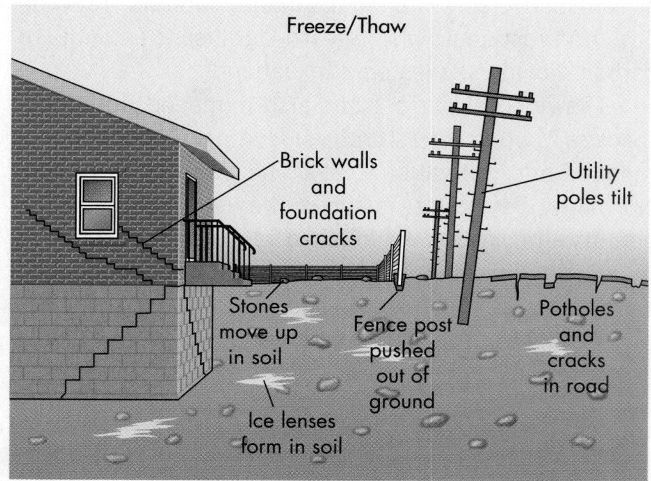

▲ **FIGURE 8.16 SOIL MOVEMENT EFFECTS** (a) Shrinking and swelling of expansive soils and (b) freezing and thawing of frost-susceptible soils have similar effects. They crack and shift foundations and pavement, and tilt utility poles, road signs, and fence posts. Soil contraction can produce deep cracks in the ground, and soil expansion can produce an irregular land surface. Freeze-thaw activity moves large objects such as stones upward in the soil to the surface.

8.4 Links between Subsidence, Soil Volume Changes, and Other Natural Hazards

As mentioned previously, subsidence can be a side effect of earthquakes, volcanic activity, and climate change. In addition, subsidence can amplify other hazardous processes. The link between subsidence and flooding, for example, is a common one.

Flooding can be a severe problem in areas that are rapidly subsiding, especially where subsidence is being caused by the over-pumping of groundwater (see the opening to this chapter). In many cities, the demand for clean drinking water leads to depletion of the resource and subsidence. Overexploitation of groundwater is partly implicated in the increase in the number and severity of coastal and river floods in some coastal cities such as Bangkok, Thailand. Subsidence induced by groundwater extraction has also exacerbated flooding in parts of Mexico City.[29]

Permafrost thaw, frost heaving, and shrinking and swelling of clay soils drive mass wasting in some landscapes. They increase the rate of gravity-driven creep, cause landslides, and produce uneven land surfaces.

Subsidence can also have links to climate change. Drought conditions in arid areas commonly lower the groundwater table, causing unconsolidated sediments to compact. In the deserts of the southwestern United States, drought is contributing to regional subsidence and to the formation of large, polygonal *desiccation cracks*. These cracks are the same size and depth as linear cracks, called *earth fissures*, produced by over-pumping of groundwater.[30,31]

Climate warming is the primary cause of thawing of permafrost in the Arctic and of the current global rise in sea level. Thawing permafrost, in turn, accelerates wave erosion of ice-rich sediments along Arctic Ocean shorelines. In coastal areas, such as Venice, New Orleans, and the Nile

8.2 CASE STUDY

Loss of Wetlands on the Mississippi Delta

The Mississippi Delta supports nearly 40 percent of the coastal swamps and marshes in the United States[16] and is the largest wetland complex in North America. The present delta is geologically young, having formed at the mouths of the Mississippi and Atchafalaya rivers over the past 7500 years.[32] As the two rivers approach the sea, they divide into a number of *distributary channels* that cross the delta plain (Figure 8.17). Each of these distributaries is building a lobe of sand and mud into the Gulf of Mexico. Floodwaters may cut through the channel banks, allowing the river to take a shorter path to the sea; the river may then partially or totally abandon its old distributary and begin to spread a new lobe of deltaic sediment seaward at the mouth of the new channel. This process has produced a complex of overlapping deltaic lobes that collectively extend for more than 500 km along the coast of Louisiana.

As the channels of the Mississippi and Atchafalaya rivers have shifted, two natural processes have kept the delta plain above sea level: deposition of sand and mud during floods, and accumulation of organic sediment in swamps and marshes. All these sediments are loose and water-saturated, much like a wet sponge. Over time, they compact and the surface subsides unless replenished with new sediment.

▲ **FIGURE 8.17 THE MISSISSIPPI RIVER DELTA** The Mississippi River transports large amounts of sediment into the Gulf of Mexico, as seen in this satellite image acquired on February 6, 2007. The watershed of the Mississippi River is the largest in North America and the third largest in the world. Over the past century, confinement of river flow to levee-bound channels has caused extensive loss of wetlands. If the current rate of loss (65 km^2 per year) continues, Louisiana could be left with less than 5 percent of the wetlands it had in the 1930s. As wetlands are lost, parts of Louisiana, such as New Orleans (the light yellow crescent just below the inland lake, Lake Pontchartrain), are exposed to storms coming in from the Gulf of Mexico. Many scientists attribute the scale of the Hurricane Katrina disaster in 2005 to the loss of surrounding wetlands. *(© European Space Agency)*

8.2 CASE STUDY *(Continued)*

Nearly three centuries ago, French settlers in New Orleans began to construct levees along the Mississippi River, restricting natural sedimentation on the delta plain and thus changing forever the balance between sedimentation and subsidence. By the early twentieth century, the U.S. Army Corps of Engineers completed levees on both sides of the river, which prevented the Mississippi from flooding its plain. Without these floods, the delta plain was no longer replenished with sediment and subsidence accelerated.

Sediment starvation, however, has not been the only cause of subsidence. Large amounts of groundwater, soil, and natural gas have been extracted from beneath the delta plain, exacerbating the problem.[33]

A consequence of increased subsidence of the Mississippi Delta has been a major loss of wetlands. According to one source, an area of wetlands the size of a football field is lost every 30 minutes; this translates to the disappearance of 65 km^2 each year.[34] Wetland loss increases the vulnerability of the coastline, including New Orleans, to hurricane storm surges and has led to major economic losses to fisheries.

Rising sea level and construction of navigation and drainage canals have worsened the loss of Mississippi Delta wetlands. Canals built primarily for oil and gas operations have exposed the delta plain to erosion and increased the salinity of marsh waters, killing many freshwater-dependent plant communities[35] (Figure 8.18).

About 25 percent of the coastal wetlands in Louisiana disappeared in the past century through subsidence, erosion, and canal construction. What will happen to the remaining wetlands, which cover an area of about 14 000 km^2? More than a decade ago, scientists, engineers, government officials, and business leaders in Louisiana concluded that a massive wetland and barrier island restoration effort was needed to reverse the loss of coastal wetlands in the state[34] (Figure 8.19). This project will likely take 50 years and cost more than U.S.$14 billion. Although $14 billion is a lot of money, it is a small fraction of the $135 billion in property damage caused by Hurricane Katrina.

▲ **FIGURE 8.18 WETLAND LOSS ON THE MISSISSIPPI DELTA** Excavation of numerous canals for oil and gas drilling, like these near Leesville, Louisiana, have contributed to wetland loss on the Mississippi Delta. Each canal drains the adjacent wetland and exposes organic soil to decomposition and erosion. Loss of this soil causes the land surface to subside. In many areas, the only land remaining is the piles of sediment that were dredged to make the canals. *(Philip Gould/CORBIS)*

▲ **FIGURE 8.19 BREACHING A NATURAL LEVEE TO RESTORE A WETLAND** The floating dredge *California* is cutting a channel through the natural levee along the west bank of the Mississippi River south of New Orleans. The channel is part of a U.S.$22 million project to divert sediment and water into West Bay. The aim of the project, which was completed in 2003, was to allow the river to establish a new wetland in the bay. *(U.S. Army Corp of Engineers, New Orleans District)*

River delta, subsidence adds to the rise of sea level caused by climate warming, increasing the resulting rate of loss of land.

8.5 Natural Service Functions of Subsidence

As we have seen, subsidence causes many environmental and economic problems, but it also provides some benefits. Karst formations are some of the world's most productive sources of clean water, and cavern systems and tower karst are important aesthetic and scientific resources (Figure 8.5). The caves of karst areas are also home to rare, specially adapted creatures.

Water Supply

About 25 percent of the world's population gets its drinking water from karst formations, and 40 percent of the U.S. population relies on water from karst terrains.[36] For example, carbonate rocks beneath the Edwards Plateau, in Texas, provide drinking water for more than 2 million people. Only a small percentage of Canadians draw their drinking water from karst, but karst rocks are common in Canada—limestones and other soluble rocks form over 10 percent of the country's landmass.

Aesthetic and Scientific Resources

Karst is an important aesthetic resource. Rolling hills, extensive cave systems, and beautiful formations of tower karst are striking landscape features found in karst terrains. Caves are popular destinations for both spelunkers and tourists. Mammoth Cave National Park, Kentucky, contains the world's longest cave system and attracts about 2 million visitors each year.

Aesthetics aside, karst regions provide scientists with a natural laboratory in which to study climate change. Stalactites, stalagmites, and flowstone contain records of changes in the ratios of two important stable isotopes of oxygen (^{16}O and ^{18}O), from which changes in past cave temperatures and there-

▲ **FIGURE 8.20 CAVES ARE UNIQUE ECOSYSTEMS** Unique species, such as this endangered Texas blind salamander from Ezell's Cave National Natural Landmark in central Texas, have evolved to live in the total darkness of caves. *(Robert Mitchell Photography)*

fore climate can be reconstructed. The caves also provide an ideal environment for preserving animal remains, making them important resources to palaeontologists and archaeologists.

Unique Ecosystems

Caves are home to rare creatures that are adapted to live only in this environment. Caves beneath the Edwards Plateau in Texas, for example, are home to more than 40 unique species, eight of which are legally designated as endangered (Figure 8.20). Karst-dependent species known as *troglobites* have evolved to live in the total darkness of caves; they include flatworms, beetles, eyeless fish, shrimp, and salamanders. Other species, such as bats, rely on caves for shelter.

8.6 Human Interaction with Subsidence

People living in areas of karst, subsiding ground, permafrost, or expansive soils can compound existing problems and create new ones. They contribute to problems caused by subsidence by withdrawing subsurface fluids, excavating underground mines, thawing frozen ground, restricting deltaic sedimentation, draining wetlands, and using poor landscaping practices.

Withdrawal of Fluids

Withdrawal of subsurface fluids such as oil, gas, and groundwater can induce or increase subsidence.[37] Fluid pressures in sediment and rock help support the material above, much as buoyancy makes a heavy object at the bottom of a swimming pool seem lighter. Removal of the fluid by pumping reduces support and causes surface subsidence.

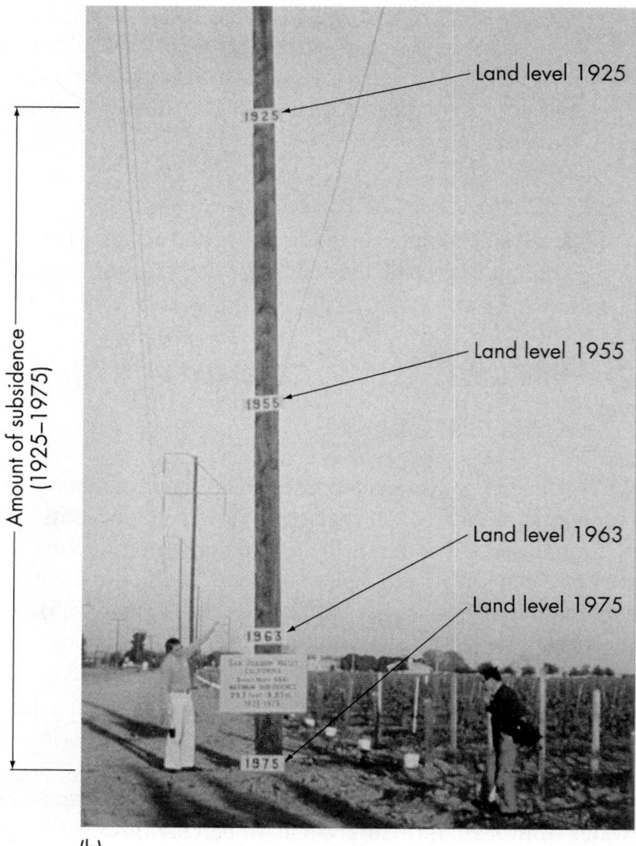

▲ **FIGURE 8.21 LAND SUBSIDENCE CAUSED BY GROUNDWATER EXTRACTION** (a) Areas of land subsidence in California resulting from groundwater removal. (b) Subsidence in the San Joaquin Valley, California. The marks on the telephone pole are the positions of the ground surface in recent decades. The total amount of subsidence is nearly 8 m. *((a) From The Geological Society of America; (b) Ray Kenny PhD.)*

A classic example of this phenomenon can be found in the Central Valley of California, where thousands of square kilometres of land have subsided due to the removal of groundwater for irrigation and other uses (Figure 8.21a). One of the areas of greatest subsidence is centred around Los Banos and Kettleman City, where most of the land has subsided more than 0.3 m and the maximum subsidence is about 9 m (Figure 8.21b). As the water was mined, fluid pressure was reduced, sedimentary grains compacted, and the surface subsided (Figure 8.22).[38,39] Subsidence caused by over-pumping has also been documented near Phoenix, Arizona; Las Vegas, Nevada; the Houston-Galveston area in Texas; San Jose, California; and Mexico City, Mexico.

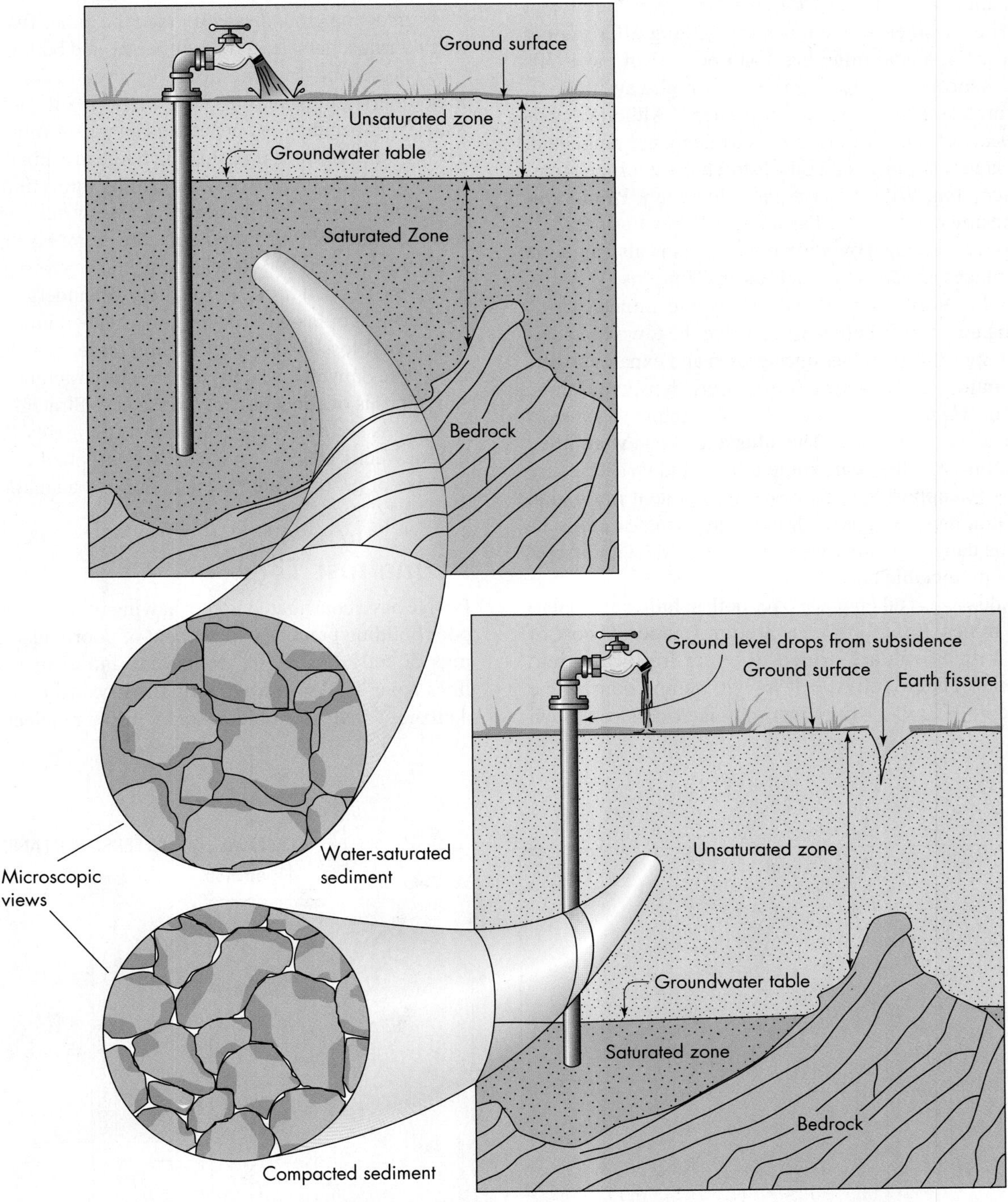

▲ **FIGURE 8.22 SUBSIDENCE FROM PUMPING OF FLUIDS** Idealized diagram showing how groundwater pumping can cause ground subsidence. The unsaturated zone is above the groundwater table where the space between sediment grains contains both air and water. In the saturated zone beneath the groundwater table, the pores are completely filled with water. When groundwater is removed, the pores become smaller as grains become more tightly packed. This compaction causes the land surface to subside. *(Based on Kenny, R. 1992. "Fissures."* Earth *2(3):34–41)*

Underground Mining

Sudden surface subsidence has accompanied or followed underground mining of coal, salt, and phosphate. Most subsidence over coal mines is caused by the failure of pillars of coal left behind to support the mine roof. With time, the pillars weather, weaken, and collapse. The roof then caves in and the land surface above the mine subsides. In the United States more than 8000 km^2 of land, an area twice the size of Rhode Island, have subsided because of underground coal mining. The subsidence continues today, long after mining ended. In 1995, a coal mine that last operated in the 1930s collapsed beneath a 600-m section of a highway in Ohio; repairs were completed three months later.[40] Although coal-mine subsidence most often affects farmland and rangeland, it has damaged buildings and other structures in urban areas, such as Scranton, Wilkes-Barre, and Pittsburgh, Pennsylvania; Youngstown, Ohio; and Farmington, West Virginia.[40,41]

Ground subsidence over mine workings is also an issue in some places in Canada, such as in Timmins, Ontario, where gold is still being mined. Many old mines, which closed long ago, were shallow but outside the town boundaries at the time. The town has since grown and expanded over these old mines. The collapse of old *adits* (horizontal mine tunnels) and *shafts* caused considerable damage to some of the newer parts of the town. The adits and shafts were filled with sand, but over the years some of the sand washed away and ceased to support the adit roofs. Subsequent mitigation involved bracing the adits with concrete. This example illustrates the danger of allowing urban expansion without recognizing a predictable hazard.

Subsidence has taken place over both solution and open-shaft salt mines. Solution mines are the source of most of our table salt. Shafts are drilled to inject freshwater into salt deposits. The dissolved salt is then pumped out of the shaft, leaving behind a cavity that may later collapse. Such collapses and subsequent surface subsidence have occurred at solution mines in Kansas, Michigan, and Texas.

Open-shaft mining is used to extract rock salt beneath Goderich, Ontario; Detroit, Michigan; and Cleveland, Ohio; and potash from sedimentary rocks beneath Saskatchewan. In the past 35 years, two underground salt mines in the United States have flooded catastrophically, causing subsidence and surface damage. One, the Retsof Mine near Geneseo, New York, was once the largest salt mine in the world. Its roof collapsed in 1994, allowing groundwater to flood the mine. Two large sinkholes formed, damaging roads, utilities, and buildings.[42]

The second flooding event occurred at Jefferson Island in southern Louisiana in 1980, when an oil rig drilling for natural gas accidentally penetrated an underground salt mine. The rig was mounted on a floating barge in a small lake above a salt dome (Figure 8.23). After drilling into a mine shaft, the rig toppled over and disappeared as lake water began to drain into the mine. Within three hours, the entire lake had drained and a 90-m-deep, 800-m-wide subsidence crater had formed above the flooded mine. Fortunately, 50 underground miners and seven people on the drilling rig escaped injury. The mine was a total loss, and buildings and gardens on Jefferson Island were damaged by subsidence. The structural integrity of salt mines is of concern because the U.S. Strategic Petroleum Reserve is stored in four Gulf Coast salt mines. A fifth mine below Weeks Island, Louisiana, was emptied of oil in 1999 because of groundwater seepage through a sinkhole.

Permafrost Thaw

People have contributed to the thawing of permafrost through poor building practices. Placement of poorly insulated buildings directly on frozen ground and burial of warm utility lines have locally thawed permafrost, causing considerable damage.[8,43] Most problems arose in the nineteenth century

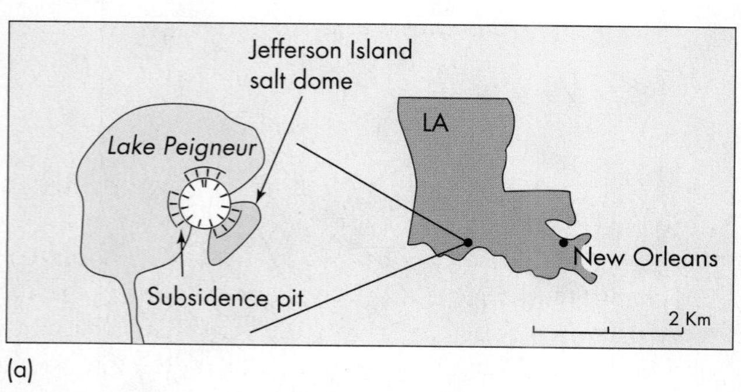

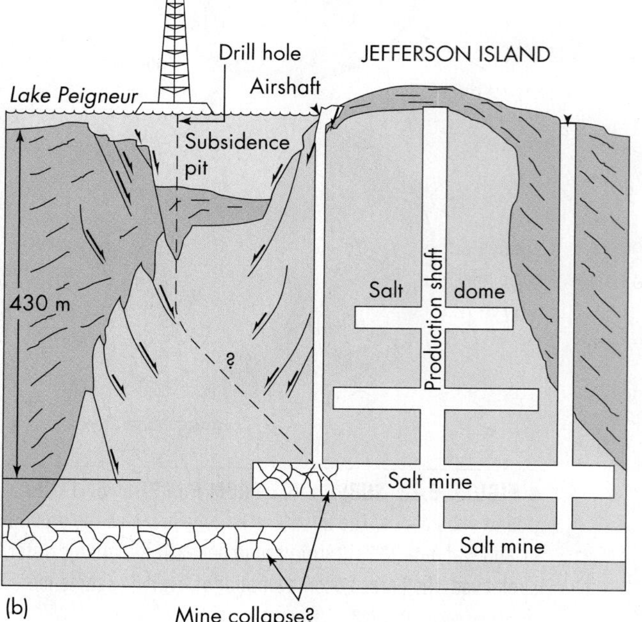

▲ **FIGURE 8.23 FLOODING AND COLLAPSE OF A SALT MINE** (a) The location of Lake Peigneur, Louisiana, and (b) an idealized diagram showing the Jefferson Island Salt Dome collapse. About 15 million m^3 of water flooded a salt mine when a shaft was penetrated during drilling.

and the first half of the twentieth century when permafrost was not understood. Notable examples include problems that arose from poor building practices in Dawson City during the Klondike gold rush at the end of the nineteenth century, and problems that beset the U.S. Army when it built the Alaska Highway in northern British Columbia, southwest Yukon Territory, and Alaska during the Second World War.

Restricting Deltaic Sedimentation

Marine deltas require continual addition of sediment to their surface to remain at or above sea level. People have reduced or stopped sedimentation on most delta plains by constructing upstream dams, by building levees at the sides of distributary channels, and by diverting sediment-laden river water into canals. All these practices can contribute to subsidence of a delta plain.

Draining Wetlands

People have drained wetlands for agriculture and settlement for centuries. Most of the western part of the Netherlands was drained for agriculture between the ninth and fourteenth centuries.[6] Draining of wetlands in the United States has caused or increased subsidence in the Florida Everglades and the Sacramento–San Joaquin Delta in California.

Some wetlands in Canada are drained to extract peat for horticultural use. The industry generates revenues of nearly $200 million per year, mainly in Alberta, Quebec, and New Brunswick. The total area of peat mined, about 17 000 ha, is small relative to the total area of wetlands in Canada (roughly 100 million ha), but the conventional mining practice is damaging and leads to a lowering of the wetland surface. A wetland that is about to be mined is first drained. Then peat is sucked from the dry surface with a vacuum harvester. Lowering of the wetland surface results from both removal of peat and compaction of the remaining organic sediments after mining is finished.

Landscaping on Expansive Soils

Problems related to shrinking and swelling of expansive soils can be amplified by poor landscaping practices. Planting trees and large shrubs close to buildings may cause damage from soil shrinkage during dry periods when plant roots pull moisture from the soil.[19] At the other extreme, planting a garden or grass that needs frequent watering close to a building can cause damage from soil swelling. Rather than maintaining the soil at a constant moisture level, watering systems commonly leave excess water in the soil. Excess water is the principal cause of damage from swelling soils.[19]

8.7 Minimizing Subsidence Hazards

It is difficult to prevent natural subsidence. However, people can take steps to minimize the risk of damage from this phenomenon.

Restricting Fluid Withdrawal

We will always be faced with subsidence problems in areas where carbonate bedrock is being aggressively dissolved or where groundwater levels are falling because of drought. We can, however, prevent human-caused subsidence associated with extraction of groundwater, oil, and gas.

Groundwater mining is the unsustainable extraction of groundwater—removal of more groundwater by wells and springs than is replenished through infiltration of rain and surface water. The practice commonly causes irreversible subsidence. Further subsidence can be minimized only by reducing fluid withdrawals.

From the early 1900s to the mid-1970s, groundwater mining in the Houston–Galveston area of Texas was the primary cause of up to 3 m of subsidence over an area of 8300 km^2.[44] In 1975, the problem prompted the Texas Legislature to create a regulatory district to issue well permits. Groundwater pumping was reduced and subsidence has greatly decreased or stopped altogether. The same cannot be said for parts of Florida, Arizona, and Nevada, where groundwater mining is continuing to produce subsidence, sinkholes, and earth fissures.

Injection wells are sometimes used to minimize or stop subsidence caused by fluid withdrawal. They were used with some success in the 1950s when water was injected at the same time that oil was being pumped at Long Beach, California. However, this method is not practical for **aquifers**, or subsurface water bodies, in porous fine-grained sediments because the sediment particles compact after the groundwater is removed and it is difficult to push them apart by injecting fluid.[45] Further, it seems pointless to replace groundwater with water injected from the surface.

Regulating Mining

The best way to prevent damage from mining-induced subsidence is to not allow mining in settled areas. Such laws are currently in place in many countries, but old abandoned mines still pose a threat.

Preventing Damage from Thawing Permafrost

Most engineering practices for building on permafrost presume the ground will remain frozen if heat from a building, pipeline, or other structure does not reach it. With recent thawing of permafrost caused by climate warming, new and more costly practices are being developed, such as placing buildings on adjustable screw jacks or lattice-like foundations to accommodate recurrent freezing and thawing of ground ice[8] and the use of heat pipes to transfer heat from the ground to the atmosphere. The heat pipes typically contain anhydrous ammonia, which turns into a vapour as it absorbs heat from the ground. The vapour flows through the pipe and condenses back into a liquid when it reaches the surface and comes into contact with cold above-ground air. As it condenses, the ammonia releases heat to the air.

Another strategy was used in the construction of a new $92 million K-12 school in Inuvik, which opened in 2012. The building is elevated above ground level, allowing heat to escape without warming the ground. Its foundation is seated on pipes filled with sand and sunk 20 m into the ground.[27] The pipes remain frozen and stable through the summer, providing a stable foundation for the school. Ground temperatures are monitored continuously. The architect and engineers also completed snow and wind studies to determine whether and where snow would build up around the building. Northwest Territories officials hope that the lessons learned from this facility can be applied to construction of other public buildings in the future.

Special precautions were taken to prevent thawing of permafrost during construction of the pipeline from oil fields on the north slope of Alaska to tidewater at Valdez. The oil had to be heated to lower its viscosity sufficiently to be conveyed hundreds of kilometres through pipes; thus the pipeline had to be elevated or insulated to prevent the transfer of the heat to the frozen ground (Figure 3.31).[46]

The Geological Survey of Canada has conducted research on permafrost in northern Canada for nearly 40 years in response to the need for knowledge to manage hydrocarbon exploration, development, and transportation.[47] More recently, additional permafrost research has been stimulated by mineral exploration in the North, especially for diamonds. Permafrost was a critical consideration in the design and management of tailings (crushed waste rock) piles at the Ekati and Diavik diamond mines in the Northwest Territories. Dams with frozen cores have been used both for tailings containment and drainage diversion.

Reducing Damage from Deltaic Subsidence

Preventing subsidence of delta plains and restoring deltas to a natural state are not feasible plans. Levees will continue to prevent sediment-laden floodwaters from entering urbanized areas, and pumping systems will remove excess surface water from the areas protected by the levees. However, in undeveloped areas, levees could be removed or breached to restore the supply of sediment and freshwater necessary to reestablish marshes. Restored marshes may help protect subsided urban areas from storms and rising seas.

Stopping the Draining of Wetlands

It is not possible to restore organic sediments that have dried out through the drainage of wetlands. Only proper water management of existing marshes and swamps will minimize their subsidence in the future.

Preventing Damage from Expansive Soils

Proper design of subsurface drains, rain gutters, and foundations can minimize damage caused by expansive soils. These techniques improve drainage and allow foundations to accommodate some shrinking and swelling of the soil.[19] Another preventive method is to construct buildings on a layer of compacted fill that forms a barrier between the structure above and the expansive soil below. The fill helps stabilize moisture levels in the soil and provides a stable base on which to build. For larger buildings, roads, and airports, it may be cost-effective to excavate the upper part of an expansive soil or to add lime to bind soil particles together. The Edmonton Convention Centre is an example of a large structure that was designed to accommodate swelling clays.

8.8 Perception of and Adjustments to Subsidence and Soil Hazards

Perception of Subsidence and Soil Hazards

Few people living in Canada are concerned about subsidence or soil expansion and contraction. However, people living in areas of karst or permafrost, on deltas, or where groundwater is being mined are more likely to understand the hazard. People living in areas where sinkholes are common are generally well aware of the hazard and perceive it to pose a real risk to property.

Adjustments to Subsidence and Soil Hazards

The most appropriate adjustment to subsidence and soil hazards is to avoid building in areas that are prone to the hazards. This approach is clearly not possible everywhere, because large areas of the Canadian Prairies and the western United States are underlain by swelling soils, much of Canada and Alaska has permafrost and organic sediments, and a significant portion of the eastern United States is underlain by karst. The best we can do is to identify high-risk areas in which construction should be prohibited or limited, or proceed with appropriate engineering. Unfortunately, in many areas, subsidence is difficult or impossible to predict. Some methods are available, however, to identify areas of potential subsidence and soil expansion and contraction.

Geologic and Soil Mapping Detailed geologic and soil maps can be made to accurately identify hazards. An understanding of the geology, soils, and surface and groundwater systems greatly helps in predicting and avoiding areas where subsidence and soil expansion and contraction might occur.

Surface Features Features such as cracks, hummocky ground, and closed depressions in areas underlain by limestone, rock salt, and gypsum are tell-tale signs of subsidence. The sudden appearance of cracks in the ground may indicate that collapse is imminent, and appropriate steps should be taken to avoid damage and injury. Cracks in the ground in arid areas may provide warning of damage from expansive soils or a falling water table.

Geophysical Surveys Knowledge of the subsurface environment is essential when decisions are made about where to build structures in karst terrain. The subsurface can be explored with *ground penetrating radar* (GPR) and other geophysical methods, and by drilling boreholes before construction begins. Additional geologic surveys may be needed to assess high-risk areas encountered during construction. In areas of expansive soils or permafrost, geotechnical borings and soil testing might be needed to properly design foundations.

Some American states, such as Colorado, require disclosure of the presence of expansive soils when houses are sold. Disclosure requirements apply to homebuilders, homeowners, and real estate brokers.[48] Homeowners who live in areas where subsidence has occurred in the past or is currently occurring should check the hazard coverage in their insurance policies; many insurers do not cover damage from sinkholes or mine subsidence.

REVISITING THE FUNDAMENTAL CONCEPTS

Subsidence and Soil Expansion and Contraction

❶ Hazards can be understood through scientific investigation and analysis.

❷ An understanding of hazardous processes is vital to evaluating risk.

❸ Hazards are commonly linked to each other and to the environment in which they occur.

❹ Population growth and socioeconomic changes increase the risk from natural hazards.

❺ Damage and loss of life from natural disasters can be reduced.

1. We can predict where subsidence is likely to occur based on geology and hydrology. Subsurface geophysical surveys may reveal large subsurface openings that could initiate subsidence, for example a collapse sinkhole. Regions where large areas of land would subside during a subduction earthquake can be identified. Areas where clay-rich soils will expand and contract can be mapped, and appropriate steps then taken to prevent damage to buildings and other structures. As a rule, however, predicting exactly when subsidence will actually happen is not possible.
2. Risk assessment for subsidence is based on the magnitude and probability of subsidence and the consequences. In the case of subsidence, consequences are easy to estimate; estimating the magnitude of the subsidence event is also possible, with some uncertainty. As mentioned above, however, the probability that subsidence will occur is more difficult to determine. Risk assessment for expansive soils is straightforward and is standard engineering practice before construction begins in regions where such soils occur.
3. Links exist between subsidence and other hazards. For example, subsidence in coastal areas can increase the possibility of coastal erosion and flooding (e.g., Venice, New Orleans, Mexico City). Prolonged drought and pumping of groundwater may lower the water table in an area, causing the ground to subside. Thawing of permafrost is closely linked to climate warming, which itself is linked to other hazards such as sea-level rise and, in some areas, more frequent landslides and debris flows.
4. Due to expanded urbanization related to population growth, more areas with subsidence hazards are being developed. As a result, the consequences of these hazards are increasing. In areas with extensive limestone, such as the southeastern United States, sinkhole collapses are increasingly damaging property because there are more people and buildings in these areas than ever before and more groundwater is being extracted. Expansive soils are an increasing issue in some areas of western Canada and the southwestern United States.
5. Although permafrost thaw cannot be easily prevented in a warming climate, roads and buildings can be designed and built to minimize or prevent the transfer of heat from the structures to frozen ground and thus not make the problem worse. Expansive soils cause billions of dollars of damage every year. These costs can be greatly reduced with appropriate engineering. We know where expansive soils occur, and we can develop plans to minimize their impact.

Summary

Ground subsidence is caused by natural processes, human activities, or a combination of the two. Natural causes include dissolution of limestone, rock salt, or gypsum; lowering of the groundwater table or fluid pressures; thawing of permafrost; a reduction or cessation of sediment deposition on delta plains; draining of wetlands; shrinking of expansive soils; earthquakes; and deflation of magma chambers.

Dissolution of limestone by acidic groundwater creates a landscape of caves and sinkholes known as karst topography. Related features include disappearing streams, springs, and tower karst. Most sinkholes form by collapse of cavern roofs, often triggered by a falling water table or by groundwater mining.

During the past several decades, permafrost thaw has become a major hazard in the Arctic. Most of the thaw has resulted from recent climate warming. Thawing permafrost causes subsidence and can produce thermokarst, a terrain consisting of uneven ground with mounds, ponds, and lakes.

Saturated, loose sediment compacts when the water table falls or fluid pressure is reduced. A lowering of the water table may be natural or it may result from human activities such as groundwater mining. Subsidence can be irreversible if the sediment compacts or dries out. Surface features associated with compaction include large earth fissures and desiccation cracks.

Compaction and subsidence occur naturally on deltas. Sediment deposition during floods keeps pace with sediment compaction on deltas that are in their natural state. Construction of dams, levees, and canals interferes with sedimentation and can cause the delta surface to subside, aggravating flooding in urban areas that are not properly protected.

Soil expansion can cause the ground surface to move upward, whereas soil contraction can lower the surface. Upward movements occur in expansive and frost-susceptible soils; sinking takes place in collapsible or thawing soils.

Expansive soils contain smectite, which swells when it becomes wet and shrinks when it dries. Wetting and drying of this clay cause the soil to expand and contract, which can cause extensive structural damage. Factors that affect the moisture content of an expansive soil include climate, vegetation, topography, and drainage.

Most frost-susceptible sediments are silts that accumulate ice in pores or lenses. Growth of ice lenses displaces the surrounding soil and produces frost heaving, which causes damage to structures and roads in cold environments.

Collapsible sediments comprise loosely packed or weakly cemented particles that are normally dry. These sediments are susceptible to subsidence when wet.

Karst is common in many parts of Canada and the United States. Hazards include sinkhole collapse and groundwater pollution. Subsidence and soil expansion and contraction are common problems in many areas. Permafrost underlies more than half of Canada, and soils susceptible to seasonal frost occur over most of the remainder of the country. Organic sediments are abundant in all but the driest parts of Canada. Collapsible sediments occur in arid and semiarid regions, such as the southwest United States. Expansive soils are a problem primarily in southern Canada and the western United States, where they are responsible for significant economic damage to highways, buildings, pipelines, and other structures.

Although subsidence causes many problems, it also has benefits. About 25 percent of the world's population gets its drinking water from karst, and karst regions have important aesthetic and scientific values. Limestone caves, for example, are home to rare animals that are specially adapted to live underground.

Humans can exacerbate subsidence problems by extracting subsurface fluids and rock, placing poorly insulated structures on frozen ground, preventing sedimentation on deltas, draining wetlands, and using poor landscaping and drainage practices on expansive soils. Natural subsidence is difficult to prevent, but human-induced subsidence can be minimized or prevented by regulating groundwater pumping and underground mining. Damage from expansive soils can be minimized by using sound construction and landscaping practices. An understanding of the local geologic and hydrologic systems can help prevent groundwater contamination in karst areas.

Adjustments to subsidence hazards include identification of problem areas through geologic, soil, and subsurface mapping. Homeowners can protect themselves with insurance, but they must make sure that policies cover the subsidence and soil hazards in their area.

Key Terms

aquifer (p. 229)
cave system (p. 212)
collapsible sediment (p. 216)
expansive soil (p. 216)
frost heaving (p. 215)
frost-susceptible sediment (p. 214)
groundwater mining (p. 229)
karst (p. 210)
karst plain (p. 211)
organic sediment (p. 216)
permafrost (p. 213)
piping (p. 215)
sinkhole (p. 211)
soil (p. 210)
spring (p. 213)
subsidence (p. 210)

Did You Learn?

1. Explain how limestone dissolves.
2. Name the rock types that are especially susceptible to dissolution.
3. Explain how caves and cave systems form.
4. Name the features found in karst areas.
5. Describe the two processes by which sinkholes form.
6. What is permafrost and how has climate change affected it?
7. What keeps a delta plain from subsiding?
8. Describe what happens to wetlands when they are drained.
9. Explain why expansive soils shrink and swell.
10. Name the features that might indicate the presence of expansive soils.

11. What factors influence the moisture content of expansive soils?
12. Identify the types of subsidence and soil hazards that are likely to be found in the (a) Canadian Prairies, (b) eastern United States, and (c) northern Canada and Alaska.
13. What factors contribute to the formation of sinkholes?
14. Explain why groundwater is sometimes polluted in karst terrains.
15. What conditions were responsible for the catastrophe caused by Hurricane Katrina in New Orleans?
16. How is subsidence linked to changes in climate?
17. Explain how fluid withdrawal and mining can increase subsidence.
18. How can we minimize and adjust to subsidence and soil hazards?

Critical Thinking Questions

1. You are considering building a home in an area in Quebec underlain by limestone and are concerned about possible karst hazards. What are some of your concerns? What might you do to determine where to build your home?
2. A new community is being planned in the Northwest Territories to exploit a diamond mine. The site of the community is in permafrost. What advice would you give planners to ensure that the ground does not subside during and after development due to permafrost thaw?
3. You have bought a house built on a concrete slab in southern Saskatchewan. The soil outside the house contains smectite. What could you do to minimize damage from the shrinking and swelling of the soil?
4. You are a member of the municipal council of a small town in Ontario. Your council has been asked to approve a permit to develop a property that is partly underlain by silty river sediments and partly by a marsh. What questions should you ask the developer about his plan to place a subdivision on the river silts and his proposal to drain the marsh and place a school and recreational facilities on it?

MasteringGeology

MasteringGeology **www.masteringgeology.com**. Looking for additional review and test prep materials? Visit the Study Area in MasteringGeology to enhance your understanding of this chapter's content by accessing a variety of resources, including **Self-Study Quizzes, Geoscience Animations, GEODe Tutorials, RSS feeds, flashcards,** web links and an optional **Pearson eText.**

CHAPTER 9

► **FLOODING IN CALGARY IN 2013** Calgary experienced the worst flooding in its history in June 2013 when the Bow and Elbow rivers, swollen from heavy rain, spilled over their banks and into the city. About 75 000 people were evacuated, which is about 7 percent of the population of the city. This photo shows the centre of Calgary, including the Saddledome and the Calgary Stampede grounds, at the height of the flood. *(Thomson Reuters (Markets) LLC)*

River Flooding

Learning Objectives

Water is essential to life on Earth. However, it can injure and kill people and destroy property in certain situations, such as during a flood. Flooding is the most universally experienced natural hazard. Floodwaters have killed more than 10 000 people in North America since 1900, and during the past decade, property damage from flooding averaged more than $4 billion per year. Your goals in reading this chapter should be to

- Understand basic river processes
- Understand the process of river flooding and know the difference between upstream and downstream floods
- Know what geographic regions are at risk from river flooding
- Know the effects of flooding and the links with other natural hazards
- Know the benefits of periodic river flooding
- Understand how people affect the flood hazard and flood risk
- Become familiar with adjustments people can make to minimize injury and damage from river flooding

The Alberta Floods of 2013

Before 2013, many people would have been surprised to learn that Calgary, Alberta, is vulnerable to flooding, because the Bow River, which flows past the city, is normally calm and well within its banks. However, the Bow River has flooded Calgary in the past, most recently in June 2005. Awareness of Calgary's flood hazard changed dramatically in late June 2013, when the Bow and Elbow rivers spilled over their banks and flooded part of the downtown core and thousands of homes in nearby areas. The flood was the worst natural disaster in the city's history. Other communities in southwest Alberta also experienced severe flooding.

Southern Alberta is situated east of the Canadian Rocky Mountains and is a relatively dry region that rarely receives high amounts of rainfall. In June 2013, however, a high-pressure system in northern Alberta blocked the passage of a low-pressure area to the south. With circulation blocked, winds from the east, opposite the prevailing direction, pushed the humid air mass into the Rocky Mountain Foothills and onto the eastern slopes of the Rocky Mountains, triggering heavy rainfall. More than 100 mm of rain fell in less than two days in many areas, particularly west and southwest of Calgary. In Canmore, over 220 mm fell in 36 hours, nearly half of the city's average annual precipitation.[1] The rain fell on already saturated ground and melted remnants of the winter snowpack. The result was near-instantaneous runoff and a rapid increase in the size of several rivers.

At the peak of the flood, the Bow and Elbow rivers were flowing through Calgary at three times their peaks in the 2005 flood. That flood caused $400 million in insured damages. Within 48 hours, by the morning of June 21, the flow of the Bow River had reached 1740 m^3/s, about six times its normal discharge rate for that time of year. The Elbow River, which flows into the Bow River within the city of Calgary, had a peak discharge of more than 500 m^3/s—about ten times its June average.

About 75 000 people from 26 Calgary neighbourhoods were evacuated as the rivers spilled over their banks (Figure 9.1). All schools in the city were closed and officials urged residents to avoid unnecessary travel. The city's largest indoor arena, the Saddledome, was among the facilities damaged; floodwaters covered the first ten rows of the lower

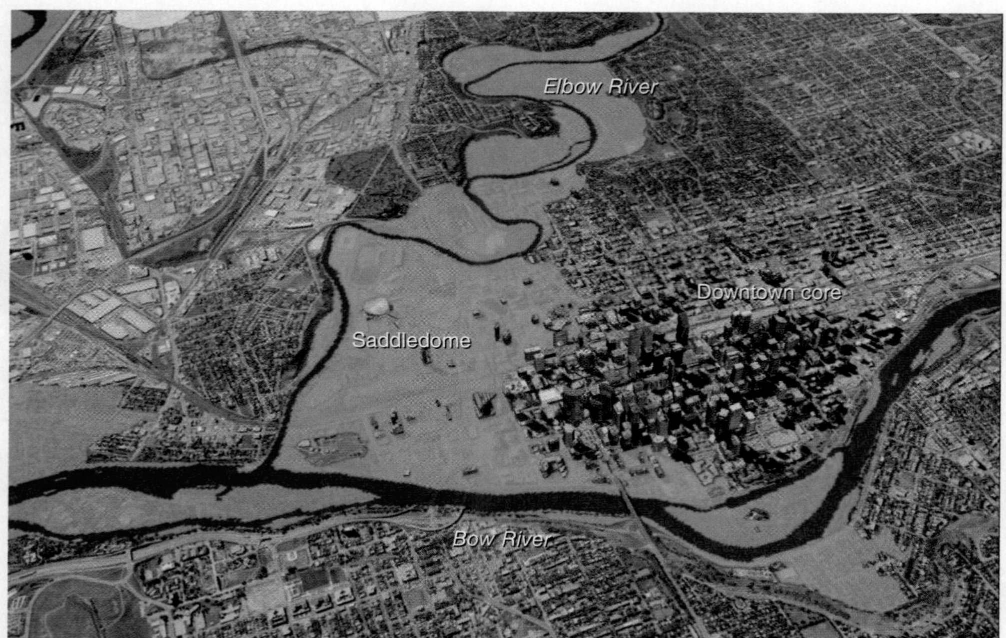

◄ **FIGURE 9.1 AREA FLOODED IN CALGARY IN 2013** This map shows the area in Calgary, Alberta, that was flooded in June 2013. Thousands of homes located on the floodplains of the Bow and Elbow rivers were affected. *(Reproduced or adapted with the permission of Natural Resources Canada 2013, courtesy of the Geological Survey of Canada (Bulletin 548))*

seating area. The Calgary Stampede grounds, adjacent to the arena, also were flooded less than two weeks before the opening of the annual exhibition and rodeo.

Damage was not limited to Calgary. The town of High River was evacuated as the Highwood River inundated the town, and the mountain towns of Banff and Canmore, west of Calgary, were cut off from neighbouring communities after the Trans-Canada Highway was severed. The city of Medicine Hat, located on the South Saskatchewan River, also experienced significant flooding. Nearly 10 000 of its residents were evacuated ahead of the flooding, and some residential areas and facilities, including the Medicine Hat Arena, were flooded late on the evening of June 23.[1]

Thanks to a timely and coordinated emergency response by the Alberta government and the city of Calgary, only four lives were lost in the floods—one in Calgary and the other three along the Highwood River. Flood damage, however, was extraordinarily large. At the time of writing of this book (in the week following the disaster), estimated damage to homes, businesses, vehicles, and other private property was more than $5 billion, making it, by far, the worst natural disaster in Alberta's history.

A significant part of Calgary is built on the flat floodplains of the Bow and Elbow rivers. In hindsight, the municipal and provincial governments should have recognized that a flood disaster was possible. Three major floods occurred during the period between 1875 and 1902; early Calgary residents would have been well aware of the flood danger.[2] Between 1932 and 2005, however, not a single major flood occurred on the Bow or Elbow rivers, perhaps creating a false sense of security. During that period, thousands of homes were built on the floodplains of the Bow and Elbow rivers, despite the fact that a series of reports commissioned by the city of Calgary in the 1970s recommended that land along the two rivers subject to the 70-year flood be officially designated as floodplain for management purposes.[2] The report recommended a floodplain management scheme in which hazardous areas would be officially delineated. New development would be prevented or discouraged in hazardous areas, and existing structures would have to meet certain flood-proofing standards. Communities on the floodplain opposed these management schemes. Most residents felt that property values would decrease if the hazard was officially recognized.[2] In addition, most residents had not personally experienced a flood and thus did not believe a hazard existed, in spite of the reports' conclusions. Largely as a result of opposition by community groups, a committee struck by the city and province recommended that most of the plan for floodplain management be rejected. Similar flood management recommendations were made in the most recent flood hazard report, released in 2005.

The 2013 Alberta floods have great significance for Canada. Flooding is our country's most common natural hazard, yet we have been slow to recognize the benefits of restricting development on **floodplains**. Changes in land use, especially urbanization, have increased flood risk along thousands of streams and rivers in towns and cities across Canada. The best way to reduce and minimize this risk is simple: Do not build on floodplains.

9.1 Introduction to Rivers

Streams and rivers are part of the *hydrologic cycle*. Water evaporates from Earth's surface, primarily from the oceans; it exists as a gas in the atmosphere; and it precipitates on oceans and on the 30 percent of Earth's surface that is land. Some of the water that falls on land as rain and snow infiltrates the ground or returns to the atmosphere through evaporation and transpiration by plants. Much of the water that falls on land, however, returns to the oceans via surface flow along paths determined by the local topography.

Surface flow, referred to as *runoff*, finds its way to small **streams**, which are *tributaries* of larger streams, or **rivers**. The region drained by a single stream is variously called a **drainage basin**, *watershed*, river basin, or *catchment* (Figure 9.2a). Each stream thus has its own drainage basin that collects rain and snow. A large river basin, such as the Red River basin, is made up of hundreds of small watersheds drained by smaller tributary streams.

One important characteristic of a river is the slope of the surface over which it flows, or its *gradient*. The gradient of a river is determined by calculating the drop in elevation of the channel over some horizontal distance and is commonly expressed in metres per kilometre or as a dimensionless number (elevation drop in metres divided by horizontal distance in metres). In general, the gradient of a river is greatest in its *headwaters*, decreases downstream, and is lowest at the river mouth, which is its *base level*. The base level of a river is the lowest elevation to which it may erode. Generally, this elevation is at or near sea level, although a lake may serve as a temporary base level. A graph showing downstream changes in a river's elevation is called a *longitudinal profile* (Figure 9.2b).

The valley of a river is steeper-sided and narrower in its headwaters than near the river mouth, where a wide floodplain may be present (Figure 9.2c and 9.2d). At higher elevations,

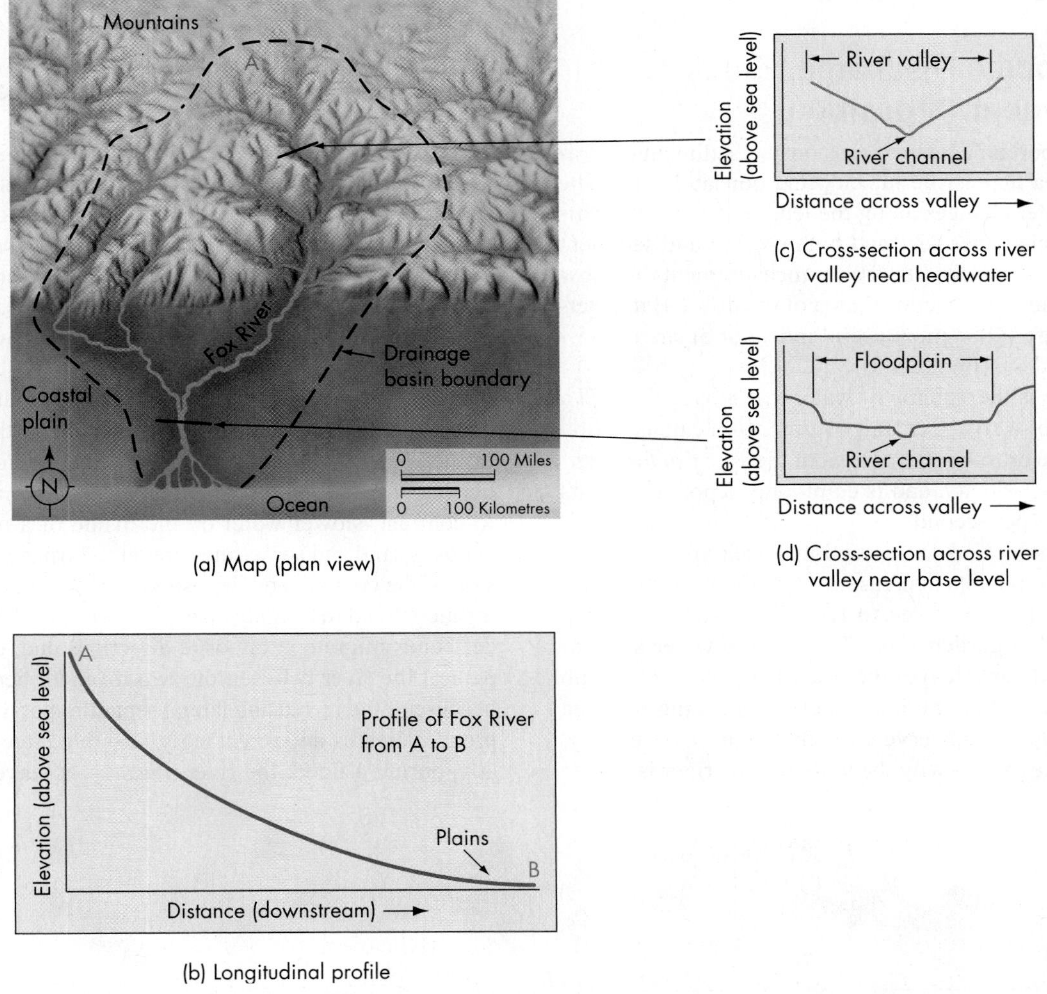

▲ **FIGURE 9.2 A DRAINAGE BASIN AND RIVER PROFILE** (a) The drainage basin, or watershed, of the hypothetical Fox River is delineated by a black dashed line. (b) The longitudinal profile of Fox River from point A at the head of the river to point B at its mouth; note that the vertical scale is greatly exaggerated. (c) A cross-section of the river valley near its headwaters; the valley is steep-sided and the valley floor is narrow. (d) A cross-section of the valley on the coastal plain near the river mouth; the valley floor is broad and the valley walls more gentle than in the headwaters.

the steeper gradient of the river facilitates erosion and downstream transfer of sediment.

Earth Material Transported by Rivers

Rivers move not only water but also a tremendous amount of material. This material, called the *total load*, consists of bed load, suspended load, and dissolved load. *Bed load* comprises particles of sand and gravel that slide, roll, and bounce along the river channel in rapidly moving water. It constitutes less than 10 percent of the total load of most rivers. *Suspended load* comprises mainly silt and clay particles carried in suspension above the riverbed. It accounts for nearly 90 percent of the total load of most rivers and gives them a muddy appearance during periods of high flow. *Dissolved load* comprises electrically charged atoms or molecules, called *ions*, that are carried in solution in the water. Most dissolved load is derived from chemical weathering of rock and sediment in the drainage basin. Ions in discharging underground springs, sewage, and chemical effluent can be a significant part of the dissolved load of some rivers.

River Velocity, Discharge, Erosion, and Sediment Deposition

Rivers are important agents of erosion and sediment deposition and play a major role in sculpting our landscape. The velocity of water changes along the length of a river and affects channel characteristics and both erosion and sediment deposition. Hydrologists combine measurements of flow velocity (V) and cross-sectional area of the flow (A) to determine **discharge** (Q), a more useful indicator of stream flow than velocity alone (Figure 9.3).

Discharge is the volume of water that moves through a cross-section of a river per unit of time. It is calculated by multiplying the cross-sectional area of the water in the channel by the flow velocity, and is commonly reported in units of cubic metres per second.

Flow velocity and cross-sectional area are related. If the cross-sectional area decreases, the velocity of the water must increase for discharge to remain constant. You can prove this with a garden hose. Turn on the water and observe its velocity as it leaves the hose. Then put your thumb partly over the end of the hose, thereby reducing the area of outflow. You will observe an increase in flow velocity. This principle explains why the velocity of a river is higher in a deep, narrow canyon than in an area where the flow is less confined.

The gradient of a river decreases where it flows from mountains onto a plain or into an ocean or lake. In these places, the river is no longer able to transport its load; consequently it builds a fan-shaped body of sediment on land referred to as an *alluvial fan* (Figure 9.4), or a triangular or irregularly shaped deposit in water called a *delta* (Figure 9.5).

Channel Patterns and Floodplain Formation

Streams and rivers flow in channels, and these channels can have different patterns. Three **channel patterns** are common: *braided*, with a large number of intersecting active channels; *anastomosing*, with two or more channels and intervening stable islands or bars where sediment is temporarily stored; and *meandering*, with a single channel shaped like a snake.

Braided floodplains (Figure 9.6) have numerous unvegetated sand and gravel bars that divide and reunite the main channel, especially during low flow. Overall, braided channels tend to be wide and shallow in comparison to meandering channels. A river is likely to have a braided pattern if it has an abundant coarse bed load and large short-term variations in discharge. These conditions are found in areas where the land is rising because of tectonic processes and where rivers receive abundant water and sediment from glaciers.

Many rivers have *meanders*, curving channel bends that migrate back and forth across a floodplain over years or centuries (Figure 9.7). Although no one is certain why rivers meander, we know a great deal about how water flows in meandering channels.

Canoeists and rafters have long known that water moves faster along the outside of a meander bend than along the inside. The fast-moving water erodes the riverbank on the outside of the bend to form a steep slope known as a *cutbank*. In contrast, slower water on the inside of a meander bend deposits sand and sometimes gravel to form a *point bar*. Erosion of the cutbank and deposition on the point bar cause the meander bend to migrate laterally over time. Adjacent meander bends migrate in opposite directions and, over time, the path of the river between the two meander bends lengthens because of the increasing lateral separation of the bends. This process creates an increasingly unstable situation. Eventually, during a flood, the river *avulses*—it leaves its channel

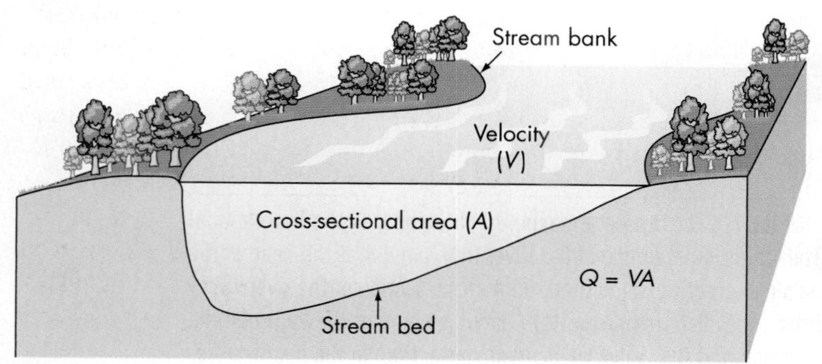

◄ **FIGURE 9.3 CALCULATING STREAM DISCHARGE** The discharge of a stream (Q) is determined by multiplying the velocity of the flow (V) times the cross-sectional area (A) of the water in the channel.

◄ **FIGURE 9.4 ALLUVIAL FAN IN DEATH VALLEY** This alluvial fan formed where a stream leaves a steep canyon and flows onto the floor of Death Valley, California. Infrequent floods discharge water onto the white salt flats seen in the upper left. *(Michael Collier)*

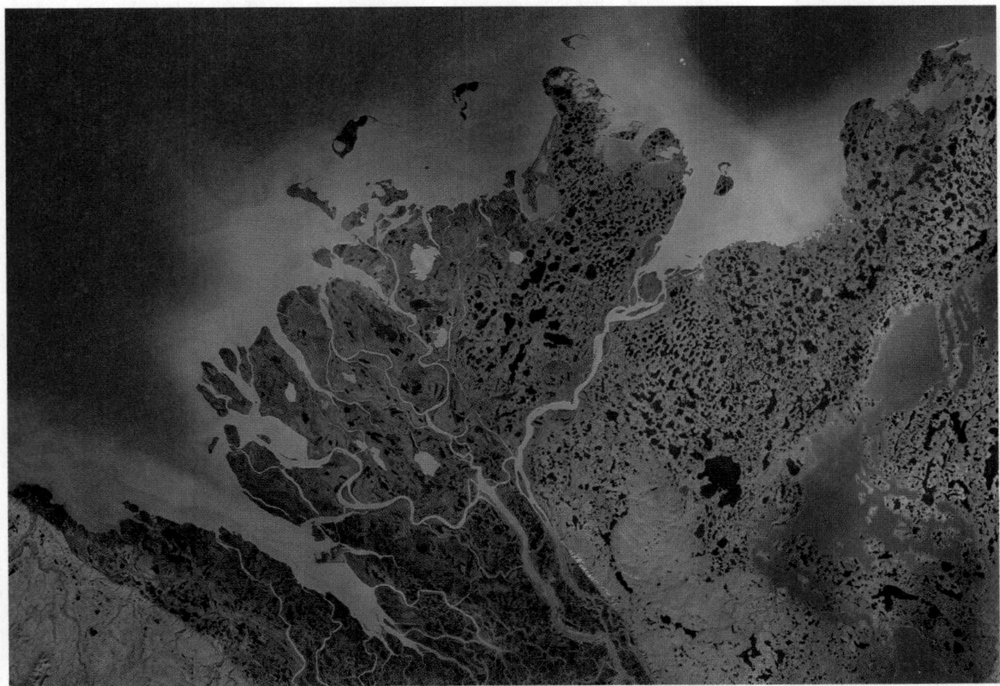

◄ **FIGURE 9.5 MACKENZIE RIVER DELTA** The Mackenzie River has built a large delta into the Beaufort Sea during the Holocene. In this true-colour satellite image, a myriad of distributary channels carry the flow of the river to the sea. The width of the image is about 150 km. *(Planet Observer/Science Source)*

directly downstream of the first bend and cuts a new channel along the much shorter route to the upstream end of the second bend. During avulsion, the river abandons a looping section of the old channel, which is left as a *meander cutoff*. Lakes within the old abandoned channel are termed *oxbow lakes*. This process is important in constructing and maintaining some floodplains.

Floodplains are also built at times of *overbank flow*, when rising waters spill over the riverbank and onto the floodplain. The escaping waters deposit fine sand, silt, and clay, thus building up the floodplain.

Meandering and anastomosing channels commonly contain a series of regularly spaced pools and riffles (Figure 9.8). *Pools* are deep areas produced by scour at high flow, and *riffles* are shallow areas formed by sediment eroded from the pools. At low flow, pools contain relatively deep, slow-moving water, and riffles have shallow, fast-moving water. Such changes in water depth and velocity along a stream create different *habitats* for organisms, increasing the diversity of aquatic life.[3] For example, fish may feed in riffles and seek shelter in pools, and pools have different types of insects from those found in riffles.

(a)

(b)

◄ **FIGURE 9.6 BRAIDED RIVER** (a) The braided pattern of Hopkins River, near Lake Ohau on the South Island of New Zealand, is formed by shallow channels flowing around and across numerous gravel bars. The river's coarse bed load is derived from the Southern Alps. (b) Surface view of a braided channel system below the toe of Fox Glacier in New Zealand, with intertwined gravelly channels. The distance across the channel at the bottom of the photo is about 10 m. *((a) © davidwallphoto.com; (b) John J. Clague)*

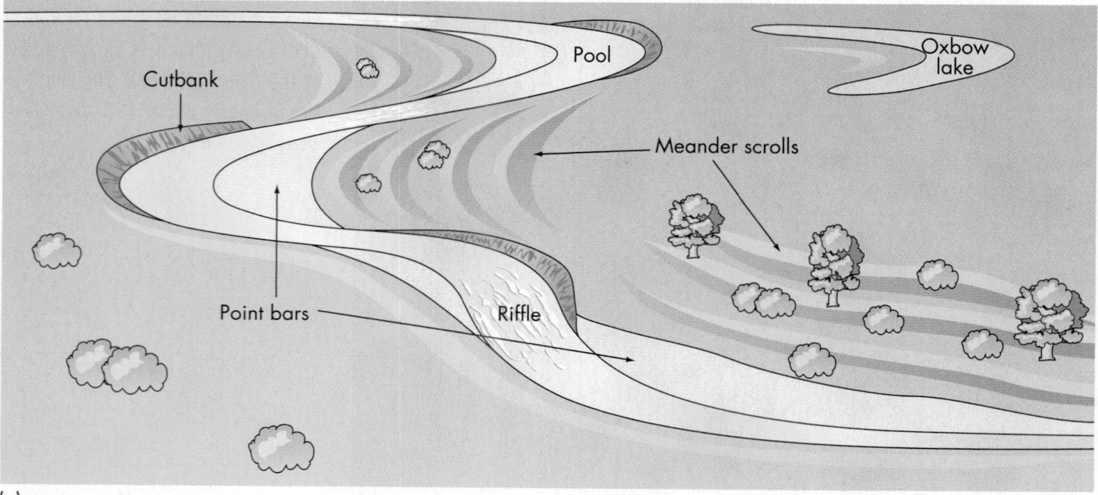

(a)

(b)

◄ **FIGURE 9.7 MEANDERING RIVER** (a) This schematic shows the important features of a meandering river. Migration of meander bends is commonly indicated by low, curving, vegetation-covered ridges called meander scrolls. The scrolls adjacent to the point bar in the left centre of the figure indicate that the stream migrated from right to left. (b) The Waiau River, South Island, New Zealand. Note the unvegetated point bars directly downstream of meander bends. *(© davidwallphoto.com)*

▲ **FIGURE 9.8 POOLS AND RIFFLES** A pool–riffle sequence in Sims Creek near Blowing Rock, North Carolina. Deep pools lie under the smooth, reflective water surface in the centre and lower right of the photograph; shallow riffles lie under the rough, non-reflective water in the far distance and left foreground. *(Edward A. Keller)*

▲ **FIGURE 9.9 THE 1894 FRASER RIVER FLOOD** The town of Chilliwack, British Columbia, at the peak of the Fraser River flood in May 1894. The Fraser River has flooded many times, but the 1894 flood is the largest on record. It was caused by the melt of an unusually heavy snowpack in southern British Columbia during a hot, wet spring. *(Image A-03933, courtesy of the Royal BC Museum, BC Archives)*

Having presented some of the characteristics and processes of water flow and sediment transport in rivers, we will now discuss the process of flooding in greater detail.

9.2 Flooding

Overbank flow is called **flooding** (Figure 9.9; Case Study 9.1). The magnitude and duration of a flood are determined by the following factors: the amount, distribution, and duration of precipitation in the drainage basin; the rate at which the precipitation soaks into the ground; the presence or absence of a snowpack; air temperature; and the speed at which surface runoff reaches the river. The amount of moisture in the soil at the time precipitation starts also plays an important role in flooding. Water-saturated soil is like a wet sponge that cannot hold additional moisture. Flooding will probably occur if heavy rains fall on ground that is already saturated. Dry soil might be able to absorb considerable moisture and thus reduce or prevent flooding. Flooding can also result from the build-up of water behind ice jams on rivers, the damming of rivers by landslides, and the sudden draining of lakes impounded behind moraines, glaciers, and landslide deposits[4] (Figure 9.10).

Floods happen at different times of the year. Their timing depends mainly on the size of the watershed and on regional climate. Many large rivers in North America—for example the Mississippi and Fraser—flood only in late spring or summer, following winters marked by abnormally heavy snowfall. Thick snowpacks melt rapidly during extended periods of unusually warm weather, sometimes accompanied by heavy rain. The watersheds of large rivers are too big to be significantly influenced by local thunderstorms or by single cyclonic storms that occur at other times of the year. Mid-size rivers also can carry peak flows during late spring,

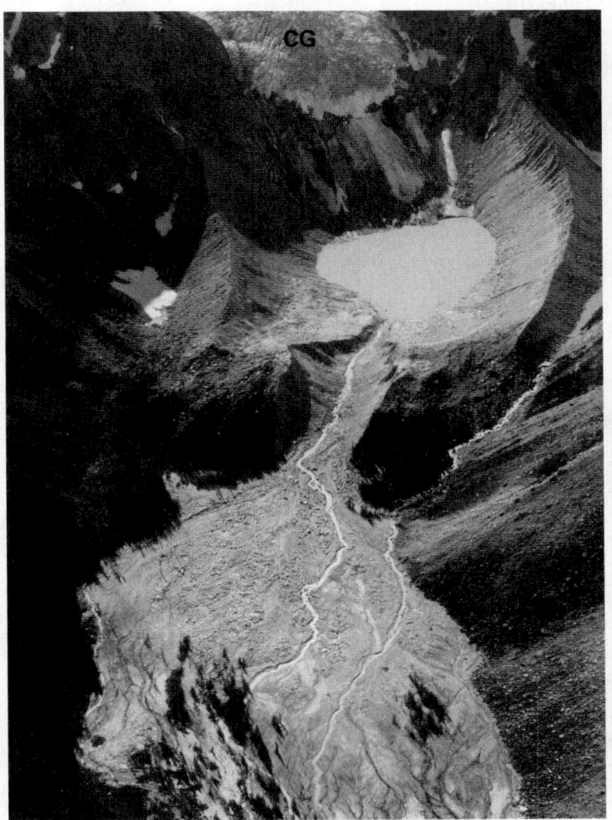

▲ **FIGURE 9.10 OUTBURST FLOOD FROM A MORAINE-DAMMED LAKE** Oblique aerial view of the breached moraine at Nostetuko Lake in the southern Coast Mountains of British Columbia. In 1983, an ice avalanche from the retreating toe of Cumberland Glacier (CG) plunged into Nostetuko Lake, generating waves that overtopped and breached the moraine. The sudden release of 6 million m^3 of water produced a short-lived, but huge flood in the valley below. *(S. G. Evans/Geological Survey of Canada)*

9.1 CASE STUDY

Mississippi River Floods of 1973 and 1993

The Mississippi River flooded thousands of square kilometres of farmland in the spring of 1973, forcing the evacuation of tens of thousands of people. Fortunately, there were few deaths, but the flooding caused about U.S.$1.2 billion in property damage.[5] The 1973 flood occurred despite a tremendous investment in upstream flood-control dams on the Missouri River, the largest tributary of the Mississippi River. Reservoirs behind these dams inundated some of the most valuable farmland in North and South Dakota; despite these structures, the flood on the Mississippi near St. Louis was record-breaking.[6] Impressive as this flood was at the time, it did not compare either in magnitude or in the suffering it caused with flooding that occurred 20 years later.

During the summer of 1993, the Mississippi River and its tributaries experienced one of the largest floods of the century. The flood lasted from late June to early August and caused 50 deaths and more than U.S.$15 billion in property damage. In all, about 55 000 km^2, including numerous towns and large tracts of farmland, were inundated (Figure 9.11).[7,8]

The 1993 flood resulted from a major weather anomaly that affected the entire U.S. Midwest—the Mississippi and Lower Missouri river watersheds.[9] It was preceded by a wet autumn and heavy spring snowmelt that saturated the ground in the upper Mississippi drainage basin. Then, early in June, a high-pressure centre developed over the U.S. east coast, drawing moist, unstable air into the Midwest. The high-pressure centre prevented storm systems in the Midwest from moving east. At the same time, air moving in over the Rocky Mountains initiated unusually heavy rainstorms.[9] The summer of 1993 was the wettest on record in Illinois, Iowa, and Minnesota. Cedar Rapids, Iowa, for example, received about 90 cm of rain from April through July—the equivalent of a normal year's rainfall in just four months![8] Intense precipitation falling on saturated ground led to a tremendous amount of runoff and an unusually large flood during the summer. The floodwaters remained high for a prolonged time, putting pressure on the flood defences of the Mississippi River, particularly **levees**, which are earthen embankments built parallel to the river to contain the floodwaters (Figure 9.12).

Before construction of the levees, the Mississippi floodplain (flat land adjacent to the river that periodically floods) was much wider and contained extensive wetlands. Since the first levees were built in 1718, approximately 60 percent of the wetlands in Wisconsin, Illinois, Iowa, Missouri, and Minnesota—all particularly hard hit by the flooding in 1993—have been lost. In some urban areas, such as St. Louis, levees have been replaced with floodwalls designed to protect the city against floods. The effects of these floodwalls can be seen in a satellite image taken in mid-July 1993, at the peak of the flood (Figure 9.13). The image shows that the river is narrow at St. Louis, where it is contained by the floodwalls, and broad upstream near Alton, Illinois, where extensive flooding occurred. The floodwalls produced a bottleneck effect—they forced floodwaters through a narrow channel between the walls, causing it to back up. This effect contributed to the flooding upstream of St. Louis in 1993.

◄ **FIGURE 9.11 MISSISSIPPI RIVER FLOODS FARMLAND AND TOWNS** In one of the largest floods of the twentieth century, the Mississippi and Missouri rivers in 1993 broke through levees and flooded vast areas. A flooded town, inundated farmland, and a flooded access bridge over the river are visible in this photograph. A long line of trees surrounded by brown floodwaters at the upper left marks one of the river banks at low flow. *(Andrea Booher/FEMA photo)*

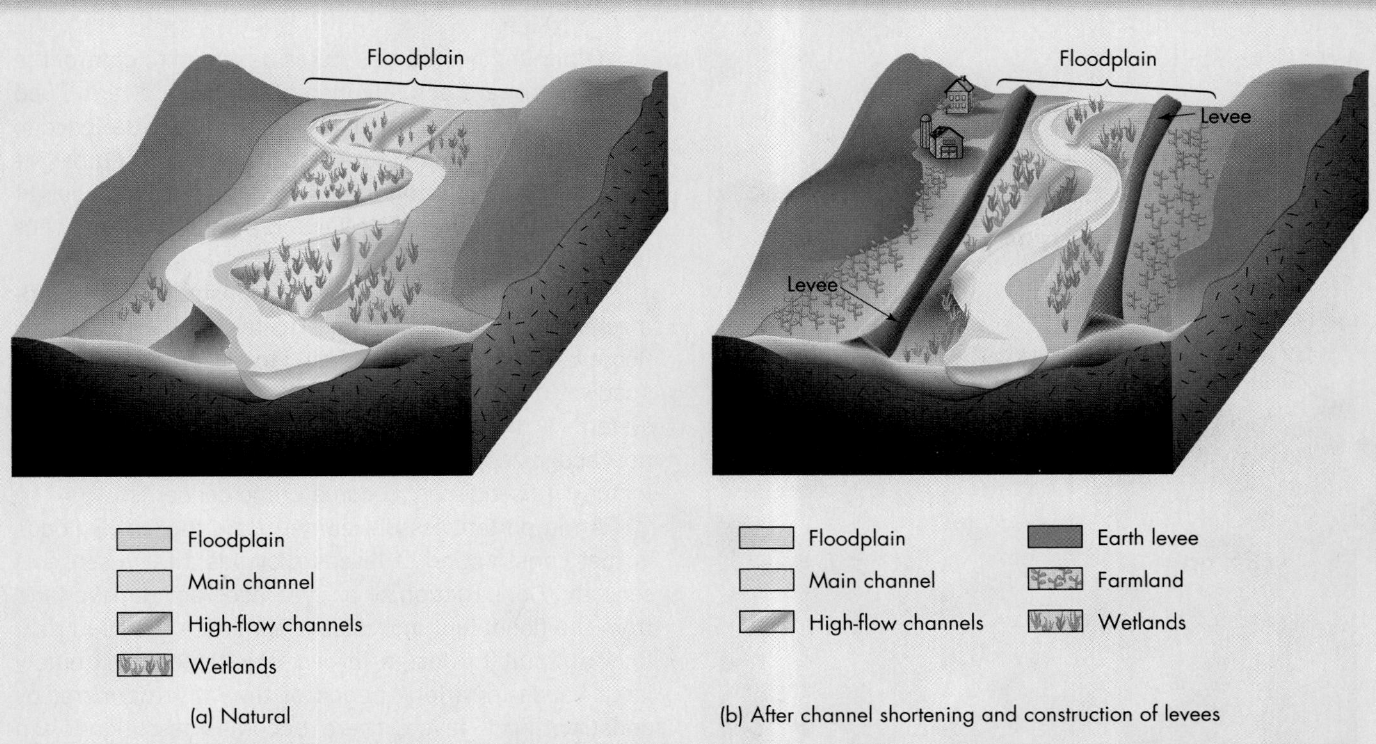

▲ **FIGURE 9.12 FLOODPLAIN WITH AND WITHOUT LEVEES** Idealized diagrams of (a) a natural floodplain with wetlands and (b) the same floodplain after the river channel has been shortened and confined between levees. Land behind the levees is farmed. *(Edward A. Keller)*

The floodwaters reached to within 0.6 m of the top of the floodwalls in St. Louis, averting a disaster in the city. However, as the floodwaters rose to record levels, they leaked through the floodwalls, requiring round-the-clock efforts to keep them from failing. Failure of earthen levees downstream from St. Louis partially relieved the pressure, possibly saving the city from flooding.

Levee failures were common during the flood (Figure 9.14).[10] Almost 80 percent of the levees built by

◀ **FIGURE 9.13 MISSISSIPPI RIVER FLOOD OF 1993** Satellite view of the extent of flooding in 1993 where the Illinois and Missouri rivers join the Mississippi. Areas covered by 1993 floodwaters are orange; dark blue is normal river flow. A series of floodwalls built to protect St. Louis constricts the Mississippi River at the lower right. This bottleneck caused widespread flooding upstream, including the town of Alton on the east bank of the river near the right centre edge of the image. *(© Cindy Brown/Corbis)*

9.1 CASE STUDY (Continued)

▲ **FIGURE 9.14 LEVEE FAILURE** A breach of this levee in Illinois in 1993 flooded the town of Valmeyer. Water can be seen rushing through the breach in the grass-covered earthen levee. *(Jupiter Images)*

farmers and homeowners failed.[8,9] In contrast, most of the levees built by the U.S. government survived the flooding and undoubtedly saved lives and property. Unfortunately, there is no uniform building code for the levees, so some areas have levees that are higher or lower than others. Failures occurred as a result of overtopping and breaching, or collapse of water-saturated material.[8]

Other engineered structures designed to control the Mississippi have actually increased the long-term flood hazard.[11] For example, river-control structures designed to optimize the movement of loaded river boats and barges during normal flow conditions can increase flood height similar to the bottleneck effect created by the floodwalls in St. Louis.

Of course, building houses, industry, public buildings, and farms on a floodplain invites disaster. Too many floodplain residents have refused to recognize the natural floodway of the river for what it is: part of the natural river system. If the floodplain and its relation to the river are not recognized, flood control and drainage of wetlands, including floodplains, become prime concerns.

An important lesson learned from the 1993 floods is that construction of levees provides a false sense of security. Construction of a levee does not remove land from the floodplain, nor does it eliminate the flood risk. It is difficult to design levees to withstand extremely large floods for a long period of time. Furthermore, by constructing a levee, there becomes less floodplain space to "soak up" the floodwaters.[7] The 1993 flood caused extensive damage and property loss; in 1995, Mississippi floodwaters inundated floodplain communities once again.

In the first six to seven years after the 1993 flood, the United States government bought out many homeowners living on the floodplain upstream from St. Louis.[12] Now real estate agents have reversed the trend and local governments are allowing development on the floodplain behind levees.[12] Weak floodplain regulations and government subsidies have allowed construction of over 25 000 new homes and other buildings on the floodplain behind new, higher levees.[13] Fortunately, not all communities along the river are short-sighted; some have re-evaluated their strategies concerning the flood hazard—they have moved to higher ground! Of course, this is exactly the adjustment that is appropriate.

although those in wet temperate areas, such as the Pacific coast of northwest North America, commonly flood in the fall during periods of heavy rain after the first snow has fallen (Figures 9.15 and 9.16).

Ice-jam floods are common in northern areas of Canada and occur when rivers freeze in the fall or, more commonly, during *break up* in spring (Figure 9.17). An example is the particularly destructive ice-jam flood that destroyed the Cree community of Winisk in northern Ontario in May 1986. An ice jam on the Winisk River caused flooding upstream over a distance of 6 km; as a result, every structure in Winisk except two were washed into Hudson Bay. Winisk was abandoned, and the community was later rebuilt 30 km upriver on higher ground.

Small streams can flood at any time of the year: in early spring, during mid-winter thaws, or during summer thunderstorms. Small streams reach flood stage very quickly if heavy, warm rain falls on frozen ground or snow. Severe

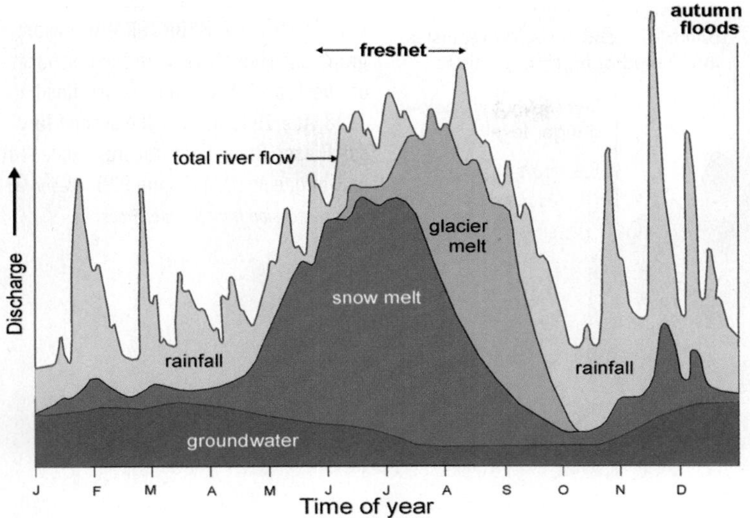

◄ **FIGURE 9.15 ANNUAL STREAM DISCHARGE VARIATIONS IN WESTERN CANADA** This graph shows variations in the four sources of water supplying a hypothetical small to mid-size stream draining a watershed containing glaciers in western Canada. The four sources are rainfall, snow melt, glacier melt, and groundwater. Discharge is commonly greatest during the spring and early summer freshet, when snow and glacier melt are greatest. Some mid-size rivers have peak flows during heavy rains in the autumn. Small streams can flood during rainstorms at any time of the year.

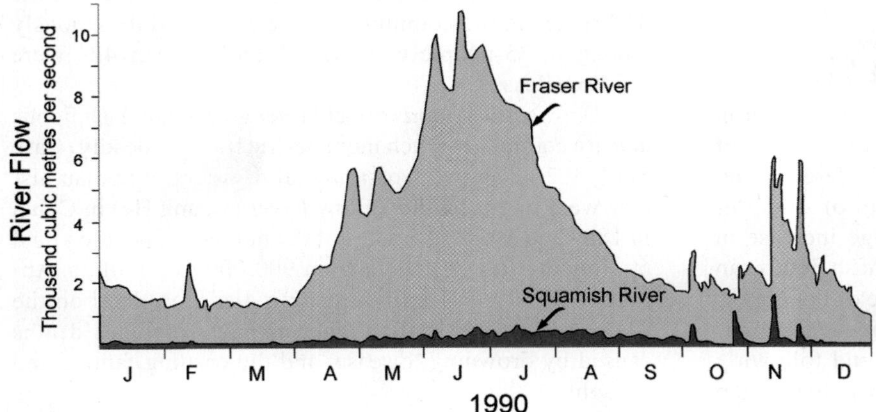

◄ **FIGURE 9.16 COMPARISON OF ANNUAL FLOW OF TWO BRITISH COLUMBIA RIVERS** The Fraser River has a much larger watershed, and thus a much larger discharge, than the Squamish River. Fraser River peak flow always occurs in May or June when the seasonal snowpack melts. The Squamish River also has high flows in late spring and summer due to snow and ice melt, but its peak discharge in some years, such as 1990, is during the autumn due to heavy rains in its watershed. *(Reprinted with permission from Tricouni Press)*

◄ **FIGURE 9.17 ICE-JAM FLOODING** An upstream view of an ice jam along the Mackenzie River at Fort Simpson, Northwest Territories, on May 3, 1989. The ice jam is 1200 m wide at this location. It extended more than 30 km upstream along a tributary of the Mackenzie River. Note the flooding adjacent to the buildings along the riverbank. *(Water Survey of Canada, Fort Simpson, NWT)*

flooding can also result from the intense rainfall that accompanies hurricanes and cyclones and from the surges created by these severe storms (Survivor Story and Chapter 11). Human modification of rivers and floodplains, discussed in Section 9.7, can also affect river processes and flooding.

A flood begins when a stream achieves *bankfull discharge*—the discharge at which water first flows out of the channel. *Flood discharge* can also be defined as the level of the river surface at a point, or its *stage*. A graph showing changes in discharge, water depth, or stage over time is called a *hydrograph* (Figure 9.18).

The term *flood stage* is frequently used to indicate that a river has reached a level likely to cause property damage. This definition is subjective—the elevation of a river at flood stage depends on human use of the floodplain.[14] The relations among stage, discharge, and recurrence interval of floods are described and illustrated in the Closer Look feature on page 251. The **recurrence interval** (or *return period*) of

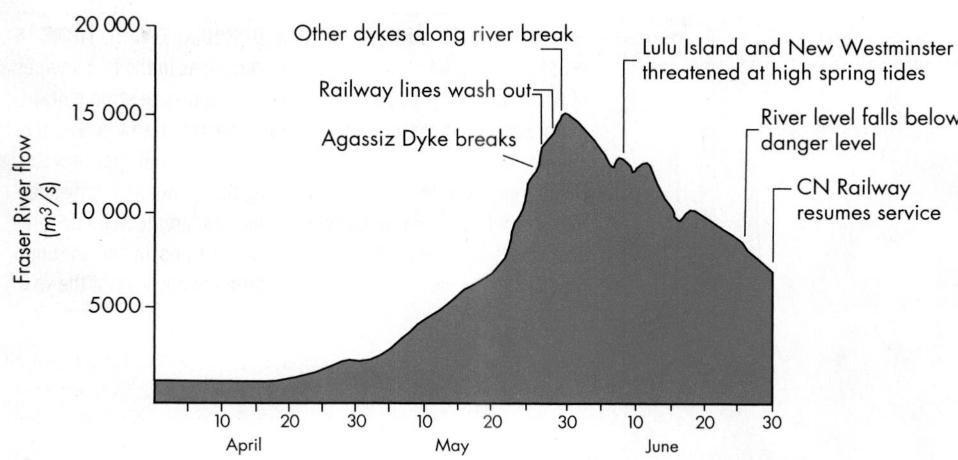

◀ **FIGURE 9.18 HYDROGRAPH** A hydrograph showing changes in the discharge of the Fraser River during its flood in 1948. The 1948 flood is the second largest Fraser River flood record, exceeded only by one in 1894 (Figure 9.9). *(Reprinted with permission from Tricouni Press)*

a flood of a particular magnitude is the average time between events of that magnitude.

Flash Floods and Downstream Floods

Floods can be further characterized by where they occur in a drainage basin. **Flash floods** occur in the upper parts of watersheds and in small tributary basins of a larger river. They are generally caused by intense rainfall of short duration over a small area, and involve a large increase in discharge in a short time. Peak discharge of flash floods can be reached in less than 10 minutes. This type of flooding is most common in arid and semi-arid environments, in areas with steep topography or little vegetation, and following breaks of dams, levees, and ice jams. Most people who die during flash floods are in automobiles. Deaths occur when people attempt to drive through shallow, fast-moving floodwaters. A combination of buoyancy and the strong lateral force of the rushing water sweeps automobiles off the road into deeper water, trapping people in sinking or overturned vehicles.

Flash floods can be very damaging, as shown by the events in the Colorado Front Range in July 1976. A system of thunderstorms swept through several canyons west of Loveland, Colorado, and delivered up to 250 mm of rain in a few hours. The floods killed 139 people and caused more than U.S.$35 million in damage to highways, roads, bridges, homes, and small businesses. Most of the damage and all the loss of life was in Big Thompson Canyon, where hundreds of residents, campers, and tourists were caught with little or no warning (Figure 9.20). Although the storms and flood were rare events in these Front Range canyons, comparable floods have occurred in the past and others can be expected in the future.[15,16,17]

Another example of a sudden-onset flood in a small watershed is the 1952 Lynmouth flood in southwest England. In mid-August of that year, an intense storm dumped 23 cm of rain in 24 hours on an already waterlogged moor near Lynmouth. Debris-laden floodwaters cascaded down the northern escarpment of the moor. A dam formed by fallen trees and boulders in the upper West Lyn valley gave way, sending a huge wave of water and debris into Lynmouth. Overnight, over 100 buildings and 28 of the 31 bridges in the community were destroyed or seriously damaged, 35 people were killed, and another 420 were left homeless.[18]

Downstream floods affect larger areas than flash floods and are commonly much more destructive and deadly (Case Study 9.2). The two worst natural disasters in human history were floods on the Yellow River (Huang He) in China in 1887 and 1931. Estimates of the number of people killed by flooding in 1887 range from 900 000 to 2 million. Another 850 000 to 4 million people died in a flood on the same river in 1931; the higher estimate includes deaths caused by drowning, disease, and the ensuing famine and drought.

The root cause of the severe flood problem on the Yellow River is geologic.[19] The river carries an average of 37 kg of sediment per cubic metre of water in its lower reaches, which is a very large sediment load. A substantial amount of this silt is deposited in the lower reaches of the river, where the channel becomes wider and the river's velocity decreases. Silting of the river channel has led to the construction of high dykes and levees to protect one of the most populous regions in China. Today, in some areas, the river bottom is 3 to 5 m higher than the surrounding floodplain. Its bed is more than 10 m above the city of Kaifeng in Henan Province; there the bed of the river lies above the rooftops of the houses behind the levees. The Chinese government has made an effort to reduce severe erosion in upstream areas, but the problem persists and the elevation of the riverbed continues to rise along the lower reaches of river. The consequences of a breach in the Yellow River dykes during the wet season would be disastrous.

Destructive downstream floods are common in other parts of the world. For example, India suffered some of the worst floods in its history in 2005, and rivers draining the southern Rocky Mountains in Alberta flooded during prolonged heavy rains in the same year. In 2004, heavy rains from a series of hurricanes and tropical storms in the eastern United States caused record or near-record flooding. In Pennsylvania, the Susquehanna River crested 2.5 m above flood

SURVIVOR STORY

Flooding from Hurricane Hazel

The worst flood disaster in Canadian history occurred in 1954, when Hurricane Hazel struck Toronto.

As Hurricane Hazel approached Ontario from the Caribbean on October 13, 1954, it showed signs of weakening. It had crossed the Allegheny Mountains and its wind velocities were rapidly falling off. However, the warm, moisture-laden air came in contact with a cold front lying over southern Ontario, producing record rainfall. From the morning of October 14 to midnight on October 15, about 210 mm of rain fell on the watersheds of several streams in Toronto.

Streams in Toronto are characterized by steep slopes with little natural storage capacity. Thus, even under the best of conditions, the intense rainfall from Hurricane Hazel would have caused flooding. In this case, the situation was made worse by already saturated soils. Autumn rainfall had been unusually heavy and had soaked the soils, thus preventing infiltration of any portion of the storm's downpour. Ninety percent of the rain that fell on the Humber River watershed during the storm left as runoff.

Hurricane Hazel caused the most severe flooding in Toronto in more than 200 years. Most of the affected floodplains had been developed by 1954, so the flood damage was high. More than 20 bridges were destroyed or damaged beyond repair, 81 lives were lost, and nearly 1900 families were left homeless. The Humber River swept away a full block of houses on one drive alone, killing 32 residents in one hour (Figure 9.19). After the disaster,

(a)

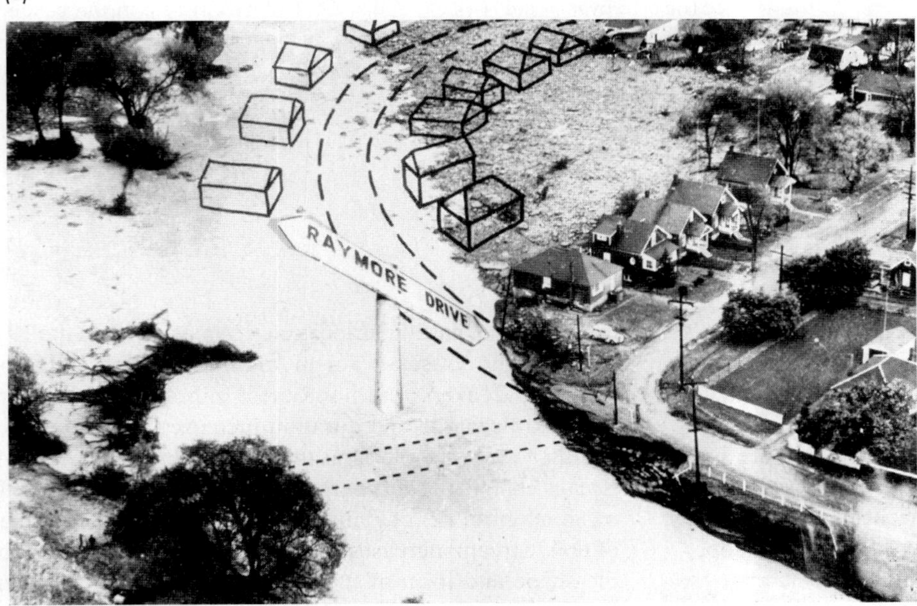

(b)

◄ **FIGURE 9.19 FLOOD DAMAGE FROM HURRICANE HAZEL** (a) Flood damage to a residential neighbourhood in suburban Toronto from Hurricane Hazel in October 1954. The floodwaters swept away houses on Raymore Drive in the upper centre of the photo. (b) Houses (box-shaped symbols) on Raymore Drive (curved dashed lines) that were destroyed by the floodwaters; compare with (a). *(Courtesy of Toronto and Region Conservation Authority)*

(continued)

SURVIVOR STORY *(Continued)*

the Metropolitan Toronto and Region Conservation Authority undertook a comprehensive program of management of the watersheds under its jurisdiction. It acquired floodplain property and converted it to parkland.

The following account is one family's harrowing experience during the flood, as told by Betty Kennedy in 1979. The farm family lived at Holland Marsh, a vegetable-growing area north of Toronto.

> Early in the evening of Hurricane Hazel the farm's foreman dropped in to warn them that the Marsh might flood, but that if so, they would be looked after. Like the rest of the family, 15-year-old Harry was not overly worried. "After all, we were 1500 miles away from the ocean, and the ocean was where the floods came from."
>
> Close to 9 o'clock, however, Harry, his elder brother, and a visiting friend were sent off to the local store to get candles in case the storm caused a power failure. When they reached the store they were astonished to find the doors wide open and everyone gone. They helped themselves to the candles (planning to pay later) and set out back home.
>
> At this point things became serious. The water was now running over the road, and the wind was so strong that the boys had to hold hands as they walked. They noticed that the only lights around came from their house, and realized that everyone else had cleared out.
>
> Back in the house they held a family conference and decided that with only five adults and twelve children they were not in a position to make a break for higher ground. Now the water was flowing into the house, and floating onion crates began to bang into the back door, forcing it open. Harry's father finally decided to nail the door shut with a board, which was our first big mistake, since it meant that the water level rose higher outside the house than inside.
>
> There was two feet of water in the house and the furniture had been stacked on tables when from upstairs the family saw a neighbouring house start to move. Then as they felt a "terrific jolt" they realized that *they* were moving, floating off the pilings that formed the house's base. "The amazing part was," as Harry recalls, "all the lights in the house stayed on, because we were moving towards the power lines, and they were slackening."
>
> Then the house hit the road and a hydro pole, shearing it right off and dropping the wires on the roof, where they lay shooting sparks all over the place. For a while it seemed that the weight of the wires over one corner might tip the house over, but then, with another jolt, the house was free.
>
> The house just took off like a boat, a real Noah's Ark. From 11:30 till 6:30 we floated aimless through the Marsh, bumping into houses, greenhouses, barns, hydro poles, everything. The area over by the Holland River had a faster current and somehow our house got caught in that current and starting spinning like a top, faster and faster, and rocking to and fro. We all would run from one side of the house to the other when it tilted, trying to balance it out. One of my younger brothers, Bastina, actually got violently sea-sick.
>
> Until then we had been too busy to really worry and then one of the younger ones asked if we were all going to die. My mother said only one person knew that, the Lord, and we all knelt down and prayed, the Lord's Prayer. And we did get out of the current and finally came to rest against a service road near the [Highway] 400, where a complete field of carrots had floated up to the surface and helped hold us in place. We were two and half miles away from where we started, with lots of side trips that had often taken us near our original place.
>
> At that time there were still cars going along the 400 and we shouted and waved to attract their attention. I even fired off a .22, but with the noise of the wind and the water, it couldn't be heard. Then we waved bedsheets and motorists saw us, and soon an amphibious truck from Camp Borden came along. One man got out, tied a rope around his waist, and plunged in to swim towards our house. We were about 250 feet away and the water was pretty wild and cold but he made it. We knocked a window out downstairs and pulled him in. Then another man came along the rope in a canoe which kept tipping but he told us we'd be okay with the extra weight of two people in the canoe. So we made it out to the truck in seven trips, and were taken to Bradford Town Hall.

(From* Hurricane Hazel *by Betty Kennedy. Reprinted with permission.)

stage, making it one of the five greatest floods in the river's history. Downstream flooding on the Ohio River in Marietta, Ohio, was the worst in 40 years, and flooding in Atlanta, Georgia, set all-time records.

Downstream floods inundate large areas and are produced by storms of long duration or by rapid melting of snowpacks. Flooding in small tributary basins is generally limited, but the combined runoff from thousands of slopes in tributary basins produces a large flood downstream. A flood of this kind is characterized by a large, slow rise in discharge at a particular location.[20] An example is the 1972 flood on the Fraser River in southern British Columbia (Figure 9.21a). As the flood crest migrated downstream, its peak discharge and duration increased (Figure 9.21b). Another way of looking at the flood is to examine downstream changes in discharge per unit area of the drainage basin (Figure 9.21c). This approach eliminates the effect of downstream increases in discharge and better illustrates the shape and form of the flood peak as it moves downstream.[21]

◄ **FIGURE 9.20 FLASH FLOOD IN BIG THOMPSON CANYON** Heavy rains in the Colorado Front Range in July 1976 caused a flash flood in Big Thompson Canyon that claimed 139 lives. *(R. R. Shroba, U.S. Geological Survey Photographic Library)*

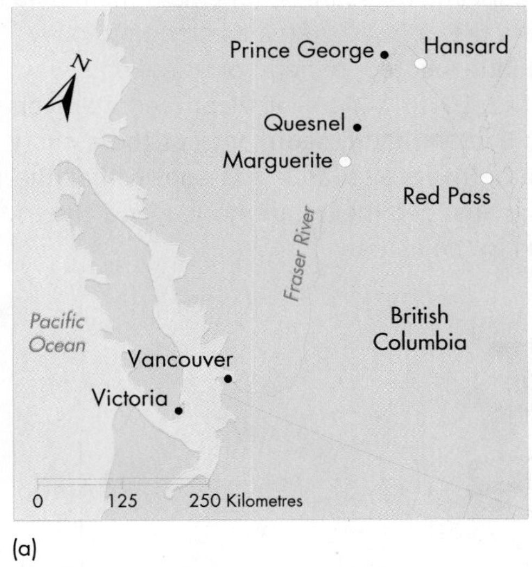

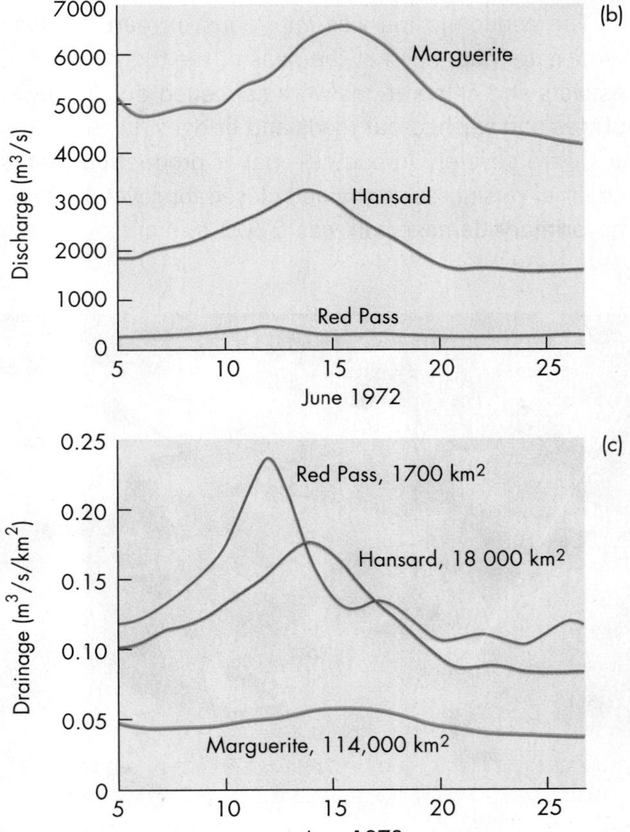

▲ **FIGURE 9.21 DOWNSTREAM MOVEMENT OF A FLOOD CREST** Changes in discharge as a flood wave moved downstream along the Fraser River in southern British Columbia in June 1972. (a) A map of the area. (b) The volume of water passing Red Pass, Hansard, and Marguerite; the volume and duration of the flood increased as tributaries added more water. (c) The volume of water per unit area at the same points; flooding at Marguerite lasted much longer than flooding at Red Pass in the headwaters of the basin.

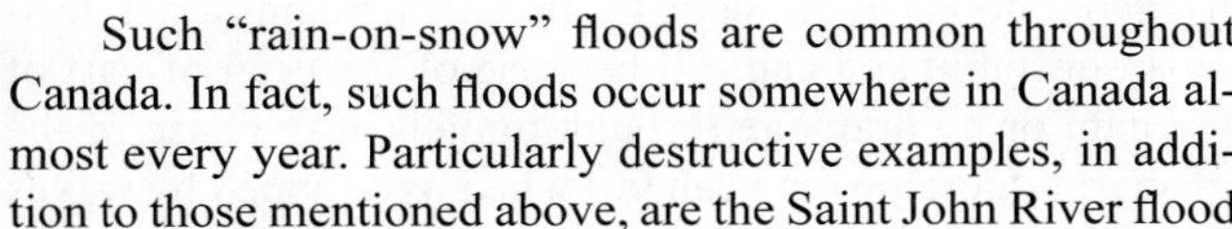

Such "rain-on-snow" floods are common throughout Canada. In fact, such floods occur somewhere in Canada almost every year. Particularly destructive examples, in addition to those mentioned above, are the Saint John River flood in Fredericton in May 2008 (about $12 million damage) and flooding along the Assiniboine River in Manitoba in May 2011 (more than $1 billion damage).

9.2 CASE STUDY

The Saguenay Flood

The Saguenay flood (in French, *le déluge du Saguenay*) occurred on July 19 and 20, 1996, in the Saguenay–Lac-Saint-Jean region of southern Quebec.[22,23] Two weeks of heavy rain filled reservoirs and raised rivers to flood stage. On July 19, about 270 mm of rain fell on the region within a few hours, an amount equal to the total normal July rainfall. Reservoirs filled to capacity and raging rivers eroded their banks, destroying bridges and buildings. Residents were forced to flee their homes as the waters rapidly rose and water poured into their basements. Over 2.5 m of water ran through parts of Chicoutimi and La Baie, destroying an entire neighbourhood (Figure 9.22). About 16 000 people were evacuated from the affected area and ten people were killed by landslides triggered by the heavy rain.

The economic toll was enormous, in part because the disaster happened at the peak of the tourist season. Festivals and celebrations were cancelled due to power outages and washed-out roads and bridges. Local industries were severely impacted—paper production plants and other businesses remained closed for several weeks. The primary damage toll was $700 million, but when both insured and uninsured losses are included, as well as indirect costs to the economy, the total losses likely exceeded $1.5 billion.[24]

A commission of enquiry into the flood found that the region's system of dams and dikes had been poorly maintained. Recommendations were made to strengthen floodgates, lower reservoir levels, and ensure the integrity of dams and dikes. The commission also suggested stopping construction in flood-prone areas. Faced with climate change and the possibility of more extreme weather in the future, there is doubt that these measures will be enough to prevent future catastrophes.

A small white house, referred to in French as *La petite maison blanche* ("the little white house") that stood while torrents of water flowed around it became the symbol of the flood. It has been preserved as a historical park and museum commemorating the flood and, despite a fire in 2002, remains standing today.

An unexpected benefit of the flood was that it deposited 10 to 50 cm of clean sediment on top of heavily contaminated sediments in the estuary of the Saguenay River. Research has shown that the buried contaminated sediments are no longer a threat to ecosystems in the estuary.[24]

◄ **FIGURE 9.22 FLOODING IN CHICOUTIMI** Flooding in downtown Chicoutimi, Quebec, triggered by heavy rainfall in July 1996. The floodwaters overtopped a small dam and spilled through the city. *(Jacques Boissinot/CP Images)*

Outburst Floods

Very large, short-lived floods result from the sudden draining of glacier-, moraine-, and landslide-dammed lakes. These "outburst" floods commonly have peak discharges many times larger than normal rainfall- or snowmelt-triggered floods in the same basin. Glacier dams are notoriously unstable and can fail because of flotation of part of the dam or by drainage through tunnels at the base of the glacier. A moraine dam can fail when overtopped by waves triggered by a landslide or ice avalanche (Figure 9.10).

A CLOSER LOOK 9.1

Magnitude and Frequency of Floods

The catastrophic floods that are reported on television and in newspapers are produced by infrequent, large, intense storms. Smaller floods may be produced by less intense storms that occur more frequently. All floods can be measured or estimated from data collected at stream-gauging stations (Figure 9.23). Such data show that there is a relation between the peak discharge of a flood and its average recurrence. Flood peak discharge at any station can be compared with discharges measured over time at that station. Values of annual peak flow, which is the largest flow of the year, are calculated from the station records. An average recurrence interval for each peak flow value is then determined by using the following equation and plotted to create a discharge-frequency curve:

$$R = (N + 1)\,M$$

where R is the recurrence interval in years, N is the number of years of record, and M is the rank of the individual flow within the recorded years.[25] Turning to Figure 9.24, for example, we see that the highest flow for nine years of data for the Patrick River is approximately 280 m^3/s, and that flow has a rank M equal to 1.[26] The recurrence interval of this flood is:

$$R = (N + 1)\,M = (9 + 1)\,1 = 10,$$

which means that a flood with a magnitude equal to or exceeding 280 m^3/s can be expected, on average, once every 10 years. We call this a 10-year flood. The probability that the 10-year flood will occur in any one year is 1/10 or 10 percent. Likewise, the probability that a *100-year flood* will occur in any year is 1 percent.

Extrapolating discharge-frequency curves is risky. A curve shouldn't be extended much beyond twice the number of years for which there are discharge records. For example, the discharge-frequency curve for the Red River shown in Figure 9.25 is based on more than 105 years of records, including the largest flood of the twentieth century in 1997. It has been extended to predict that the 200-year flood should have a peak discharge of about 5500 m^3/s, which is more than 20 percent larger than the peak discharge of the 1997 event.[22]

Data from many streams and rivers show that channels are formed and maintained by

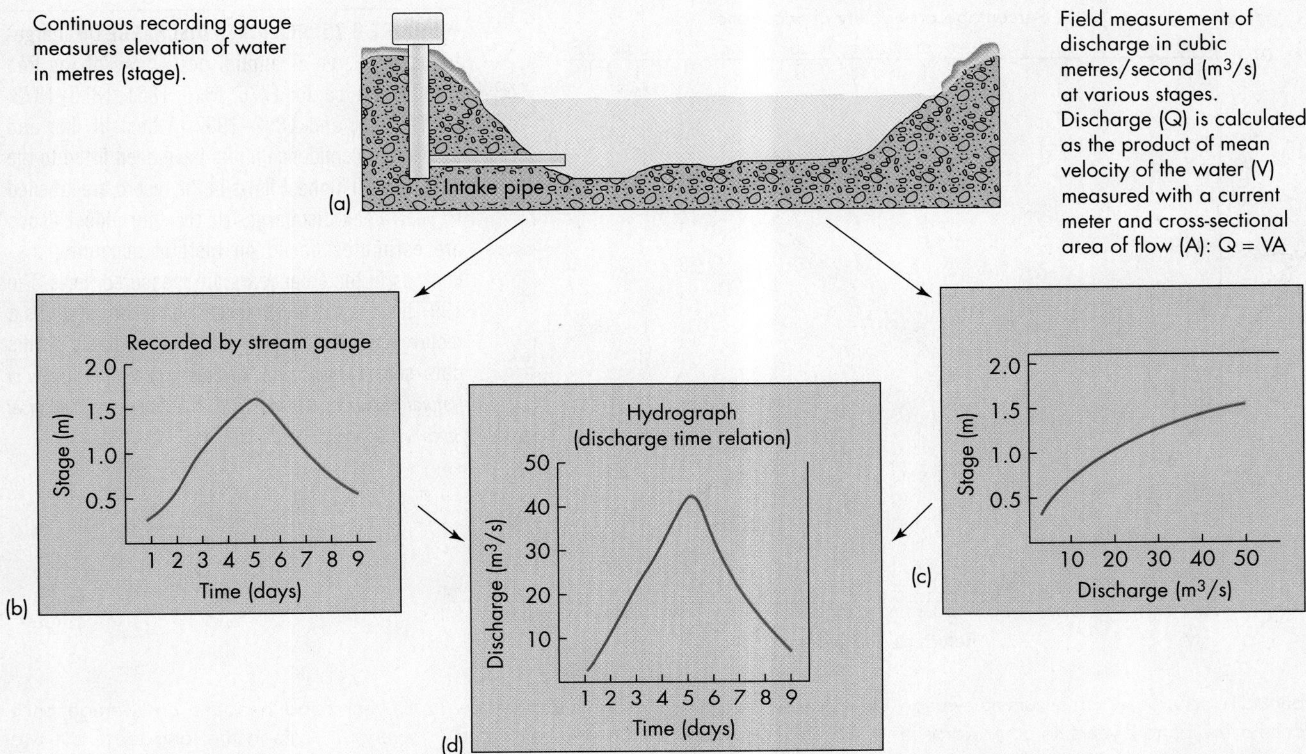

▲ **FIGURE 9.23 HOW A HYDROGRAPH IS PRODUCED** A hydrograph is made by installing a recording gauge (a) to obtain a continuous record of the water level, or stage. This record is then used to produce a stage-time graph (b). Field measurements at various flows provide a stage-discharge graph (c). Graphs (b) and (c) are then combined to make the final hydrograph (d).

(continued)

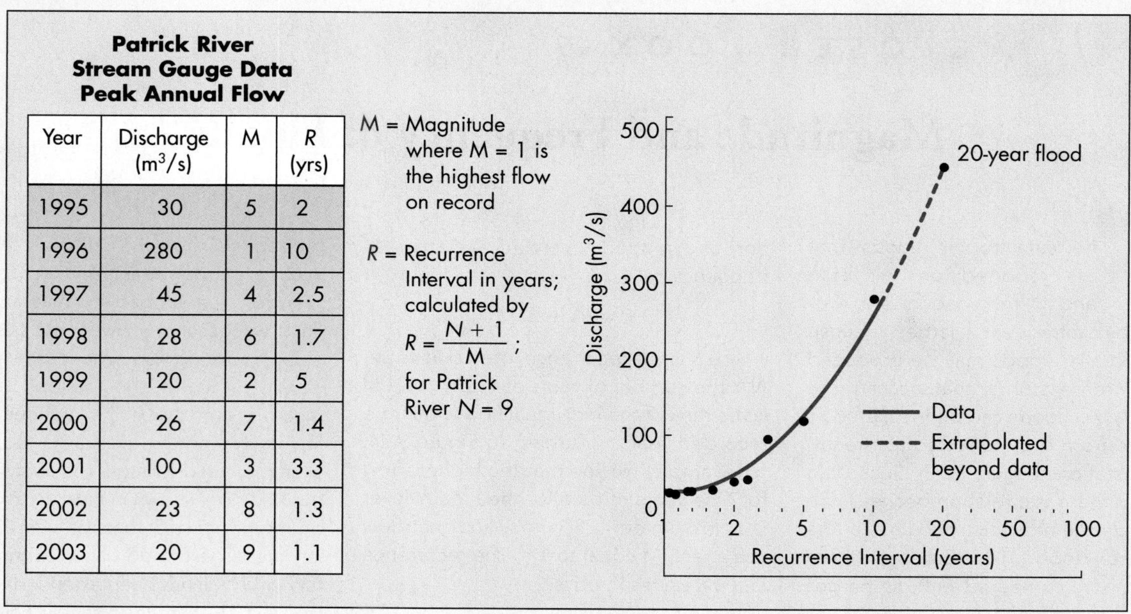

Patrick River Stream Gauge Data Peak Annual Flow

Year	Discharge (m^3/s)	M	R (yrs)
1995	30	5	2
1996	280	1	10
1997	45	4	2.5
1998	28	6	1.7
1999	120	2	5
2000	26	7	1.4
2001	100	3	3.3
2002	23	8	1.3
2003	20	9	1.1

▲ **FIGURE 9.24 DISCHARGE-FREQUENCY CURVE** We can make a discharge-frequency graph for the Patrick River by ranking (1 through 9) the largest flow for each of nine successive years of discharge measurements in m^3/s. We then calculate the recurrence interval, or frequency, of the largest annual flow by using the formula shown above and tabulated in the table on the left. Finally, we plot the discharges as a function of the recurrence interval to produce the graph on the right. The curve can be extended or extrapolated to estimate a peak discharge of 450 m^3 per second for a 20-year flood. *(After Leopold, L. B. 1968.* Hydrology for Urban Land Planning: A Guidebook on the Hydrologic Effects of Urban Land Use. *U.S. Geological Survey Circular 554)*

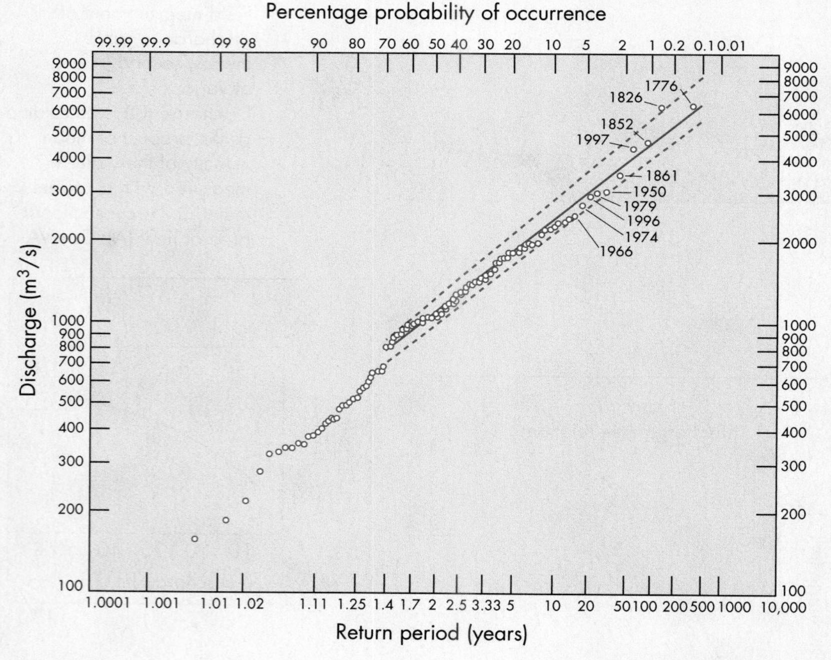

◀ **FIGURE 9.25 RED RIVER DISCHARGE** Discharge-frequency curve of annual peak flows of the Red River, Manitoba, for 1776, 1826, 1851, 1875–1878, 1880–1885, and 1892–1997. A best-fit line and 95 percent confidence limits have been fitted to the data. The 10 highest flows in the record are labelled by year. Peak discharges of the four oldest floods are estimates based on historic accounts; discharges in the other years are measured flows. The 1997 flood is the fourth largest on record and has a recurrence interval of 62 years on the basis of this data set. *(Reproduced or adapted with the permission of Natural Resources Canada 2013, courtesy of the Geological Survey of Canada (Bulletin 548))*

bankfull discharge, which occurs on average once every 1.5 to 2 years. In other words, a stream in a natural state overflows its banks and covers part of the floodplain with water and sediment once every year or two.

As more river flow data are collected, we can more accurately predict floods. However, designing structures for any flow with a long return period is a calculated risk because the predictions are commonly based on extrapolation of data. For many streams, the flow record is far too short to accurately predict the magnitude and frequency of large floods. Furthermore, a 50-year flood happens on average once every 50 years in the long term, but two 50-year floods could occur in successive years, as could two 100-year floods![27] As long as people build highways, bridges, homes, and other structures on floodplains, we should expect loss of life and property.

Outburst floods and associated debris flows can be deadly when people live along their paths. A flood of debris and water caused by the failure of a moraine dam in the Cordillera Blanca of Peru in 1941 killed about 5000 people in the city of Huaraz.[28] Perhaps the largest outburst flood from a glacier-dammed lake in the twentieth century occurred in 1929 in the Himalayas. The sudden emptying of Lake Shyok in 1929 produced a flood that raised the level of the Indus River 29 m at Attock, 700 km downstream from the lake.[29]

9.3 Geographic Regions at Risk for Flooding

Flooding can occur along any stream or river and thus is the most widespread natural hazard (Figure 9.26). A single flood can cause billions of dollars of property damage and large numbers of deaths (Table 9.1). Developing countries suffer much greater loss of life than developed ones because of the larger numbers of people at risk, the lack of monitoring and warning capabilities, poor infrastructure and transportation systems, and inadequate resources available for effective disaster relief.[14,21]

The disproportionate impact of disasters on developing countries is illustrated by the 2010 flood in Pakistan (Figure 9.27). The flooding—the worst in the country in the last 80 years—was caused by persistent heavy monsoon rains in the Karakoram Himalaya of northern Pakistan. Floodwaters killed over 2000 people and destroyed over a million homes. More than 21 million people, nearly one-third of the country's population, were left homeless, which is more than the combined total number of persons affected by the 2004 Indian Ocean tsunami, the 2005 Pakistan earthquake, and the 2010 Haiti earthquake.[30] At one point, about 20 percent of Pakistan's land area was under water (Figure 9.28).

The flood was a catastrophe that unfolded in slow motion. The flood wave moved southward along the Indus River through the month of August, from northern regions through west Punjab and into the southern province of Sindh. More than 7 million ha of Pakistan's most fertile cropland was submerged and 200 000 livestock were killed.[30]

An estimated 4000 km of highway and 6000 km of railway were damaged. The power infrastructure of Pakistan also took a heavy blow—10 000 electrical transmission lines were damaged by the floods, as well as many power-generating facilities, causing a temporary shortfall in electricity.

No country can cope with such as tragedy, least of all one with limited resources such as Pakistan. Recovery will take many years and will require large amounts of assistance from wealthy countries.

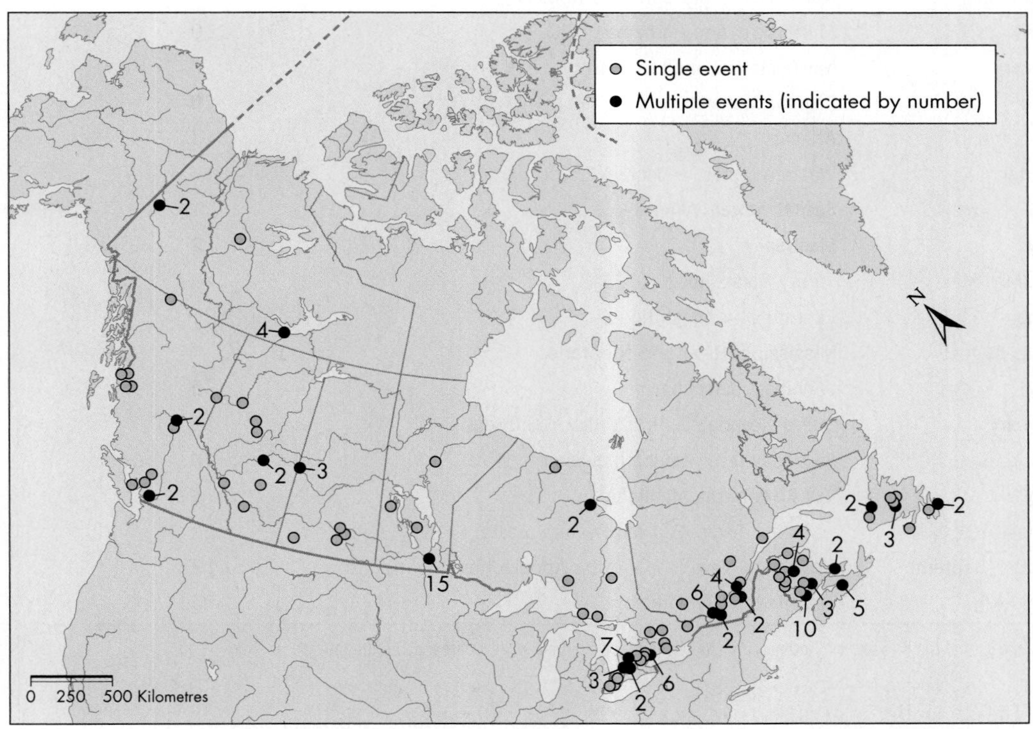

▲ **FIGURE 9.26 FLOOD DISASTERS IN CANADA IN THE TWENTIETH CENTURY** Locations of major historic floods in Canada. Symbols depict the general location of floods; most affected areas, however, are much larger than the symbol itself. *(Reproduced or adapted with the permission of Natural Resources Canada 2013, courtesy of the Geological Survey of Canada (Bulletin 548))*

TABLE 9.1 Selected Floods in the United States and Canada

Year	Month	Location	Lives Lost	Property Damage ($millions)*
1933	**May**	**Southern New Brunswick**	**7**	**Unknown**
1937	January–February	Ohio and lower Mississippi River basins	137	418
1938	March	Southern California	79	25
1940	August	Southern Virginia and Carolinas and eastern Tennessee	40	12
1947	May–July	Lower Missouri and middle Mississippi River basins	29	235
1948	**May–June**	**Lower Fraser River**	**10**	**15**
1950	**May**	**Red River basin, Manitoba**	**1**	**>100**
1951	June–July	Kansas and Missouri	28	923
1954	**October**	**Southern Ontario**	**81**	**100**
1955	December	West coast of United States	61	155
1963	March	Ohio River basin	26	98
1964	June	Montana	31	54
1964	December	California and Oregon	40	416
1965	June	Sanderson, Texas (flash flood)	26	3
1969	January–February	California	60	399
1969	August	James River basin, Virginia	154	116
1971	August	New Jersey	3	139
1972	June	Rapid City, South Dakota (flash flood)	242	163
1972	June	Eastern United States	113	3000
1973	March–June	Mississippi River	0	1200
1974	**January**	**Southern Quebec**	**0**	**60**
1976	July	Big Thompson River, Colorado (flash flood)	139	35
1977	July	Johnstown, Pennsylvania	76	330
1977	September	Kansas City, Missouri, and Kansas	25	80
1979	April	Mississippi and Alabama	10	500
1983	**January**	**Newfoundland and Labrador**	**0**	**34**
1983	**June**	**Regina, Saskatchewan**	**0**	**60**
1983	September	Arizona	13	416
1986	Winter	Western states, especially California	17	270
1986	**July**	**Saskatchewan, Alberta**	**1**	**28**
1987	**July**	**Montreal**	**2**	**94**
1990	January–May	Trinity River, Texas	0	1000
1990	June	Eastern Ohio (flash flood)	21	Several
1993	June–August	Mississippi River and tributaries		16 000
1993	**July**	**Winnipeg, Manitoba**	**0**	**>500**
1997	January	Sierra Nevada, Central Valley, California	23	Several hundred
1996	**July**	**Saguenay River, southern Quebec**	**10**	**1500**
1997	**May**	**Red River basin, Manitoba**	**3**	**4000**
2001	June	Houston, Texas, Buffalo Bayou (coastal river)	22	2000
2004	August–September	Georgia to New York and the Appalachian Mountains	ca.13	400
2006	June–July	Virginia to New York	16	1000

*Damages relate to the year of the disaster and are not adjusted for inflation. Canadian floods are indicated in bold.

◀ **FIGURE 9.27 CATASTROPHIC FLOODING IN PAKISTAN** An aerial view of flooded areas in Dera Alayar in Balochistan province, Pakistan, on August 16, 2010. *(Waheed Khan/Landov)*

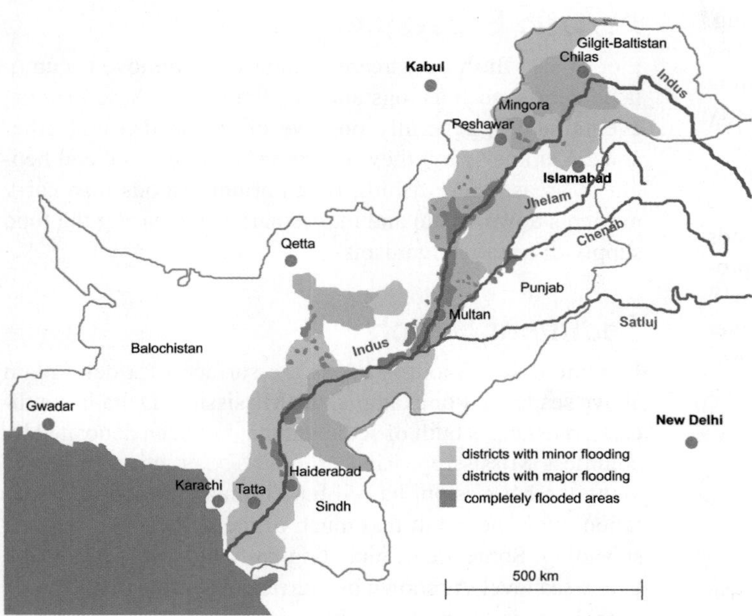

◀ **FIGURE 9.28 2010 PAKISTAN FLOOD** A map showing the area in Pakistan that was flooded in August 2010. At the peak of the flood, nearly 20 percent of the country was under water. About 2000 people drowned in this disaster. *(Wikipedia. 2010. "Pakistan floods." Used under Creative Commons Attribution 3.0 Unported licence; http://en.wikipedia.org/wiki/File:Indus_flooding_2010_en.svg)*

9.4 Effects of Flooding and Links between Floods and Other Hazards

Flood damage may be primary (that is, caused directly by the flood) or secondary (resulting from disruption of services and systems). Primary effects include loss of life; injury; and damage to farms, homes, buildings, railroads, bridges, roads, and other engineered works from flowing water, debris, sediment, and inundation. Floods can also remove or bury soil and vegetation. Secondary effects include pollution, hunger, disease, displacement of people, and losses of services and income. Failure of wastewater ponds, treatment plants, sanitary sewers, and septic systems may contaminate floodwaters with disease-causing microorganisms. For instance, record rainfall in southern New England in June 1998 caused partially treated sewage to float into Boston Harbour, forcing closure of many areas of Rhode Island's Narragansett Bay to swimming and shellfish harvesting.

Several factors affect the damage caused by floods:

- Land use on the floodplain
- Extent, height, and strength of dykes
- Depth and velocity of floodwaters
- Rate of rise and duration of flooding
- Season of the flooding
- Amount and type of sediment deposited by floodwaters
- Effectiveness of flood forecasting, warning, and evacuation

Damage is far greater on floodplains that have commercial and residential development than on floodplains that are used for farming, ranching, and recreation. Flooding during the growing season may damage or destroy crops, whereas

◄ **FIGURE 9.29 FLOODED CITY** These buildings in Grand Forks, North Dakota, were burned by a fire caused by the Red River flood in 1997. The flood forced the evacuation of 50 000 people in Grand Forks. *(Eric Hylden/ Grand Forks Herald)*

the same flooding during winter months is less damaging. Along large rivers, flood forecasts by Environment Canada and the U.S. National Weather Service provide the warning time needed to build temporary levees or remove property from vulnerable areas.

As discussed in other chapters, floods can be a primary effect of hurricanes and a secondary effect of earthquakes and landslides. Although seemingly counterintuitive, floods can also cause fires in populated areas. Floodwaters can produce shorts in electric circuits or erode and break gas lines, sparking dangerous fires. For example, the 1997 Red River flood caused a fire in Grand Forks, North Dakota, that burned part of the city centre (Figure 9.29). River erosion during floods can also trigger landslides.

9.5 Natural Service Functions

Flooding is a risk only when people live or have property on a floodplain or when they try to cross a flood-swollen river. In fact, periodic flooding has many benefits. Floods provide fertile sediment for farming; they benefit aquatic ecosystems; and in some areas they keep the surface of the land above sea level.

Fertile Land

When a river overflows its banks, the velocity of flow decreases, and the suspended load of fine sand, silt, clay, and organic matter is deposited on the floodplain. These periodic additions of nutrient-rich sediment are the reason that floodplains are some of the most fertile and productive agricultural areas in the world. Ancient Egyptians planned their farming around regular flooding of the Nile River. They understood that good harvests followed large floods, and they referred to flooding as "The Gift of the Nile." Unfortunately, with completion of the Aswan Dam in 1970, Egypt's annual floods have been effectively stopped. Now farmers must use fertilizer to successfully grow crops on what was once naturally fertile land.

Aquatic Ecosystems

Floods also flush out stream channels and remove accumulated debris, such as logs and tree branches. These flushing events have a generally positive effect on fish and other aquatic animals, and they may translate into a societal benefit in areas where fishing is important. Floods also carry nutrients downstream and into estuaries, increasing the food supply of aquatic organisms.

Sediment Supply

In some cases, flooding keeps the surface of a delta plain above sea level. For example, the Mississippi Delta in southeast Louisiana is built of sediment that has been deposited by countless Mississippi River floods. Construction of levees along the Mississippi has all but eliminated flood sedimentation, with the result that much of the delta is now slowly subsiding. Some areas, including parts of New Orleans, are below sea level. As shown by Hurricane Katrina in 2005, this subsidence is a very big problem indeed!

9.6 Human Interaction with Flooding

Human activity can significantly affect river processes and alter the magnitude and frequency of flooding. Land-use changes can increase or decrease sediment supply to a stream, which can, in turn, change the gradient and shape of the channel. Urbanization, especially increases in the number of paved areas, buildings, and storm sewers, changes runoff and, in some places, the risk of local flooding. Dyking of rivers in order to lower flood risk may, ironically, increase it because floodwaters can no longer spread out over the floodplain and are confined to a restricted space.

Land-Use Changes

Rivers are open systems that generally maintain a *dynamic equilibrium*—that is, an overall balance between the work

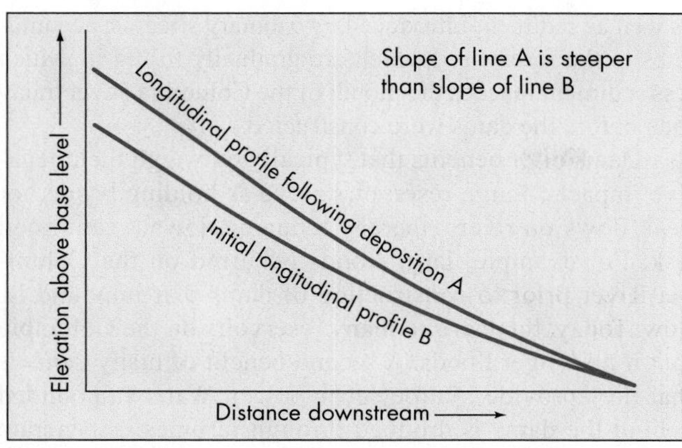

▲ **FIGURE 9.30 EFFECT OF DEPOSITION ON RIVER SLOPE** An idealized diagram illustrating that deposition in a stream channel increases the channel gradient; the slope of longitudinal profile A is steeper than that of longitudinal profile B.

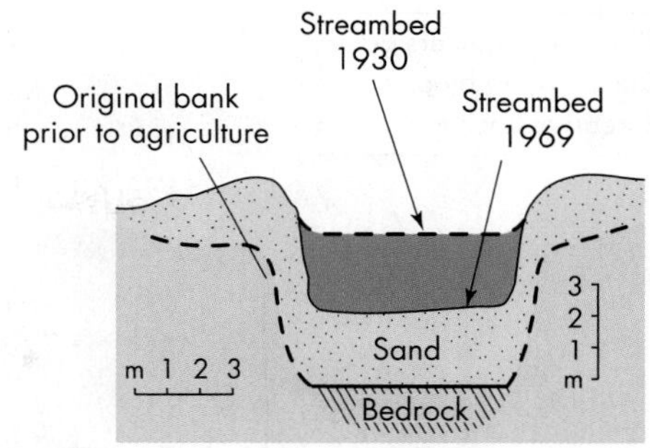

▲ **FIGURE 9.31 RESPONSE OF A STREAM TO CHANGES IN LAND USE** A cross-section of a stream at the Mauldin Millsite, Georgia, showing channel changes through time. Clearing of forest for agriculture increased sediment deposition in the channel until 1930. A return to woodland after 1930 increased stream erosion, thus lowering the channel. *(Based on Trimble, S. W. 1969.* Culturally Accelerated Sedimentation on the Middle Georgia Piedmont. *Master's thesis, Athens, Georgia: University of Georgia. Reproduced by permission)*

the river does and the sediment load it receives. Sediment is supplied by tributaries, landslides, and erosion of bank materials. A river tends to maintain the gradient and cross-sectional shape that provide the flow velocity it needs to move its sediment load.[31]

A change in the amount of water or sediment carried by a river brings about changes in channel gradient or cross-sectional shape, effectively changing the velocity of the water. The change in velocity may, in turn, increase or decrease the amount of transported sediment. Land-use changes that alter a stream's sediment or water supply may set into motion a series of events that bring about a new dynamic equilibrium.

Consider, for example, a land-use change from forest to an agricultural row crop, such as corn, that increases the amount of sediment delivered to a stream. At first, the stream will be unable to transport the additional load and will deposit some of the sediment, increasing the channel gradient. As the channel steepens, flow velocity increases, allowing the stream to move more sediment. If we assume that base level remains constant, this process will continue until the stream is flowing fast enough to carry the new load. If it seems counterintuitive that sediment deposition increases channel gradient, study the longitudinal profiles in Figure 9.30. A new dynamic equilibrium can be reached, provided the increase in sediment supply levels off and the channel gradient and shape can adjust before another change occurs.

Suppose now that the reverse situation occurs; that is, farmland is converted to forest. Surface erosion will decrease and less sediment will be deposited in the stream channel. Erosion of the channel will eventually lower the gradient, which in turn will lower the velocity of the water. Erosion will continue until a new equilibrium is achieved between sediment supply and work done.

The sequence of events just described occurred in parts of the southeast United States over the past 250 years. Most forests between the Appalachian Mountains and the Atlantic coastal plain were cleared for farming by the 1800s. The change from forest to farming accelerated soil erosion and triggered deposition of sediment in local streams (Figure 9.31). Channels that existed before farming began to fill with sediment. After 1930, the land reverted to pine forest. This change, in conjunction with soil-conservation measures, reduced the quantity of sediment delivered to streams. By 1969, formerly muddy, sediment-choked streams had cleared and eroded their channels (Figure 9.31).

Dam Construction

Consider the effects of building a dam on a river. Upstream of the dam, the river enters the reservoir and deposits much of its sediment there (Figure 9.32). Downstream of the dam,

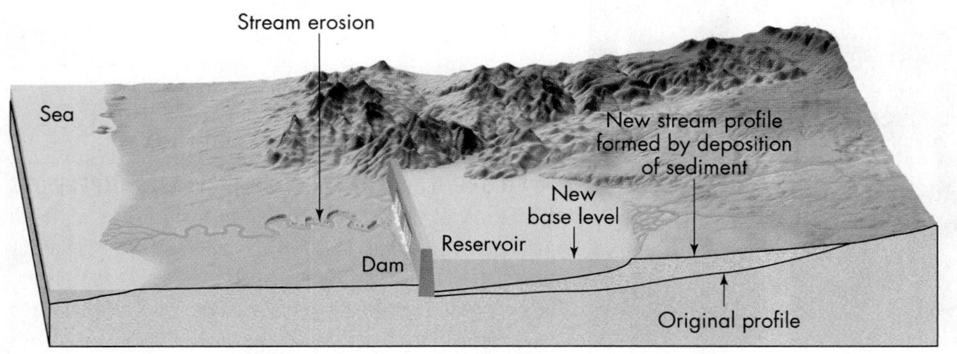

◄ **FIGURE 9.32 EROSION AND DEPOSITION CAUSED BY A DAM** Following dam construction, sediment accumulates within and upstream of the reservoir shown in this figure. Erosion occurs downstream of the dam because water leaving the reservoir carries much less sediment than the stream is capable of transporting. *(Based on Tasa, D. 2005. In Tarbuck E. J., and F. K. Lutgens.* Earth: An Introduction to Physical Geology, *8th ed. Upper Saddle River, NJ: Pearson Prentice Hall)*

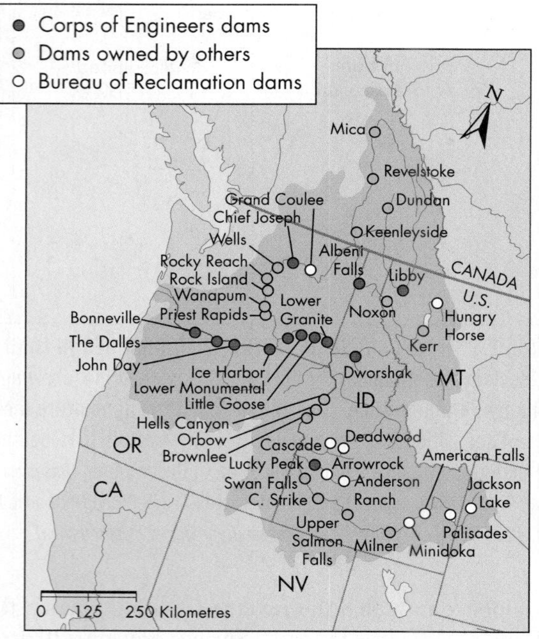

▲ **FIGURE 9.33 TAMING OF A RIVER** Much of the Columbia River upstream of Portland, Oregon, is a series of reservoirs impounded by dams built between 1933 and 1984. Sediment transported to the Pacific Ocean before dam construction now accumulates in the reservoirs.

the river is transporting little sediment and thus erodes its channel more aggressively. The gradient of the river downstream of the dam will decrease until a new equilibrium is attained.

Dam construction has radically altered sediment and water flow on the Columbia River in British Columbia, Washington, and Oregon. A series of dams built in the past century for irrigation, electrical power generation, and flood control has left the Columbia with few free-flowing reaches (Figure 9.33). Most of the sediment transported by the river, as well as sediment introduced by tributary streams, accumulates in the reservoirs, which are gradually filling in. Much less sediment reaches the mouth of the Columbia River today than before the dams were constructed.

Dams offer benefits that typically outweigh their negative impacts. Large reservoirs serve as holding basins for peak flows on rivers, thereby reducing downstream flood risk. For example, large floods occurred on the Columbia River prior to construction of dams that impound its flow. Today, there are so many reservoirs on the Columbia that it no longer floods. A second benefit of many dams is that they provide hydroelectric power. Water impounded behind the dams is dropped through turbines to generate electricity. About two-thirds of the electricity generated in Canada currently comes from hydroelectric facilities.[32] A third important benefit of dams is irrigation. Agriculture would not be possible on the Columbia Plateau of eastern Washington and in southern Saskatchewan, for example, without irrigation waters derived from reservoirs. Finally, some artificial reservoirs provide recreational opportunities that would not otherwise exist. Franklin D. Roosevelt Lake, which is impounded behind Grand Coulee Dam in Washington State, is the centrepiece of a U.S. National Recreation Area that has 27 public campgrounds and 22 public boat launches. The 220-km-long lake is a popular boating, fishing, hunting, camping, and swimming destination.

Urbanization and Flooding

Urbanization can increase the magnitude and frequency of floods in small drainage basins. Changes in runoff in such situations depend on the amount of land that is covered with roofs, pavement, and cement (elements referred to as *impervious cover*) (Figure 9.34), and the percentage of the area served by storm sewers. In most urban areas, storm sewers start at drains at the sides of streets. They carry runoff to

◄ **FIGURE 9.34 URBANIZATION INCREASES IMPERVIOUS COVER** Aerial view of Toronto, Ontario, which, like all North American cities, has much of its land surface covered by paved streets, sidewalks, parking lots, and buildings. This impervious land cover greatly reduces infiltration of rainwater and increases surface runoff. *(Pavel Baudis/Czech Republic)*

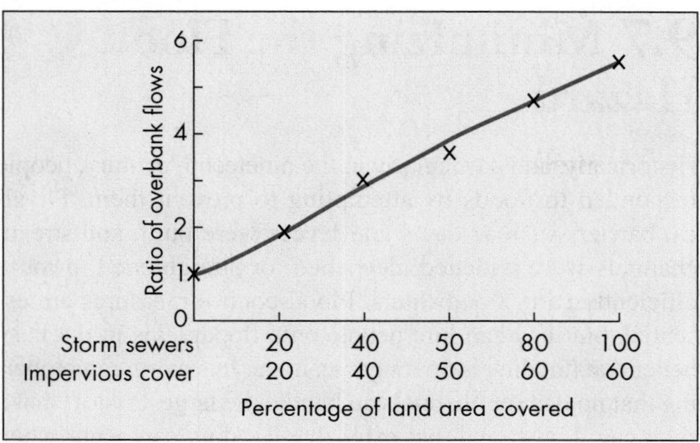

▲ **FIGURE 9.35 EFFECT OF URBANIZATION ON THE FREQUENCY OF FLOODS** The relation between the increase in overbank flows and the percentage of area with impervious cover and with storm sewers. A ratio of three, for example, means that three floods occurred after urbanization for every one that took place before urbanization, or that flooding was three times more frequent after urbanization. The graph shows that the frequency of floods increases with increasing urbanization. *(After Leopold, L. B. 1968.* Hydrology for Urban Land Planning: A Guidebook on the Hydrologic Effects of Urban Land Use. *U.S. Geological Survey Circular 554)*

stream channels much more quickly than surface flow in natural settings. An urban area with 40 percent impervious cover and 40 percent of its area served by storm sewers may experience up to three times as many floods of a given magnitude as before urbanization (Figure 9.35). However, floods are less affected by urbanization as the size of the drainage basin increases.

Urban flooding can also result from poorly constructed or maintained drains. Long periods of only moderate rainfall can cause flooding if storm drains become blocked with sediment and storm debris. Water begins to pond behind debris in the drains, overflowing into low areas.

Urbanization also changes how rapidly floods develop. Comparison of hydrographs before and after urbanization shows a significant reduction in the *lag time* between peak rainfall and the flood crest after urbanization (Figure 9.36). A short lag time, referred to as *flashy discharge*, is characterized by a rapid rise and fall in discharge.

Urbanization also greatly reduces stream flow during the dry season. Normally, small streams continue to flow during dry periods because groundwater discharges into channels. However, because urbanization significantly reduces infiltration, less groundwater is available to recharge streams. The reduced flow affects both water quality and the appearance of a stream. Urban streams with low discharges may carry heavy loads of pollutants.[26] Some of the pollutants, such as nitrogen and phosphorus derived from fertilizer, can stimulate the growth of algae, which reduces the dissolved oxygen content of the water and harms aquatic life.

Impervious cover and storm sewers are not the only types of construction that increase flooding. Some flash floods occur because bridges built across small streams

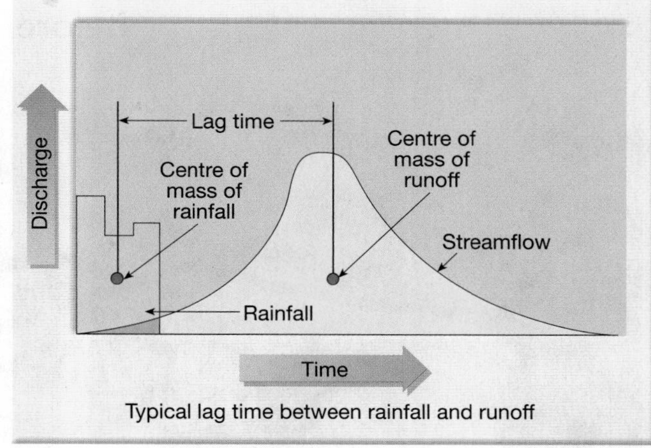

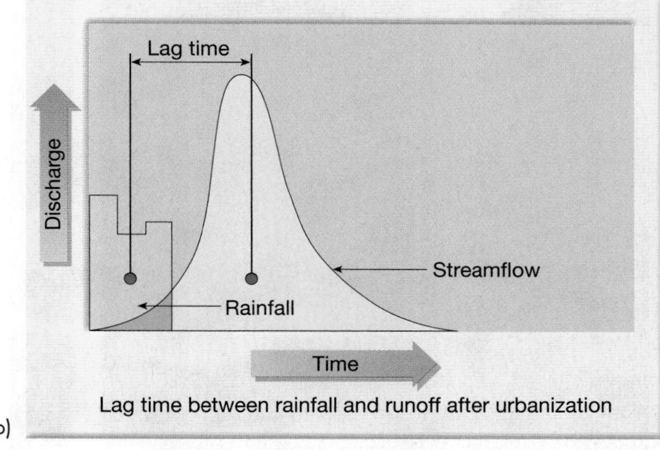

▲ **FIGURE 9.36 URBANIZATION SHORTENS LAG TIME** (a) A generalized hydrograph showing the lag between the time of peak rainfall and the time of flooding. (b) A hydrograph showing the decrease in lag time and more rapid rise and fall of discharge caused by urbanization. *(Based on Tarbuck, E. J., and F. K. Lutgens. 2005.* Earth: An Introduction to Physical Geology, *8th ed. Upper Saddle River, NJ: Pearson Prentice Hall)*

block the passage of floating debris, which then forms a temporary dam. If the dam fails, a destructive wave of water and debris moves downstream. One of Canada's worst natural disasters, and an example of such an occurrence, happened at Britannia, British Columbia, in October 1921. Britannia was a bustling mining town at the mouth of Britannia Creek on Howe Sound during the early decades of the past century. When the town was built, no consideration was given to the possibility that Britannia Creek might flood, but flood it did on October 28, 1921. Heavy rains had fallen in the preceding days, and Britannia Creek ran high. Culverts beneath a large earthen railway embankment that crossed Britannia Creek directly above the town became plugged with debris and a lake began to form. As the lake grew larger, the trapped waters exerted more and more pressure on the embankment. Eventually, the dam failed and a torrent of debris-laden water cut a swath through the community (Figure 9.37). The flood killed 37 people and destroyed about half the 170 houses in the town.[33]

▲ **FIGURE 9.37 DISASTER AT BRITANNIA** The mining community of Britannia, built on the floodplain of Britannia Creek, before and after the catastrophic flood of October 1921. *(Britannia Mine Museum)*

9.7 Minimizing the Flood Hazard

Historically, and particularly in the nineteenth century, people responded to floods by attempting to prevent them. Physical barriers such as dams and levees were built, and stream channels were widened, deepened, or straightened to more efficiently carry floodwaters. Flood-control measures are essential, but they can lure people onto floodplains in the false belief that flooding is no longer an issue. It is worth remembering that no dam or channel can handle the largest runoff that a river can deliver and that extensive flooding can occur when the flow ultimately exceeds the capacity of the structure.

The Structural Approach

Physical Barriers Engineered structures built to reduce the risk of flooding include earthen levees, or *dykes* (Figure 9.38), concrete floodwalls, dams to store water for later release, floodways that bypass populated areas, and stormwater-retention basins. Unfortunately, the benefits of these physical barriers are often lost because they encourage development on the floodplains they are intended to protect. For example, the winters of 1986 and 1997 brought tremendous storms and flooding to the western American states, particularly California, Nevada, and Utah. In all, damage exceeded several hundred million dollars and several people died. During one of the floods in 1986, a levee on the Yuba River in California broke, causing more than 20 000 people to flee their homes. An important lesson from this flood is that levees constructed long ago might be in poor condition and might fail during floods.

In Canada, dykes are widely used to protect people and property from river flooding. A good example is the dyking system along the lower Fraser River in British Columbia. Before European settlement, the mighty Fraser River was untamed—it regularly overtopped its banks and covered parts of the floodplain from Hope to the river mouth downstream of New Westminster. Beginning in 1864, dykes were built along

◄ **FIGURE 9.38 MISSISSIPPI RIVER LEVEE** Earthen or concrete levees border the lower Mississippi River in Louisiana. A road on the top of this levee appears as a curving white line. This levee protects farms, homes, and businesses along the highway on the left. *(Jupiter Images)*

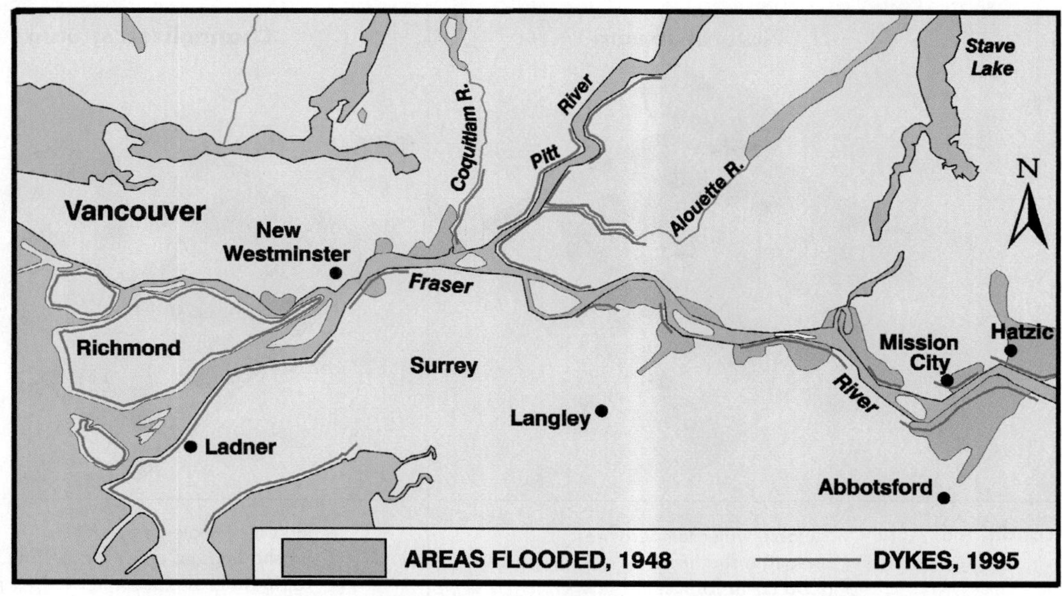

▲ **FIGURE 9.39 RIVER AND SEA DYKES PROTECT PEOPLE AND PROPERTY ON THE FRASER RIVER FLOODPLAIN** A map of the lower Fraser Valley showing locations of sea and river dykes and areas inundated during the 1948 Fraser River flood. *(Reprinted with permission from Tricouni Press)*

the Fraser River and its tributaries to protect people and property from flooding. Following the last big Fraser River flood in 1948, more than 260 km of dykes were repaired or rebuilt by the Fraser River Dyking Board, and by 1960, some 375 km of river and sea dykes provided protection to most of the vulnerable floodplain and reclaimed tidal lands[33] (Figure 9.39). In 1968, the federal and provincial governments established the Fraser River Flood Control Program to further rehabilitate and improve the system of dykes in the Fraser Valley. By 1994, the two governments had spent almost $300 million on flood-control structures and programs. A good investment, most would argue, because a recurrence of a flood similar to that of 1948 would cause at least $2 billion damage. The present dyke system should withstand a flood as large as any that has happened in the last 150 years.

The Fraser Valley dyking system includes sea dykes as well as river dykes. Dykes at the western and southern fronts of the Fraser River delta protect Richmond and Delta from incursions of the sea during severe storms and when tides are high. Richmond, a city of 250 000 people, is located below the upper limit of tides and would be inundated by waters of the Strait of Georgia were it not for the sea dykes ringing the city.

Channelization Straightening, deepening, widening, clearing, and lining existing stream channels are all methods of **channelization** (Figure 9.40). Channelization is used to control floods, drain wetlands, and maintain navigable river channels.[34] Thousands of kilometres of streams in Canada and the United States have been channelized without adequate consideration of the adverse effects of the practice. Thousands of additional kilometres of channelization projects are planned or in progress.

◄ **FIGURE 9.40 STREAM CHANNELIZATION** An extreme case of stream channelization: a concrete-lined channel in Los Angeles, California. *(Edward A. Keller)*

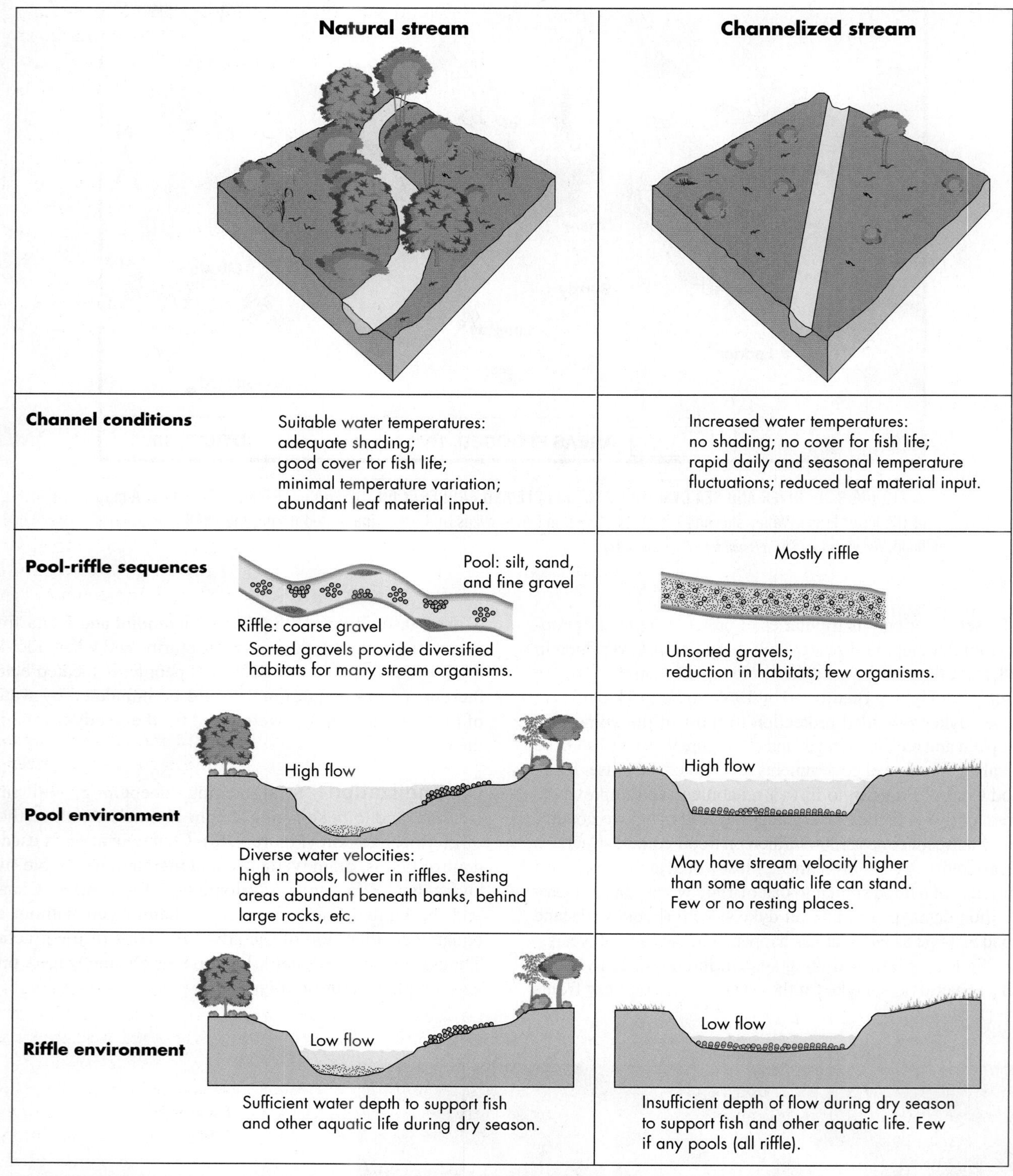

▲ **FIGURE 9.41 COMPARISON OF NATURAL AND CHANNELIZED STREAMS** The channelization of a stream significantly alters flow conditions, pool and riffle development, and aquatic ecosystems. *(Based on Corning,* Virginia Wildlife, *February 1975)*

Opponents of channelizing streams emphasize that the practice degrades river and wetland ecosystems (Figure 9.41). Their arguments are as follows:

- Draining wetlands adversely affects plants and animals by eliminating habitats necessary for their survival.
- Removal of trees along riverbanks eliminates shade and cover for fish and exposes the stream to the sun, which can damage plant life and heat-sensitive aquatic organisms.
- Cutting trees on floodplains eliminates many animal habitats, while increasing sediment delivery to the stream.

▲ **FIGURE 9.42 RED RIVER FLOODWAY** The Red River Floodway (channel leading to the top of the photo) diverts some of the flow of the swollen Red River (bottom left) during the 1997 flood, reducing the discharge of the river downstream in Winnipeg. *(Greg Brooks, Geological Survey of Canada)*

- Straightening a river channel destroys the diversity of natural flow conditions and feeding and breeding areas for aquatic life.
- Conversion of a meandering stream to a straight channel seriously degrades the aesthetic value of a natural area.[34]

Some drainage projects are beneficial. The benefits are probably greatest in urban areas with a high flood risk and in rural areas where previous land use has created drainage problems. In some areas, channel modification has improved navigation or reduced flooding without significant environmental degradation.

The largest flood channelization project in Canadian history was construction of the Red River Floodway between 1962 and 1968.[35] The floodway consists of four elements: the floodway channel, an inlet control structure, dykes, and an outlet structure (Figure 9.42). The floodway channel is 48 km long and has a designated flow depth of 8 m. The width of the top of the channel ranges from 213 m to 305 m. Material excavated from the channel was deposited along the channel sides to form an embankment 6 m high. The inlet control structure is located on the Red River just downstream from the floodway inlet near St. Norbert, Manitoba. Its purpose is to regulate the flow between the natural channel of the Red River and the floodway channel during periods of high flow. Dykes upstream of the inlet control structure prevent floodwaters from bypassing it. The vertical drop over the entire length of the floodway is 5 m, only half the corresponding drop of the Red River. The purpose of the outlet structure is to dissipate the potential energy in the water at its point of re-entry into the Red River near Lockport, thereby preventing damage to the floodway.

The Red River Floodway was extensively criticized when it was planned and constructed, but its benefits have been considerable. Had the floodway not existed during the great flood of 1997, Winnipeg would have been flooded. Damage to Winnipeg that might have occurred during Red River floods in 1969, 1970, 1974, 1979, and 1987 would have been in excess of $2 billion. The total cost of the Red River Floodway and two other flood-control structures was $94 million at the time of construction, equivalent to about $500 million today.

Channel Restoration: An Alternative to Channelization

Most streams in urban areas are no longer natural. **Channel restoration** is a suite of measures that attempts to return severely modified streams to a more natural state. These measures include (1) removing urban waste from stream channels, (2) protecting existing channel banks by planting native trees and other vegetation, and (3) re-establishing deeper pools and shallower riffles within channels.[36] Trees are important because they provide shade for a stream and their root systems protect stream banks from erosion.

The objective of channel restoration is to create a more natural channel by allowing the stream to meander and develop variable water-flow conditions. Where lateral bank erosion must be controlled, the outsides of bends can be defended with large stones known as *riprap* or with wire baskets filled with rocks known as *gabions* (Figure 9.43).

Kissimmee River Restoration Restoration of the Kissimmee River in Florida may be the most ambitious project of its kind attempted in the United States. Channelization of the river began in 1960 and took 10 years to complete at a cost of U.S.$32 million. The Kissimmee was changed from a 165 km meandering river into an 83 km straight ditch. As a result of the channelization, water quality decreased and numbers of waterfowl and fish declined. More than 800 km^2 of floodplain wetlands were drained during the project. Ironically, channelization increased the flood hazard because the floodplain wetlands no longer stored runoff.

Within a year after the project was completed, the State of Florida called for the river to be restored. The U.S. Congress in 1991 mandated that the U.S. Army Corps of Engineers, which was responsible for the original channelization, begin restoring about one-third of the river at a cost of U.S.$400 million, more than 10 times the cost of the original project. Although restoration is certainly the right thing to do, more careful environmental evaluation before the original

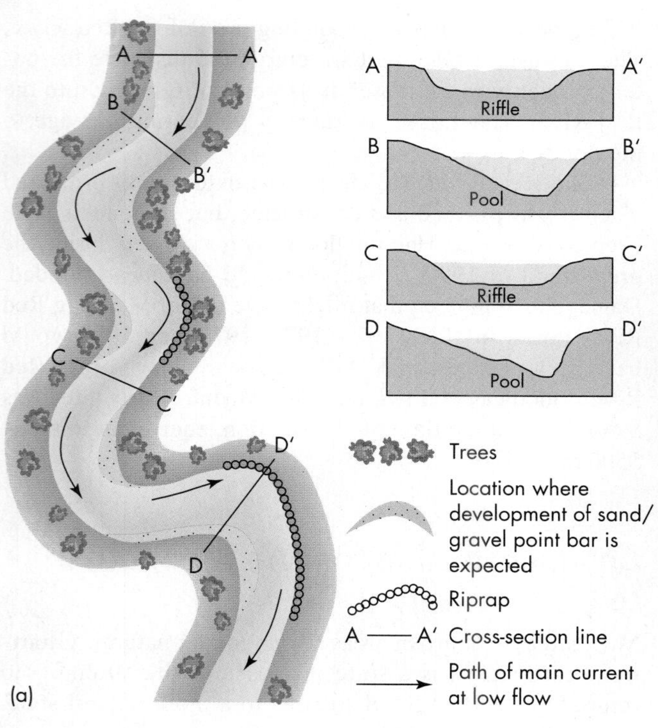

◀ **FIGURE 9.43 URBAN STREAM RESTORATION** (a) A channel restoration strategy that uses changes in channel shape to cause scour and deposition at desired locations. (b) Dump truck placing riprap to defend the bank of Briar Creek in Charlotte, North Carolina. Grass was planted and covered with straw mulch (to the left of men) to help stabilize the stream bank. *((a) Modified after Keller, E. A., and E. K. Hoffman. 1977.* Journal of Soil and Water Conservation *32:237–240; (b) Edward A. Keller)*

project began would have revealed the potential damage that necessitated it.[37]

Flood Forecasts and Advisories

The specific dates and magnitudes of floods cannot be predicted over long periods. For some watersheds, however, spring discharges can be estimated several days in advance by using models that incorporate snowpack depths, stream and lake levels, ground moisture, temperature, wind, evaporation, and weather forecasts. *Outlooks* of peak spring discharge can be made weeks in advance by using similar models with a variety of scenarios of snow melt and rainfall. Potential peak discharges of floods are expressed in terms of probability of occurrence, but the actual peak runoff is dependent on climatic events that are yet to occur and are thus inherently uncertain. Nevertheless, worst-case scenarios can be identified and planned for well in advance of the actual flooding.

Flood advisories or warnings are issued in anticipation of severe weather conditions, such as convective storms, thunderstorms, and hurricanes. Although the advisories are generally issued only hours in advance of the storm, they can provide sufficient time to mobilize resources for flood control and, in extreme cases, to evacuate people, livestock, or property from areas likely to be affected.

Northward-flowing rivers in Canada, such as the Yukon and Mackenzie rivers, routinely produce ice-jam flooding in the spring. Residents know when the flooding is likely to occur and are generally well prepared for it. Historical data can be used to relate the time and magnitude of the peak water level during breakup to current conditions. Further, once an ice jam has formed somewhere along a river, empirical methods can be used to predict the rise in water level.

9.8 Perception of and Adjustment to Flood Hazards

Perception of Flood Hazards

Most government agencies, planners, and policymakers have an adequate perception and understanding of flooding (see Professional Profile), but many individuals do not. Public knowledge of floods, an understanding of future flooding risk, and a willingness to adjust to the hazard are highly variable.

Progress in reducing flood risk at the institutional level requires preparation of hazard maps of flood-prone areas. Flood-hazard maps show areas that are susceptible to flooding along streams, lakes, and coastlines, and areas where urbanization is likely to cause problems in the future. In addition, federal, provincial, territorial, and state governments have encouraged local communities to adopt floodplain management plans.[38] Still, the idea of restricting or prohibiting development on floodplains or of relocating present development to sites off the floodplain is problematic; extensive community discussion is needed before the general population will accept such measures. This need was tragically shown in the 2006 flood on the U.S. Atlantic seaboard. Over 200 000 residents of Pennsylvania alone were evacuated from the Susquehanna River floodplain, and damages of approximately U.S.$1 billion were incurred. The Susquehanna River crested about 4.2 m above flood level near Binghamton, New York. Sixteen people lost their lives in flood-swollen rivers and creeks.

Adjustments to the Flood Hazard

In recent decades, scientists and planners have increasingly recognized the advantages of alternatives to structural flood-control measures. Alternatives include flood insurance and land-use controls on floodplains. Planners, policymakers, and hydrologists generally agree that no single adjustment is best in all cases. An integrated approach that incorporates adjustments appropriate for a particular situation is a more effective strategy.

Floodplain Regulation From an environmental perspective, the best adjustment that can be made to flood hazards is through **floodplain regulation**. The goal of floodplain regulation is to maximize the benefits that floodplains offer while minimizing flood damage and the cost of flood protection.[39] This approach is a compromise between the indiscriminate development of floodplains, which results in loss of life and tremendous property damage, and the complete abandonment of floodplains, which gives up a valuable natural resource.

Engineered structures are necessary to protect lives and property on floodplains that have extensive development. We must recognize, however, that the floodplain is part of the river system and that any encroachment that reduces the cross-sectional area of the floodplain increases flood risk. One approach to reducing flood risk is to disallow new development that would lessen a river's access to its floodplain; in other words, to design with—rather than against—nature. Realistically, the most practical approach is a combination of physical barriers and floodplain regulations that minimize physical modification of the river system. For example, reasonable floodplain zoning may reduce the size of a floodwater diversion channel or an upstream reservoir required to produce a prescribed level of flood protection.

A preliminary step in floodplain regulation is flood-hazard mapping.[40] Flood-hazard maps may delineate past floods or floods of a particular return period, for example the 100-year flood. They are useful in regulating development, purchasing land for parks and other public use, and creating guidelines for future land use on floodplains.

Flood-hazard evaluation can be accomplished in a general way by direct observation and measurement. For example, extensive flooding in the Red River Valley in 1997 was clearly mapped using satellite imagery and aerial photographs. The flood hazard can also be assessed from field measurement of high-water lines, flood sediments, scour marks, and the distribution of woody debris on the floodplain after the water has receded.[38] Once flood-hazard maps have been produced, planners can modify zoning maps, regulations, and building codes (Figure 9.44).

Flood-Proofing Several methods of flood-proofing are currently available:[41]

- Raising the foundation of a building above the anticipated level of flooding by using piles or columns or by extending foundation walls or earth fill

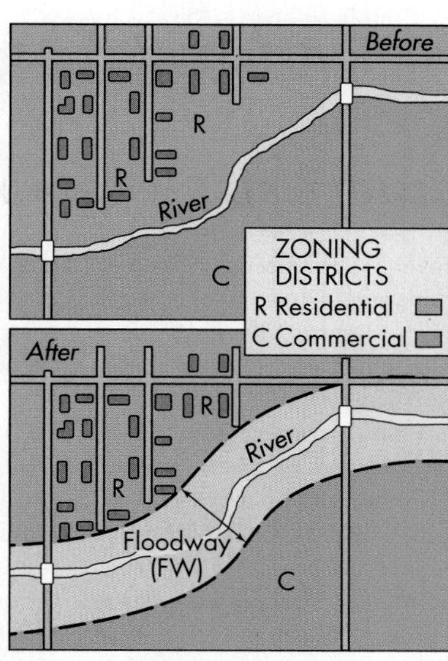

▲ **FIGURE 9.44 FLOODPLAIN ZONING** Zoning map before and after the implementation of flood regulations. *(From Water Resources Council. 1971.* Regulation of Flood Hazard Areas, *vol. 1)*

- Constructing flood walls or earthen mounds around buildings to isolate them from floodwaters
- Using waterproofed doors, basement walls, and windows
- Installing drains with pumps to remove water
- Installing ground-level windows that open to allow floodwaters to pass through the building without washing it away.

Other structural modifications can reduce damage when floodwaters enter a building. For example, ground floors of expensive riverfront properties in some communities in Germany are designed to withstand the forces of floodwaters and can be easily cleaned following a flood.[41]

Different Approaches Canada and the United States have adopted different approaches to managing their flood hazards. The Canadian approach involves planning, regulations to discourage new development on floodplains, and government flood damage compensation.[42] The U.S. approach includes some of these elements but places importance on private and public flood insurance.

The *Canada Water Conservation Assistance Act*, enacted in 1953, was the first federal legislation directly concerned with water resource management. It enabled the federal government to provide financial assistance to the provinces and territories for construction of works to conserve and control water. Under the act, the federal government contributed 37.5 percent of the cost of such works.

The *Canada Water Act*, passed in 1970, superseded the *Canada Water Conservation Assistance Act*. It allowed the federal government to participate with the provinces and territories in water-management programs. The philosophy behind the act was that:

PROFESSIONAL PROFILE

Eve Gruntfest, Geographer

It was early evening on July 31, 1976, and a summer thunderstorm was massing over Colorado's Front Range. The clouds cast a shadow over the steep ravines of the Big Thompson Canyon. When a hard rain began to fall, residents, accustomed to such weather, didn't think twice. But this was no ordinary storm. Within a few hours, it drenched the upper reaches of the Big Thompson Canyon with more than 30 cm of rain. By morning, 143 people were dead, drowned in the roiling waters.

Since that day, geographer Eve Gruntfest of the University of Colorado in Colorado Springs has dedicated her career to studying how those in the path of the Big Thompson Flood saved or lost their lives (Figure 9.45). Gruntfest applies social science methods to analyze the impacts of weather on society.

Trying to find out what people were doing during the flood, says Gruntfest, "was like a detective story." She researched newspaper accounts, tracked down eyewitnesses, handed out questionnaires, and even took out ads in local papers asking survivors to contact her for interviews. The story she pieced together surprised even herself.

For one thing, only a quarter of the victims were tourists. "Experts thought most local people knew what to do," Gruntfest says. In fact, many residents had tragically flawed ideas about how to escape a flood. Some fled to the roofs of motels, but the flash flood swept the structures away. Others tried to drive through the rising water. Gruntfest showed that people who left the shelter of their vehicles and scrambled to higher ground fared far better.

"The more I learned about the Big Thompson Flood, the more amazed I was that more people didn't die," Gruntfest says. "There were many incredible close calls—people driving with four flat tires, a bus that had just crossed a bridge seconds before it washed out."

The largest Front Range cities—Denver, Colorado Springs, and Boulder—"were just really lucky that the storm had situated in the canyon and not over them," Gruntfest says. Each has a history of flash flooding in both historical time and the geologic past.

Gruntfest's work led to the signs now posted in Front Range canyons warning drivers to "climb to safety" in the event of a flash flood. "If people are driving along, and the environmental cues suggest this storm might be unusually severe—the river is louder than normal, it's raining very hard, there are lots of fallen trees on the road—they might say, 'I know what to do, I've seen the sign before,'" Gruntfest says.

To encourage physical scientists to examine how weather impacts society, Gruntfest founded the Weather and Society Integrated Studies workshops, or WAS*IS. The purpose of the workshops is to train scientists, engineers, and others in how to use social science research methods to identify what kind of weather information is most useful to the public.

▲ **FIGURE 9.45 DR. EVE GRUNTFEST** Professor Gruntfest is an internationally recognized expert in natural hazard warning systems and flash floods. She has worked on natural hazard mitigation for nearly 30 years and teaches in the Department of Geography and Environmental Studies at the University of Colorado. *(Dr. Eve Gruntfest)*

For example, the U.S. National Weather Service often agonizes over the decision to declare a tornado warning for fear of frustrating people with false alarms. Gruntfest has found, however, that most people would rather endure extra warnings and be safe than risk being caught in a tornado. Says Gruntfest, "You have to listen to find out what will prompt them to respond appropriately to warnings."

—Kathleen Wong

- Planning should be comprehensive and include all water uses and their economic, social, and environmental importance.
- Views of the people affected by the programs should be sought.
- Non-structural flood-control alternatives should be considered.
- Planning should take place on a watershed scale.

Flooding was recognized as a problem that required a new approach, beyond the traditional structural approach and providing disaster assistance. This approach was born of dissatisfaction with many aspects of the structural approach, the issue of "income transfer" from the general public to the minority of floodplain residents, changing social values, increased urbanization, economic conditions, and the seemingly endless escalation in flood damage costs, even with structure controls. The new approach was embodied in the federal Flood Damage Reduction Program established in 1975.

The Flood Damage Reduction Program operates under a series of federal-provincial and federal-territorial cost-sharing agreements. The two levels of government agree to the following policies:

- They will not build, approve, or finance flood-prone development in designated flood-risk areas.
- They will not provide flood disaster assistance for any development built after an area is designated to have a flood risk, except for flood-proofed structures.

- They will encourage local authorities to zone on the basis of flood risk.

Agreements between the two levels of government differ across provinces and territories, but normally a general agreement outlining policies of the Flood Damage Reduction Program is supplemented by a subsidiary agreement on mapping. Additional agreements may be made on flood forecasting, structure controls, and research. The general agreement sets out the basic approach, which is to consider all applicable structural and non-structural measures to reduce flood risk. Effectiveness, costs, benefits, and environmental impacts are considered. The mapping agreements provide for a program to delineate and designate areas at risk from flooding. The federal criterion for defining the flood-risk area is the 100-year flood. (As discussed above, a 100-year flood is one that has a 1 percent chance of occurring in any given year.) This criterion, however, is not applied across the country. For example, the *Canada–British Columbia Flood Damage Reduction Agreement* uses the 200-year, rather than the 100-year, flood.

Areas in Canada at risk from flooding are commonly subdivided into two zones: the *floodway* and the *flood fringe*. The floodway is the portion of a river's floodplain with the deepest, fastest, and most destructive waters. Flood-vulnerable infrastructure is discouraged in the floodway because of the danger to life and of potential property damage. New development may be permitted within the flood fringe, where water is shallower and slower, provided that structures are adequately flood-proofed.

Local governments play an important role in floodplain management in Canada because they are generally responsible for land-use planning and regulation of new development. The federal–provincial and federal–territorial agreements require that local authorities be encouraged to zone designated areas on floodplains according to the flood risk. In some provinces and territories, local governments are required to incorporate flood hazard information into municipal planning through official plans, zoning bylaws, subdivision plans, and flood regulations.

Provinces and territories have primary jurisdiction for responding to disasters, but the federal government will provide assistance if the cost of a disaster is larger than a province or territory can reasonably be expected to bear on its own. The federal financial contribution is determined by a formula based on provincial/territorial population and federal guidelines for defining eligible costs. Public Safety and Emergency Preparedness Canada (PSEPC) administers the *Disaster Financial Arrangements* on behalf of the Government of Canada. Not all damages are eligible for cost-sharing. For example, the program does not cover damage to large businesses, industries, crops, or summer cottages.

As mentioned above, an important element of the United States' flood-management program is private and public insurance. In 1968 the U.S. Congress established the *U.S. National Flood Insurance Program* to make flood insurance available at subsidized rates. This program is administered by the Federal Emergency Management Agency (FEMA) and requires mapping of special Flood Hazard Areas, defined as areas that would be inundated by a 100-year flood. Flood Hazard Areas are designated along streams, rivers, lakes, alluvial fans, deltas, and low-lying coastal areas.

New property owners in areas that might be flooded must buy insurance at rates determined by the risk they face. In this case, risk is evaluated by using the flood-hazard maps. The insurance program is intended to provide short-term financial aid to victims of floods and to establish long-term land-use regulations that discourage development of floodplains. Part of this program involves revising building codes to limit new construction in a Flood Hazard Area to flood-proofed buildings (Figure 9.46) and to prohibit all new construction in areas that would be inundated by a 20-year flood. Before joining the National Flood Insurance Program, a community must have FEMA prepare maps of the 100-year floodplain, and the community must adopt minimum standards of land-use regulation within identified Flood Hazard Areas. Nearly all U.S. communities with a significant flood risk have basic flood-hazard maps and have initiated some form of floodplain regulation. Several million property

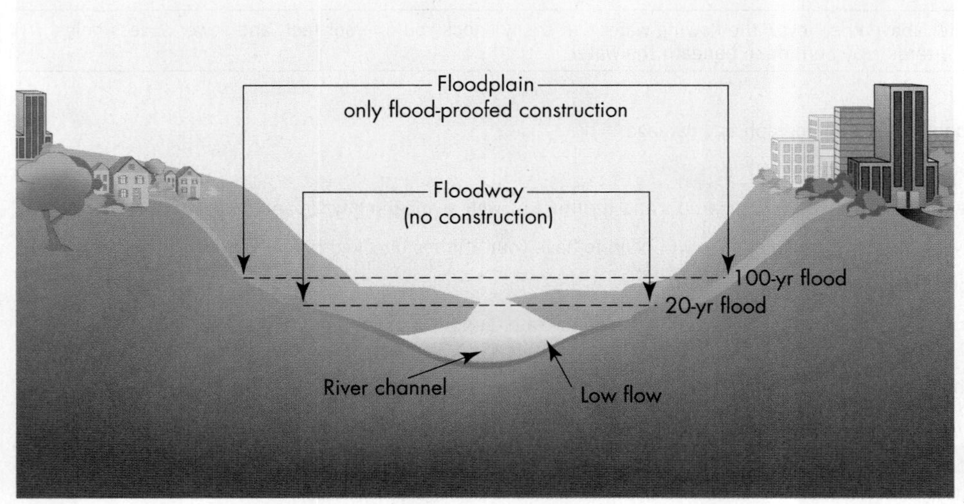

◄ **FIGURE 9.46 FLOODPLAIN REGULATION** Schematic diagram showing areas inundated by 100- and 20-year floods, which are the floods used for regulation under the U.S. National Flood Insurance Program.

owners in the United States currently have flood insurance policies.[41]

By the early 1990s, policymakers and flood-control professionals recognized that the flood insurance program was in need of reform. In response, in 1994, the U.S. Congress passed the *National Flood Insurance Reform Act*, with provisions to further mitigate flood risk through flood-proofing, relocation, and buyouts of properties likely to be frequently flooded.[41]

Relocating People from Floodplains

To reduce future losses from flooding in the United States, local, state, and federal governments have been selectively purchasing and removing homes damaged by floodwaters. The following are two examples of voluntary relocation. In September 1999, nearly 50 cm of rain from Hurricane Floyd flooded many areas in North Carolina. Floodwaters damaged approximately 700 homes in Rocky Mount, a city with a population of 60 000. The state and federal governments later decided to spend nearly U.S.$50 million to remove 430 of these homes, the largest single home buyout ever approved. Following purchase, the homes were demolished and the land was preserved as open space.

A wet cycle that began in 1993 caused Devils Lake in North Dakota to rise 8 m. With no natural outlet and flat land surrounding its shore, the lake more than doubled in size and inundated the land around the town of Churchs Ferry. By late June 2000, the town was all but deserted, the population having dropped to approximately 7 from 100 people. Most residents took advantage of a

TABLE 9.2 What to Do and What Not to Do before and after a Flood

What to Do	**Preparing for a Flood**
	• Enquire about the risk of flooding where you live.
	• If your property is at risk, purchase flood insurance if you can and be sure that you know how to file a claim.
	• Buy sandbags or flood boards to block doors.
	• Assemble a flood kit that includes a flashlight, blankets, raingear, battery-powered radio, first-aid kit, rubber gloves, key personal documents, and medication. Keep the kit upstairs if possible.
	• Find out where to turn off your gas and electricity. If you are not sure, ask your utility company.
	• Talk to your family or housemates about the possibility of flooding. Consider writing a flood plan and storing it with your flood kit.
What Not to Do	• Never underestimate the damage a flood can do.
What to Do	**When a Flood Warning Has Been Issued**
	• Be prepared to evacuate.
	• Observe water levels and monitor radio and television news and weather reports.
	• Move people, pets, and valuables upstairs or to higher ground.
	• Move your car to higher ground. It takes only 0.6 m of fast-flowing water to wash it away.
	• Check on your neighbours. Do they need help? They might not be able to escape upstairs or might need help moving furniture.
	• Do as much as you can in daylight. If the electricity fails, it will be hard to do anything at night.
	• Keep warm and dry. A flood can last longer than you think, and you can get cold. If you are forced out of your home, take warm clothes, blankets, a Thermos, and food supplies.
What Not to Do	• Do not walk in floodwater above knee level; the flowing water can easily knock you off your feet, and sewer-access holes, road works, and other hazards may be hidden beneath the water.
What to Do	**After a Flood**
	• Check your home for damage and photograph any damage.
	• If insured, file a claim for the damage.
	• Obtain professional help in removing or drying carpets and furniture as well as cleaning walls and floors.
	• Contact gas, electricity, and water companies. You will need to have your utilities checked before you turn them back on.
	• Open doors and windows to ventilate your home.
	• Wash water taps and run them for a few minutes before use. Your water supply could be contaminated; check if you are concerned.
	• Disinfect and thoroughly clean everything that became wet.
What Not to Do	• Do not touch items that have been in contact with the water. Floodwater could be contaminated.

Source: Modified from the Environment Agency, United Kingdom. 2004. "Floodline: Prepare for flooding." http://www.environment-agency.gov.uk/homeandleisure/floods/default.aspx. Accessed August 9, 2013.

voluntary federal buyout plan and moved to higher ground, many to the town of Leeds about 24 km away. They were offered as compensation the appraised value of their homes plus an incentive; most considered the offer too good to turn down. They also recognized that the town eventually would have been destroyed by flooding anyway. Nevertheless, there was some bitterness among the town's population and not everyone participated. The mayor and the fire chief were among the seven people who decided to stay. The buyout program demonstrated how emotional such a process can be. It is difficult for people to leave their homes, even though they know floodwaters will eventually destroy them.

Personal Adjustments: What to Do and What Not to Do

We cannot prevent floods, but we can prepare for them by learning what to do and what not to do before, during, and after floods (Table 9.2).

REVISITING THE FUNDAMENTAL CONCEPTS

River Flooding

❶ **Hazards can be understood through scientific investigation and analysis.**

❷ **An understanding of hazardous processes is vital to evaluating risk.**

❸ **Hazards are commonly linked to each other and to the environment in which they occur.**

❹ **Population growth and socioeconomic changes increase the risk from natural hazards.**

❺ **Damage and loss of life from natural disasters can be reduced.**

1. Flooding is an extensively studied and well understood natural hazard. Floods of different magnitudes and their return periods can be predicted based on past flow records. In the case of large rivers, such as the Mississippi, Red, and Fraser rivers, the progress of flood waves from headwaters to river mouth can be predicted and orderly evacuations conducted in advance of the arrival of the flood wave. Rainfall-triggered floods on small streams are more difficult to predict, although forecasts of heavy rainfall provide some warning. Weather conditions that lead to ice-jam floods are also known, as are locations where such flooding is a problem.
2. Risk assessments for flooding are based on the probability of a flood multiplied by the consequences. These factors are relatively easy to quantify, allowing flood risk to be reliably estimated. These estimates are important because floods are the most widely experienced natural hazard. When considering the hazard, however, it is important to take into account the possible effects of climate change on the future frequency or magnitude of floods.
3. Links exist between flooding and other natural hazards. Flooding is triggered by hurricanes and extratropical storms. Landslides may block streams, forming up-valley lakes that can overtop the dams, causing downstream flooding. Floodwaters may also erode river banks and valley walls, triggering landslides. Finally, floodwaters might carry dead animals, chemicals, and raw sewage that can contaminate drinking water and spread disease.
4. More people than ever are living on floodplains, and development on floodplains is increasing. Although dyking and other measures are reducing the hazard in some areas, exposure, which is an element of the risk equation, is increasing. Furthermore, urbanization and other land-use changes, such as deforestation, are increasing the frequency and peak discharges of floods.
5. Careful land-use planning and engineered structures can reduce flood risk. A strong case can be made for disallowing certain forms of development on floodplains. Unfortunately, in many parts of the world, significant development on floodplains has resulted from past unwise land-use decisions. As a result, large expenditures will be required to build and improve levees and floodwalls along rivers to protect existing development. Of particular importance in reducing damage and loss of life from floods is disaster preparedness. We know where flooding will occur, and we can develop plans so that the impact of flooding is minimized and recovery is expedited.

Summary

The region drained by a river and its tributaries is called a drainage basin or watershed. Streams carry chemicals in their dissolved load and sediment in their suspended and bed loads. Discharge refers to the volume of water moving past a particular location per unit time.

Sediment accumulates on the floodplain when a stream migrates laterally or when it overtops its banks during floods. The configuration of the stream channel is called the channel pattern and can be braided, anastomosing, or meandering.

Flooding happens when a stream overtops its banks. Upstream floods and flash floods in small drainage basins are commonly produced by intense rain falling over a small area. Downstream floods are produced by rapid snowmelt and by storms of long duration or high intensity that affect a large area.

Flood magnitude and frequency are difficult to predict for many streams because of changing land use and short historical records. This difficulty is especially acute for extreme events, such as the 100-year flood. The probability that a 100-year or greater flood will take place each year is the same, regardless of when the last 100-year flood occurred.

River flooding is the most universally experienced natural hazard. Floods can occur anywhere there are streams, and all regions of Canada and the United States face some threat of flooding. Flooding causes a great deal of death and damage, but it also maintains fertile lands, provides benefits to aquatic ecosystems, and deposits sediment on delta plains, countering the subsidence that may occur there.

Urbanization has increased flooding in small drainage basins by covering much of the ground with buildings, parking lots, and roads, thereby increasing runoff of storm water.

Loss of life from flooding is relatively low in developed countries, which have monitoring and warning systems and adequate resources for recovery. On the other hand, property damage is much greater in developed countries than in developing ones because floodplains are often extensively developed and have more infrastructure at risk.

The best strategy for minimizing flood damage is floodplain regulation. Engineered structures are required, however, to protect existing development on floodplains. These structures include physical barriers, such as levees and flood walls, and dams that regulate the release of water from reservoirs.

Channelization is the straightening, deepening, widening, cleaning, or lining of existing streams, with the goal of controlling floods or improving drainage. Many channelization schemes have caused serious environmental damage, and thus new projects must be carefully evaluated. New approaches to channel modification mimic natural processes, and some channelized streams are being restored to a more natural state.

Flood hazards and risk are generally understood at the institutional level but not necessarily at the individual level. More educational programs are needed to help people understand the risk of living in flood-prone areas.

Adjustments to flood hazards include flood insurance, floodproofing, and floodplain regulation. Dykes and other protective structures tend to encourage further development of floodplains by providing a false sense of security, thereby increasing risk. The first step in floodplain regulation is flood-hazard mapping. Planners use flood-hazard maps to zone flood-prone areas for appropriate uses. In some cases, homes in flood-prone areas have been purchased and demolished by governments and people have relocated to safe ground.

Key Terms

channel pattern (p. 238)
channel restoration (p. 263)
channelization (p. 261)
discharge (p. 238)
drainage basin (p. 237)
flash flood (p. 246)
flooding (p. 241)
floodplain (p. 236)
floodplain regulation (p. 265)
levee (p. 242)
recurrence interval (p. 245)
river (p. 237)
stream (p. 237)

Did You Learn?

1. Describe what a drainage basin is.
2. Name the three components that constitute the total load of a stream.
3. What lessons were learned from the 1993 flood on the Mississippi River?
4. How do braided, anastomosing, and meandering rivers differ?
5. Explain the difference between pools and riffles.
6. How do upstream and downstream floods differ?
7. List the major factors that determine the amount of damage a flood causes.
8. Explain how urbanization affects flood hazard.
9. What is meant by floodplain regulation?
10. Explain how levees and floodwalls can worsen flooding.
11. Explain channel restoration.
12. Describe the techniques used to flood-proof structures.
13. What do we mean when we say that a 100-year flood has occurred?
14. How do Canadian and American flood-damage-reduction programs differ?

Critical Thinking Questions

1. You are a planner working for a community that is planning to put new subdivisions in the headwaters of a watershed. You are aware of the effects of urbanization on flooding and want to make recommendations to avoid some of these effects. Outline a plan of action.
2. You work for a municipality that has been channelizing streams for many years. Bulldozers are used to straighten and widen channels, and the agency has been criticized for causing extensive environmental damage. You have been asked to develop new plans for channel restoration as part of a stream-maintenance program. Devise a plan that will convince the official in charge of the channelization program that your ideas will improve the urban stream environment while reducing flood risk.
3. What is your personal opinion about development on floodplains given competing uses for valuable land? What type of development, if any, should be allowed? What measures should be taken to protect structures on or near the floodplain? Formulate an argument to support your views.

MasteringGeology

MasteringGeology **www.masteringgeology.com**. Looking for additional review and test prep materials? Visit the Study Area in MasteringGeology to enhance your understanding of this chapter's content by accessing a variety of resources, including **Self-Study Quizzes, Geoscience Animations, GEODe Tutorials, RSS feeds, flashcards,** web links and an optional **Pearson eText.**

CHAPTER 10

▶ **ICE STORM CRIPPLES CENTRAL CANADA** Montrealers negotiate their way through a maze of broken branches and downed power lines in January 1998 after an ice storm plunged much of the city into darkness and cold. Parts of Quebec and Ontario had no electricity for weeks following the crippling storm. *(©Christopher Morris/Corbis)*

Atmosphere and Severe Weather

Learning Objectives

Atmospheric processes and energy exchanges are driven mainly by energy reaching Earth from the sun and are linked to climate and weather. Thunderstorms, tornadoes, blizzards, ice storms, dust storms, heat waves, drought, and floods resulting from intense precipitation are natural processes that are hazardous to people and are responsible for significant destruction and many deaths each year throughout the world. Your goals in reading this chapter should be to

- Understand Earth's energy balance and energy exchanges that produce climate and weather
- Know the different types of severe weather events and how they differ in terms of energy and size
- Know the main effects of severe weather events and their relations to other natural hazards
- Understand how people interact with severe weather
- Understand how to minimize the adverse effects of severe weather
- Know the natural service functions of severe weather

The 1998 Ice Storm

The worst North American ice storm of the twentieth century began on January 5, 1998. By January 10, parts of Quebec, Ontario, New Brunswick, and the northeastern United States were gripped by up to 10 cm of accumulated ice. Fifty-seven communities in eastern Ontario and 200 in Quebec were declared disaster areas. About 1000 steel electrical pylons (Figure 10.1), 35 000 utility poles, and millions of trees collapsed under the weight of the ice. More than 3 million people in Quebec and 1.5 million in eastern Ontario were without power.[1,2] At least 45 people were killed as a result of house fires, falling ice, carbon monoxide poisoning, and hypothermia, and almost 1000 were injured.[1] About 100 000 people were evacuated to shelters. The estimated cost of the ice storm was U.S.$6.2 billion, with more than 80 percent of the losses in Canada.[3,4]

On January 8, the military was brought in to help clear debris, provide medical assistance, evacuate residents, and canvass door to door to make sure people were safe. They also helped to restore power. Power was restored in most urban areas in a matter of days, but in Quebec alone, 150 000 persons were without electricity three weeks after the beginning of the storm.

The economies of Ontario and Quebec were hit hard. Nearly one-quarter of Canada's dairy cows, one-third of the cropland in Quebec, and one-quarter of the farmland in Ontario were in the affected area. Milk-processing plants were closed, and about 10 million L of milk had to be dumped. Most of the maple trees tended by Quebec maple syrup producers were killed. It was estimated that syrup production in the province would take several decades to return to normal levels.

The 1998 ice storm was the most expensive natural disaster in Canadian history. According to Environment Canada, the storm directly affected more people than any previous weather event in Canada.

Meteorological Conditions

What was the cause of this disaster? The meteorological set-up for the ice storm involved a persistent flow of warm, moist air from the Gulf of Mexico that rose up over a thin wedge of cold Arctic air (Figure 10.2). Cold air was driven south by northeasterly winds circulating around an

◄ **FIGURE 10.1 ICE STORM** Supercooled rain can turn to clear ice when it falls on a surface with a temperature at or below freezing. The ice storm in January 1998 downed electrical transmission lines that provide electricity to southern Quebec and Ontario. *(Jacques Boissinot/ CP Images)*

Arctic high-pressure system centred over Hudson Bay. The cold air settled over southern river and mountain valleys, including the St. Lawrence and Ottawa valleys. Warm air was driven northward by a high-pressure system centred over Bermuda and a low-pressure trough located over the Mississippi River valley.[5]

Weather conditions in the region remained essentially unchanged for almost a week. The flow of moist air from the Gulf of Mexico acted like a conveyor belt, delivering water vapour for precipitation both north and south of the stationary front. Heavy rains south of the front caused severe flooding in North Carolina and Tennessee, killing nine people.[6] North of the front, most precipitation fell in the form of freezing rain, with total water-equivalent accumulations ranging from 20 to 100 mm.[2]

Lessons Learned

Why was this storm so catastrophic? The answer lies in both the physical and social sciences. For many Canadian communities, including Ottawa and Montreal, the ice storm of 1998 was the longest period of freezing rain on record, lasting more than 80 hours.[2,5] On the human side, the ice storm demonstrated how dependent our society has become on electricity, especially the electrical grid that links large areas. The storm destroyed more than 120 000 km of power lines. In many communities, emergency backup electrical generators were absent, failed, or ran out of fuel. Also apparent was a lack of resiliency in the system. For example, it became difficult to repair the electrical grid when essential supplies, such as utility poles, had to be shipped long distances on roads that were barely passable.

10.1 Energy

Atmospheric processes involve huge amounts of energy, amounts so large that they can be expressed in terawatts (tW). One tW equals 1 million megawatts (mW) or 1 trillion watts (W). Typical household incandescent light bulbs have power ratings of 40 to 100 watts. The total solar energy absorbed at Earth's surface is approximately 120 000 tW per year, and it is mainly this energy that heats our planet,

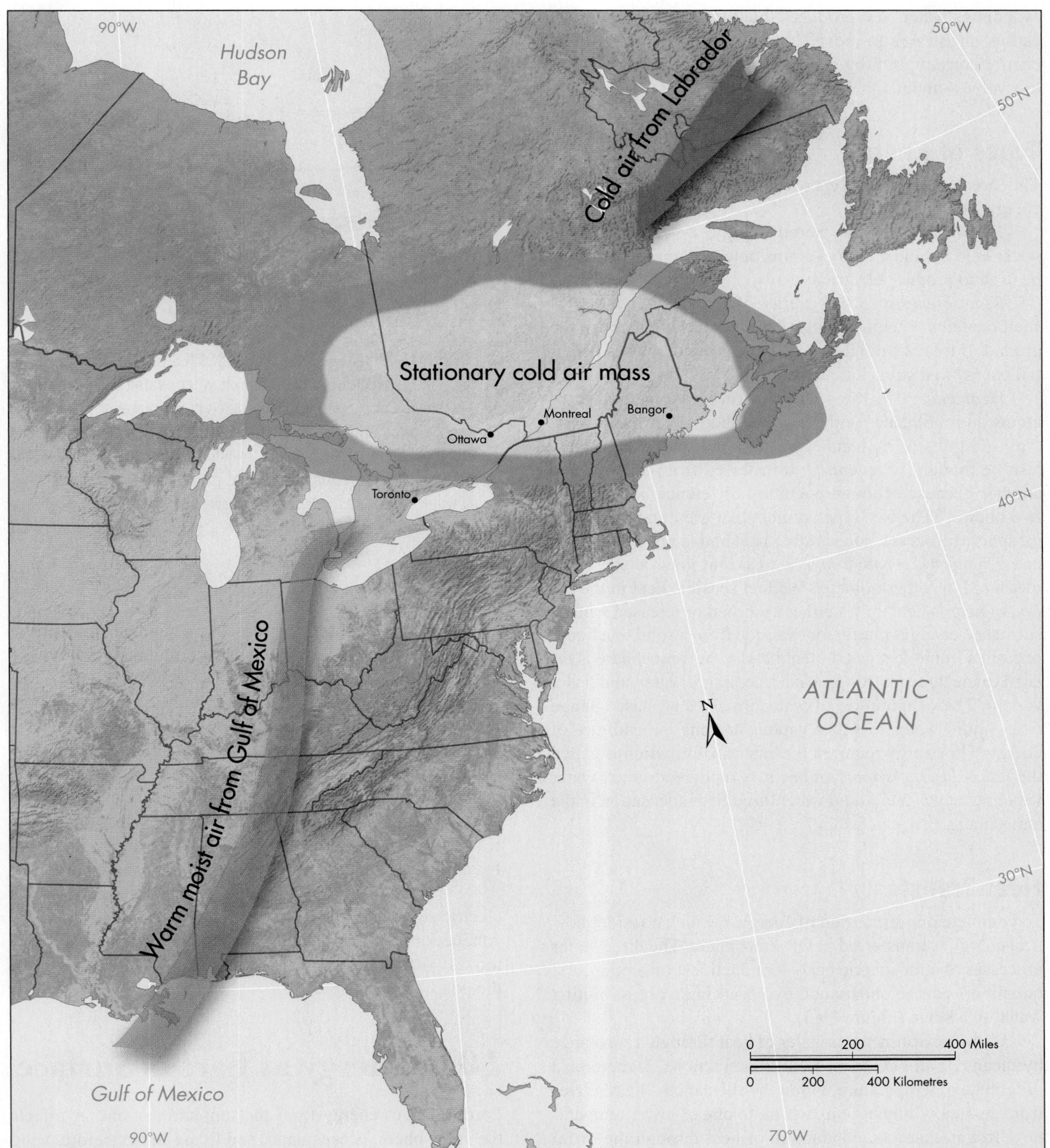

▲ **FIGURE 10.2 WEATHER CONDITIONS THAT CAUSED THE 1998 ICE STORM** A continuous flow of warm, moist air from the Gulf of Mexico overrode a stationary cold air mass in southern Quebec, eastern Ontario, New Brunswick, Nova Scotia, Prince Edward Island, and the northeastern United States during early January 1998. The system remained in place for nearly a week, causing the worst natural disaster in Canadian history. *(After Environment Canada, http://www.weatherof?ce.gc.ca/canada_e.html)*

evaporates water, and produces the differential heating that causes air masses to move. In comparison, current global energy consumption by people is less than 0.01 percent of this value—about 13 tW per year.[7]

Types of Energy

The three main types of energy are potential energy, kinetic energy, and heat energy.

Potential energy is stored energy. For example, the water held behind a dam contains potential energy that may be used to produce electricity.

Kinetic energy is the energy of motion. A book on a shelf contains potential energy based on its height above the ground. If it falls from the shelf to the ground, it loses potential energy and gains kinetic energy.

Heat energy is the energy of the random motion of atoms and molecules and can be defined as the kinetic energy of atoms or molecules within a substance. Heat may also be thought of as energy transferred from one body to another because of the temperature difference between the two bodies.[8] The two types of heat that are important in atmospheric processes are sensible heat and latent heat. As the name suggests, *sensible heat* is heat that may be sensed or measured by a thermometer. We feel sensible heat in the air. *Latent heat* is heat that is either absorbed or released when a substance changes phase, for example from a solid to a liquid or from a liquid to a gas. Latent heat in the atmosphere is related to the three phases of water: ice, liquid water, and water vapour. The evaporation of water involves a phase change from liquid water to water vapour and the expenditure of energy. The energy required for this transformation is called the *latent heat of vaporization*. It is recovered when water vapour changes into liquid water through condensation in the atmosphere, producing rain.[8]

Heat Transfer

To complete our discussion of energy, we now consider how heat energy is transferred in the atmosphere. The three major processes of atmospheric heat—conduction, convection, and radiation—can be understood by observing a pot of boiling water on a stove (Figure 10.3).

Conduction is the transfer of heat through a substance by means of atomic or molecular interactions. It requires a difference in temperature within the substance—heat moves from an area of higher temperature to one of lower temperature. In our example, conduction of heat through the metal pot causes the handle to heat up. Conduction also occurs in the atmosphere, on land, and in bodies of water, such as the ocean. For example, warm surface ocean water may lose heat by conduction to the cooler air above.

Convection is the transfer of heat by the movement of a fluid, such as water or air. In our example, water at the bottom of the pot warms and rises upward to displace the cooler water at the surface. The cooler water sinks to the bottom of the pot. This process physically mixes the water by creating a circulation loop known as a *convection cell*.

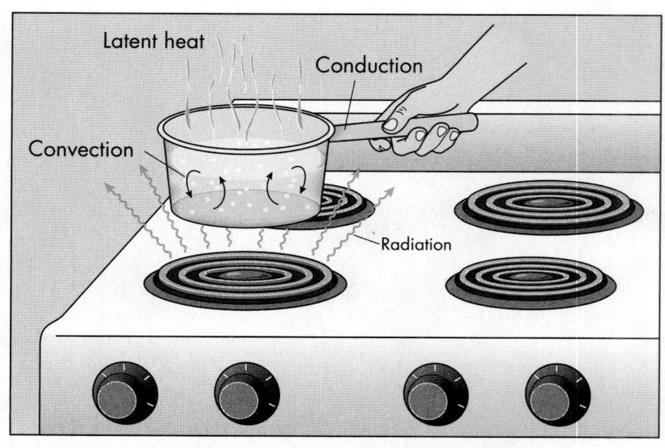

▲ **FIGURE 10.3 HEAT TRANSFER PROCESSES** Heat is transferred by conduction, convection, and radiation when water is boiled on an electric stove. Conduction is taking place in the metal of the pan and its handle; circulating water is transferring heat by convection; and electromagnetic radiation is being transferred through the air from the glowing coil of the stovetop. Latent heat is absorbed by water vapour as liquid water evaporates. *(Based on Christopherson, R. W. 2006.* Geosystems: An Introduction to Physical Geography, *6th ed. Upper Saddle River, NJ: Pearson Prentice Hall)*

Convection is an important process in transferring heat in thunderstorms and in the large-scale circulation of air away from the equator.

Radiation is wave-like energy emitted by a substance that possesses heat. The transfer of energy by radiation occurs by oscillations in electric and magnetic fields; thus the waves are called *electromagnetic waves*. In our example, heat energy radiates from the heating element in the electric stove to the pot on the stove. Some of these electromagnetic waves are visible to the eye—for example, light emanating from the glowing coil on the stove—but most are not.

Summarizing our example, heat is transferred from the electric stove by radiation from the glowing heating coil, by conduction through the metal pot, and by convection, which moves warm water upward from the bottom to the top of the pot.

10.2 Energy at Earth's Surface

Earth receives energy from the sun, and this energy affects the atmosphere, oceans, land, and living things before being radiated back into space. *Earth's energy balance* is the equilibrium between incoming and outgoing energy (Figure 10.4). Energy changes form repeatedly and in a complex manner from the time it reaches Earth to the time it leaves, but as stated in the First Law of Thermodynamics, it is neither created nor destroyed.

Earth intercepts only a tiny fraction of the total energy emitted by the sun, but the intercepted energy is adequate to sustain life. Solar energy also drives the hydrologic cycle (Figure 1.9), ocean waves and currents, and global atmospheric

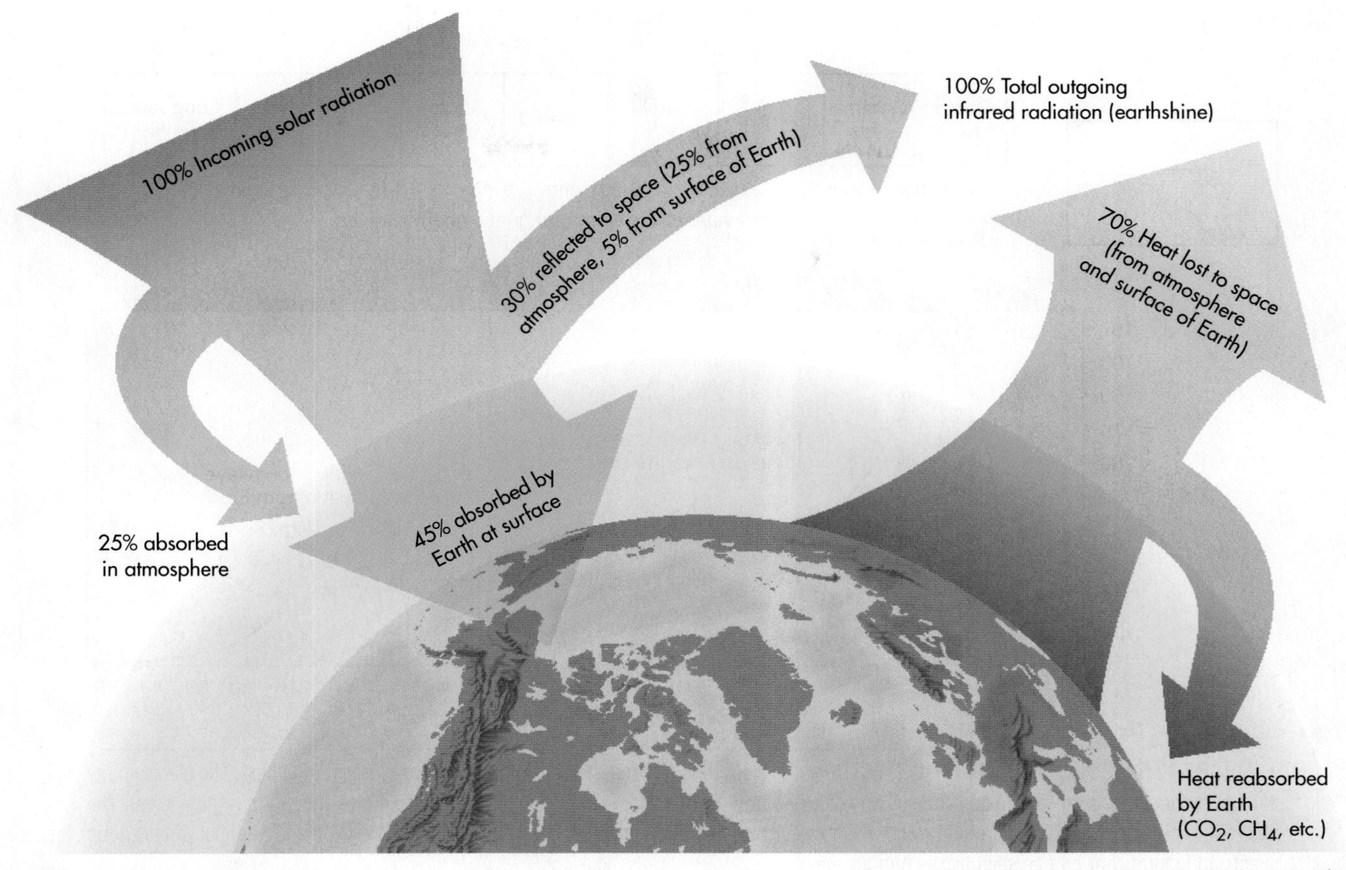

▲ **FIGURE 10.4 EARTH'S ENERGY BALANCE** Most of the energy flow to Earth from the sun is either reflected or radiated back into space. Only a small proportion (0.01 percent) of the heat at Earth's surface comes from the interior of the planet. *(Modified after Pruitt, N. L., L. S. Underwood, and W. Surver. 1999.* Bioinquiry Learning System 1.0: Making Connections in Biology. *Toronto: John Wiley & Sons)*

circulation. Earth's energy balance has several important components, but nearly all the energy available at Earth's surface comes from the sun.

Electromagnetic Energy

Much of the energy emitted by the sun is *electromagnetic energy*. This energy—a type of radiation—travels through space at the speed of light, nearly 300 000 km/s. Electromagnetic radiation is commonly described as having a wave form, and the distance between the crests or troughs of two successive waves is referred to as the *wavelength*. The *electromagnetic spectrum*, which encompasses all possible wavelengths, is large, and the different types of electromagnetic radiation are defined on the basis of their wavelengths (Figure 10.5). Electromagnetic energy with long wavelengths—greater than 1 m—includes radio waves and microwaves, whereas the types with the shortest wavelengths—less than about 0.0001 mm—are X-rays, gamma rays, and cosmic rays. Visible electromagnetic radiation, which we know as light, makes up only a very small part of the electromagnetic spectrum, with wavelengths ranging from about 0.0004 to 0.0007 mm. Other types of radiation that have environmental significance include infrared (IR) and ultraviolet (UV) radiation. Infrared radiation is involved in global warming, and levels of UV radiation at Earth's surface are influenced by the amount of ozone in the upper atmosphere (Chapter 14).

Energy Behaviour

Once electromagnetic energy from the sun reaches Earth, it is redirected, transmitted, or absorbed by the atmosphere, oceans, and land (Figure 10.4). In the case of *redirection*, the energy is either reflected like a light bouncing off of a mirror or scattered in different directions. Reflection from the tops of clouds, the water, and the land is one of two ways that solar energy returns to outer space (see Chapter 14); the other way is emission (see below). Scattering disperses energy in many directions, and scattered energy is generally weaker than reflected or transmitted energy. *Transmission* involves the passage of energy through the atmosphere, like light passing through window glass. On a clear day, the atmosphere transmits most of the energy it receives from the sun. *Absorption* of electromagnetic energy either alters the structure of molecules it comes into contact with or causes them to vibrate and emit energy. A change in molecular structure can take place with penetrating short-wave radiation, such as UV radiation. The energy that is emitted by this process is in

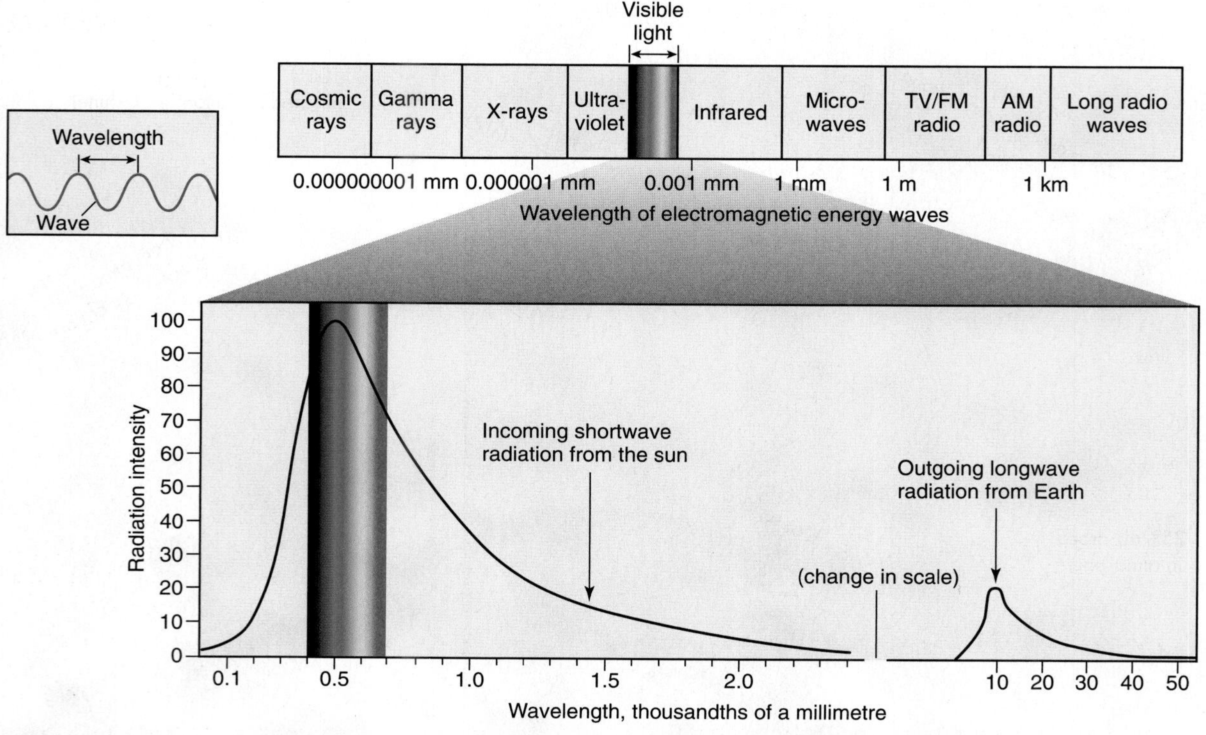

▲ **FIGURE 10.5 THE ELECTROMAGNETIC SPECTRUM** Wavelength is the distance between one wave crest and the next. The electromagnetic spectrum spans an enormous range of wavelengths from trillionths of a metre (cosmic rays) to kilometres (long radio waves). The incoming energy from the sun has shorter wavelengths than radiation emitted from Earth. Only a small part of the electromagnetic spectrum is light that can be seen by the human eye.

the form of heat or electromagnetic energy. For many gases, the emitted electromagnetic energy has a longer wavelength than the incoming short-wavelength radiation.

The relative amount of energy reflected and absorbed is important in determining the temperature of the air and land. For example, the colour of the land surface determines its reflectivity to visible light. This reflectivity is referred to as the *albedo* of the surface. In general, dark-coloured surfaces reflect less energy than light-coloured ones and have lower albedos. Dark coniferous forest, for example, reflects 5 to 15 percent of incoming solar radiation, whereas lighter-coloured grassland reflects up to 25 percent of this radiation.[9] Much of the energy that is not reflected is absorbed, causing darker surfaces to heat up more than lighter ones. Many atmospheric gases are selective absorbers—that is, they absorb some wavelengths of energy but are transparent to others. For example, carbon dioxide and water vapour selectively absorb IR waves, which contributes to the warming of the lowest layer of the atmosphere (see Chapter 14).

Absorbed energy is emitted as heat and electromagnetic radiation. The temperature of the object emitting the radiation affects the wavelength of the emitted radiation. Hotter objects radiate energy at shorter wavelengths, whereas cool objects radiate longer-wavelength radiation. This fact explains why the sun emits mainly short-wavelength radiation—gamma rays, X-rays, UV rays, and visible light. In contrast, Earth's land surface, oceans, and clouds are so cool that they emit mainly longer-wavelength IR radiation.

10.3 The Atmosphere

Having completed our brief discussion of energy and Earth's energy balance, we turn now to Earth's atmosphere and atmospheric circulation.

The **atmosphere** is the thin envelope of gases that surrounds Earth. It is made up of gas molecules, suspended solid and liquid particles, and falling rain and snow. The atmosphere is responsible for the weather we experience every day, and it keeps Earth warm enough to be habitable. Knowledge of the structure and dynamics of the atmosphere is critical to understanding severe weather, as well as the mechanism and causes of global warming, which we will cover in Chapter 14.

Composition of the Atmosphere

Earth's atmosphere is mainly made up of nitrogen and oxygen, but it contains smaller amounts of argon, water vapour, and carbon dioxide, as well as traces of a host of other elements and compounds. With the exception of water vapour, we will discuss these gases in greater detail in Chapter 14.

Water vapour plays an important role in cloud formation and atmospheric circulation. The term *humidity* describes the amount of water vapour, or moisture, in the atmosphere at a specific temperature. Humidity commonly changes with temperature, because warm air can hold more moisture than

◄ **FIGURE 10.6 EARTH'S THIN ATMOSPHERE** Viewed from space, the atmosphere appears as a thin layer surrounding Earth. *(NASA)*

cold air. The amount of moisture in the air is expressed as **relative humidity**—the ratio, or percentage, of the water vapour present in the atmosphere to the maximum amount of water vapour that could be there. Values range from a few percent to 100 percent.

Almost all water vapour in the atmosphere is evaporated from water at Earth's surface. Water is constantly being exchanged between the atmosphere on the one hand, and oceans, lakes, and continents on the other. Sleet, snow, hail, and rain remove water from the atmosphere and deposit it on Earth, where it can enter rivers, lakes, oceans, and Earth's crust, or be added as snow and ice to glaciers. Eventually this water evaporates and returns to the atmosphere to begin the cycle again. This constant cycling of water between the atmosphere and Earth's surface is a major part of the hydrologic cycle (see Chapter 1).

Structure of the Atmosphere

Images from orbiting satellites show that Earth's atmosphere is very thin compared with the diameter of the planet and that its upper limit is indistinct (Figure 10.6). Essentially the entire atmosphere lies below an altitude of 100 km.

Earth's atmosphere consists of four major layers or spheres (Figure 10.7). In this chapter, we focus on the lowest

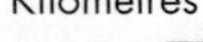

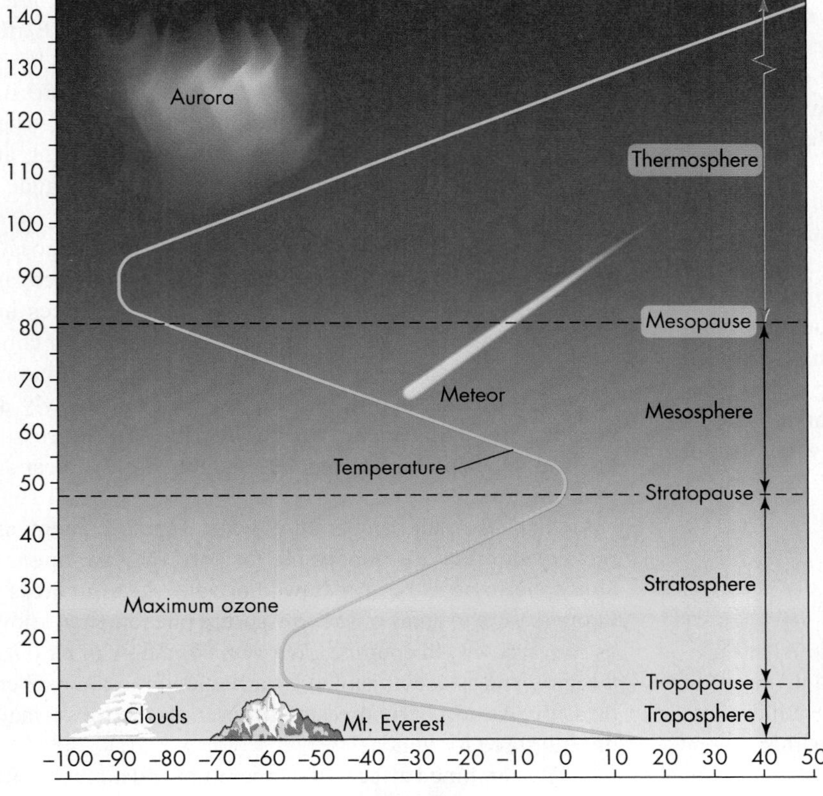

◄ **FIGURE 10.7 STRUCTURE OF EARTH'S ATMOSPHERE** Four atmospheric layers are defined on the basis of vertical changes in air temperature. The yellow line shows the change in air temperature with height above the surface of Earth. Weather develops in the lowest layer, the troposphere. *(Based on Lutgens, F. K., and E. J. Tarbuck. 2004.* The Atmosphere: An Introduction to Meteorology, *9th ed. Upper Saddle River, NJ: Pearson Prentice Hall, with data from the National Weather Service)*

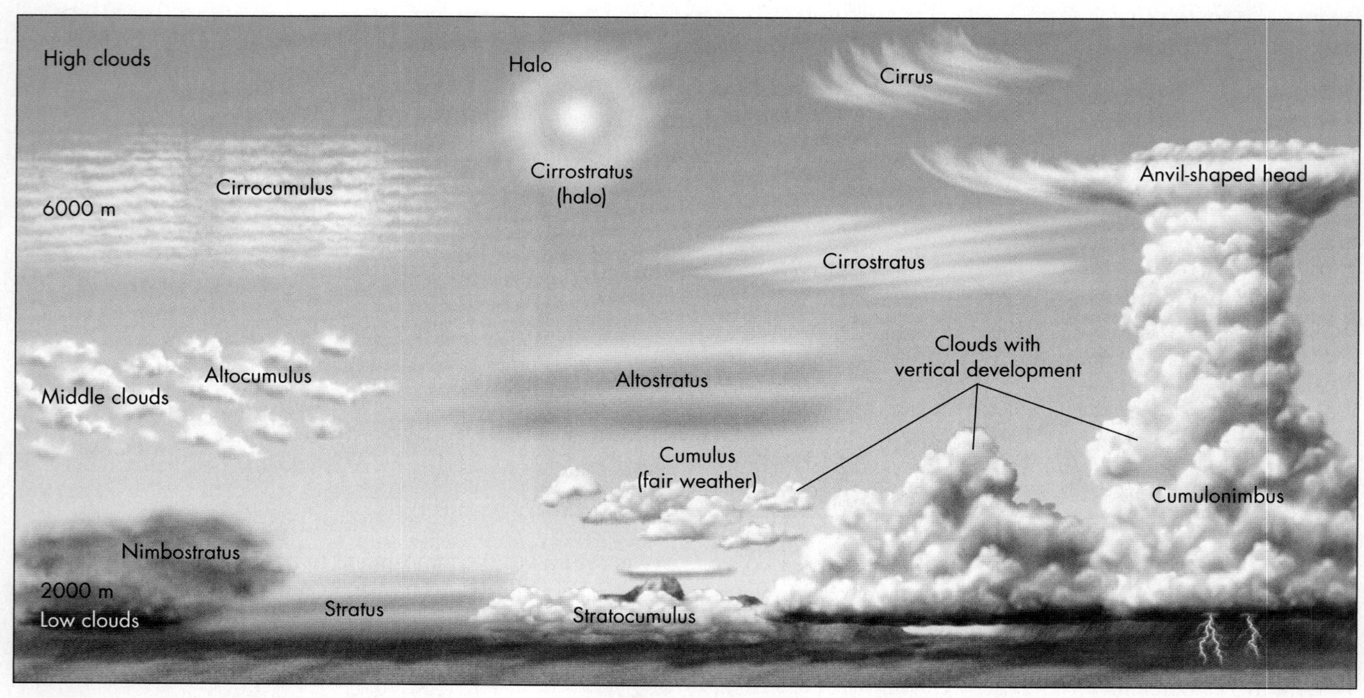

▲ **FIGURE 10.8 COMMON CLOUD TYPES** Clouds consist of very small water droplets and are classified on the basis of altitude (low, middle, and high) and form (cirriform, cumuliform, and stratiform). Cumulonimbus clouds are associated with severe thunderstorms and tornadoes. *(Based on Christopherson, R. W. 2003.* Geosystems, *5th ed. Upper Saddle River, NJ: Prentice Hall)*

layer, the **troposphere**, which extends from the ground to an altitude of between 8 and 16 km. Except during some jet airplane flights, we spend our entire lives in the troposphere. The defining characteristic of the troposphere is a rapid decrease in temperature upward, but the most visible feature is abundant condensed water vapour in the form of clouds. Clouds and the weather that directly affects us are in the troposphere.

The formation and development of clouds are particularly important to our discussion of weather. Clouds comprise small water droplets or ice crystals; without them, there would be no rain, snow, thunder, lightning, or rainbows. You are probably familiar with two of the most common types of clouds: puffy, fair-weather *cumulus* clouds that resemble pieces of floating cotton, and towering *cumulonimbus* clouds, which release tremendous amounts of energy during thunderstorms through condensation of water vapour (Figure 10.8).

10.4 Weather Processes

A complete discussion of atmospheric conditions and processes associated with severe weather is beyond the scope of this book. Here we focus on four aspects of the atmosphere that are important for understanding severe weather: atmospheric pressure, vertical stability of the atmosphere, the Coriolis effect (see A Closer Look), and fronts.

Atmospheric Pressure

Atmospheric pressure, also known as *barometric pressure*, is the weight of a column of air at a point on or above Earth's surface. It can also be thought of as the force exerted by gas molecules on a surface (Figure 10.9a). Atmospheric pressure is greater at sea level than at the top of a mountain, where there is less air (Figure 10.9b). Nearly all the weight, and thus the pressure, of the atmosphere is below an altitude of 50 km (Figure 10.9b).

Atmospheric pressure also differs over distances of tens to hundreds of kilometres across Earth's surface, and these differences strongly affect weather. Air rises and cools in areas of low atmospheric pressure; as the air cools, water vapour condenses to form clouds and precipitation. In contrast, in areas of high pressure, drier air slowly descends, producing clear skies. The air in high-pressure areas may be cold or hot depending on the time of year and other factors.

Changes in the temperature, water vapour content, and movement of air are responsible for horizontal variations in atmospheric pressure. Air flows horizontally from areas of high pressure to areas of low pressure, a phenomenon known as *convergence*. In contrast, *divergence* is a flow of air out of a region and is accompanied by a reduction in atmospheric pressure. Atmospheric pressure differences are thus a major driving force for wind.

The combined effects of temperature and air movement produce the *low-pressure centres* (L) and *high-pressure*

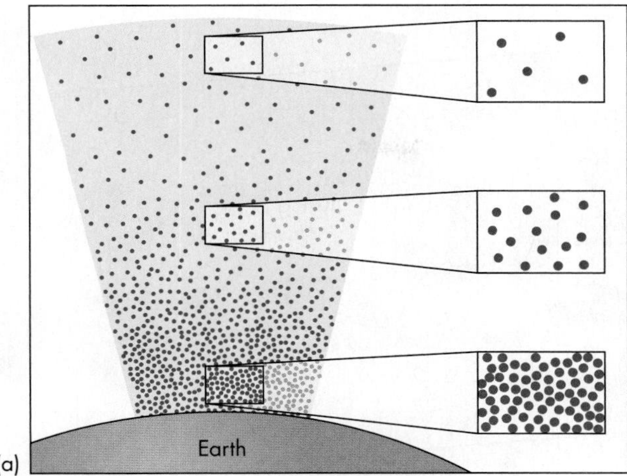

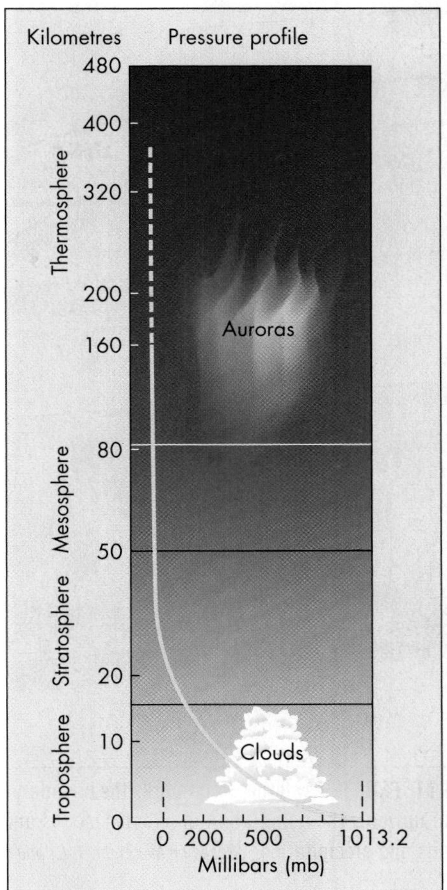

▲ **FIGURE 10.9 AIR PRESSURE** Both the density (a) and the pressure (b) of Earth's atmosphere decrease with altitude. The dots in (a) represent gas molecules in the air. Air pressure in (b) is measured in millibars; one atmosphere of pressure equals 1 bar or 1000 millibars. *(Based on Christopherson, R. W. 2003.* Geosystems, *5th ed. Upper Saddle River, NJ: Prentice Hall)*

centres (H) that you see on weather maps (Figure 10.10). At the surface, air flows from areas of high pressure to areas of low pressure. Less dense air rises in areas of low pressure and diverges in the upper troposphere. A common result is an area of high pressure above a near-surface area of lower pressure. Air can also flow from a high-pressure centre aloft to an area of lower pressure aloft, where it sinks to create a near-surface high-pressure centre. Thus a high-pressure centre at the surface is generally overlain by a low-pressure centre aloft, and vice versa (Figure 10.10).

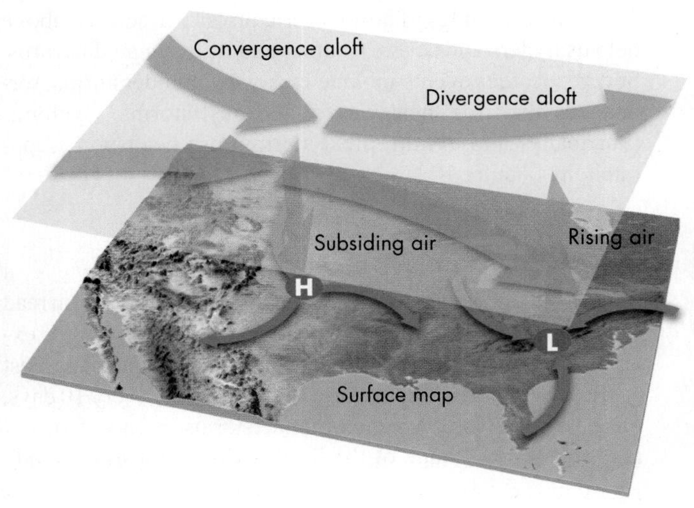

◀ **FIGURE 10.10 LOW- AND HIGH-PRESSURE CENTRES** Variations in atmospheric pressure and temperature cause air to circulate in the troposphere. Airflow at the surface converges (purple arrows pointing inward) in low-pressure centres and diverges (purple arrows pointing outward) in high-pressure centres. *(Lutgens, F. K., and E. J. Tarbuck. 2007.* The Atmosphere: An Introduction to Meteorology, *10th ed. Upper Saddle River, NJ: Pearson Prentice Hall. Reprinted and Electronically reproduced by permission of Pearson Education, Inc., Upper Saddle River, New Jersey.)*

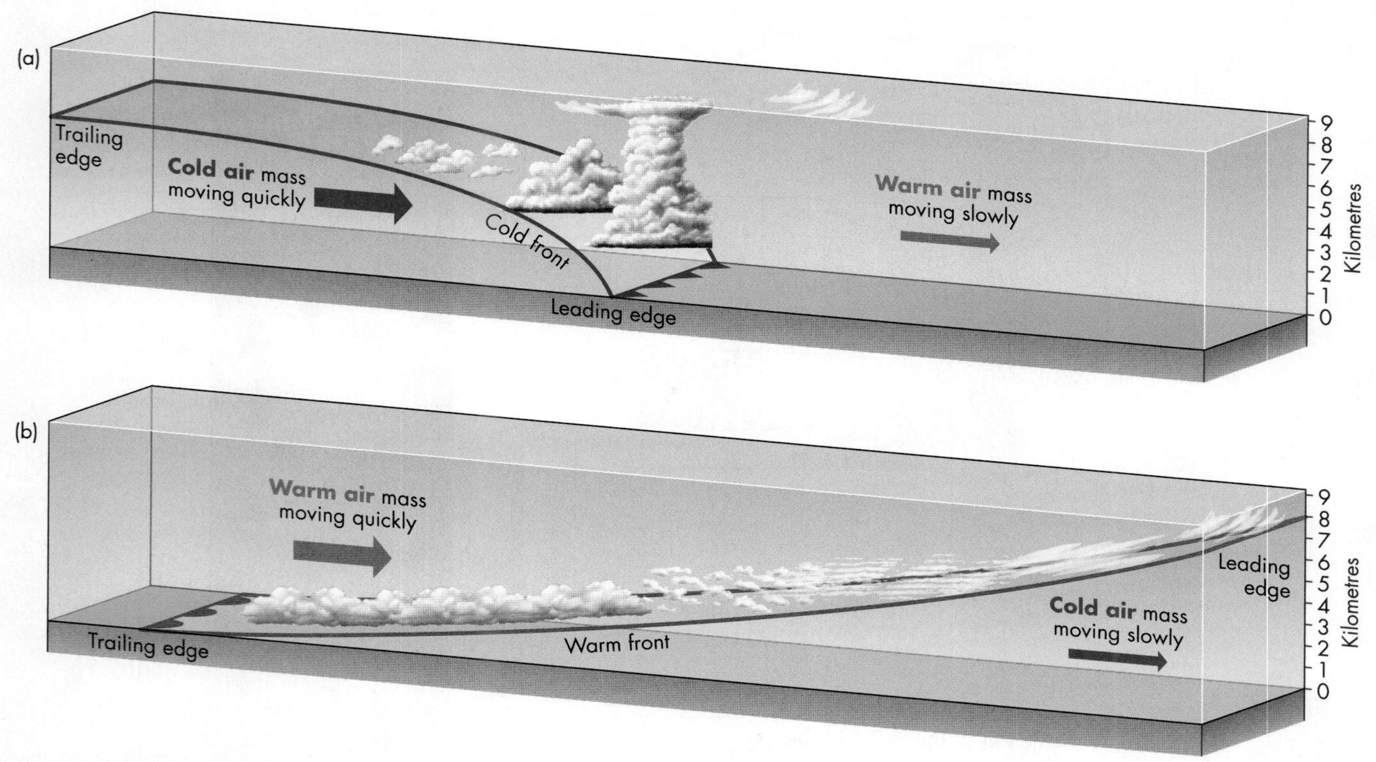

▲ **FIGURE 10.11 FRONTS** A weather front marks the boundary between two air masses of different density and temperature. (a) An advancing cold front forces warm air upward. The rising warm air creates clouds and heavy precipitation. (b) An advancing warm front forces warm air over cooler air, again producing clouds and precipitation. *(Based on McKnight, T. L., and D. Hess. 2004.* Physical Geography, *8th ed. Upper Saddle River, NJ: Pearson Prentice Hall)*

Vertical Stability of the Atmosphere

Air also moves because of the vertical heterogeneity of the troposphere. We can understand these movements by examining the behaviour of a small volume or *parcel* of air. The tendency of a parcel of air to remain in place or to change its vertical position is referred to as *atmospheric stability*. An air mass is stable if its parcels resist vertical movement or return to their original position after they have moved. In contrast, an air mass is said to be unstable if its parcels rise until they reach air of similar temperature and density.[10] The atmosphere becomes unstable when lighter warm air is overlain by denser cold air. Under these conditions, the instability causes some parcels of air to sink and others to rise like hot air balloons. Air turbulence and severe weather are associated with unstable atmospheric conditions.

Fronts

Meteorologists refer to the boundary between cool and warm air masses as a *front*. A cold front forms when cool air moves into a mass of warm air; the opposite is true for a warm front (Figure 10.11). A *stationary front* has a boundary that does not move much. All three types of fronts can cause severe weather. A storm typically develops when a fast-moving cold front comes into contact with a slow-moving warm front. As the storm intensifies, the cold front may overtake the warm front and come into contact with another cool air mass, forming an *occluded front*. Figure 10.12 is a weather map that shows several types of fronts as well as areas of high and low pressure.

10.5 Hazardous Weather

The basic principles of atmospheric physics described above help us understand severe weather and its associated hazards. Severe weather events include cyclones, thunderstorms, tornadoes, blizzards, heat waves, and dust storms. Cyclones (hurricanes and extratropical storms) are dealt with separately in Chapter 11.

Thunderstorms

Thousands of thunderstorms are occurring on Earth as you read this chapter. Most of them are in equatorial regions. For example, the city of Kampala, Uganda, near the equator in East Africa experiences thunderstorms nearly 7 out of every 10 days, on average. In North America, thunderstorms are most frequent along the Front Range of the Rocky Mountains in Colorado

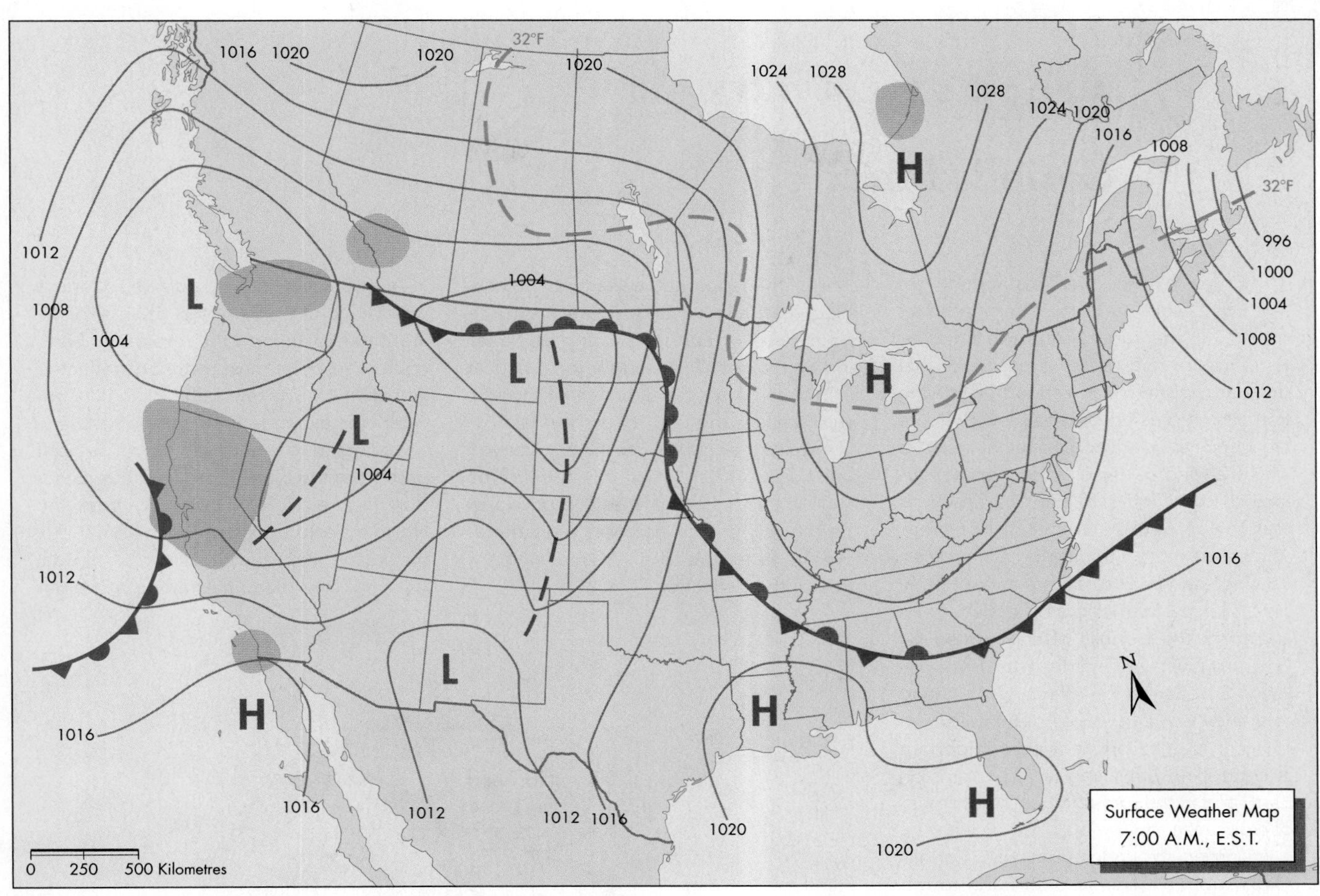

▲ **FIGURE 10.12 WEATHER MAP** This map depicts surface weather over the United States and southern Canada on April 13, 2003. It shows positions of high- and low-pressure centres (H and L), two low-pressure troughs (reddish-brown dashed lines) extending from low-pressure centres in southern Idaho and eastern Montana, a cold front extending east into the Atlantic Ocean (blue line with triangles), a warm front in North Dakota (red line with half circles), a stationary front in the lower Mississippi Valley (alternating cold and warm front symbols on opposite sides of the same line), an occluded front off the California coast (alternating cold and warm front symbols on the same side of a purple line), a line delineating a surface temperature of 0°C (curving dashed blue line over central and eastern Canada), and areas of precipitation (solid green). The green contour lines show atmospheric pressure in millibars (e.g., 1016). *(U.S. National Weather Service)*

and New Mexico and in a belt encompassing all of Florida and the southern parts of Georgia, Alabama, Mississippi, and Louisiana (Figure 10.16). However, most readers are likely to have experienced at least one thunderstorm because they occur in almost every part of Canada and the United States.

Although rain falls anywhere that clouds become oversaturated and are forced to release water, a thunderstorm requires a special set of atmospheric conditions. First, water vapour must be present in the troposphere to feed clouds as the storm forms. Second, a temperature gradient must exist in the troposphere, so that rising air can rapidly cool and become emplaced over warmer, moist air. Third, an updraft must force moist air upward to cooler levels of the atmosphere.

As moist air is forced upward, it cools and water vapour condenses to form a cumulus cloud. The cloud continues to grow upward as long as the moisture supply and updraft persist. This upward growth marks the beginning of the *cumulus stage* of thunderstorm development (Figure 10.17). The cumulus cloud evolves into a cumulonimbus cloud with the upward growth of domes and towers that look like a head of cauliflower. This growth involves a continuous release of latent heat due to water vapour condensation, which warms the surrounding air and causes it to rise farther.

As a cumulus cloud grows, precipitation starts by two mechanisms. First, the cloud expands into colder air, causing water droplets to freeze into ice crystals and snowflakes. The larger crystals and snowflakes fall until they enter warmer air and melt to form raindrops. Second, in warm air in the lower part of the cloud, large droplets collide with smaller droplets and coalesce to become raindrops. Once raindrops are too large to be supported by the updraft, they begin to fall, creating a downdraft.

A CLOSER LOOK 10.1

Coriolis Effect

The unequal distribution of solar energy that reaches the surface of Earth leads to temperature and pressure gradients that drive atmospheric circulation. Air moving from a high-pressure area to a low-pressure area tends to flow along a straight path. So why are wind patterns across Earth's surface curved (Figure 10.13)? The answer is that the planet, our frame of reference, rotates beneath the flowing air masses, causing the winds to take a curved path. This change in motion or deflection is known as the **Coriolis effect**, named for Gaspard-Gustave Coriolis, who first published a paper on the effects of a rotating reference frame in 1835. As you will see in Chapter 11, the Coriolis effect is important in controlling prevailing wind patterns on Earth and in the formation and movement of hurricanes.

In order to understand the Coriolis effect, we must first step off Earth to get a perspective on our rotating planet. As Figure 10.14 shows, Earth rotates from west to east and, when viewed from the Northern Hemisphere, rotates in an anticlockwise direction. Conversely when viewed from above the South Pole, the west to east rotation of Earth results in a clockwise motion. We recommend that you try this with a globe or basketball. Continuously rotate the ball in one direction and view it from above and from below. Because of the different rotation directions, the Coriolis effect behaves differently in the Northern Hemisphere than in the Southern Hemisphere. Specifically, winds are deflected to the right in the Northern Hemisphere and to the left in the Southern Hemisphere (Figure 10.13). Winds along the equator are not affected by the Coriolis effect.

Let's now examine what causes the Coriolis deflection. Notice in Figure 10.14 that the rotation speed of Earth differs with latitude, increasing from zero at the poles to 1675 km/h at the equator. It is this difference in rotation speed that causes the apparent deflection of anything that flows or flies above the rotating surface of Earth, including airplanes and air masses. The Coriolis force acts at a right angle to the direction of motion, which causes the path of an object moving above the surface to have a curved path (Figure 10.15). The magnitude of the Coriolis effect increases as the speed of the moving object increases, and therefore the faster the air mass is moving the greater the deflection.

A good way to visualize the Coriolis effect is to use an analogy of an airplane flight to explain how the rotation of Earth could cause air masses to be deflected across the surface of the planet. Consider, for example, the case of a plane travelling due south along a straight path from the North Pole to Quito, Ecuador (Figure 10.14). Departing from the North Pole (the axis of Earth's rotation), the plane has zero eastward rotation. Although the plane flies along a straight

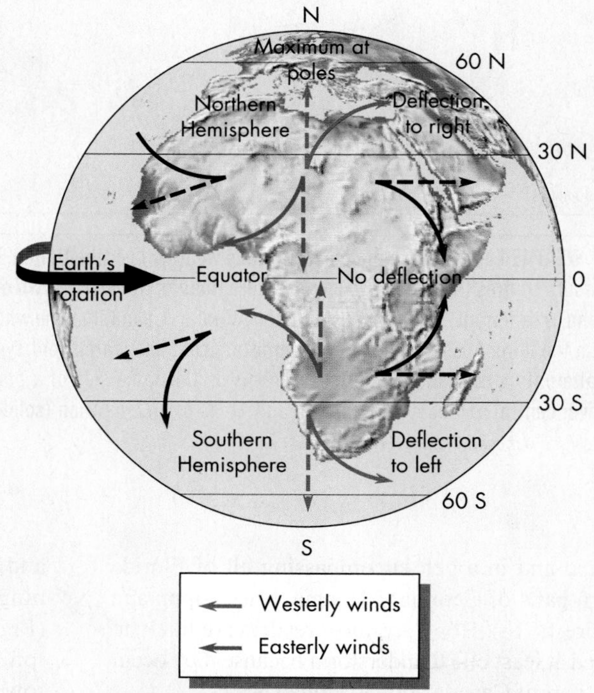

▲ **FIGURE 10.13 SURFACE WIND PATTERNS THAT RESULT FROM THE CORIOLIS EFFECT** Air masses moving across the surface of Earth are deflected to the right in the Northern Hemisphere and to the left in the Southern Hemisphere due to the Coriolis effect. The deflection influences the prevailing wind directions, creating westerly (red arrows) and easterly (blue arrows) winds from subtropical high-pressure zones. *(Based on Christopherson, R. W. 2009.* Geosystems, *7th ed. Upper Saddle River, NJ: Pearson Prentice Hall)*

(continued)

path toward Quito, the greater rotation speed of Earth near the equator carries that city eastward. Upon arrival at the latitude of Quito, the pilot would find herself over the Pacific Ocean (Figure 10.14). To an observer in Quito watching the airplane, the land surface would feel stationary, yet the airplane would appear to veer toward the west and head over the sea, even though its path was straight. On the return flight from Quito, the greater eastward rotation at the equator compared to more northern latitudes would cause the plane to be deflected toward the east. In both cases, when viewed from above, the plane appears to have been deflected to the right.

Let's now look at an example of how the Coriolis effect can be used to explain the predominant wind directions on Earth. Air travelling southward from the Northern Hemisphere's subtropical high-pressure zone (30°N latitude) will be deflected toward the west, producing easterly winds (Figure 10.13), whereas air moving toward the north will produce westerly winds. In contrast, air moving southward from the Southern Hemisphere's subtropical high-pressure zone (30°S latitude) will be deflected toward the west, producing easterly winds, whereas air moving toward the north will produce westerly winds.

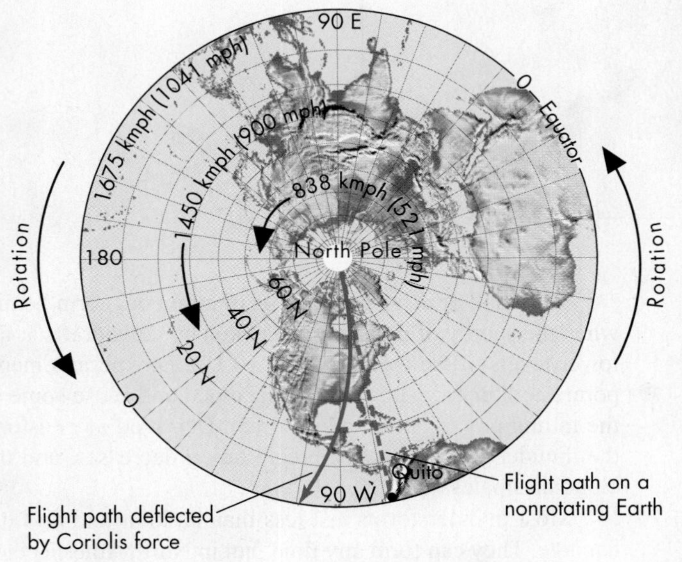

◄ **FIGURE 10.14 EARTH'S ROTATION AND THE CORIOLIS EFFECT** This example illustrates the Coriolis effect on an airplane travelling along a straight path from the North Pole to Quito, Ecuador, which is on the equator. *(Based on Christopherson, R. W. 2009.* Geosystems, *7th ed. Upper Saddle River, NJ: Pearson Prentice Hall)*

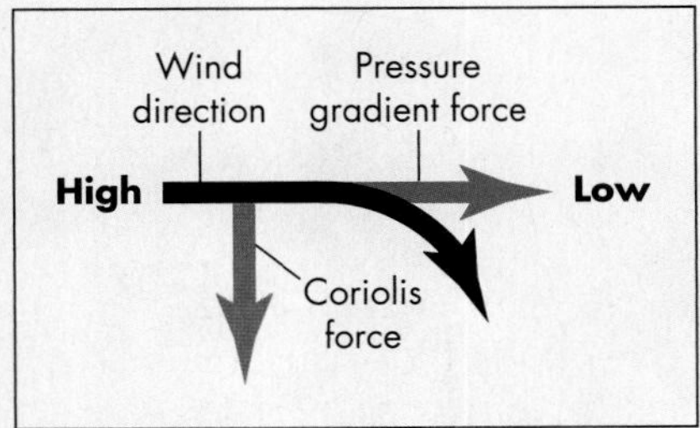

◄ **FIGURE 10.15 FORCES INVOLVED IN WIND PATTERNS** The path of an air mass is controlled, in part, by the pressure gradient and Coriolis forces. On a non-rotating Earth, air masses would flow along a gradient from areas of high pressure to areas of low pressure (blue arrow). However, on a rotating Earth, the Coriolis force (red arrow) is perpendicular to a horizontal wind direction in this example, which causes the path of the air mass to curve to the right (black arrow).

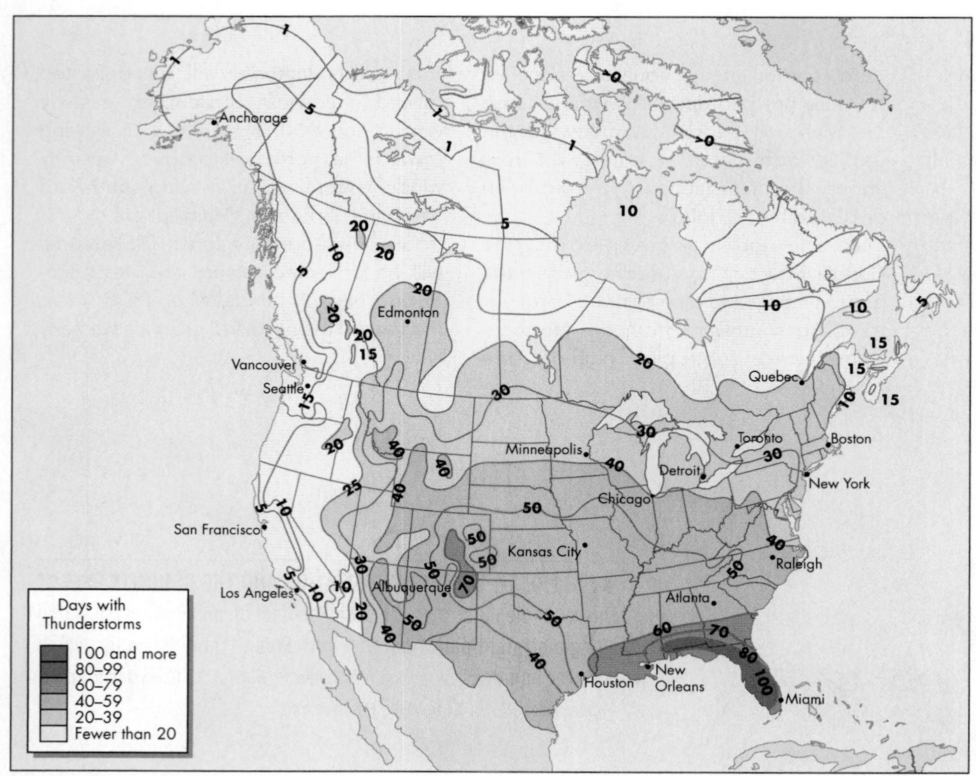

◄ **FIGURE 10.16 THUNDERSTORM OCCURRENCE IN CANADA AND THE UNITED STATES** This map of Canada and the United States shows the average number of days per year with thunderstorms. South-central Florida has the largest number of days with thunderstorms, and Baffin Island in the Canadian Arctic has the least. *(From Christopherson, R. W. 2003.* Geosystems, *5th ed. Upper Saddle River, NJ: Prentice Hall, with data from the National Weather Service and Map Series 3, Climate Atlas of Canada)*

The *mature stage* of thunderstorm development begins when the downdraft and falling precipitation leave the base of the cloud (Figure 10.17). At this time, the storm has both updrafts and downdrafts, and it continues to grow until it reaches the top of the unstable atmosphere near the *tropopause*. Updrafts may continue to build the cloud outward to form a characteristic anvil shape. The storm produces its most intense precipitation, thunder, and lightning during the mature stage (see Case Study 10.1).

The final, or *dissipative*, stage of a thunderstorm begins when the supply of moist air is blocked by downdrafts at the lower levels of the cloud (Figure 10.17). Downdrafts incorporate cool dry air surrounding the cloud and cause some of the falling precipitation to evaporate. Deprived of moisture, the thunderstorm weakens, precipitation decreases, and the cloud dissipates.

Most thunderstorms last less than an hour and do little damage. They can form any time, but in North America they

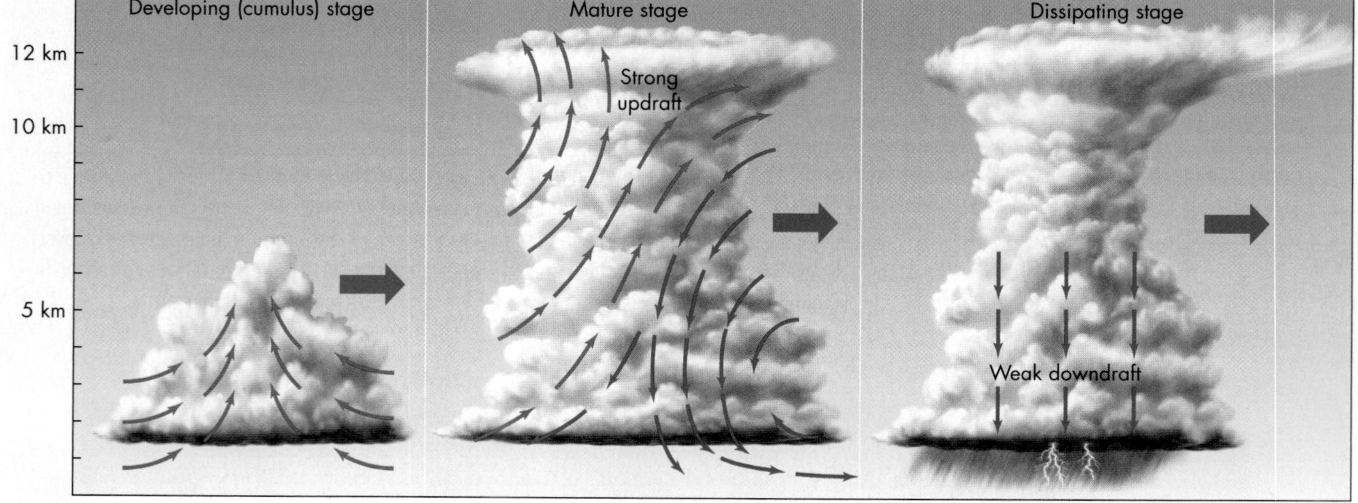

▲ **FIGURE 10.17 LIFE CYCLE OF A THUNDERSTORM** An idealized diagram showing stages in the development and dissipation of a thunderstorm. Red arrows show updrafts of warm air and blue arrows show downdrafts of cold air. The large blue arrows show the direction that the storm is moving. *(Based on Aguado, E., and J. E. Burt. 2001.* Understanding Weather and Climate, *2nd ed. Upper Saddle River, NJ: Prentice Hall)*

10.1 CASE STUDY 10.1

Lightning

Lightning consists of flashes of light produced by the discharge of millions of volts of electricity (Figure 10.18) and is common during thunderstorms. The discharges momentarily heat the air in their paths to temperatures as high as 30 000°C, five times hotter than the surface of the sun.[11] This rapid heating causes the surrounding air to expand rapidly, producing *thunder*.

Most lightning bolts travel from cloud to cloud—that is, they start and end in clouds. Cloud-to-ground lightning is less common, but many tens of millions of lightning bolts strike Canada and the United States each year.[12]

Cloud-to-ground lightning is more complex than it appears to an observer (Figure 10.19). Most lightning comes from cumulonimbus clouds that have grown high enough for ice crystals and pellets to form. Rising ice crystals interact with falling ice pellets, creating an electrical field strong enough to produce lightning.[13] Electrons moving between the two forms of ice build up a positive electrical charge in the upper part of the cloud and a negative charge in the lower part (Figure 10.19). Because like electrical charges of the same polarity repel, the negative charge in the lower part of the cloud drives away any negative charge on the Earth surface below, leaving the ground beneath a thunderstorm with a net positive charge. A strong difference in electrical charge between clouds and the ground is required for lightning to overcome the natural insulating property of air.

Each lightning strike begins when a column of electrically charged air, called a *stepped leader,* advances downward from the base of a cloud. Within milliseconds, this column branches downward until it is close to the ground (Figure 10.19). As the stepped leader approaches the ground, a spark surges upward from a tall object, such as a tree or building, and attaches to the leader. Once it makes contact, the spark completes a conductive channel of ionized air molecules that results in an upward surge of positive electrical charge. This surge heats the air to create a brilliant *return stroke* of lightning to the cloud. Within milliseconds, additional leaders and return strokes move along the same path, often causing the lightning flash to appear to flicker to the human eye. Thus, although a lightning strike appears to come down from the base of the cloud, the electrical discharge actually moves upward from the ground.[11]

Lightning strikes are a serious hazard, although the number of annual lightning deaths in North America has decreased as more people leave rural areas to live in cities.

▲ **FIGURE 10.18 LIGHTNING** A night-time photograph of cloud-to-ground and cloud-to-cloud lightning strokes. *(NOAA)*

(continued)

10.1 CASE STUDY (Continued)

Lightning kills about seven people in Canada each year and injures 60 to 70 more; in the United States, it kills about 100 people and injures more than 300 each year.[13,14,15] Seventy percent of survivors of a lightning strike suffer serious long-term health effects (see Survivor Story).[16] The chances of being struck by lightning in the United States are estimated to be 1 in 240 000 each year.[12] Over an average lifespan of 80 years, an individual's risk is approximately 1 in 3000, but much depends on where he or she lives and works.

◄ **FIGURE 10.19 DEVELOPMENT OF CLOUD-TO-GROUND LIGHTNING** (a) Electrical charge separation develops in a cumulonimbus cloud, most likely by the interaction between rising ice crystals and falling ice pellets. The build-up of a negative electrical charge at the base of the cloud locally induces a positive electrical charge below the thunderstorm. (b) An invisible stepped leader forms a column of electrically charged air that branches downward toward the ground. (c) When the stepped leader approaches the ground, a spark jumps from an object on the ground, such as a tree, and attaches to the leader. (d) Flowing electrons heat the channel to an extremely high temperature and create a return stroke back up to the cloud to produce the lightning strike.

are most common during afternoon and evening hours in the spring and summer.

Severe Thunderstorms The scenario described above is typical of most thunderstorms, which never become severe. However, under certain conditions, some thunderstorms do become severe. Canada's Atmospheric Environment Service classifies a thunderstorm as severe if it has wind speeds in excess of 90 km/h or hailstones larger than 1.5 cm, or if it generates a tornado.[11] Severe thunderstorms require favourable atmospheric conditions over a large area and the ability to maintain themselves. They commonly appear in groups and can last from several hours to several days.

Conditions required to form a severe thunderstorm include *wind shear* produced by winds blowing in different directions, high water-vapour content in the lower troposphere, uplift of air, and the existence of a dry air mass above a moist air mass.[11] Of these four conditions, wind shear in a vertical direction is especially important—in general, the greater the vertical wind shear, the more severe the thunderstorm.[17]

Three types of severe thunderstorms have been identified on the basis of their organization, shape, and size: roughly circular clusters of storm cells called mesoscale convective complexes, linear belts of thunderstorms called squall lines, and large cells with single updrafts called supercells.

Mesoscale convective complexes are the most common of the three types of severe thunderstorms. They are very large clusters of self-propagating storms in which the downdraft of one cell leads to the formation of a new cell nearby. Unlike many single-cell thunderstorms that last for less than an hour, these complexes can continue to grow and move for periods of 12 hours or more. Their downdrafts can come together to form *outflow boundaries*—curved lines of thunderstorms that travel long distances.

Squall lines average 500 km in length and are lines of individual storm cells.[11] They commonly develop parallel to, and 300 km to 500 km ahead of, cold fronts. Updrafts in the advancing squall line typically form anvil-shaped clouds whose tops extend ahead of the line. Downdrafts originating on the backside of the storms may surge forward as a *gust front* of cold air in advance of precipitation. Squall lines can also develop along *drylines*, an air mass boundary similar to a front, but along which the air masses differ in moisture content rather than in temperature. Drylines develop in the southwestern United States during the spring and summer, sometimes producing daily squall lines.

The most damaging of all severe thunderstorms are *supercell storms*. A supercell storm is a thunderstorm with a deeply rotating updraft flanked by smaller updraft elements. The flanking rising air masses usually merge with and augment the main rotating updraft, rather than developing into separate and competing thunderstorm cells. Although smaller than mesoscale convective complexes and squall lines, supercell storms are extremely violent and are the breeding ground for most large tornadoes. They are typically 20 km to 50 km in diameter and last from two to four hours.

Downbursts of air from severe thunderstorms, especially mesoscale convective complexes, can also generate strong, straight-line windstorms. The largest of these storms, called *derechos*, produce severe, tornado-strength wind gusts along a line that is at least 400 km long.[18] In general, derecho winds exceed 90 km/h and can topple trees and cause widespread power outages, serious injuries, and fatalities. The strongest derechos generate winds approaching 210 km/h and produce the same damage as a medium-sized tornado.[19] More than a dozen derechos strike North America every year, with most occurring in the eastern two-thirds of the United States and southern Canada.[18] Smaller thunderstorm downbursts, called *microbursts*, are more common than derechos and are a serious hazard to aviation.

Hail Many large thunderstorms produce hard, rounded, or irregular pieces of ice called *hailstones*. Large hailstones from severe thunderstorms can cause much damage. You can infer how hailstones form by cutting one in half to reveal its bull's-eye pattern of concentric rings of ice. The rings form when a hailstone moves up and down in the turbulent air of a thunderstorm. Starting with a small ice pellet as a nucleus, a hailstone gets a coating of liquid water in the lower part of the storm; the liquid freezes when a strong updraft carries the stone upward into cold air. This process is repeated many times to form a large piece of hail. The largest authenticated hailstone in North America fell from a severe thunderstorm in Aurora, Nebraska, in June 2003. It was 18 cm in diameter, nearly the size of a volleyball, weighed 800 g, and was estimated to have hit the ground at a velocity of more than 160 km/h![20]

Damage caused by hail in Canada and the United States averages more than $1 billion per year.[20] In North America, damaging hailstorms are most common on the Great Plains of the United States, particularly in northeastern Colorado and southeastern Wyoming, and on the Prairies of southern Alberta. North-central India, Bangladesh, Kenya, and Australia also experience frequent damaging hailstorms. Although rare in North America, deaths from hailstones are common in areas of high population density and poorly constructed dwellings, notably in Bangladesh and India.[20]

Tornadoes

A **tornado**, or "twister," is one of nature's most violent natural processes. Tornadoes kill an average of 50 people per year in the United States, and parts of Canada also experience deadly tornadoes: Barrie, Ontario (1985), Edmonton (1987, Case Study 10.2), and Pine Lake, Alberta (2000). These spinning columns of wind range in shape from ropes to funnels and are capable of inflicting tremendous damage. To be called a tornado, a spinning column of wind or *vortex* must extend downward from a cloud and touch the ground. Funnel-shaped vortices that do not touch the ground are called *funnel clouds* (Figure 10.21b). The number of funnel clouds that develop each year far exceeds the number of tornadoes.

Tornadoes form where there are large differences in atmospheric pressure over short distances, most commonly during a severe thunderstorm. Although meteorologists do not completely understand how tornadoes form, they know that most tornadoes go through similar stages in development (Figure 10.21).

In the initial, *organizational stage* of a tornado, wind shear causes air to begin to rotate within the thunderstorm (Figure 10.22a). Strong updrafts in front of the advancing cold front, and upwelling due to latent heat of condensation, tilt the horizontally rotating air vertically (Figure 10.22b). If the updraft and the wind shear are strong enough, a large vertically rotating column of air known as a *mesocyclone* forms (Figure 10.22c). Updrafts, commonly in the rear of the storm, lower a portion of the cumulonimbus cloud to form a *wall cloud* (Figure 10.21a). The wall cloud might itself begin to slowly rotate, and a short funnel cloud can descend from

SURVIVOR STORY

Struck by Lightning

Michael Utley came within inches of death when he was struck by lightning on a Cape Cod golf course.

Like most of us, Michael Utley didn't worry much about being struck by lightning. After all, the odds are very low. And so at a charity golf tournament to benefit a local YMCA near his Cape Cod home, he was not concerned about a looming thunderstorm. "I didn't pay attention to it," he says.

Four holes into the game, a horn blasted, warning golfers to seek shelter. Utley replaced the flag in the hole and was several yards behind his three companions when he was struck by lightning.

Utley's friends heard a thunderous crack behind them, and when they turned around, they saw smoke coming from Utley's body, his shoes torn from his feet, and his zipper blown open.

Luckily, Utley remembers none of this. In fact, he remembered nothing for the next 38 days, which he spent in an intensive-care unit. Fortunately for Utley, one of his companions had recently been retrained in CPR, which may have saved his life.

Utley comments that some of the ideas we have about people struck by lightning are accurate and some are not. "The hair standing up, that's real. When that happens, you're pretty close to being dead," he says. But "I didn't see a white light at the end of the tunnel. I don't burn clocks off the wall when I touch them."

Utley, like many victims of lightning, has had, in his words, "a variety of ups and downs" after the accident. His wife, Tamara, recalls that he did not slip into a coma-like state for several days. "It was very, very strange," she says. "When he first got to the ICU, he was able to move and he was very lucid." But within a few days he began to lose consciousness for extended periods. Utley remembers none of this.

The road to recovery is not over for Utley. He still bumps into walls when he walks and has had to relearn many basic activities. Since the accident, he has spent a great deal of time working to educate the community, including the Professional Golfers' Association (PGA) (Figure 10.20), about the perils of lightning. He sums up his message to golfers with this slogan: "If you see it, flee it; if you hear it, clear it."

—Chris Wilson

▲ **FIGURE 10.20 LIGHTNING SAFETY** This U.S. National Weather Service poster featuring professional golfer Vijay Singh reminds people to be careful outdoors during a thunderstorm. *(NOAA)*

▲ **FIGURE 10.21 STAGES IN TORNADO DEVELOPMENT** (a) A wall cloud near Miami, Texas. Wall clouds extend downward during a severe thunderstorm and are commonly where tornadoes form. (b) A funnel cloud over the state capitol buildings in Austin, Texas. (c) A tornado in the organizational stage of development near Enid, Oklahoma. The funnel cloud extends downward from the thunderstorm and debris is beginning to rotate on the ground below it. (d) A tornado in the mature stage of development near Seymour, Texas. (e) A tornado in the shrinking, rope stage near Cordell, Oklahoma. *((a) NOAA; (b) Austin Public Library; (c), (d), and (e) NOAA)*

it (Figure 10.21b). At the same time, dust and debris on the ground begin to swirl below the funnel (Figure 10.21c). If the two rotating columns of air join, the funnel cloud becomes a tornado. Not all wall clouds and mesocyclones produce tornadoes, and a tornado can develop without the presence of a wall cloud or mesocyclone.

In the second, *mature stage* of a tornado, a visible funnel extends from the thunderstorm cloud to the ground and moist air is drawn upward (Figure 10.21d). Small intense whirls, called *suction vortices*, may form within larger tornadoes. They orbit the centre of the vortex and can be responsible for the tornado's greatest damage.[21]

The tornado enters the *shrinking stage* when the supply of warm moist air is reduced. The funnel begins to tilt and becomes narrower. Winds can increase at this stage, making the tornado even more dangerous.

In the final, *rope stage* (Figure 10.21e), the upward-spiralling air comes into contact with downdrafts, and the tornado begins to move erratically. This behaviour marks the beginning of the end for the tornado, but it can still be extremely dangerous. Tornadoes can go through all the stages described above or they can skip stages, and new tornadoes can form nearby as older tornadoes dissipate.

Tornadoes pick up soil, vegetation, and debris as they move across land. This material gives the tornado cloud its characteristic dark appearance (Figure 10.21d). Tornadoes typically have diameters of tens of metres and wind speeds of 65 km/h to over 400 km/h.[11] Most tornadoes travel fewer than 10 km and last only a few minutes before weakening and disappearing. However, the largest and most damaging tornadoes can have widths of hundreds of metres and paths a few hundred kilometres long.

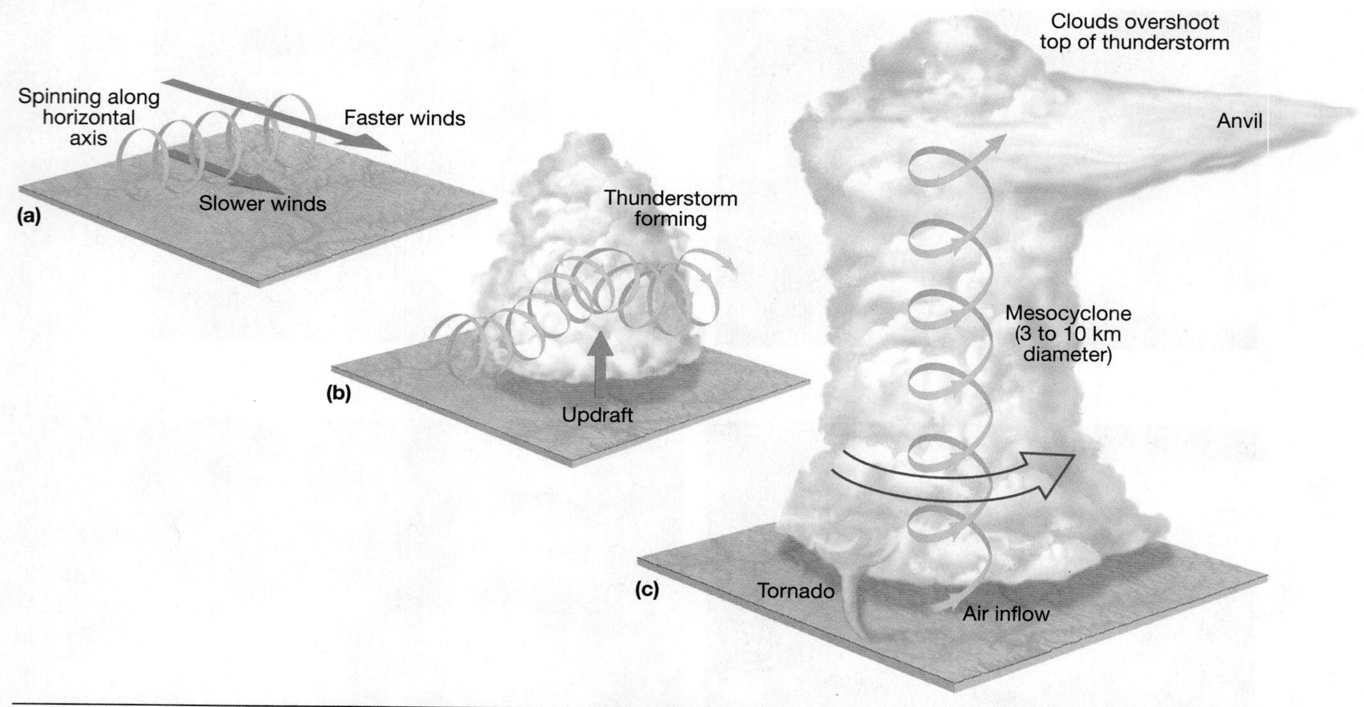

▲ **FIGURE 10.22 TORNADO FORMATION** (a) Vertical wind shear creates a horizontal rotation or rolling of air. (b) Updrafts produced by rising warm air lift the horizontally rolling air into a vertical column. (c) A vertical rotating mesocyclone develops. If conditions are right, a tornado will develop from the slowly rotating wall cloud toward the rear of the storm. *(McNight, T. L., and D. Hess. 2004.* Physical Geography, *10th ed. Upper Saddle River, NJ: Pearson Prentice Hall)*

Classification of Tornadoes Tornadoes are classified according to their maximum wind speeds and the damage they produce. They are assigned values on the *Enhanced Fujita Scale* or **EF Scale** (Table 10.1) on the basis of post-storm damage surveys. The surveys determine the levels of damage experienced by 26 types of buildings, towers, and poles, and by hardwood and softwood trees. Wind speeds are estimates of the maximum three-second wind gust in the tornado, and range from 105 to 137 km/h for an EF0 tornado to over 322 km/h for an EF5 tornado.[22] The EF Scale replaces the F Scale developed by T. Theodore Fujita in 1971 and is analogous to the Modified Mercalli Scale for assessing earthquake intensity (Chapter 3).

Tornadoes that develop over water are called *waterspouts*. Most waterspouts develop beneath fair-weather cumulus clouds and appear to be associated with wind shear along the boundary of contrasting air masses.[21] Small wind-shear-generated tornadoes can develop on land and, like waterspouts, are typically weak EF1 or EF0 tornadoes.[21]

TABLE 10.1 Enhanced Fujita Scale of Tornado Intensity

Scale	Wind Speeds[1]	Typical Damage[2]
EF0	105–137 km/h	Light damage
EF1	138–177 km/h	Moderate damage
EF2	178–218 km/h	Considerable damage
EF3	219–266 km/h	Severe damage
EF4	267–322 km/h	Devastating damage
EF5	> 322 km/h	Incredible damage

[1]Wind speeds are maxima estimated for a three-second gust.

[2]Accurate placement on the EF Scale involves expert judgment of eight levels of damage to 28 indicators, including homes and other types of buildings, towers, poles, and trees. A report explaining the EF Scale and its basis is available on the Texas Tech University website: www.depts.ttu.edu/weweb/EFScale.pdf.

Source: Data from NOAA, http://www.spc.noaa.gov/faq/tornado/ef-scale.html. Accessed September 21, 2010.

10.2 CASE STUDY

The Edmonton Tornado

On the afternoon of July 31, 1987, a tornado rated EF4 on the Enhanced Fujita Scale ripped through Edmonton (Figure 10.23). Twenty-seven people were killed, more than 300 were injured, and property damage amounted to more than $330 million. The tornado remained on the ground for an hour, cutting a path of destruction 40 km long and up to 1 km wide. It achieved wind speeds of up to 460 km/h. In terms of loss of life and injuries, this tornado was the worst natural disaster in Alberta's history and one of the worst in Canada's history.[23]

Weather forecasts issued for Edmonton by Environment Canada during the morning and early afternoon of July 31, 1987, warned of a high potential for severe thunderstorms. Environment Canada responded swiftly on receipt of the first report of a tornado touching down adjacent to Edmonton's southern boundary—it issued a tornado warning over weather radio five minutes after the report.

While municipal emergency agencies, fire departments, ambulance, and police were responding, the Canadian Department of National Defence placed helicopters and ambulances on standby at Canadian Forces Base Edmonton and provided reconnaissance flights for the City of Edmonton and the deputy prime minister. During the storm, Emergency Preparedness Canada established contact with the Alberta Government Emergency Response Centre and established a liaison office at the response centre later that day.

▲ **FIGURE 10.23 EDMONTON TORNADO** On July 31, 1987, a tornado sliced through Edmonton, Alberta, killing 27 people and injuring more than 300 others. *(Robert den Hartigh)*

The post-disaster response period lasted approximately three weeks, during which immediate disaster assistance was provided to victims. At the end of August 1987, the Government of Alberta announced a disaster-recovery program with assistance from the Government of Canada.

The Alberta Emergency Public Warning System was developed as a result of the disaster. It interrupts private and public broadcasts on radio, television, and cable systems, alerting the public to imminent potential disasters.

Occurrence of Tornadoes Although tornadoes occur throughout the world, they are much more common in the United States than in any other country (Table 10.2). The United States has just the right combination of weather, topography, and geographic location to make it the perfect tornado spawning ground.[21] Most U.S. tornadoes happen in spring and summer in the region extending from Florida to Texas and north to the Dakotas, Indiana, and Ohio. The area at highest risk on the Great Plains is called "Tornado Alley."

A recent example of the severity of the hazard is the Joplin tornado, a catastrophic EF5 multiple-vortex tornado that struck Joplin, Missouri, in the late afternoon of May 22, 2011. It was part of a large tornado outbreak in late May that caused widespread damage and loss of life in Tornado Alley. The tornado had an unusually wide path of destruction (its maximum width was 1.6 km) as it tracked through the southern part of the city. It intensified as it tracked eastward across the city and into rural portions of two counties. One hundred fifty-eight people were directly killed by the tornado, 1100 were injured, and damage totalled U.S.$3 billion.[24] The maximum wind speed was about 400 km/hr. It was the deadliest tornado to strike the United States since one in Texas and Oklahoma in 1947, and it ranks as the seventh most deadly and costly tornado in U.S. history. The winds were so strong that a semi-truck was thrown hundreds of metres and wrapped completely around a debarked tree.[24] The Joplin tornado occurred one month after a devastating outbreak of tornadoes in the central and southern U.S.—between April 25 and 28, 2011, tornadoes in nine states claimed 324 lives.

Canada experiences far fewer and less deadly tornadoes than the United States, but it has several tornado-prone regions, including southern Alberta, southern Ontario, southeastern Quebec, and an area extending from southern Saskatchewan through southern Manitoba to Thunder Bay, Ontario (Figure 10.24, Table 10.2).[25] The deadliest tornado in Canadian history is the so-called Regina Cyclone, an EF4 tornado that devastated the city of Regina, Saskatchewan, and killed 28 people on June 30, 1912.[26] The tornado tore a swath through two residential areas, the downtown business district, rail yards, and warehouse district. The most recent major tornado in Canada happened in Goderich, Ontario, on August 21, 2011. Rated EF3 on the Enhanced Fujita Scale, the

TABLE 10.2 Tornadoes Causing More than 100 Deaths in the United States and More than Five Deaths in Canada

Name (state or province)	Date	Deaths
Tri-State (Missouri, Illinois, Indiana)	March 18, 1925	747
Natchez (Mississippi)	May 7, 1840	317
St. Louis and East St. Louis (Missouri, Illinois)	May 27, 1896	255
Gainesville (Georgia)	April 5, 1936	203
Woodward (Oklahoma)	April 9, 1947	181
Joplin (Missouri)	May 22, 2011	158
New Richmond (Wisconsin)	June 11–12, 1899	117
Flint (Michigan)	June 8, 1953	116
Goliad (Texas)	May 18, 1902	114
Regina (Saskatchewan)	June 30, 1912	ca. 28
Edmonton (Alberta)	July 31, 1987	27
Windsor-Tecumseh (Ontario)	June 17, 1946	17
Pine Lake (Alberta)	July 14, 2000	12
Salaberry-de-Valleyfield (Quebec)	August 16, 1888	9
Windsor (Ontario)	April 3, 1974	9
Barrie (Ontario)	May 31, 1985	8
Sudbury (Ontario)	August 20, 1970	6
Sainte-Rose (Quebec)	June 8, 1953	6
Bouctouche (New Brunswick)	August 6, 1879	5
Portage la Prairie (Manitoba)	June 22, 1922	5

Note: Table does not include tornado outbreaks, which are multiple tornadoes affecting large areas; several tornado outbreaks have claimed more than 400 lives.

Source: Data from Wikipedia. 2013. "List of North American tornadoes and tornado outbreaks." http://en.wikipedia.org/wiki/List_of_North_American_tornadoes_and_tornado_outbreaks. Accessed May 9, 2013.

tornado developed from an isolated supercell that unexpectedly crossed Huron County on the afternoon of August 21. It began as a waterspout over Lake Huron and ripped through the lakeside town of Goderich, severely damaging the historic downtown area and homes in the surrounding area. One person died and 37 were injured.[27] The Goderich tornado was the strongest and most damaging tornado to strike Ontario since 1996.

Tornadoes are also common in Bangladesh, Australia, New Zealand, northern India, South Africa, Argentina, Japan, eastern China, and central Europe from France and Great Britain eastward to Russia and Ukraine. However, violent tornadoes (EF4 and EF5) are rare or nonexistent outside North America, with the possible exception of Bangladesh.[21]

Most waterspouts occur in tropical and subtropical waters, but they have been reported off the New England and California coasts, on the Great Lakes, and on Lake Winnipeg.[21] They are especially common along the Gulf Coast of the United States, in the Caribbean Sea, in the Bay of Bengal, and in the South Atlantic Ocean. A study conducted in the Florida Keys reported 390 waterspouts within 80 km of Key West over a five-month period.[21]

Blizzards, Extreme Cold, and Ice Storms

Blizzards are severe winter storms characterized by high winds, blowing snow, low visibility, and low temperatures for an extended period. The official threshold for blizzard conditions differs in Canada and the United States. In Canada, winds must exceed 40 km/h with visibility less than 1 km for at least three hours, whereas in the United States winds must exceed 56 km/h with visibility less than 400 m for at least three hours.[28,29]

Blizzards can produce heavy snowfall, wind damage, and large snowdrifts (Figure 10.25). The Saskatchewan blizzard of 1947 lasted for 10 days and buried an entire train in a snowdrift 8 m deep and 1 km long.[30] Another famous storm, the "Blizzard of 1888," killed more than 400 people and paralyzed the northeastern United States for three days with snow drifts that reportedly covered the first floors of buildings.

Storms that produce heavy snowfall and blizzards form when upper-level winds associated with a low-pressure trough interact with a surface low-pressure system.[31] Blizzards typically occur on the northwest side of these storms as they move north along the east side of a low-pressure trough.[31]

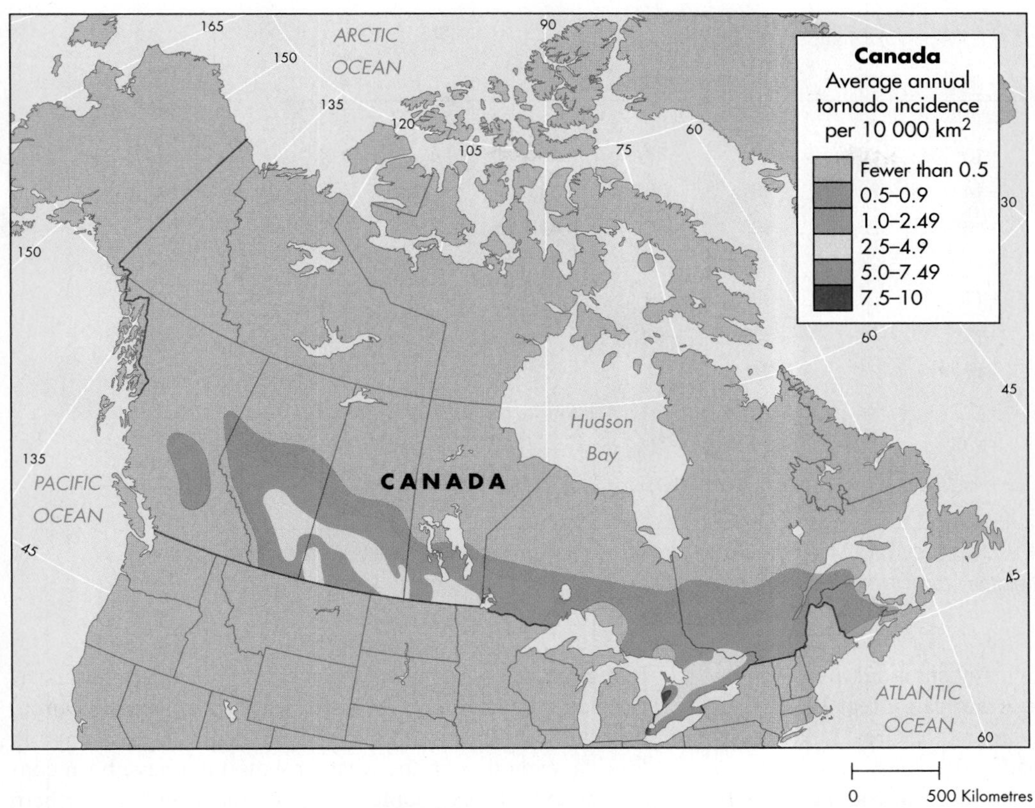

▲ **FIGURE 10.24 TORNADO OCCURRENCE IN CANADA** Average annual number of tornadoes per 10 000 km^2 in Canada based on data from 1950 to 1997. *(Reprinted by permission of David Eitken)*

▲ **FIGURE 10.25 AFTERMATH OF A BLIZZARD** Residents of Buffalo, New York, recover cars buried in deep snow following a blizzard. *(Darren Pittman/CP Images)*

Wind speed, km/h	Actual Air Temperature in °C								
Calm	4°	−1°	−7°	−12°	−18°	−23°	−29°	−34°	−40°
8	2°	−4°	−11°	−17°	−24°	−30°	−37°	−43°	−49°
16	1°	−6°	−13°	−20°	−27°	−33°	−41°	−47°	−54°
24	0°	−7°	−14°	−22°	−28°	−36°	−43°	−50°	−57°
32	−1°	−8°	−16°	−23°	−30°	−37°	−44°	−52°	−59°
40	−2°	−9°	−16°	−24°	−31°	−38°	−46°	−53°	−61°
48	−2°	−9°	−17°	−24°	−32°	−39°	−47°	−55°	−62°
56	−2°	−10°	−18°	−26°	−33°	−41°	−48°	−56°	−63°
64	−3°	−11°	−18°	−26°	−34°	−42°	−49°	−57°	−64°
72	−3°	−11°	−19°	−27°	−34°	−42°	−50°	−58°	−66°
80	−3°	−11°	−19°	−27°	−35°	−43°	−51°	−59°	−67°

Frostbite times: 30 min. 10 min. 5 min.

▲ **FIGURE 10.26 WIND CHILL CHART** Table of equivalent temperatures felt by the body for different wind speeds. *(Data from the National Weather Service and Meteorological Services of Canada, version 11/01/01)*

Blizzards associated with heavy snowfall are common through much of Canada, the American Great Plains and northeastern states, and in areas around the Great Lakes. Fast-moving storms east of the Canadian Rockies, called *Alberta Clippers*, are generally drier, with less snow and extremely cold temperatures. Blizzards on the east coast of Canada and the United States commonly occur during *nor'easters*. Nor'easters wreak havoc with hurricane-force winds, heavy snowfall, and high waves that damage coastal infrastructure. These storms are most common between September and April and can create blizzard conditions in Halifax, New York, Boston, and other east coast cities. In March 1993, a severe nor'easter paralyzed the east coast of the United States, causing snow, tornadoes, and flooding from Alabama to Maine. More than 240 people were killed in the "Blizzard of '93," and damage was more than U.S.$1 billion.

Less than three years later, in January 1996, a strong winter storm brought another massive blizzard to the east coast of the United States. The storm crippled the east coast for several days and produced record-breaking snowfall in Philadelphia and parts of New Jersey; it dropped 51 cm of snow in New York City's Central Park. The blizzard killed at least 100 people and caused an estimated U.S.$2 billion in damage.

A blizzard can also occur without snowfall. *Ground blizzards* in Antarctica, Alaska, Canada, and on the American Great Plains rework existing snow to produce whiteout conditions with visibility limited to a few metres or less.

The **wind chill** effect makes blizzards more dangerous than other snowstorms. Moving air rapidly cools exposed skin by evaporating moisture and removing warm air next to the body. This chilling reduces the time it takes for *frostbite* to develop. In blizzards, the wind chill temperature (WCT) is a more important measure of possible effects on the body than is the air temperature (Figure 10.26).

Blizzards and outbreaks of Arctic air can produce *extreme cold*. What constitutes extreme cold depends on climate averages and on personal and community preparedness. In regions relatively unaccustomed to winter weather, near-freezing temperatures are considered "extreme cold." During the winter of 1989–1990, 26 people died of hypothermia in Florida, even though the weather would not have been considered extreme by people living in Canada and the northern United States. Because of normally mild temperatures, many Florida homes lack adequate heating and insulation, and the outdoor lifestyle in the state leads to danger for those unprepared for freezing temperatures. In the far north of Canada, where people are accustomed to cold winters, temperatures below about −45°C are considered extreme cold. The lowest recorded temperature in North America was −63°C at Snag, Yukon Territory, in 1947. In general, people rapidly lose body heat whenever temperatures drop markedly below normal, especially during blizzard conditions. Extreme cold is a dangerous situation that can bring on health emergencies in susceptible people, such as those without shelter, stranded motorists, and people who live in homes that are poorly insulated or without heat. Environment Canada and the U.S. Department of Health and Human Services provide suggestions and information to help people minimize the risk of death or injury from extreme cold.

Ice storms are prolonged periods of freezing rain. They can be more damaging than blizzards and just as dangerous. During an ice storm, rain at a temperature near 0°C freezes when it comes into contact with cold surfaces. Heavy accumulations of ice are especially harmful to utility lines and trees, and can make driving treacherous. Ice storms typically occur during the winter on the north side of a stationary or warm front when three conditions are met: (1) ample moisture is present in the warm air mass south of the front, (2) warm air overlies a shallow layer of cold air, and (3) objects on the ground are at or below the freezing temperature. Snow begins to fall from the cooled top of the warm air mass and melts as it passes through. The resulting raindrops become supercooled as they descend through near-surface cold air (Figure 10.27). When they strike cold objects on the ground,

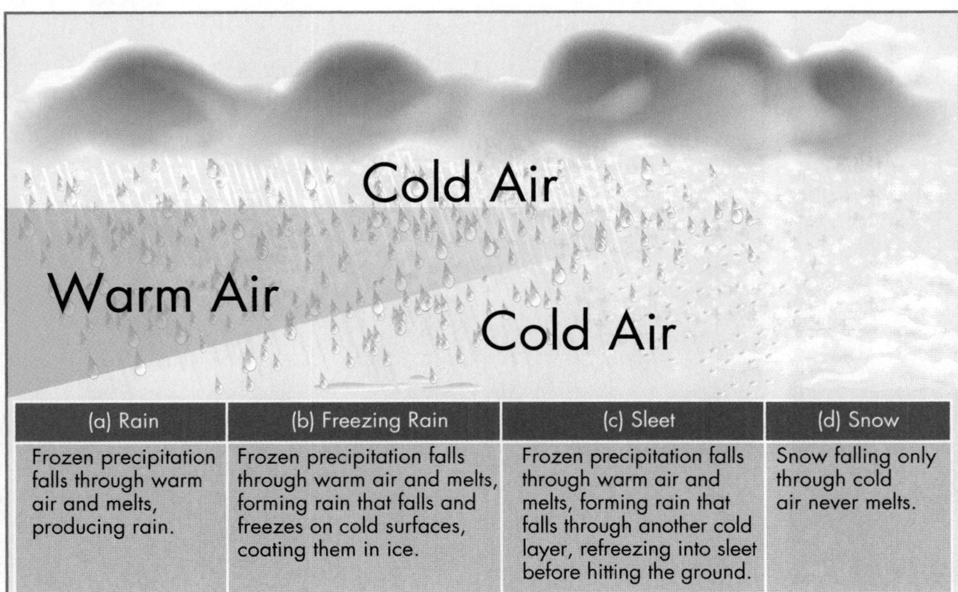

◄ **FIGURE 10.27 THE THICKNESS OF COLD AIR DETERMINES TYPE OF PRECIPITATION** In winter storms, the thickness of cold air at the surface determines the type of precipitation. (a) Snowflakes melt to produce rain where there is a thick warm air layer above the surface. (b) Raindrops become supercooled and freeze immediately upon contact with a cold surface if there is only a thin layer of cold air at the surface. (c) Falling snowflakes melt if they pass through warm air aloft; the resulting rain may refreeze to form ice pellets (sleet) if there is sufficient near-surface cold air. (d) Precipitation falls as snow if it does not pass through warm air before reaching the ground. *(Linda Scott/Austin American-Statesman)*

such as roads, trees, and utility lines, the raindrops immediately freeze, forming a coating of ice. Layers of ice up to 20 cm thick have been produced by prolonged ice storms. Areas in North America most prone to ice storms include southern Ontario, southern Quebec, the mid-Atlantic and New England states, the Ohio River Valley, the south-central Great Plains, and the Columbia River Valley in the Pacific Northwest (see opening story in this chapter).[1]

Fog

Fog is cloud in contact with the ground. It is hazardous only when it obscures visibility for travel and when it contains high levels of air pollutants, forming smog or vog (see Chapter 5). Fog contributed to the worst aircraft accident in history when 583 people died in the collision of two Boeing 747 aircraft on a runway on the Canary Islands in 1977.[32] Dense fog was also responsible for the worst maritime accident in Canada—in 1914 two ships collided on the St. Lawrence River, with the loss of 1014 lives.[33]

Fog forms either by condensation as air cools or by the evaporative addition of water vapour to already cool air. Cooling mechanisms include radiation of heat from land at night, the passage of moist air over cold coastal waters, and the upward movement of humid air along mountain slopes. Evaporative mechanisms include the flow of cold air over river, lake, or coastal waters that are warmer than the air, and precipitation of warm rain through cool air along a frontal boundary.

Drought

Drought is defined as an extended period of unusually low precipitation that produces a shortage of water for people, animals, and plants. More than 1 billion people live in semiarid regions where droughts are common, and more than 100 million people can suffer malnutrition or death if drought causes crops to fail. In addition to diminished crop growth and yield, drought can result in: (1) dust storms and related erosion; (2) famine; (3) ecosystem damage; (4) hunger; (5) malnutrition, dehydration, and related diseases; (6) population migrations, producing international refugees; (7) reduced electricity production; (8) shortages of water for industrial users; (9) social unrest; and (10) war.

Drought can be defined in three main ways:[34]

1. *Meteorological drought* is a lengthy period with lower than average precipitation. Meteorological drought usually precedes the other two types of drought.
2. *Agricultural drought* affects crop production. This condition can arise either from low precipitation or when soil conditions and erosion caused by poor agricultural practices cause a shortfall in water available to crops.
3. *Hydrological drought* occurs when the amount of water in aquifers, lakes, and reservoirs is below average for a lengthy period. It commonly materializes more slowly than other types of drought because it involves stored water that is used but not replenished. Like agricultural drought, it can be triggered by other than just a reduction in rainfall. For example, Kazakhstan recently received money from the World Bank to restore water that had been diverted to other nations from the Aral Sea under Soviet rule.

Recurring meteorological droughts in the Horn of Africa have led to massive food shortages. The Darfur conflict in neighbouring Sudan, and also affecting Chad, was fuelled by decades of drought and desertification.[35] In 2005, parts of the Amazon basin experienced the worst drought in 100 years. Researchers at Woods Hole Research Center suggest that the Amazon rainforest in its present form could survive only three years of severe drought.[35] Prolonged drought could push the rainforest toward a "tipping point" where it could it start to die and turn into savannah, with catastrophic consequences for the world's climate.

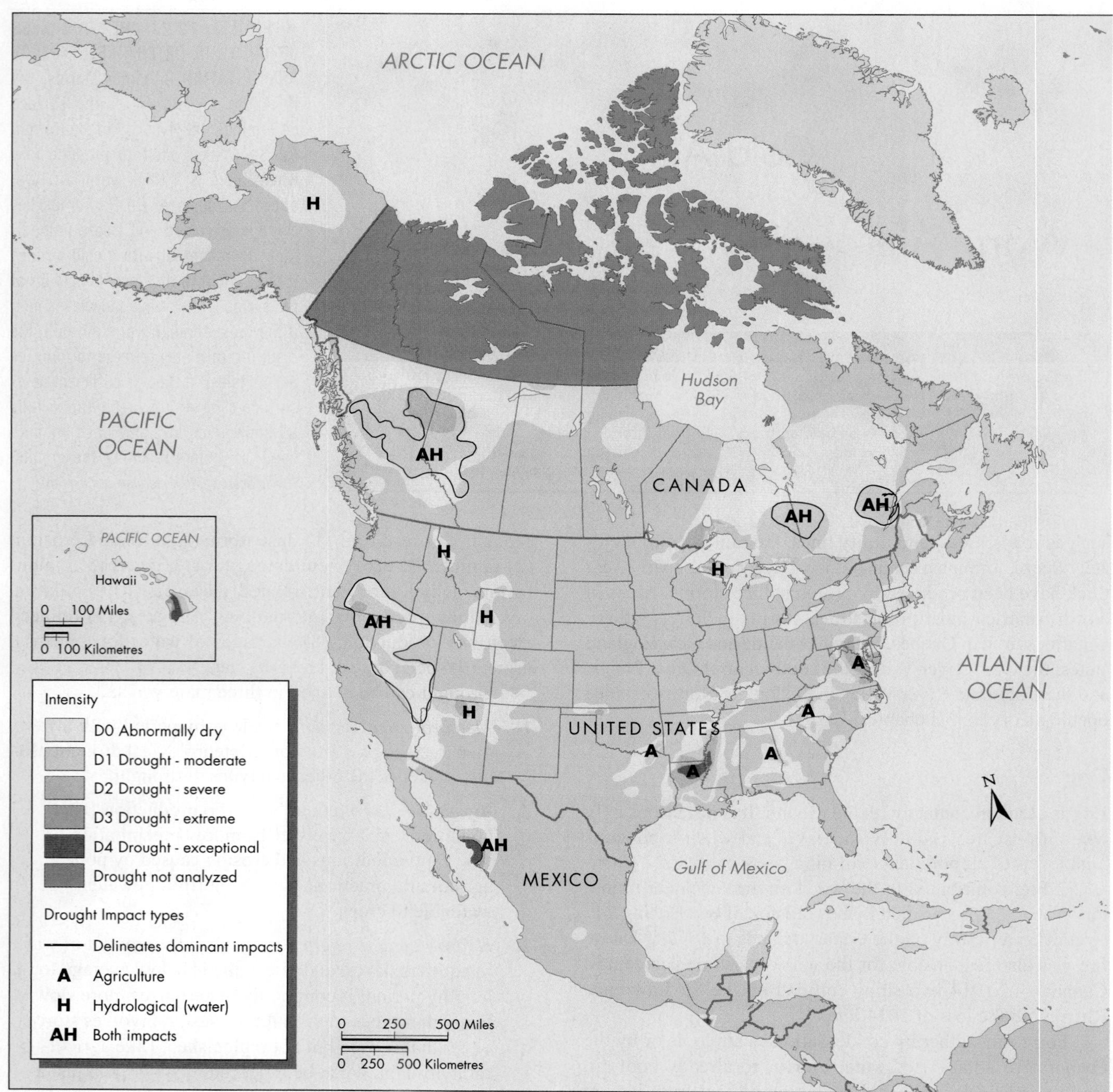

▲ **FIGURE 10.28 NORTH AMERICA DROUGHT CONDITIONS AND EFFECTS** A map showing the severity and effects of drought in the United States, Canada, and Mexico in October 2004. *(Courtesy NOAA National Climate Data Center)*

Droughts affect more people in Canada and the United States than any other natural hazard. Southern Alberta and southwestern Saskatchewan are semiarid areas that are prone to drought. They are also some of the most important grain-producing areas in Canada. Droughts on the Canadian Prairies from 1999 to 2002 caused losses in grain production of \$3.6 billion. Losses in the United States from drought typically total U.S.\$6 billion to U.S.\$8 billion each year.[36] Droughts can cause serious water and power shortages, as well as agricultural problems (Figure 10.28).

Droughts are linked to global and regional weather patterns and ocean circulation. There might be a relation between drought in some areas of North America and ocean circulation in the equatorial Pacific (see Chapter 14). La Niña events, during which cool waters well up off the west coast of South America and flow westward along the equator toward Australia and New Guinea, can lead to dry warm weather throughout much of the United States.

Whatever the cause, droughts are a normal part of the climate system. The droughts of the Dust Bowl period in the

1930s and droughts during the past several decades should not be thought of as unusual. In fact, historic droughts are not the most severe that are possible. Geologic studies of past vegetation and climate suggest that severe droughts lasting several decades have occurred at times during the past millennium in North America.

In the future, droughts will increasingly threaten our rapidly growing cities. As of 2011, about 20 cities had populations in excess of 10 million people; at least six of them have more than 20 million inhabitants. The number and size of megacities will continue to increase, stressing the available supply of clean freshwater. Severe droughts will challenge our ability to meet the basic water and food needs of large cities.

Dust and Sand Storms

Dust storms are strong windstorms that transport large amounts of fine sediment. Wind velocities in these storms exceed 48 km/h and visibility is less than 800 m.[8] A typical dust storm is several hundred kilometres wide and may carry more than 100 million tonnes of dust. Most dust particles are mineral and rock grains less than about 60 μm (0.6 mm) in diameter, but natural dust also contains small amounts of fine biological material, notably spores and pollen. Dust storms remove valuable topsoil and disrupt transportation and commerce. Airborne dust particles can also affect human health. Fine dust particles can be carried long distances in the upper atmosphere. Satellite images have shown dust storms from West Africa crossing the Atlantic Ocean to Florida, and dust from Asia occasionally reaches western North America.

Sand storms, like dust storms, are strong windstorms that move large amounts of sediment. They differ from dust storms in transporting sand rather than silt. During a sand storm, countless grains of sand are moved near the ground surface. Sand, unlike silt, is not transported long distances and is concentrated in a layer close to the ground.

Dust and sand storms occur mostly in mid-latitude semiarid and arid regions. Huge dust storms in the southern U.S. Plains during the 1930s produced conditions known as the "Dust Bowl." A combination of drought and poor agricultural practices during the Great Depression caused severe soil erosion in parts of five states. Frequent, sometimes daily, dust storms in this area destroyed crops and pastureland. Drought extended into Alberta and Saskatchewan, causing severe hardship for farmers and ranchers.

Heat Waves

Heat waves—periods of heat that are longer and hotter than normal—are one of the deadliest of natural hazards, yet they get little publicity and are not long remembered. In recent years, heat waves have killed an average of 200 people per year in Canada and the United States, about the same as the combined number of deaths from flooding, lightning, tornadoes, and hurricanes.

The threshold ambient temperature at which people are at risk from heat differs with location, but when summer temperatures reach about 5°C above the norm for a lengthy period, incidences of heat-related illness increase dramatically. High humidity compounds the effects of the heat by reducing evaporation, rendering perspiration less effective as a cooling mechanism. Under normal circumstances, people maintain a body temperature of about 37°C. When subject to extreme heat, the body tries to maintain this temperature by changing blood circulation and perspiring. At a temperature of 40°C, vital organs are at risk and, if the body's temperature is not reduced, death follows.

Heat waves take their greatest toll in cities. Urban centres, where the area of heat-absorbing dark roofs and pavement exceeds the vegetated area, are *heat islands* and can be as much as 5°C warmer than the surrounding countryside.[37] People in rural areas generally get some relief from the heat when temperatures fall at night, but cities remain warm because of radiation from warm buildings and roads. Air pollution is generally worse in cities than in rural areas, exacerbating the effects of high temperatures by further stressing the body's respiratory and circulatory systems.

A record heat wave claimed about 35 000 lives in Europe in the summer of 2003. Switzerland experienced its hottest June in 250 years.[38] Extreme daily high temperatures of 35° to 40°C continued through July in central and southern Europe (Figure 10.29). That August was the warmest on record in the Northern Hemisphere. In France alone, nearly 15 000 people died from temperatures that reached 40°C.[38] Germany lost some 7000 people, and Spain and Italy each suffered about 4200 heat-related deaths. The large number of deaths was due to the extreme temperatures that developed in poorly ventilated apartments and houses without central air-conditioning or even window units. Most of the deaths from the heat wave were in cities, which were warmer than surrounding areas and suffered from stagnant polluted air. The heat wave led to over 25 000 fires, which burned a total area of about 647 000 ha of forest.[38] Drought also had a major impact on power production; France, Europe's major energy producer, had to cut its power exports by 50 percent because of a lack of cooling water for its power plants.

Heat-related fatalities are common in parts of India because of high temperatures that occur before the monsoon, the large population, and the lack of means of protecting people. In May 2003, peak temperatures between 45°C and 49°C claimed more than 1600 lives throughout the country. In the state of Andhra Pradesh alone, 1200 people died from the heat. A year earlier, a one-week heat wave with temperatures topping 50°C claimed 1000 lives.

Several of the worst heat waves of the twentieth century occurred in Canadian and American cities. A heat wave in July 1936 set all-time records in Manitoba and Ontario: Temperatures reached 40°C in Toronto and 42°C in Winnipeg. Nearly 800 Canadians died during this heat wave. In 1972 New York City experienced a two-week heat wave that claimed 891 lives, and in 1995 an extreme heat wave in Chicago killed 739 people.

Most heat waves in Canada and the United States are associated with elongated areas of high pressure called *ridges*.

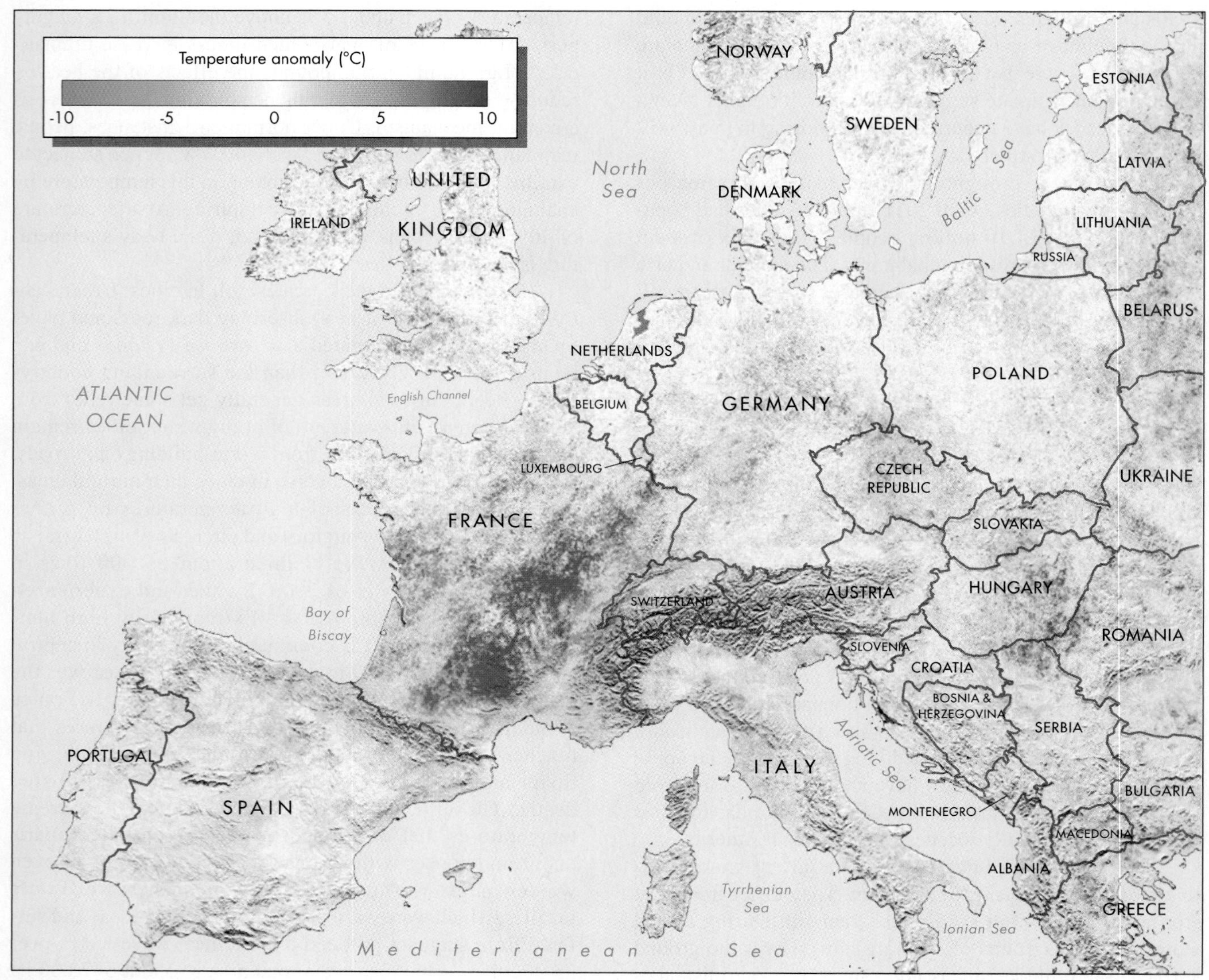

▲ **FIGURE 10.29 THE 2003 HEAT WAVE IN EUROPE** Difference between daytime surface temperature in Europe in July 2003 and the cooler, more normal temperatures of July 2001. Temperatures were measured by the Moderate Resolution Imaging Spectroradiometer (MODIS) sensor on NASA's Terra satellite. The dark red areas were at least 10°C warmer in 2003 than in 2001. *(After image by Reto Stockli and Robert Simmon, NASA Earth Observatory Team; courtesy of the National Aeronautics and Space Administration)*

Conditions west of the ridge are generally wet, whereas sunny and dry weather prevails to the east. If the ridge remains stationary for several days, air temperatures below the ridge can rise, triggering a heat wave.

The air can be either humid or extremely dry during heat waves. In either case, it is important to monitor the **humidex index**, or **heat index** (Figure 10.30). This index measures the body's perception of air temperature, which is greatly influenced by humidity. For example, a temperature of 35°C will feel significantly hotter in Toronto, where the relative humidity might be greater than 70 percent, than in Las Vegas, where the humidity is much lower. In the Toronto example, the combination of high temperature and high humidity produces a humidex temperature of 51°C, which is dangerous.

Heat waves cause other problems. In summer 2006, a heat wave strained electrical grids in Ontario to their limits. Electricity suppliers were barely able to meet the increased demand from people cooling homes and workplaces with fans and air conditioners. The province of Ontario was not able to generate enough power to meet its demand during the heat wave and had to ask people to cut back. It also had to import electricity from Quebec, New York, and Michigan. However, power imports cannot not be assured during such a heat wave. The electricity that Quebec sells to Ontario is hydroelectric power produced from dammed lakes and rivers. Drought preceding or accompanying heat waves lowers reservoirs in Quebec and reduces power production and the amount of energy it can export. Ontario's only other option would have been to cut power to certain areas. But power cuts would have had a considerable impact on the province's economy, forcing work slowdowns at Ontario's big steel plants and automakers.

RH (%) T (°C)	100	95	90	85	80	75	70	65	60	55	50	45	40	35	30	25	20
21	29	29	28	27	27	26	26	24	24	23	23	22					
22	31	29	29	58	28	27	26	26	24	24	23	23					
23	33	32	32	31	30	29	28	27	27	26	25	24	23				
24	35	34	33	33	32	31	30	29	28	28	27	26	26	25			
25	37	36	35	34	33	33	32	31	30	29	28	27	27	26			
26	39	38	37	36	35	34	33	32	31	31	29	28	28	27			
27	41	40	39	38	37	36	35	34	33	32	31	30	29	28	28		
28	43	42	41	41	39	38	37	36	35	34	33	32	31	29	28		
29	46	45	44	43	42	41	39	38	37	36	34	33	32	31	30		
30	48	47	46	44	43	42	41	40	38	37	36	35	34	33	31	31	
31	50	49	48	46	45	44	43	41	40	39	38	36	35	34	33	31	
32	52	51	50	49	47	46	45	43	42	41	39	38	37	36	34	33	
33	55	54	52	51	50	48	47	46	44	43	42	40	38	37	36	34	
34	58	57	55	53	52	51	49	48	47	45	43	42	41	39	37	36	
35		58	57	56	54	52	51	49	48	47	45	43	42	41	38	37	
36			58	57	56	54	53	51	50	48	47	45	43	42	40	38	
37					58	57	55	53	51	50	49	47	45	43	42	40	
38							57	56	54	52	51	49	47	46	43	42	40
									56	54	53	51	49	47	45	43	41
										57	54	52	51	49	47	44	43
											56	54	52	50	48	46	44
												56	54	52	50	48	46
													56	54	51	49	47

RH = relative humidity

Humidex (°C)	Degree of comfort
20–29	No discomfort
30–39	Some discomfort
40–45	Great discomfort; avoid exertion
46 and over	Dangerous; probable heat stroke

FIGURE 10.30 HUMIDEX CHART The humidex index combines air temperature and relative humidity to provide a numerical measure of the body's perception of air temperature. The National Weather Service in the United States uses a similar heat index chart. *(Data from http://accuracyproject.org/heatindexchart.html, used with permission of the Internet Accuracy Project, www.accuracyproject.org)*

10.6 Human Interaction with Weather

Human activities can have an impact on weather events. For example, the practice of plowing cropland after the fall harvest and leaving the topsoil exposed to wind significantly increased the size of the dust storms in the Dust Bowl during the 1930s. Also, locating mobile homes in areas subject to frequent high winds and tornadoes greatly increases damage and loss of life from this type of severe weather. Finally, land-use practices in cities can intensify the effects of heat waves. Large areas of pavement and sparse parkland contribute to higher temperatures than in the surrounding countryside.

Scientists have expressed concern that global warming is increasing the size and frequency of severe weather events. Global warming is likely to increase the number of heat waves in continental interiors, as well as the intensity of precipitation in many areas.[39] It also is likely to increase the frequency of drought in mid-latitude continental interiors and could increase the strength of hurricanes, typhoons, and other tropical cyclones.[39,40]

10.7 Links with Other Hazards

Severe weather is directly linked to flooding, landslides, and wildfires. Links related to tropical and extratropical cyclones are discussed in Chapter 11.

Intense thunderstorms can produce flooding. Light prevailing winds can cause a mesoscale convective system to remain fixed over one area, or they can result in a stationary front. Because thunderstorms commonly move parallel to a stationary front, stagnation of the front can cause storms to track over the same area, dropping large amounts of rainfall. Stagnation of thunderstorms can also occur along mountain fronts, resulting in deadly flash floods.

Thunderstorms also produce lightning, the primary, natural ignition mechanism for wildfires (see Chapter 13). However, not all lightning strikes trigger wildfires—by one estimate, only one in four cloud-to-ground lightning strikes has the continuous current necessary to start a fire, and then the fire will spread only if the fuel is dry.[41]

Global warming could make drought, dust and sand storms, and heat waves more common in some regions. High temperatures in the Arctic have already contributed to thawing of permafrost, which in turn has caused ground subsidence and the release of greenhouse gases (Chapter 8).

10.8 Natural Service Functions of Severe Weather

It might be difficult to envision any benefits from severe weather, possibly because most of the natural service functions are long-term and not obvious. For example, lightning is the primary ignition source of natural wildfires, and wildfires are a vital process in prairie and forest ecosystems (see Chapter 13). Windstorms also help maintain the health of forests. They topple dead and diseased trees, which then are recycled into the soil. Fallen trees also create clearings that become new habitat for some plants and animals. Even ice storms, which appear to be so destructive to trees, are part of a natural ecological cycle that increases plant diversity in forests.

Blizzards and other snowstorms and thunderstorms are important sources of water. Water derived from snowmelt and seasonal storms reduces a region's vulnerability to drought.

Finally, there are some uniquely human benefits to severe weather. Snowfall, storm clouds, and lightning have aesthetic value. Severe weather also excites many people. Thanks to movies and television, tornado chasing has become a new form of tourism in the United States. Tourists are taken in specially equipped vehicles into Tornado Alley to chase and photograph tornadoes. Tornado chasing, however, can be dangerous—serious injury or death can occur if a vehicle is struck by a tornado.

10.9 Minimizing Severe Weather Hazards

Thunderstorms, tornadoes, droughts, and other severe weather events cannot be prevented and will continue to claim lives and destroy property. However, we can take steps to minimize loss of life and damage from severe weather events if we can better predict and prepare for them.

Forecasting and Predicting Weather Hazards

Timely and accurate prediction of severe weather events is essential if human lives are to be saved. Even with improvements in satellite sensors and computer modelling, severe weather events are still difficult to forecast, and their behaviour is unpredictable.

Installation of a network of *Doppler radar* stations across North America has significantly improved our ability to predict the paths of tornadoes and other severe storms. Doppler radar antennas send out electromagnetic waves with a wavelength a little longer than microwaves. Clouds, raindrops, ice particles, and other objects in the sky reflect the electromagnetic waves. The wavelength of the reflected waves differs depending on whether the objects are moving toward or away from the antenna. The change in wavelength, called the *Doppler effect*, is similar to the difference in pitch of a siren as an ambulance approaches you and then goes past. The changes in radar wavelength are analyzed and can be used to make short-term predictions about weather, on the scale of hours. For example, Doppler radar can detect a mesocyclone within a thunderstorm, allowing meteorologists to issue some warnings up to 30 minutes in advance of the touchdown of a tornado.

Tornado Watches and Warnings A tornado **watch** is a public advisory that one or more tornadoes might develop in a specified area in the near future. A typical tornado watch covers an area of 50 000 to 100 000 km^2 and lasts four to six hours. A watch does not guarantee that a tornado will happen; rather, it alerts the public to the possibility of a tornado and suggests they monitor local radio or television stations for more information.

When a tornado has been sighted or detected by Doppler radar, the watch is upgraded to a **warning**. A warning indicates that the affected area is in danger and that people should take immediate action to protect themselves and others. Watches can be upgraded to warnings, or warnings can be issued for an area not previously under a watch. Both watches and warnings can be issued not only for tornadoes but also for any type of severe weather: thunderstorms, tropical storms, hurricanes, heat waves, and blizzards, with some differences in the area covered and the duration of the watch or warning.

People's perception of the risk they face from severe weather hazards differs according to their experience. Someone who has survived a tornado will more likely perceive the hazard as real than will a person who has lived in a tornado-prone region but has never experienced one. Incorrect predictions of where or when a severe weather event will strike can also affect people's perception of risk. For example, if people are repeatedly warned of severe thunderstorms that never materialize, they might become complacent and ignore future warnings. An accurate understanding and perception of risk by planners and the public alike is key to reducing injury and death from severe weather events.

Our current theoretical understanding of severe weather and our monitoring technology do not always allow accurate prediction of the intensity and extent of many events. For example, even though fewer than 1 percent of tornadoes—EF4 and EF5 storms—are responsible for half of all tornado-caused deaths, meteorologists are not yet able to include predictions of intensity in official tornado watches and warnings.[42] Likewise, it is difficult to forecast the amount of icing and depths of snow accumulation in winter storms. However, researchers are developing computer-based models for making real-time predictions of the intensity of some severe weather events.[43] These models use information from weather radar, satellites, and automated weather stations to predict the path and development of severe storms after they form.

Adjustment to Severe Weather Hazards

Actions taken to prevent or minimize death and damage are referred to as *mitigation*. They include the engineering and building of structures with safety in mind and the installation of warning systems. *Preparedness* involves the preparation of community and individual plans and procedures to deal with a possible natural disaster.[44]

Mitigation Mitigation methods differ for each weather hazard, but some general statements can be made. Building new wind-resistant structures and retrofitting existing ones can save lives and lessen property damage during severe storms such as tornadoes and thunderstorms.

Ensuring that electric, gas, water, and wastewater systems can function following severe storms is an important part of weather-hazard mitigation. This need was painfully apparent after the 1998 ice storm in Canada and the United States, when millions of people were left in the cold and dark for periods ranging from days to weeks.[45] Without backup power sources that can function for days until the power grid is restored, water and wastewater systems are vulnerable to power interruptions caused by lightning strikes, wind, and ice storms. Redundancy inherent in modern communication systems—a combination of landline telephones, cell phones, voice-over-internet, multi-frequency radio, and satellite communication links—ensures that they are able to survive severe weather events.

Another mitigation method involves development and installation of warning systems. The purpose of warning systems is to give the public the earliest possible notification of impending severe weather. Announcements can be made by radio, television, and the internet. Sirens can be installed in communities and used to issue tornado warnings.

Mitigation strategies for droughts and heat waves are more complicated and much more difficult to implement. Droughts lower reservoir levels and the capacity to generate hydroelectric power; they also reduce stream flow. Mitigation and adaptation measures that can be taken to deal with drought include increasing the capacity of reservoirs, exploiting groundwater resources, desalinization, and water conservation. With the exception of conservation, these measures are expensive and they take time to implement. Governments can find it difficult to convince the public to spend funds in anticipation of droughts that might occur at some unknown time in the future.

Preparedness Individuals can take steps to prepare for severe weather. Some of the steps should be taken before a watch or warning is issued, whereas others are more appropriate when the danger is imminent. Information about how to prepare for weather-related disasters is available from Environment Canada and its subsidiary, the Meteorological Service of Canada; from the U.S. National Oceanic and Atmospheric Administration (NOAA) and its subsidiary, the National Weather Service; and from the U.S. Federal Emergency Management Agency (FEMA).

Wearing proper clothing and modifying travel plans are prudent adjustments to many types of severe weather. The need for insulated clothing in blizzards is intuitive for most adults. However, many people are not prepared for cool (5° to 15°C), wet, windy weather, in which the body can rapidly lose heat. Rapid loss of core body heat causes loss of muscle coordination, mental sluggishness, and confusion, a medical condition known as *hypothermia*. In such weather, an improperly dressed person can become hypothermic within several hours. People can become hypothermic in 25°C water, or, if improperly dressed and wet, in 20°C air temperature.[46,47] Unless body heat is restored, a hypothermic person will become unconscious and eventually die.

Low-visibility conditions can develop rapidly in dust storms, blizzards, and fog and are especially hazardous for travel on freeways and other roads where vehicles normally move at high speed. When travelling by automobile, "Onward through the Fog" is not the best course of action.

REVISITING THE FUNDAMENTAL CONCEPTS

Atmosphere and Severe Weather

1 Hazards can be understood through scientific investigation and analysis.

2 An understanding of hazardous processes is vital to evaluating risk.

3 Hazards are commonly linked to each other and to the environment in which they occur.

4 Population growth and socio-economic changes increase the risk from natural hazards.

5 Damage and loss of life from natural disasters can be reduced.

1. Atmospheric conditions can be forecast, and short-term weather predictions made, with the use of radar, satellite sensors, and computer models. However, because severe weather, such as tornadoes and blizzards, involves many variables, the accuracy of predictions can be limited. For example, short-term tornado watches can be issued over large areas several hours in advance of the detection of large-scale mesocycles, but the storm might never produce a tornado or, if it does, the exact location cannot be predicted. When radar does detect a tornado, a tornado watch can be issued, but there may be little time to seek shelter.
2. Risk assessment for severe weather is based on the probability that an event will occur and the possible

consequences of the event. Historic records of tornadoes, blizzards, and ice storms provide information on probabilities of damaging events in a given area, and property values and population numbers and distribution provide a measure of possible consequences. Many homeowners in the U.S. Midwest, for example, have tornado insurance; the cost of the insurance policy is based on the location of the house with respect to past tornadoes and the value of the house and property should they be damaged or destroyed.

3. Links exist between severe weather and several other hazards. Heavy rainfall from severe weather can cause flooding in low-lying areas and flash floods in mountainous areas (see Chapter 9). Lightning can ignite fires that can become wildfires if conditions are right (see Chapter 13). Over a longer time, changes in global climate (see Chapter 14) will result in a reorganization of Earth's weather systems, which will change the incidence of drought, flooding, heat waves, and blizzards in some areas.
4. Severe weather events that previously caused little damage are more commonly producing disasters. More and more people are living in regions where severe weather events are common, resulting in an increase in loss of life and property when such events occur. Additionally, population growth has forced large numbers of people to migrate to areas where water and food supplies are limited or temperatures more extreme; thus the effects of events such as drought and heat waves are greater.
5. The consequences of severe weather can be minimized through improvements in infrastructure, warning systems, and education. Buildings can be designed to withstand the effects of high winds and constructed with better insulation to protect against extreme heat and cold. Basements and safe zones protect life even if a structure is destroyed by a tornado. Advances in weather forecasting provide greater lead time for warnings, which, if combined with a well informed and prepared population, will greatly reduce the loss of life in severe weather.

Summary

Energy emitted by the sun sustains life and drives many processes at or near Earth's surface, such as atmospheric and oceanic circulation.

Potential, kinetic, and heat energy are the three primary forms of energy. In the atmosphere, heat energy occurs as sensible heat that can be measured and latent heat that is stored. Evaporation absorbs heat and cools the air, whereas condensation, which forms clouds, releases latent heat and warms the air.

Energy is transferred in the atmosphere by convection, conduction, and radiation. Of these three mechanisms, convection is the most dynamic and significant in producing severe weather.

Earth receives primarily short-wavelength electromagnetic energy from the sun. This energy is reflected, scattered, transmitted, or absorbed on Earth. Dark-coloured surfaces reflect less and absorb more energy than light-coloured surfaces. Most absorbed solar energy is radiated back into the atmosphere as long-wavelength infrared radiation.

Most weather occurs in the troposphere, the lowest of the four major layers of the atmosphere. Clouds in the troposphere consist of water droplets and ice crystals. Changes in atmospheric pressure and temperature are responsible for air movement. Air flows from areas of high pressure to areas of low pressure.

Winds blowing over long distances curve because Earth rotates beneath the atmosphere, resulting in what is termed the Coriolis effect. Wind patterns are controlled by the Coriolis effect, horizontal differences in atmospheric pressure, and, in the case of surface winds, friction.

Boundaries between cooler and warmer air masses, called fronts, can be cold, warm, stationary, or occluded. Many thunderstorms, tornadoes, snowstorms, ice storms, and dust storms are associated with fronts. Thunderstorms form when there is moist air in the lower troposphere, rapid cooling of rising air, and updrafts to create cumulonimbus clouds. Most thunderstorms form during maximum daytime heating, either as individual storms or as lines of storms associated with fronts. Hail forms by ice added in layers to ice pellets that rise and fall within a thunderstorm.

Severe thunderstorms have winds over 93 km/h, commonly produce destructive hail, and can spawn tornadoes. Downdrafts in severe thunderstorms create potentially damaging gust fronts and microbursts. Three major types of severe thunderstorms are mesoscale convective complexes, linear squall lines, and supercells. Mesoscale convective complexes can produce derechos, which are straight-line windstorms that can be as damaging as tornadoes. Squall lines develop ahead of cold fronts and along drylines.

Tornadoes are funnel-like columns of rotating air with winds ranging from 65 km/h to over 400 km/h that extend downward from a cloud to the ground. They form by wind shear in severe thunderstorms, especially supercells. Tornado damage is rated on the Enhanced Fujita (EF) Scale, with EF5 being the most severe. Waterspouts form under fair-weather conditions along ocean coastlines and over lakes, and are generally weaker than tornadoes.

Lightning develops when there are large differences in electrical charge within a thunderstorm cloud or between the cloud and the ground. The flow of electrical charges within the cloud, or between the ground and the cloud, produces lightning.

Blizzards are severe winter storms in which large amounts of falling or blowing snow reduce visibility for extended periods. Most blizzards are produced by extratropical cyclones that have travelled eastward across the Rocky Mountains or have moved along the Atlantic coast as nor'easters.

Ice storms result from prolonged freezing rain along a stationary or warm front. Freezing rain requires sub-zero surface temperatures and a shallow layer of cold air at the surface.

A dust storm can greatly reduce visibility, causing deadly accidents. Other hazardous weather conditions include sand storms, fog, drought, and heat waves. Drought and heat waves are commonly associated with high-pressure centres that stagnate over regions for extended periods. Heat waves are intensified in cities because of the urban heat island effect. Safety in hot weather requires knowledge of the humidex index or heat index, which is a combination of air temperature and relative humidity.

People interact with hazardous weather in numerous ways. At the local level, land-use practices such as housing and agricultural processes may increase the impact of severe weather. On the global scale, atmospheric warming caused by burning fossil fuels and clearing land may be changing climate and weather. Warming of both the atmosphere and the oceans can feed more energy into storms, potentially increasing the incidence of severe weather events.

Minimizing injury and damage from thunderstorms, tornadoes, heat waves, and blizzards requires: (1) better forecasting and warning of severe weather events; (2) construction practices that prevent or minimize death and loss of property; (3) hazard preparedness, including actions that individuals and communities can take once they have been warned of severe weather; (4) education; and (5) insurance programs to reduce risk.

Key Terms

atmosphere (p. 278)
blizzard (p. 294)
conduction (p. 276)
convection (p. 276)
Coriolis effect (p. 284)
drought (p. 297)
dust storm (p. 299)
EF Scale (p. 292)
fog (p. 297)
heat index (p. 300)
heat wave (p. 299)
humidex index (p. 300)
ice storm (p. 296)
lightning (p. 287)
radiation (p. 276)
relative humidity (p. 279)
tornado (p. 289)
troposphere (p. 280)
warning (p. 302)
watch (p. 302)
wind chill (p. 296)

Did You Learn?

1. Name the three types of energy and explain how they differ.
2. Name the three types of heat transfer. How do they differ from one another?
3. Describe how Earth's energy balance works.
4. List the following types of electromagnetic energy in order, from shortest wavelength to longest wavelength: radio waves, ultraviolet radiation, gamma radiation, visible light, infrared radiation, X-rays, and microwaves.
5. Describe the characteristics of the troposphere. How do meteorologists identify the top of the troposphere?
6. Explain why atmospheric pressure decreases with increasing altitude.
7. What is the difference between stable and unstable air?
8. Explain the Coriolis effect. How does it influence weather?
9. Name the different types of severe weather.
10. Describe the three stages of thunderstorm development.
11. Define supercells, mesoscale convective complexes, and squall lines. How do these severe thunderstorms differ?
12. What is hail and how does it form?
13. Characterize a tornado in terms of wind speed, size, typical speed of movement, duration, and length of travel.
14. Describe the stages of tornado development.
15. What is a blizzard and how does it develop?
16. What is a nor'easter? How is it related to blizzards?
17. Describe the weather conditions that cause an ice storm.
18. How are the humidex index and wind chill temperature alike? How are they different?
19. Explain the difference between a severe weather watch and a severe weather warning.
20. Explain how drought, soil moisture, and a heat wave can be related.
21. Explain how a change in Earth's climate can contribute to severe weather hazards.
22. What measures can be taken to minimize the adverse impacts of heat waves?
23. Explain how Doppler radar is used to lessen the potential injury and loss of life from a tornado.
24. What are some of the natural service functions of severe weather?

Critical Thinking Questions

1. Assume that you live in a rural area in Tornado Alley in the United States. What factors would influence whether you buy tornado insurance?
2. What measures should be taken to prevent a repeat of the 1998 ice storm disaster in Ontario and Quebec? Consider measures that could be taken by individuals, municipal and provincial governments, the federal government, and companies.
3. Parts of southern Alberta and Saskatchewan are drought-prone. These areas produce much of Canada's grains. What effect do you think a severe drought in this region would have on the Canadian economy? What measures could be taken to reduce the economic damage of such an event?
4. What severe weather events are potential hazards in the area where you live? What are some steps you might take to protect yourself from these hazards? Which of these hazards is your community least prepared for?
5. Tornadoes can generally be spotted on weather radar, whereas many clouds cannot. What makes tornadoes visible?
6. Has your community ever experienced a heat wave? If so, when did it occur and how were people affected? Does your community have a heat health warning system? If so, what actions do local officials and emergency personnel take when the system is activated? If not, what type of system would you recommend?

MasteringGeology

MasteringGeology **www.masteringgeology.com**. Looking for additional review and test prep materials? Visit the Study Area in MasteringGeology to enhance your understanding of this chapter's content by accessing a variety of resources, including **Self-Study Quizzes, Geoscience Animations, GEODe Tutorials, RSS feeds, flashcards,** web links and an optional **Pearson eText.**

CHAPTER 11

▶ **HURRICANE KATRINA** Satellite image of Hurricane Katrina heading north toward landfall on the Mississippi Delta near Buras, Louisiana. Spiralling rain bands of the storm fill the entire eastern Gulf of Mexico from southwestern Louisiana on the left to the west coast of Florida on the right. A large, well-developed eye marks the centre of the hurricane, which had a Category 5 intensity at the time this infrared image was taken by NOAA's GOES-12 satellite on Sunday, August 25, 2005. *(NASA/Goddard Space Flight Center Scientific Visualization Studio)*

Hurricanes and Extratropical Cyclones

Cyclones, including hurricanes and their mid-latitude relatives—extratropical cyclones—are the most powerful storms on Earth and among the most deadly and costly of natural hazards. Hurricane Katrina alone was one of the most expensive natural disasters in history. Understanding how these storms work helps us appreciate the threat they pose to our highly technological society. As our climate changes and sea level rises, the threat these storms pose will increase. Yet, coastal populations in areas vulnerable to hurricanes continue to grow. As Hurricane Katrina and Hurricane Sandy demonstrated, choices made by individuals and governments can turn an "act of God" into an "act of man." Your goals in reading this chapter should be to

- Understand the weather conditions that create and sustain cyclones
- Know what geographic regions are at risk from hurricanes and extratropical cyclones
- Recognize links between cyclones and other natural hazards
- Know the benefits that cyclones provide
- Understand adjustments that can be taken to minimize damage and injury from cyclones
- Know the prudent actions to take when a hurricane watch or warning is issued

Hurricane Katrina

Unlike the huge lateral blast from Mount St. Helens volcano in 1980, which surprised many scientists and public officials, scientists had predicted the flooding of New Orleans by a major hurricane for decades. As David Brooks of the *New York Times* later wrote, "Katrina was the most anticipated natural disaster in American history."[1]

Why was this disaster so anticipated? First, hurricanes are common along the coastlines of the southeastern United States. Second, 95 percent of New Orleans is below sea level, and the city relied on a patchwork of levees, floodwalls, and pumps for protection from flooding.[2] Parts of the flood protection and drainage system are over 100 years old. Third, in the past century, coastal Louisiana has lost wetlands and barrier islands that lessen the magnitude of hurricane storm surges (see Chapter 12). Finally, many coastal areas have experienced large increases in population, and one-third to one-half of the residents living in these areas today have never experienced a major hurricane.

Hurricane Katrina was a Category 3 hurricane when it made landfall on the Mississippi Delta about 80 km southeast of New Orleans.[3] The storm's intensity dropped as it moved inland, and most of New Orleans experienced a Category 1 or weak Category 2 hurricane.[3] As bad as Katrina was, its drop in intensity was fortunate for New Orleans—less than 24 hours before landfall, it had been a Category 5 storm, one of the most powerful ever observed in the Gulf of Mexico.[3]

The U.S. National Weather Service (NWS) did a good job of forecasting the hurricane's landfall and alerting government officials and the public. Forecasters began monitoring the hurricane from its inception as a tropical depression over the Bahamas almost a week before it came ashore in Louisiana. Air Force and NOAA (U.S. National Oceanic and Atmospheric Administration) reconnaissance aircraft, satellites, and land-based radars tracked the progress of the storm as it intensified, crossed Florida, and then doubled in size over the Gulf of Mexico. A day and a half before Katrina made landfall, Dr. Max Mayfield, then director of the NWS National Hurricane Center, called the governors of Louisiana and Mississippi and the mayor of New Orleans, urging evacuations.[2] The National Weather Service accurately forecast storm surge

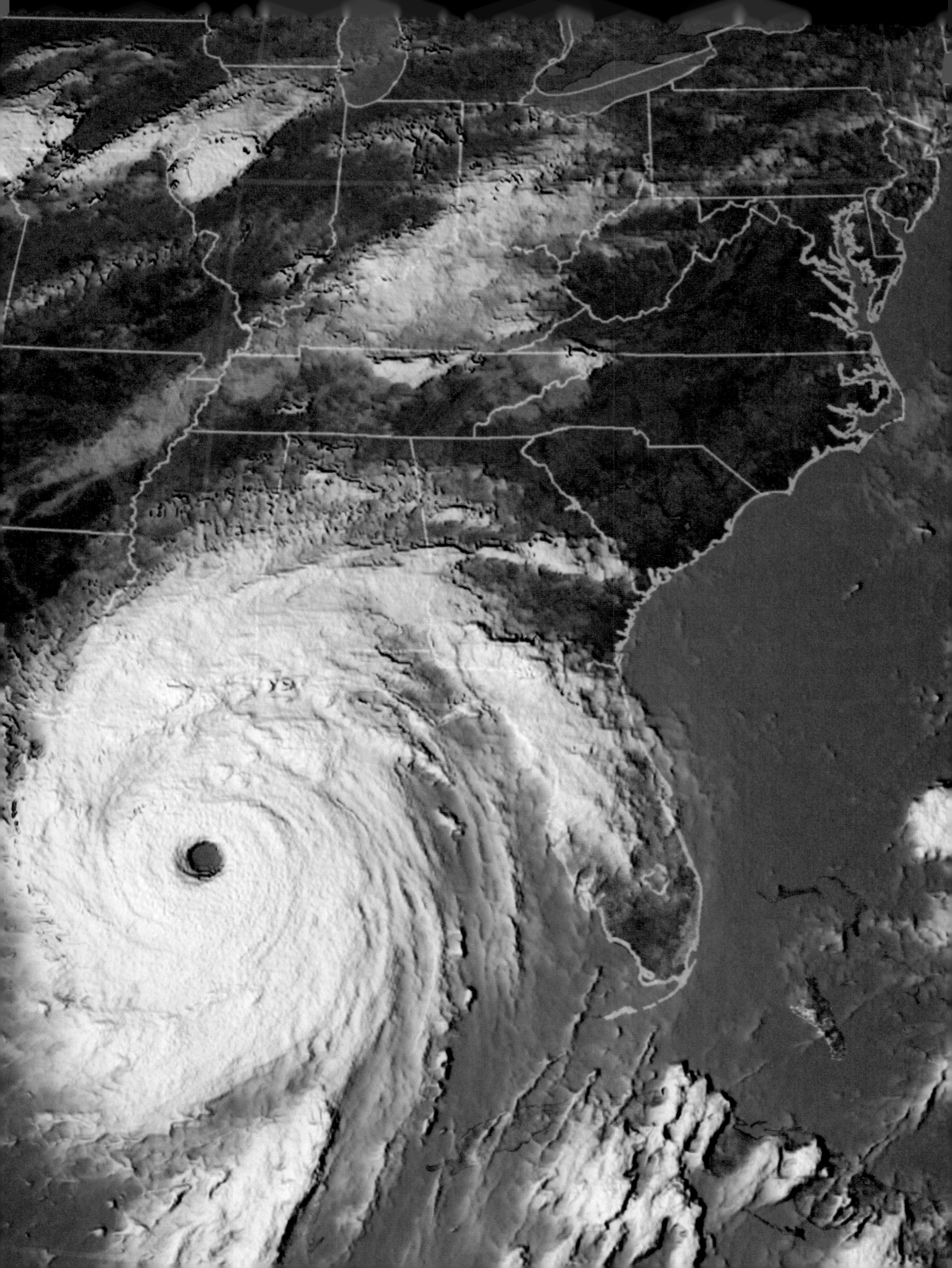

◄ **FIGURE 11.1 NAVIGATION CANAL LEVEE BREACHED BY STORM SURGE** Hurricane Katrina's storm surge breached the Inner Harbor Navigation Canal levee, inundating New Orleans's lower Ninth Ward. Poor foundation design contributed to the failure. *(Vincent Laforet/AP Images)*

heights of up to 8.5 m, and Katrina's landfall was within 35 km of the forecast location.[3]

The failure of New Orleans's flood protection system was the most costly failure of an engineered system in history.[4] Flood walls and levees were breached in at least eight locations, inundating 85 percent of the greater New Orleans metropolitan area to depths of 0.3 to 6.1 m (Figure 11.1).[4] Along the east side of New Orleans, navigation channels funnelled and focused the storm surge, contributing to the failure of adjacent levees and flood walls (Figure 11.2).[2] Much of the city remained flooded for two to three weeks before levees were temporarily repaired and floodwaters pumped out. Hurricane Katrina and its storm surge were also responsible for the collapse of the I-10 causeway east of New Orleans; damage to the roof of the Superdome; loss of cell phone

▲ **FIGURE 11.2 NEW ORLEANS INUNDATED AFTER ITS FLOOD PROTECTION SYSTEM FAILED** Levees and floodwalls designed to protect New Orleans from a Category 3 hurricane were breached in at least eight places by Hurricane Katrina. This photo, taken one day after landfall, shows cars abandoned on freeway overpasses. *(David J. Phillip/AP Images)*

communications and power for residents, businesses, and hospitals; and the collapse of several offshore oil and gas production platforms.

The hurricane was also a catastrophe from a social standpoint. More than 1 million people were evacuated from the affected area. Evacuation was especially difficult in New Orleans, where close to 250 000 people had no private transport.[5] Those who had no means to evacuate, or chose not to, had few places to go. In the first 10 days of the disaster, Coast Guard boats, ships, and aircraft rescued over 23 000 people along the Gulf Coast, and other rescuers saved many tens of thousands (see Survivor Story).[6]

Was Hurricane Katrina an unusual storm for the Gulf Coast? It was not. Hurricanes of equal or greater strength had struck the Texas, Louisiana, and Mississippi coasts several times during the previous 50 years. In 1965, Hurricane Betsy, a Category 4 storm, came within 60 km of New Orleans, breached its levees with a 2.4- to 3.0-m storm surge, and flooded parts of the city. Four years later, Category 5 Hurricane Camille just missed New Orleans and devastated the Mississippi coast with sustained winds of 300 km/h. Camille's winds were over three times stronger than Katrina's.[7] Its storm surge was nearly as great, but the population of the coastal zone in 1969 was much smaller than it is today. In 2004, Hurricane Ivan, a Category 4 storm, headed straight for New Orleans, only to veer to the right and slam into the Alabama coast. As it moved inland, Hurricane Ivan spawned more than 100 tornadoes and produced heavy rainfall that caused river flooding as far north as New York State.[8]

The storm surge from Hurricane Katrina was devastating, with water levels normally associated with a Category 5 hurricane. Katrina inundated the Mississippi coastline in water 5.5 to 7.3 m deep, pushed salt water up to 10 km inland, and levelled entire neighbourhoods (Figure 11.3), washing houses onto levees and stranding large ships and barges on land.[3] The storm surge was responsible for most of the hurricane-related deaths in Mississippi.[3]

If Hurricane Katrina was not an extraordinary storm, and there was ample advance warning, why did over 1600 people die? The answer to this question is complex and multifaceted, involving failure in the design, construction, and maintenance of levees and flood walls; an over-reliance on technology to protect life and property; social and psychological denial of the hazard; poverty and limited education of many residents in the affected area; diversion of military and government resources to the conflict in Iraq; and failures in political leadership, communication, and public policy at all levels. The policy failures include a system of flood insurance and post-disaster aid that encouraged people to live in hazardous coastal areas and that rewarded developers, businesses, and individuals for rebuilding in previously flooded zones.[9]

▲ **FIGURE 11.3 STORM SURGE FLATTENS GULFPORT** A 7- to 8.5-m storm surge washed away most of the houses in this neighbourhood in Gulfport, Mississippi, during Hurricane Katrina. *(© Tyrone Turner/National Geographic)*

SURVIVOR STORY

Hurricane Katrina

Abdulrahman Zeitoun's Family Evacuation—He Stayed to "Mind the Damage"

Contractor Abdulrahman Zeitoun (Figure 11.4) has seen his share of tropical storms in the more than 30 years he has lived in New Orleans. So when forecasters began advising Gulf Coast residents to evacuate as Hurricane Katrina approached in late August 2005, Zeitoun chose not to leave. "I say look, I'm not going. Someone has to stay behind and mind the damage."

On Saturday, August 27, Zeitoun waved goodbye to his wife and four children as they drove to Baton Rouge to stay with relatives, and settled in to wait out the storm. The next day, the wind and rain grew stronger and stronger, the power went out, and the roof of his house began to leak. But by late Monday, the weather had calmed and the danger seemed over. "I see water on the street a foot and a half deep, a few trees down. I call my wife, I say everything's done, if you want to come back."

In fact, the worst was yet to come. Ocean waters surging inland were overtopping city levees, bursting through in several places to inundate entire neighbourhoods.

Zeitoun heard the floodwaters before he saw them. "There were noises like when you sit by a river. I stick my head out the window and see water in the backyard rising very fast." He moved family photographs, the first-aid kit, flashlights, and small items upstairs, and then stacked the furniture on the first floor of the house in tall piles. "I sacrificed one thing to save another one; anything I could do to minimize the damage."

The water rose for another 24 hours. Zeitoun called his wife and told her not to return, as the city would not be habitable for a long time. Then he climbed in his canoe to check on his tenants elsewhere in town.

What he saw as he paddled the silent streets was beyond his worst imaginings. Water on his block reached the bottom of a stop sign 2 m high. A neighbour asked for a lift to check on his truck. They found the vehicle in water up to its roof. An older couple waving white flags asked to be rescued. Unable to fit them in the canoe, Zeitoun promised he would find help. Farther along, they heard a voice from a one-storey house. Jumping in the water, Zeitoun managed to open the door. An elderly woman, skirts ballooning around her, stood in water up to her shoulders. To rescue her, Zeitoun and his neighbour flagged down one military boat after another, but to no avail. Finally, a few civilians with a hunting boat stopped. Zeitoun positioned a ladder beneath her legs, and together the men levered the woman into the boat. They went on to rescue an older couple and delivered all three seniors to a local hospital.

At his rental property, Zeitoun found his tenant healthy and, miraculously, in possession of a working phone. Neighbours began to converge to call friends and relatives.

Zeitoun helped several more people the next day, obtaining a boat to rescue an older man in a wheelchair. In the evening, Zeitoun, his tenant, a friend, and a man using the phone were gathered at the rental property when a military boat floated by. "They jump into the house with machine guns and say, 'What are you doing here?' I say I own the place, it's my house." But the soldiers arrested all four men and took them to a bus station being used as a temporary jail. They were thrown into a makeshift cage made of chicken wire, interrogated for three days, and moved to a nearby prison. After spending a month behind bars, Zeitoun was released on bail. Charges for suspected looting were later dropped.

Despite his ordeal, Zeitoun plans to stay in New Orleans. "My business is here. I have good relations here with my associates, friends, and customers. I feel like I have a family in this city," he says.

—Kathleen Wong

▲ **FIGURE 11.4 HURRICANE KATRINA SURVIVOR** New Orleans resident Abdulrahman Zeitoun evacuated his family but chose to stay behind in the face of danger from Hurricane Katrina. *(Kathy Zeitoun)*

11.1 Introduction to Cyclones

In meteorological terminology, Hurricane Katrina was a **cyclone**, a general term applied to large cells of moisture-laden air that rotate around an area of low pressure. Because of the Coriolis effect, the winds in a cyclone blow anticlockwise in the Northern Hemisphere and clockwise in the Southern Hemisphere (Chapter 10). Cyclones are classified as tropical or extratropical based on their place of origin and the temperature of their centre or core region.

Tropical cyclones form over warm tropical or subtropical oceans, typically between 5° and 20° latitude. They are not associated with fronts and have warm centres. In contrast, **extratropical cyclones** develop over land or water in temperate regions, typically between 30° and 70° latitude. These mid-latitude cyclones are generally associated with fronts and have cool centres. Both types of cyclones are characterized by their intensity, which is indicated by their sustained wind speeds and lowest atmospheric pressure.

Cyclones are associated with most severe weather in North America. Tropical cyclones create high winds, heavy rain, surges of seawater, and tornadoes. Extratropical cyclones can produce the same effects, as well as snowstorms and blizzards during winter. Not all extratropical cyclones produce severe weather, whereas hurricanes, by definition, are severe storms.

Although the destructive effects of tropical and extratropical cyclones are similar, the two types of storms differ in their source of energy and their structure. Tropical cyclones derive energy from warm ocean water and the latent heat that is released as rising air condenses to form clouds. Extratropical cyclones obtain their energy from the contrast between air masses on opposite sides of a front. In hurricanes, warm air rises to form a spiralling pattern of clouds. The rising and warming air surrounding the centre of a hurricane heats the entire core of the storm. In contrast, most extratropical cyclones are fed by cold air at the surface and a flow of cool, dry air aloft. The resulting storm has a cool core from bottom to top. Tropical cyclones that move over land or cooler water lose their source of heat and either dissipate or become extratropical cyclones moving along a front.

Classification

A variety of terms are used to describe cyclones in different parts of the world. For example, both forecasters and residents use the term *nor'easter* for an extratropical cyclone that tracks northward along the east coast of the United States and Canada. Onshore winds from these storms blow from the northeast and can sometimes reach hurricane strength.

In the Atlantic and eastern Pacific oceans, strong tropical storms are called **hurricanes** after a native Caribbean word for an evil god of winds and destruction.[10] In the western Pacific Ocean and north of the equator, hurricanes are referred to as **typhoons** after a Chinese word meaning "scary wind" or "wind from four directions."[10] Hurricanes in the Pacific Ocean south of the equator and in the Indian Ocean are referred to as cyclones, a term coined by a curator of the Calcutta Museum in India and based on the Greek "coil of a snake."[10] For simplicity, we will refer to all of these strong tropical cyclones as hurricanes.

Hurricanes are classified by wind speed on a damage-potential scale developed in the 1970s by Herbert Saffir, a consulting engineer, and Robert Simpson, a meteorologist (Table 11.1). The Saffir-Simpson Scale is divided into five categories based on the highest one-minute average wind speed in the storm; the scale was first used for public advisories in the United States in 1975.[7] The category of a hurricane changes as it intensifies or weakens, and all but the weakest hurricanes will have more than one category assigned to them over their duration. Meteorologists describe categories 3 through 5 hurricanes as major hurricanes. As a hurricane's wind speed and thus its category increases, the atmospheric pressure in the centre of the storm drops. Category 5 hurricanes generally have a central atmospheric pressure of less than 920 millibars. Hurricane Wilma set the record low pressure for an Atlantic hurricane—882 millibars—in 2005.[11]

Naming

A small percentage of cyclones are given names, either to identify where they form or to track their movement. Some extratropical cyclones, especially those that become snowstorms, are named for the geographic area where they form. In contrast, all hurricanes are given individual names by government forecasting centres. The names are established by international agreement through the World Meteorological Organization. A name is assigned once the maximum sustained winds of a tropical depression exceed 63 km/h. Initially, Pacific cyclones were given only female names, but since 1978, both male and female names have been applied to these storms. In 1979, this practice was extended to hurricanes in the Atlantic Ocean and the Gulf of Mexico. Names are assigned alphabetically each year from a previously agreed-upon list for the region in which the storm forms. For example, in the Atlantic Ocean, the first three names for 2011 were Arlene, Bret, and Cindy.

The names of hurricanes come from one of six standard lists of alternating male and female names. Each list has 21 names derived from English, Spanish, or French. The six lists are used in rotation; therefore the 2013 list will be used again in 2019 and 2025. The names of some particularly intense and destructive Atlantic hurricanes have been retired and replaced with new names. For example, the name "Mitch" was retired after a Category 5 hurricane of that name devastated several countries in Central America and killed more than 11 000 people.[7] Katrina, Rita, Ivan, and Charley have also been retired from the rotating lists.

TABLE 11.1 The Saffir-Simpson Hurricane Scale

The Saffir-Simpson Hurricane Scale is a five-point scale based on hurricane wind speed and the size of the associated storm surge. It is used to give an estimate of the potential property damage and flooding along the coast when a hurricane makes landfall.

Category 1 Hurricane

Winds 119 km/h to 153 km/h. Storm surge generally 1.2 m to 1.5 m above normal. No major damage to structures. Damage primarily to unanchored mobile homes, shrubs, and trees. Some damage to poorly constructed signs. Some flooding of coastal roads and minor pier damage. At their peak, hurricanes Allison (1995) and Danny (1997) were Category 1 hurricanes.

Category 2 Hurricane

Winds 154 km/h to 177 km/h. Storm surge generally 1.8 m to 2.4 m above normal. Some roof, door, and window damage. Considerable damage to shrubs and trees; some trees blown down. Considerable damage to mobile homes, poorly constructed signs, and piers. Coastal and low-lying escape routes flooded two to four hours before arrival of the eye of the hurricane. Small craft in unprotected anchorages break moorings. Hurricanes Bonnie (1998), George (1998), and Juan (2003) are examples of Category 2 hurricanes.

Category 3 Hurricane

Winds 178 km/h to 209 km/h. Storm surge generally 2.7 m to 3.7 m above normal. Some structural damage to small residences and utility buildings, including wall failures. Foliage blown off trees and large trees blown down. Mobile homes and poorly constructed signs destroyed. Low-lying escape routes cut off by rising water three to five hours before arrival of the eye of the hurricane. Flooding near the coast destroys smaller structures, and larger structures are damaged by floating debris. Land lower than 1.5 m above mean sea level may be flooded as far as 13 km inland. Evacuation of low-lying residences within several blocks of the shoreline may be required. Hurricanes Roxanne (1995) and Fran (1996) were Category 3 hurricanes at landfall.

Category 4 Hurricane

Winds 210 km/h to 249 km/h. Storm surge generally 4.0 m to 5.5 m above normal. Some roofs blown off residences, and extensive wall failures. Shrubs, trees, and all signs are blown down. Complete destruction of mobile homes. Extensive damage to doors and windows. Low-lying escape routes may be cut off by rising water three to five hours before arrival of the eye of the hurricane. Major damage to lower floors of structures near the shore. Terrain lower than 3.1 m above sea level may be flooded, requiring evacuation of residential areas as far as 10 km inland. Hurricanes Luis, Felix, and Opal (1995) were Category 4 hurricanes.

Category 5 Hurricane

Winds greater than 249 km/h. Storm surge generally greater than 5.5 m above normal. Complete roof failure on many residential and industrial buildings. Many complete building failures. All shrubs, trees, and signs blown down. Complete destruction of mobile homes. Severe and extensive window and door damage. Low-lying escape routes are cut off by rising water three to five hours before arrival of the eye of the hurricane. Major damage to lower floors of all structures located less than 4.6 m above sea level within many hundreds of metres of the shoreline. Evacuation of residential areas on low ground within 16 km of the shoreline may be required. Hurricane Mitch (1998) was a Category 5 hurricane at its peak over the western Caribbean. Hurricane Gilbert (1988) was a Category 5 hurricane at its peak and is the strongest Atlantic tropical cyclone of record. Hurricanes Rita and Katrina (2005) were also Category 5 hurricanes at their peak; they were, respectively, the fourth- and sixth-strongest Atlantic hurricanes ever recorded.

Source: Modified after Spindler, T., and J. Beven. 1999. "NOAA, Saffir-Simpson Hurricane Scale." NOAA, www.nhc.noaa.gov/aboutsshs.shtml. Accessed February 17, 2010.

11.2 Cyclone Development

Tropical and extratropical cyclones differ not only in their characteristics, but also in their development. Most tropical and extratropical cyclones form, mature, and dissipate independently. Some tropical cyclones, however, transform into extratropical cyclones if they encounter an upper-level, low-pressure trough as they weaken over land or over cooler seawater at higher latitudes.

Tropical Cyclones

A tropical cyclone is a general term for large thunderstorm complexes rotating around an area of low pressure that has formed over warm tropical or subtropical ocean water. A variety of names are applied to these complexes, depending on their intensity and location. Low-intensity tropical cyclones are called tropical depressions and tropical storms. High-intensity tropical cyclones are hurricanes. To be termed a hurricane, a tropical cyclone must have sustained winds of at least 119 km/h.[7] Hurricanes require tremendous amounts of energy, which they acquire from rising warm moist air that condenses at altitude. They form only in oceans that are warmer than 26°C.

Most hurricanes start out as a **tropical disturbance**, which is a large area of unsettled weather that is typically 200 to 600 km in diameter, has organized thunderstorms, and persists for more than 24 hours. A tropical disturbance is associated with an elongated area of low pressure called a *trough*. Air in the disturbance rotates weakly because of the Coriolis effect (see A Closer Look in Chapter 10).

Tropical disturbances form in a variety of ways—along lines of convergence similar to squall lines, in upper-level troughs of low pressure, from remnants of cold fronts, and from easterly waves of converging and diverging winds that develop in the tropics. In the Atlantic Ocean, most tropical cyclones develop from easterly waves that form over western Africa.

A tropical disturbance may develop into a **tropical depression** if winds increase and rotate around the area of disturbed weather. Warm moist air is drawn into the depression and begins to spin faster, much as spinning

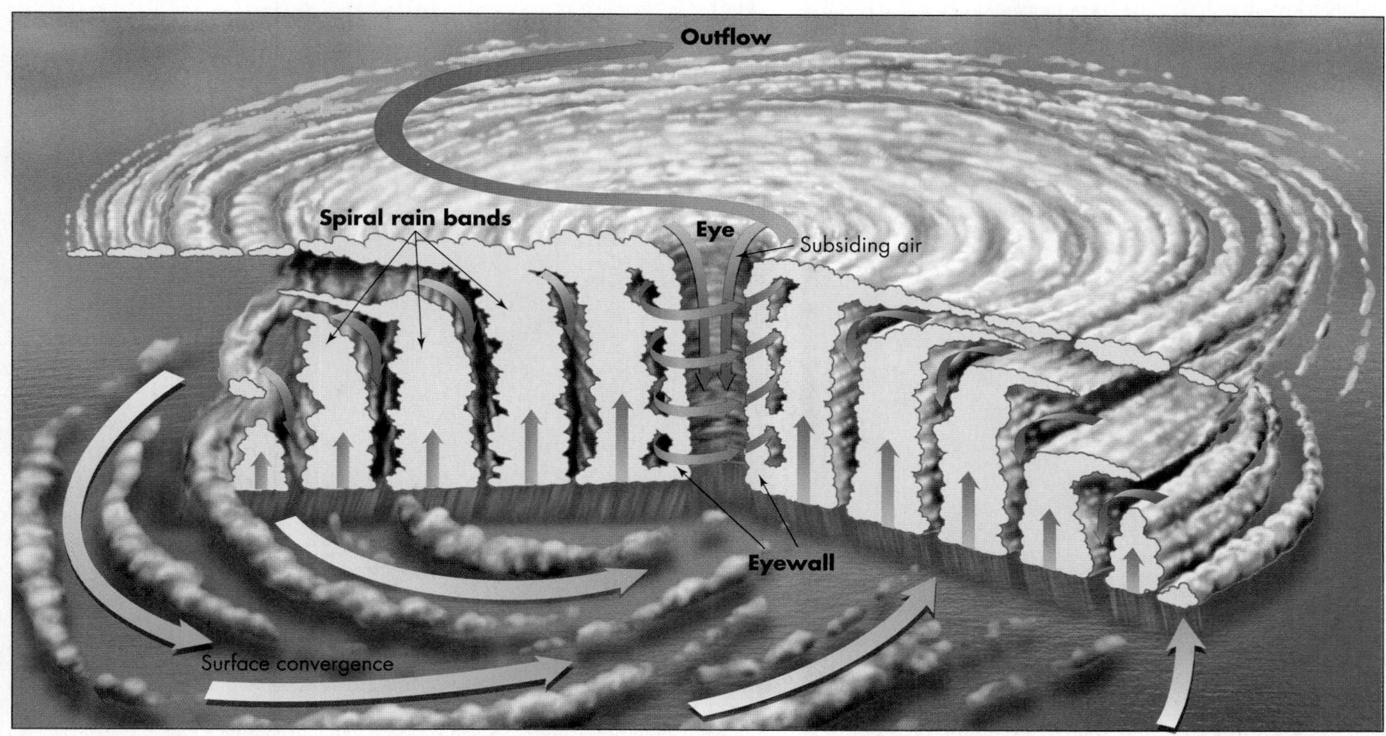

▲ **FIGURE 11.5 CROSS-SECTION OF A HURRICANE** The cloudless eye of a hurricane is surrounded by upward-spiralling winds and rain. Warm, moist air at the surface spirals toward the centre of the storm and rises to form rain bands. Sinking dry air creates spaces with fewer clouds between the bands. The innermost rain band is the eyewall cloud that surrounds the centre of the storm. Rising warm air from the eyewall either leaves the storm in one of several outflow jets aloft or loses moisture and sinks back into the eye of the storm. Subsiding dry air in the eye warms by compression, giving the storm its characteristic warm core. The vertical dimension of this diagram is greatly exaggerated. *(NOAA)*

ice skaters increase their rate of rotation by bringing in their arms. Once maximum sustained wind speeds reach 63 km/h, the depression is upgraded to a **tropical storm** and receives a name. Although the winds are not as strong in tropical storms as in hurricanes, the rainfall amounts can be as large. If the winds continue to increase, a tropical storm may become a hurricane.

Just as not all tropical depressions develop into tropical storms, few tropical storms develop into hurricanes. Several conditions must be met for a hurricane to form. First, warm ocean waters (at least 26°C) must extend to a depth of several tens of metres.[12] Warm surface temperatures alone are not sufficient to form a hurricane; there has to be a thick pool of warm water to provide energy for the storm. Second, the atmosphere must cool fast enough upward from the surface for moist air to continue to be unstable and convect. Layers of warm air aloft stop or cap hurricane development. Finally, there must be little vertical wind shear—that is a change in wind speed—between the surface and the top of the troposphere. Strong winds aloft prevent a hurricane from developing.

Once developed, a mature hurricane averages about 500 km in diameter and consists of anticlockwise spiralling clouds that swirl toward the storm's centre.[13] This rotation gives hurricanes their characteristic circular appearance in satellite images (see chapter opener photograph). Hurricane-strength winds (119 km/h or more) are limited to the innermost 160 km of an average hurricane. The outer area of the storm has gale-force winds, defined as winds with speeds of more than 50 km/h.

The clouds that spiral around a hurricane are referred to as **rain bands** and contain numerous thunderstorms (Figure 11.5). Both the thunderstorms and the surface winds increase in intensity toward the centre of the storm. The most intense winds and rainfall occur in the innermost bands of clouds, known as the **eyewall**. Rainfall rates directly beneath the eyewall may reach 250 mm/h.[13] A hurricane's eyewall is constantly changing as the storm progresses, and some strong hurricanes develop double eyewalls. In these storms, the inner eyewall may gradually dissipate and be replaced with a strengthening outer eyewall at the time the hurricane intensifies. NASA scientists discovered that very tall clouds up to 12 km high, called *hot towers*, can develop within an eyewall six hours before a storm intensifies (Figure 11.6).[14]

The eyewall surrounds a circular area of calm conditions and broken clouds that ranges from 5 km to over 60 km in diameter, known as the **eye**.[15] Most eyes are smaller at the surface and widen upward to form a large, amphitheatre-like area of nearly cloud-free blue sky surrounded by the white clouds of the eyewall. When a hurricane's eye passes directly overhead, there is "calm before the storm," because the fierce winds of the other side of the eyewall will soon blow from the opposite direction.

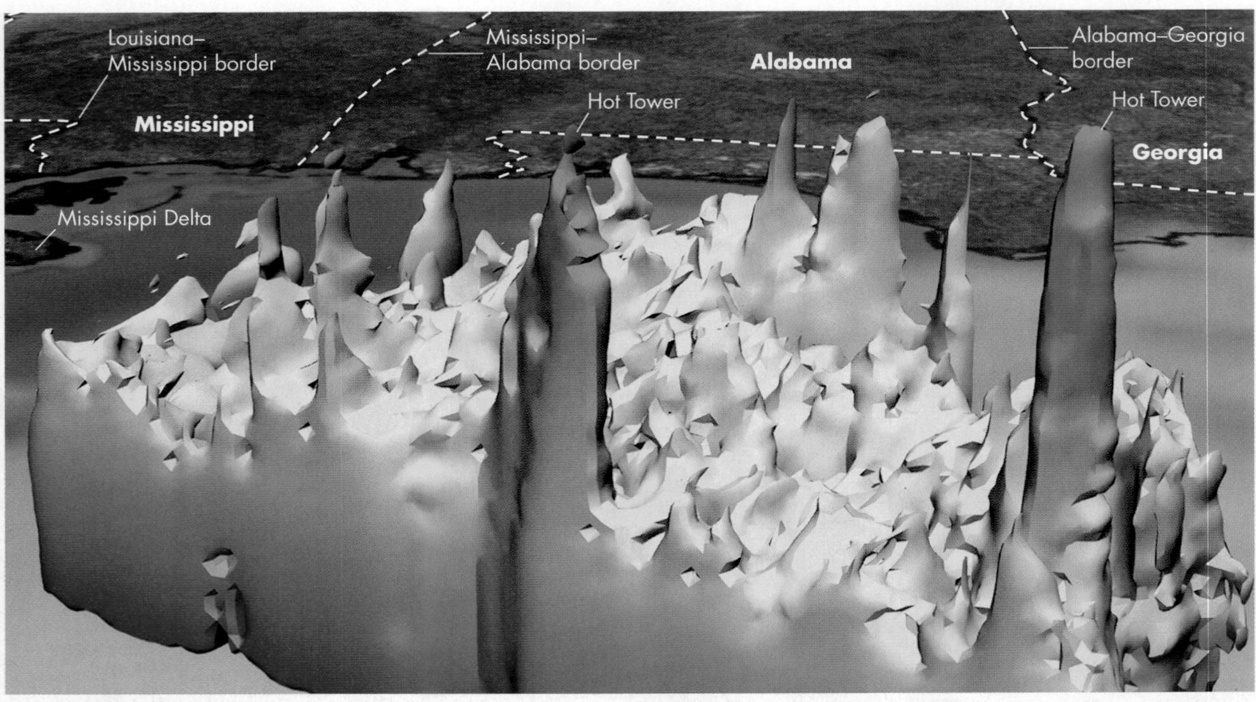

▲ **FIGURE 11.6 HOT TOWERS IN HURRICANE KATRINA** NASA's Tropical Rainfall Measuring Mission (TRMM) satellite spotted a pair of extremely tall thunderstorms in the eyewall and outer rain band of Hurricane Katrina shortly before it intensified to Category 5. Referred to as "hot towers," these clouds were more than 15 km high. The towers released huge amounts of heat and contributed to the rising winds in the storm. Detection of hot towers could help predict changes in hurricane intensity. *(Courtesy NASA/JAXA)*

In a typical hurricane, rising warm and moist air spirals upward around the eyewall. As the air rises, it loses moisture, clouds form, and the air becomes drier. This upward-spiralling rotation draws air from the eye and causes some of the drier air aloft to move downward into the centre of the storm. The upward-moving air also flows out of the top of the storm and builds a large cloud platform. Outward-flowing air at the top of the storm is concentrated in one or more outflow jets (Figure 11.6).[10] This exhaust system is critical for survival of the hurricane because it allows additional warm, moist air to converge inward at the lower level of the storm. Outflow of warm air at the top of hurricanes is also a mechanism for the transfer of heat from the tropics into polar regions.

Movement of a hurricane is controlled by the Coriolis effect and by steering winds that are 8 km to 11 km above the surface. In the Northern Hemisphere, hurricanes track westward in the trade winds across the Atlantic Ocean and curve to the right (Figure 11.7). The storms first curve to the northwest, then to the north and northeast. Deviations

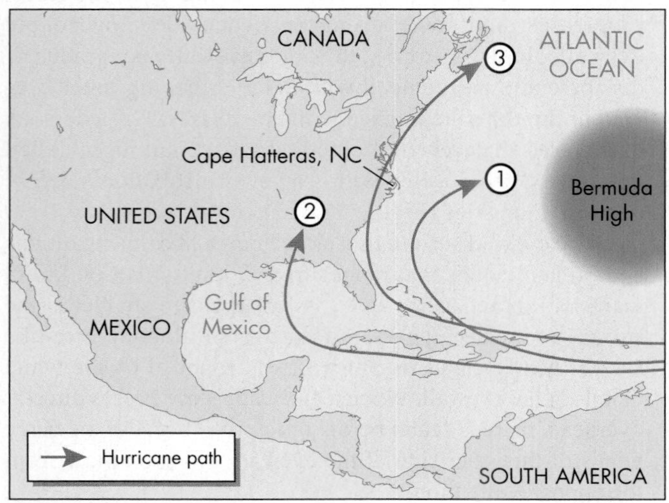

◄ **FIGURE 11.7 HURRICANE PATHS** The three common paths of hurricanes generated in the Atlantic Ocean. Each path, marked by a curving red arrow, starts in the central Atlantic Ocean to the east (right) of the map. Hurricanes that follow all three paths threaten Caribbean islands; hurricanes that take path 2 threaten the southeastern United States; and those that take path 3 threaten the Atlantic coasts of the United States and Canada.

from this track are common when steering currents are weak. Some hurricanes have even reversed course and made an entire loop in their track. In the North Atlantic, hurricane tracks are also influenced by the location and size of the *Bermuda High*, a persistent high-pressure anticyclone that remains anchored in the North Atlantic during the summer and early fall. Many hurricane tracks curve around the west side of the Bermuda High (Figure 11.7).

A mature hurricane will move forward at a speed of 19 to 27 km/h.[15] In most cases, this slow forward speed means that a hurricane is a two-day event for communities in the path of the storm. Toward the end of its life, however, a hurricane's forward speed can suddenly increase to 74 to 93 km/h.[15] This increase in speed characterizes many hurricanes that move up the east coast toward Atlantic Canada.

In summary, tropical cyclones develop in several stages from tropical disturbances. Each stage represents an increase in sustained winds and a decrease in atmospheric pressure. A tropical disturbance may become a tropical depression, then a tropical storm, and finally a hurricane.

In the Atlantic Ocean, most hurricanes start out as a tropical disturbance that forms where trade winds converge and diverge off the west coast of Africa. Hurricane development is favoured where (1) there is a thick layer of surface warm water, (2) warm moist air is free to rise upward toward the top of the troposphere, and (3) upper-level winds are relatively weak. Hurricanes obtain their energy from the evaporation and subsequent condensation of warm tropical or subtropical seawater, and they generally lose strength when they move over land.

Extratropical Cyclones

Two main factors contribute to the formation of an extratropical cyclone: a strong temperature gradient in the air near the surface, and strong winds in the upper troposphere. Surface temperature gradients are generally strongest along a cold, warm, or stationary front; therefore, most extratropical cyclones develop along fronts.

Strong winds in the upper troposphere occur in a concentrated flow of air called a **jet stream**. The Northern Hemisphere has two jet streams: one at an average altitude of 10 km called the *polar jet stream*, and the other at an average altitude of 13 km referred to as the *subtropical jet stream* (Figure 11.8).[13] These jet streams range from less than 100 km to over 500 km wide and are typically several kilometres thick.[16] Jet streams cross North America from west to east at an average speed of 100 km/h in winter and 90 km/h in summer.[14] Peak wind speeds in the jet streams can be twice the average wind speed.

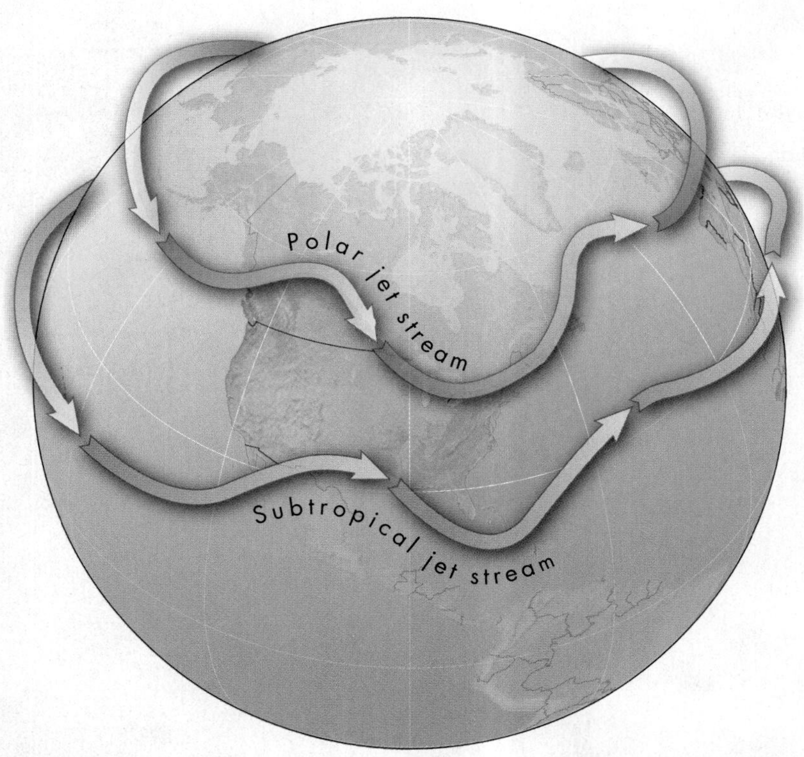

▲ **FIGURE 11.8 POLAR AND SUBTROPICAL JET STREAMS** Strong winds near the boundary between the troposphere and stratosphere are concentrated in jet streams. The strongest winds are in the polar jet stream at an altitude of about 10 km, and generally less intense winds form the subtropical jet stream at about 13 km. *(Lutgens, F. K., and E. J. Tarbuck. 2007.* The Atmosphere: An Introduction to Meteorology, *10th ed. Upper Saddle River, NJ: Pearson Prentice Hall)*

The polar jet stream shifts from a path crossing the conterminous United States in winter to one crossing southern Canada in summer. This migration in path causes a shift in the location of severe thunderstorm and tornado activity during late winter, spring, and early summer. The subtropical jet stream normally crosses Mexico and Florida and is strongest in winter.

Large high-pressure ridges and low-pressure troughs in the upper troposphere cause jet streams to bend north or south of their normal paths, producing long meanders or waves in their flow. A jet stream may also split in two around isolated high-pressure centres and reunite down-flow. Extratropical cyclones commonly develop in an area where jet stream winds curve cyclonically or diverge, such as on the east side of a low-pressure trough.

Bending or splitting can cause the polar jet stream to dip southward, and the subtropical jet stream to flow northeastward, into the conterminous United States. The southern branch of a split polar jet stream in the Pacific Ocean brings warm moist air out of the tropics and can be recognized on infrared satellite images as a band of clouds extending northeastward from the equatorial Pacific Ocean (Figure 11.9). During some winters, a series of extratropical cyclones track northeastward along this southern branch. West coast weather forecasters refer to this flow of warm moist air as an "atmospheric river" or the *Pineapple Express* because of its origin near Hawaii. Extratropical storms called nor'easters often form when bends of the polar and subtropical jet streams begin to merge off the southeast coast of the United States.

Extratropical cyclones can intensify when they cross areas with strong low-level temperature gradients. For example, extratropical cyclones that hit the west coast commonly weaken as they cross the Rocky Mountains and then deepen and strengthen when they reach the Great Plains.

Most extratropical cyclones start as a low-pressure centre along a frontal boundary, with a cold front developing on the southwest side of the cyclone and a warm front on the east (Figure 11.10). A conveyor-like flow of cold air circulates anticlockwise around the cyclone, wedging beneath the warm air to the east. Lighter, warm, moist air rises in a conveyor-like flow on the southeast side of the cyclone, creating a comma-like pattern of clouds (Figure 11.11). A conveyer-like flow of dry air aloft can feed the cyclone; the dry air sinks behind the cold front, forming a *dry slot* that is sometimes visible on satellite images. The cyclone matures and merges with the warm front to become an *occluded front* with warm air trapped aloft (Figure 11.10). At first, the storm intensifies; then, as the cold air completely displaces the warm air on all sides of the cyclone, the pressure gradient weakens and the storm dissipates within a day or two.

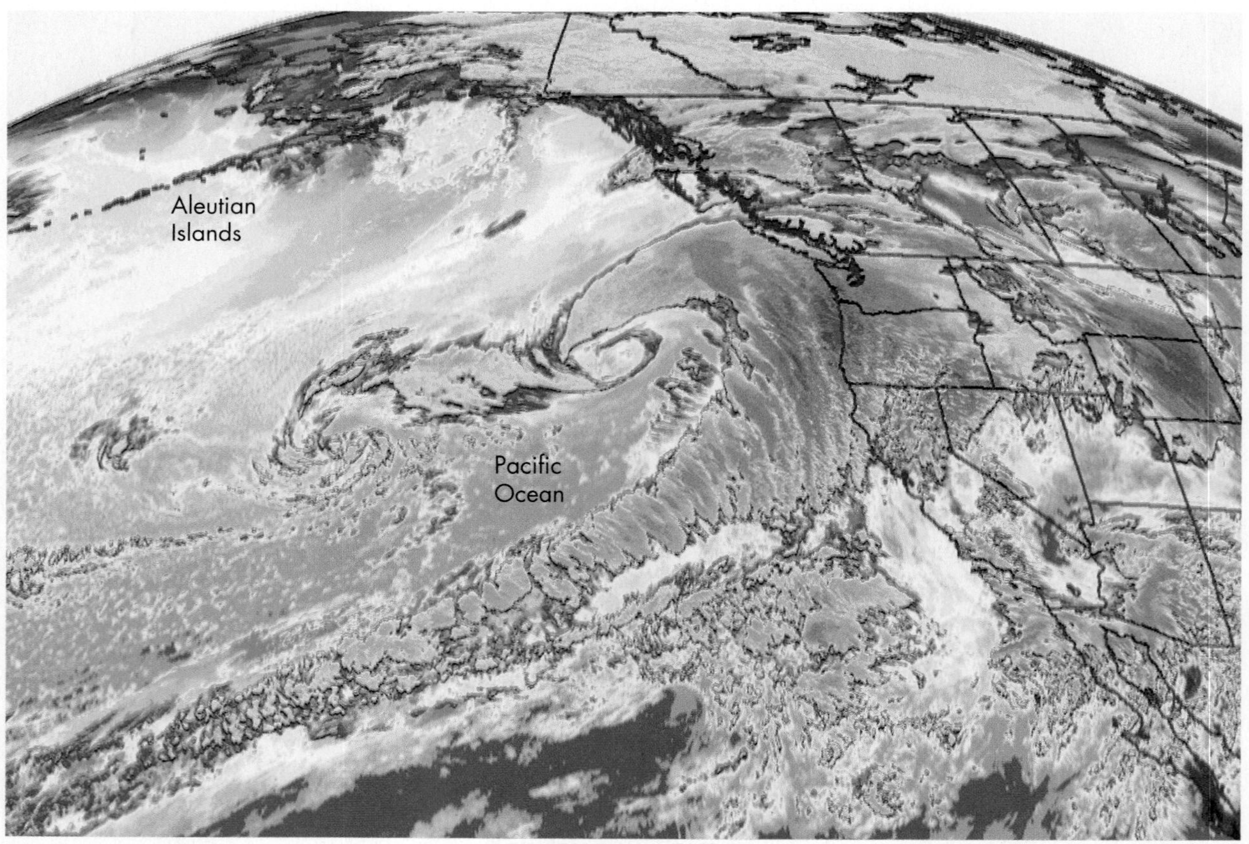

▲ **FIGURE 11.9 THE PINEAPPLE EXPRESS** In this satellite image, a stream of warm moist air is carried to the northeast across the North Pacific Ocean by a southern branch of the polar jet stream. Storms following this stream of tropical air bring heavy precipitation to the west coast of North America, causing flooding, landslides, and coastal erosion. This colour-enhanced infrared image was taken by the NOAA GOES-9 satellite. *(NOAA)*

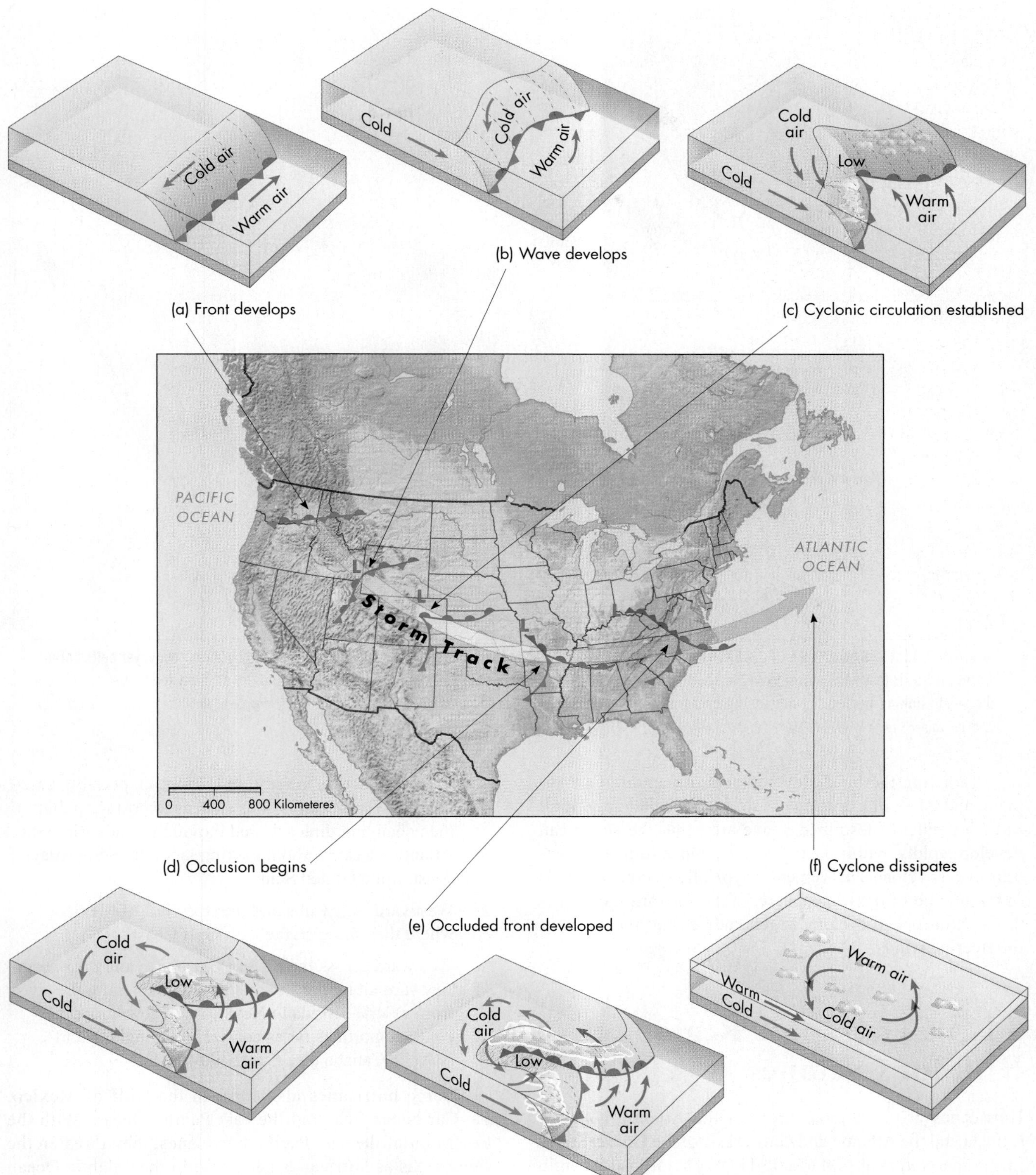

▲ **FIGURE 11.10 DEVELOPMENT OF AN EXTRATROPICAL CYCLONE** Stages in the development of an extratropical cyclone: (a) Air masses, moving in opposite directions along a front such as the stationary front shown here, provide the temperature contrast for cyclone development; (b) a wave forms along the front at a place where upper-level winds diverge, such as a bend in the jet stream; (c) a surface low-pressure centre develops as cold air pushes south on the west side of the front and warm air pushes north on the east side of the front; (d) the cold front advances faster than the warm front, catches up with the warm front, and displaces warm air upward; (e) an occluded front develops, with warm air held aloft by cold air masses at the surface; (f) the temperature contrast at the surface disappears, friction slows winds, and atmospheric pressure rises as the cyclone dissipates. *(Lutgens, F. K., and E. J. Tarbuck. 2007.* The Atmosphere: An Introduction to Meteorology, *10th ed. Upper Saddle River, NJ: Pearson Prentice Hall)*

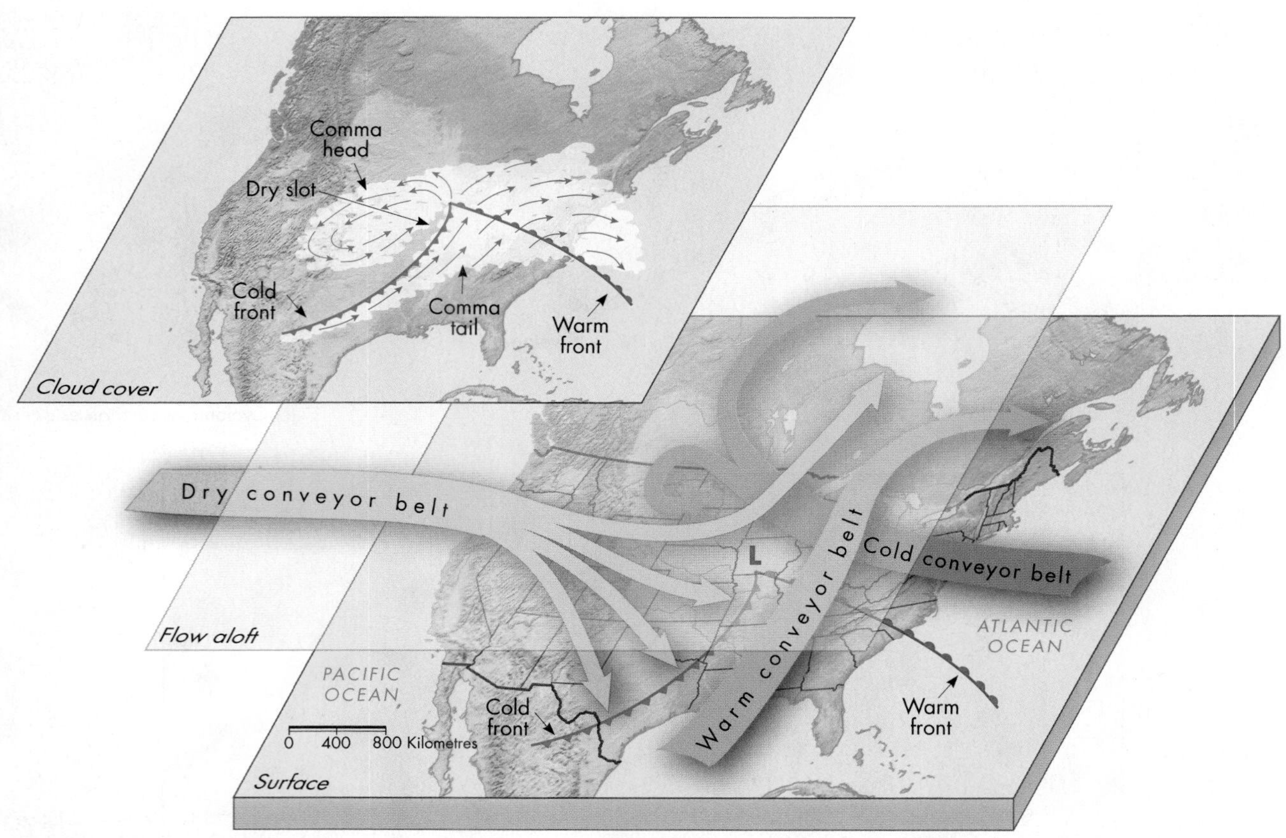

▲ **FIGURE 11.11 STRUCTURE OF AN EXTRATROPICAL CYCLONE** Once an extratropical cyclone develops, it is fed by three "conveyer belts" of air: warm moist air feeds the storm from the south; it rises up over a stream of cold air coming from the east; and dry air aloft feeds the storm from the west, sinking behind the advancing cold front at the surface. *(Lutgens, F. K., and E. J. Tarbuck. 2007.* The Atmosphere: An Introduction to Meteorology, *10th ed. Upper Saddle River, NJ: Pearson Prentice Hall)*

Predicting the birth, development, direction of movement, and death of extratropical storms is challenging. If all three ingredients described above are right, the storm can develop rapidly, within 12 to 24 hours. Once formed, an extratropical cyclone's movement is typically steered by winds in the middle of the troposphere.[13] The forward movement of an extratropical cyclone is generally at half the speed of the steering winds.[17]

11.3 Geographic Regions at Risk for Cyclones

Hurricanes are a serious threat to the Atlantic coast of Canada and the Atlantic and Gulf coasts of the United States in summer and early fall (Table 11.2). The official Atlantic hurricane season starts on June 1 and ends on November 30. Most Atlantic hurricanes occur in August, September, and October, when the sea surface is warmest. In contrast, the season for tropical storms in the Southern Hemisphere is from January to April.

The U.S. Atlantic and Gulf coasts experience five hurricanes each year, on average. These hurricanes form off the west coast of Africa and take one of three tracks (Figure 11.7):

1. Westward across the eastern Caribbean, possibly passing over Caribbean islands such as Puerto Rico, but then changing direction and moving northeast into the Atlantic Ocean without making landfall on the Atlantic coast of the United States.
2. Westward over Cuba and into the Gulf of Mexico, where they threaten the U.S. Gulf Coast.
3. Westward across the eastern Caribbean and then northeastward along the Atlantic coast, threatening the coast from central Florida to New York. A few hurricanes continue north as far as coastal New England and Atlantic Canada (see Case Study 11.1).

Some hurricanes also form in the Gulf of Mexico, the Caribbean Sea, and the east Pacific Ocean. With the exception of the east Pacific hurricanes, they threaten the same areas as hurricanes generated in the Atlantic Ocean (Figure 11.12). Hurricanes that form in the east Pacific Ocean are a serious hazard to the Pacific coast of Mexico as far north as the city of Guaymas. Moist, unstable air from the remnants of Pacific tropical storms and hurricanes move inland over Mexico and may cause torrential rains as far east as Texas.

Hurricane risk can be assessed from maps that show the probability that a hurricane will strike a particular coastal

TABLE 11.2 Hurricanes That Have Killed More than 200 People in North America

Name, State/Province	Year	Category	Deaths[1]
Great Galveston Hurricane, Texas	1900	4	8000
Newfoundland Hurricane, Newfoundland	1775	?	4000
Lake Okeechobee Hurricane, Florida	1928	4	2500
Hurricane Katrina, Louisiana, Mississippi	2005	3	1200
Cheniere Caminanda, Louisiana	1893	4	1100–1400
Sea Islands, South Carolina, Georgia	1893	3	1000–2000
Great New England Hurricane, New England states	1938	3	680–800
Georgia, South Carolina	1881	2	700
Nova Scotia Hurricane, Nova Scotia, Newfoundland	1873	3	>600
Hurricane Audrey, Louisiana, Texas	1957	4	416
Great Labor Day Hurricane, Florida	1935	5	408
Last Island, Louisiana	1856	4	400
Miami Hurricane, Florida, Alabama	1926	4	372
Grand Isle, Louisiana	1909	4	350
Florida Keys, Florida	1919	4	287[2]
New Orleans, Louisiana	1915	4	275
Galveston, Texas	1915	4	275
Hurricane Camille, Mississippi, Louisiana	1969	5	256

[1]Loss of life is approximate.

[2]An additional 500 lives were lost on ships at sea.

Sources: U.S. hurricanes—Rappaport, E. N., and J. Fernandez-Partagas. 1995. The Deadliest Atlantic Tropical Cyclones, 1492–1996. *NOAA Technical Memorandum NWS NHC 47 (updated 22 April 1997 by J. Beven). http://www.nhc.noaa.gov/pastdeadly.shtml. Accessed May 10, 2013.*

Canadian hurricanes—Wikipedia. 2013. "List of Canadian hurricanes." http://en.wikipedia.org/wiki/List_of_Canada_hurricanes. Accessed May 13, 2013.

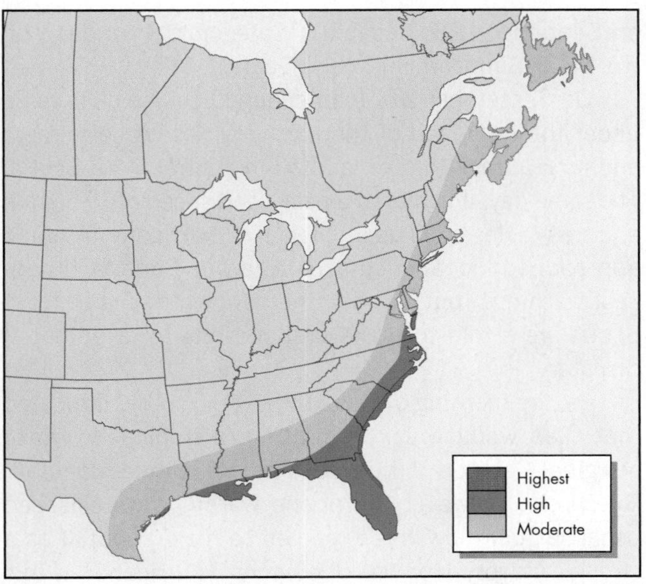

▲ **FIGURE 11.12 NORTH AMERICAN REGIONS AT RISK FROM HURRICANES** The highest-risk area (red band) is likely to experience 60 hurricanes in a 100-year period (the word "experience" here means that the hurricanes will pass within 160 km of the indicated area). The high-risk area (dark blue) can expect 40 to 60 hurricanes in 100 years, and the moderate risk area (light blue) will likely experience fewer than 40 hurricanes in 100 years. The map is based on observations from 1888 to 1998. *(Based on data from U.S. Geological Survey and Atlas of Canada)*

reach in a single year. Hurricane-strike probabilities are highest along the coasts of North Carolina, South Carolina, southern Florida, Louisiana, and eastern Texas.

The hurricane threat to New England and Atlantic Canada is often underestimated. Yet in 1815, a major hurricane made landfall on Long Island and crossed into Massachusetts and New Hampshire. Six years later, a hurricane passed very close to New York City and flooded parts of it with a 4-m storm surge.[10] The Great Hurricane of 1938, one of the fastest-moving hurricanes on record, had an intensity and path nearly identical to the 1815 hurricane. It struck Long Island without warning and killed at least 680 people in New York and several New England states. Winds were clocked at 130 km/h in New York, and about 80 percent of New England lost electrical power during the storm.[10] If the 1938 storm were to hit the same area today, the damage would likely exceed U.S.$40 billion. More recently, in 2003, Hurricane Juan caused $200 million damage in Nova Scotia and Prince Edward Island and claimed eight lives. The most recent damaging hurricane to reach Atlantic Canada was Hurricane Earl in September 2010. It claimed one life and cut power to more than 250 000 people in Nova Scotia.

Our emphasis on North America might lead one to believe that the North Atlantic Ocean is the most hazardous area for hurricanes and tropical storms. In fact, the Pacific and Indian oceans have far more hurricanes than the North Atlantic (Figure 11.13).[7] The South Atlantic and southeast

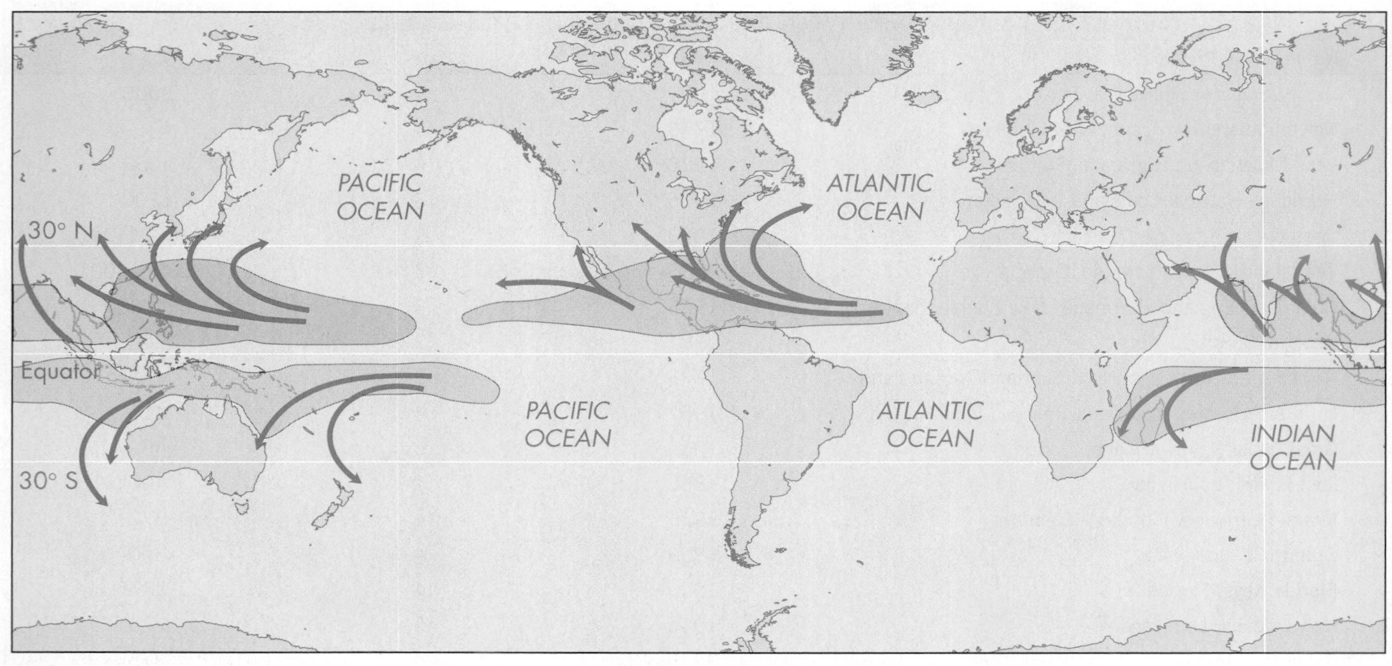

▲ **FIGURE 11.13 TYPICAL TROPICAL CYCLONE PATHS AND REGIONS WHERE THEY FORM** Most tropical cyclones develop between 5° and 20° latitude. Intense tropical cyclones are called hurricanes in the green region, typhoons in the blue region, and severe tropical cyclones or severe cyclonic storms in the yellow region. Red arrows show the typical directions and paths of these storms.

11.1 CASE STUDY

Hurricane Juan

Most Canadians do not perceive hurricanes as a threat, but many parts of Canada have experienced hurricanes in the past and will again. Hurricane Hazel caused 81 fatalities in Toronto in 1954 (see Chapter 9), and Hurricane Frieda claimed six lives in southwest British Columbia in 1962. The worst recent hurricane to hit a Canadian city was Juan in 2003.

Hurricane Juan was the sixth hurricane of the 2003 Atlantic hurricane season. It struck Nova Scotia and Prince Edward Island as a Category 2 hurricane, causing major damage to trees and property, especially within the urban core of Halifax (Figure 11.14). The storm killed eight people and caused more than $200 million in damage. It was the region's most powerful hurricane since 1873.

The tropical depression that would become Juan formed on September 24, 2003, northeast of the Bahamas. On September 25, the depression began to organize and it turned to the northwest. As the storm crossed the waters of the Gulf Stream, which were unusually warm for that time of year, it intensified and, on September 26, became a Category 1 hurricane on the Saffir-Simpson Hurricane Scale.

On September 27, Juan changed course as it came under the influence of the northerly flowing jet stream and took a track toward landfall in central Nova Scotia. The same day, it achieved a peak wind speed of 175 km/h and was reclassified as a Category 2 hurricane. As Juan approached, local media in Atlantic Canada broadcast warnings; public and emergency officials in Nova Scotia were told to make preparations for a potential disaster.

On the morning of September 28, reports indicated that Juan would weaken to either a tropical storm or marginal Category 1 hurricane before it made landfall. By 6 p.m., however, hurricane warnings were issued because Juan now was expected to make landfall as a strong Category 1 or weak Category 2 hurricane. At this point, it was too late for much of the general public to make necessary preparations.

The storm made landfall on September 29 with peak winds of 170 km/h and moved directly north toward Halifax. It continued north, crossing Nova Scotia and Prince Edward Island within hours, before diminishing to a strong Category 1 hurricane. By midday, the

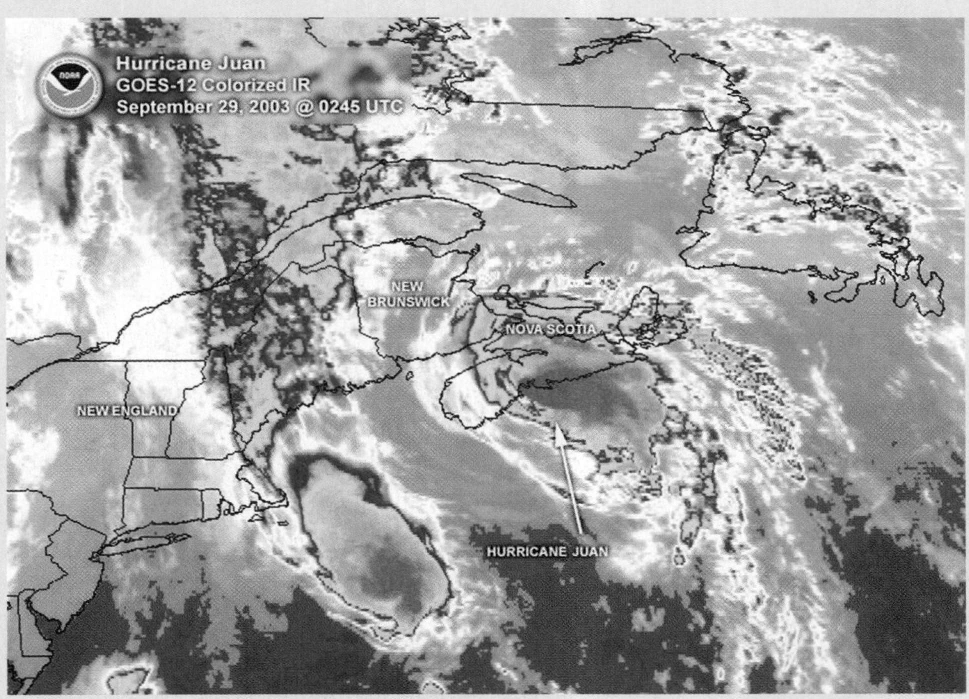

▲ **FIGURE 11.14 DAMAGE FROM HURRICANE JUAN** Hurricane Juan caused more than $200 million in damage when it crossed Nova Scotia and Prince Edward Island on September 29, 2003. The damage was caused by winds of up to 170 km/h and a storm surge. The photo shows damage to the coastal community of Prospect, Nova Scotia. *(NOAA;Tim Krochak/CP Images)*

(continued)

(Continued)

storm was centred over the Gulf of St. Lawrence and was downgraded to a tropical storm.

Hurricane Juan caused widespread damage in central Nova Scotia and Prince Edward Island. Most property damage was in the western urban core of Halifax. Truro, Nova Scotia, and Charlottetown, Prince Edward Island, also experienced significant property damage. Wave-rider weather buoys off the entrance to Halifax Harbour snapped their moorings after recording waves in excess of 20 m. Some of the harbour's populated shorelines, especially in Bedford Basin, were severely eroded by waves. The hurricane left more than 300 000 people without power, and it took a week and a half to restore power to the hardest hit rural areas of Nova Scotia's Eastern Shore and the Musquodoboit River valley. Seventy percent of the trees in Halifax's Point Pleasant Park were destroyed, fundamentally changing the character of the park. The Halifax Public Gardens were also badly damaged. Boats and seawalls along Charlottetown's waterfront sustained heavy wave damage, and the forest in that city's downtown core was heavily damaged.

Hurricane Juan resulted in several changes to the Meteorological Service of the Canadian Hurricane Centre (CHC). The centre was relocated from a vulnerable and exposed office building in Dartmouth, Nova Scotia, to a location that can withstand hurricane winds. The CHC's hurricane warning system has also been improved. Traditionally, the CHC issued only high wind and heavy rainfall warnings, not standard hurricane or tropical storm watches or warnings. Wind and heavy rainfall warnings were not heeded by local residents. Since 2004, the CHC has issued standard hurricane warnings for storms that could potentially affect Canada.

The name Juan was retired from the hurricane list in April 2004 and will never again be used for an Atlantic hurricane. It was the first time that the Meteorological Service of Canada specifically requested that a hurricane name be retired. The name Juan was replaced by Joaquin.

Pacific have few hurricanes because of their cold surface waters. Hurricanes do not form close to the equator because of the absence of the Coriolis effect and, once formed, do not cross the equator.

The geographic region at risk from extratropical cyclones in North America is far larger than that at risk from tropical cyclones. Severe weather from extratropical cyclones is greatest in the continental interior, but may also occur in coastal areas. Extratropical cyclones create strong windstorms in winter months along the Pacific and Atlantic coasts. They also produce heavy snowstorms and blizzards, especially on the Prairies east of the Rocky Mountains. During spring and summer, extratropical cyclones are responsible for severe thunderstorms and tornadoes east of the Rocky Mountains.

11.4 Effects of Cyclones

Tropical and extratropical cyclones claim many lives and cause enormous amounts of property damage every year. Damage, injuries, and death from tropical and extratropical cyclones are caused by winds, storm surges, and freshwater flooding, and by the thunderstorms and tornadoes that the cyclones generate (see Chapter 10). Extratropical cyclones can also create snowstorms and blizzards. Storm surges cause the greatest damage and contribute to 90 percent of all hurricane-related fatalities.[18]

Storm Surge

A **storm surge** is a rapid local rise in water level that happens when hurricane winds push water onto a shoreline (see Case Study 11.2). Tropical cyclones commonly generate storm surges of more than 3 m, and surges of 12 m or more have been recorded in Bangladesh and Australia.[19] Larger and faster-moving hurricanes create higher storm surges than do smaller, slower-moving hurricanes. High storm surges also develop on broad, shallow coastlines where the forward motion of wind-driven water is impeded because of friction. Storm surge and wind damage in the Northern Hemisphere are greatest in the forward right quadrant of the storm as it makes landfall—this is the direction in which the storm is both travelling and rotating, and thus the wind speed at the ground is higher there (Figure 11.15). The height of the surge will also be greater if the hurricane comes onto land at high tide.

Two mechanisms cause the storm surge. The first, and by far the more significant one, is stress exerted by wind on the sea surface. The larger the area over which the wind blows (*fetch*), the higher the water will rise. The second mechanism is the low atmospheric pressure in the storm, which raises the water level. The sea surface rises 1 cm for every millibar that atmospheric pressure drops.[10] In most intense hurricanes, the drop in atmospheric pressure is about 100 millibars; thus the height of the surge is raised about 1 m by this mechanism.

The storm surge is also affected by the shape of the coastline. In a narrow bay, lagoon, or lake, the height of

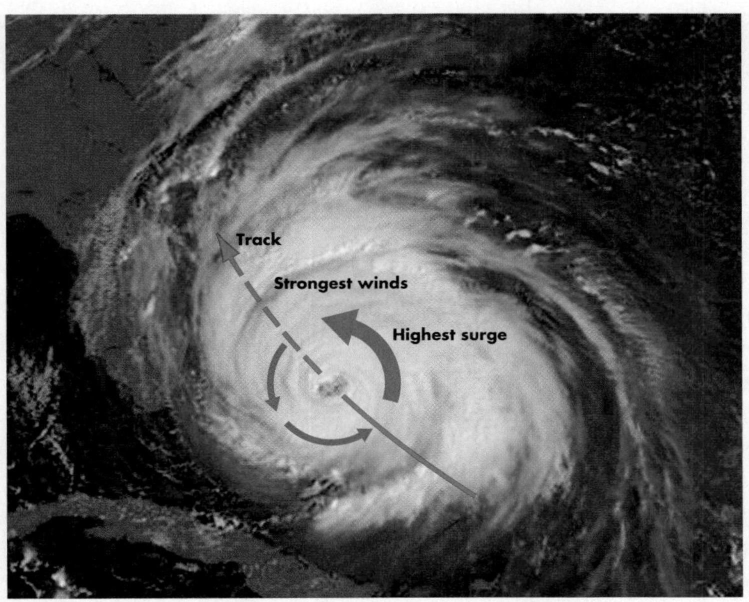

◄ **FIGURE 11.15 HAZARD IS GREATEST IN THE RIGHT FORWARD QUADRANT OF ATLANTIC HURRICANES** For hurricanes in the Northern Hemisphere, like Hurricane Floyd shown here, the right forward quadrant of the storm has the highest storm surge, the strongest winds, and the heaviest rainfall. This image was processed from the NOAA GOES satellite. *(NASA Goddard Space Flight Center/Dennis Chester)*

the surge may increase as water sloshes back and forth in the enclosed or partially enclosed body of water. This phenomenon is termed *resonance* and occurs when a wave reflecting from one shore is superimposed on a wave moving in another direction. The net effect is an amplification of the surge, similar to the amplified shaking that occurs when some seismic waves move through a sedimentary basin (Chapter 3).

A storm surge is not an advancing wall of water, but rather a steady increase in water level. Large storm waves are superimposed on the surge and, together with the water itself, are responsible for much of the damage of a hurricane. The storm waves also erode beaches, islands, and roads.

Sand eroded from beaches and coastal dunes is carried landward by the storm surge and deposited as **overwash**. Overwash may occur in *washover channels* cut through the beach, sand dunes, or a barrier island. Washover channels can form during a single hurricane and, on islands and peninsulas, isolate one area from another (Figure 11.16). Most of the channels are naturally filled by *littoral transport* and wave action in the months following the storm.

High Winds

To most people, the damage caused by wind is more obvious than that caused by a storm surge, in part because of the larger affected area. The highest sustained hurricane wind speed that has been measured, 310 km/h, occurred during three Category 5 hurricanes: Camille in the Gulf of Mexico in 1969, Tip in the west Pacific in 1979, and Allen in the Caribbean Sea in 1980. Even higher gusts occurred during

◄ **FIGURE 11.16 HATTERAS ISLAND BREACHED BY HURRICANE ISABEL STORM SURGE** Three channels, totalling 400 m in width, were cut through Hatteras Island, North Carolina, by the 2.3-m storm surge in the right forward quadrant of Category 2 Hurricane Isabel in September 2003. *(U.S. Geological Survey)*

11.2 CASE STUDY

Hurricane Sandy

Hurricane Sandy devastated portions of the Caribbean and the mid-Atlantic and northeastern United States during late October 2012. It had lesser impacts in the southeastern and midwestern U.S. and central Canada. Sandy was the tenth hurricane of the 2012 Atlantic hurricane season and was a Category 2 hurricane at its peak intensity. Although far from the strongest hurricane in U.S. history, Sandy was the largest in terms of the area affected, with a diameter of 1800 km.[20] An estimate of direct damage is still preliminary at the time this edition of the textbook was prepared, but probably is greater than U.S.$75 billion. If this figure proves to be correct, Hurricane Sandy would be the second-costliest Atlantic hurricane in history, behind only Hurricane Katrina. At least 253 people were killed in seven countries along the path of the storm. The severe and widespread damage the storm caused in the United States, combined with the fact that it merged with a frontal system in the northeastern U.S., led the media and several government agencies to nickname the hurricane "Superstorm Sandy."

Hurricane Sandy developed from a tropical depression in the western Caribbean Sea on October 22, 2012 (Figure 11.17). It quickly strengthened and was upgraded to Tropical Storm Sandy six hours later. The cyclone moved slowly northward and gradually intensified; by October 24, Sandy became a hurricane and made landfall near Kingston, Jamaica. A few hours later, it re-entered the Caribbean Sea and strengthened into a Category 2 hurricane. On October 25, Sandy hit Cuba, and a day later passed through the Bahamas. On October 27, Sandy briefly weakened to a tropical storm, but strengthened again to a Category 1 hurricane. Early on October 29, Sandy curved north-northwest before moving ashore near Atlantic City, New Jersey, with hurricane-force winds.

The hurricane left a long trail of devastation. Seventy percent of the residents of Jamaica were left without electricity, and strong winds blew roofs off buildings, killed one person, and caused about U.S.$55 million in damage in that country. Although Sandy passed offshore of Haiti, heavy rains brought flooding that killed at least 54 people and left about 200 000 homeless. Cuba experienced extensive coastal flooding and wind damage. Some 15 000 homes were destroyed and 11 people were killed there; total damage was about U.S.$2 billion. Two

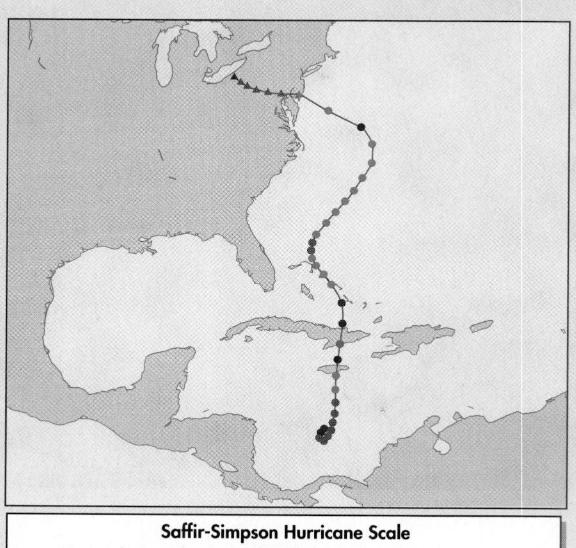

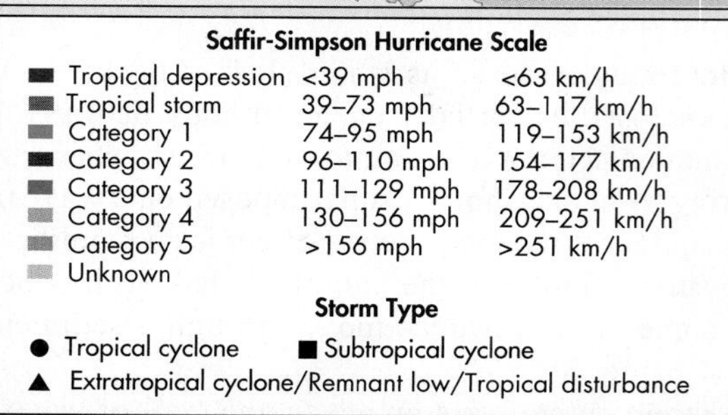

▲ **FIGURE 11.17 PATH OF HURRICANE SANDY** Hurricane Sandy formed from a tropical depression in the western Caribbean Sea in late October 2012. This diagram shows its track over a one-week period from October 22 to 29. The points show the location of the eye of the storm at six-hour intervals. The shape of the points represents the size of the storm according to the Saffir-Simpson Hurricane Scale, with the colours corresponding to the maximum sustained wind speeds. *(Wikipedia Commons; background image courtesy of NASA)*

people died in the Bahamas and damage totalled about U.S.$300 million. In the United States, Hurricane Sandy affected 24 states, including the entire eastern seaboard from Florida to Maine and west to Michigan and Wisconsin. The greatest damage was in New Jersey and New York (Figure 11.18). Sandy's storm surge hit New York City on October 28, flooding streets, tunnels, and subway lines, and severing power in and around the city.[20] Part of the Atlantic City boardwalk in New Jersey was washed away; half of the city of Hoboken was flooded; and several other towns were devastated by the storm surge. New Jersey suffered about half of the total damage from the hurricane. Heavy wet snow blanketed the Appalachian Mountains when Arctic air pushed south through the region. Two people died in Ontario and total damage in Ontario and Quebec amounted to about $100 million.

Hurricane Sandy was unusual in more ways than its size. As Atlantic hurricanes move north, they typically are forced east and out to sea by the prevailing winds of the jet stream. In Sandy's case, however, this typical pattern was blocked by a ridge of high pressure over Greenland that formed a kink in the jet stream and caused the cyclone to turn back on itself off the northeast coast of the United States. The blocking high over Greenland also stalled an Arctic front that merged with the storm.

The oft-asked question of whether a weather disaster is related to climate warming quickly surfaced in the case of Hurricane Sandy. Kevin Trenberth, the senior climatologist at NCAR (National Center for Atmospheric Research) responded that this is the wrong question: "All weather events are affected by climate change because the environment in which they occur is warmer and moister than it used to be."[20] Trenberth agrees that the storm was caused by natural variability, but also thinks it was enhanced by global warming. One factor contributing to the storm's strength was abnormally warm sea surface temperatures off the east coast of the U.S., which is probably linked to global warming. Also, as the temperature of the atmosphere increases, its capacity to hold water increases, leading to stronger storms and more rainfall. Harvard geologist Daniel Schrag said that Hurricane Sandy's 4-m storm surge is an example of what will, by mid-century, be the "new norm on the Eastern Seaboard."[20]

◄ **FIGURE 11.18 HURRICANE SANDY DAMAGE** Aerial view of the damage caused by Hurricane Sandy to the New Jersey coast. This photograph was taken during a search and rescue mission. *(© Archive Image/ Alamy)*

three hurricanes that made landfall in the United States: one in the Florida Keys in 1935; Camille in Mississippi in 1969; and Andrew near Homestead, Florida, in 1992 (Figure 11.19).[19] The destruction wrought by 310 km/h winds is nearly total. In the 1935 hurricane, a train carrying veterans working in the Florida Keys was blown entirely off its tracks (Figure 11.20). Some of the 423 people who died during this hurricane appeared to have been sandblasted to death by sand blown from the beach.[19]

Wind speeds in most hurricanes diminish rapidly once they make landfall; they generally drop by half within about seven hours after reaching land.[10] However, a few hurricanes do not lose all their strength as they move inland and instead transition into extratropical cyclones. During the transition, some of these storms maintain or even increase their wind speeds.[10] In 1954, Hurricane Hazel made landfall in North Carolina and maintained 160 km/h winds and caused extensive damage as far north

▲ **FIGURE 11.19 AERIAL VIEW OF HURRICANE WIND DAMAGE** This mobile home community in southern Florida was devastated by wind gusts up to 320 km/h during Hurricane Andrew in 1992. About 250 000 people were left homeless by the hurricane. *(Corbis Images)*

as Toronto (Chapter 9). The transition to an extratropical storm may occur if a dying hurricane merges with an upper-level extratropical cyclone or cold front, or if it rapidly moves from warm to cold water.[10]

Strong winds and heavy rains from extratropical cyclones cause most of the severe weather on the Pacific coast of North America. One of the most destructive of these storms was the "Big Blow of 1962," which produced wind gusts of nearly 300 km/h along the British Columbia, Washington, and Oregon coasts and killed 46 people.[10]

Extratropical storms are responsible for the high winds in blizzards and tornadoes in Canada and the United States and for deadly windstorms on the Great Lakes. A storm in 1913 sank or grounded 17 ore ships on Lakes Huron and Erie.[19] Another cyclone in 1975 sank the S.S. *Edmund Fitzgerald*, one of the largest ore ships ever to navigate the Great Lakes, drowning the entire crew.[17]

Heavy Rains

A third hazard associated with hurricanes is inland flooding. An average hurricane releases about four trillion litres of water each day, nearly 30 times the amount of freshwater consumed in Canada each year.[10] The heaviest rainfalls in history have occurred where hurricanes passed over, or close to, mountainous islands. La Reunion Island in the Indian Ocean has several rainfall records from tropical cyclones, including 1144 mm in 12 hours and 2467 mm in 48 hours.[10] Tropical storms, and even weaker tropical depressions, can produce heavy rains and extensive flooding. For example, Tropical Storm Allison, which never became a hurricane, dumped up to 940 mm of rain on Houston, Texas, in 2001.[21] The storm killed 41 people and caused over U.S.$5 billion in damage.[21] Rainfall from tropical storms and remnants of hurricanes is commonly most intense when the storms encounter topographic barriers such as mountain ranges. In some hurricanes, the damage from inland flooding exceeds that of wind in the coastal zone. The flooding can last for days as waters slowly return to the sea.

Four factors affect the extent of inland flooding: the storm's speed; the terrain over which the storm moves; interaction with other weather systems; and the amount of water in the soil, streams, and lakes before the storm arrives.[22] Inland flooding from Hurricane Ivan affected a huge area from Georgia north to New York State. The flooding was caused in part by the slow movement of the storm, the mountainous terrain of the Appalachians, and saturation of the soil by heavy rains from the remnants of Hurricane Frances less than 10 days earlier.

11.5 Links between Cyclones and Other Natural Hazards

Intense precipitation associated with cyclones causes flooding, erosion, and landslides. Severe coastal erosion may occur when cyclones make landfall. Wind-driven waves superimposed on a storm surge actively erode beaches, coastal dunes, and sea cliffs. A cyclone's storm surge and heavy rainfall also produce widespread coastal flooding. Coastal areas that have subsided because of an earthquake,

▲ **FIGURE 11.20 HURRICANE BLOWS TRAIN OFF TRACKS IN FLORIDA** Wind gusts of over 320 km/h blew the cars of this train off its tracks in the Florida Keys in 1935. Hundreds of World War I veterans working in the area were killed on the train as they were being evacuated from the storm. *(History Miami)*

soil compaction, or groundwater withdrawal are highly vulnerable to storm surges.

At landfall, severe hurricanes cause both saltwater flooding from the storm surge and freshwater flooding from heavy rains. Steering winds allow the storms to stagnate over the same area for many days, prolonging the period of flooding.

In mountainous areas, heavy rains associated with cyclones can trigger devastating landslides and debris flows. For example, in 1998 heavy rains from Hurricane Mitch in Guatemala, Honduras, El Salvador, and Nicaragua caused widespread flooding, landslides, and debris flows that killed at least 11 000 people.[23]

Wind damage from cyclones is not restricted to the winds circulating about the eye of the storm. Cyclones can generate *downbursts* with speeds of over 160 km/h and numerous tornadoes.[7] Approximately half the hurricanes that come ashore in the United States produce tornadoes, and most of these tornadoes occur within 24 hours after a hurricane makes landfall.[24]

Finally, as global sea level rises because of global warming, the effects of cyclones will become more severe, especially along coastlines with relatively flat coastal plains. In coming decades, storm surges from cyclones will be able to penetrate farther inland than ever before.

11.6 Natural Service Functions of Severe Weather

Cyclones are the primary source of precipitation in most areas of the United States and Canada. In the eastern and southern United States, hurricanes and tropical storms may provide much-needed precipitation to moisture-starved areas. And along the Pacific coast, extratropical storms are the major source of rain and snowfall during the winter rainy season.

Cyclones also serve an important function in equalizing the temperatures of our planet. They elevate warm air from the tropics and distribute it toward higher latitudes.

Cyclones benefit ecosystems. Hurricane-generated waves can stir up deeper, nutrient-rich waters, producing plankton blooms in the open ocean and in estuaries. Cyclones rejuvenate ecosystems ranging from old-growth forests to tropical reefs. High winds topple weak and diseased trees, and strong waves break apart some types of corals, creating new surfaces for organisms to colonize. Fallen trees also create clearings that become new habitats for many plants and animals. Overall, these storms help maintain species diversity in many ecosystems. Lightning starts wildfires, which are important in maintaining prairie and forest ecosystems (see Chapter 13).

11.7 Human Interaction with Weather

In the past 50 years, population growth in the United States has been greatest in coastal areas, and today about 53 percent of the people in the U.S. live in coastal counties.[25] More and more people choose to live along coastlines that have been devastated by hurricanes in historic time, including, for example, Galveston, Texas, and the greater Miami–Ft. Lauderdale and Tampa–St. Petersburg areas of Florida.

Construction along the shoreline can potentially affect the severity of a hurricane. Removal of coastal dunes during development increases the vulnerability of a shore to storm surges. Seawalls and bulkheads that are built to protect property can contribute to erosion farther along the shoreline (see Chapter 12). Improperly attached roofing and other building materials commonly become dangerous projectiles during hurricanes, damaging other buildings and injuring—and sometimes killing—people.

Many scientists have become concerned about the effects our changing climate could have on tropical cyclones. Global warming, caused by gases that humans are putting into the atmosphere, is raising the temperature of the sea surface and contributing to rising sea level (see Chapter 14). Warm sea-surface temperature is a major factor in tropical cyclone development, and it is possible that warmer oceans will increase hurricane intensity.[26] Rising sea level will also increase the reach of storm surges and extend the effects of the large waves that ride the surge.

11.8 Minimizing the Effects of Cyclones

We cannot prevent cyclones, so the primary way of reducing property damage and preventing loss of life is to accurately forecast these storms and issue advisories to warn people in their path. Other ways to minimize damage, such as improved building codes and evacuation procedures, are adjustments to living with the hazard and are discussed later (Section 11.9). Although the following discussion focuses on hurricanes, many of the challenges faced in forecasting their behaviour, such as changes in path and intensity, also apply to intense extratropical cyclones.

Forecasts and Warnings

Timely and accurate prediction of hurricanes is essential if lives are to be saved. The public must be warned in time to prepare or evacuate. Hurricanes can be difficult to forecast because they encompass many different weather processes and they develop far from shore where there are few, if any, observers. Once a hurricane has formed, meteorologists must predict if it will reach land, where and when it will strike, how strong the winds will be, how large an area will be affected, and how much rainfall and how high a storm surge will accompany the storm.

Hurricane forecast centres are located across the globe and include the Canadian Hurricane Centre (CHC) in Dartmouth, Nova Scotia, and the U.S. National Hurricane Center (NHC) in Miami, Florida. The NHC watches the Atlantic Ocean, the Caribbean Sea, the Gulf of Mexico, and the eastern Pacific between May 15 and November 30 each year, and the CHC monitors cyclones that are likely to affect Atlantic Canada.[27,28] Both centres issue **hurricane watches** and **hurricane warnings** to the public and support hurricane research. They use information from weather satellites, hurricane-hunter aircraft, Doppler radar, weather buoys, reports from ships, and computer models to detect and forecast hurricanes.

Hurricane Forecasting Tools Weather satellites are probably the most valuable tool for hurricane detection and tracking, because these great storms form over the open ocean. Satellites can detect storms that may become hurricanes, thus alerting meteorologists to areas that should be watched closely.[16] Satellites cannot, however, provide accurate information about wind speed and other conditions within the storms. Once a hurricane is detected, other tools are used to refine predictions about the storm's behaviour.

Aircraft are an invaluable tool, and special planes are flown directly into a hurricane to gather data. U.S. Air Force hurricane-hunter aircraft perform most of this work, although NOAA research aircraft also fly missions through hurricanes. During winter months, Air Force and NOAA aircraft also collect meteorological data on extratropical cyclones in the Pacific Ocean and along the Atlantic coast of the United States.

Doppler radar is another tool for collecting data about hurricanes that come within about 320 km of the United States mainland or some islands, such as Puerto Rico and Key West. Radar can provide information about rainfall, wind speed, and the direction the storm is moving (Figure 11.21).

Weather buoys floating along the Atlantic and Gulf coasts are also used to forecast hurricanes. These buoys are automated weather stations that continuously record weather conditions. They transmit the information to the Meteorological Service of Canada and the U.S. National Weather Service. Some information is also obtained from ships in the vicinity of a hurricane.

Meteorologists use computers to process all available data and make predictions about hurricane tracks and strength (Figure 11.22). Although these models have greatly improved our ability to predict where and when a storm will strike, they are still unable to accurately predict the intensity of the storm; therefore reconnaissance flights remain an integral part of hurricane prediction.

Storm Surge Predictions Forecasts of the path and intensity of cyclones are used to predict the height and extent of the storm surge. Forecasters can use wind speed, fetch, and water depth to get a general idea of the height of the storm surge, but they must also predict the arrival time because of tidal effects. The tidal range along shores of the Gulf of Mexico is only about 1 m, but it is 3 m or more in Atlantic Canada, and thus it is more important to take tides into account when predicting storm surges in this region.

Predicting the extent of the storm surge requires detailed information about land elevations along the coast. An airborne laser surveying technique, known as LIDAR (Light Detection and Ranging), is used to prepare detailed digital elevation models of coastal areas. The survey data can be processed with advanced computer programs to predict both the height and extent of the storm surge (Figure 11.23). The

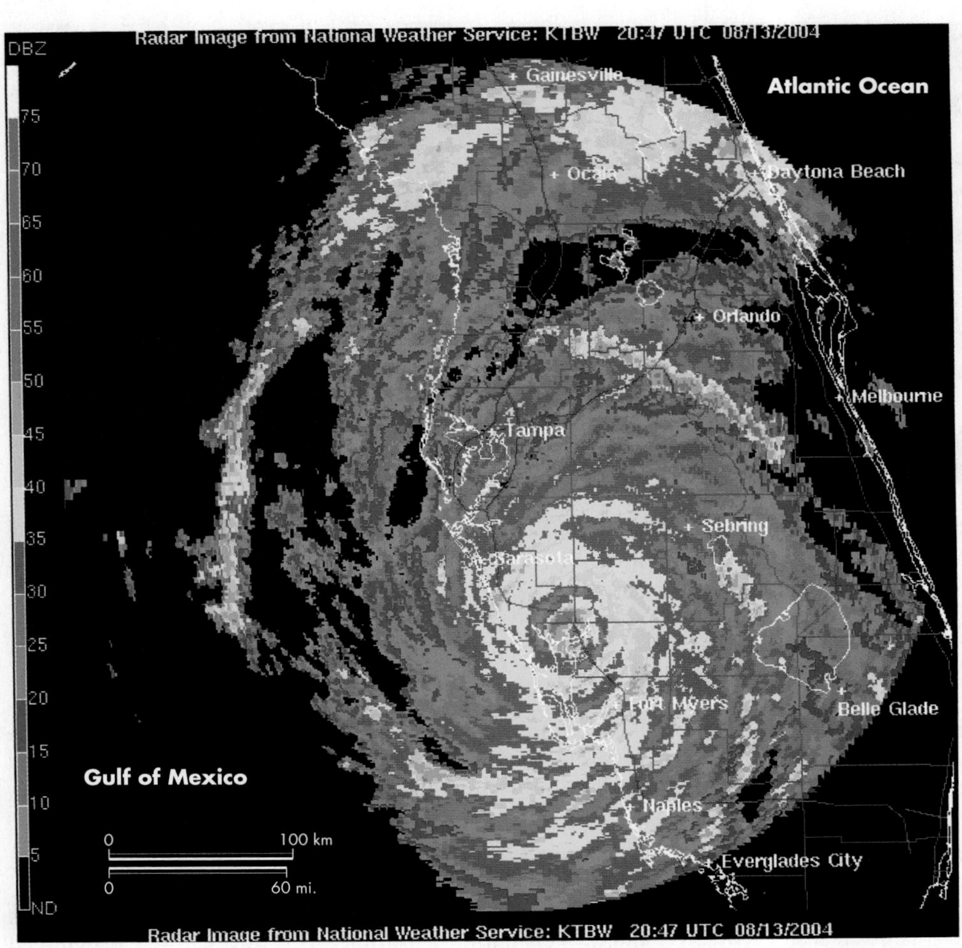

◄ **FIGURE 11.21 RADAR IMAGE OF HURRICANE CHARLEY** Image from weather radar after Hurricane Charley, a Category 4 storm, made landfall between Fort Myers and Sarasota on the west coast of Florida in 2004. Red and orange colours in the spiralling rain bands have the highest winds. *(National Weather Service)*

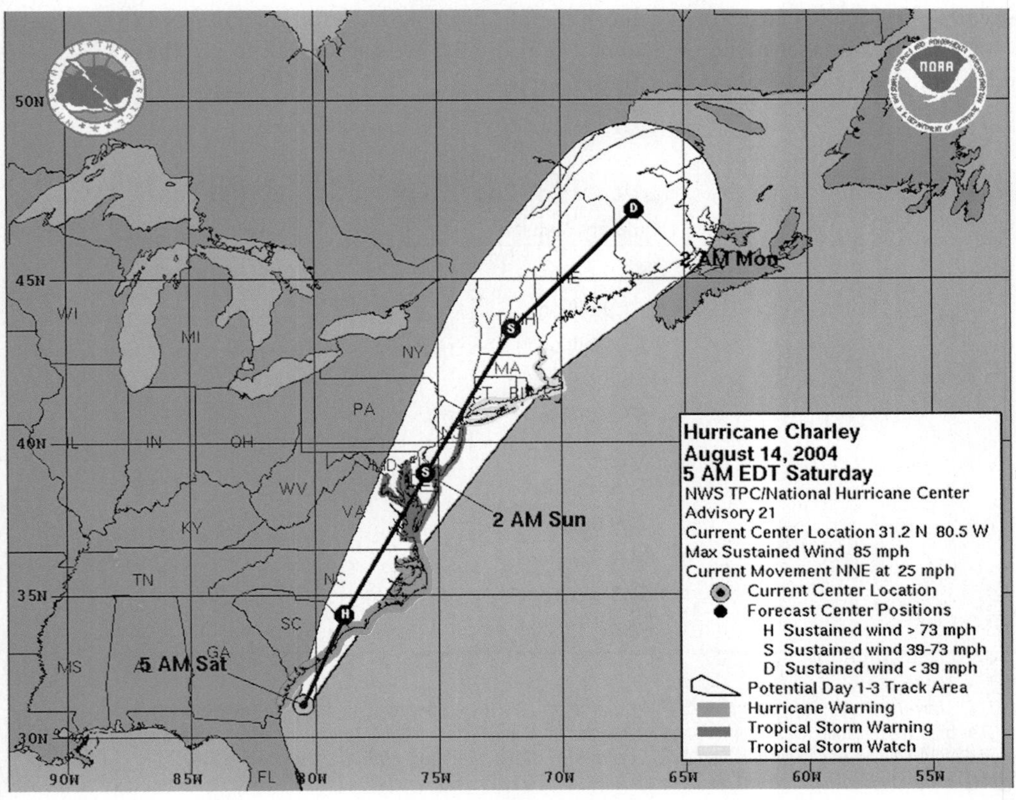

◄ **FIGURE 11.22 HURRICANE TRACK MAP** The predicted path of Hurricane Charley over a period of three days beginning on August 14, 2004. The width of the white area indicates the uncertainty in the predicted path. Severe weather watches and warnings are shown—red indicates a hurricane warning, blue a tropical storm warning, and yellow a tropical storm watch. *(National Weather Service)*

▲ **FIGURE 11.23 HURRICANE STORM SURGE FLOODING OF GALVESTON, TEXAS** About 65 000 people live on Galveston Island, a barrier bar south of Houston, Texas. (a) Photograph of part of Galveston. (b) Image showing area that would be flooded by a 5.8-m storm surge when a Category 5 hurricane hits the island. In 1900, catastrophic storm surge flooding by a Category 4 hurricane drowned more than 6000 people in the most deadly natural disaster in U.S. history. *(University of Austin Bureau of Economic Geology)*

programs take into account the forecast atmospheric pressure of the storm; its size, forward speed, direction, and wind speed; wind and wave stress; tides; and seafloor topography (bathymetry). The forecasts are generally accurate to 20 percent or better; thus if the computer model calculates a 31-m storm surge, you can expect the actual surge to be between 2.4 and 3.6 m.

Hurricane Prediction and the Future Until Hurricane Katrina in 2005, hurricane deaths in North America had dropped dramatically, largely because of better forecasting, improved evacuation, and greater public awareness.[7] However, Hurricane Katrina, which killed 2140 people in New Orleans, Biloxi, and other communities in southern Louisiana and Mississippi, highlighted the extreme vulnerability of people on the Atlantic and Gulf coasts of the United States to severe tropical storms. Unfortunately, coastal populations are skyrocketing, which may contribute to more hurricane-related deaths if inexperienced residents are not adequately prepared, if residents ignore evacuation warnings, or if evacuations are slowed by increased traffic.[7]

Property damage costs have increased as more and more people build homes and businesses in the coastal zone (Table 11.3). We can expect to see these costs continue to rise.

TABLE 11.3 Property Damage Costs from Hurricanes on the U.S. Atlantic and Gulf Coasts

Decade	Property Damage[1]
1900–1909	$2.0 billion
1910–1919	$4.9 billion
1920–1929	$3.1 billion
1930–1939	$8.3 billion
1940–1949	$7.6 billion
1950–1959	$18.5 billion
1960–1969	$34.5 billion
1970–1979	$29.6 billion
1980–1989	$29.9 billion
1990–1999	$78.5 billion
2000–2006	$182.7 billion

[1] In 2006 U.S. dollars.

Source: E. S. Blake, J. D. Jarrell, E. N. Rappaport, and C. W. Landsea. 2006. The Deadliest, Costliest, and Most Intense United States Tropical Cyclones from 1851 to 2005 (and Other Frequently Requested Hurricane Facts). *NOAA Technical Memorandum NWS TPC-4, Miami, FL: NOAA/National Weather Service, Tropical Prediction Center, National Hurricane Center. http:www.nhc.noaa.gov/Deadliest_Costliest.shtml. Accessed July 14, 2013.*

11.9 Perception of and Adjustment to Cyclones

Perception of Cyclones

Residents of regions at risk from hurricanes or coastal extratropical cyclones might have significant experience with them, but they do not always perceive the danger. During every hurricane, many people choose not to evacuate. People's perception of the risk of hurricanes differs according to their experience. Many people living in coastal areas are relatively new residents, and although they might have experienced a few hurricanes, they do not comprehend the threat of a major hurricane, and they might underestimate the hazard. Incorrect predictions of where or when a hurricane will strike can also lower the perception of risk. For example, if people are repeatedly warned of storms that never arrive, they might become complacent and ignore future warnings. As with any other hazard, accurate risk perception by planners and the public is key to reducing threats associated with cyclones.

Adjustments to Cyclones

Warning systems, evacuation plans and shelters, insurance, and building design are key adjustments to hurricanes. Emergency warning systems are designed to give people as much possible advance notice that a hurricane is headed their way. Warning methods include media broadcasts of hurricane watches and warnings and, in the case of immediate danger, the local use of sirens. Evacuation plans must be developed prior to the hurricane season to ensure the most well organized evacuation possible. At the time of evacuation, public transportation must be provided, shelters opened, and the number of outbound traffic lanes on evacuation routes increased. As was apparent with Hurricanes Katrina and Rita in 2005, it can take days to evacuate heavily populated areas. Lastly, property insurance should be available to people living in hurricane-prone regions.

Homes and other buildings can be constructed to withstand hurricane-force winds and also elevated to allow the passage of a storm surge. People living on hurricane-prone coastlines must also be aware of the hurricane season. Before the season begins, they should prepare their homes and property by trimming dead or dying branches from trees and installing heavy-duty shutters that can be closed to protect windows. They should be aware of evacuation routes and discuss emergency plans with family members. Disaster preparedness for a hurricane is similar to that for an earthquake and includes having at the ready flashlights, spare batteries, a radio, a first-aid kit, emergency food and water, a can opener, cash and credit cards, essential medicines, and sturdy shoes.[29] Tune in to a local radio or television station once a hurricane or tornado watch has been issued. If an evacuation order is given, gather emergency supplies and follow instructions.

REVISITING THE FUNDAMENTAL CONCEPTS

Hurricanes and Extratropical Cyclones

❶ Hazards can be understood through scientific investigation and analysis.

❷ An understanding of hazardous processes is vital to evaluating risk.

❸ Hazards are commonly linked to each other and to the environment in which they occur.

❹ Population growth and socio-economic changes are increasing risk from natural hazards.

❺ Damage and loss of life from natural disasters can be reduced.

1. Atmospheric conditions and sea surface temperatures can be monitored with radar, satellite and ocean sensors, and aircraft. Computer modelling of these data provides short-term predictions of the paths that cyclones will follow. As in the case of severe weather, however, cyclone formation and intensity are affected by numerous atmospheric and oceanic variables that continuously change, limiting the accuracy of the predictions over periods of more than a few days. When a hurricane approaches a coastal area, aircraft provide accurate information on wind speeds that, when combined with modelling, can improve forecasts of the intensity of the storm and its path and likely time and location of landfall.
2. Risk assessment for hurricanes is based on the probability of events of different intensities and their possible consequences. Hurricane probabilities for different areas of the Gulf and Atlantic coasts of the United States and the coasts of Atlantic Canada are known from the record of past hurricanes that have made landfall. Populations and infrastructure along these coasts are also known; therefore the risk for any area can be calculated. However, how the risk might change in the future due to climate change is much more difficult to predict.
3. Cyclones are linked to coastal erosion, flooding, mass wasting, tornadoes, and blizzards, some of which can be more destructive than the direct effects of the cyclones. For example, Hurricane Katrina and Hurricane Sandy were the most expensive hurricanes in U.S. history. The enormous property damage was largely the result of flooding caused by the storm surges generated by the hurricanes. In the case of Hurricane Katrina, levee failures resulting from the storm surge resulted in the deaths of 1300 people.

(continued)

4. In the past few decades, growth in the United States has been concentrated in coastal states. Populations in cities along the hurricane-prone Gulf and Atlantic coasts have grown and thus the risk has increased. Hurricanes Katrina and Sandy illustrate the increasing risk from tropical storms in the U.S. In the coming decades, continued population growth and development along the Atlantic coast will likely lead to ever-greater damage from hurricanes. Likewise, more extratropical storm disasters can be expected in other areas of North America, notably the Pacific coast where the population continues to grow.
5. Minimizing damage and loss of life from cyclones requires accurate forecasting, better construction practices, and more judicious land use in the coastal environment. Advances in remote sensing, computer modelling, and warnings reduced loss of life from hurricanes in the United States prior to Hurricane Katrina. Yet, as the Katrina example showed, poorly constructed levees, which failed due to the cyclone's storm surge, led to a catastrophe. Had the levees in New Orleans been designed to withstand a Category 5 hurricane, as they should have been, Katrina, which was a Category 4 storm at landfall, would have been far less costly.

Summary

Cyclones are large areas of low atmospheric pressure with wind converging toward the centre. Because of the Coriolis effect, winds blow anticlockwise in cyclones in the Northern Hemisphere.

Cyclones can be either tropical or extratropical based on their characteristics and origin. Tropical cyclones have warm cores, are not associated with weather fronts, and form between 5° and 20° latitude over tropical and subtropical oceans. These storms are referred to as hurricanes in this book. Extratropical cyclones have cool cores, develop along weather fronts, and form between 30° and 70° latitude over either land or ocean. They produce windstorms, snowstorms, blizzards, and severe thunderstorms. Both types of cyclones can produce tornadoes, heavy precipitation, and coastal storm surges.

A hurricane is assigned to a category on the Saffir-Simpson Hurricane Scale based on measured or inferred sustained wind speeds. Tropical storms and hurricanes are given male and female names from internationally agreed-upon lists for the region where they form. Extratropical storms can be named for the geographic area of origin or prevailing wind direction.

Most tropical cyclones begin as tropical disturbances—large thunderstorm complexes associated with low-pressure troughs. In the Atlantic Ocean, these disturbances form off the west coast of Africa as waves in the trade winds. A tropical disturbance that develops a rotational wind circulation pattern becomes a tropical depression and can become a tropical storm if its winds exceed 63 km/h. Only a few tropical storms become hurricanes. Conditions needed to form a hurricane include waters that are at least 26°C to a depth of several tens of metres, uninhibited convection of moist air evaporated from the sea surface, and weak winds aloft that keep vertical wind shear to a minimum.

Hurricanes develop spiralling bands of clouds around a calm central eye. The eye is bordered by an eyewall cloud that has the storm's strongest winds and most intense rainfall. Warm, moist air condensing in rain bands and the eyewall releases latent heat and provides a supply of energy for the hurricane. Excess heat is vented out of the top of the storm. Most hurricanes move forward at 19 to 27 km/h and are steered by winds in the middle and upper troposphere. They generally lose intensity over land or cool water.

Extratropical cyclones form along fronts where there are strong, diverging winds in the upper troposphere. Strong upper-level winds occur in polar and subtropical jet streams. Many extratropical cyclones intensify as they move east from the Rocky Mountains or along a coastline. In North America, most extratropical cyclones have a cold front trailing to the southwest and a warm front to the east. Cold air circulating around these cyclones collides with warm air rising in the eastern part of the storm.

In North America, the Gulf and Atlantic coasts are at greatest risk from hurricanes. Hurricane-strike probabilities are greatest in southern Florida and the northern Gulf coast. Extratropical cyclones are the primary severe weather hazard on the Pacific coast. Nor'easters, named for their strong northeasterly winds, are a hazard in Atlantic Canada and in the eastern United States.

Cyclones produce coastal storm surges, high winds, and heavy rains. In the Northern Hemisphere, the greatest storm surges, highest winds, and most tornadoes are in the right forward quadrant of cyclones. Strong currents from storm surges can cut channels through barrier islands and deposit sand as overwash. Wind damage from hurricanes is more widespread but less deadly than the storm surge. Coastal cyclones cause both saltwater and freshwater flooding. Inland flooding occurs from hurricane remnants, especially if the storms move slowly, encounter hills or mountains, interact with other weather systems, or track over already saturated ground.

Cyclones are closely linked to other severe weather, flooding, landslides, and debris flows. The hazards posed by storm surges and erosion from coastal cyclones will increase as sea level rises due to global warming. Hurricane intensity is also forecast to increase in a warmer climate. Damage from both hurricanes and extratropical cyclones will increase as coastal populations and per capita wealth grow. On the positive side, cyclones are important sources of precipitation, help redistribute heat on Earth, and contribute to long-term ecosystem health.

Perception of cyclones depends on individual experience and proximity to the hazard. Community adjustments to hurricanes involve building protective structures or modifying people's behaviour through land-use zoning, developing evacuation procedures, and using warning systems. Individual adjustments to hurricanes involve having emergency supplies on hand, preparing in advance of the arrival of the hurricane, and, if required, evacuating before the storm strikes. Homes can be constructed to withstand hurricane-force winds and elevated to allow passage of small storm surges.

Hurricane forecasting relies on weather satellites, aircraft flights, Doppler radar, and automated weather buoys. Computer models predict hurricane tracks more accurately than hurricane intensities. Storm surge predictions are based on wind speed, fetch, average water depth, and timing of landfall in relation to tidal levels.

Key Terms

cyclone (p. 311)
extratropical cyclone (p. 311)
eye (p. 313)
eyewall (p. 313)
hurricane (p. 311)
hurricane warning (p. 328)
hurricane watch (p. 328)
jet stream (p. 315)
overwash (p. 323)
rain band (p. 313)
storm surge (p. 322)
tropical cyclone (p. 311)
tropical depression (p. 312)
tropical disturbance (p. 312)
tropical storm (p. 313)
typhoon (p. 311)

Did You Learn?

1. Explain the difference between a tropical cyclone and an extratropical cyclone.
2. How do tropical cyclones, tropical storms, tropical depressions, and tropical disturbances differ?
3. Explain how and where hurricanes form.
4. Describe the conditions necessary for hurricane formation.
5. Explain how the Coriolis effect influences hurricane winds and movement.
6. How are hurricanes categorized? What category is the most intense?
7. Which areas of the United States and Canada are at greatest risk from hurricanes?
8. Where on Earth are tropical cyclones most common?
9. Sketch the three common paths taken by hurricanes spawned in the North Atlantic.
10. Name the three most important causes of hurricane damage. Which is typically the most deadly?
11. Explain the causes and effects of storm surges.
12. Describe the changes that occur when a hurricane moves inland and becomes an extratropical cyclone.
13. Why is risk from cyclones likely to increase in the future?
14. Explain how cyclones are linked to other natural hazards.
15. What are the natural service functions of cyclones?
16. Describe the tools used in making hurricane forecasts.
17. Describe the adjustments that people can make to survive hurricanes.

Critical Thinking Questions

1. If you had to evacuate your home due to an approaching hurricane, where would you go? If you had to evacuate your home and travel at least 160 km away from it, where would you go? In either case, what would you take with you? What problems might you or your neighbours face during an evacuation?
2. Obtain a topographic map for an urban area on the coast of Nova Scotia. Shade or colour the area that would be affected by a 5-m storm surge. Examine your completed map and assess the damage that would be caused by such a surge, the routes people would take for evacuation, and restrictions you would recommend for future development.
3. Hurricane Juan, which struck Nova Scotia and Prince Edward Island in 2003, caused more than $200 million damage. At some time in the future, Atlantic Canada will experience another hurricane. What measures do you think can be taken to reduce the damage from that inevitable storm?
4. Should everyone except emergency response personnel be forced to evacuate an area in the path of an approaching hurricane? Why or why not?
5. Some scientists have argued that extratropical storms pose a greater risk than earthquakes to Vancouver and Victoria. Do you agree?

MasteringGeology

MasteringGeology **www.masteringgeology.com**. Looking for additional review and test prep materials? Visit the Study Area in MasteringGeology to enhance your understanding of this chapter's content by accessing a variety of resources, including **Self-Study Quizzes, Geoscience Animations, GEODe Tutorials, RSS feeds, flashcards,** web links and an optional **Pearson eText.**

CHAPTER 12

▶ **SURGING OCEAN** Key West residents Brian Goss (left), George Wallace, and Michael Mooney (right) flee the Florida coast as Hurricane George's 3.6-m storm surge washes away houses, cars, roads, and beaches in September 1998. *(Dave Martin/AP Images)*

Waves, Currents, and Coastlines

Learning Objectives

In this chapter we focus on one of the most dynamic environments on Earth—the coast, where the sea meets the land. Large numbers of people live on and visit beaches and rocky coastlines, but most of us have little understanding of how ocean waves and currents form and change coastlines. This chapter explains the processes at work in coastal areas and the hazards that result from wind and waves. Your goals in reading this chapter should be to

- Understand waves, currents, and beach processes
- Understand coastal hazards
- Know what coastal regions are at risk from these hazards
- Recognize the links between coastal processes and other natural hazards
- Know the benefits of coastal processes
- Understand how our use of the coastal zone affects coastal processes
- Understand the measures that can be taken to avoid damage and injury from coastal hazards

Harris Meisner's Farm by the Sea

Harris Meisner was a true Canadian Maritimer. He was born and spent his early years on Meisner Island on the east coast of Nova Scotia (Figure 12.1). The island, which was named after Harris's ancestors, was home to three families. It had the same number of working farms, several animals, and three dug water wells. But during Harris Meisner's lifetime, the island was eroded away by the sea. Meisner wryly noted that "there's a lot of water in the family well these days." The disappearance of Meisner Island demonstrates that the coastline, regardless of human intervention, is always changing.

To understand why and when Meisner Island disappeared, we need to look at the local geology. Many farms in Nova Scotia, as in other parts of Canada, are located on *drumlins*—glacially sculpted ridges that have the shape of a whale back. Drumlins in Nova Scotia consist of till, which is a mixture of boulders, gravel, sand, and mud deposited by a glacier. The three families on Meisner Island made a living off the land by mixed farming. Regrettably, this situation was forever changed by the 30-cm rise in sea level along the east coast of Nova Scotia during the past century, a change that sealed the fate of Meisner Island.

Geologists have tracked the destruction of the island using aerial photographs extending back to 1945 and historical records (Figure 12.2). The drumlin on which Mr. Meisner lived was relatively stable from 1766, when Captain James Cook mapped the coast, until 1945, when erosion accelerated. Then, within only three decades, the island disappeared.

Between 1954 and 1966, the part of the drumlin most exposed to ocean waves retreated 7 m/y; between 1966 and 1974 the rate of erosion was 9.4 m/y. At that rate it didn't take long to lose the island. Severe Atlantic storms, including Hurricane Edna in September 1954, made the problem worse. The number and violence of storms increased during the early 1950s, and some sand and gravel was removed from the shore for construction purposes, compounding the problem. In one generation, Meisner Island was consumed by the sea and its sediments recycled, leaving behind a boulder shoal (Figure 12.2).

What Harris Meisner perhaps did not realize is that change is the norm along coasts. In Nova Scotia, the sea typically destroys land in some

▲ **FIGURE 12.1 MEISNER ISLAND** Archival photograph of Meisner Island taken in 1928. *(Harold Meisner)*

places, but builds new land in others. Changes can be seen elsewhere, as in the high Arctic, where the remains of house rings and fire pits preserved on raised beaches are a testament to seasonal visits by nomadic humans as far back as 4000 years ago. The camps become progressively younger toward the present shoreline due to the inexorable rise of the land. Native oral traditions on the west coast of Canada speak of sudden changes of the shoreline there, caused by great earthquakes and tsunamis that wiped out entire villages. Today, coastal landowners are aware of changes that happen suddenly, such as during a hurricane. But people tend to think of these changes as "acts of God" rather than as normal events that shape the coast.

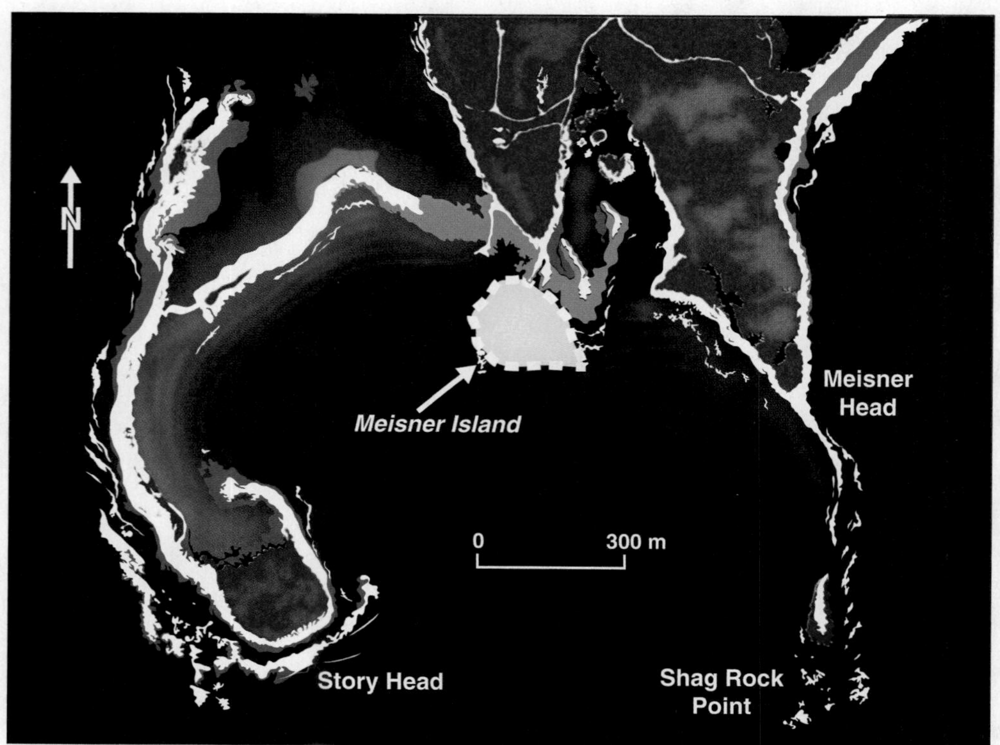

▲ **FIGURE 12.2 MEISNER ISLAND**, located off the east coast of Nova Scotia, was destroyed by erosion during the second half of the twentieth century.

Geologists determine shoreline changes over periods ranging from years to millennia. They track the movement of sediment by waves and currents, and determine the effects of changing sea level on the coast. They have observed that in natural settings, where shorelines have no hard protective structures, such as seawalls and groins, beaches generally do not disappear, but rather change seasonally or shift location. As we shall see, interference with natural shore processes can significantly alter shoreline processes and cause severe erosion. And even in natural environments, shorelines will respond to gradually rising sea level. Sea level has been rising globally at an average rate of about 2 to 3 mm/y over the past century due to glacier melt and warming and expansion of upper ocean waters (see Chapter 14).[1] Although this rate of change may seem small, sea-level rise may accelerate in coming decades and cumulatively result in a total rise of many tens of centimetres or more by the end of the century. The effects for hundreds of millions of people who live at or close to the shore will be catastrophic.

12.1 Introduction to Coastal Hazards

Coastal areas are among the most dynamic environments on Earth. In these areas, terrestrial and oceanic processes converge to produce landscapes that can change rapidly.

On a continental scale, coastal topography is influenced by plate tectonics. The Atlantic coast of Canada and the United States is located on a tectonically *passive margin*, because it is distant from a convergent plate boundary (see Chapter 2). Tectonically passive margins commonly have wide continental shelves and extensive sandy beaches and barrier islands. Rocky shorelines are less common, except where ancient mountain ranges, such as the Appalachians, meet the coast. Some shorelines along the Great Lakes are sandy, whereas others, especially around Lake Superior and Lake Huron, cross the Canadian Shield and are rocky.

In contrast, the Pacific coast of Canada and the United States is located on a tectonically *active margin* close to plate boundaries (see Chapter 2). The southern British Columbia coast and the coasts of Washington and Oregon border the convergent boundary between the North American and Juan de Fuca plates. The northern British Columbia coast and most of the California coast are adjacent to transform boundaries between the North American and Pacific plates. Southern Alaska and the Big Island of Hawaii are also tectonically active. Mountain building in tectonically active areas has produced coasts with sea cliffs and rock shorelines. Although long sandy beaches are present in these regions, they are not as abundant as, for example, along the Atlantic and Gulf coasts of the United States.

At a local level, coastal topography is strongly influenced by geology—specifically the type and structure of the rock at the shoreline. Coastlines formed on granite, for example, differ from those developed in flat-lying sandstone and shale, and the latter differ from shorelines developed in dipping sandstone and shale, or in limestone. Coastlines formed in glacial sediments commonly have rapidly eroding sea cliffs and sand and gravel beaches.

Shorelines are also affected by climate and by plants and animals. Many shorelines in southeast Alaska, for example, owe their form to sediment-laden rivers flowing from glaciers, and Arctic shorelines are affected by seasonal sea ice and permafrost. Coastlines in temperate regions have intertidal marshes that support a wide variety of plants, birds, and burrowing organisms, and many coastlines in the tropics and subtropics are bordered by mangrove swamps or offshore coral reefs.

The impact of hazardous coastal processes is considerable, because many cities and towns are located near the sea or lakes. Millions of Canadians live along the Pacific and Atlantic coasts and along Lake Ontario. Shoreline problems will increase as more people move to or visit coastal areas, and these problems will be compounded by rising sea levels caused by climate warming.

The main coastal hazards are the following:

- Strong coastal currents, including rip currents generated in the surf zone and tidal currents in narrow bays and channels
- Coastal erosion
- Sea-level rise
- Storm surges from tropical and extratropical cyclones (see Chapter 11)
- Tsunamis (Chapter 4)

12.2 Coastal Processes

Waves

Waves are generated by wind, in some cases thousands of kilometres from the coast. Wind blowing over the ocean or a lake transfers some of its energy to the water, producing waves. The waves, in turn, travel through the water and eventually expend their energy at the shoreline.

Waves have a range of sizes and shapes. The size of waves depends on a combination of the following:

- The velocity of the wind. The stronger the wind, the larger the waves.
- The duration of the wind. Winds that last longer, such as those associated with storms, have more time to impart energy to the water, thereby producing larger waves.
- The distance the wind blows across the water, which is referred to as the *fetch*. A longer fetch allows larger waves to form; thus waves in the ocean are generally larger than those in a lake.

12.1 CASE STUDY

Rogue Waves

Rogue waves form when a series of similar-size waves meet and coalesce to produce a much larger wave. This process is known as *constructive interference*. The new wave may be as high as the sum of the waves that coalesced. Seafloor irregularities and currents may also be important in forming rogue waves.

Rogue waves can be extremely dangerous. They can appear out of nowhere, crashing over a pier or a rocky headland and sweeping unsuspecting people to their deaths. Each summer several beachgoers in Canada and the United States are swept into the ocean by rogue waves. Shores, however, are not the only areas at risk from rogue waves. These waves can appear in the open ocean in otherwise fairly calm water (Figure 12.3) and can be large enough to break in open water, threatening ships or offshore drilling platforms, such as those on the continental shelf off Atlantic Canada. In stormy seas, where all waves are large, rogue waves can reach 30 m in height. A rogue wave 20 m high struck the cruise ship *Norwegian Dawn* about 400 km off the coast of Georgia in April 2005, damaging the ship and ending the cruise. Large rogue waves were apparently common during the October 1991 "Perfect Storm," an intense extratropical cyclone featured in a book and a movie, which sunk the fishing boat *Andrea Gail* and killed its entire six-person crew. Ten separate rogue waves more than 25 m high were identified in the world's oceans in a study of three weeks of global satellite radar data.[2] Such information is important to the shipping industry because most ships are designed to withstand waves only 15 m high.

Rogue waves can develop almost anywhere in the world's oceans, but they are most common where local or regional conditions favour the constructive interference that permits them to form. One such place is off the southern tip of Africa where the southwest-flowing Agulhas Current, the eastward-flowing Antarctic Circumpolar Current, and the swell from Antarctic storms interact.

◄ **FIGURE 12.3 ROGUE WAVE** This huge rogue wave is approaching the bow of the *JOIDES Resolution*, the scientific drilling ship of the Ocean Drilling Program. Rogue waves have sunk supertankers and large cargo ships. *(Ocean Drilling Program, Texas A&M University)*

As waves move away from their source, they become organized into groups, or *sets*, of similar size and shape. Sets of waves may travel long distances across the ocean and arrive at distant shores with little loss of energy. Unexpected, unusually large **rogue waves** occasionally arrive at the shore, sometimes with disastrous results (see Case Study 12.1).

Waves moving across deep water have a similar basic shape, or wave form (Figure 12.4a). Three parameters describe the size and movement of a wave: *wave height* (H), which is the difference in height between the trough and the crest of a wave; *wavelength* (L), the distance between successive wave crests; and *wave period* (P), the time in seconds for successive waves to pass a reference point. The reference point used to determine wave period could be a pier or another object anchored on the seafloor or lake bottom.

By studying the motion of an object at the surface of the water and one below the surface, you can understand how waves transmit energy through the water. When a wave

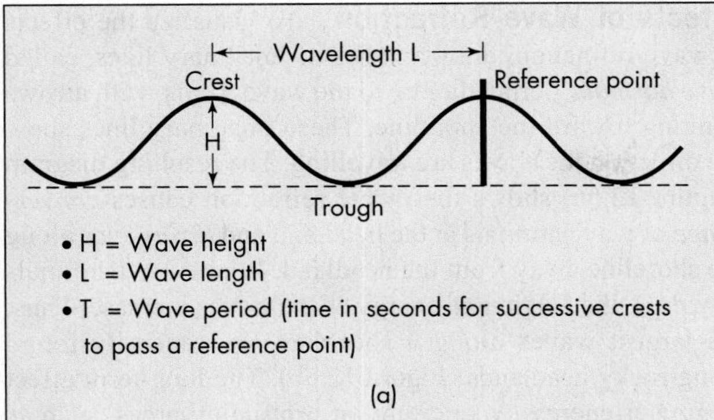

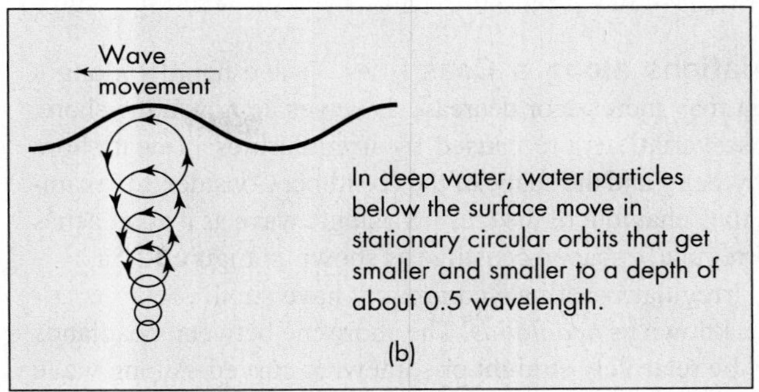

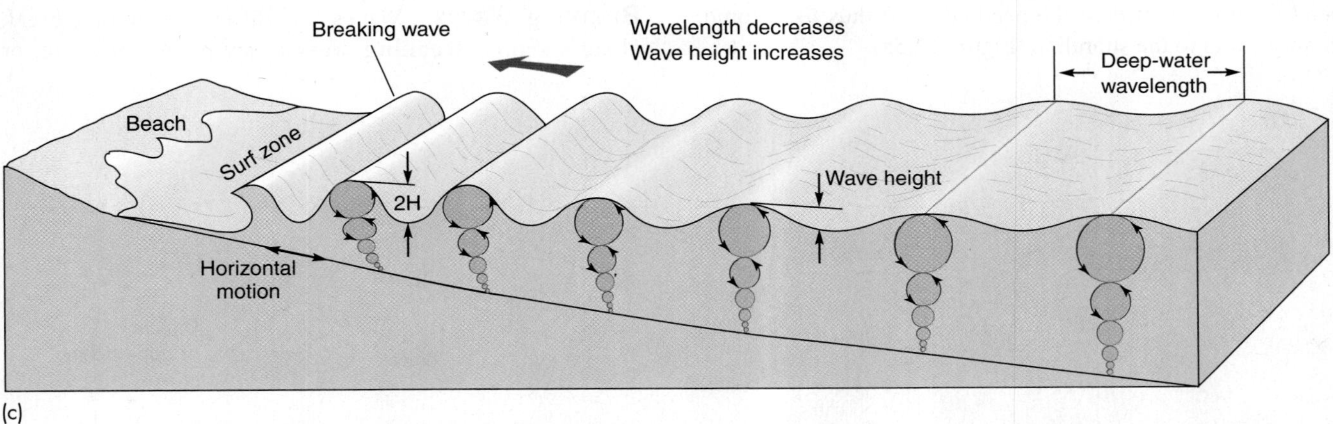

▲ **FIGURE 12.4 WAVES AND BEACHES** (a) Waves form in deep water where the water depth is greater than one-half of the wavelength (L). The curving black line is the water surface and the thick vertical black line is a pier or some other reference point for determining the wave period (T). The black dashed line connecting the bottom of the troughs is the reference line for calculating wave height (H). (b) Motion of water particles associated with the movement of a wave in deep water. The water particles follow the path of the arrows in the black circles. (c) The motion of water particles in shallow water at a depth less than 0.5 L. The water particles follow the path of the arrows in the circles. The waves are approaching the shore from right to left in (b) and (c).

moves through the area, an object 20 m below the surface has a circular orbit, moving up, forward, down, and back, always returning to the same place (Figure 12.4b). An object nearer the water surface also moves in circles, but the circles are larger.

The shape of the orbital motions changes as waves enter shallow water. At a depth of less than about one-half their wavelength (i.e., depth 0.5 L), the waves begin to "feel the bottom," causing the circular orbits to become ellipses (Figure 12.4c). In very shallow water, motion at the bottom may be nearly horizontal, with only a small vertical component. You probably have experienced this while standing or swimming in shallow water; the water repeatedly pushes you toward the shore and then away from it.

Wave sets generated by storms far out at sea are called *swell*. Wave height, period, and velocity of the swell can be predicted by using equations based on the fetch, wind velocity, and the length of time the wind blows over the water. These predictions are important: By knowing the velocity and height of the waves generated by a distant storm, we can estimate when the waves will strike the shore and how erosive they will be.

As waves approach the coast, they become unstable and break. Wavelength and velocity decrease and wave height increases; only the wave period remains constant. The waves also change shape—the rounded crests and troughs found in deep water are replaced by peaked crests with relatively flat troughs in shallow water. Perhaps the most dramatic change is the rapid increase in wave height. As waves approach their breaking point in shallow water, their height may increase to twice that in deep water (Figure 12.4c). The wave crest becomes unstable when the water depth is about 1.3 times the wave height. It keeps moving forward while the lower part of the wave slows down, and the wave collapses, or breaks, toward the shore.

Waves release a large amount of energy as they rush ashore. Wave energy is approximately proportional to the square of wave height. Thus, if wave height increases from 1 m to 2 m, the wave energy, or wave power, increases by a factor of 4. Waves 5 m high, which are common in large storms, expend about 25 times the energy of waves 1 m high. The energy spent on a 400-km length of open coastline by waves about 1 m high is approximately the same as the energy produced by an average nuclear power plant over the same period.[3] The energy released by larger storm waves along the same stretch of coast is commonly many times this amount.

Variations along a Coastline Wave heights along a coast may increase or decrease as waves approach the shore. These variations are caused by irregularities in near-shore bathymetry and the shape of the coastline. Consider, for example, the behaviour of the crest of a single wave as it approaches an irregularly shaped coastline as shown in Figure 12.5a.

Irregular coastlines commonly have small rocky peninsulas known as *headlands*. The shoreline between headlands may be relatively straight or somewhat curved. A long wave approaching the coast will first slow down in the shallow water off the headland. The slowdown causes the wave to bend, or *refract*, around the headland and thus to become more parallel to the shoreline (Figure 12.5a).

Effects of Wave Refraction To visualize the effects of wave refraction, draw a series of imaginary lines, called *wave normals*, perpendicular to the wave fronts, with arrows pointing toward the shoreline. These imaginary lines show the direction the waves are travelling. The resulting diagram (Figure 12.5a) shows that wave refraction causes *convergence* of wave normals at the headland and *divergence* along the shoreline away from the headland. Where wave normals converge, the height and energy of the waves increase. Thus, the largest waves along a shoreline are generally found along rocky headlands (Figure 12.5b). The long-term effect of greater energy expenditure on protruding areas, such as headlands, is that the shoreline tends to become straightened.

Breaking Waves Waves also differ in how they break along a shore. **Breaking waves** may plunge or surge, or

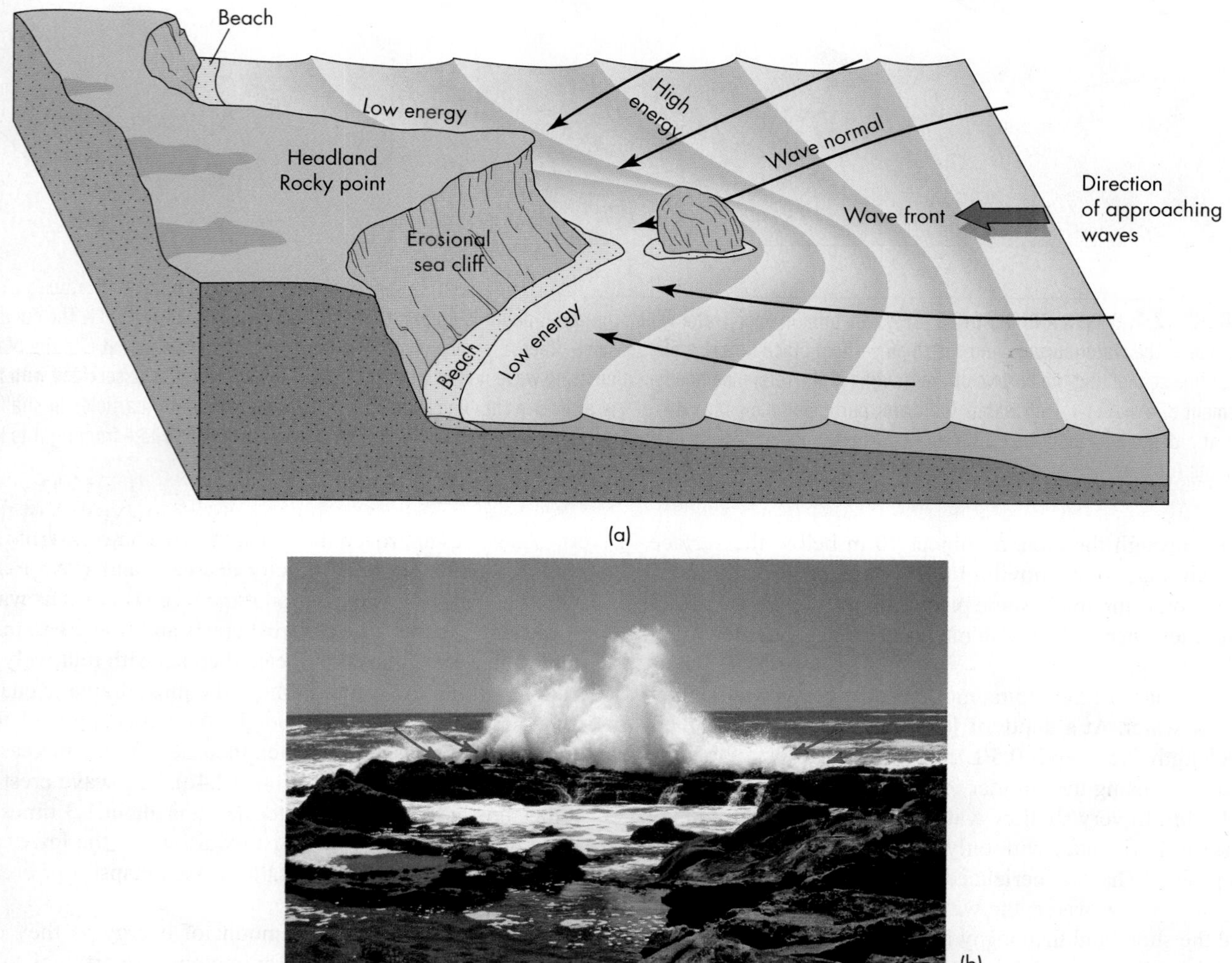

▲ **FIGURE 12.5 CONVERGENCE AND DIVERGENCE OF WAVE ENERGY** (a) A schematic diagram showing wave refraction and concentration of wave energy at a headland. Refraction, the bending of wave fronts, causes convergence of *wave normals* (long curving black arrows) on the headland and divergence along indented sections of the coast. (b) Large waves striking a rocky headland along the Pacific Coast at Pebble Beach, California. *(Robert H. Blodgett)*

▲ **FIGURE 12.6 TYPES OF BREAKERS** A schematic diagram and photographs showing (a) plunging breakers on a steep beach and (b) spilling breakers on a gently sloping beach. *((a) Peter Cade/Stone/Getty Images; (b) Patti Altridge)*

they may gently spill onto the shore, depending on local conditions. *Plunging breakers* (Figure 12.6a) typically form on steep beaches and can be very erosive. *Spilling breakers* (Figure 12.6b) commonly develop on wide, gently sloping, sandy beaches and are less erosive than plunging breakers.

In estuaries subject to large tides, inflowing tidal waters may be slowed by outflowing river water to produce **tidal bores**.[4] Tidal bores have very steep fronts, in rare cases up to several metres high, and surge forward like broken waves (Figure 12.7). They occur at the mouths of the Amazon, Hooghly, Meghna, Indus, Severn, and Yangtze rivers; within the Bay of Fundy in Nova Scotia; and in Ungava Bay in Quebec. Breakers along a particular reach of the coast can also change seasonally and with changes in underwater slope and topography.

◄ **FIGURE 12.7 TIDAL BORE** Steep landward-moving wave produced by retardation of inflowing tidal water by outward-flowing river water in upper Cook Inlet, Alaska. *(NOAA)*

Beach Form and Processes

A **beach** consists of loose material, typically sand or gravel, that has accumulated by wave action at the shoreline. Sand and gravel are derived from a wide variety of rocks and therefore can differ in colour and composition. The white beaches of South Pacific islands, for example, are made of broken bits of shell and coral, and Hawaii's black sand beaches consist of sand-size fragments of basalt. Wave energy and beach shape are other factors affecting beach materials. Most steep, high-energy beaches are gravel, whereas gently sloping, lower-energy beaches commonly consist of sand.

The Beach Onshore An understanding of beach processes requires knowledge of beach forms (Figure 12.8). The landward boundary of the beach can be a cliff, called a **sea cliff** along a seashore and a **bluff** along a lakeshore; one or more sand dunes; or terrestrial vegetation. Sea cliffs and lakeside bluffs are erosional landforms produced by waves, currents, and, in some cases, landslides. In contrast, coastal sand dunes form by deposition of wind-blown beach sand.

The onshore portion of many beaches can be divided into two zones that are parallel to the shoreline—one that is flat or slopes gently landward, called the *berm*; and another that slopes seaward, called the *beach face*. The berm is located at the backshore of a beach and consists of sand deposited by waves as they rush up and expend the last of their energy. Beaches may have more than one berm or none at all. The beach face is located seaward of the berm, where the beach slope steepens toward the water. It lies in part within the **swash zone**, which experiences the repeated uprush and backwash of waves. The swash zone shifts seaward or landward because of changes in water level resulting from storms or tides.

The Beach Offshore The surf zone and the breaker zone lie offshore of the swash zone (Figure 12.8). The **surf zone** is located just seaward of the swash zone and is the place where waves move turbulently toward the shore after they break. Beyond the surf zone is the **breaker zone**, where incoming waves peak and break. A low ridge consisting of sand or gravel, called a *longshore bar*, occurs on the seafloor or lake floor in the breaker zone (Figure 12.8). A *longshore trough* may form by wave and current action landward of the longshore bar. Both the bar and the trough are elongated parallel to the crests of the breaking waves. Wide and gently sloping beaches may have several lines of breakers and longshore bars.[5]

Sand Transport Beach sand is not static; wave action constantly moves the sand in the surf and swash zones. Storms erode sand from the beach and redeposit it either offshore or landward in dunes. Most of the sand moved offshore during a storm returns to the beach later, during fair weather.

Sand also is carried parallel to the shore in the swash and surf zones by a process termed **littoral transport** (Figure 12.9). Littoral transport consists of two processes—beach drift and longshore drift. In *beach drift*, the repeated shoreward and seaward movement of sand in the swash zone produces a sinuous or zigzag transport path (Figure 12.9). The shoreward component of the movement is at an angle to the shoreline, whereas the seaward component is nearly perpendicular to the shoreline. *Longshore drift* is the transport of sediment by currents that flow parallel to the shoreline, called *longshore currents*. Both types of drift occur where waves strike the coast at an angle other than 90 degrees (Figure 12.9). The terms *updrift* and *downdrift* refer to the direction that sediment is moving or accumulating relative to the direction of the incoming waves. For example, updrift in Figure 12.9 is toward the upper right of the diagram.

The direction of littoral transport differs markedly along the Pacific, Atlantic, and Arctic coasts of Canada, primarily because of the complex shape of the coastline. Significant littoral transport is restricted to those parts of the Canadian coast with broad sandy beaches, largely on the Pacific and Atlantic coasts. Littoral transport is limited or does not occur on the steep rocky shorelines that form much of the Canadian coast and also on those parts of the Arctic coast where sea ice limits wave formation during

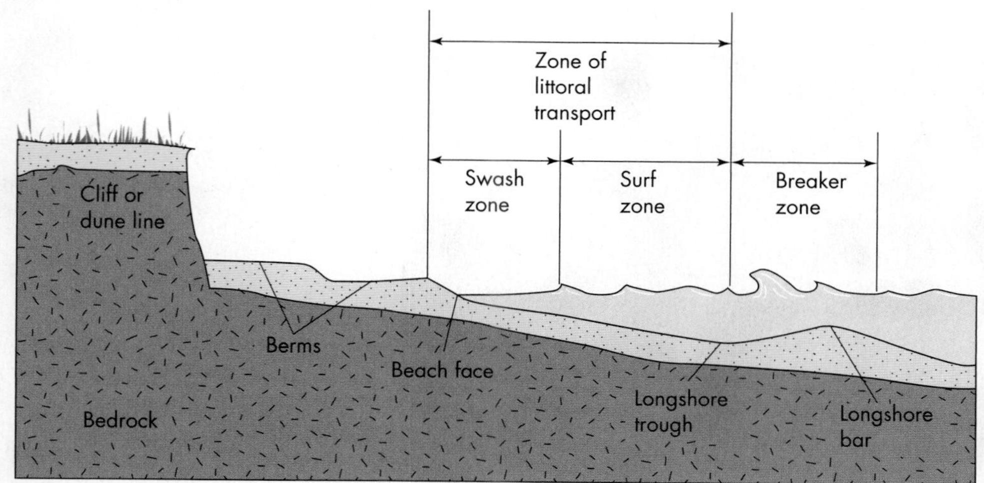

◄ **FIGURE 12.8 BEACH TERMINOLOGY** Basic terms for landforms and wave action in the beach and nearshore environment. A sea cliff or line of coastal sand dunes marks the landward limit of the beach. Two berms slope gently toward the cliff. The beach face slopes toward the water. A longshore trough and longshore bar are seaward of the beach face; the longshore bar forms within the breaker zone. The zone of littoral transport includes the swash and surf zones.

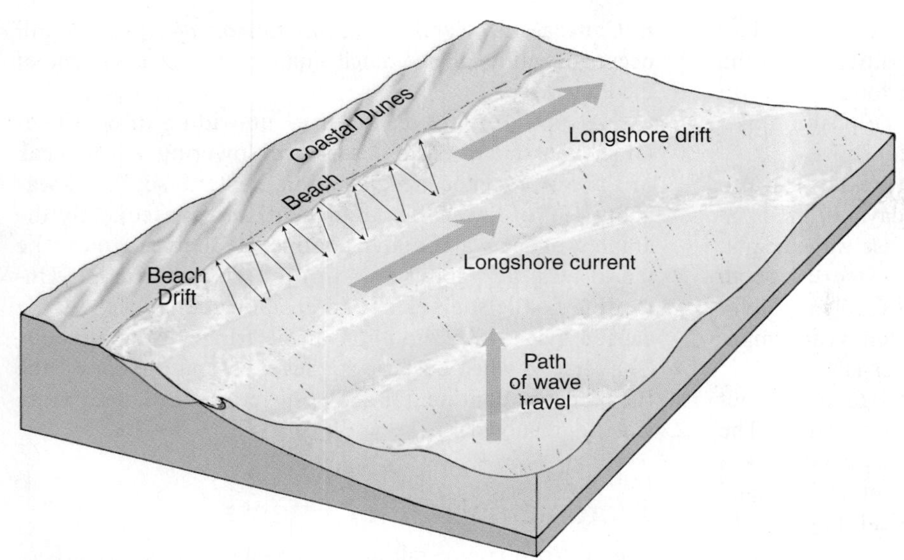

◄ **FIGURE 12.9 TRANSPORT OF SEDIMENT ALONG THE SHORE** A block diagram illustrating the processes of beach drift and longshore drift. In this diagram, waves approach the coast from the lower right. The direction of beach drift is shown by the many thin black arrows in the swash zone. The direction of long drift is shown by the straight thick dark-blue arrow in the surf zone. Collectively, these two types of drift move sand along the coast in a process termed littoral transport.

much of the year. Broad sandy beaches are less common in Canada than in the United States. Rates of longshore drift for some U.S. beaches range from 150 000 m^3 to 300 000 m^3 of sediment per year.[6] Rates of longshore drift for sandy shorelines of the Great Lakes are much less—about 6000 to 69 000 m^3 per year. Even these rates, however, are substantial when you consider that a typical dump truck carries a meagre 8 m^3 of sand.

Landforms Produced by Littoral Drift Beaches are not the only features produced by waves, currents, and longshore drift. Other landforms produced by these processes include spits (Figure 12.10), tombolos, and barrier islands. **Spits** are narrow and low, finger-like ridges of sand or gravel that are attached to the coast and extend out into the sea or a lake. **Tombolos** are similar to spits except they are connected at both ends to the shore. **Barrier islands** are long and relatively narrow islands that lie offshore of a coast and are parallel to it. Except for the tidal and storm channels that breach them, barrier islands isolate bays and coastal waterways from the open ocean. Barrier islands are common features on the Atlantic coast of the United States.

12.3 Sea-Level Change

The level of the sea at the shoreline is constantly changing due to tidal fluctuations. The maximum tidal range on Earth is 17 m and occurs at two places in Canada—the Bay of

◄ **FIGURE 12.10 GOOSE SPIT** A spit on the east coast of Vancouver Island, British Columbia. Goose Spit has formed over the past several thousand years from sand and gravel carried by south-flowing longshore currents. The source of the sediment is eroding sea cliffs to the north (top-centre of the photo). The log boom and pier are located on the landward side of the spit, where they are protected from waves. *(Waite Air Photos Inc.)*

Fundy, Nova Scotia, and Ungava Bay, Quebec.[7] Tidal fluctuations are produced by the gravitational attraction of the moon and, to a lesser extent, the sun. These forces cause sea level to fluctuate daily and seasonally as the position of a coast relative to the moon and sun constantly shifts.

Changes in wind speed and atmospheric pressure can also change sea level over periods of hours or days. In the open ocean, strong winds pile up water and increase wave height, producing the swell described earlier. The swell increases both water level and wave heights when it reaches the shore. Storm surges from hurricanes and extratropical cyclones can temporarily raise sea level many metres (see Chapter 11).

Other processes change sea level more gradually—not cyclically, as tides do, or episodically, as weather does. The position of the sea at the shore, referred to as **relative sea level**, is controlled by the vertical movement of both the land and sea water. These movements can be local, regional, or global in extent.

Eustatic Sea-Level Change

Sea level rises or falls when the amount of water in the world's oceans increases or decreases, or when there is a change in the shape of ocean basins. These global changes in sea level are referred to as **eustasy**. The shapes of oceans change due to plate movements over timescales of millions of years and are not important in the short term. Climate is the dominant control on eustatic sea level today, exerting this control in two ways. First, the average air temperature influences the average temperature of the oceans. As sea water warms it expands, and as it cools it contracts. Sea level is currently rising at a rate of about 3 mm/y, and about half of this rise is thought to be caused by *thermal expansion* driven by atmospheric warming.[1] Second, changes in atmospheric temperature control the amount of snow and glacier ice on land. The volume of water locked up in alpine glaciers and ice sheets on Greenland and Antarctica is closely related to the average air temperature over years to decades. Atmospheric warming over the past century has caused alpine glaciers to melt and thus increased the amount of water in the oceans. About half of the eustatic sea-level rise in recent years is due to glacier melt.

Isostatic Sea-Level Change

Local and regional processes also affect relative sea level. The general equilibrium of the forces tending to elevate or depress Earth's crust is referred to as **isostasy**. Isostatic sea-level changes are caused by an increase or decrease in the weight of ice or water on the crust or by a change in the thickness of the lithosphere. The land can rise or fall slowly as Earth's crust responds to changes in glacier cover on the continents. The disappearance of large ice sheets in North America and Eurasia at the end of the Pleistocene Epoch removed a huge load on the crust. This unloading was accompanied and followed by hundreds of metres of uplift of the land in some areas and smaller amounts in others. In some areas, such as Hudson Bay and the Arctic Islands in Canada and Scandinavia, *glacio-isostatic uplift* is still occurring, although at a much smaller rate than at the end of the Pleistocene.

A similar effect results from unloading of continental shelves due to eustatic sea-level lowering. At the peak of the last Pleistocene glaciation, eustatic sea level was about 130 m lower than at present, and consequently the amount of sea water above continental shelves around the world was only about one-third of what it is today. Unloading of the shelves due to eustatic sea-level lowering caused *hydro-isostatic uplift* at the fringes of continents. Conversely, when eustatic sea level rose at the end of the Pleistocene, continental shelves were loaded with more water and subsided.

Tectonic and Other Effects

The land can rise or fall due to tectonic movements. These movements can be slow, of the order of a few millimetres per year or less, or, in the case of earthquakes, rapid and up to many metres. The shoreline of Montague Island, southeast of Anchorage, Alaska, rose nearly 10 m within perhaps a minute during the giant Alaska earthquake in 1964.[8]

Land can subside due to deposition of large amounts of sediment along the coast. Deltas are loci of sedimentation and thus commonly are areas of subsidence. The surface of the Mississippi Delta, for example, is subsiding due to the weight of the large body of sediment deposited on the crust along the Louisiana coast. Subsidence of deltas is commonly compounded by compaction of the sediments that results from a loss of pore water.

12.4 Coastal Hazards

Coastal processes create hazards for both individuals and communities. Individuals face hazards from strong coastal currents when they swim, as well as hazards to themselves and their property during storm surges and tsunamis. Coastal communities may be threatened by coastal erosion and sea-level rise. Coastal erosion affects beachfront homes and businesses, lighthouses, docks, and roads and utility lines close to the shore. Over periods of decades, erosion can be exacerbated by rising sea level.

Our discussion of coastal hazards focuses on strong nearshore currents (called rip currents), beach and cliff erosion, and sea-level rise. Tsunamis (Chapter 4) and cyclones (Chapter 11) are discussed elsewhere.

Rip Currents

When large waves strike a shoreline, powerful **rip currents** carry large amounts of water directly away from the shore. They develop when waves pile up water between the longshore bar and the swash zone. This water moves seaward in narrow zones, commonly metres to tens of metres wide. Rip currents can extend seaward perpendicular to the shoreline for distances of hundreds of metres. They widen and dissipate once they have passed the line of breaking waves. Many beachgoers and lifeguards incorrectly call these currents "riptides" or "undertow." The currents are not tidal and they do not pull people under water; however, they do carry people away from shore into deeper water.

On average, 100 people drown in rip currents each year in the United States. There are few fatalities in Canada, not because rip currents are any less common here, but because Canadian waters are too cold for most swimmers. On an annual basis, the number of deaths in the United States from rip currents is about the same as from river flooding and is greater than deaths from hurricanes or earthquakes. People drown in rip currents because they panic and struggle to swim directly back to shore. Rip currents can have velocities in excess of 6 km/h, which even strong swimmers cannot withstand for long. A swimmer trying to fight a rip current soon becomes exhausted and might not have the energy to stay afloat.

Coastal Erosion

Coastal erosion is becoming a serious global problem because of sea-level rise and increasing development on shorelines (see Professional Profile). A great deal of money is spent to control erosion, but many of the benefits are temporary and much of the money is wasted.

PROFESSIONAL PROFILE

Phil Hill, Coastal Geologist

Dr. Philip Hill (Figure 12.11) is a marine geologist with 30 years of experience working on Canada's Atlantic, Pacific, and Arctic coasts, as well as in Hudson Bay. "I have seen more of Canada's coast than almost anyone, and understand how it will be impacted by climate change," says Hill.

Hill spent part of his professional career as a professor of oceanography at the Université du Québec à Rimouski, but moved to British Columbia in 2001 and currently works for the Geological Survey of Canada in Sidney on Vancouver Island. Hill has been the leader of a multidisciplinary research program focused on issues relating to ocean management and climate change in the Strait of Georgia.

Hill manages a program designed to provide geoscience information to assist Canadians in understanding, preparing for, and adapting to the effects of changing climate. "There's so much research being done on climate change," says Hill; "the challenge for scientists is to provide the right kind of information to planners and policymakers so that they can make science-based decisions for their communities."

Hill is especially interested in marine deltas, which are vulnerable to erosion and flooding from sea-level rise. The Fraser Delta, south of Vancouver, is Canada's most densely populated coastal area and a critically important wetland ecosystem and habitat for migratory bird species. "More than 250 000 people currently live less than 1 m above the high tide line on the Fraser delta," says Hill, "and the number is growing every day."

Hill and his colleagues have studied the impact of sea-level rise on the Fraser Delta. They use a diverse set of tools, including multi-beam swath mapping and high-resolution seismic profiling of the delta front, LIDAR, and sediment transport modelling. The team mapped the delta's tidal flats, modelled wave propagation and the evolution of marshes at the delta front, and used satellite and GPS technologies to document rates of delta subsidence. The results of their research have been used by the Corporation of Delta to develop a climate change action plan to reduce the community's vulnerability to sea-level rise. Hill says that this project is one of the most rewarding that he has done: "The whole community, from the mayor down to individual citizens, has really engaged us to address the climate change issue. I see our work making a real contribution."

—*John J. Clague*

▲ **FIGURE 12.11 PHIL HILL, COASTAL GEOLOGIST** Dr. Hill (upper left, background) describes the geology of the tidal flats of the Fraser Delta, south of Vancouver, to a community group. *(Dr. Phil Hill)*

Most Canadian provinces and territories and the 30 U.S. states that border oceans or the Great Lakes have problems with coastal erosion. Average local erosion rates along some barrier islands in the United States approach 8 m/y, and some Great Lakes shorelines have retreated at rates up to 15 m/y![9] However, these are extreme examples; generally, rates of coastal erosion are much lower, ranging from less than 1 cm/y to about 50 cm/y. Erosion rates depend primarily on the strength of the cliff materials and wave energy.[10]

Steep rocky shorelines that are stable or experience only limited erosion are common along the Pacific coast north of southern California; they form about 85 percent of the British Columbia coastline.[11] The coastal zone of central and northern British Columbia and Alaska is studded with rocky islands, some as large as many U.S. states. Steep rocky shorelines are also common in other tectonically active areas around the world.

Shorelines can shift considerably on barrier islands and spits. Entire islands or the inlets between them can move. An extreme example is Hog Island, east of Chesapeake Bay in Virginia, which has shifted up to 2.5 km in historic time.[12] Aransas Pass, the tidal inlet that provides ships access to Corpus Christi, Texas, moved over 3 km between 1859 and 1911, when its location was fixed with the construction of jetties.[13] Bayocean spit near Tillamook, Oregon, and Nauset Beach spit near Chatham, Massachusetts, were breached by storms in the twentieth century and separated from the mainland. The possible migration of barrier islands and spits makes them poor places for coastal development.

Beach Erosion An easy way to visualize beach erosion is to consider its **sediment budget**. This budget is similar to a bank account. In the case of a bank account, deposits made to the account are input, the account balance is how much money is in the account at a given time, and withdrawals are output. Similarly, we can partition the sediment budget of a beach in terms of input, storage, and output of sand and gravel (see A Closer Look). Beaches both gain and lose material. Sand and gravel are added to a beach by the coastal processes that move sediment along the shore (Figures 12.8 and 12.9) or by local wave erosion. Beach drift and longshore drift bring sediment from updrift sources, such as rivers. Local wave erosion of sand dunes or sea cliffs also adds sediment to a beach. Wave erosion is greatest during storms when waves are largest and reach farthest inland. Sediment in storage on a beach is what you see when you visit a site. Output of sediment occurs when coastal processes move sand and gravel away from the beach. These coastal processes include littoral drift, a return flow to deep water during storms, and wind erosion.

When input exceeds output, the beach grows—more sediment is stored and the beach widens. When input and output are about the same, the beach is in equilibrium and remains fairly constant in width. If output of sediment exceeds input, the beach erodes and becomes narrower.

Storms affect sediment supply, although generally only temporarily. Along the west coast of North America, for example, many beaches lose sediment during winter when storms are common. This sediment is replenished during the summer when storms are few. On the Atlantic coast, much erosion is caused by hurricanes and by other severe storms known as nor'easters (see Chapter 11).[3]

Long-term changes in the sediment budget of beaches can be caused by climate change or by human interference with natural shore processes.[14] One long-term change on some beaches in the United States has been a reduction in sand supply from rivers. Construction of dams on rivers flowing to the coast, such as the Savannah River in Georgia and the Rio Grande in Texas, has decreased the sand supply to beaches and increased the coastal erosion problem.

Cliff Erosion Sea cliffs and lakeshore bluffs can be eroded by waves and landslides. The problem can be aggravated when people alter cliffs during development.

Storm waves erode many sea cliffs and lakeshore bluffs, thus moving the shoreline landward (Figure 12.13). The possibility of erosion might not be apparent during periods of fair weather when a beach is present at the base of the cliff. Beach sediment protects the cliff from wave erosion during periods between storms. During fall and winter, however, successive storms remove much sand from beaches, leaving cliffs vulnerable to erosion (Figure 12.14). Most sea-cliff erosion on the west coast of North America happens during severe winter storms. Sea cliffs on the Atlantic coast of the northeast United States and Canada may be severely eroded during nor'easters. Storm waves also are responsible for most of the erosion of lakeshore bluffs along the Great Lakes.

Bluff erosion is a periodic problem along the coasts of the Great Lakes and is most severe during extended periods of high water levels caused by above-normal precipitation. Measurements by the U.S. Army Corps of Engineers show that the level of Lake Michigan has fluctuated about 2 m since 1860. During periods of below-average lake level, wide beaches dissipate energy from storm waves and protect the shore. However, as lake levels rise, beaches become narrower and storm waves expend considerable energy against the shoreline. Even a small rise in water level will inundate a wide section of a gently sloping beach.[15] During periods of high water level, storm waves have eroded some lakeshore bluffs at an average rate of 0.4 m/y,[16] destroying many buildings, roads, retaining walls, and other structures (Figure 12.15).[15]

Sea ice limits fetch and therefore wave erosion throughout much of the central and northern Arctic. However, erosion rates of up to 2 m/y have been observed along parts of the Beaufort Sea shoreline in the western Arctic.[17] Thawing permafrost is a problem in the Arctic that is not an issue elsewhere (see Chapter 8). The Arctic climate has warmed significantly over the past several decades. This warming, together with wave erosion and sea-level rise, has eroded ice-rich coastal sediments (Figure 12.16). Some Inuit villages are threatened by the erosion and might have to be relocated at considerable cost.

Sea-cliff and lakeshore erosion is a natural process that cannot be completely prevented, even with investments of large amounts of money. Therefore, we must learn to live

A CLOSER LOOK 12.1

Beach Budget

For a given segment of a beach, the total volume of sand added to the beach can be compared to the volume of sand lost to produce a "beach budget"[18]:

- If losses are greater than gains, the beach erodes.
- If losses are less than gains, the beach grows by accretion of sand.
- We can evaluate the beach budget over a specific period, such as 1 year or 10 years.

As an example, imagine a simple coast with rivers supplying sand, a sea cliff, beach, and a submarine canyon (Figure 12.12). This coast defines a *littoral cell*, which is a segment of the coast that includes sediment sources and the littoral transport system that moves sand along the beach. In this simple case, rivers deliver sand, as does erosion of the sea cliff. Sand is transported along the shoreline; some sand moves away from the near-shore environment and is transported into deep water via a *submarine canyon*—an offshore canyon that heads near the coast.

In this example, we determine the budget before and after a dam was constructed near the mouth of the river supplying sand to the beach, which explains why erosion occurred on beaches south of the submarine canyon.

Sources of Sand for Our Beach Budget Example (+ = gain; − = loss)		
Littoral transport (+)	South	200 000 m^3/y
	North	50 000 m^3/y
	Net	150 000 m^3/y to the south
Sea-cliff erosion (Scf) (+)	Erosion rate is 0.5 m/y; average height of sea cliff is 6 m; total length is 3000 m. Assume 50 percent of eroded material remains on beach.	
Scf = 0.5 m/y × 6 m × 3000 m × 0.5 = 4500 m^3/y		
River source (Sr) (+)	Assume drainage area of 800 km^2	
A dam was constructed 15 years ago, reducing the drainage basin delivering sediment to the coast to 500 km^2.		

To estimate the sediment delivered to the coast by the river, we can use equations (numerical models) or regional graphs.[18] The area of the drainage basin contributing sediment to the coast before the dam was constructed is 800 km^2. From our analysis of sediment production, we estimate that the sediment delivered to the beach from the river prior to dam construction was 300 m^3 for every square kilometre of the basin per year ($m^3/km^2/y$). Following dam construction, the sediment-contributing area is 500 km^2, with a yield of 400 $m^3/km^2/y$. The increase per unit area results because smaller tributaries below the dam have a higher sediment yield.

Assume that 30 percent of sediment delivered from the river will remain on the beach and that it is sand-size and larger:

Before dam:

$$Sr(+) = 300\ m^3/km^2/y \times 800\ km^2 \times 0.3 = 72{,}000\ m^3/y$$

After dam:

$$Sr(+) = 400\ m^3/km^2/y \times 500\ km^2 \times 0.3 = 60{,}000\ m^3/y$$

Loss to submarine canyon (Scn) (−):

220 000 m^3/y (estimated from offshore observations)

(continued)

Budget before dam:		
Longshore drift (Sl)	+	150 000 m^3/y
Cliff erosion (Scf)	+	4500 m^3/y
River (Sr)	+	72 000 m^3/y
Submarine canyon (Scn)	–	220 000 m^3/y
Net	+	6500 m^3/y

More sand is arriving than is leaving; therefore the beach is growing.

Budget after dam:		
Longshore drift (Sl)	+	150 000 m^3/y
Cliff erosion (Scf)	+	4500 m^3/y
River (Sr)	+	60 000 m^3/y
Submarine canyon (Scn)	–	220 000 m^3/y
Net	–	5500 m^3/y

More sand is leaving than is being supplied; therefore the beach is eroding.

Assume there are several beach homes near point X. What advice would you give the homeowners?

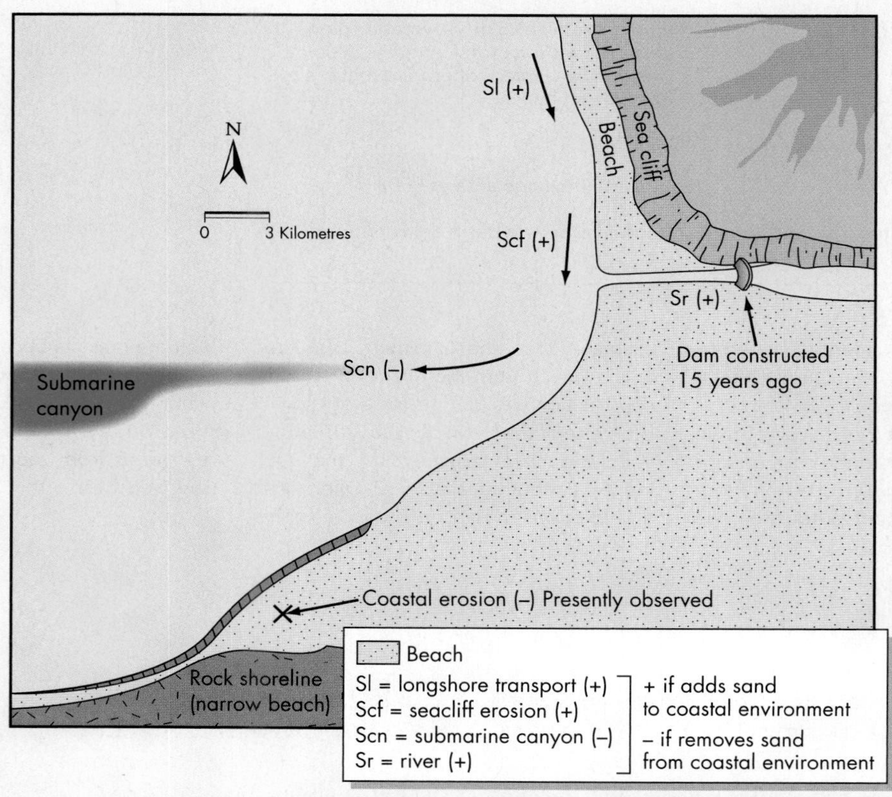

▲ **FIGURE 12.12 EXAMPLE OF A BEACH BUDGET** Beach erosion is occurring at point X. Arrows indicate the direction of sediment transport. See text for calculation of the beach budget before and after construction of a dam on the river to the north.

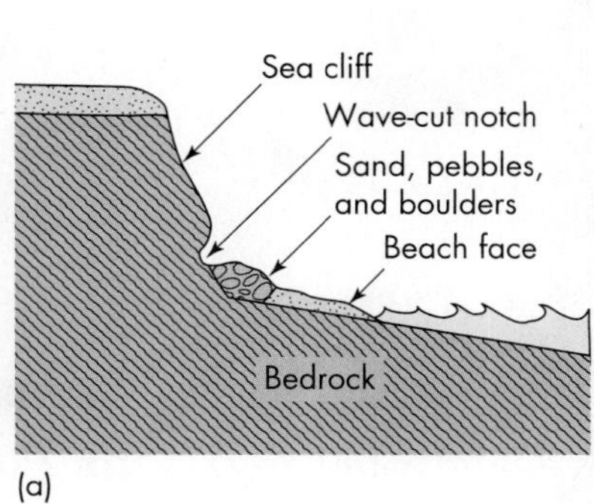

▲ **FIGURE 12.13 SEA CLIFF AND BEACH** (a) A generalized cross-section and (b) photograph of a sea cliff, beach, and intertidal rock platform at Santa Barbara, California. *(Donald Weaver)*

◄ **FIGURE 12.14 SEA CLIFF EROSION** A steep, eroded sea cliff at Vancouver, British Columbia. Loose Pleistocene sands, which form the cliff, are easily eroded by waves during storms. Groundwater seepage from the lower part of the cliff (dark zone) contributes to the problem. The dotted line near the top of the sea cliff marks the contact between Pleistocene sands and an overlying thin layer of coarser sediment deposited by a glacier. *(John J. Clague)*

with some erosion, while not exacerbating the problem by allowing uncontrolled runoff on cliffs or by placing homes and other structures close to the top edge of a cliff.

Sea-Level Rise

Rising sea level contributes to coastal erosion because storm waves can erode farther inland. Some shorelines in Atlantic Canada, where sea level is rising more rapidly than elsewhere in North America,[19] are retreating at rates of up to 10 m/y. Even at lower rates of erosion, entire islands off Nova Scotia have disappeared in the recent geological past (see story about Meisner Island at beginning of this chapter).

Sea-level rise in this region is due, in part, to continuing glacio-isostatic subsidence associated with the disappearance of the *Laurentide ice sheet*, which covered central and eastern Canada as recently as about 12 000 years ago. The ice sheet depressed the land on which it lay by several

▲ **FIGURE 12.15 LAKESHORE EROSION** Coastal erosion destroyed this home at the shore of Lake Michigan. *(Steve Leonard/Stone/Getty Images)*

▲ **FIGURE 12.16 COASTAL EROSION IN THE ARCTIC** Waves have attacked ice-rich sediments that form the shore along this part of Canada's Arctic coast. *(S. A. Wolfe/Geological Survey of Canada)*

hundred metres, displacing viscous mantle material laterally within Earth to and beyond the ice margins. This lateral displacement of mantle material created an uplifted "bulge" beneath Atlantic Canada. When the ice sheet melted, the uplifted bulge collapsed and the land beneath Atlantic Canada subsided. This process continues today at a low rate and is augmenting the rise in sea level caused by melting of glaciers and thermal expansion of ocean waters.

Estimates of just how much sea level will rise over the remainder of this century differ, from as little as 20 cm to more than 1 m.[1] Although precise estimates are not possible, even a modest 40-cm rise in sea level would have huge environmental, economic, and social impacts. A rise of this amount would increase coastal erosion and increase the vulnerability of buildings and other structures to storm surges. It would flood low-lying island nations in the Pacific and Indian oceans (Figure 12.17), as well as large areas of Vietnam, Bangladesh, and other Asian countries. A rise in sea level of more than 1 m would threaten parts of many North American cities, including Vancouver, Victoria, Miami, New Orleans, New York, Boston, and Washington, D.C. Expensive measures would be needed to protect property in the coastal zone; governments would have to choose between making these investments or allowing beaches and estuaries to migrate landward.

12.5 Links between Coastal Processes and Other Natural Hazards

Coastal processes have links to other natural hazards, including earthquakes, tsunamis, volcanoes, landslides, subsidence, flooding, cyclones, and tornadoes.

Intense precipitation and storm surges associated with hurricanes and other cyclones cause flooding, coastal erosion, and landslides. A hurricane's storm surge and heavy rainfall combine to produce widespread coastal erosion and flooding (Figure 12.18). Coastal areas that have subsided because of an earthquake, soil compaction, or groundwater withdrawal are highly vulnerable to both river flooding and storm surges.

Another common hazard associated with coastal processes is landslides. Wave erosion undercuts or steepens sea cliffs and lakeside bluffs, triggering landslides (Figure 12.19).

Coastal erosion is also linked to climatic conditions that change over periods of decades. For example, trade winds control sea surface temperatures in the equatorial

◄ **FIGURE 12.17 AERIAL VIEW OF MALÉ IN THE MALDIVE ISLANDS** A seawall surrounds most of Malé, the capital of the Maldives, an island nation of 300 000 people. The seawall was constructed to protect the island from waves up to 2 m high, at a cost of about $4000/m^2$. Seawalls have been built on only a few of the 1192 islands that make up this country. Approximately 80 percent of the land area of the Maldives is less than 1 m above mean sea level. *(Peter Essick/Aurora & Quanta Productions Inc.)*

◄ **FIGURE 12.18 STORMS CAUSE COASTAL EROSION** Two hurricanes, 21 days apart, eroded sand from this shoreline in Vero Beach on Florida's east coast. The top photograph, taken in August 1997, before the hurricanes, shows two homes located a few tens of metres from the shore; a sand dune provided some protection. In the middle photograph, taken on September 8, 2004, after Hurricane Frances, the shoreline has retreated to the edge of the houses. The third photograph, taken on September 29, 2004, after Hurricane Jeanne, shows one of the houses undermined and severely damaged by storm waves. The red arrow points to the same place in all three photographs. *(Courtesy of Coastal and Marine Geology Program, U.S. Geological Survey)*

▲ **FIGURE 12.19 ROOM WITH A VIEW** These homes, perched on a sea cliff along Puget Sound west of Port Townsend, Washington, were not damaged by this landslide in 1997, but they could be destroyed in a future landslide. *(Gerald W. Thorsen, Washington Division of Geology and Earth Resources)*

Pacific Ocean. Prolonged slackening of these winds leads to warming across the equatorial Pacific, a condition known as El Niño (see Chapter 14). The warmer seas, in turn, affect climate throughout North and South America. A very strong El Niño in 1982 and 1983 resulted in an increase in the intensity and frequency of extratropical cyclones reaching the coasts of California and Oregon, causing severe coastal erosion there.[14]

The giant oil spill in the Gulf of Mexico, from a blowout at the *Deepwater Horizon* well drilled from a platform in deep water off the coast of Louisiana, highlights the fact that coastal processes are linked to some environmental problems. The spill was the most serious in U.S. history. As oil moved with the wind and currents to the Gulf coast, coastal processes became an important factor (Figure 12.20). Oil was carried along beaches by the tides and by longshore currents driven by waves. Oil contaminated the beaches of barrier islands and moved through inlets to tidal marshes. Predicting the movement of the oil along the beaches and in tidal marshes required detailed information on:

- Wave height and frequency
- Direction and rate of longshore transport
- Strength of tidal flow into and out of barriers and island inlets
- How far inland tidal flow inundates marshes
- How marsh vegetation and sediment interact with oil
- The effects of the oil on plants and animals on beaches and in marshes

12.6 Natural Service Functions of Coastal Processes

As is true of many of the other hazards we have discussed, it is difficult to imagine any benefits of coastal erosion and sea-level rise. But some coastal processes do provide benefits. They contribute to the ecological health and aesthetic value of the coastal zone and provide a variety of recreational activities.

Erosion is a problem for property owners in coastal areas, but the beauty of the coastal zone largely results from wave action and erosion (Figure 12.21). The stunning cliffs, rocky headlands, and sandy beaches of British Columbia and Atlantic Canada are the result of erosion and are important aesthetic resources.

Erosion of dunes, sea cliffs, and lakeshore bluffs provides much of the sand that forms beaches. Longshore drift and other coastal processes redistribute this sand along sea coasts and lakeshores.

Coastal processes have also concentrated valuable deposits of diamonds and placer gold along some sea coasts. Examples are diamonds off the coast of Namibia and gold offshore Nome, Alaska.

12.7 Human Interaction with Coastal Processes

Human interference with natural shore processes is a major cause of coastal erosion. Most problems arise in populated areas where efforts to stop coastal erosion have interrupted

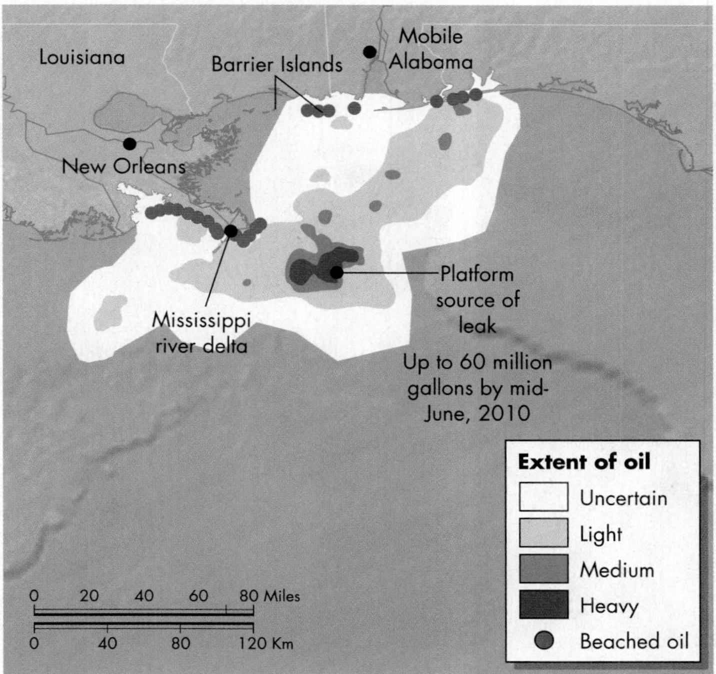

(a)

(b)

▲ **FIGURE 12.20 GULF OF MEXICO OIL SPILL 2010** (a) Map of the extent of the spill in mid-June, two months after the blowout at the *Deepwater Horizon* well. (b) Cleaning oil from a Louisiana beach. *(a) NOAA with modification; (b) (Newscom)*

the natural longshore movement of sand. Engineered structures may help beaches to grow in some areas, but they can cause downdrift erosion that damages beachfront property. This type of interference is common on all coasts of North America, with the exception of the Arctic coast.

The Atlantic coast of the United States is fringed by barrier islands that are separated from the mainland by a lagoon or bay. Most of the barrier islands have been altered by humans.

The history of barrier islands along the Maryland coast of the United States illustrates the interplay between people and coastal processes in this area. Demand for the 50 kilometres of oceanfront in Maryland is very high. The barrier islands are used seasonally by residents of the Washington

◄ **FIGURE 12.21 SCENIC HEADLAND** This beautiful rock formation on the west coast of the South Island of New Zealand was formed by wave erosion. It is a popular tourist attraction and an important aesthetic resource. *(John J. Clague)*

and Baltimore metropolitan areas, and Ocean City, located on Fenwick Island, has experienced particularly rapid growth (Figure 12.22). Since the early 1970s, high-rise condominiums and hotels have been built on Ocean City's waterfront. Coastal dunes were removed to make space for new construction. This activity and serious beach erosion have increased the vulnerability of the island to hurricanes. Coastal dunes act as natural barriers to storm waves and partially protect

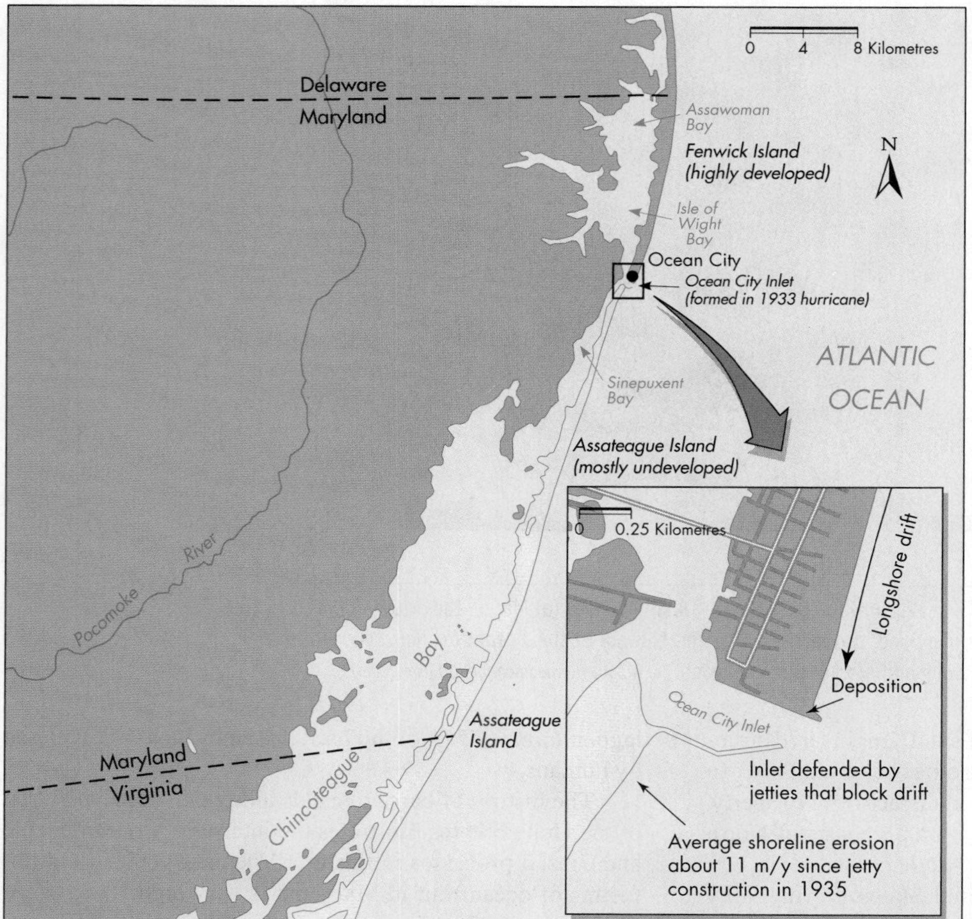

◄ **FIGURE 12.22 JETTY INCREASES BEACH EROSION** Fenwick Island, a barrier island on the Maryland coast in the United States, has experienced rapid urban development. Potential hurricane damage is a concern. The inset is a map of Ocean City and a washover channel that formed during a hurricane in 1933. The north end of Assateague Island, shown in beige, is relatively undeveloped and has experienced rapid shoreline erosion since a jetty was constructed in 1935.

structures built behind them. Ocean City experienced a hurricane in 1933. The hurricane's storm surge formed Ocean City inlet directly south of the city (Figure 12.22). This event indicates that Ocean City could be damaged or destroyed by the storm surge of a future hurricane.[20]

Assateague Island is located south of Ocean City, just across the inlet. It spans two-thirds of the Maryland coastline and is mostly undeveloped, in contrast to highly urbanized Fenwick Island. Both islands, however, share the same sand supply. At least that was the case until 1935, when jetties were constructed to stabilize Ocean City inlet (Figure 12.22). Since construction of the jetties, beaches at the north end of Assateague Island have receded about 11 m/y, nearly 20 times the long-term rate of shoreline retreat for the Maryland coast. During the same period, beaches directly north of the inlet, in Ocean City, became wider, requiring lengthening of a recreational pier.[21]

Historic changes to Maryland's outer coast are clearly linked to human interference with the natural movement of sand along the shore. Construction of the Ocean City inlet jetties interfered with the southward drift of sand. The jetties diverted sand offshore, preventing it from continuing southward to nourish the beaches on Assateague Island.

Erosion is also a problem along the densely populated and extensively modified Gulf of Mexico coastline, which, like the Atlantic coast, has numerous barrier islands. A study of the Texas coastal zone suggests that, in the past 100 years, human modification of the coastline has accelerated erosion by 30 to 40 percent.[22] Much of the accelerated erosion stems from coastal development, subsidence because of groundwater and hydrocarbon withdrawal, and the damming of rivers that supply sand to beaches.

The coastline of southern California is heavily populated and has been extensively altered by people. Winter storms during El Niño years have eroded beaches and sea cliffs, causing large amounts of property damage. Jetties and groins, built to improve beaches and to prevent infilling of the mouths of rivers with sediment, have in many instances caused severe downdrift erosion. Much of the California shoreline is developed in Tertiary mudstone and sandstone. These rocks erode more slowly than sandy shorelines, but structures built too close to cliff tops are at risk (Figure 12.23).

Some coastal residents of Washington and British Columbia live too close to the tops of sea cliffs. The problem is particularly acute in parts of the Strait of Georgia and

◄ **FIGURE 12.23 RETREAT OF SEA CLIFF IN ROCK** Apartment buildings at the edge of a sea cliff in Isla Vista, California. The outdoor deck might have been "open," but it was unsafe because it overhung the cliff by at least 1 m. Note the exposed cement pilings that provide support for the buildings. The decks and apartment buildings were in imminent danger of collapse and were condemned in 2004. *(Edward A. Keller)*

Puget Sound (Figure 12.19), where homes are located at the tops of cliffs formed of loose Pleistocene sediments. Storm waves that would not be considered noteworthy on the exposed Pacific Coast can severely erode these cliffs.[11,23]

12.8 Minimizing Damage from Coastal Hazards

Coastal erosion might appear easier to control than most other natural hazards. For example, we can do nothing to prevent earthquakes or volcanic eruptions; we can only prepare for them. There are things we can do, however, to control coastal erosion:[24]

- Shoreline stabilization with such structures as groins and seawalls—the "hard" solution
- Beach nourishment and drainage—the "soft" solutions
- Land-use changes—avoid the problem by not building in hazardous areas or by relocating threatened buildings

One of the first tasks in minimizing and managing coastal erosion is to estimate future erosion. Estimates are based on historic shoreline change or analysis of waves, winds, and sediment supply. Setback recommendations can then be made. A *setback* is the distance from the shoreline, beyond which development, such as of houses and roads, is allowed. The setback concept is at the heart of coastal zone erosion management (see A Closer Look).

Hard Stabilization

Structures built along the shoreline to slow erosion and protect buildings from damage include seawalls, groins, breakwaters, and jetties. **Seawalls** (Figure 12.24) are built at the shore and parallel to it; they are made of concrete, large stone blocks called *riprap,* wood or steel pilings, cemented sandbags, or other materials. Although seawalls help retard erosion and protect buildings from damage, they have been criticized because they redirect the energy of incoming storm waves back onto the beach (Figure 12.26). They can thus promote beach erosion and produce a narrower beach with less sand. Furthermore, seawalls are frequently damaged during severe storms and require costly repairs. Design and construction of seawalls must be carefully tailored to specific sites and needs. Because of these problems, some geologists believe that seawalls cause more problems than they solve and should rarely, if ever, be used.

Groins are linear structures placed perpendicular to the shore, commonly in groups called *groin fields* (Figure 12.27). Each groin in a groin field traps a portion of the sand carried by longshore drift. A small amount of sand accumulates updrift of each groin, contributing to an irregular but wider beach. The wider beach protects the shoreline from erosion.

However, groins, like seawalls, can create problems—sand is deposited updrift of the structure, but erosion can occur downdrift of it. Thus a groin or groin field will produce a wider, more protected beach in a desired area, but at a cost to the adjacent shoreline. Once a groin has trapped all the sediment it can hold, sand is transported around its offshore end to continue its journey along the beach. Therefore, erosion may be minimized by artificially filling the space in front of each groin with sand. Despite such precautions, groins may cause undesirable erosion in the downdrift area; therefore, their use should be carefully evaluated.

Breakwaters and **jetties** are linear structures of riprap or concrete that protect limited stretches of the shoreline from waves. The purpose of a breakwater is to intercept waves and provide a protected area, or harbour, for mooring boats. They may be attached to or separated from the beach (Figure 12.28a and 12.28b). Jetties extend perpendicular to the shore at the mouth of a river or at the entrance to a bay or lagoon (Figure 12.28c). They are commonly built in pairs and are designed to prevent sediment from accumulating at

◄ **FIGURE 12.24 SEAWALL** The shoreline along part of West Vancouver, British Columbia, is protected from wave erosion by a seawall and riprap. A popular pedestrian path is located on the seawall. *(Bob Turner)*

A CLOSER LOOK 12.2

E-Lines and E-Zones

The U.S. National Research Council (NRC), at the request of the U.S. Federal Emergency Management Agency (FEMA), developed coastal zone management recommendations, some of which are listed below:[24]

- Estimates of future erosion should be based on historic changes to the shoreline or on statistical analysis of local wave and wind conditions and sediment supply.
- After average local erosion rates have been determined, maps should be made showing erosion lines and zones, which are referred to, respectively, as E-lines and E-zones (Figure 12.25). An **E-line** is the expected position of the shoreline after a specified number of years; for example, the E-10 line is the expected shoreline in 10 years. **E-zones** are the areas between the present shoreline and the E-line; the E-10 zone, for example, is the area between the shoreline and the E-10 line.
- No new habitable structures should be allowed in the E-10 zone.
- Movable structures are allowed in the E-10 to E-60 zones, which are deemed to be at intermediate and long-term risk.
- Permanent large structures must have setbacks to beyond the E-60 line.
- All new structures built seaward of the E-60 line, with the exception of those on high bluffs or sea cliffs, should be built on pilings. They should be designed to withstand erosion that could be caused by a storm that recurs, on average, once every 100 years.

NRC recommendations on setbacks are considered to be minimum standards for state and local coastal erosion management programs. At present, only a small number of states use a setback based on erosion rates.

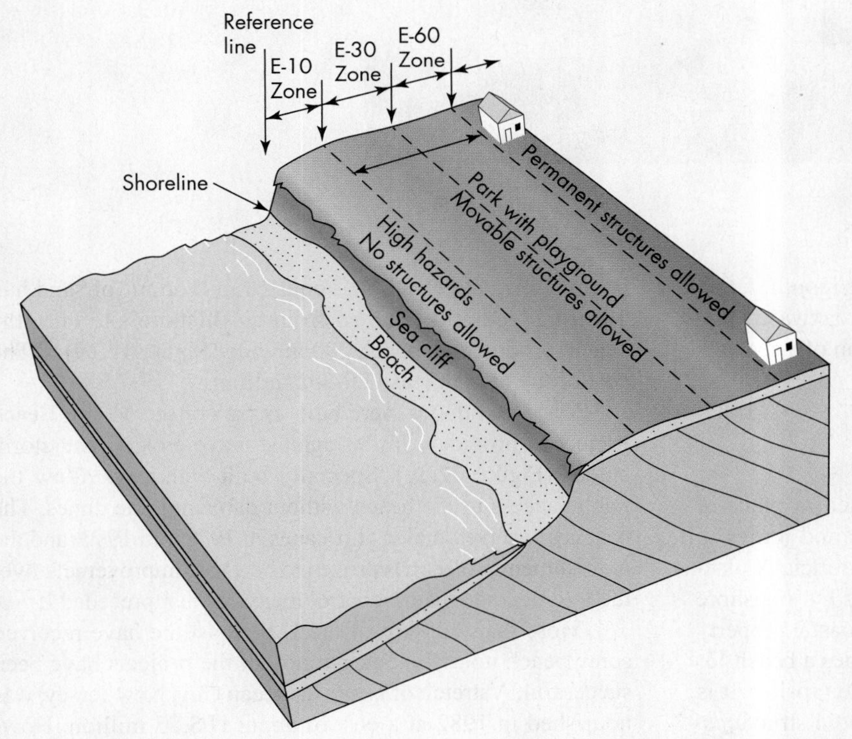

◄ **FIGURE 12.25 E-LINES AND E-ZONES** A conceptual diagram illustrating E-lines and E-zones. The dashed lines are E-lines. Widths of E-zones depend on the rate of erosion and can be used to define setback distances. *(Based on National Research Council. 1990.* Managing Coastal Erosion. *Washington, DC: National Academy Press)*

the river mouth or bay entrance. Jetties also shelter the channel from large waves. Like groins, jetties block littoral transport and thus cause the updrift beach adjacent to one jetty to widen and downdrift beaches to erode.

Breakwaters and jetties block or interfere with littoral transport and alter the shape of the shoreline as new areas of deposition and erosion develop. The result can be serious erosion, or the harbour entrance can be blocked with newly deposited sand. Therefore, these structures must be carefully designed and constructed to minimize adverse effects. The sand may have to be moved by *artificial bypass*—that is, by dredging and pumping the sand adjacent to the breakwater

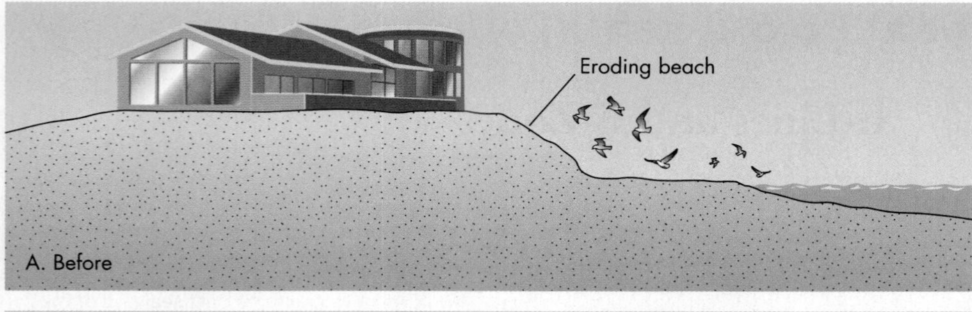

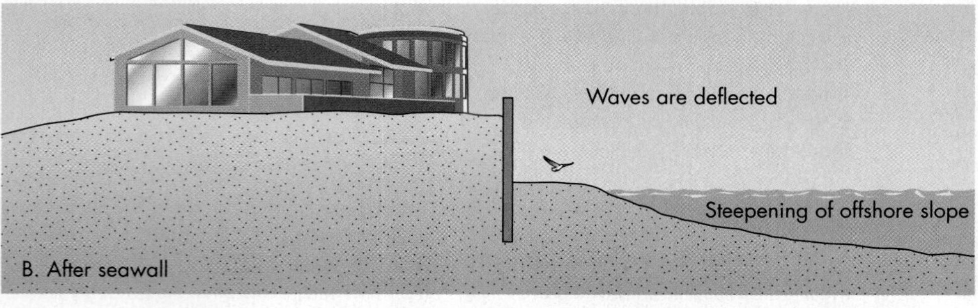

◄ **FIGURE 12.26 SEAWALL INCREASES BEACH EROSION** Construction of seawalls to prevent coastal erosion and protect buildings commonly causes beach erosion. A seawall reflects storm waves and reduces or eliminates the supply sediment to the beach.

and redepositing it downdrift of the harbour mouth. Other measures that minimize adverse effects of breakwaters and jetties include beach nourishment, construction of a seawall, and emplacement of riprap.[5]

Soft Solutions

Sometimes **beach nourishment** can be used in place of "hard" engineered structures such as groins and jetties to control erosion. Beach nourishment involves artificially placing sand on beaches to compensate for losses by longshore drift. Ideally, a nourished beach will protect coastal property from wave attack.[5] Beach nourishment provides a beach for recreation and some protection from shoreline erosion. It is also aesthetically preferable to many engineered structures. Unfortunately, beach nourishment is expensive and must be periodically repeated to stabilize a beach. Considerable care must be taken when selecting the appropriate range of sand sizes for a beach replenishment project, and locating a nearby source of this sand can be difficult.

In the mid-1970s, the City of Miami Beach in Florida and the U.S. Army Corps of Engineers began an ambitious beach-nourishment program to reverse the serious erosion that had plagued the area since the 1950s. The program aimed to widen the beach and protect coastal resorts from storm damage.[25] By 1980, about 18 million m^3 of sand had been dredged and pumped from an offshore site onto the beach, producing a beach 200 m wide (Figure 12.29).[26] The cost of the project was U.S.$62 million.

Vegetated dunes were built as part of the Miami Beach project to provide a buffer against wave erosion and storm surges (Figure 12.30). Specially built walkways allow the public access to the beach without damaging the dunes. The beaches survived major hurricanes in 1979 and 1992, and the nourishment project has proven to be a vast improvement over the fragmented erosion-control measures that preceded it.[5]

More than 600 km of the U.S. coastline have received some beach nourishment, but not all the projects have been successful. A stretch of beach at Ocean City, New Jersey, was nourished in 1982 at a cost of about U.S.$5 million. However, the sand that was placed on the beach was removed in less than three months by a series of storms. Beach nourishment remains controversial; some scientists consider sediment added during beach nourishment projects to be nothing more than "sacrificial sand" that will eventually be removed by coastal erosion.[26] Most beaches require frequent replenishment to remain intact.[27] Nevertheless, beach nourishment has become a commonly used method of restoring or creating beaches and protecting shorelines from erosion around the world.

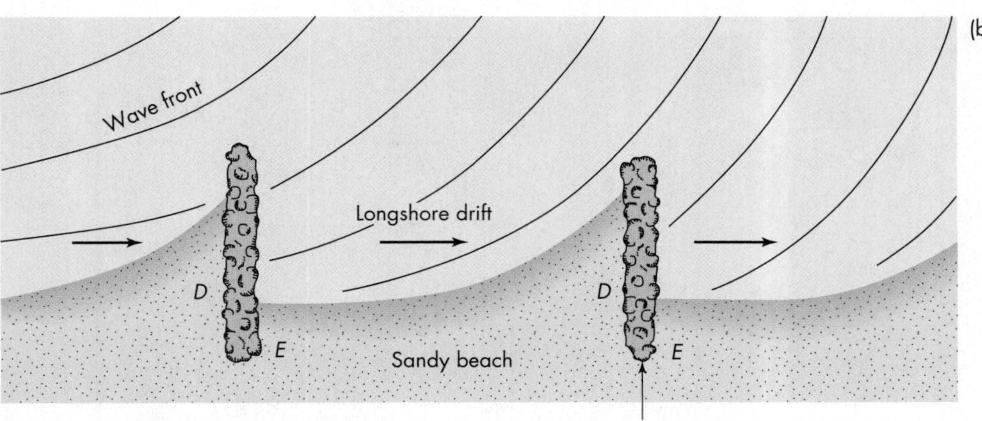

◄ **FIGURE 12.27 BEACH GROINS** (a) Several groins and breakwaters were built along the shoreline at Vancouver, British Columbia, to minimize the loss of beach sand through erosion. (b) Two groins trap sand at D, updrift of the structures. Erosion occurs at E. *(John J. Clague)*

Coastal erosion can be triggered or aggravated by removing vegetation from the shore. Removal of vegetation may destabilize sea cliffs, causing landslides. It may also reduce the resistance of the shore to erosion by storm waves, thus increasing the rate of shoreline retreat. Intertidal plants and plants on the beach berm give shore materials greater resistance to erosion.

Another "soft solution" to coastal erosion is **beach drainage**.[28] The principle of beach drainage is that by lowering the water table beneath a sand beach, erosion is reduced. On a beach with an artificially lowered water table, water in wave backwash infiltrates into the sand; thus less water and sand return to the sea. Beach drainage is accomplished by installing one or more pipes beneath and parallel to the beach. The pipes collect seawater that infiltrates into the sand. The collected water may be discharged back to the sea or, alternatively, used in desalinization plants, swimming pools, and in other applications. More than 30 beach drainage systems have been installed in Denmark, the United States, the United Kingdom, Japan, Spain, Sweden, France, Italy, and Malaysia.

12.9 Perception of Coastal Hazards

Past experience, proximity to the coastline, and the possibility of property damage affect a person's perception of risk from hazardous coastal processes. People perceive land as stable, and most societies view it as a permanent commodity that can be bought and sold. In most cases, this perception is appropriate because the land does not change for millennia. However, coastlines are dynamic environments and shorelines can shift over periods of months and years.

One study of sea-cliff erosion in California found that shoreline residents in an area likely to experience damage in the near future were generally well informed and saw the erosion as a threat.[29] People living a few hundred metres from the shore, although aware of the erosion problem, knew little about its severity. Still farther inland, people were aware that the coast was eroding but did not consider it a hazard.

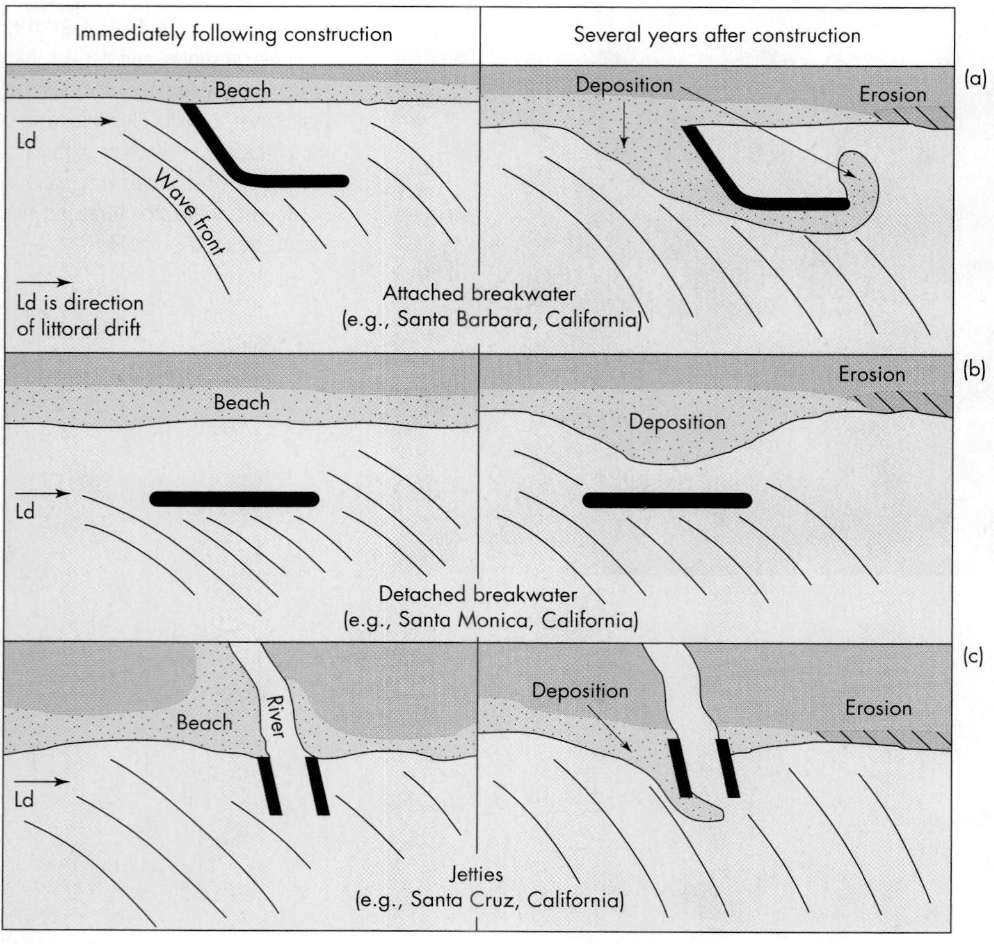

◄ **FIGURE 12.28 EFFECTS OF ENGINEERED COASTAL STRUCTURES ON SEDIMENTATION AND EROSION** Breakwaters, the thick black lines in (a) and (b), and jetties, the thick black lines in (c), alter deposition (stippled tan) and erosion (lined brown). Each of the three rows shows conditions immediately after construction of a structure (left) and later, following a period of deposition and erosion (right). Thin, curved black lines show wave fronts, and the arrows indicate the direction of littoral transport.

(a)

(b)

▲ **FIGURE 12.29 BEACH NOURISHMENT** Miami Beach (a) before and (b) after beach nourishment in the mid-1970s. *(Courtesy of U.S. Army Corps of Engineers, Headquarters)*

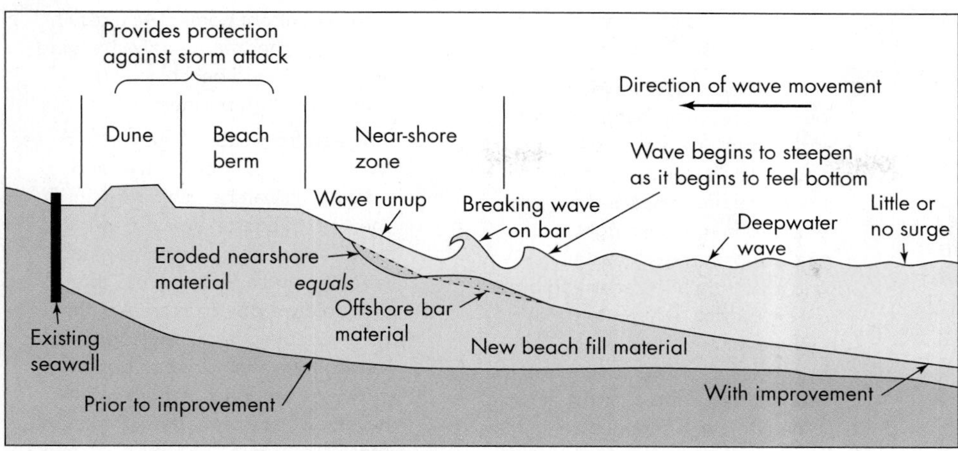

◄ **FIGURE 12.30 MIAMI BEACH AFTER NOURISHMENT** A diagrammatic cross-section of Miami Beach after the beach nourishment project in the mid-1970s. The constructed dune and the beach berm protect the shore from erosion by waves during severe storms. A pre-existing seawall is indicated by the thick black vertical line on the left. Added beach sand is light brown. The orange areas are sand that may be eroded and deposited during a storm. The line at the top of the darker brown area shows the pre-storm profile of the beach. *(Courtesy of U.S. Army Corps of Engineers, Headquarters)*

Many people do not perceive of waves and currents as hazards. Their first visits to beaches were as children, and their parents closely limited their activities. Yet waves and currents can be dangerous for wading and swimming. Rip currents are responsible for about 80 percent of all lifeguard rescues on seashores.[30]

12.10 Future Coastal Zone Management

We are at a crossroads today with respect to coastal zone management. One path leads to the increasing use of coastal defences such as seawalls to control erosion. The other path requires that we live with coastal erosion and involves controlled and appropriate use of land in the coastal zone.[24,27]

Most structures in the coastal zone should be considered temporary and expendable; few critical facilities should be placed near a shore that is retreating because of erosion. Development in the coastal zone must be in the best interest of the public rather than the few who profit from it. In other words, the shoreline belongs to all people, not just those fortunate enough to purchase beachfront property. This concept is at odds with the attitude of developers who consider the coastal zone too valuable not to develop. In Canada, all beaches are public property and local property owners cannot deny others access to the beach. Likewise in some U.S. states, such as Texas and Oregon, almost all beaches are public property and coastal zoning now requires avenues for public access.

The idea that, with minor exceptions, coastal zone development is temporary and expendable and that the general public should be given first consideration is founded on five principles:[27]

1. *Coastal erosion is a natural process.*
2. *Any shore construction causes change.* Shore structures interfere with natural processes and may produce adverse effects.
3. *Stabilization of beaches with dykes, groins, jetties, and other engineered structures protects property, not the beach itself.* Most protected property belongs to a small number of people, whereas the costs commonly are borne by the general public.
4. *Engineered structures designed to protect a beach may eventually destroy it.*
5. *Once built, defensive structures lead to further coastal development.* This trend is difficult, if not impossible, to reverse.

Remember these guidelines if you plan to purchase land in the coastal zone: (1) allow for a sufficient setback from the beach, sea cliff, or lakeshore bluff; (2) ensure that the property is above the possible limit of flooding; (3) construct buildings to withstand high winds; and (4) if hurricanes are a possibility, be sure that the property has an adequate evacuation route.[27]

REVISITING THE FUNDAMENTAL CONCEPTS

Coastal Hazards

1 Hazards can be understood through scientific investigation and analysis.

2 An understanding of hazardous processes is vital to evaluating risk.

3 Hazards are commonly linked to each other and to the environment in which they occur.

4 Population growth and socio-economic changes are increasing the risk from natural hazards.

5 Damage and loss of life from natural disasters can be reduced.

1. Scientists have an advanced understanding of coastal processes. We have instruments that record weather, waves, tides, and

(continued)

land use, and the geology of the coastal zone is well known. Much is also known about beach form and processes, as well as the processes that operate in associated coastal environments, such as sea cliffs, tidal marshes, and barrier islands. As a result, it is possible to predict coastal erosion problems that might be caused by land-use changes, such as jetty or groin construction. One potential area of uncertainty is the future impact of sea-level rise on coastlines.

2. Establishing the risk that accompanies coastal erosion is fairly straightforward. The magnitude and likelihood of erosion can be predicted and potential economic losses can be estimated. Risk is the product of these two variables. It is important that we document coastal erosion in Canada because this information will allow us to identify areas that should be developed with caution, as well as areas where we will have to adjust to, or protect against, coastal erosion.
3. Several links exist between coastal erosion and other natural hazards, including tsunamis, severe storms, and flooding. Oil spills are linked to coastal processes and represent a hazard in their own right. Some of the most serious coastal erosion occurs during hurricanes and other severe storms. A hurricane may change long reaches of coastline—beaches and barrier islands can be eroded, new inlets formed, and old inlets filled with sediment.
4. Coasts are among the most desirable areas to live, and the rate of population increase of coastal communities has exceeded the average rate of population growth around the world. As a result, coastal erosion is becoming a more serious issue, with increasing economic losses. What used to be a natural coastline may now be lined with oceanfront homes. As a result, erosion, rather than simply changing an undeveloped beach, may damage homes and businesses.
5. Damage from coastal erosion and loss of life from rogue waves and rip currents can be minimized. In the case of coastal erosion, the environmentally preferable solution is zoning and avoiding areas that are rapidly eroding. In areas with considerable coastal development, managed retreat is not an option and large amounts of money will have to be spent in efforts to control erosion problems. In the future, minimizing the effects of coastal erosion will become more difficult as sea level continues to rise. Thus today we are at a crossroads. Many people want to protect the coastline at all cost. Others would rather maintain beaches for future generations and choose a softer path of minimizing coastal erosion, such as managed retreat or beach nourishment.

Summary

Waves are generated by storms at sea or on large lakes; they expend their energy at the shoreline. The size of the waves in open water depends on a combination of wind speed, duration of the wind, and fetch. Waves approaching the shore are slowed by friction and increase in height. Waves become higher in storms—wave energy is proportional to the square of wave height. Irregularities in the shoreline and differences in coastal geology account for local differences in wave erosion and are largely responsible for determining the shape of the coast.

Most beaches consist of sand or gravel that has been carried to the coast by rivers or eroded from rocky headlands. Some beaches consist of locally derived broken shells and coral fragments. Most waves strike a beach at an angle, producing a longshore current parallel to the shore. This current, combined with beach drift, produces littoral transport of sediment parallel to the coast. Hazardous rogue waves can form in the open ocean by the constructive interference of waves of similar size and wavelength.

Almost all shorelines, including those of large inland lakes, are at risk from one or more coastal hazards, including rip currents generated in the surf zone, erosion, rogue waves, rising sea level, tsunamis, and cyclones. Tidal currents can be hazardous in narrow bays and channels, especially on coasts with large tidal ranges such as Alaska, British Columbia, and Atlantic Canada. Beaches experiencing waves from storm swells or plunging breakers commonly have hazardous rip currents that can carry even the strongest swimmers offshore. Coastal erosion can be caused by river damming; construction of dykes, jetties, and other shoreline defences; severe storms; or a rise in sea level. Although less damaging than river flooding, earthquakes, and tropical cyclones, coastal erosion is a serious problem along parts of Canadian and U.S. coasts and Great Lakes shorelines.

Sea level is controlled by eustasy, isostasy, tectonics, and sediment loading. Global, or eustatic, changes in sea level are driven mainly by changes in the amount of glacier ice on land and by thermal expansion or contraction of upper ocean waters. Global sea level is rising today because upper ocean waters are warming and glaciers are shrinking.

Most coastal erosion problems occur in populated areas. Seawalls, groins, breakwaters, jetties, and, more recently, beach nourishment have been used to combat beach erosion where property is at risk. These approaches have had mixed success and can cause additional problems in adjacent areas. Sand tends to accumulate on the updrift side of groins and jetties, and is eroded on the downdrift side. Seawalls reflect storm waves and cause beach erosion. Such engineering structures are expensive, require maintenance, and, once in place, are difficult to remove. The cost of engineered structures may eventually exceed the value of the properties they protect. Some structures may even destroy the beaches they were intended to save. Beach nourishment has been successfully used to restore and widen beaches, but it remains to be seen whether this "soft solution" to coastal erosion will be effective in the long term.

Perception of coastal hazards depends mainly on an individual's experience with and proximity to the hazard. Coastal erosion and flooding risks in wealthy countries are reduced through protective structures, land-use zoning, warning systems, education, and government and individual emergency planning.

Key Terms

barrier island (p. 343)
beach (p. 342)
beach drainage (p. 359)
beach nourishment (p. 357)
bluff (p. 342)
breaker zone (p. 342)
breaking wave (p. 340)
breakwater (p. 356)
E-line (p. 358)
E-zone (p. 358)
eustasy (p. 344)
groin (p. 356)
isostasy (p. 344)
jetty (p. 356)
littoral transport (p. 342)
relative sea level (p. 344)
rip current (p. 344)
rogue wave (p. 338)
sea cliff (p. 342)
seawall (p. 356)
sediment budget (p. 346)
spit (p. 343)
surf zone (p. 342)
swash zone (p. 342)
tidal bore (p. 341)
tombolo (p. 343)

Did You Learn?

1. Explain how waves form on the ocean or a lake.
2. What factors affect the size of waves on the ocean or a lake?
3. Describe how water particles behave in the open ocean when a wind-driven wave passes. How does this behaviour change in shallow water?
4. Explain why waves converge at a coastal headland.
5. Name the two main types of breaking waves. Where do they form?
6. Explain how littoral transport takes place. How is longshore drift related to the direction of wave approach?
7. Name the most serious coastal hazards.
8. Describe how rip currents form.
9. Name the main causes of relative sea-level change.
10. What are the main causes of coastal erosion? Why is coastal erosion becoming more of a problem?
11. Describe the processes that cause sea cliffs and lakeshore bluffs to retreat.
12. What coastal regions are most vulnerable to erosion and sea-level rise?
13. Explain how coastal processes are related to flooding, landslides, subsidence, and tornadoes.
14. What are the natural service functions of coastal processes?
15. Explain the purpose of seawalls, groins, breakwaters, and jetties.
16. List arguments for and against beach nourishment.
17. Name the five principles that should be accepted if we choose to live with, rather than control, coastal erosion.

Critical Thinking Questions

1. Has human activity increased coastal erosion? Outline a research program that could address this question.
2. Do you agree or disagree with the following statements? (1) All structures in the coastal zone, with the exception of critical facilities, should be considered temporary and expendable. (2) A balance of public and private interests is required for any development in the coastal zone. Explain your position on both statements.
3. What are the specific factors that must be known or estimated to predict the future effects of coastal erosion?
4. Assume you live in a community on the shores of the Beaufort Sea in Arctic Canada. Part of the town is now threatened by erosion because permafrost is thawing and waves have strengthened in summer due to the earlier loss of Arctic sea ice. The problem is likely to get worse in the future as the climate continues to warm. You have been hired by the community to advise on a land-use plan and to consider the potential severity of the erosion problem. What sorts of data would you need in order to advise on coastal setbacks for future development? In addition, people in the community wish to live as close to the shore as possible. How would you talk to them about the erosion problem?
5. Do you think a person has a right to build a home on a barrier island on the U.S. Atlantic coast if it could be destroyed by coastal erosion during a hurricane?

MasteringGeology

MasteringGeology **www.masteringgeology.com**. Looking for additional review and test prep materials? Visit the Study Area in MasteringGeology to enhance your understanding of this chapter's content by accessing a variety of resources, including **Self-Study Quizzes, Geoscience Animations, GEODe Tutorials, RSS feeds, flashcards,** web links and an optional **Pearson eText.**

CHAPTER 13

▶ **OKANAGAN MOUNTAIN WILDFIRE** This image from the Advanced Spaceborne Thermal Emission and Reflection (ASTER) radiometer on the Terra satellite was acquired on September 2, 2003. It shows the widespread devastation left by the Okanagan Mountain wildfire in southern British Columbia. The pink area to the east (right) of the bend in Okanagan Lake is the burned area. Smoke (light blue) and areas of active burning (pinkish-white) are visible at the northeast perimeter of the fire. The city of Kelowna is north of the fire. *(NASA)*

Wildfires

Learning Objectives

As Earth's population grows, more people are living in and near scrublands and forests where wildfires occur. This trend increases the risk of property damage and loss of life from fires. Your goals in reading this chapter should be to

- Understand wildfire as a natural process
- Understand the effects of fires and the link to climate
- Know how wildfires are linked to other natural hazards
- Know the potential benefits of wildfires
- Know the methods employed to reduce wildfire hazard
- Understand how people adjust to the threat of wildfires

Wildfires in British Columbia in 2003

Hundreds of thousands of North Americans, seeking a more relaxed lifestyle, have moved into the forested fringes of cities in Canada and the United States. The lives of thousands of such Canadians changed abruptly in July and August 2003, when wildfires—ignited by careless smokers and lightning strikes—exploded out of control in southern British Columbia (Figure 13.1). These wildfires were the most damaging the region had ever endured.

The worst of the wildfires was the Okanagan Mountain fire, which burned approximately 250 km^2 of forest over a three-week period in August and September 2003 (Figure 13.2). It began on August 16 with a lightning strike in Okanagan Mountain Provincial Park south of Kelowna. The fire rapidly spread and became an unstoppable firestorm. In just four days, it destroyed the park, jumped a 50-m-wide fireguard, and approached the outskirts of Kelowna, a city of about 150 000 people. On one night at the height of the fire, a tongue of flames moved into a forested suburb of Kelowna and destroyed 239 homes. More than 45 000 people had to be evacuated in Kelowna and many other towns in southern British Columbia that summer. Property losses totalled more than $250 million, making it the second most costly wildfire in Canadian history.[1]

The Kelowna fire was only one of many wildfires in British Columbia in 2003; other fires were even larger. In total, more than 2500 km^2 of the province's forests burned that summer. The total area burned was not of unprecedented size—more forest land was burned during the 1971 and 1983 fire seasons than in 2003. What made 2003 different from any other year were the number of people and amount of property in the paths of the fires.

Kelowna was again threatened by wildfires in 2009. The summer of 2009 in southern British Columbia was exceptionally dry and hot, and forest fires raged on the outskirts of Kelowna. Aggressive fire suppression, and perhaps lessons learned from the 2003 wildfire, paid off—although many residents were evacuated during the fires, no homes were lost in Kelowna. However, the total cost of fighting wildfires in British Columbia during the summer of 2009 was huge—about $400 million, equal to almost $100 for every man, woman, and child living in the province.

Two factors contributed to the severity of the fires in southern British Columbia in 2003 and 2009: (1) prolonged droughts combined with

Kelowna
Westbank
Okanagan Lake
Okanagan
Mountain Park

◄ **FIGURE 13.1 WILDFIRES AND SMOKE IN BRITISH COLUMBIA** A satellite image of southern British Columbia, showing major forest fires (red squares) and their plumes of smoke in the summer of 2003. The area shown in this image is 600 km across. *(MODIS Rapid Response Project at NASA/GSFC)*

◄ **FIGURE 13.2 THE OKANAGAN MOUNTAIN FIRE** The Okanagan Mountain fire in southern British Columbia in August and September 2003 was the most damaging of nearly 2500 wildfires that burned in southern British Columbia that summer. Dry fuel, high winds, and low humidity hampered efforts to contain the fire, which burned for three weeks, destroyed hundreds of homes, and caused more than $250 million in damage. Nearly 45 000 people, one-third of Kelowna's residents, were evacuated from their homes, and in a single night, 239 houses in the city were lost. *(CP Photo/Richard Lam)*

hot, very dry summers left the forests and soils tinder-dry; and (2) the forest floor had an overabundance of deadfall and other fuel, the result of decades of effective fire suppression.

The wildfires in British Columbia in 2003 pointed out some fundamental principles involved in living with this natural hazard: First, large wildfires require abundant fuel; second, wildfires are becoming more disastrous because more and more people are moving into scrublands and forest lands at the *rural–urban interface*; and third, our fire management policies need to be re-evaluated to minimize future damage by wildfires.

13.1 Introduction to Wildfire

Wildfire is an ancient phenomenon, dating back more than 350 million years to the time trees evolved and spread across the land. Grasses appeared about 20 million years ago, providing fuel for other types of wildfire. Much more recently, at the end of Pleistocene Epoch, the climate became warmer and perhaps drier than it is today, leading to a marked increase in wildfires, at least in the Northern Hemisphere.[2] The geologic record shows a significant increase in the amount of charcoal in sediment dating from the first several thousand years of the Holocene Epoch in western North America, indicating high wildfire frequency at that time. Part of this increase might have been due to humans' use of fire to clear land and assist in hunting.[3,4,5]

After a fire, colonizing plants become established on the burned landscape. The vegetation goes through a post-fire cycle from early colonization to a mature ecosystem adapted to the climate at that particular location and particular time. This cycle, which operates today, is so ancient that some plants have evolved to rely on and use fire to their advantage. For example, oak and redwood trees have bark that resists fire damage, and some species of pine trees have seed cones that open only after a fire. Native eucalyptus trees in southeast Australia, which is regularly swept by bush fires in summer, sprout new leaves after the trees are charred. Often within a year or two, the "bush" has grown back to such an extent that it is not easy to see where it had burned.

Natural fires started by lightning strikes and volcanic eruptions allowed early humans to harness fire for heat, light, and cooking. The benefits of fire allowed humans to broaden their diet, settle in colder areas, and spread across all continents except Antarctica.[4] Aboriginal peoples used fire as a tool in hunting, warfare, and agriculture. Early Europeans commented on the many fires set by Aboriginal peoples for a variety of purposes. This practice continues today in parts of the world, occasionally with serious negative consequences.

13.2 Wildfire as a Process

Wildfire is a self-sustaining, rapid, high-temperature biochemical oxidation reaction that releases heat, light, carbon dioxide, and other products.[4,5] It requires three things—fuel, oxygen, and heat (Figure 13.3). If any of these three things is removed, the fire goes out.

▲ **FIGURE 13.3 THE FIRE TRIANGLE** Wildfire requires fuel, oxygen, and heat. If one of them is missing, a fire cannot start. Likewise, if a leg of the triangle is removed, a fire goes out. *(Based on Contrell, W. H., Jr. 2004.* The Book of Fire, *2nd ed. Missoula, MT: Mountain Press)*

During a wildfire, plant tissue and other organic material is rapidly oxidized by combustion, or burning. Grasslands, scrublands, and forest lands burn because, over long periods, these systems establish a balance between carbon production and decomposition. Plants remove carbon dioxide from the atmosphere by photosynthesis and temporarily sequester carbon in their tissues. Microbes alone do not decompose plant material fast enough to balance the addition of carbon through continued plant growth. Wildfires help to restore this balance. In this simplified view of wildfire, carbon dioxide, water vapour, and heat are released when plants burn. The combustion process, however, is complex—numerous chemical compounds are released in solid, liquid, and gaseous form. Common trace gases include nitrogen oxides, carbonyl sulphide, carbon monoxide, methyl chloride, and hydrocarbons, such as methane.[6] These gases, along with solid particles of ash and soot, form the smoke observed during a wildfire. Both ash and soot are powdery residues that accumulate after burning. Ash consists primarily of mineral compounds and soot is made of unburned carbon.

Flames and smoke provide an accurate mental picture of wildfire, but they do not capture fire's complexity. Wildfires have three phases, all of which operate continuously:[5] (1) pre-ignition, (2) combustion, and (3) extinction.

In the first or **pre-ignition** phase, vegetation is brought to a temperature and water content at which it can ignite and burn. Pre-ignition involves two processes, preheating and pyrolysis.

As vegetation is *preheated*, it loses a great deal of water and other volatile chemical compounds. A volatile compound is one that is easily vaporized. Gasoline, for example, contains many volatile chemical compounds and, when spilled, evaporates quickly to a gas that you can smell.

The other important process of pre-ignition is **pyrolysis**, which literally means "heat divided." Pyrolysis is actually a group of processes that chemically degrade the preheated fuel. Degradation takes place as heat splits large hydrocarbon molecules into smaller ones. Products of pyrolysis include volatile gases, mineral ash, tars, and carbonaceous char. Pyrolysis takes place when you scorch a piece of toast, turning it black. The burnt toast is covered with char, and smoke coming out of the toaster contains small black droplets of tar. A similar thing happens when you scorch cotton fabric with a hot iron.

The two processes of pre-ignition—preheating and pyrolysis—operate continuously in a wildfire. Heat radiating from flames causes both preheating and pyrolysis in advance of the fire. These processes produce the first fuel gases that ignite in the next phase of the fire.

The second phase of a wildfire, **combustion**, begins with ignition and marks the start of a set of processes that are completely different from those related to pre-ignition. Pre-ignition processes absorb energy, whereas combustion involves external reactions that liberate energy in the form of heat and light.[5] Ignition does not necessarily lead to a wildfire. In fact, ignitions, both natural and human-caused, are far more numerous than full-blown wildfires. Wildfires develop only when the vegetation is dry and has accumulated in sufficient quantities to carry fire across the land.

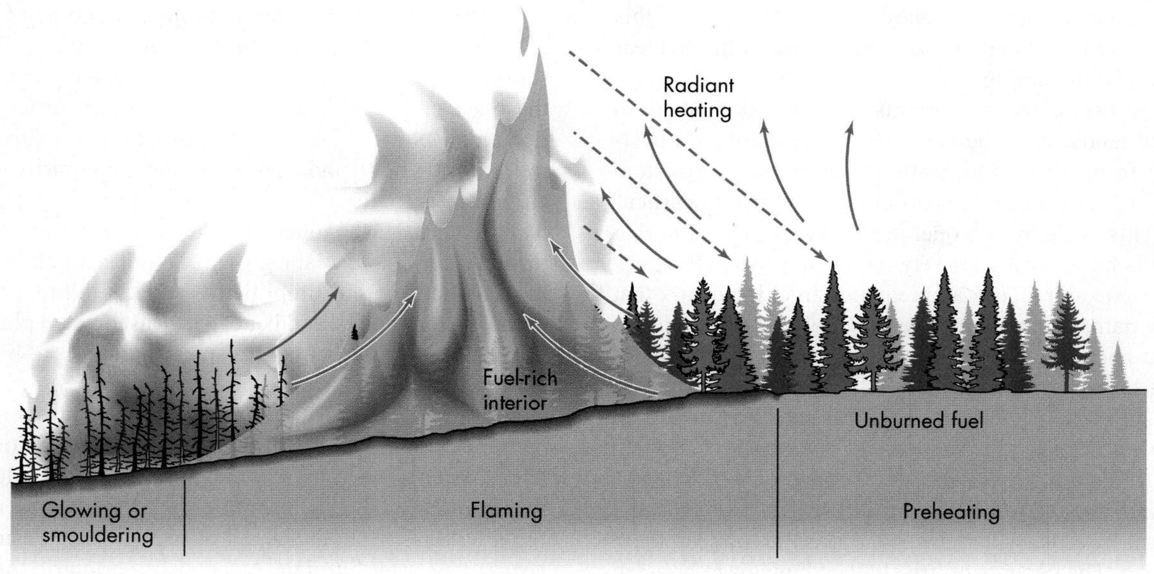

▲ **FIGURE 13.4 PARTS OF A WILDFIRE** Schematic diagram of an advancing wildfire showing its three phases: (1) an unburned area where vegetation is being preheated, (2) an area of flaming combustion with rising air currents, and (3) an area of glowing or smouldering combustion. The fire is advancing from left to right. Solid arrows indicate the motion of the air, and dashed lines show radiant heating from the fire. *(Based on Ward, D. 2001. "Combustion chemistry and smoke." In E. A. Johnson and K. Miyanishi (eds). 2001.* Forest Fires: Behavior and Ecological Effects, *pp. 55–77. San Diego, CA: Academic Press)*

Once an area has burned, the low fuel supply prevents future wildfires. Fire will not threaten again until there is sufficient new fuel. These facts are contrary to past fire-management philosophy that only people can prevent forest fires. On ecologically important timescales—several decades to centuries—fires will occur whether people start them or not; the role of humans in triggering large wildfires is secondary to that of lightning.

Most people think that a wildfire involves a single ignition. Actually, the process is more complex. Ignition repeats time and time again as a fire moves across the land, like sparks and embers popping out of a fireplace, campfire, or barbecue grill. The dominant types of combustion are flaming combustion and glowing or smouldering combustion (Figure 13.4). These two types of combustion differ in that they proceed by different chemical reactions and have different appearances. Flaming combustion is the rapid, high-temperature conversion of fuel to thermal energy by oxidation reactions. It leaves a large amount of residual unburned material. This form of combustion sustains flames as the fire advances across the landscape. As volatile gases are removed from the fuel, woody material continues to burn, but the amount of fuel is less and ash begins to cover new fuel. The ash is non-combustible and may hinder flaming combustion. These processes lead to glowing or smouldering combustion, which can take place at lower temperatures and does not require rapid pyrolysis.

Three primary processes control the transfer of heat as a wildfire moves across the land: *conduction*, which is the transmission of heat through molecular contact; *radiation*, heat transfer by electromagnetic waves; and *convection*, heat transfer by the movement of heated gases driven by temperature differences in a liquid or gas (see Chapter 10). Wildfires transfer heat mainly by convection, although radiation also plays a role. Convective and radiant heating by the fire increases the surface temperature of the fuel. As heat is released, air and other gases become less dense and rise. The rising air removes both heat and combustion products from the zone of flaming. This process shapes the fire as it pulls in the fresh air required to sustain combustion.[5]

Finally, **extinction** is the point at which combustion, including smouldering, ceases. A fire is considered extinct when it no longer has sufficient heat and fuel to sustain it.

Fire Environment

The behaviour of a large wildfire can be explained by three factors: fuel, topography, and weather. With sufficient information about these three factors, we can understand and predict wildfire behaviour.[7]

Fuel Wildfire fuels are complex and differ in type, size, quantity, arrangement, and moisture content. Types of fuel include leaves, woody debris and decaying material on the forest floor, grasses, mosses, ferns, shrubs, small and large living trees, and standing dead or dying trees. Smaller, finer fuels burn most readily and most vigorously, but large woody materials also contribute to fires. If disease or a storm downs a large number of trees, the wood will eventually dry and decay, allowing it to more easily burn during a wildfire. Fuel arrangement can also be important. Forests may consist of trees that are relatively uniformly spaced or that differ considerably in spacing. Most forests have one or more canopies of trees at different heights, as well as shrubs and other plants on the ground. In contrast, some forests have little ground

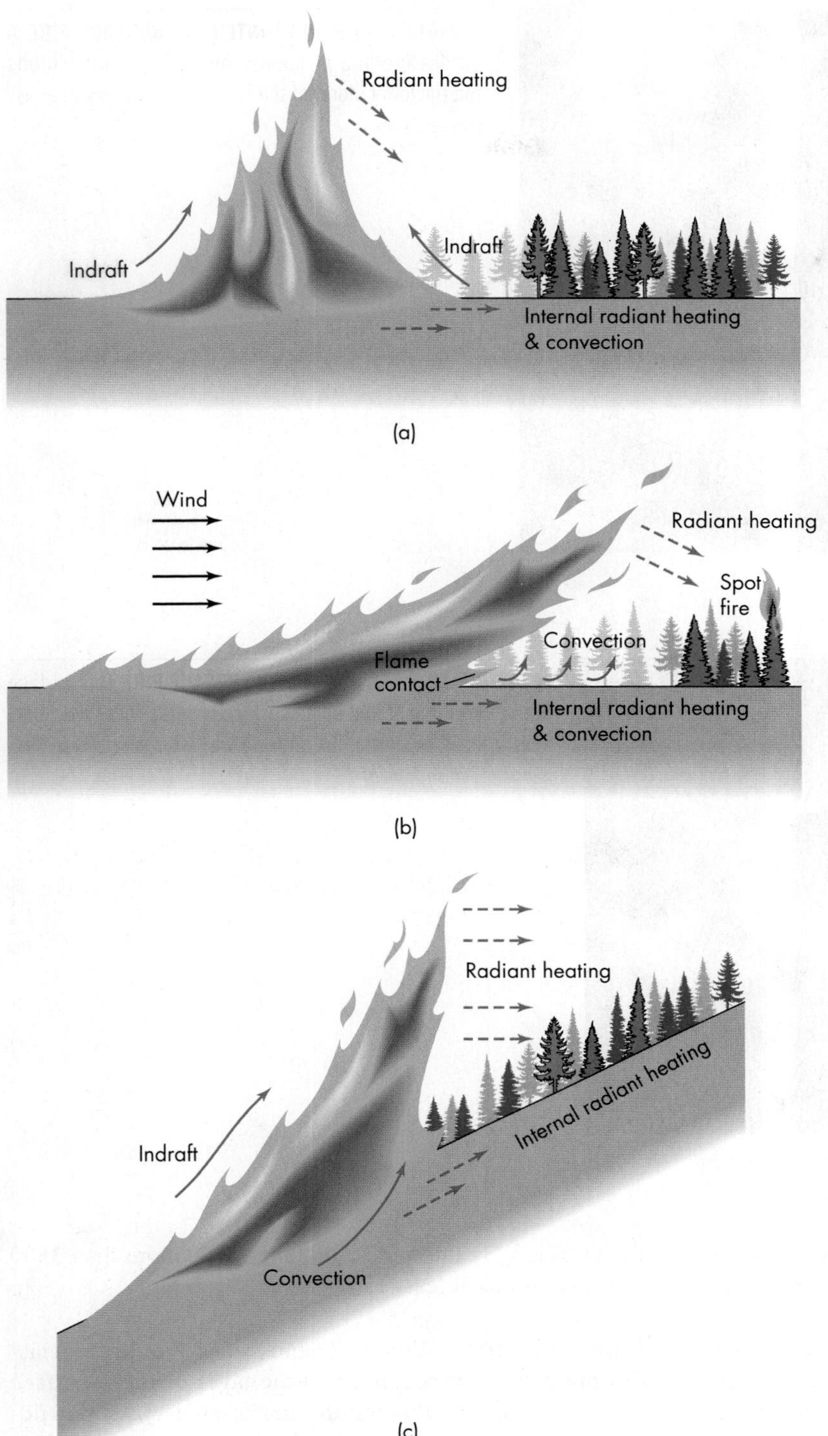

◀ **FIGURE 13.5 WILDFIRES ARE INFLUENCED BY WIND AND TOPOGRAPHY** Idealized diagrams of flame shape and processes associated with a spreading fire under three conditions: (a) flat ground and no wind, (b) flat ground with wind, and (c) a slope with wind driving the fire uphill. *(Modified after Rothermel, R. C. 1972.* A Mathematical Model for Predicting Fire Spread in Wildland Fuels. *U.S. Forest Service Research Paper INT-115)*

cover. The density of the trees in a forest is also important. Dense stands of lodgepole and jack pine in parts of temperate North America, and stands of spruce forming the vast boreal forest of Canada and Eurasia, contain abundant fuel and easily sustain wildfires once they start.

Topography Topography affects fire in several ways. First, in the Northern Hemisphere, south-facing slopes are relatively warm and dry, and fuel on these slopes has a lower moisture content and burns more readily than fuel on north-facing slopes. Slopes exposed to prevailing winds also tend to have drier vegetation than do slopes sheltered from the wind. In mountainous areas, winds tend to move up or down canyons, providing easy paths for wildfires. Wildfires burning on steep slopes preheat fuel upslope from the flames (Figure 13.5c). This preheating increases the rate of movement and thus spreading of a fire moving upslope. Fires may also advance downslope, especially if they are driven by wind.[7]

◄ **FIGURE 13.6 HIGH-INTENSITY SURFACE FIRE** A surface fire in a ponderosa pine forest burns shrubs and the lower storey of the forest. *(Kent and Donna Dennen/ Science Source)*

◄ **FIGURE 13.7 LOW-INTENSITY SURFACE FIRE** This surface fire is burning soil and some surface vegetation and is characterized by smouldering and glowing combustion. *(Remote Sensing Solutions GmbH)*

Weather Weather exerts a strong influence on wildfires. Large wildfires are particularly common following droughts that reduce fuel moisture content. Fires spread more rapidly under hot, dry conditions when humidity is low than when the air is cool and moist. Winds greatly influence the spread, intensity, and form of a wildfire. Strong and changing winds help the fire preheat adjacent unburned fuel. Winds can also carry burning embers far ahead of the flaming front to ignite *spot fires* (Figure 13.5b).[7]

The influence of weather on fires was dramatically illustrated during the drought in Florida between 1998 and 2002, which was one of the worst in the state's history.[8] Hot, dry conditions during the drought reduced water levels in swamps and wetlands, and wildfires occurred in normally wet bald cypress swamps and grassy marshes. One fire in 1998 charred 700 km^2 of marsh in the Everglades and surrounding upland near Miami, and another in 2001 burned over 250 km^2 of swamp and commercial woodland west of Tallahassee.[9] By the middle of June 2001, more than 3500 fires had burned at least 1500 km^2.

Types of Fires Wildfire scientists and firefighters classify fire behaviour according to the layer of fuel—surface or crown—that is allowing the fire to spread. Most wildfires involve both types of fire. **Surface fires** move close to the ground and involve shrubs, dead leaves, twigs, fallen trees, and grass (Figure 13.6). Some surface fires also burn the soil just under the ground surface (Figure 13.7). Surface fires differ greatly in their intensity. Low-intensity surface fires advance relatively slowly with glowing or smouldering combustion and limited flaming. Other surface fires, such as those in the scrublands of southern California and in the grasslands of Africa, can be extremely intense and release large amounts of heat as they move swiftly across the land (Figure 13.8). Some grass fires have been clocked at more than 300 m/min.[10]

◄ **FIGURE 13.8 SURFACE FIRE IN A GRASSLAND** A surface fire in the western United States. Wildfires can spread rapidly in grasslands when driven by strong winds. *(Mark Thiessen/National Geographic Stock Collection)*

◄ **FIGURE 13.9 CONTINUOUS CROWN FIRE** Crown fires spread rapidly in the upper canopy of coniferous forests. They may move ahead of the fire on the surface. In 1995, high winds turned this crown fire, on Mount Lemmon outside Tucson, Arizona, into a firestorm. *(A. T. Willett/Alamy Images)*

Crown fires move rapidly through the forest canopy by flaming combustion (Figure 13.9). *Intermittent crown fires* consume the tops of some trees, whereas *continuous crown fires* consume the tops of all or most of the trees. Crown fires are common in the mixed coniferous forests of western North America and the vast boreal forests of Canada and Siberia. They can be fed by surface fires that move up stems into the limbs of trees, or they can spread independently of surface fires. Large crown fires are generally driven by strong winds and can be aided by steep slopes. Such fires will grow and expand as long as conditions for combustion are favourable. They can advance at rates of up to 100 m/min and can also spread by throwing out fire brands that can be carried by the wind to start new fires up to several kilometres from the fire front.[7] Large wind-driven crown fires are nearly impossible to stop; people and animals must evacuate threatened areas.

13.3 Geographic Regions at Risk from Wildfires

An estimated 70 percent of all forest fires occur in the tropics—half of them in Africa. Dry conditions caused by recent El Ninõ events may have expanded the area burned in tropical Africa.[11] Hotspots in sub-Saharan Africa include northern Angola, the southern Congo, southern Sudan, and the Central African Republic. In these areas, grasslands and fields that have replaced tropical forests decades ago readily burn.[11]

Most areas of Canada and the United States are at risk of wildfires during prolonged dry weather (Case Studies 13.1 and 13.2). Even the Sonoran desert in Arizona, rainforests of coastal British Columbia, and normally wet marshes and swamps in Canada, Alaska, and the southeast United States

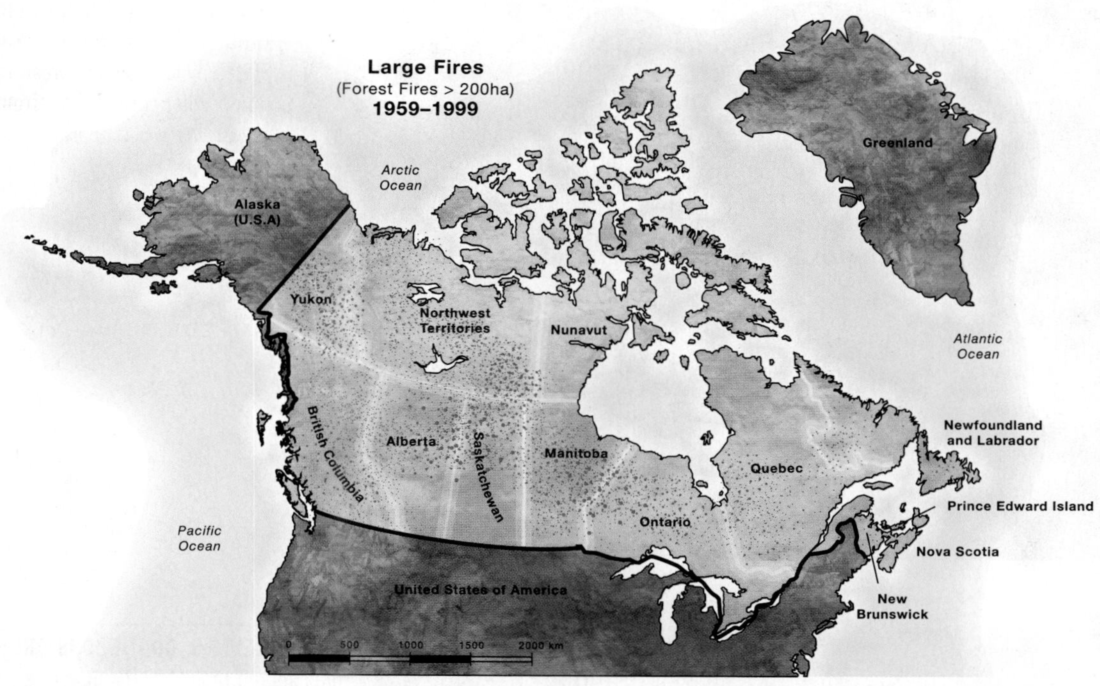

▲ **FIGURE 13.10 FOREST FIRES IN CANADA** This map shows the locations of large forest fires in Canada between 1959 and 1999. Most of the fires are in the boreal forest, which extends in a broad band across the country from Yukon Territory to Newfoundland and Labrador. Forest fires are also common in the mixed coniferous forests of southern and central British Columbia. *(Adapted from Natural Resources Canada, Canadian Forest Service. Reprinted with permission.)*

can experience wildfire. In Canada, the hazard is greatest in British Columbia and in the boreal forest, which extends in a belt almost 5000 km wide across the country from Yukon Territory to Newfoundland and Labrador (Figure 13.10).[10,12]

The specific geographic region at greatest risk from wildfires shifts from year to year, depending on the weather. In some years, the largest area burned is in Ontario or Quebec, whereas in other years, as for example in 2009, British Columbia is Canada's hotspot. Texas recorded the greatest

13.1 CASE STUDY

Wildfires in Canada

Wildfires are a significant agent of change in the vast Canadian forest. They have played a major role in forming and maintaining the boreal forest ecosystem. Large fires in Canada have high rates of fuel consumption, spread rapidly, and release large amounts of energy. The standard measure of fire intensity in the Canadian Forest Fire Behaviour Prediction System is the energy release per unit length of fire front, measured in kilowatts per metre (kW/m). Typical fire intensities are 100 kW/m for surface fires, 2000 kW/m for intermittent crown fires, and 10 000 kW/m or more for continuous crown fires.[13] By one estimate, the amount of energy released annually by fires in Canada is enough to supply the electricity needs of the entire country for six months.[13]

On average, 8500 fires burn 2.5 million ha of forest in Canada each year. Only 3 percent of wildfires in Canada are larger than 200 ha, but these fires are responsible for 97 percent of the total area burned. The number of fires and the area burned differ widely across the country and from year to year; the largest total area burned, some 7.5 million ha, was in 1989 (Table 13.1). In comparison, Canada has about 400 million ha of forested land, which is about 10 percent of the world's forests.[13] Close to 1 million ha are harvested each year.

On average, lightning strikes cause about 45 percent of the forest fires in Canada, but they result in 85 percent of the total area burned.[13] Typically, lightning-caused fires occur in remote areas where detection is more difficult and wildfires are not fought as aggressively as in populated areas. People are responsible for the remaining 55 percent of forest fires in Canada. Most of these fires result from recreational, residential, and industrial activities, but more than 10 percent of human-caused fires are intentionally set.

TABLE 13.1 Total Forest Land Burned in Canada, 1970–2009 (in km^2)

Year	Atlantic Canada	Quebec	Ontario	Man.	Sask.	Alberta	B.C.	YT and NWT	Canada[1]
1970	112	312	227	371	5318	524	1056	1885	10 588
1971	45	2621	420	85	862	631	3513	7054	16 951
1972	504	1053	320	187	2068	493	261	2911	7802
1973	117	30	36	234	2386	107	334	8572	11 844
1974	561	30	5239	1612	260	184	210	387	8486
1975	1802	171	169	233	950	58	243	6690	10 317
1976	2187	569	5441	1281	910	225	570	6952	18 138
1977	42	319	4163	2316	1301	157	38	5989	14 379
1978	83	65	75	246	927	78	501	864	2892
1979	346	32	637	824	2297	1946	294	19 965	27 007
1980	46	316	5603	5143	13 488	6272	656	13 452	47 767
1981	140	25	1795	3762	16 480	13 656	1066	10 201	53 931
1982	117	80	39	154	646	6884	3487	5570	17 063
1983	194	2389	4437	992	522	28	674	2698	11 942
1984	91	31	1204	1302	3212	790	199	610	7654
1985	1566	27	10	118	1101	129	2346	2199	7552
1986	1493	1972	1456	103	132	27	169	4123	9501
1987	190	368	756	1695	2265	361	345	4874	10 856
1988	41	2757	3907	4857	811	145	115	724	13 361
1989	692	21 095	4039	35 679	4705	64	254	9059	75 596
1990	546	833	1837	164	1873	340	758	2742	9344
1991	700	4383	3188	1273	2394	62	247	3582	15 847
1992	81	271	1760	4575	970	35	305	671	8687
1993	280	1282	1047	673	6606	261	52	9447	19 677
1994	1114	1160	835	14 288	9949	296	303	34 251	62 960
1995	17	7277	6124	8892	13 870	3361	481	30 858	70 951
1996	849	6916	4488	1253	97	20	150	4777	18 549
1997	98	3931	385	418	39	47	19	1367	6307
1998	410	4183	1583	4506	9610	7270	437	18 145	46 143
1999	424	977	3283	1105	1737	1205	117	7399	6246
2000	1496	392	67	1109	1410	147	177	1855	6653
2001	24	331	107	789	1843	1541	97	1290	16 264
2002	361	10 138	1721	946	8790	4965	86	628	27 704
2003	380	879	3142	5479	1266	745	2647	1766	17 434
2004	30	30	16	260	2589	2361	2205	22 359	31 838
2005	237	8001	423	700	2138	608	347	3888	16 715
2006	54	1363	1495	1574	12 037	1188	1392	3131	22 236
2007	121	3427	407	3178	2129	1037	294	4828	15 420
2008	79	15	132	1571	11 302	208	132	3669	16 989
2009	—	940	—	29	376	—	—	21	—

Note: Areas in hectares can be calculated by multiplying values in table by 100.
[1]Includes areas burned in national parks, which are not listed separately under provinces and territories.

Source: Data are based on Canadian Council of Forest Ministers, National Forestry Database Program. 2010. "3.1 forest fire statistics by province/territory/agency, 1970–2009, Table B. Area burned." http://nfdp.ccfm.org/data/comp_31e.html. Accessed August 30, 2010. © Canadian Council of Forest Ministers 2008.

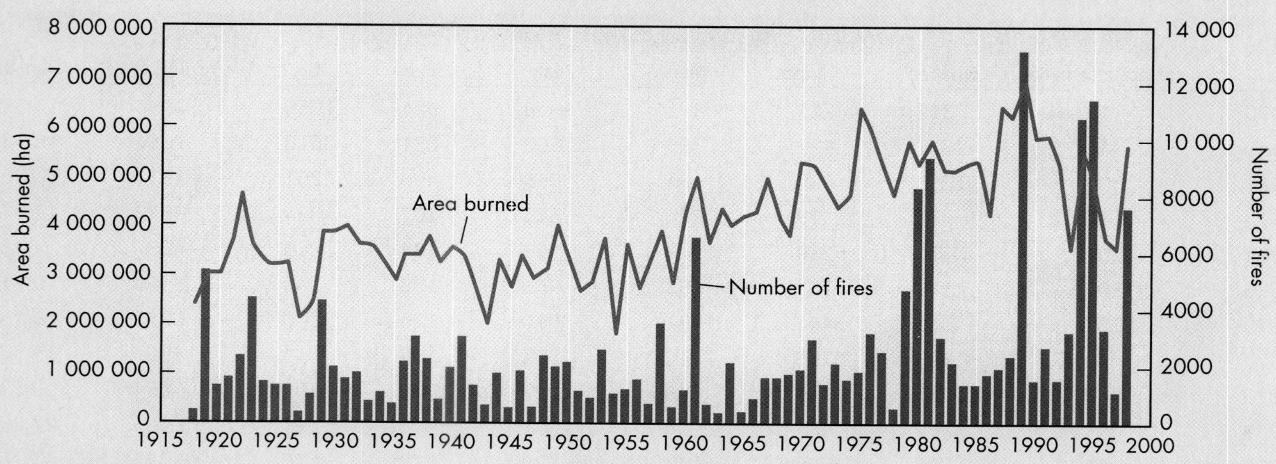

▲ **FIGURE 13.11 ANNUAL VARIABILITY OF FOREST FIRES IN CANADA** The total area burned in Canada increased around 1950. *(Based on Canadian Council of Forest Ministers. 2005.* Canadian Wildland Fire Strategy: A Vision for an Innovative and Integrated Approach to Managing the Risks. *A Report of the Canadian Wildland Fire Strategy ADMs Task Group, p. 5, CIFFC data for 2005 and 2006; and Weber, M. G., and B. J. Stocks. 1998. "Forest fires and sustainability in the boreal forests of Canada."* Ambio *27(7):545–550)*

The Canadian Forest Service has nearly a century of data on forest fires. The data indicate that both the number of forest fires and the total area burned were greater in the second half of the twentieth century than in the first half (Figure 13.11). The total area burned began to increase in the late 1950s but has decreased slightly since about 1990. In comparison, in six separate years since 1980, the number of fires has exceeded the highest number in any year during the first half of the twentieth century. The reasons for these differences are not clear, although climate change, population growth, and increased recreational and industrial use of wildlands could be factors.

Fire-suppression costs in Canada average approximately $400 million per year. Property damage averages more than $10 million per year. In comparison, forestry in Canada is a $60 billion per year industry.[13]

13.2 CASE STUDY

The 2011 Slave Lake Wildfire

Slave Lake is a town with a population of about 7000 people situated in the boreal forest, about 250 km north of Edmonton. The community had been threatened by fires in the past. Residents on the outskirts of Slave Lake were evacuated in 2001 during a nearby fire, and the town escaped a fire in 1968 when the winds shifted.

Wildfires are common in Canada's boreal forest, and conditions preceding the Slave Lake disaster are typical of those that trigger these fires. Alberta faced unusually dry conditions and high winds throughout the spring of 2011, and by May the fire hazard was extreme across much of the province. By mid-May, over 100 fires were burning including 23 that were out of control. As a consequence, the Alberta government enacted a complete fire ban ahead of the Victoria Day weekend; it was only the second time that Alberta had issued such an order.

The fire that devastated Slave Lake was set by an arsonist about 15 km east of town on the afternoon of Saturday, May 14.[14] It grew at an astonishing rate—to 500 ha within three hours. By the afternoon of May 15, the province had declared the fire situation around Slave Lake to be a level 3 emergency, but local officials felt that firefighters were gaining control of the fire and remained hopeful that the town would be spared. However, a change in the winds and gusts up to 100 km/h pushed the flames past firebreaks and into town a short time later.[14]

Residents began to evacuate Slave Lake late in the afternoon of May 15. At 9:30 that evening, the province raised the emergency to level 4—the highest on its scale—and ordered a mandatory evacuation of all residents. Over 7000 people fled east and south to neighbouring towns,

▲ **FIGURE 13.12 AFTERMATH OF SLAVE LAKE FIRE** Aerial photograph of a residential area in the town of Slave Lake two days after a wildfire swept through the community in May 2011. The fire gutted all of the homes visible in this image. *(Newscom)*

as well as to Edmonton. The evacuation was the largest in Alberta's history.

As residents fled Slave Lake, the winds sent burning embers overhead, igniting homes and businesses. Firefighters attempted a last stand in the east part of town, but they were forced to retreat due to the intense heat and burning embers thrown forward by the blaze.[14]

The fire destroyed 374 properties in Slave Lake, about one-third of the buildings in the town, leaving over 700 residents homeless (Figure 13.12). Fifty-nine other properties were destroyed in the surrounding area.[14] The town hall, library, radio station, and a mall were gutted by the fire. Insured damage caused by the fire was estimated at over $700 million, making it one of the costliest disasters in Canadian history.

Fortunately, there were no casualties among Slave Lake residents as a result of the fire, but a pilot was killed when his helicopter crashed while battling the blaze.

number of hectares burned in the United States in 2006, whereas it was Alaska in 2004 and 2005, and California and the northern Rocky Mountains in 2003.[15] In 2002, the spotlight was on Georgia and Colorado. A 2340-km^2 fire in Okefenokee Swamp in Georgia was the largest wildfire in the southeast United States in over a century. The Missionary Ridge fire consumed 285 km^2 of forest in Colorado in June and July 2002. The total direct and indirect costs of this wildfire were about U.S.$152 million.[16] Even worse fires occurred in Colorado in the summer of 2012. Numerous fires in June, July, and August in tinder-dry forests were triggered by dry thunderstorms and arsonists. The fires followed an extremely dry winter, with precipitation far below the average. Temperatures in June in some parts of the state exceeded 40°C, and humidity was low. Colorado Springs was especially hard hit—the Waldo Canyon fire forced the evacuation of 34 000 residents, destroyed 346 homes, and claimed two lives in the community. Insured damage from that fire alone exceeded U.S.$350 million.[17]

On average more than $400 million is spent fighting wildfires in Canada each year, but the cost can exceed $1 billion in bad fire years. No civilian lives have been lost due to wildfires in Canada since the 1938 Dance Township fire in Ontario, which took 20 lives. However, over the past 30 years, about 40 fire management personnel have been killed fighting fires, for example in plane crashes.

13.4 Effects of Wildfires and Links with Climate

Wildfires affect many aspects of the local environment. They burn vegetation, harm wildlife, release smoke into the atmosphere, char soil, increase erosion and runoff, and cause landslides.

Effects on the Geological Environment

Wildfires affect soils differently depending on the type and moisture content of the soil and the duration and intensity of the fire. The amount and intensity of precipitation after a fire also influence how a wildfire affects soil.[18]

Hot fires that scorch dry, coarse soil may leave a near-surface, water-repellent layer called a *hydrophobic layer*. Water repellency is caused by the accumulation of chemicals derived from burning vegetation.[5] The water-repellent layer increases surface runoff and erosion because the burned surface lacks vegetation to hold the loose soil above it. Water flows over a hydrophobic soil layer much as rain runs off a water-repellent raincoat. Soil above the hydrophobic layer quickly becomes saturated during rains and may wash downslope.[19] Hydrophobic layers can persist in soils for several years following a fire.

Soil erosion and debris flows are common following wildfires, although the effects are variable. Areas that are susceptible to erosion commonly experience increased erosion for a few years after a fire, whereas areas that normally experience little erosion may show little or no effect.[5] Erosion and landslides are significantly greater on steep slopes charred by a severe burn than on gentler, less severely burned ones.

Heavy rains can significantly increase the incidence of erosion and landslides in burned areas. Wildfires in California in 1997 denuded vegetation from steep slopes just before the winter rains. Of 25 burned areas mapped by the U.S. Geological Survey, 10 produced debris flows during the first winter storm.[20] Sediment washed from the burned slopes choked swollen streams and rivers. The wildfires in southern British Columbia in 2003 set the stage for increased debris flow activity, erosion, and sedimentation in severely burned areas.[21]

Effects on the Atmosphere

Wildfires significantly increase the concentration of particulates in the atmosphere for weeks, even months (Figure 13.13). Increases of airborne particulates can be observed thousands of kilometres downwind of large, long-lasting fires. During the past decade, fires in Mexico, Guatemala, and other countries in Central America raised atmospheric particulates in some cities in the southern United States to levels exceeding Clean Air standards. Fires burning in Indonesia and Malaysia in 1997 and 1998 affected people in many other countries in the region.

Links with Climate

A possible consequence of climate warming expected later in this century is an increase in the intensity and frequency of wildfires brought about by changes in air temperature, precipitation, and the frequency of severe storms. Related biological changes may affect the type and quantity of fuel available for wildfires. Some of these changes may already be underway.[22,23]

In late October 2003, wildfires in southern California burned about 6000 km^2 of forest and scrubland, destroyed

▲ **FIGURE 13.13 SMOKE, GASES, AND PARTICULATES FROM WILDFIRES** Wildfires introduce large amounts of ash, soot, carbon dioxide, water vapour, and other gases into the atmosphere. *(Kip Frasz/The Canadian Press/Kelowna Daily Courier)*

PROFESSIONAL PROFILE

Bob Krans

The Cedar fire in southern California began on October 25, 2003, when a lost hunter lit a signal fire in the Cleveland National Forest. By the time it was extinguished in November, the Cedar fire had become the worst fire in California history.

Bob Krans, fire division chief in Poway, California, had been in the firefighting business for 30 years and had never seen anything like it. "I saw conditions that I'd never seen before," Krans says. "It was extremely frustrating to deal with."

The smoke appeared orange because the blaze was lighting the smoke from within. Krans describes the fire as sounding like a freight train because of the strong accompanying wind, some of which was being generated by the fire itself because it was consuming huge amounts of oxygen.

Fighting a wildfire has hazards beyond the obvious ones of intense heat and quickly changing conditions. Krans says that firefighters near the blaze can suffer third-degree "steam burns" when the heat vaporizes the sweat inside their protective suits.

Krans recalls that the Cedar fire at times was consuming up to 5000 ha/h, whipped by winds blowing 105 km/h. At that rate, it is easy to observe the movement of the fire as it draws closer (Figure 13.14).

◄ **FIGURE 13.14 FIRE DIVISION CHIEF BOB KRANS** Chief Krans examines the Cedar fire as it moves rapidly toward him. The fire reached the spot where Krans is standing in three minutes, destroying the house directly behind him. *(Chris Thompson, Poway Reserve firefighter)*

Although fire departments take extensive safety precautions, one firefighter perished and several others were badly injured while battling the Cedar fire when a second blaze closed off their escape route. The smoke can be so thick that vehicles require headlights and firefighters can become disoriented. "Daytime literally turns to night time," Krans notes.

—Chris Wilson

several thousand homes, and killed 24 people. Twelve major fires started in a single week. Catastrophic "runaway" fires could not initially be contained by firefighters and spread rapidly to damage large areas (see Professional Profile). Autumn and winter are particularly hazardous seasons for wildfires in southern California. Hot, dry *Santa Ana winds*, which result from large-scale atmospheric circulation, produce dangerous conditions over the entire region.[24,25]

Although the number of large wildfires in southern California hasn't changed significantly in the past few hundred years, their size and intensity and their effect on people certainly have. The population of southern California has doubled in the past 50 years, and fire suppression has increased natural fuels in forests and scrublands. As a result, the severity and intensity of fires are increasing, as are their human and ecological consequences.[24] The association of wildfire and Santa Ana winds is well known, but the exact timing of fires is controlled by rainfall during winter and the weather the summer before. The relation of the seasonal Santa Ana winds to projected climate warming has not been studied and is not understood. However, if the trend toward drier, windier conditions in southern California continues, wildfires will likely become larger and more intense.[25,26]

The concentration of carbon dioxide in the atmosphere is expected to be twice the pre-Industrial Revolution value before the end of the twenty-first century, leading to an increase in atmospheric temperatures. Increases in temperature in most areas may make the wildfire hazard worse. For example, the climate of western North America from California to Alaska is likely to become warmer, drier, and windier—three factors that promote wildfires. Both the number and the intensity of wildfires will probably increase in this region. More property will be destroyed, more lives lost, and more floods and landslides will follow the fires. In addition, the costs of suppressing fires and of fire insurance will increase.[25] Similarly, warmer springs and summers over much of sub-Arctic Canada may increase the incidence of large wildfires in the boreal forest that covers much of this huge region.

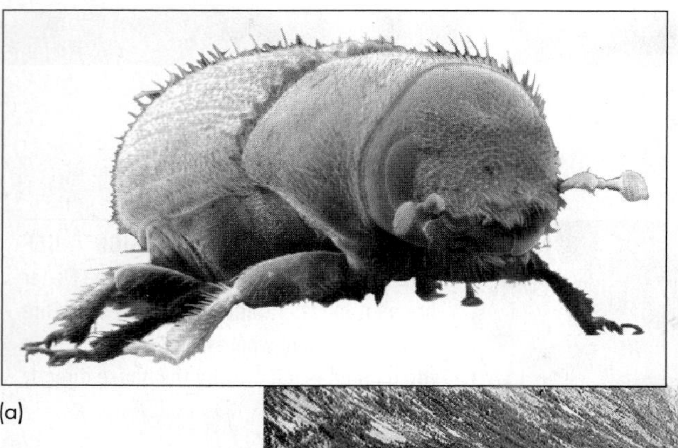

(a)

(b)

◀ **FIGURE 13.15 MOUNTAIN PINE BEETLE** (a) The mountain pine beetle has devastated lodgepole pine forests in British Columbia. (b) During the past 15 years, the beetle has infested 15 million ha of forest. Dead stands of pine like this pose a wildfire hazard if not removed. *((a) Leslie Manning, Canadian Forest Service, Natural Resources Canada, 2010; (b) Natural Resources Canada, 2010)*

Increases in fire frequency and intensity may also change the vegetation of semi-arid areas that currently have limited moisture. Grasslands will expand at the expense of dry coniferous forests in places like southern British Columbia and California. The boundary between the boreal forest and prairie grasslands in Canada may move north as climate warms and wildfires become more common.

Warmer and drier conditions can make trees more susceptible to beetle infestations, weakening them and rendering them more vulnerable to fire. An example is the current mountain pine beetle infestation in western Canada. Larvae of mountain pine beetle (*Dendroctonus ponderosae*) mature just inside the bark of lodgepole pine, an important forest tree in interior British Columbia, western Alberta, and southern Yukon Territory.[27,28] The adult beetles (Figure 13.15a) emerge from pine trees in early summer. After mating, the females lay their eggs in healthy trees. As the larvae develop, they mine the living part of the tree beneath the bark and eventually cut off the tree's supply of water and nutrients. The beetles also introduce a bluestain fungus that defeats the tree's natural defences against the attack by killing living cells in the inner bark and sapwood. The combination of larvae and fungus girdles and thus kills most affected trees. The beetle larvae can be killed by extreme and prolonged winter temperatures (−40°C) or by cold snaps of −25°C in early fall or late spring. However, over the past century, western Canada has warmed considerably, and large areas of lodgepole pine forest have not experienced such cold in nearly two decades. Partly as a result of this warming, the mountain pine beetle has affected more than 15 million ha of forest; about 80 percent of all mature pines in British Columbia are now dead. The beetle has now spread into the Rocky Mountains of Alberta and has attacked yellow (ponderosa) pine forests in southern B.C. The infestation might well be the largest ever recorded, and the extensive stands of dead trees (Figure 13.15b) are a serious fire hazard.

Other damaging insect infestations have occurred in western North America in recent years. Warmer temperatures on Alaska's Kenai Peninsula have led to a large infestation of another species of beetle there. An estimated 40 million trees have been killed in an area nearly three times the size of Prince Edward Island. As the trees die and desiccate, catastrophic wildfire will be much more likely.[25] The spruce budworm (*Choristoneura fumiferana*) has caused extensive damage to eastern Canada's coniferous forests. The preferred host for the insect is mature balsam fir.

An increase in wildfires because of climate warming will affect human health. As mentioned previously, increases in airborne particulates from wildfires lower air quality and can harm people with breathing disorders. For example, wildfires in Florida in 1998 greatly increased the number of people seeking emergency treatment for asthma, bronchitis, and chest pain.[25]

13.5 Impacts of Wildfires on Plants and Animals

Fire affects people, property, companies, and the environment in a variety of ways. The level of impact depends largely on fire intensity and size. A large fire can result in town evacuations, road and airport closures, property loss, timber loss, habitat loss for some plants and animals, an increase in habitat for others, removal of the vegetation, and consumption of the forest floor.

Timely evacuations have greatly reduced the number of deaths from wildfires in North America, but several large fires caused by land clearing and slash burning killed many people in the nineteenth century. A wildfire in 1825 burned more than 1.2 million ha of forest in New Brunswick and Maine, killing 160 people. Another fire in 1871 in Wisconsin and Michigan charred more than 1 million ha and claimed 1500 lives. Ten years later, in 1881, a fire in eastern Michigan burned more than 400 000 ha and killed 169 people.[29]

Rains have a much greater erosive impact on an area after a fire. For example, in May 1996, a 48-km^2 fire burned two of the watersheds that supply drinking water for Denver, Colorado. Two months after the fire, a storm caused flooding that delivered large amounts of floating debris and high concentrations of manganese to the reservoir from which Denver draws its water. Two years after the fire, water quality was still poorer than it had been before it.[20]

Smoke and haze produced by fires can harm human health. As mentioned previously, exposure to smoke and haze produces eye, respiratory, and skin problems. Firefighters and others who have experienced prolonged exposure to smoke can experience chronic, sometimes fatal, respiratory problems.

Wildfire can, of course, destroy personal property. More and more people are living in densely vegetated areas at the fringes of cities. As a result, property losses because of wildfires are increasing dramatically. When a wildfire occurs in scrubland or forest near Victoria, Los Angeles, or Sydney, Australia, thousands of homes might be at risk.

Large intense wildfires can kill even the fastest-running animals, but animals generally are able to escape advancing fire. Even rodents have a good chance of survival because the ground in which they take refuge is a good insulator. Fish and other aquatic species may suffer from increased sedimentation, and stream temperatures may increase because plants along their banks have been destroyed.[18] However, many species of birds, insects, reptiles, and mammals benefit from wildfire (Case Study 13.3). Forest fires produce open areas suited for grazing mammals, including deer, moose, and bison, while maintaining grasslands required by some birds and rodents.

13.6 Natural Service Functions of Wildfires

Wildfire is an important element of terrestrial ecosystems and, as mentioned earlier, has been for hundreds of millions of years. It temporarily reduces competition for moisture, nutrients, and light, allowing both the surviving and new species to thrive, thus maintaining biodiversity (Figure 13.16). It maintains stand age diversity in forests by opening gaps in the canopy, allowing more sunlight to reach the forest floor. Fire also acts a regulator; where a fire has burned, the amount of fuel is reduced and the fire risk is decreased.

Fire triggers the release of seeds and stimulates flowering of species that depend on it for reproduction. Lodgepole pine, aspen, and fireweed are examples of plants that require abundant sunlight and are common pioneer species after forest fires in North America. Fire stimulates the release of an enzyme in aspen that can cause charred trees to produce huge numbers of new sprouts.[30] These sprouts are an important food for elk and moose. Fire melts the resinous coating on lodgepole pine cones, releasing the seeds.[5] The seeds fall to a nutrient-rich, ash-covered forest floor that is devoid of competition from other plants. Species with these reproductive adaptations exist mainly in environments where fire is frequent. If wildfires were eliminated from their environment, these species might eventually become extinct.

Fire increases the nutrient content of a soil, leaving an accumulation of carbon on the surface in the form of ash. If the ash is not removed by erosion, it forms a nutrient reservoir that is beneficial to local plants. Fires can also reduce populations of soil micro-organisms, benefiting the plants that compete with them for nutrients. Some micro-organisms are parasites or carry diseases.[5]

The pre-settlement Canadian Prairie, for example, was an ecosystem well adapted to wildfires. Grasses are especially

▲ **FIGURE 13.16 RENEWAL AFTER FIRE** Fireweed and young conifers have become established on a burned landscape in southwest Yukon Territory. *(John J. Clague)*

13.3 CASE STUDY

Yellowstone Fires of 1988

Because wildfires benefit ecosystems, many scientists believe that natural fires in forested areas should not be suppressed. In 1976 officials in Yellowstone National Park, which covers 9000 km^2 in Wyoming and Montana, instituted a policy of allowing natural fires to burn in areas of the park managed as wilderness, provided they did not endanger human life, threaten visitor areas such as Old Faithful, or spread to areas outside the park. Any human-caused fire, however, would be extinguished immediately. Before the summer of 1988, the worst fire in the park's history had burned only about 100 km^2 of forest. This figure stands in stark contrast to over 3200 km^2 of forest consumed during the 1988 fire season, one-third of Yellowstone's total area. The 1988 fires led to a major controversy over the National Park Service's natural-burn policy.

The problem started when lightning strikes ignited 50 fires in the park in the early summer of 1988. Twenty-eight of the 50 fires were allowed to burn according to the natural-burn policy. These fires quickly expanded under the hot, dry, summer conditions, fuelled by high winds in mid-July (Figure 13.17). On July 17, Yellowstone officials bowed to political pressure and sent fire crews in to fight one of the natural fires. Within four days, many natural fires in and around Yellowstone were being fought, but it was clear that they were beyond the control of the crews. The fires did not slow until September 10, when rains fell throughout the area. Snows in November finally extinguished the flames.

Yellowstone's natural-burn policy was severely criticized during and after the 1988 fires. Although most scientists agree that fire is good for the natural environment, it was difficult for people to sit and watch the park burn. Eventually, 9500 firefighters were deployed at a cost of U.S.$120 million. Critics claimed that the fires would not have been so big if park officials had fought them from the beginning. Others argued, however, that the fires would not have been so severe if prior fire-suppression policies had not allowed fuel to accumulate to dangerous levels in the area. Post-fire studies in Yellowstone and elsewhere have shown that natural wildfires are beneficial to the environment. Yellowstone park officials still adhere to a natural-burn policy. This policy is correct because Yellowstone's ecosystems have, through geologic time, adapted to and become dependent on wildfire. The fires of 1988 did not destroy the park; on the contrary, they revitalized ecosystems through natural transformations that cycle energy and nutrients through solids, plants, and animals.

◄ **FIGURE 13.17 YELLOWSTONE FIRE** Firefighters watch a forest fire in Yellowstone Park, Wyoming, in 1988. *(Ted Wood/Stone/Getty Images)*

suited for growth after a fire, because they grow from their base rather than leaf tips and because 90 percent of their biomass is underground and generally not damaged by fire.[31] Fire removes surface litter and allows nutrients and moisture to reach the roots of grasses. Native North Americans long ago understood the importance of fire for the prairie ecosystem.[31]

Wildfire also is indispensible to the health of the boreal forest that covers large areas of Canada and Russia. It removes stands of diseased, dying trees and limits the spread of beetle infestations. It also facilitates the cycling of nutrients and provides clearings in the boreal forest that are required by some of the animals that inhabit it.

Many species of birds, insects, reptiles, and mammals benefit from wildfire. For example, deer and elk find new forage in shrubs, tree seedlings, and grasses that proliferate following fire. Burned, decaying logs provide homes for insects and food for a wide variety of animals.

13.7 Fire Management

The danger posed by wildfires to humans and property is minimized through *fire management*. The elimination of wildfire is not possible, nor is it economically feasible or ecologically desirable. A wildfire, however, cannot be allowed to run its natural course when it threatens lives, property, or valuable resources. The aim of fire management is to control wildfires for the benefit of ecosystems while preventing them from harming people and destroying property.

The fire season in Canada extends from April through October, whereas in parts of the United States, it is nearly year-round. In Canada, there typically is no winter wildfire activity, a flurry of spring fires after the snow melts, followed by a decline as forests green up in early summer, a peak of fires in mid-summer, and then decreasing activity in the fall. Characteristics of Canada's environment, society, and geography that have shaped fire management include (1) a sparse population concentrated along the U.S. border, (2) large areas with few roads, (3) both high- and low-value property and resources to protect, (4) limited high-cost labour, (5) excellent technological infrastructure, (6) a short fire season, and (7) typical stand-replacing fires.[32]

Fire management in Canada is a provincial and territorial responsibility, except on federal lands. Thirteen provincial and territorial agencies set their own fire policies, depending on local and regional fire regimes, forest types, and infrastructure at risk. Funding for fire management in Canada comes from general tax revenues, timber sale revenues, or levies based on land ownership.[32]

The main elements of fire management in North America are scientific research, data collection, and fire suppression, including the use of prescribed burns.

Scientific Research

Scientific research on wildfires and especially the role of fire in ecosystems is critical to fire management: we cannot manage what we do not understand. Further research is needed to better understand the **fire regime**, or pattern of fire activity of an area, including, for example: (1) the types of fuel present in plant communities; (2) typical fire behaviour, characterized by the size and intensity of the fire and the amount of biomass removed; and (3) the fire history of the area, including fire frequency and size. Reconstruction of natural fire histories is difficult in many areas because fires have been suppressed for nearly 100 years.[33,34] Nevertheless, fire management is more likely to be successful if fire regimes can be defined for specific ecosystems.

Data Collection

Remote sensing has become an important tool for fire management. Satellite imagery is now routinely used to map vegetation and determine fire potential (Figure 13.18). Since the early 1990s, the U.S. Geological Survey, in co-operation with the National Oceanic and Atmospheric Administration (NOAA), has prepared weekly and biweekly maps of parts of the contiguous United States and Alaska illustrating vegetation cover and biomass production. The maps, combined with field measurements of the moisture content of vegetation, are an invaluable resource for fire managers. A management tool called the Fire Potential Index was developed in the United States to characterize the fire potential of forests, rangelands, and grasslands.[35] It takes into account the total amount of burnable plant material, or fuel load; the water content of dead vegetation; and the percentage of the fuel load that is living vegetation. The Fire Potential Index is based on weekly satellite images that portray the approximate time of green-up in spring and the relative amount and condition of growing plants. Regional and local fire-potential maps are prepared daily in a geographic information system (GIS) to help land managers develop plans to minimize the threat of fires.

Fire Suppression

For the past century, fire management in Canada and the United States has been guided by a policy of wildfire suppression, mainly to protect human interests. Four principles underlie fire suppression in Canada: (1) reliance on information about fire danger, risk assessment, and resource allocation; (2) reliance on equipment, including vehicles, pumps and hoses, and communications; (3) reliance on aircraft to detect and suppress fires and to transport firefighters; and (4) rapid mobilization, both within and across agencies. Fire suppression is a provincial and territorial responsibility in Canada. A typical provincial or territorial fire management agency has four components.[32] An executive unit prepares plans and budgets, and evaluates fire programs, such as for prevention, suppression, and training. A strategic unit monitors fire danger and fire activity, deploys resources across the agency, and authorizes interagency movement of resources. A management unit is responsible for day-to-day activities, including readiness, detection, and initial attack dispatches. An operational unit transports crews and equipment to fires,

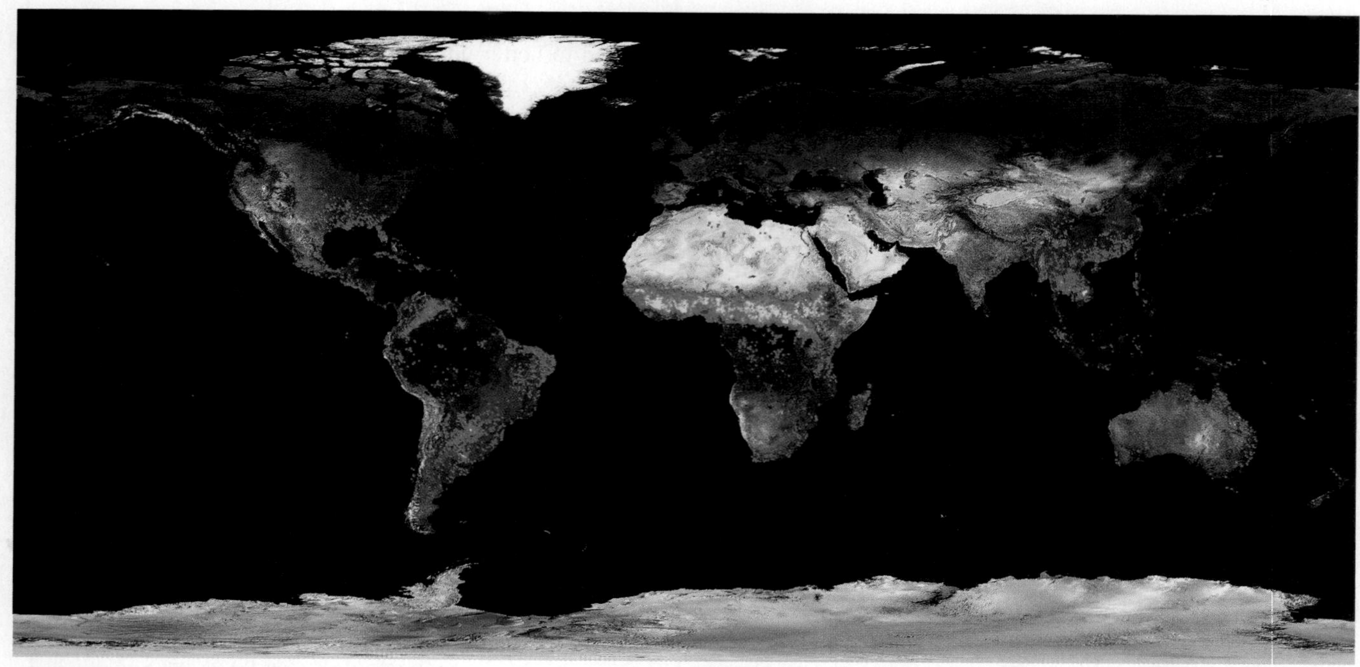

▲ **FIGURE 13.18 GLOBAL FIRE MAP** A map showing fires detected by the MODIS sensor on the Terra satellite from January 1 to 10, 2006. Each coloured dot is a location where MODIS detected at least one fire during the 10-day period. Colour ranges from red where the number of fires is small to yellow where the count is large. NASA has produced 10-day composite global fire maps like this since 2000. *(NASA)*

provides logistical support, and is directly responsible for fire suppression. A common operational practice is to steer the fire into an area with no fuel, called a *fire break*. Natural fire breaks include rivers, lakes, and roads. Where no natural fire breaks exist, one can be made by using bulldozers to create a corridor through scrub or forest or by burning ahead of the fire.

Reliance on fire suppression in North America has led to a build-up of fuel in forests and an increased potential for large, high-intensity fires. One way to counter the build-up of fuel is to ignite controlled fires, also known as **prescribed burns**. The use of prescribed burns for forest management is not new; such fires have been used for decades as an alternative to total fire suppression.[34]

The purpose of controlled burning is to reduce the amount of fuel in forests and thus the likelihood of a catastrophic wildfire.[20] Fire ecologists have found that more frequent, smaller fires result in fewer large, dangerous fires.

Each prescribed burn has a written plan that outlines the objectives—where and how it will be carried out, under whose authority, and how the burn will be monitored and evaluated. Those in charge of a prescribed burn take on a great deal of responsibility: they have to predict the natural behaviour of the fire and keep it under control. They face the difficult task of predicting the fuel and weather conditions under which they can safely control the fire. Such factors as temperature, humidity, and wind all must be taken into account.

Changes in winds during a prescribed burn can have disastrous results. On May 4, 2000, fire managers lost control of a prescribed burn in Bandelier National Monument in New Mexico, when the winds unexpectedly changed direction. After two weeks, the fire had burned 190 km^2 of forest and was still uncontrolled. Fed by drought-parched pine forest and grassland, the fire eventually destroyed 280 homes and forced the evacuation of 25 000 people in the town of Los Alamos. The question asked after the flames had been extinguished was, How could a controlled burn, ordered by the National Park Service itself, end up nearly destroying a town?

A formal review of the disaster called for changes in the Park Service's prescribed burning policy. The changes included a more careful analysis and review of burn plans, better coordination and co-operation among federal agencies in developing burn plans, and use of a standard checklist before setting a prescribed fire. These and other new policies have been instituted. The 1995 U.S. Federal Fire Policy was also reviewed and updated. Specific changes include emphasizing the role of science in developing and implementing fire-management programs.

Prescribed burn policies in populated parts of Canada are similar to those in the United States and have been similarly updated because of damaging wildfires in British Columbia, Ontario, Quebec, and New Brunswick. The situation is slightly different in forested areas in northern Canada, where the population is small and concentrated in widely spaced towns. Prescribed burns are generally not done in these areas

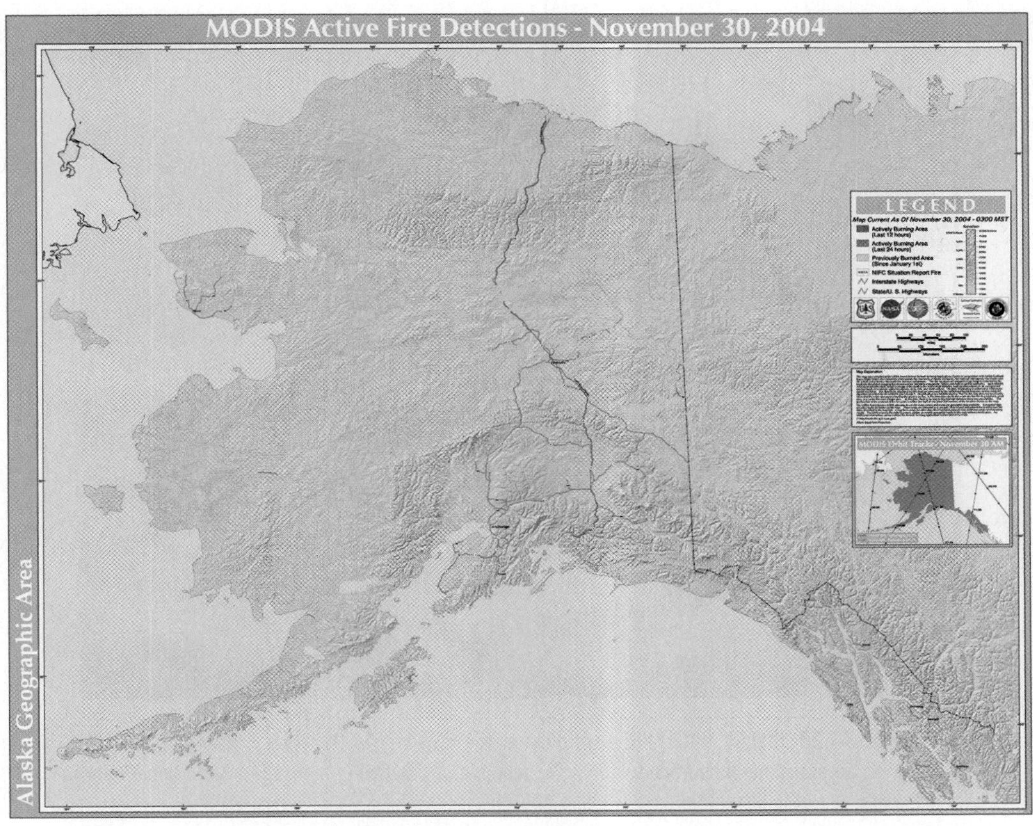

▲ **FIGURE 13.19 FIRES IN YUKON AND ALASKA** This map shows (in yellow) areas in Yukon Territory and Alaska that burned during the summer of 2004. Most of the wildfires occurred in wilderness areas and were allowed to burn. *(U.S. Department of Agriculture, Forest Service)*

because wildfires caused by lightning strikes are allowed to burn unless towns or highways are threatened or valuable forest resources are at risk. Natural fires in these areas can be very large. In 2004, more than 4500 km^2 of forest in Yukon Territory and eastern Alaska burned (Figure 13.19). Some of these fires were allowed to burn themselves out; others, however, threatened small towns and mining operations and were vigorously fought.

In 2003, the U.S. Congress passed the *Healthy Forests Restoration Act*, a forest management plan with the stated objective of reducing large damaging wildfires by thinning trees on federal lands. The act reduces or limits the environmental review that is normally required for new logging projects. Those in support of the act state that the new management procedures will reduce the risk of catastrophic fires to towns in and around national forests, will save the lives of forest residents and firefighters, and will protect wildlife, including threatened and endangered species. Those opposing the act counter that large-scale logging will occur far from communities at risk and will damage forests. They further argue that the best way to minimize risk from wildfires is by selectively removing vegetation around communities and homes and through education and planning.

13.8 Perception of and Adjustment to the Wildfire Hazard

Perception of Wildfire Hazard

In general, people who live in seasonally hot and dry forested areas and scrublands do not fully appreciate the risk they face from wildfires. Included in this group are hundreds of thousands of California residents. Wildfires occur every year in California, yet development continues on brush-covered hills. The demand for hillside property has raised property values in these areas, which means that the people whose property is most at risk from fire have paid a premium for the "privilege" of living there. Fire insurance and disaster assistance may worsen the situation—if people know their fire losses will be reimbursed by insurers or governments, they may conclude that there is no reason not to live where they choose, regardless of the risk. In the past century, hundreds of fires have burned scrublands and forests in California, and tens of thousands of homes and other structures have been destroyed. Yet, population and property values continue to increase in high-risk areas.

▲ **FIGURE 13.20 KILLER WILDFIRE** A wildfire in the hills above Oakland, California, in 1991 devastated an entire neighbourhood, killing 25 people and destroying more than 3800 homes and apartments. *(Tom Benoit/Stone/Getty Images)*

The result of this lack of understanding of wildfire risk was tragically illustrated in October 1991, when a wildfire destroyed about 3800 homes and apartments in the cities of Oakland and Berkeley (Figure 13.20). The fire killed 25 people and caused more than U.S.$1.68 billion damage, making it one of the worst urban disasters in United States history.

The fire started on the evening of October 19, when flames escaped from a cooking fire in a camp of homeless people. Firefighters thought they had extinguished the fire and left the site around midnight. However, the next day was hot and windy, and embers left from the night before reignited the fire. Urbanization had reduced open land on the slopes above Berkeley and Oakland from 47 percent in 1939 to barely 20 percent in 1988. It also added fuel to the slopes, which previously were grass-covered with scattered oak and redwood trees. The additional fuel included numerous homes and non-native trees, mainly eucalyptus. When the fire reignited, it quickly became uncontrollable and firefighters could only evacuate residents. The fire moved through the urban landscape quickly; during the first hour it consumed a home every five seconds!

The Oakland wildfire was started by people, but other factors proved more important in making it a disaster: an ample fuel supply of buildings, brush, and trees, and hot, windy weather. Furthermore, from a fire-hazard perspective, land-use planning in the area was inadequate. Houses were not fire resistant, the density of buildings and the placement of utilities were not regulated, and no rules required removal of excess vegetation from around buildings.[5]

Reducing Wildfire Risk

Wildfire risk can be reduced through fire danger warnings, education, codes and regulation, insurance, evacuation, and structural protection.

Fire Danger Warnings Federal agencies in Canada and the United States have developed rating systems to alert land managers, residents, and visitors to fire danger. The ratings are based on information about fuel conditions, topography, weather, and risk of ignition. Weather forecast offices also issue warnings about extreme fire conditions to put citizens, public officials, and firefighters on alert.

Education Community awareness programs and presentations on fire safety in schools may help reduce the fire risk. Unfortunately, even with fire education programs, the risk might not seem real to many people who have never experienced a wildfire.

Codes and Regulations One way to reduce risk in fire-prone areas is to enact and enforce building codes that require structures be built with fire-resistant materials.

For example, roofs of houses can be constructed with clay tiles instead of flammable cedar shingles or asphalt tiles. In seismically active areas, however, homes made with heavy clay tile roofs and masonry walls might not be safe during earthquakes. Nevertheless, making appropriate fire-resistant materials mandatory for new structures would significantly reduce the amount of damage caused by fires.

Fire Insurance Fire insurance allows people whose property has been destroyed by a fire to be reimbursed for part or all of their losses. However, as mentioned above, insurance may provide a false sense of security and prompt more people to live in fire-prone areas.

Evacuation Temporary evacuation of people from threatened areas is the most common response to wildfires. Evacuation ensures personal safety but does not protect homes and other fixed structures (see Survivor Story).

Structural Protection You can take actions to protect your home from a wildfire (Table 13.2).[36] If you live or work in a forested area, you might want to acquire a book that describes how to prepare for a wildfire.[37]

TABLE 13.2 Reducing Your Fire Risk at Home

Make Your Home Fire Resistant

- Consider using fire-resistant roofing and building materials, such as stone, brick, concrete, and metal, when building or renovating a home.
- Use 3-mm mesh screen beneath porches, decks, floor areas, and the home itself. Also screen openings to floors, roof, and attic.
- Install multi-pane windows, tempered safety glass, or fireproof shutters to protect large windows from radiant heat.
- Install a smoke alarm on each level of your home; test monthly and change the batteries at least once a year.
- Install freeze-proof exterior water outlets on at least two sides of the house and near other structures on the property. Install additional outlets at least 15 m from the house.
- Consider installing heavy fire-resistant draperies in the home.
- Regularly clean the roof and gutters.
- Inspect chimneys at least twice a year and clean them at least once a year.
- Teach each family member how to use a fire extinguisher and show them where it is kept.
- Consider obtaining a portable gasoline-powered pump in case the electrical power is cut off.

Let the Landscape around Your Home Defend You

- Keep tools for fire protection nearby: a garden hose that is long enough to reach any area of the house and other structures on the property, a shovel, rake, axe, and buckets.
- Keep a ladder that will reach the roof.
- Mow grass on a regular basis.
- Rake leaves, dead limbs, and twigs. Clear all flammable vegetation.
- Remove leaves and rubbish from under structures.
- Remove dead branches that extend over the roof.
- Prune tree branches and shrubs within 5 m of a stovepipe or chimney.
- Ask your power company to clear branches from power lines.
- Remove vines from the walls of the home.
- Clear a 3-m area around propane tanks and the barbecue. Place a screen made of non-flammable material over the grill.
- Landscape your property with fire-resistant grasses and shrubs to prevent fire from spreading quickly.
- Stack firewood at least 30 m from your home and all secondary structures.
- Store flammable materials, liquids, and solvents in approved safety cans and place cans in a safe location away from the base of buildings.
- Identify and maintain an adequate outside source of water such as a small pond, cistern, well, swimming pool, or hydrant.
- Make sure that water sources, such as hydrants and ponds, are accessible to the fire department.

Source: Adapted from Federal Emergency Management Agency, US Fire Administration. 2004. Wildfire. Are You Prepared? *FEMA Publication FA-287/August 2004, http://www.usfa.fema.gov/citizens/home_fire_prev/wildfire/index.shtm.*

SURVIVOR STORY

The Cedar Fire

Lisza Pontes knew that her home near San Diego, California, was vulnerable to wildfire, so she made plans in case fire threatened her neighbourhood. But when the devastating Cedar Fire reached the Pontes home in October 2003, none of her plans seemed to work.

"We had all these plans for what to do, and nothing happened that way," Pontes says. "We went through Plan B, Plan C, Plan D."

To begin with, Lisza and her husband did not awaken until the fire was close. None of the nine smoke alarms in the house activated because the house was well insulated against outside smoke.

Lisza's 24-year-old daughter was the first member of the family to awaken when she noticed the "odd orange glow" outside and the noise of the flames. "The sound is unbelievable," Pontes says. "It stays with me. It's surround sound, what I imagine a war zone is like." Outside, trees were exploding.

Strangely, Pontes does not remember either the heat or the smoke. "I'm asthmatic, and I don't remember coughing."

Once they had been awakened by their daughter, Pontes and her husband hastily prepared to evacuate. They rounded up their three dogs and wrapped themselves in wet towels.

Among her recollections of the fire, Pontes vividly recalls the wind that accompanied the blaze. "These fires create their own storm and their own wind," she says. Later she would learn that nearby gusts had been clocked at 130 km/h.

Pontes describes the air as having the "same strange, static feeling" that a friend of hers described feeling shortly before being struck by lightning. "The air was very electric," she remarks.

The Pontes got into their truck and began to flee, but it was knocked off the road when a neighbour's propane tank exploded. They had to retreat back up the hill, which was already burning, to reach their second vehicle. "It was really strange when we were running uphill, and the wind was whipping like crazy," she says. "At some point I remember wondering if my shoes would hold together." Pontes also recalls the odour of burning hair from the dog she carried in her arms.

They reached Lisza's old Mercedes and drove to the bottom of the driveway, which bordered a heavily wooded area. "That was an inferno," Pontes says. "That was when we first felt the heat."

They now were faced with the difficult decision of which way to go. "We had no idea where the fire had come from," she recalls. In the end, her husband decided to take the shortest road out.

The smoke had reduced visibility to almost zero, and they had to rely on occasional gusts of wind and familiar bumps in the road to pick their way out of the blaze. Pontes describes the light as a "*Twilight-Zone* orange. Not a normal light. We passed cars that were dead on the road. We were some of the last to get out of this area."

They were able to reach safety, but many others in the area did not. Twelve of their neighbours perished in the blaze. Of the 11 homes in their neighbourhood, only theirs survived (Figure 13.21). Their efforts to protect their home against such a disaster, including applying a fire-resistant coat of paint to the exterior of the house, paid off.

But their entire property, some 1.4 ha, was completely burned. And as Pontes notes, the damage was indiscriminate. "That's Mother Nature," she says. "She doesn't care."

—Chris Wilson

▲ **FIGURE 13.21 HOMES DESTROYED BY THE 2003 CEDAR FIRE** Many large homes in suburban areas of San Diego were burned in the Cedar Fire, one of the worst wildfires in California's history. Fanned by Santa Ana winds, the fire claimed 14 lives, destroyed nearly 2300 homes, and burned an estimated 1100 km^2 of land. *(Dernis Poroy/AP Images)*

REVISITING THE FUNDAMENTAL CONCEPTS

Wildfires

1. **Hazards can be understood through scientific investigation and analysis.**
2. **An understanding of hazardous processes is vital to evaluating risk.**
3. **Hazards are commonly linked to each other and to the environment in which they occur.**
4. **Population growth and socio-economic changes increase the risk from natural hazards.**
5. **Damage and loss of life from natural disasters can be reduced.**

1. The science of wildfires is a mature area of research. The frequency and recurrence of wildfires are well known for many regions of the world. We can also estimate the potential for wildfire under different weather conditions and antecedent soil moisture, and by documenting the amount of fuel in the forest and on the forest floor. What is more difficult to forecast is the effect of future climate change on wildfire frequency and magnitude. It is likely, however, that increasing warmth and drought in parts of Canada and the U.S. will lead to more frequent and larger wildfires and may extend the fire season.
2. The risk from wildfire is defined as the product of the probability of a fire and the consequences. Much research has been done on the increased risk from wildfires that have resulted from regional demographic changes and fire-suppression techniques. More and more people are living in areas at high risk from wildfire, and successful fire suppression practices have led to a dangerous build-up of fuels in many inhabited forested areas.
3. Wildfires are linked to several other natural hazards. Runoff and soil erosion increase after fire, and, as a result, the flood hazard increases. Debris flows and other types of landslides can also increase in steep terrain following wildfires. Large amounts of sediment eroded from burned slopes can choke streams, affecting aquatic ecosystems.
4. The consequences of wildfire have increased dramatically in North America and some other parts of the world in recent decades as populations in areas of high fire risk have grown. Large numbers of people now live in forested areas near cities. In addition, wildfires have been vigorously suppressed in the wildland–urban interface, leading to a build-up of combustible fuel in many areas. As a result, fires are becoming larger and more intense.
5. Damage and injury from wildfire can be reduced. Homeowners can create a defendable space around their homes and be prepared to evacuate. Emergency planning at the community level is also important. Evacuating a large number of people is difficult and requires advanced planning at the municipal level. The severity of wildfire can be reduced by managing forest fuels in high-risk areas. Selective forest thinning and removal of dead, woody material from the forest floor are effective, albeit costly, strategies for reducing fire fuel. Fire breaks can also be created in high-risk areas.

Summary

Wildfire is a high-temperature biochemical oxidation reaction that is rapid, self-sustaining, and releases heat, light, carbon dioxide, and other products. Wildfires in natural ecosystems maintain a rough balance between plant productivity and decomposition. The two main processes associated with wildfires are pre-ignition and combustion.

Two types of wildfires are distinguished by the part of the vegetation cover that burns. Surface fires burn in grasslands and along the forest floor. Fast-moving crown fires spread in the upper canopy of the forest.

Wildfire behaviour is influenced by fuel, weather, topography, and the fire itself. Predictions of fire behaviour require an understanding of ecosystem fire regimes.

Fire can increase runoff, erosion, flooding, and landslides. Natural service functions of fires include enriching soils with nutrients, initiating regeneration of plant communities, creating new habitat for animals, and reducing the risk of future large fires.

Fire management involves scientific research, data collection, education, and the use of prescribed burns. Large wildfires in hot, dry, windy weather are difficult to prevent and generally cannot be suppressed. Education, building restrictions, fire insurance, and evacuation are the main adjustments to wildfire hazard.

Key Terms

combustion (p. 367)
crown fire (p. 371)
extinction (p. 368)
fire regime (p. 381)
pre-ignition (p. 367)
prescribed burn (p. 382)
pyrolysis (p. 367)
surface fire (p. 370)
wildfire (p. 366)

Did You Learn?

1. Describe how the relationship between wildfires and people has changed over time.
2. How are wildfires related to photosynthesis and plant decomposition?
3. Name the major gases and solid particles that are produced by a wildfire.
4. Name and describe the three phases of a wildfire.
5. Explain how the two processes of the pre-ignition phase of a wildfire prepare plant material for combustion.
6. Explain how the processes of combustion differ from those of ignition.
7. What are the two types of combustion and how do they differ?
8. Describe the three processes of heat transfer in a wildfire. Rank them in order of their importance.
9. What factors control the behaviour of a wildfire?
10. Describe the two types of fire.
11. Describe the effects of wildfires on the geological environment and on the atmosphere.
12. What are the links between wildfires and climate?
13. How are wildfires linked to other natural hazards?
14. Explain how wildfires can increase erosion of the land.
15. Describe how climate change could change the frequency and intensity of wildfires.
16. Describe the natural service functions of wildfires.
17. What are the methods that can be used to reduce wildfire hazard and risk?
18. Name the four main components of fire management.
19. What are the difficulties associated with prescribed burns?
20. Explain how people can adjust to the wildfire hazard.

Critical Thinking Questions

1. You live in an area where wildfires are possible. Make a list of actions you should take to protect your family and home.
2. The staff of a national park is reviewing its wildfire policy. They have called on you for advice. Their current policy is to suppress all fires as soon as they start and not to use controlled burns. They are considering switching to a policy of allowing natural fires to burn without human intervention. What would you suggest? List the pros and cons of each policy before making your decision.
3. Describe the features in and around your home that place it at risk in the event of a wildfire.
4. Do you think development should be restricted in areas with a high fire hazard? Why or why not?
5. Do you think a person has a right to enter public forest lands during a period of extreme fire hazard? Why or why not?

MasteringGeology

MasteringGeology **www.masteringgeology.com**. Looking for additional review and test prep materials? Visit the Study Area in MasteringGeology to enhance your understanding of this chapter's content by accessing a variety of resources, including **Self-Study Quizzes, Geoscience Animations, GEODe Tutorials, RSS feeds, flashcards,** web links and an optional **Pearson eText.**

CHAPTER 14

▶ **GLACIERS RESPOND TO CLIMATE CHANGE** Glaciers around the world have thinned and receded due to climate warming over the past century. Water produced by the melting of glaciers is contributing to current sea-level rise of over 3 mm/y. Glaciers flowing into the sea shed, or *calve*, large blocks of ice, which also contributes to sea-level rise. This photograph of the calving front of Dawes Glacier in southeast Alaska shows large blocks toppling into Yakutat Bay. *(© Paul Souders/Corbis)*

Climate Change

Learning Objectives

Climate has changed throughout geologic time, and some of the warming we have experienced in the past century is normal. Evidence is mounting, however, that the current warming is unprecedented and is probably due, at least in part, to human modification of Earth's atmosphere. Further warming may change the frequency or severity of some of the hazardous natural processes discussed in this book. A basic understanding of climate science is necessary to understand these changes. Your goals in reading this chapter should be to

- Understand the difference between climate and weather
- Know the basic concepts of atmospheric science, specifically the structure, composition, and dynamics of the atmosphere
- Understand the causes of climate change
- Understand how climate has changed in the recent geologic past and how human activity may be altering climate
- Know how climate change can affect the frequency or severity of some natural hazards
- Know how we can adjust to the problems climate change may cause

Arctic Threatened by Climate Change

Although people living in the mid-latitudes of North America argue about whether the climate is warming, to the Inuit and Dene in Arctic Canada climate change is a fact. In recent decades, the Aboriginal peoples of northern Canada have witnessed a marked warming of climate, a decrease in Arctic sea-ice cover, increased permafrost thaw, and sea-level rise caused by melting glaciers and ice sheets.

Arctic Sea Ice

Satellite microwave sensors indicate that the area of summer sea ice in the Arctic Ocean has decreased by 9 percent per decade since the late 1970s. If this trend continues, the Arctic Ocean will be ice-free in summer by the middle of the twenty-first century (Figure 14.1). The Arctic ice pack changes in size with the seasons. It grows each winter as the sun sets for several months and intense cold sets in. Some of that ice is pushed out of the Arctic by winds, but much of it melts in place during summer. Arctic sea ice reaches its minimum extent in August or September. Thicker, older ice that survives one or more summers is more likely to persist through the next summer.

Scientists have observed a marked loss of Arctic sea ice over the past decade. In September 2002, the Arctic ice pack was the least extensive it had been since 1979. Since then, there have been a series of record or near-record lows in the area of sea ice. Only 4.2 million km^2 of sea ice were left in the Arctic in August 2012, the smallest amount ever recorded and less than 60 percent of the sea ice at summer's end in the late twentieth century;[1] the melt-back in 2008 was nearly as great. At the same time, average air temperatures across most of the Arctic region were as much as several degrees warmer than average.

Even more ominous than the reduced area of sea ice is its thinning. As the area of sea ice has decreased, the remaining ice has thinned. Seasonal sea ice averages about 2 m in thickness, whereas ice that has lasted through more than one summer averages about 3 m, although it can grow much thicker in some locations near the coast.

Recent data show that the Arctic ice pack is continuing to thin. Using two years of data from NASA's Ice, Cloud, and Land Elevation Satellite

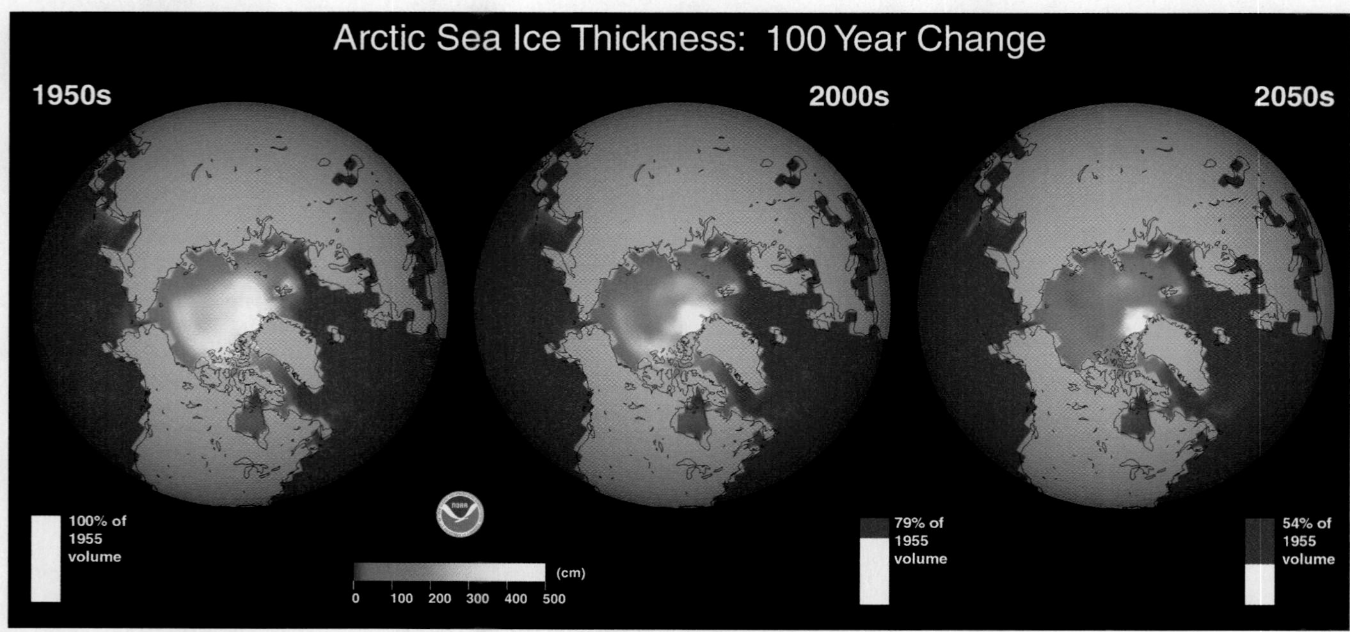

▲ **FIGURE 14.1 DECREASE IN ARCTIC SEA ICE COVER** These three images show minimum Arctic sea ice cover in the 1950s (left); 2000s (middle); and, projected, 2050s (right). The volume of sea ice in the middle of this century may be only half what it was 100 years earlier. *(NOAA, National Geophysical Data Center)*

(ICESat), a team of researchers from NASA's Jet Propulsion Research Laboratory produced the first map of sea ice thickness over the entire Arctic basin for 2005 and 2006. Thin seasonal ice—ice that melts and re-freezes every year—made up about 70 percent of the Arctic sea ice that winter, up from 40 to 50 percent in the 1980s and 1990s.[2] Thicker ice, which survives two or more years, now constitutes less than 30 percent of the wintertime ice cover.

The reduction in the extent and thickness of Arctic sea ice provides a good example of a *positive feedback*—enhancement of an effect by the process that gives rise to it. The melting may contribute to even higher Arctic temperatures in the future. White ice reflects more of the sun's energy than dark ocean water, so with more of the ocean exposed, the seawater absorbs more heat, reducing the amount of solar energy reflected back into space. The resulting higher Arctic temperatures further increase the melt of the ice pack.

The reduction in Arctic ice cover is harmful to some northern animals. Polar bears, for example, use sea ice to hunt for marine mammals, such as seal. The bears now have a shorter season during which they can access the ice and, as a consequence, are experiencing problems finding their most important food source. Biologists have suggested that recent declines in the size and health of polar bears are due, in part, to this problem.

Another important consequence of the thinning of Arctic sea ice is that a Northwest Passage will open year-round between Baffin Bay and the Chukchi and Bering seas. In August 2000, open water was found at the North Pole, and for a short time in 2006 an open channel extended across the Arctic Ocean to the pole. A commercial shipping passage through the Arctic Ocean, sought by Sir John Franklin during the fateful voyage on the *Erebus* and *Terror* in 1845, could become a reality before the middle of this century. The economic advantages over the Panama Canal of a northern shipping route are considerable, but commercial shipping through the high Arctic raises environmental and sovereignty issues that Canada has not yet addressed.

Ice Sheets and Ice Shelves

While the Arctic sea ice has been thinning and retreating, the Greenland Ice Sheet has also been showing signs of change. The ice sheet is about 2 km thick on average, covers about 80 percent of Greenland, and is the second largest ice body in the world, after the Antarctic Ice Sheet.

The Greenland Ice Sheet has experienced record melting in recent years and is likely to contribute substantially to sea-level rise and possibly to changes in ocean circulation in the future. The area of the ice sheet that experiences some summer melting increased by about 16 percent between 1979, when measurements were first made, and 2002 (Figure 14.1). More recent research, which uses data from 1996 to 2005, shows that the ice sheet is thinning at progressively faster rates.[3] In 1996 the ice sheet was losing about 96 km^3 of ice per year. In 2005, the loss increased to about 239 km^3 per year because of rapid thinning near the ice-sheet margin.[4] The most recent study, based on reprocessed

▲ **FIGURE 14.2 MELTING OF GREENLAND ICE SHEET HAS INCREASED** The portion of the Greenland Ice Sheet that experienced surface melting during the summer increased considerably between 1992 (pink-coloured) and 2005 (pink and red colours combined). The area of greatest increase was the southern third of Greenland, where the entire ice sheet experienced surface melting during summer months. *(Courtesy Russell Huff and Konrad Steffan/University of Colorado CIRES)*

and improved data between 2003 and 2008, confirms the earlier work, reporting an average annual loss of 195 km^2.[5] Some of the ice lost at the margins of the ice sheet is being offset by increases in snowfall at its centre (Figure 14.2). Although seemingly counterintuitive, the increased snowfall at high elevations in Greenland is a consequence of climate warming. Warm air masses hold more moisture than cold ones. Relatively warm, moist air masses that reach the centre of Greenland release more snow than the cooler ones that prevailed before climate started to warm early in the twentieth century.

Yet another recent effect of ongoing climate warming at high latitudes is the anomalous calving of Canadian, Greenland, and Antarctic ice shelves.[6] An *ice shelf* is a thick, floating sheet of ice fronting a glacier or an ice sheet that reaches the coast. Ice shelves are found only in Arctic Canada, Greenland, and Antarctica. In the past several decades, glaciologists have observed that some Arctic ice shelves are decreasing in extent due to melting, calving, and disintegration. The Ellesmere Ice Shelf in Arctic Canada has decreased by 90 percent in size in the twentieth century, leaving the separate Alfred Ernest, Ayles, Milne, Ward Hunt, and Markham ice shelves. A report on Canadian ice shelves in 1986 concluded that 48 km^2 of ice calved from the Milne and Ayles ice shelves between 1959 and 1974.[7] The Ayles Ice Shelf completely broke apart on August 13, 2005. The Ward Hunt Ice Shelf, the largest remaining section of thick land-fast sea ice along the northern coast of Ellesmere Island, lost 600 km^2 of ice in a massive calving event in 1961–1962,[8] and experienced another major breakup in 2002.[9]

Two sections of Antarctica's Larsen Ice Shelf broke apart into hundreds of small fragments in 1995 and 2002. In April 2009, about 700 km^2 of the Wilkins Ice Shelf, which originally was about 15 000 km^2 in size, collapsed in a massive calving event.[10] An even larger break-up, involving 1000 km^2 of the ice shelf, occurred in 1998. Other Antarctic ice shelves that experienced similar large losses of ice over the past 30 years include the Prince Gustav Channel, Larsen Inlet, Larsen A, Wordie, Müller, and Jones ice shelves, as well as Larsen B, which disappeared in 2002.

Looking Ahead

Climate models project that local warming in the Canadian Arctic will exceed 3°C during this century. Ice sheet models indicate that such a warming would initiate the long-term melting of the Greenland Ice Sheet and lead to its disappearance within centuries.[11] If the entire 2.85 million km^3 of ice were to melt, global sea level would rise by 7 m, inundating almost every major coastal city on the planet and drowning several small island nations.[11]

How fast the melt would occur is a matter of debate. According to the Intergovernmental Panel on Climate Change (IPCC), the expected 3°C warming would, if temperature did not rise further, result in about 1 m of sea-level rise by the end of this century.[11] Some scientists have cautioned that this rate of melting is too low as it assumes a linear progression. James Hansen, an atmospheric scientist formerly at NASA's Goddard Institute for Space Studies, has argued that positive feedbacks could lead to nonlinear ice sheet disintegration much faster than claimed by the IPCC. In a 2007 paper, Hansen and his co-authors stated, "We find no evidence of millennial lags between forcing and ice sheet response in paleoclimate data. An ice sheet response time of centuries seems probable, and we cannot rule out large changes on decadal timescales once wide-scale surface melt is underway."[12]

14.1 Global Change and Earth System Science: An Overview

Preston Cloud, a famous earth scientist who studied the history of life on Earth, human impact on the environment, and the use of Earth resources, proposed two central goals for Earth science:[13]

1. Understand how Earth works and how it and life have evolved over the 4.6 billion years of its existence.
2. Apply that understanding to better manage our environment.

Until recently, it was thought that human activity caused only local or, at most, regional environmental change. Now, however, we recognize that the effects of humans on Earth are so extensive that we are conducting an unplanned planetary experiment. To recognize and perhaps modify the changes we have initiated, we need to understand how the Earth works as a system. The aim of the discipline called **Earth system science** is to further this understanding by studying how the components of the system—the atmosphere, oceans, land, and biosphere—are linked on a global scale and how these complex links affect life on Earth.[14] In this chapter, we examine the factors that affect climate and the role that humans are beginning to play in modifying our climate.

14.2 Climate and Weather

Many people form opinions about **global warming** based on day-to-day or week-to-week changes in the weather. They do not appreciate the important distinction between climate and weather. **Climate** refers to the characteristic atmospheric conditions of a region over years or decades (Figure 14.3). **Weather** is the atmospheric conditions of a region for much shorter periods, typically days or weeks. For example, we associate the coast of British Columbia with mild temperatures, high humidity, and lots of rain during fall and winter. Thus, when travelling to Vancouver or Victoria in February, we would probably take an umbrella. But we might also enjoy bright, sunny, dry weather during a weeklong stay in Vancouver or Seattle in winter, because weather can vary greatly on a daily basis.

If you were asked to characterize the climate of a region, you might do so in terms of its average temperature and precipitation. However, climate is more than averages. Two locales might have the same average annual temperature but very different climates. Halifax, Nova Scotia, and Kelowna, British Columbia, have nearly the same average temperature of 7°C to 8°C, but the annual temperature range in Halifax is much less than in Kelowna, and Halifax receives much more precipitation. Although the two cities have the same average annual temperature, they clearly have different climates.

In our discussion of climate, it is important to consider scale. We will discuss climate change on regional, hemispheric, and global scales. Although climate has warmed over most of Earth during the past century, the magnitude of this warming has differed considerably from place to place; the greatest warming has occurred at high latitudes and high elevations. Furthermore, some areas have become wetter, whereas others have become drier. In this context, the term "global climate warming" can be misleading.

14.3 The Atmosphere

Composition of the Atmosphere

Earth's atmosphere comprises nitrogen, oxygen, and smaller amounts of other gases. Atmospheric gases can be divided into two groups: *permanent gases*, notably nitrogen and oxygen, concentrations of which do not change; and *variable gases*, such as carbon dioxide, concentrations of which vary in time and space.

Permanent Gases The major permanent gases, which constitute about 99 percent by volume of all atmospheric gases, are nitrogen, oxygen, and argon (Table 14.1). Nitrogen generally occurs as molecules of two nitrogen atoms (N_2) and composes about 78 percent of the volume of all permanent gases. Although abundant, elemental nitrogen is relatively unimportant in atmospheric dynamics. However, it plays an important role in climate when it combines with other gases, notably oxygen.

The second-largest component of Earth's atmosphere is oxygen, which forms 21 percent of atmospheric gases by volume. Like nitrogen, oxygen molecules consist mainly of two atoms (O_2). All animals, including humans, require oxygen gas. Like N_2, diatomic oxygen is relatively unimportant in atmospheric dynamics. However, as discussed below, some oxygen compounds have important effects on Earth's climate.

Argon makes up most of the remaining 1 percent of the permanent gases. The atmosphere also contains small amounts of neon, helium, krypton, xenon, and hydrogen.

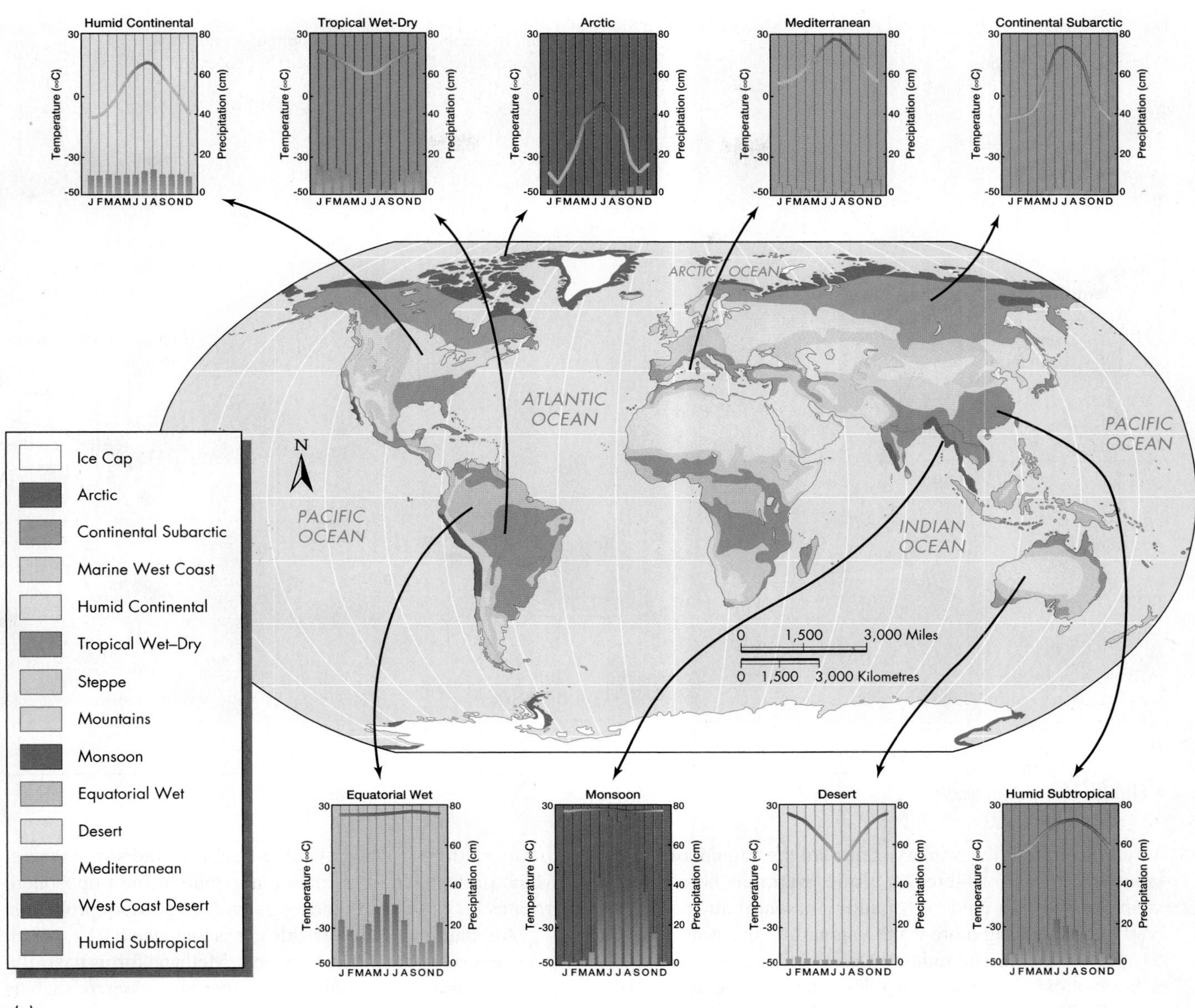

▲ **FIGURE 14.3 CLIMATES OF THE WORLD** (a) Characteristic temperature and precipitation conditions. The red lines in each plot represent temperature, and vertical bars show precipitation. (b) Schematic diagram showing the complex, linked components and changes of the climate system that produce, maintain, and change the climates of the world. *((a) Based on Marsh, W. M., and J. Dozier. 1981.* Landscape: An Introduction to Physical Geography. *New York, NY: John Wiley & Sons; (b) IPCC. 2007.* The Physical Science Basis: Working Group 1. Contribution to the Fourth Assessment Report. Intergovernmental Panel on Climate Change. *New York, NY: Cambridge University Press)*

Variable Gases The variable gases (Table 14.1) constitute only a small percentage of the total mass of the atmosphere, but some of them have important roles in atmospheric dynamics. The variable gases include carbon dioxide, water vapour, ozone, methane, nitrogen oxides, and halocarbons.

Carbon dioxide (CO_2) makes up a very small percentage of the atmosphere—currently about 400 ppm. It is a component of the carbon cycle, the biogeochemical cycle by which carbon is exchanged among Earth's biosphere, lithosphere, geosphere, hydrosphere, and atmosphere (see Chapter 1). Carbon dioxide is released into the atmosphere naturally through volcanic activity, plant and animal respiration, wildfires, and decay of organic material. It is removed from the atmosphere through photosynthesis by green plants, chemical weathering, and absorption by sea water. Carbon dioxide also enters the atmosphere through the burning of fossil fuels by people. Since the start of the Industrial Revolution around 1750, the amount of carbon dioxide added to the atmosphere from human, or **anthropogenic**, sources—mainly the burning of fossil fuels and deforestation—has greatly increased.

As mentioned in Chapter 10, water vapour (H_2O) is produced by the evaporation of water at Earth's surface. It condenses to form clouds and eventually returns to the surface as precipitation as part of the hydrologic cycle (Chapter 1).

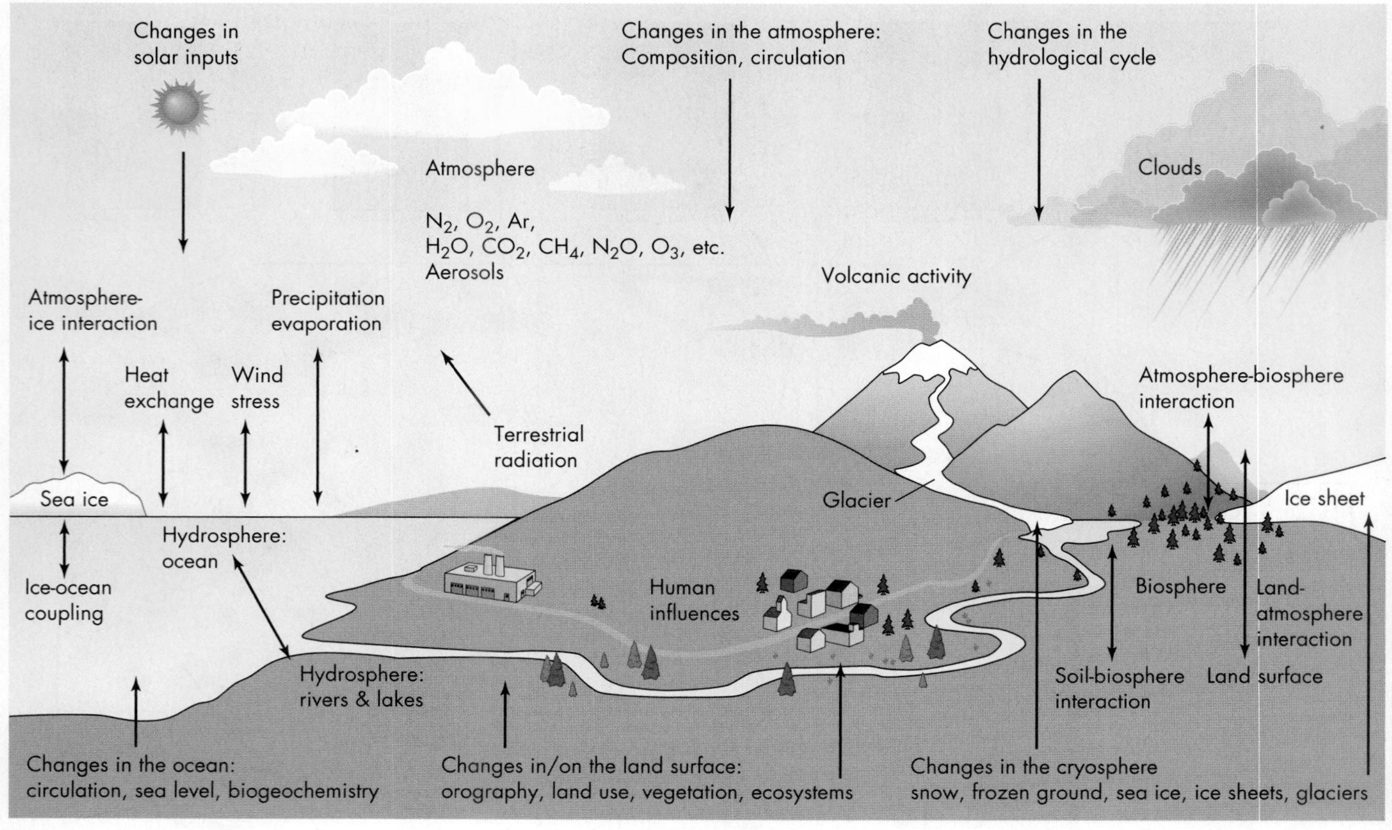

▲ **FIGURE 14.3** *(Continued)*

Air temperature is the primary control of the amount of water vapour in the atmosphere. As a rule, warm air holds more water vapour than cold air. In addition, warm air increases evaporation, putting more water vapour into the atmosphere.

Ozone (O_3) is triatomic oxygen found in small amounts in the stratosphere, the layer of the atmosphere above the troposphere (see Chapter 10). It partially shields Earth from ultraviolet (UV) radiation from the sun. Ozone is neither the cause nor the result of global warming. Smaller concentrations of ozone are also found in the troposphere near Earth's surface, where it is produced by chemical reactions during the formation of smog. Unlike ozone in the stratosphere, which protects us from radiation, ozone in the troposphere irritates lungs and eyes and aggravates respiratory problems.

TABLE 14.1 Composition of the Atmosphere

Permanent Gases		Variable Gases	
Nitrogen	78.08%	Water Vapour	0.2–4%
Oxygen	20.95%	Carbon Dioxide	0.038%
Argon	0.93%	Methane	0.00017%
Neon	0.0018%	Nitrous Oxides	0.000032%
Helium	0.00052%	Ozone	0.000004%
Krypton	0.00011%	Halocarbons	0.00000002%
Xenon	0.00009%		
Hydrogen	0.00005%		

Note: Percentage by volume.

Source: Data in part from Bryant, E. 1997. Climate Process and Change. *New York, NY: Cambridge University Press.*

Another important variable gas is methane (CH_4), which is a major component of natural gas. Methane forms naturally by bacterial decay in moist places that lack oxygen, such as marshes, swamps, bogs, and the intestinal tracts of termites, cows, and sheep. Anthropogenic sources of methane include coal mines, oil wells, leaking natural gas pipelines, rice paddies, landfills, and livestock. The concentration of methane in the atmosphere is 1.7 ppm, which is about twice its value in 1700.[15,16]

Nitrogen oxide gases, sometimes referred to as nitrous oxides or NO_x because they include both nitrogen oxide and nitrogen dioxide, are variable gases present in the troposphere and the stratosphere. Natural sources of these gases include microbiological processes in soils and oceans, wildfires, and lightning strikes. Anthropogenic sources include automobiles, power plants, jet aircraft, and fertilizers. The concentration of nitrous oxides in the atmosphere is about 0.3 ppm.[11]

Variable gases, or *halocarbons*, are chemical compounds that contain carbon and halogen elements such as chlorine and bromine. These gases include CFCs and are almost entirely anthropogenic. Halocarbons are used in industrial processes, firefighting, and as fumigants, refrigerants, and propellants. Halocarbons contribute to warming of the troposphere and ozone depletion in the stratosphere.

Aerosols Aerosols are not gases, but rather microscopic liquid and solid particles in the atmosphere. Aerosols are the nuclei around which water droplets condense to form clouds. The concentration of aerosols in the air influences the size of water droplets in clouds.[17] Natural sources of aerosols include desert dust, wildfires, sea spray, and volcanic eruptions.[17] Anthropogenic sources include the burning of forests to clear land and the consumption of fossil fuels.

Structure of the Atmosphere

Much of the heat radiating from Earth's surface is trapped in the troposphere, the lowest layer of the atmosphere (see Chapter 10). The stratosphere, which is above the troposphere, is a dry layer of the atmosphere where water vapour occurs as ice crystals. Aerosols rapidly disperse around the globe in the stratosphere. The uppermost layers of the atmosphere—the mesosphere and thermosphere—are far from Earth's surface and are not important to our discussion of climate and natural hazards.

Atmospheric Circulation

Warm moist air rises in the troposphere in equatorial areas and moves north and south toward the poles (Figure 14.4a). As this air rises, it cools and loses moisture as rain. The cooler and drier air then descends between 15° and 30° latitude (Figure 14.4b).[18] The descending air produces semi-permanent

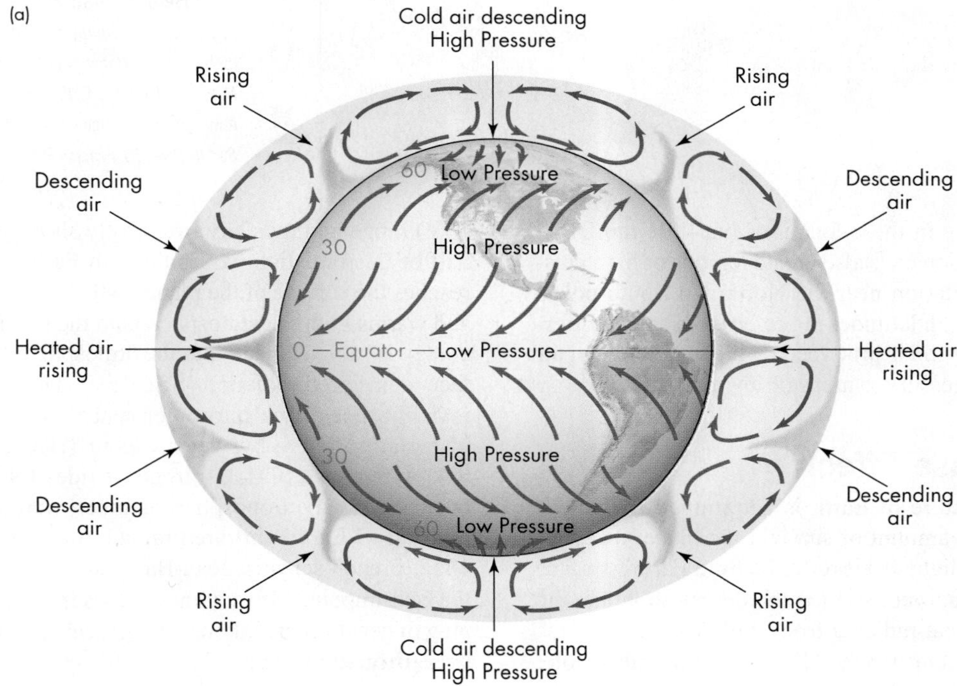

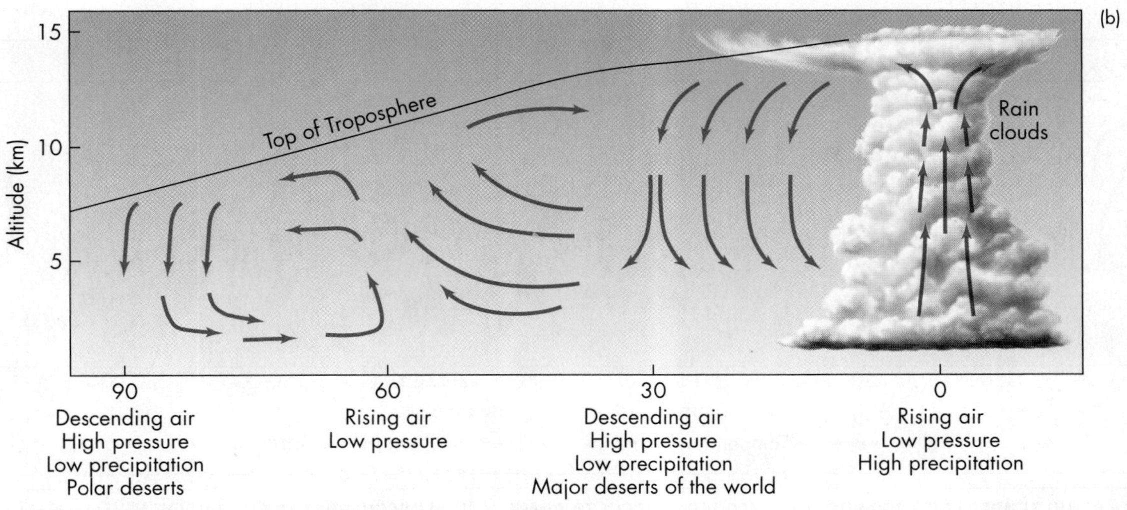

▲ **FIGURE 14.4 ATMOSPHERIC CIRCULATION** (a) The general circulation of the lower atmosphere, showing zones of rising and descending air masses and corresponding areas of low and high air pressure. (b) An idealized diagram showing atmospheric circulation along a transect from the equator to a pole.

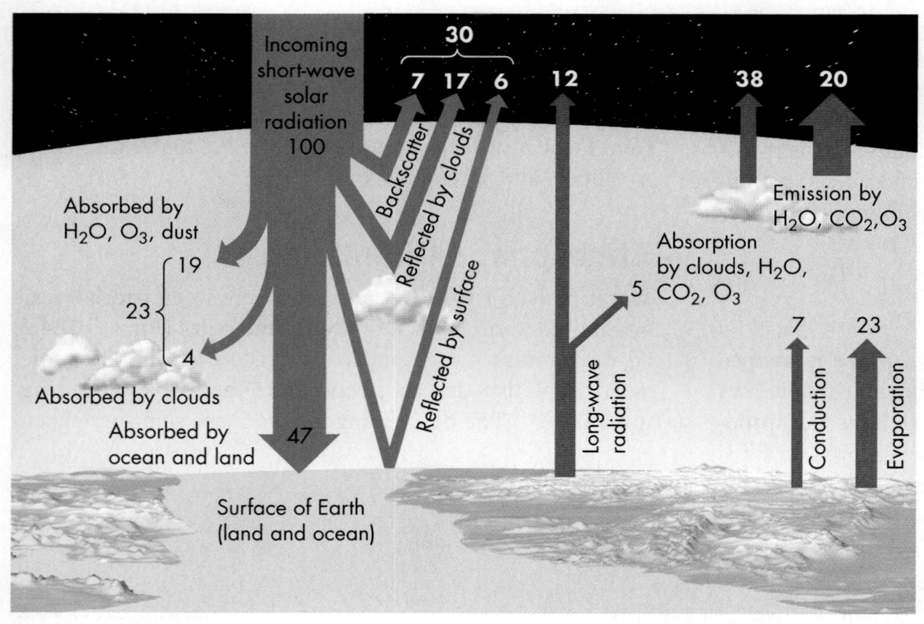

◄ **FIGURE 14.5 THE GREENHOUSE EFFECT** An idealized diagram showing Earth's energy balance. Earth absorbs approximately 47 percent of the incoming short-wave solar radiation (visible light, ultraviolet light, and some infrared radiation). Some of this energy is radiated back into the atmosphere as long-wave infrared (heat) radiation. Water vapour, water, carbon dioxide, methane, and other gases in the atmosphere absorb a portion of the long-wave radiation. These greenhouse gases reradiate some of the infrared energy to Earth and warm the atmosphere. Global warming has been attributed to increases in the concentrations of carbon dioxide and methane because of human activities. *(Based on Trujillo, A. P., and H. V. Thurman. 2005.* Essentials of Oceanography, *8th ed. Upper Saddle River, NJ: Pearson Prentice Hall; and Lutgens, F. K., and E. J. Tarbuck. 2004.* The Atmosphere: An Introduction to Meteorology, *9th ed. Upper Saddle River, NJ: Pearson Prentice Hall)*

cells of high pressure in these locations but little rainfall.[19] Atmospheric circulation cells also create regions of high pressure and low precipitation near the North and South poles. Land areas at these high latitudes are referred to as *polar deserts* (Figure 14.3a). Wind is produced by the movement of air from areas of high pressure to areas of low pressure.

The Greenhouse Effect

The surface temperature of Earth is determined mainly by three factors: (1) the amount of sunlight the planet receives, (2) the amount of sunlight that is reflected from Earth's surface and therefore not absorbed, and (3) the degree to which the atmosphere retains heat radiated from Earth (Figure 14.5).[11] Most solar radiation that reaches Earth is in the ultraviolet (UV) range and thus has a relatively short wavelength. Over half of this radiation passes through Earth's atmosphere and reaches the surface of the planet, where most of it is absorbed and warms both the atmosphere and the surface. Some of this energy is radiated back into the atmosphere from Earth's surface as infrared radiation.[11] Some of the reradiated infrared radiation goes back into outer space and some is absorbed in the atmosphere by water vapour (H_2O), carbon dioxide (CO_2), methane (CH_4), nitrogen oxides (NO_x), and several other gases. The troposphere is much warmer than it would be if all of Earth's infrared radiation escaped into space. In 1827, French scientist Jean-Baptiste Fourier recognized that the heat-trapping effect of these gases is analogous to the trapping of heat by a greenhouse for growing plants, thus the name **greenhouse effect**.

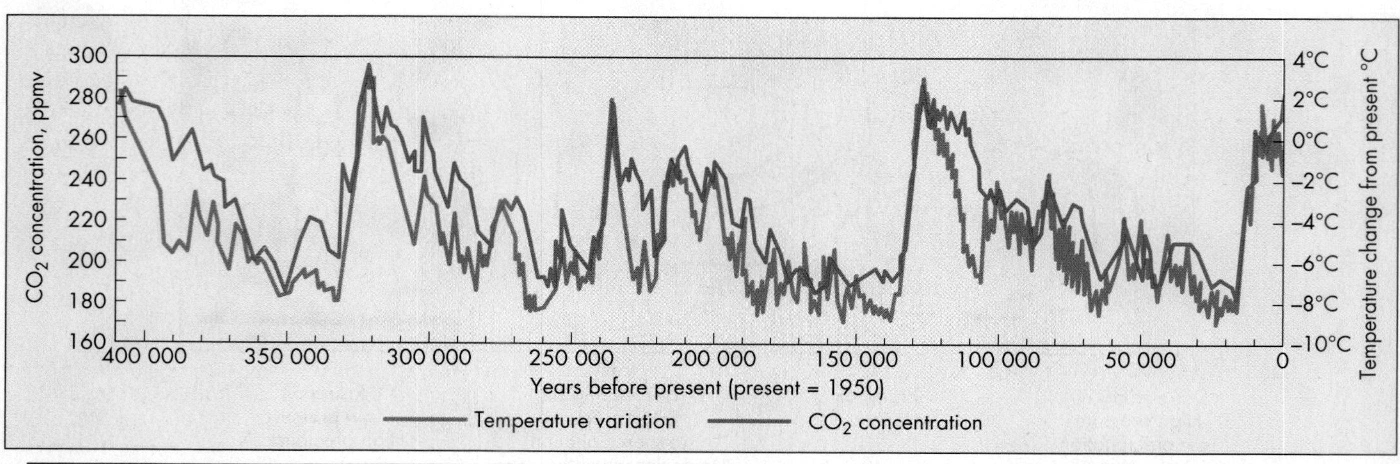

▲ **FIGURE 14.6 AIR TEMPERATURE CHANGES CORRESPOND CLOSELY TO CHANGES IN ATMOSPHERIC CARBON DIOXIDE** Measurements of the carbon dioxide content of air bubbles and oxygen isotope ratios in glacier ice from a core taken at Vostok, Antarctica, show that atmospheric CO_2 levels have corresponded closely to air temperatures for 420 000 years. Records from other ice cores show a similar pattern extending back more than 800 000 years. *(Based on Petit, J. R., J. Jouzel, et al. 1999. "Climate and atmospheric history of the past 420,000 years from the Vostok ice core in Antarctica."* Nature *399:429–436)*

Without the greenhouse effect, Earth's surface would be much cooler than it is, all surface water would be frozen, and little, if any, life would exist. Most of natural greenhouse warming is due to absorption of infrared radiation by water vapour and small particles of liquid water in the atmosphere. However, some of the warming is due to absorption of infrared radiation by carbon dioxide, methane, and nitrogen oxide gases, which are commonly termed **greenhouse gases**. In recent years the atmospheric concentration of these gases has increased through human activities.

Carbon dioxide is the most studied of the greenhouse gases. Cores of glacier ice up to one million years old have been recovered from the Antarctic and Greenland ice sheets. The cores contain atmospheric air bubbles trapped at the time the ice formed. Measurements of gas concentrations in these bubbles indicate that atmospheric concentrations of CO_2 have ranged from 180 ppm to about 400 ppm over the past 800 000 years (Figure 14.6).[20,21] The highest levels of CO_2 date to a major interglacial interval about 340 000 years ago and to ice that is forming today.[20,21] Figure 14.6 clearly shows that there is a close relationship between glaciation and CO_2 concentrations in the atmosphere. When CO_2 concentrations have been high, air temperatures have also been high; conversely, low concentrations of CO_2 correlate with periods of low temperature.

14.4 How We Study Climate Change and Make Predictions of Future Climate

The data that scientists gather to document and better understand climate change come from all regions, including the oceans. Data are available for three main time periods:[22]

- The *instrumental period.* The earliest measurements of temperature and precipitation date back to the late seventeenth and early eighteenth centuries, but it was not until about 1960 that such data were widely collected, both on land and in the oceans (Figure 14.7). About 1000 stations were reporting data in the late nineteenth century; today there are about 7000 stations around the world. The concentration of carbon dioxide in the atmosphere has been measured continuously since 1958, and accurate measurements of solar irradiation have been made over the past several decades.
- The *historical period.* A variety of historical records extend back several hundred years. They include written documents (e.g., books, newspapers, journal articles, personal journals, ship logs) and farmers' crop records. These records are not generally quantitative, but they contain useful information about past climates.
- The *paleo-proxy record.* The instrumental and historical records are short. As a result, the record needs to be extended further back in time. Paleoclimatology (the study of past climates) is part of Earth science. The paleo record of Earth's climate has provided some of the strongest data to support and test forecasts of future climate change. The term **proxy data** refers to data that are not strictly climatic but can be correlated with climate. Sources of proxy climate data include tree rings, ocean sediments, ice cores, fossils such as pollen, and carbon dioxide. The disadvantage of paleoclimate proxy data is obvious—the data are not direct measurements; thus temperature or precipitation must be inferred from the data. In spite of this limitation, paleoclimate proxy data preserved in the geologic record provide the best evidence of changes that predate the historical period.

Tree Rings

The growth of trees is influenced by climate, including rainfall and temperature. Most trees add one growth ring each year, and the width, density, and isotopic composition of annual rings provide information about past climate. By dating dead trees using the radiocarbon method or by counting rings within living trees, scientists have developed a *dendroclimatology* proxy record that extends back more than 10 000 years (Figure 14.8).

Sediments

The oceans of the world are a repository of sediments eroded from the land and delivered by rivers, winds, and volcanic eruptions. Ocean sediments contain the skeletal remains of marine plants and animals such as foraminifera and mollusks. Lakes and bogs also accumulate sediment that can be sampled for a climate signal. Sediments are recovered by coring lakes and ocean basins. Samples are taken from the cores and analyzed for their fossil content, isotopic composition, and other properties that can be related to past climates. Some of the strongest evidence for past climate change comes from these proxy records.

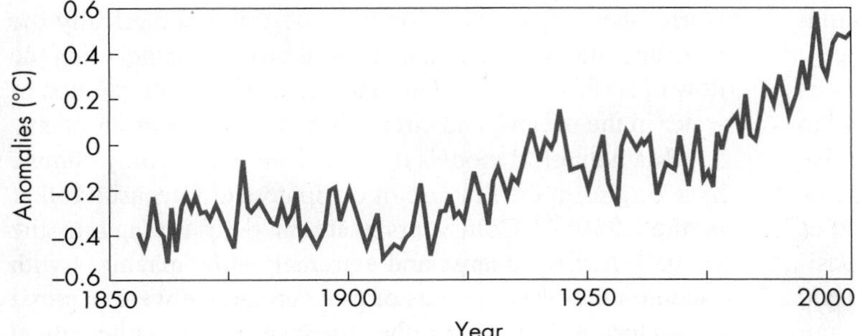

◄ **FIGURE 14.7 CHANGE IN TEMPERATURE** Diagram showing temperature changes from about 1850 to the present. The vertical scale shows departure from the average global temperature for the period 1960–1990 ("0" on the scale). The average temperature at Earth's surface has risen about 0.4°C over the past three decades. *(http://www.ncdc.noaa.gov/paleo/globalwarming/instrumental.html. Accessed August 23, 2013)*

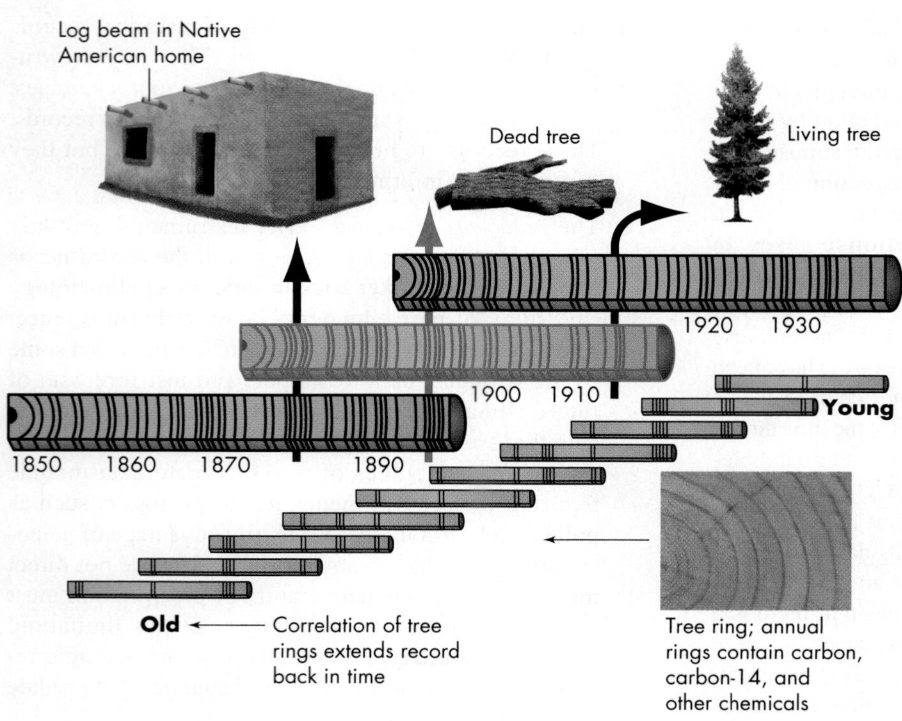

◄ **FIGURE 14.8 TREE RING CHRONOLOGY** Tree rings contain proxy information on temperature and precipitation with annual resolution. The rings in dead trees can be cross-dated with those in living trees to extend the record back in time.

▲ **FIGURE 14.9 ICE CORE** An ice core collected from the Greenland Ice Sheet is studied for the climate record it contains. *(Photo Researchers, Inc./Science Source)*

Ice Cores

Glaciers consist of ice derived from the annual accumulation of snow. The Greenland and Antarctica ice sheets are an archive of ice dating back almost 1 million years. Continuous ice cores are collected by drilling and then studied in detail to learn about past climatic conditions (Figure 14.9). Glacier ice cores contain small bubbles of air trapped at the time the snow transformed into ice. Scientists determine the composition of the gases in the air bubbles in order to infer the past composition of the atmosphere. Ice cores also contain volcanic ash and dust, which can provide proxy data that assist in evaluating climate change. The ice itself is studied to determine the oxygen isotope composition of the water, which provides proxy information on the total amount of past glacier ice on land and processes occurring in the oceans.

Pollen

Pollen from trees and other plants accumulates with sediment in a variety of environments, including oceans, lakes, and bogs. Scientists study the types and abundance of pollen in sediment in order to reconstruct past climate and environments. For example, a period of cool climate might be signalled by a change in the types of pollen in sediments, which reflects a change in vegetation. Because pollen is preserved in sediment layers that commonly can be independently dated, a chronology can be developed. Many other types of fossils, for example foraminifera, diatoms, ostracods, and beetle remains, are preserved in sediments and provide additional information on past climates.

Global Climate Models

Scientists construct mathematical models to represent real-world phenomena. These models describe numerically the links and interactions among natural processes, including the flow of surface water and groundwater, erosion, circulation of water in the oceans, and circulation of air in the atmosphere.

Mathematical models that scientists use to study climate have origins in the first use of computers to forecast weather in the 1950s.[23] Computers were in their infancy in the 1950s—they were large and extremely slow machines with vacuum tubes. Many hours of processing time were required to produce a 24-hour weather forecast. Early mathematical

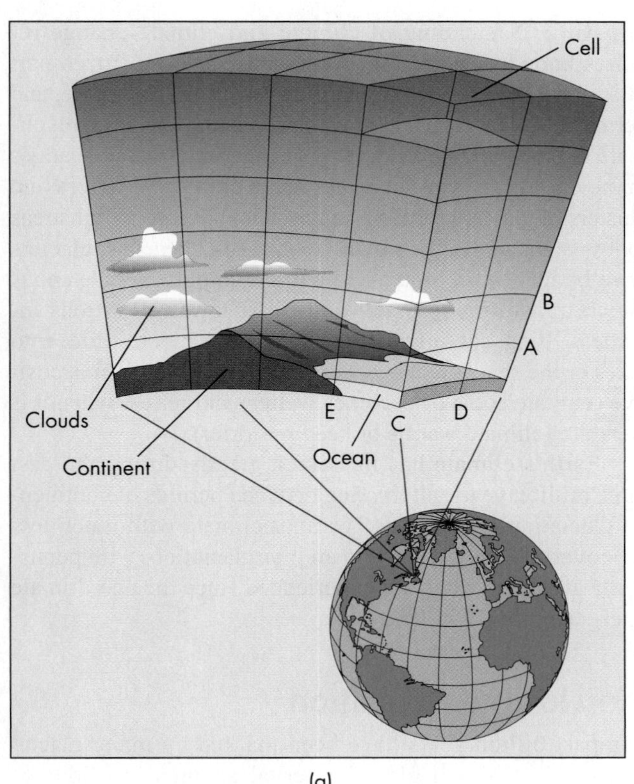

◀ **FIGURE 14.10 MODELLING CLIMATE CHANGE** (a) Idealized diagram showing cells used in global climate models. The upper layers of the cells are coupled to global circulation models that describe circulation of air in the atmosphere. The lower layers of cells are coupled to mathematical models that describe ocean circulation and land surface processes. (b) Observed (black) and predicted temperature changes from 1900 to 2005. The predicted mean temperature (red line) is the average of 58 different simulations using 14 different climate models. The yellow band is the variability in the simulations. *(Based on Climate Change 2007.* The Physical Science Basis: Working Group 1. Contribution to the Fourth Assessment Report of the Intergovernmental Panel on Climate Change. *New York, NY: Cambridge University Press)*

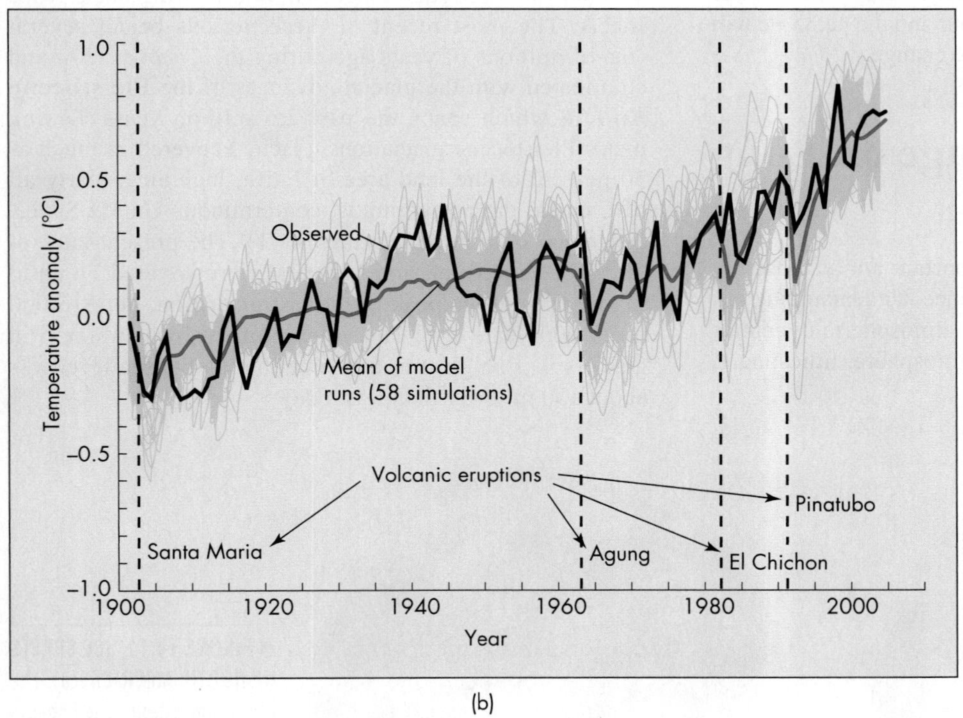

models run on computers were, of necessity, simple and regional in extent; they were used primarily to describe the general circulation of the atmosphere. These general circulation models were the foundation for more sophisticated *global climate models*. From the 1980s through the early 2000s, atmospheric general circulation models were refined and coupled with mathematical models of other Earth subsystems, such as the land surface, ocean circulation, sea ice, aerosols, and the carbon cycle.[16]

The framework of a global climate model resembles a large stack of boxes (Figure 14.10). Each box is a three-dimensional cell that covers several degrees of latitude and longitude. A typical cell has an area five times the size of the province of Nova Scotia. Each cell differs in height depending on the number of cells used to subdivide the lower atmosphere. Most models use 6 to 20 layers of cells to represent the atmosphere up to an altitude of about 30 km.[24] Data are entered into each of the cells, and boundary conditions are established for the overall model. Equations based on principles of physics describe the major atmospheric processes that interact between cells. A model is run backward ("hindcast") to test how accurately it describes historic, and even prehistoric, changes in climate. If the model accurately replicates past climate, it is then run forward to forecast future climate.

Like models used to forecast weather, global climate models are constantly being evaluated and improved to make better predictions. In 2007, an international team of climate scientists found that the global climate models used by the International Panel on Climate Change in 2001 accurately predicted or slightly underestimated global temperature, CO_2 concentration, and sea-level change for the period 1900 through 2006.[25]

Global climate models provide a great deal of information for evaluating Earth as a system and are a valuable tool for understanding climate change. The models also help scientists to identify additional data that are needed to produce better models in the future. One of the greatest difficulties in using global climate models to predict future climate is our inability to anticipate human behaviour and the paths we will take to deal with the threat of climate change.

14.5 Climate Change on Long Timescales

Climate change has become an important and extensively studied subject in many fields of science. Understanding climate change requires knowledge of atmospheric dynamics and the complex links among the atmosphere, lithosphere, hydrosphere, and biosphere.

An understanding of climate and climate change requires knowledge of Earth's **cryosphere**—the frozen part of the hydrosphere, comprising snow, sea and glacier ice, and permafrost. The cryosphere is in rough equilibrium with climate and responds quickly to a change in climate. Changes in the cryosphere can, in turn, impact climate. **Glaciers** and clusters of glaciers forming an ice sheet flow from high areas to low areas under their own weight. Like beaches, glaciers have budgets with inputs and outputs. Inputs include snow, which transforms into ice; outputs at lower elevations include melting, sublimation, and calving of blocks of ice into lakes or the sea. Glaciers advance when their budget is positive (climate cools or becomes wetter) and retreat when it is negative (climate warms or becomes drier).

Earth's climate has fluctuated greatly during the past three million years, alternating between periods of continental glaciation and times of warmer climate with much less ice cover.[26] Today we live in an interglaciation, with persistent warm conditions not experienced since the penultimate interglaciation 125 000 years ago.

Pleistocene Glaciation

The past billion years have been marked by many glacial intervals.[18,26] Each glacial interval spanned millions to tens of millions of years and was characterized by repeated *advances* and *retreats*, or growth and melting back of ice sheets. The most recent of these periods began several tens of millions of years ago during the *Cenozoic Era* and culminated with the glacial advances of the **Pleistocene Epoch**, which spans the past 2.6 million years. During major Pleistocene glaciations, glaciers covered as much as 30 percent of the land area of Earth, including nearly all of Canada, the northernmost conterminous United States, and much of Europe (Figure 14.11). The present sites of Toronto, Vancouver, and Montreal were covered by up to 2 km of ice at the peak of the last glaciation, only 16 000 to 20 000 years ago. The volume of freshwater stored in glaciers was so great at that time that global sea level was about 130 m lower than it is today.[27]

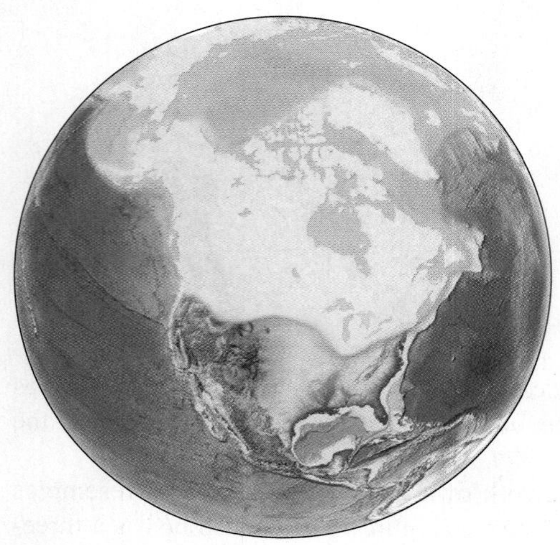

(a)

(b)

◄ **FIGURE 14.11 ICE SHEETS IN NORTH AMERICA** (a) The maximum extent of ice sheets in North America about 20 000 years ago during the late Pleistocene. Ice sheets covered almost all of Canada, northern Europe, and part of Eurasia. (b) Glacier ice in North America today. *(NOAA, http://www.ncdc.noaa.gov/)*

TABLE 14.2 Glacier Ice on Earth

	Area (km²)	Percentage	Volume (km³)	Percentage
Antarctica	12 600 000	84	31 000 000	91
Greenland	1 800 000	12	2 800 000	8
Other*	540 000	4	120 000	<1

*Ice caps and valley glaciers in Iceland, Europe, Asia, North America, South America, and New Zealand.

Today, we live in an interglaciation called the **Holocene Epoch**, which began about 11 600 years ago. Only 10 percent of Earth's land area is currently covered by glacier ice, mainly in Antarctica and Greenland (Table 14.2). Glaciers also exist in Iceland and Arctic Canada and in the high mountains of Asia, North and South America, Europe, and New Zealand.

Causes of Glaciation The causes of glaciation, and thus climate change, have long been debated. No single factor can explain glaciation. Rather, it is the result of several natural factors operating together.[26] One of the most important of these factors, however, is cyclic changes in Earth's orbital pattern, commonly referred to as **Milankovitch cycles**.

The positions of the continents significantly affect ocean and atmospheric circulation and determine how much land area lies in the mid-latitudes where continental glaciers form. When the continents are favourably situated for glaciation, small, regular changes in the amount of solar radiation reaching Earth's surface determine the advance and retreat of continental ice sheets. Milutin Milankovitch, a Serbian geophysicist, proposed in the 1920s that cyclic variations in the shape of Earth's orbit around the sun (*eccentricity*) and in the tilt (*obliquity*) and wobble (*precession of the equinoxes*) of Earth's axis of rotation lead to small changes in the distribution of solar radiation on the planet's surface (Figure 14.12).

The three Milankovitch cycles can be combined to determine net changes in radiation that any point on Earth receives through time. These changes correlate well with major advances and retreats of continental ice sheets (Figure 14.13). However, correlation is not proof of causation. Milankovitch cycles cause only small changes in seasonality and the amount of solar radiation reaching Earth. They should be considered the metronome of climate, modulating fluctuations of ice sheets on timescales of thousands of years once other factors favour global glaciation. Neither the position of the continents nor Earth's orbital cycles can explain observed fluctuations on timescales of decades to centuries. The causes of these shorter-term variations, which are most relevant to the climate change during the remainder of this century, are discussed in the next section.

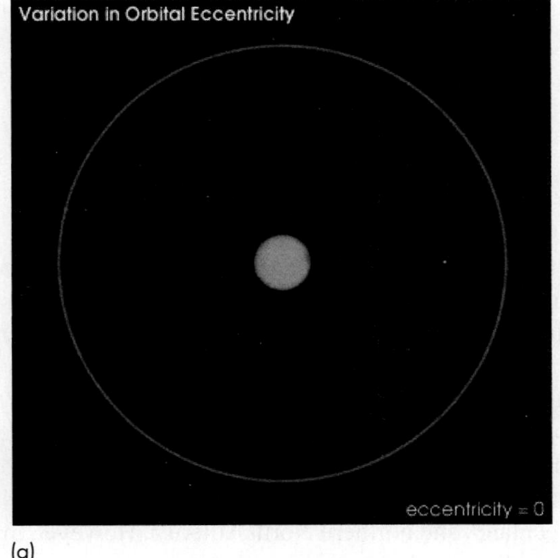

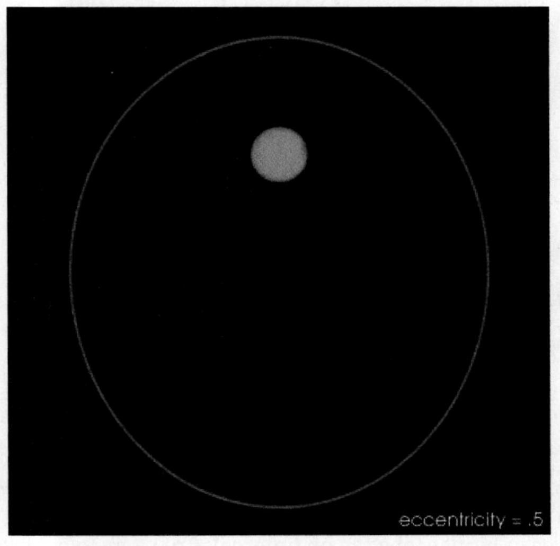

(a)

▲ **FIGURE 14.12 MILANKOVITCH CYCLES** Variations in (a) the eccentricity of Earth's orbit and (b) the wobble (precession of the equinoxes) and (c) the tilt (obliquity) of its axis of rotation lead to differences in the distribution of solar radiation on Earth's surface. (a) The eccentricity of Earth's orbit is a measure of its departure from circularity. Earth's orbit varies from nearly circular to mildly elliptical on a cycle of about 100 000 years. (b) The precession of the equinoxes is the change in the direction of Earth's axis of rotation relative to the sun at the time of *perihelion* and *aphelion*. Perihelion and aphelion are, respectively, the points in the orbit of Earth that are nearest and farthest from the sun. Earth goes through one precession cycle roughly every 20 000 years. (c) Earth's rotational axis is tilted with respect to its orbital plane; the angle of the tilt changes through 2.4° about every 40 000 years. *(NASA)*

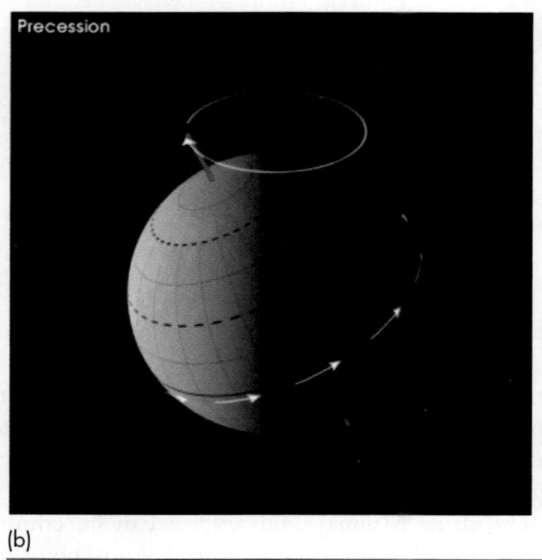

(b)

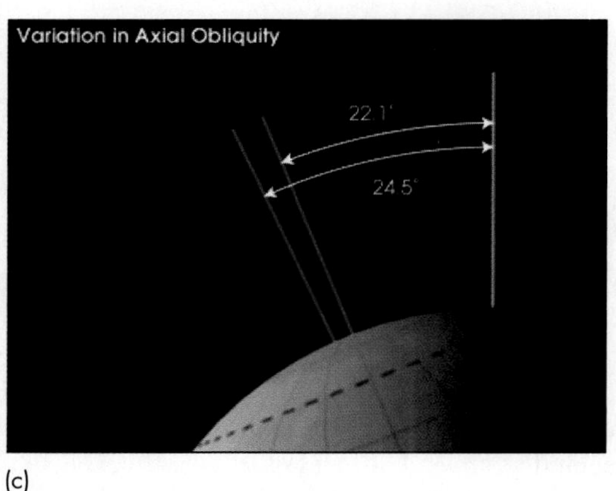

(c)

▲ **FIGURE 14.12** *(Continued)*

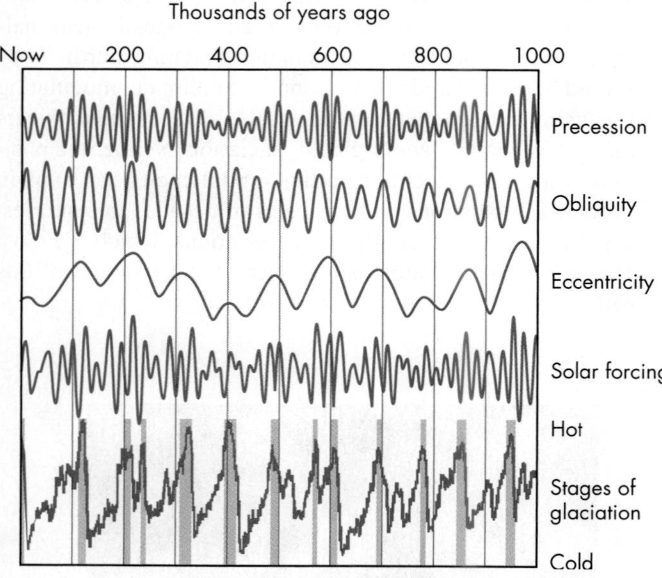

◄ **FIGURE 14.13 RELATION BETWEEN MILANKOVITCH CYCLES AND GLACIATION** Several quasi-periodic cycles occur as Earth spins and orbits the sun. The curves have a large number of sinusoidal components, but a few components are dominant. Milutin Milankovitch studied changes in eccentricity, obliquity, and precession (top three curves), which collectively change the amount and location of solar radiation reaching Earth (solar forcing, fourth curve). Changes in the Northern Hemisphere are important because of the large amount of land there; land reacts to such changes more quickly than do the oceans. The lowest curve is an independent record of global ice extent, which is a proxy for Earth temperature. It is similar to the curve of solar forcing on timescales of tens of thousands of years. *(Adapted from Robert A. Rohde, http://en.wikipedia.org/wiki/File_Milankovitch_Variations.png)*

14.6 Climate Change on Short Timescales

The above discussion highlights important drivers of climate change on timescales of thousands to hundreds of thousands of years. Clearly, however, climate changes on much shorter timescales (Figure 14.14). In this section, we examine evidence for and causes of climate change on timescales of years to centuries.

Evidence for Climate Change on Short Timescales

Historic data and proxy climate data obtained from old trees reveal complex warming and cooling trends over the past 1000 years (Figure 14.14d). Trees that are subject to moisture or temperature stress are sensitive to climate—they add thick annual rings when conditions for growth are favourable and thin rings when moisture or temperature is unfavourable. Scientists measure the thickness of long sequences of annual rings in trees in many parts of the world to determine moisture and temperature conditions back through time. These studies and historical data show that relative warmth from about A.D. 1000 to 1270 allowed the Vikings to colonize Iceland, Greenland, and northern North America. However, they abandoned their settlements in North America and parts of Greenland around A.D. 1400, early in the *Little Ice Age*.[28,29] Climate began to warm in the late 1800s, at the end of the Little Ice Age.[30] The warming was not continuous, but rather occurred in two phases, one between about 1910 and 1940, and the other from 1980 to the present. The two warming intervals were separated by a period when climate cooled slightly and many glaciers around the world advanced (Figure 14.14e).

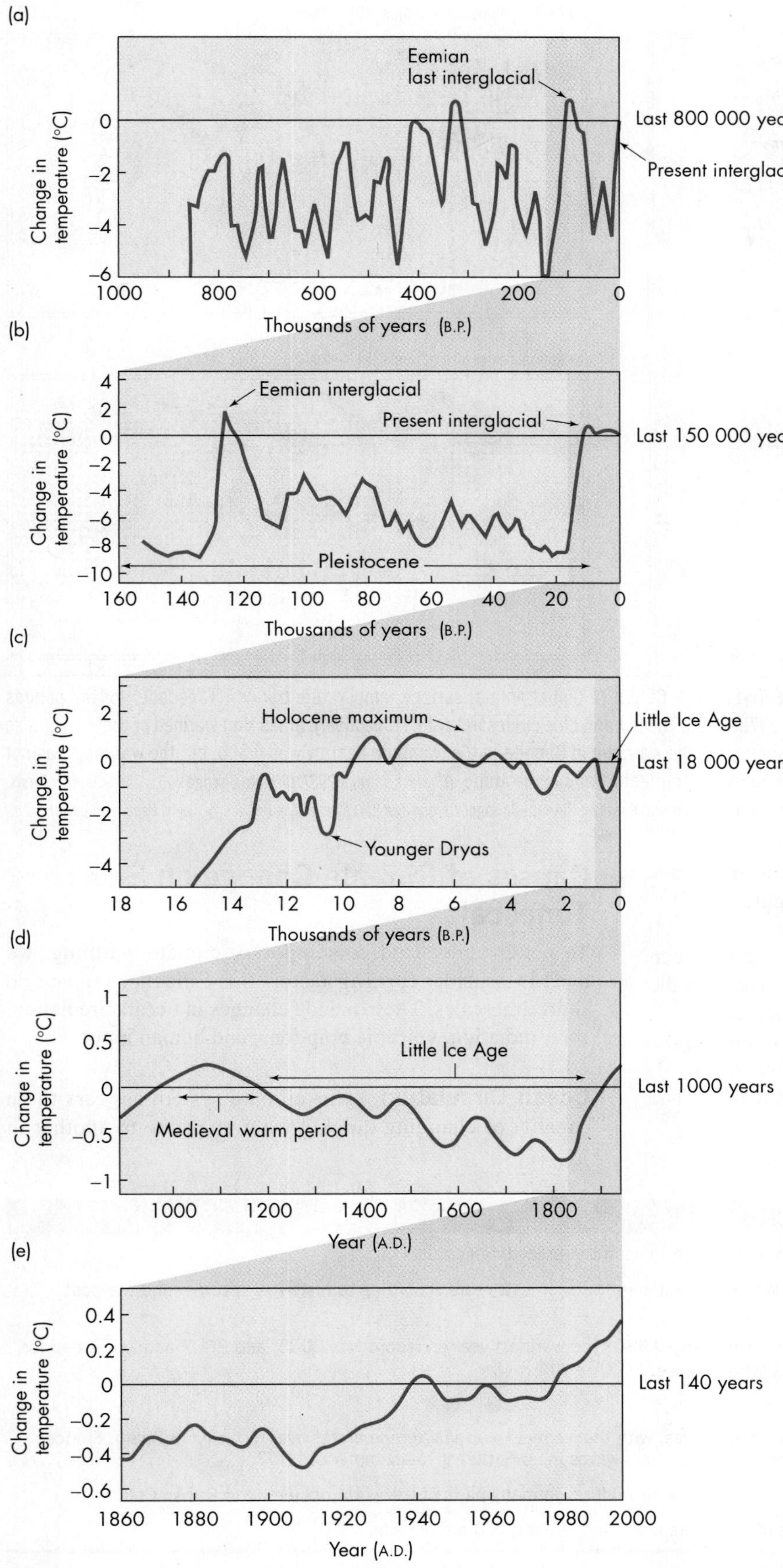

◀ **FIGURE 14.14 CHANGES IN GLOBAL TEMPERATURE** Graphs showing changes in global temperature on different timescales over the past 800 000 years. The zero line delineates the average temperature of the Holocene, which is the present interglaciation (the past 11 600 years). *(Based on University Corporation for Atmospheric Research, Office for Interdisciplinary Studies. 1991. "Science capsule: Change in time in the temperature of the Earth."* Earth-Quest *5(1); and the UK Meteorological Office. 1997.* Climate Change and Its Impacts: A Global Perspective)

(a) Annual temperature trends, 1901–2000

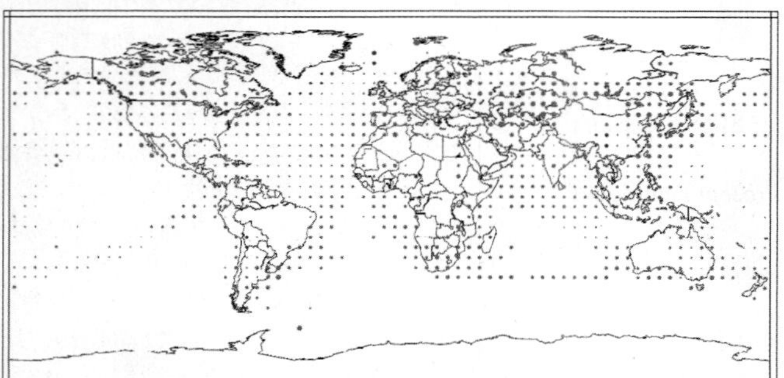

(b) Annual temperature trends, 1910–1945

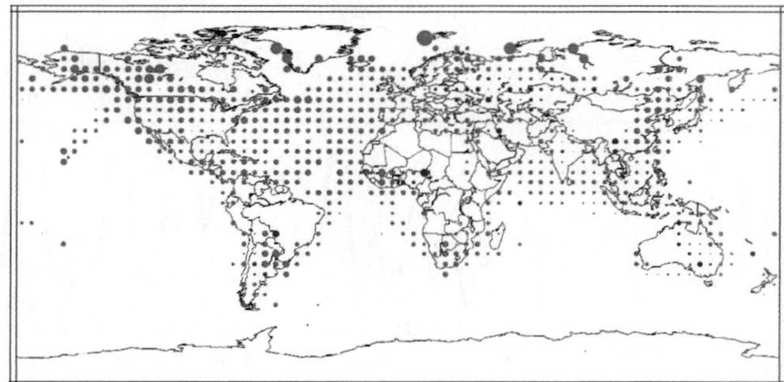

(c) Annual temperature trends, 1946–1975

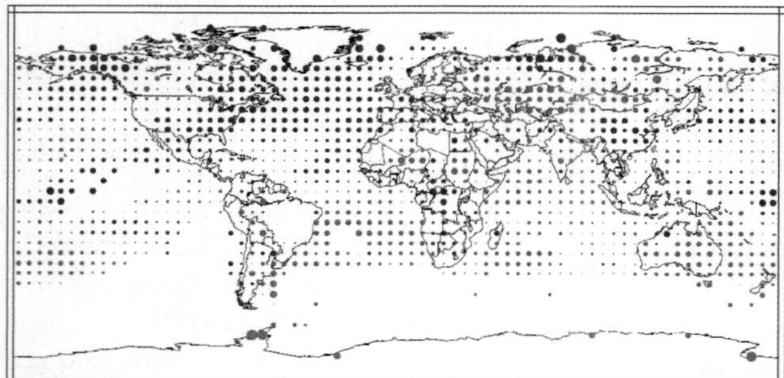

(d) Annual temperature trends, 1976–2000

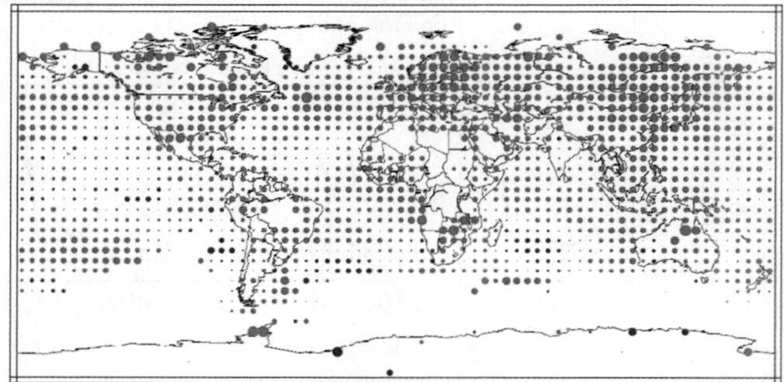

▲ **FIGURE 14.15 TEMPERATURE CHANGE DURING THE TWENTIETH CENTURY** Global annual surface temperature trends (°C/decade) for the periods (a) 1901–2000, (b) 1910–1945, (c) 1946–1975, and (d) 1976–2000. The red and blue circles indicate, respectively, areas that warmed or cooled. The size of each circle is proportional to the magnitude of the change. Average warming in Canada in the twentieth century was 0.8°C, but the warming was not uniform in either time or space. It was greatest in the last quarter of the century and in the Arctic. *(Climate Change 2001.* Climate Change 2001: The Scientific Basis. Contribution of Working Group I to the Third Assessment Report of the Intergovernmental Panel on Climate Change. *Cambridge, UK: Cambridge University Press. Used with permission.)*

The evidence that Earth's climate has warmed since the late nineteenth century is unequivocal (Figure 14.15).[11] The total increase in global mean temperature since 1900 has been about 0.8°C. About half of the increase has been in the past 30 years, and the past two decades have been the warmest since global temperatures have been monitored.[31,32] Most of the years between 1995 and 2012 were among the warmest on record; sea-surface temperatures have risen; alpine glaciers have receded; and permafrost is thawing.[11] The evidence is summarized in Table 14.3.

Causes of Climate Change on Short Timescales

To better understand contemporary climate warming, we need to consider **forcing** factors that influence climate on short timescales. They include changes in ocean circulation, solar radiation, volcanic eruptions, and human input.

Ocean Circulation Our climate system appears to be capable of changing quickly from one state to another in

TABLE 14.3 Evidence of Global Warming during the Past Century
Warming since the mid-1970s has been about three times as rapid as in the preceding century.
The first decade of the twenty-first century was the warmest decade in the past 142 years according to historical records, and the past 1000 years according to geologic data.
The 15 warmest years in recorded history have all occurred since 1990. The warmest year on record was 2005, and 2007 and 2009 tied for the second warmest years; 2002, 2003, and 2006 were very close to 2007 and 2009.
The warmest year in U.S. history was 2012.
In 2003, Europe experienced unprecedented summer heat waves, with the warmest seasonal temperatures ever recorded in Spain, France, Switzerland, and Germany. Approximately 15 000 people died in heat waves in Paris during the summer of 2003.
Warm conditions, along with drought, in 2003 contributed to severe wildfires in many parts of the world, including in British Columbia.
Since 1970, the average temperature of the surface of Earth has increased about 0.2°C per decade.

Note: A few years of anomalously high temperatures with drought, heat waves, and wildfires are not themselves an indication of long-term climate warming. The persistent trend of increasing temperatures over several decades, however, is more compelling evidence that global warming is real.

Source: Based on data from NOAA and the WMO (World Meteorological Organization).

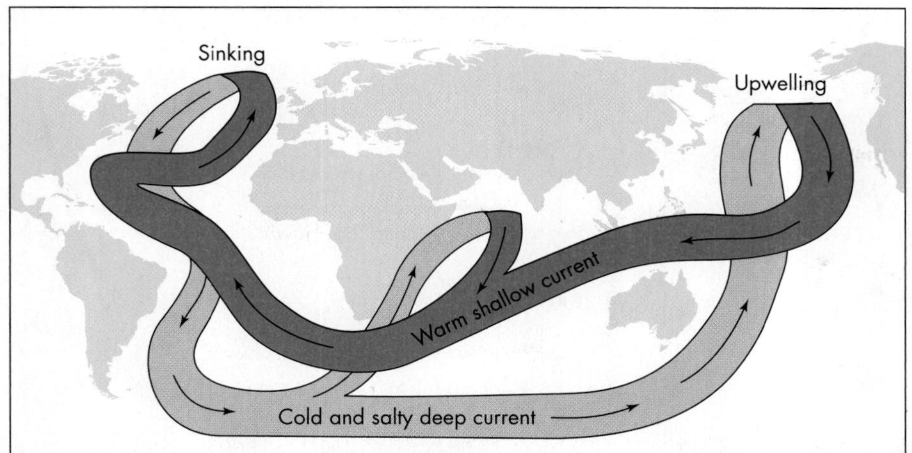

◄ **FIGURE 14.16 OCEAN CONVEYOR BELT** An idealized diagram of the oceanic conveyor belt current. The actual system is complex, but in general, warm Atlantic surface water (red) is transported northward to near Greenland where it cools from contact with cold Arctic air. As the water cools, it becomes denser, sinks to the bottom, and flows south and then east to the Pacific Ocean, where upwelling occurs. The total flow rate is about 20 million m^3 per second, and the heat released to the atmosphere from the warm water keeps northern Europe much warmer than it would be without the conveyor belt. *(Broecker, W. 1997. "Will our ride into the greenhouse future be a smooth one?"* Geology Today *7(5):1–7)*

as little as a few years. Part of what drives the climate system and allows it to change so rapidly is a global-scale circulation of ocean waters known as the **ocean conveyor belt**.

The ocean conveyor belt carries warm surface waters northward along the east side of the North Atlantic Ocean (Figure 14.16). The warm water cools and becomes more saline as it approaches Greenland. The cold salty water is denser than the surrounding water and thus sinks to the bottom, where it flows southward along the seafloor to around Africa. The flow in the conveyor belt is huge, equal to about 100 times the mean discharge of the Amazon River. The heat that it releases is sufficient to keep northern Europe 5°C to 10°C warmer than it would be if the conveyor belt were not operating. If the conveyor belt were to shut down, the planet would cool and northern Europe and eastern Canada would become much less habitable.[33]

Solar Forcing Variations in solar output should be evaluated as a cause of climate change because the sun is the source of most of the surface heat on Earth. Scientists have found a way to estimate changes in solar output over the past 10 000 years or more. The method is based on the fact that carbon-14 in tree rings is a proxy of solar output. Cosmic rays originating from fusion reactions in the sun pass through Earth's atmosphere and produce radioactive atoms of carbon-14. The larger the output of solar energy, the greater the abundance of atmospheric carbon-14 and, consequently, the higher the concentrations of the radioisotope in all living things, including trees. By measuring the concentration of carbon-14 in tree rings of known age, scientists can determine changes in solar output back through time. The record reveals that the Medieval Warm Period (A.D. 1000–1270) was a time of above-average solar radiation. Solar output was at a minimum during the major glacier advances of the Little Ice Age. It appears, therefore, that variability of energy output from the sun can partially explain climatic variability during the past 1000 years. However, the effect is small—the inferred difference between solar output during the Medieval Warm Period and the Little Ice Age is only about 0.25 percent.[30]

Volcanic Forcing Explosive volcanic eruptions introduce large amounts of aerosols into the atmosphere. Aerosol particles reflect sunlight and can cool the atmosphere. Forcing by this mechanism is believed to have contributed to Little Ice Age cooling and climatic variability.[30]

Volcanic eruptions introduce some uncertainty into predictions of future climate conditions. The eruption of Mount Tambora in Indonesia in 1815 was probably the largest explosive volcanic eruption of the past 10 000 years. Explosions, ash flows, and the tsunami from the eruption killed an estimated 90 000 people. About 50 km^3 of ash and volcanic gases were carried high into the atmosphere. The eruption caused months of cool weather, and Europe and North America suffered crop failures and famine during 1816, the "year without a summer." This event highlights the important links among geologic processes, the atmosphere, and people.[34]

Human Forcing In 2007, an international report authored by over 1200 scientists from 120 countries concluded that there is a 90 percent probability that humans are contributing to global warming, mainly through their consumption of fossil fuels.[11] The key conclusions of the report have been accepted by national academies of science, scientific associations, and government agencies in Canada, the United States, and many other nations.

Although acknowledging that uncertainties exist, the authors of the report concluded that (1) people have a discernible effect on global climate, (2) warming is now occurring, and (3) the mean surface temperature of Earth will likely increase by 1.4–5.8°C during the twenty-first century.[11] People are producing large amounts of greenhouse gases that trap heat in the atmosphere (Figure 14.17), and the evidence is strong that increases in these gases are linked to an increase in the mean global temperature of Earth (Figure 14.18). The concentration of atmospheric CO_2 was approximately 280 ppm at the beginning of the Industrial Revolution. Today the concentration is 400 ppm and is expected to reach at least 460 ppm by the year 2050—more than 1.6 times its pre-industrial level.[11]

A recent study examined climate forcing over the past 1000 years, allowing us to place late-twentieth-century

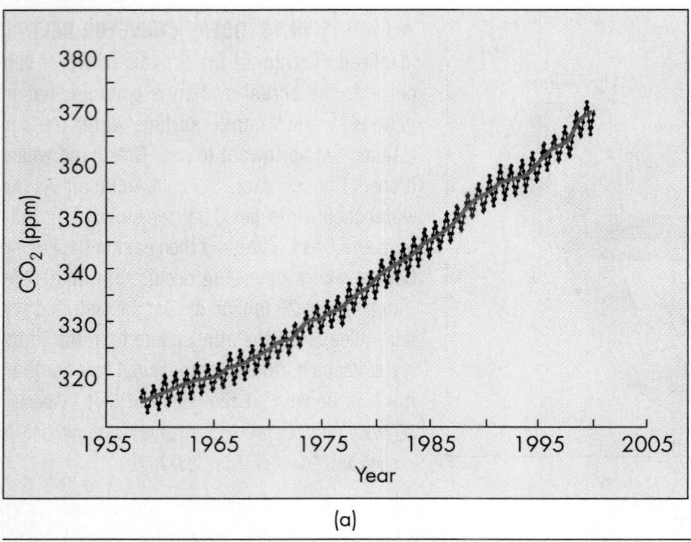

(a)

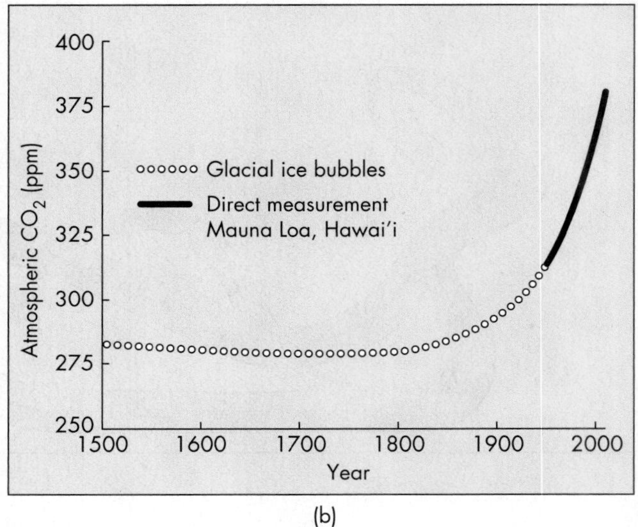

(b)

▲ **FIGURE 14.17 CARBON DIOXIDE IS INCREASING IN THE ATMOSPHERE** (a) Atmospheric concentration of carbon dioxide measured at the Mauna Loa Observatory, Hawaii. (b) Average concentration of atmospheric carbon dioxide from A.D. 1500 to 2000 based on measurements of air bubbles trapped in glacier ice and on direct measurements at the Mauna Loa Observatory. *((a) Data from Scripps Institute of Oceanography, NOAA, and C. D. Keeling, at http://www.esrl.noaa.gov/gmd/ccgg/trends/co2_data_mlo.html. Accessed December 23, 2009; (b) Data in part from Post, W. M., T.-H. Peng, W. R. Emanuel, A. W. King, V. H., Dale, et al. 1990. "The global carbon cycle."* American Scientist *78(4):310–326)*

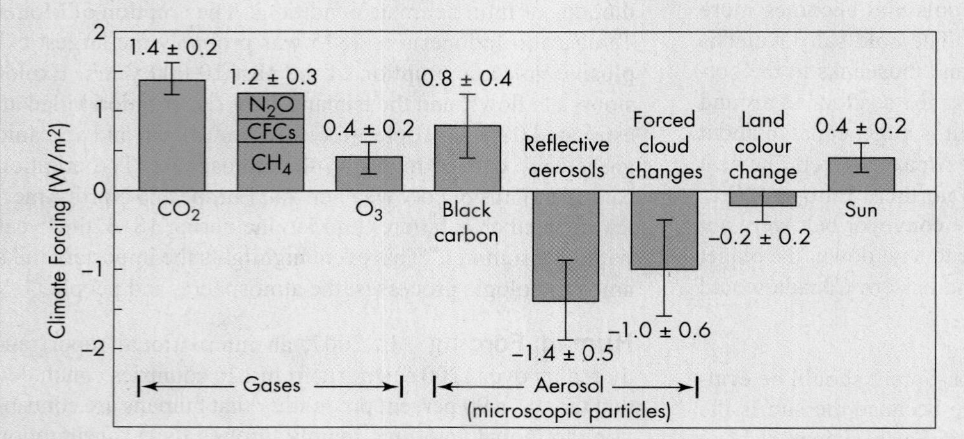

- Increases of greenhouse gases (except O_3) are known from observations and bubbles of air trapped in ice sheets. The increase of CO_2 from 285 parts per million (ppm) in 1850 to 368 ppm in 2000 is accurate to about 5 ppm. The conversion of this gas change to a climate forcing (1.4 W/m^2), from calculation of the infrared opacity, adds about 10% to the uncertainty.
- Increase of CH_4 since 1850, including its effect on stratospheric H_2O and tropospheric O_3, causes a climate forcing about half as large as that by CO_2. Main sources of CH_4 include landfills, coal mining, leaky natural gas lines, increasing ruminant (cow) population, rice cultivation, and waste management. Growth rate of CH_4 has slowed in recent years.
- Tropospheric O_3 is increasing. The U.S. and Europe have reduced O_3 precursor emissions (hydrocarbons) in recent years, but increased emissions are occurring in the developing world.
- Black carbon ("soot"), a product of incomplete combustion, is visible in the exhaust of diesel-fueled trucks. It is also produced by biofuels and outdoor biomass burning. Black carbon aerosols are not well measured, and their climate forcing is estimated from measurements of total aerosol absorption. The forcing includes the effect of soot in reducing the reflectance of snow and ice.
- Human-made reflective aerosols include sulfates, nitrates, organic carbon, and soil dust. Sources include burning fossil fuel and agricultural activities. Uncertainty in the forcing by reflective aerosols is at least 35%.
- Indirect effects of aerosols on cloud properties are difficult to compute, but satellite measurements of the correlation of aerosol and cloud properties are consistent with the estimated net forcing of −1 W/m^2, with uncertainty of at least 50%.

◄ **FIGURE 14.18 DIFFERENCES IN CLIMATE FORCINGS BETWEEN THE PRESENT AND THE MIDDLE OF THE NINETEENTH CENTURY** Positive forcings warm the atmosphere and negative forcings cool it. Human-caused forcings in recent decades have dominated other forcings. The net forcing difference is about 1.6 ± 0.1 W/m^2 (watts per square metre), consistent with the observed rise in surface temperature over the past few decades. *(Modified from Hansen, J. 2003. NASA Goddard Institute for Space Studies and Columbia University Earth Institute)*

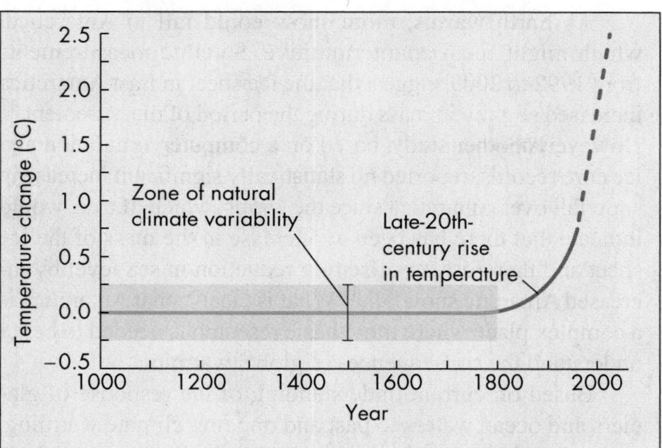

▲ **FIGURE 14.19 TEMPERATURE RISE IN THE TWENTIETH CENTURY** This graph is based on a computer model that removed the effects of solar and volcanic forcing to determine the impact of increases in greenhouse gases resulting from human activities. It suggests that human forcing caused the rise in global temperature in the twentieth century. *(Crowley, T. J. 2000. "Causes of climate change over the past 1000 years."* Science *289:270–277. Reprinted with permission.)*

warming within a longer-term context.[30] The author of the study used a mathematical model to remove the climatic effects of variations in solar output and volcanic activity and to thus isolate the effect of human forcing. He found that present warming greatly exceeds natural variability and is consistent with the warming predicted from greenhouse gas forcing (Figure 14.19). The potential effects and adjustments to this warming are described below.

14.7 Effects of Climate Change

All climate models are consistent in predicting continued warming as a result of the greenhouse gases now in the atmosphere and accelerated warming in the coming decades. Warming will continue even if we stabilize concentrations of atmospheric greenhouse gases at current levels. If carbon dioxide in the atmosphere doubles from pre-industrial levels, as expected, average global temperature will rise about 1.5°C to 4.5°C by the end of the century, with significantly greater warming in polar regions.[11] Therefore, we need to carefully examine the potential effects of such warming.

Surface processes and climate are intimately linked, and, as the term "global warming" suggests, all regions of the planet will be affected by climate change (Figure 14.20). We know where tornadoes, hurricanes, blizzards, and wildfires occur today. If climate changes, these sites may shift, forcing people to cope with new problems. In addition, climate change will melt glaciers, raise sea level, thaw permafrost, alter the location of agricultural zones, and stress plant and animal species by altering their habitat. Temperature increases might

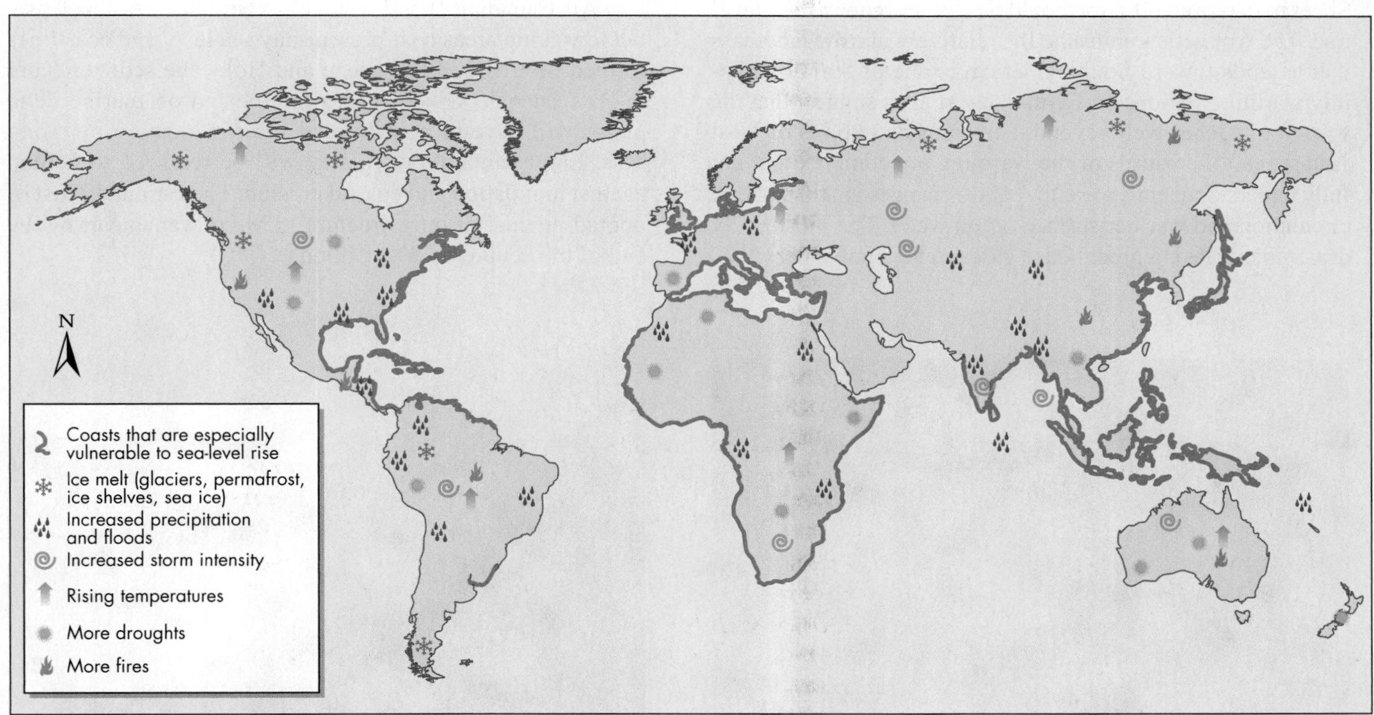

▲ **FIGURE 14.20 PROJECTED IMPACTS OF CLIMATE CHANGE** A global overview of climate-change impacts in the twenty-first century. The effects include sea-level rise, an increase in flooding, melting of glaciers and sea ice, permafrost degradation, more intense storms, rising temperatures, more droughts, and more wildfires. *(Based on Abramowitz, J. 2001.* Unnatural Disasters. *Worldwatch Paper 158, Washington, DC: Worldwatch Institute, based on data from Watson, R. T., et al. 1998.* The Regional Impacts of Climate Change: An Assessment of Vulnerability. Special Report of IPCC Working Group II. *Cambridge, UK: Cambridge University Press; McCarthy, J. J., et al., eds. 2001.* Climate Change 2001: Impacts, Adaptation, and Vulnerability. Contribution of Working Group II to the Third Assessment of the Intergovernmental Panel on Climate Change. *Cambridge, UK: Cambridge University Press; and Revenga, C., et al. 1998.* Watersheds of the World. *Washington, DC: World Resources Institute and Worldwatch Institute)*

also accelerate desertification of arid lands and increase the frequency of severe droughts and wildfires.

Glacier Ice and Sea-Level Rise

Sea-level rise related to global warming is a serious problem (see Chapter 12). Sea level is currently rising at a rate of over 3 mm/yr due to melting of glaciers and the heating and thermal expansion of upper ocean waters.[11] Of particular concern is accelerated melting of the Greenland Ice Sheet and large discharges of ice at its margin into the sea (Figure 14.2). Losses of ice from the Greenland Ice Sheet have doubled since about 1998. Surface melting produces water that flows through vertical tunnels (*moulins*) and crevasses down to the base of the ice sheet where it accelerates the movement of the glacier over its bed. Most glaciers flowing from the southern half of the ice sheet are accelerating and losing ice more rapidly than previously. In 2005 alone, over 200 km^3 of glacier ice were lost by this process.[35]

Glaciers in mountains around the world are also thinning and retreating, contributing to sea-level rise (Figure 14.21).[36,37] The changes are partly in response to a mean global temperature increase of 0.4°C over the past three decades. With the anticipated continuing increase in temperatures, most of the glaciers in Glacier National Park could be gone by A.D. 2030, and those in the European Alps could be gone by the end of the century.[38,39,40]

Scientists are increasingly scrutinizing the Antarctic Ice Sheet to determine if it too could be a contributor to sea-level rise. The Antarctic Peninsula, the relatively narrow landmass that extends toward South America, is one of the most rapidly warming regions on Earth. New studies suggest that the warming reaches well beyond the peninsula to parts of West Antarctica. The causes of the warming are complex and not fully understood, but appear to require changes in atmospheric circulation and warmer surface ocean water. The most likely driver is increasing greenhouse gases in the atmosphere.[41]

As Earth warms, more snow could fall in Antarctica, which might seem counterintuitive. Satellite measurements from 1992 to 2003 suggest that the ice sheet in East Antarctica increased slightly in mass during the period of measurement.[42] However, another study, based on a computer simulation and ice core records, reported no statistically significant increase in snowfall over Antarctica since the 1950s, which, if true, would indicate that there has been no increase in the mass of the ice sheet and therefore no offsetting reduction in sea level by increased Antarctic snowfall.[43] What is clear is that Antarctica is a complex place where more basic research is needed to better understand the consequences of global warming.

Based on current understanding of the response of glaciers and ocean waters to past and ongoing climate warming, researchers have concluded that sea level will rise somewhere between several tens of centimetres and about 1 m by the end of this century. Even sea-level rise at the lower end of this range would increase coastal erosion and flooding of low-lying coastal areas during severe storms (see Chapter 12).

Sea-level rise is already threatening some small island nations in the tropical Pacific Ocean. The nation of Tuvalu consists of about nine atolls, and the highest place in the country is only about 4.5 m above sea level. People first inhabited the island of Funafuti and other islands of Tuvalu about 200 years ago.[44] Rising sea level is threatening Funafuti, especially during high tides and storms (Figure 14.22). With continued sea-level rise, the 12 000 people on Funafuti could become the first permanent climate refugees.

All Canadian coastlines will experience sea-level rise, but low-lying areas near present-day sea level and coastlines formed of erosive Pleistocene and Holocene sediments are at greatest risk. Cities and towns located on marine delta plains will have to be protected from inundation by rising seas. Large amounts of money will be required to defend against inundation and coastal erosion. The estimated cost of upgrading coastal infrastructure in Metro Vancouver by the end of the century is $9.6 billion.[45]

(a)

(b)

▲ **FIGURE 14.21 MOUNTAIN GLACIERS ARE RETREATING** Warming of the atmosphere is the main cause of retreat of most mountain glaciers, like Muir Glacier in Glacier Bay National Park, Alaska, shown here. The photograph in (a) was taken on August 13, 1941, and the photograph in (b) was taken from the same vantage point on August 31, 2004. Between 1941 and 2004, the glacier retreated more than 12 km and thinned more than 800 m. Vegetation in the foreground of (b) is growing where there was only bare rock in 1941. *(USGS National Center)*

▲ **FIGURE 14.22 FLOODING OF FUNAFUTI** During high tides in February 2005, the sea flooded much of the island of Funafuti in Tuvalu. *(© Global Warming Images/Alamy)*

Glacier Hazards

Glaciers present hazards to people living in or visiting high mountain environments.[46] Glaciers could advance rapidly, or *surge*, and cross a valley and dam a stream, producing a temporary lake. The lake might drain catastrophically after the surge ends and the ice dam stagnates and weakens.[47,48] Similarly, a new lake could form at the margin of a retreating glacier as climate warms, and could later drain suddenly.[49] Deadly ice avalanches can occur at the terminus of a valley glacier when it retreats up a steep bedrock slope. In September 2002, a huge ice avalanche from an alpine glacier in the Caucasus Mountains in southern Russia sent a torrent of ice, rocks, and mud down a valley, killing at least 100 people in a village at the valley mouth.[50]

Glaciers that flow into the sea or a lake release large blocks of ice from their steep fronts, a process known as calving (see photograph at the beginning of this chapter). Calving produces blocks of floating ice known as *icebergs* that can float into shipping lanes and create a hazard to navigation. This danger was dramatically illustrated on April 15, 1912, when the *Titanic*, a luxury ocean liner on her maiden voyage from England, struck an iceberg and sank in the North Atlantic, claiming 1501 lives.

Thawing of Permafrost

Climate modellers suggest that the greatest warming during the remainder of this century will be at high latitudes, where the ground is frozen. The area underlain by discontinuous permafrost has been shrinking throughout most of the Northern Hemisphere, and temperatures at the top of Arctic permafrost have increased by 3°C since the 1980s.[51] Thawing of permafrost presents problems for people living in the Arctic and for many northern animals and plants (see Chapter 8). It also poses an additional problem: Very large amounts of carbon are stored in frozen peat deposits in northern Canada and Siberia. With large-scale thawing of permafrost, some of these organic sediments will decay and release carbon dioxide into the atmosphere, thus producing additional climate warming. This scenario is an example of a positive feedback, a "runaway" situation in which the output of a process amplifies or increases the process itself.

Changes in Climate Patterns

Global warming could significantly change rainfall patterns, soil moisture, and other climate factors that are important to agriculture. Some marginal agricultural areas in Canada and Eastern Europe could become more productive, whereas lands to the south will become more arid. However, such shifts do not necessarily mean that prime agricultural zones likely will also move north. Maximum production of grains and other foods depends as much on soil conditions as on climate, and suitable soils will not be present in all areas that will experience a more favourable climate for agriculture. For example, a warmer climate farther north in Ontario and Quebec will be of little advantage because the Canadian Shield has little soil capable of supporting large-scale agriculture. The uncertainty around effects of climate change on agriculture is a major concern. Increasing global production of grain is crucial for feeding the growing human population.

Global warming could also increase the frequency or intensity of violent storms, an issue just as important as which areas become wetter, drier, warmer, or cooler. Warmer oceans could feed more energy into hurricanes and typhoons. Several studies conclude that this warming is already affecting hurricane intensity in the North Atlantic Ocean.[51] Larger hurricanes would increase the risk of living in low-lying coastal areas of Mexico and the United States that are experiencing rapid population growth.

A natural phenomenon called ENSO (El Niño/Southern Oscillation), or more commonly **El Niño**, dramatically illustrates how a change in ocean circulation can affect the frequency and intensity of storms, landslides, drought, and fires. El Niño is an oceanic and atmospheric phenomenon characterized by high surface temperatures in the eastern equatorial Pacific; drought in parts of Southeast Asia, Australia, Africa, and South America; and wet weather and flooding in parts of the western and southern United States. An El Niño event probably begins with a slight reduction in the trade winds, after which warm water in the western equatorial Pacific Ocean begins to flow eastward (Figure 14.23). The eastward flow further reduces the strength of the trade winds, causing more warm water to move eastward, until surface waters across the entire equatorial Pacific from Asia to the west coast of South America are warmer than normal.[52]

During El Niño years, extra amounts of water vapour are added to the atmosphere by evaporation of warm tropical waters. The increase in water vapour and changes in the trade winds increase rainfall in Peru, Ecuador, and the southern United States; cause more frequent and severe hurricanes along the Pacific coast of Mexico; and produce severe drought in Indonesia, Papua New Guinea, Australia, and Brazil.

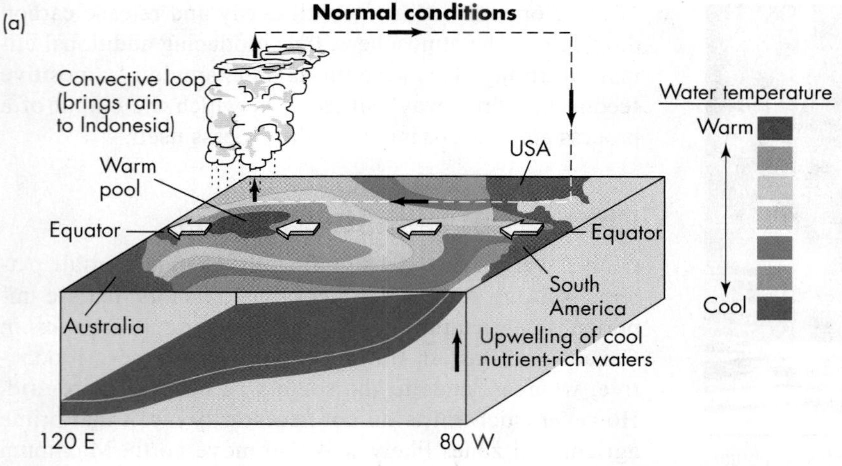

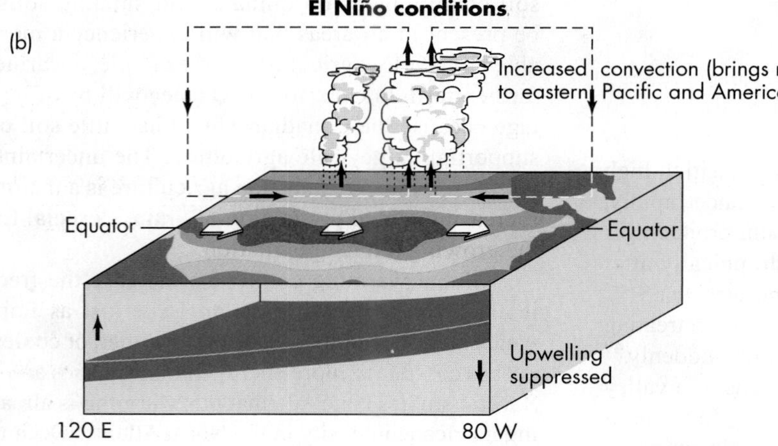

◀ **FIGURE 14.23 EL NIÑO** A schematic diagram contrasting (a) normal and (b) El Niño conditions and processes in the equatorial Pacific Ocean. Under normal conditions, trade winds blow warm surface water in the equatorial Pacific Ocean to the west toward Papua New Guinea and Indonesia. The westward flow allows colder waters to rise, or upwell, along the northwest coast of South America, and reduces rainfall in that area. During an El Niño event, winds blow the warm water east toward South America, upwelling of colder water is suppressed, and rainfall increases in parts of the Americas. *(Modified after National Oceanic and Atmospheric Administration/PMEL/TAO. 2002. "NOAA El Niño page." http://www.elnino.noaa.gov. Accessed December 23, 2009)*

A strong El Niño event in 1997 and 1998 contributed to hurricanes, floods, landslides, drought, and fires that caused widespread damage to crops, roads, buildings, and other structures (Figure 14.24). Australia, Papua New Guinea, Indonesia, the Americas, and Africa were particularly hard hit. Researchers disagree over the amount of damage the 1997–1998 El Niño caused, but all agree it was significant.[53,54]

The opposite of an El Niño event is a **La Niña**, during which eastern Pacific waters are cool and the southern United States experiences drought rather than floods. The alternation of El Niño and La Niña conditions is a natural phenomenon that was only recently recognized.[55] We do not understand the cause of these events, but scientists have voiced concern that global warming may change their frequency and strength.

Changes in the Biosphere

Global warming could be causing changes that threaten ecological systems and people. An *ecosystem* is a community of plants and animals that is adapted to the environment in which they live; energy and materials cycle through the system. Changes to ecosystems resulting from global warming and related land-use changes include a reduction in species and shifts in the range and *habitats* of plants and animals. Examples of observed changes in range are the migration of mosquitoes carrying such diseases as malaria and dengue fever to higher elevations in Africa, South America, Central America, and Mexico; the recent appearance of West Nile virus in Canada; the northward movement of some butterfly species in Europe and birds in the United Kingdom; the invasion of alpine meadows in western North America by coniferous trees; and a shift in the distribution of alpine plants in Austria to higher elevations.

Warming of several degrees Celsius will result in major changes in natural vegetation zones. For example, the boreal forest of Canada and Russia will probably expand northward at the expense of polar tundra. Similarly, grasslands and deserts may in the future extend north of their present limits. Many of the changes in vegetation assemblages might be facilitated by wildfires that remove plants no longer adapted to the area in which they occur. Other plants, rather than shifting latitudinally, will become more restricted, perhaps as isolated small populations.

Habitats of plants and animals that do not migrate are also changing. Earlier melting of sea ice in the Arctic is stressing some seabirds, walruses, and polar bears; and the warming of shallow water in the Florida Keys, Bermuda, Australia's Great Barrier Reef, and many other tropical oceans is killing corals.[11,56,57] In addition to the warmer waters, animals and algae that build coral reefs are being threatened by an increase in

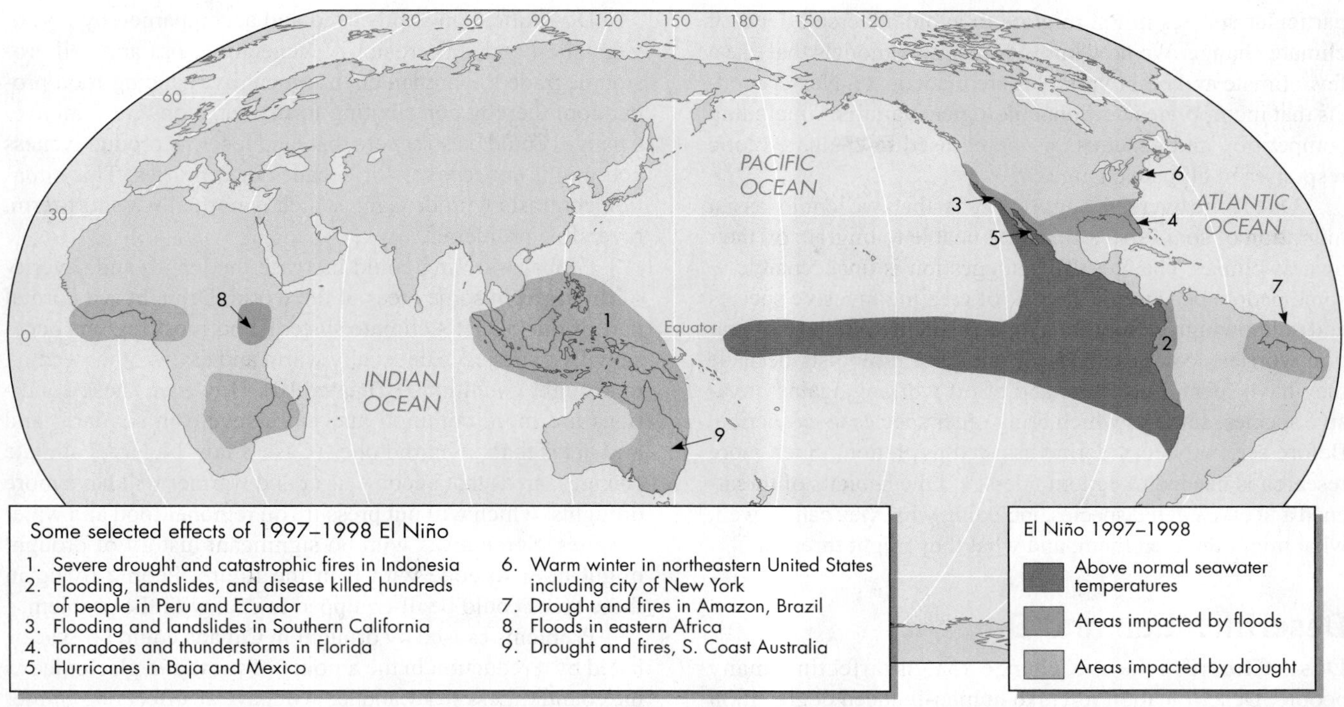

▲ **FIGURE 14.24 THE 1997–1998 EL NIÑO EVENT** The map shows areas affected by the strong El Niño event in 1997 and 1998, and the event's effects. *(Data from National Oceanic and Atmospheric Administration, 1998)*

the acidity of seawater. *Ocean acidification* results from large amounts of atmospheric carbon dioxide that are dissolved in the water (Figure 14.25). Corals and other reef invertebrates have difficulty building calcium carbonate skeletons in such waters[58] and could die on a large scale if the amount of CO_2 dissolved in seawater increases.[51] The widespread death of corals would be devastating because coral reefs are second only to rainforests in biodiversity, and reefs protect many coastlines from erosion. Overall, the projected effects of climate change on the biosphere could be tremendous, with 30 percent of all plant and animal species at risk of extinction.[51]

One problem with past evaluations of threats to species from climate change is that the evaluations were based on simple models based primarily on temperature and precipitation and, in some cases, soil type and hydrology. This information was then put into a standard climate model to predict where

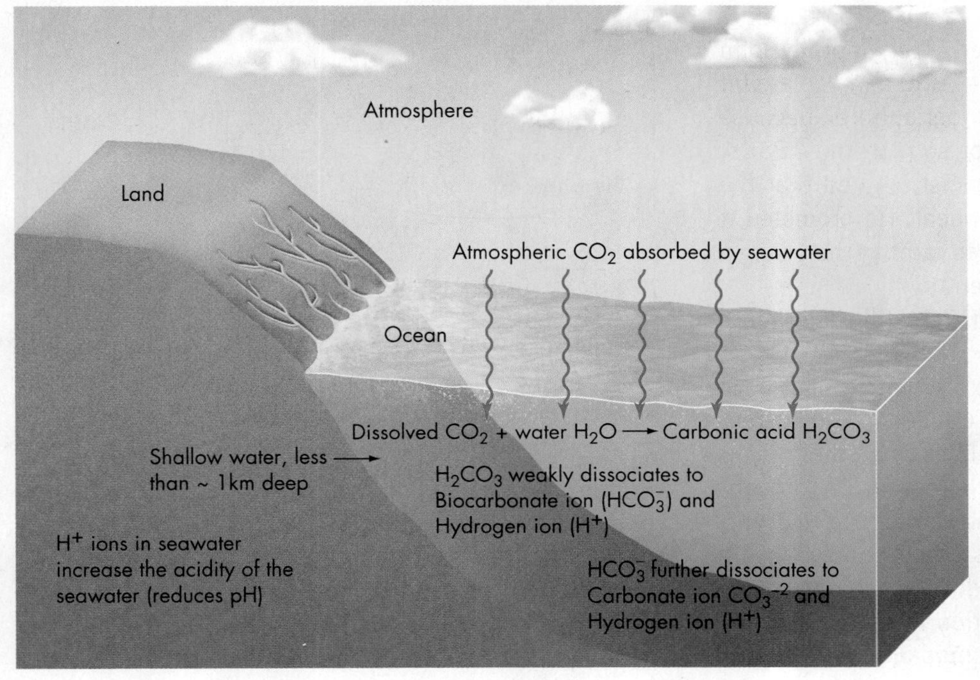

◀ **FIGURE 14.25 CHANGES IN OCEAN WATERS** Idealized diagram showing processes that change the acidity of seawater.

particular species might migrate to avoid adverse effects of climate change. We now know that simple models that use a few climate and other variables are inadequate. Newer models that incorporate additional biological variables, including competition and genetics, are being used to evaluate biotic responses to climate change.[59,60]

One controversial suggestion is that we could assist migration of some species that are unable to migrate on their own as climate changes. This suggestion is unacceptable to some people because of the risk of creating invasive species and unknowingly compromising ecosystems. Assisted migration worries ecologists and conservation biologists because they have spent much time and effort working against invasive species, some of which bring other species to extinction. Before seriously considering assisted migration, much more research is needed to better understand the habitats of threatened and endangered species, including what they can do well, what might threaten them, and what they might threaten.[59]

Desertification and Drought

Desertification is one change that is affecting many people. Desertification refers to human-induced degradation of productive land, leading ultimately to a more desert-like state.[61] Climate change may exacerbate the problem in areas that are already becoming warmer and drier.

Desertification is preceded and accompanied by loss of vegetation and soil erosion.[19] Vegetation loss and soil erosion degrade the human environment by reducing food production, thereby contributing to malnutrition and famine.[19] Changes could be so severe that land loses its productiveness and might not recover for decades or centuries. This situation contrasts with drought, which is normally a short-term, reversible problem.[61]

Global warming could increase the length and severity of **drought** in some areas of the world. Drought is a normal part of our planet's climate system, and most regions occasionally experience unusually warm and dry weather leading to drought conditions (Chapter 10). However, these conditions are more common and more severe in semiarid and arid areas than in humid ones (Case Study 14.1). As climate changes, areas that become drier and warmer will have more droughts, which will put pressure on regional food and water supplies. Some areas with no significant history of drought might have to cope with it in the future. People living in these areas could be ill-equipped to deal with the problem.

Problems caused by drought in Canada could be exacerbated by a reduction in the amount of snow and glacier ice in the country. Less snow and ice will have an effect on summer streamflow in many areas, such as British Columbia. Less streamflow will negatively impact hydroelectric power generation because snow and glacier melt currently top up many

14.1 CASE STUDY

Palliser Triangle

Nearly half of Canada's agricultural production comes from the Palliser Triangle, a semiarid grassland area in southern Alberta and southwest Saskatchewan, extending south from Red Deer to the U.S. border (Figure 14.26). The area was named after Captain John Palliser, who surveyed it in 1857–1860 and declared it "forever comparatively useless for agriculture" because of the low precipitation and poor soil. In the 1870s, John Macoun, a government botanist, argued that the land would be good for growing wheat. He promoted it to immigrant farmers who began to farm in the triangle near the end of the nineteenth century.

The homesteaders struggled from the beginning. Drought intensified in 1929 and, together with poor agricultural practices, turned the area into a dust bowl in the 1930s and helped plunge Canada into the Great Depression. Farmers watched helplessly as their livelihoods blew away in clouds of fertile topsoil. The drought lasted from 1929 until 1937 and devastated 7.3 million ha, one-quarter of Canada's arable land, causing almost 14 000 farms to be abandoned.

The Prairie Farm Rehabilitation Administration of 1935 provided financial and technical support to the

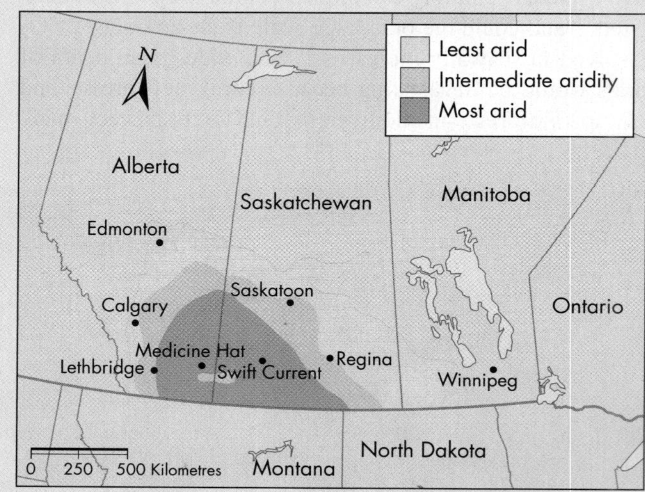

▲ **FIGURE 14.26 PALLISER TRIANGLE** The extent of the Palliser Triangle in western Canada. This drought-prone area covers parts of southern Alberta and Saskatchewan, and is notable for its lack of trees and its dry climate.

Prairie agricultural community. The agency instituted farm dugouts for watering livestock, strip farming to prevent soil drifting, the seeding of abandoned land for community pastures, and tree planting to protect the soil from wind erosion. Eventually, the new farming techniques and a series of wet years helped re-establish the Palliser Triangle as an important agricultural region, but farming has always been precarious and farmers repeatedly require government help to deal with drought conditions. During the drought of 1988, when mean annual temperature in the Palliser Triangle was 2°C to 4°C above the 30-year mean and precipitation was half the 1950–1980 average, wheat production fell by 29 percent, contributing to a $1.5-billion drop in farm receipts in Saskatchewan alone.[62]

Future sustainable agriculture in the driest parts of the region is threatened by global climate change, which is expected to result in more frequent droughts. Global circulation model simulations indicate that global warming resulting from increased greenhouse gas concentrations in the atmosphere is likely to bring warmer, drier conditions to the northern Great Plains.[63] Drought could become more frequent than it has been since the area was first settled by Canadians more than 100 years ago.[64]

Canadian researchers have gathered large amounts of evidence that allow them to reconstruct climate from thousands of years ago in the Palliser Triangle. Saline lakes in the region contain sediments that span the past 7000 years and contain microfossils and other proxy indicators of past climate. Analysis of these indicators makes it clear that the brief historical record does not capture the full range of climate variability experienced by the region. Within the past 2000 years, there have been periods, perhaps decades or more in length, that were significantly more arid than at any time in the past 100 years.[65] During these hyper-arid periods, the water table was several metres lower than today and most lakes in the region were dry. Similarly, periods of sand dune activity in the nineteenth century imply more arid conditions than exist in the region today.[66] These data highlight the vulnerability of water resources on the Canadian Prairies and raise questions about the sustainability of agriculture during droughts that are likely to occur later this century.

reservoirs during this low-flow season. Rivers and streams are likely to become warmer in summer, resulting in adverse impacts on salmon and other aquatic animals.

Wildfires

Global warming will lead to an increase in wildfires in many regions. Scientists used historical weather data and four different global circulation models to evaluate forest-fire hazard in Canada and Russia under a warmer climate. The study suggested that forest fires will be more frequent and severe and that the number of years between two successive fires will decrease.[67]

More frequent and more severe wildfires will also have ecological impacts. Large wildfires could facilitate or accelerate replacement of ecosystems that are poorly adapted to a warmer climate with ones that can tolerate more warmth and drier conditions. For example, as climate warms, grassland and scrubland ecosystems could expand at the expense of coniferous forests in the forest–grassland transition zone in Canada and the United States.

14.8 Minimizing the Effects of Global Warming

Global warming is one of the most challenging problems facing humanity because of its magnitude and global extent, the social and economic problems it creates, and the scale of the required adjustments. It directly affects our use of energy, food, and shelter; our health and safety; national security; and the survival of many species. Unlike most hazards, which produce rare, episodic events (such as earthquakes, volcanic eruptions, and floods), climate change is a slow, but continuous process that is being greatly amplified by human activities. Because of our confidence that current global warming is due to rising emissions of greenhouse gases, we can focus on reducing those emissions. Most scientists believe that we have a decade or so to significantly reduce greenhouse gas emissions if we are to avoid disastrous effects from climate warming.

A reduction in greenhouse gas emissions will not come quickly or easily. Behavioural, regulatory, and technological changes will have to be made at individual, community, national, and international levels (Table 14.4). Resistance to these changes is high, especially from individuals, businesses, and nations heavily invested in fossil fuels. But others will welcome these changes as opportunities for economic growth and development are created (Table 14.5). Economists estimate that reductions in greenhouse gases to safe levels will cost between 1 and 3 percent of the world's gross economic output in 2030.[11] To transition from fossil fuels to other types of energy, we will need binding international agreements to reduce greenhouse gases. We also must capture carbon dioxide before it enters the atmosphere and remove some of the CO_2 that is currently there. Individuals will have to change their attitudes toward fossil fuels and conservation. Only if citizens of our global community work together will we be able to reduce the inevitable warming and the problems it creates.

Most anthropogenic carbon dioxide is produced by the burning of fossil fuels. Therefore, energy planning that relies more on alternative energy sources, such as wind, solar, or geothermal power, will lead to lower CO_2 emissions into the atmosphere. A greater use of nuclear energy would also

TABLE 14.4 Individual Actions to Reduce Greenhouse Gases

- Drive less often and use mass transit or bicycles, or walk where possible.
- Make energy efficiency the top priority in purchasing vehicles, appliances, and electronics.
- Use renewable energy sources where possible.
- Buy food and other products with renewable, recyclable, or reduced packaging.
- Turn the thermostat down in the winter and up in the summer.
- Buy locally produced food and other products.
- Replace incandescent light bulbs with compact fluorescent bulbs.
- Make recycling and energy conservation priorities.
- Support businesses and politicians who act to reduce greenhouse gases and energy use.
- Turn off lights and power strips for appliances and computers when not in use.

Source: Data modified from Intergovernmental Panel on Climate Change. 2007. "Summary for policy makers." In S. Solomon, D. Qin, M. Manning, Z. Chen, M. Marquis, K. B. Averyt, M. Tignor, and H. L. Miller (eds.). *Climate Change 2007: Mitigation of Climate Change. Contribution of Working Group III to the Fourth Assessment Report of the Intergovernmental Panel on Climate Change.* New York, NY: Cambridge University Press. http://www.ipcc.ch

TABLE 14.5 Key Approaches to Reducing Greenhouse Gases

Sector	Technology or Practice
Energy	Carbon capture and storage of emissions Fuel switching from coal to natural gas and nuclear energy Greater use of renewable energy (e.g., solar, wind, geothermal, hydropower)
Transportation	Increased fuel efficiency in all forms of transportation Shift from road to mass transit Increased hybrid, electric, and non-motorized (e.g., foot, bicycle) transportation
Building	Increased energy and lighting efficiency Improved solar design and insulation Intelligent metering systems with feedback to automated controls
Industry	Improved heat and power recovery Material recycling and reuse Carbon capture and storage for cement, ammonia, and iron production
Agriculture	Reduction of methane and nitrous oxide emissions Increased soil carbon storage Improved energy efficiency and crop yields
Forest Management	Plant trees and reduce deforestation Increase biomass productivity to replace fossil fuels
Waste Management	Increased recycling and waste minimization Reduced methane emissions Improved energy recovery from landfills and incinerators

reduce the requirement for carbon-based fuels,[68] but problems with nuclear power plant safety and waste disposal have led to a great deal of resistance in many countries to adopting this alternative source of energy. Curbing greenhouse gas emissions is expensive and could be difficult or impossible for developing nations without the help of the United States, Canada, Japan, and Europe.

International Agreements

Until the late twentieth century, few international environmental agreements existed. Most governments and policymakers perceived environmental problems as being local or regional rather than global. This attitude changed in 1987, when most of the world's nations signed the Montreal Protocol on Substances that Deplete the Ozone Layer. The protocol was a successful international response to the depletion of ozone in the stratosphere by halocarbon gases, primarily CFCs such as the refrigerant Freon. Stratospheric ozone shields life on Earth from high levels of solar ultraviolet radiation. UV radiation causes sunburn and skin cancer, including melanoma, which can be fatal. The production and atmospheric level of CFCs have declined since the Montreal Protocol was implemented in 1989.[69] With time, ozone levels in the stratosphere will recover.

Could a similar international agreement be successful in reducing greenhouse gases? In 1997, the United Nations Framework Convention on Climate Change held in Kyoto, Japan, produced such an agreement, known as the *Kyoto Protocol*. This agreement established targets for nations to reduce emissions of greenhouse gases by 2012. The protocol came into effect in 2005, and by 2007, 166 countries that emit about 60 percent of the world's greenhouse gases had ratified the agreement. Canada was one of the signatories to the Kyoto Protocol, but recently withdrew after it was clear that emission targets were not being met. The United States originally agreed to reductions but in 2001 refused to honour the agreement, much to the disappointment of other nations, especially its European allies. This withdrawal has severely reduced the effectiveness of the agreement because the United States emits approximately 25 percent of the world's anthropogenic greenhouse gases. As a result, the leadership in controlling greenhouse gas emissions has shifted from the United States to the European Union.

An international summit in The Hague in 2000 focused on alternative methods of reducing greenhouse gas emissions to reach the emission targets in the Kyoto Protocol. The alternative methods include (1) clean development mechanisms by which firms and governments invest in projects or technologies that improve the "carbon balance" of developing nations, such as through reforestation to help absorb carbon dioxide; (2) joint implementation, which allows emission credits to be shared by developing states and rich countries that help the developing states achieve their goals either through investment or technology transfers; and (3) emission trading, where countries that easily meet their emissions targets can sell credits to countries that fail to meet their own. Unfortunately, The Hague conference ended without an agreement, largely because of opposition to a proposal to allow countries to earn credits against their emissions quotas for the carbon dioxide absorbed by their farmlands and forests. This proposal was supported by Canada, the United States, and Japan, but it was strongly opposed by the European Union and most environmental groups. Subsequent international meetings to address the climate change issue have produced little progress, including the most recent in Warsaw, Poland, in late 2013.

Carbon Sequestration

An important strategy for reducing greenhouse gases is to capture and store carbon dioxide before it enters the atmosphere, a process referred to as **carbon sequestration**. Carbon dioxide can be stored below ground in geologic formations, at depth in the oceans, or above ground in plants or soils. Most scientists view carbon sequestration as essential to any serious plan to address climate change. Although we face many scientific and economic hurdles in effectively sequestering carbon, none of the hurdles is serious enough to prevent the needed capture of trillions of tonnes of carbon dioxide in this century.[70]

The top priority in carbon sequestration is the removal of carbon dioxide produced by burning coal. Coal combustion produces the most CO_2 per unit energy of any fossil fuel, almost twice as much as natural gas.[70] Today, China is the greatest user of coal and the world's leading producer of CO_2.[70]

Of the three major options for sequestration—biological, oceanic, and geologic—the geologic option is the most promising. Biological sequestration, involving planting more trees, is a slower process and can remove only a small amount of the anthropogenic carbon dioxide from the atmosphere. Fertilizing phytoplankton with iron[71] also has limited potential for removing CO_2 from the atmosphere and has unknown side effects in oceans.[72] Phytoplankton are floating microscopic organisms that remove CO_2 by photosynthesis. Injecting CO_2 deep in the oceans would increase their acidity and potentially wreak havoc on marine ecosystems. In contrast, geologic sequestration is attractive because the residence time of carbon in the geologic environment is potentially hundreds of thousands of years. Furthermore, CO_2 injection has been used for decades in oil and gas fields to enhance recovery of these fossil fuels.

The general principle of geologic sequestration of carbon dioxide is straightforward. Power plants and industrial facilities must be designed to capture CO_2 before it is emitted into the atmosphere. The captured CO_2 is then compressed and injected under pressure into wells drilled into the crust. Injection may take place in depleted oil and gas fields[73] or in deep saline aquifers that have not previously been drilled for oil and gas. Injection in oil and gas fields offers the economic benefit of forcing out additional quantities of hydrocarbons. Carbon dioxide injection is currently being tested in the Weyburn oil field in Saskatchewan. The Weyburn field is receiving several thousand tonnes of carbon per day in the form of pressurized CO_2 through a 320-km-long pipeline from a coal-burning power plant in North Dakota.[74]

Statoil, Norway's largest oil company, is involved in another carbon sequestration project. Since 1996, the company has injected nearly 1 million tonnes of CO_2 per year into porous sandstone 550 to 1500 m below the sea floor. It undertook this initiative to save U.S.$40 million in carbon taxes.[75] A coalition monitoring the project estimates that the sandstone unit could store all the CO_2 produced by European power plants for the next 800 years.[76]

Safety is the primary concern with geologic sequestration. Although CO_2 is non-toxic, it can be dangerous when released in large quantities, as occurred at Lake Nyos in 1986 (see Chapter 5). Deep well injection must avoid causing earthquakes like those produced by injection wells in Denver, Colorado, in the early 1960s (see Chapter 3). Fortunately, there are many deep saline aquifers in parts of Canada and the Gulf Coast region that can be used for geologic sequestration.

Fossil Fuels and Future Climate Change

We are dependent on carbon-based fuels in part because their production and consumption are a major part of the world economy. The global economy cannot function at present without coal, natural gas, and oil. However, two possible scenarios for global warming and sea-level rise during the next 100 years demonstrate the importance of reducing our dependency on fossil fuels by transforming our global economy into one based on other energy sources. Both scenarios assume economic growth, with population peaking in the middle of the twenty-first century and declining thereafter. In the first scenario, more efficient energy technologies are introduced, but the energy system remains fossil-fuel intensive. Under this scenario, the average temperature on Earth will rise by about 4.5°C and sea level might rise as much as 1.4 m. In the second scenario, the economic system has changed, with a reduction in material consumption and the introduction of clean resource-efficient technologies. In this scenario, the average temperature will rise by about 2°C and sea level will rise less than 0.5 m.

All indications suggest that humanity is following the first scenario, with potentially grave consequences. Consumption of fossil fuels continues to rise, largely because of increasing consumption in India, China, and other countries. Efforts to bring alternative sources of energy into widespread

A CLOSER LOOK 14.1

Abrupt Climate Change

Abrupt climate change is defined as a large-scale change in the global climate system that takes place over a few decades or less. Rapid changes are thought to have occurred in the geologic past, but little is known about them or why they happened. Such changes could persist for decades or longer and substantially disrupt human and natural systems.[77] Types of abrupt climate change that are a serious risk to humans and the natural environment include:

- A rapid change in sea level
- Droughts and floods resulting from widespread rapid changes in the hydrologic cycle
- An abrupt change in the pattern of circulation of water in the Atlantic Ocean
- A rapid release of methane to the atmosphere from thawing permafrost or ocean sediments

The first issue is a rapid rise in sea level. We know that even a small rise in sea level can be very damaging. Present climate models do not adequately capture all aspects of sea-level change resulting from melting glacier ice and warming seawater. A concern exists that projections of future sea-level rise, as presented by the International Panel on Climate Change,[11] might be too low.

The second issue is the potential for abrupt changes in the hydrologic cycle. Of particular concern are changes that affect water supply, especially through protracted, severe droughts. It is now apparent that droughts can develop faster than people and society are able to adapt to them, and droughts that last several years to more than a decade have serious consequences. Scientists have pointed out that long droughts have occurred in the past and are likely to recur in the future, even in the absence of global warming.[77]

The third issue centres on changes in thermohaline circulation in the Atlantic Ocean. An ocean current carries warm water to the North Atlantic, where the water becomes cooler and sinks and moves south. This warm water makes Northern Europe habitable. The strength of the current could decrease as climate warms, although it is unlikely that the thermohaline conveyor belt will change markedly during this century.[77]

The fourth issue is whether there will be a rapid change in atmospheric methane. Methane is a potent greenhouse gas that, at higher concentrations in the atmosphere, would accelerate global warming. A very rapid change in the release of methane from thawing organic sediments or frozen methyl hydrates in marine sediments during the next 100 years is very unlikely. However, continuing warming will increase emissions of methane from wetlands at high latitudes. These wetlands are more susceptible to methane emissions because warming is greater at high latitudes than elsewhere and because, although currently frozen, these wetlands could thaw as climate warms. It thus appears that, in future decades, atmospheric methane levels will increase and cause additional warming.[77]

In conclusion, it appears unlikely that climate will change abruptly in the near future, and thus we will have some time to avert disaster. However, time is growing short and a serious response to global warming from all countries is necessary in the near future.

A better understanding of past abrupt changes in climate can be achieved by collecting and analyzing geologic data. Data obtained from sediments and glacier ice, along with monitoring, are helping scientists understand the causes of long-term changes in climate. Armed with such information, we will be better able to forecast future sea-level changes, severe droughts, and changes in ocean and atmosphere circulation patterns.

use have been resisted by some governments, some companies, and many people. The global population continues to rise and, barring some catastrophe, will probably reach 9 billion by the middle of the twenty-first century.

14.9 Adaptation

Scientists argue that, even with altered human behaviour, climate will continue to warm. This inevitability should not be used as an excuse for doing nothing, because the amount of warming will depend on how rapidly we make the transition from a carbon-based economy to one based on other sources of energy, such as hydrogen, wind, and solar power. It does, however, require that governments and people adapt to the changes in climate that are coming. **Adaptation** is an alteration in individual or group behaviour in response to a change in external conditions, such as climate change.

A major shift from fossil fuel energy to alternative sources that produce less CO_2 will not happen overnight. Rather, it will require a transition period, perhaps decades. And even after we reduce carbon dioxide emissions, there will be a considerable lag time for CO_2 levels to stabilize in the atmosphere and for global temperatures and sea level to stabilize (Figure 14.27). Yet evidence from glacier ice cores indicates that significant climate change can occur abruptly, possibly within a decade (see A Closer Look 14.1).[78] If natural or human-induced warming were to occur quickly, people would find it difficult to adapt and the consequences would be serious. It is difficult to imagine the human suffering that might result late in this century from a quick change to harsher conditions at a time when there are several billion more people on Earth to feed.[79]

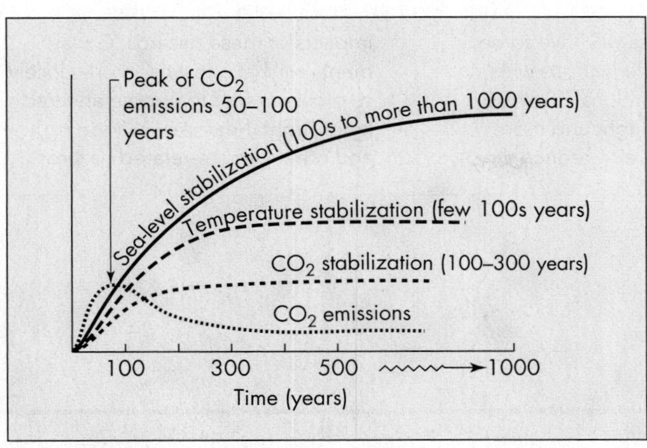

◄ **FIGURE 14.27 CLIMATE STABILIZATION WILL TAKE CENTURIES** It could take several centuries for atmospheric carbon dioxide levels; air, ocean, and land temperatures; and sea level to stabilize once CO_2 emissions are reduced. *(Based on International Panel on Climate Change. 2001. "Synthesis report: Summary for policy makers." http://www.grida.no/climate/ipcc_tar/vol4/English/pdf/spm.pdf)*

REVISITING THE FUNDAMENTAL CONCEPTS

Climate Change

1. **Hazards can be understood through scientific investigation and analysis.**
2. **An understanding of hazardous processes is vital to evaluating risk.**
3. **Hazards are commonly linked to each other and to the environment in which they occur.**
4. **Population growth and socioeconomic changes increase the risk from natural hazards.**
5. **Damage and loss of life from natural disasters can be reduced.**

1. Climate science is an area of intense research. Geologic data on past climates are continually being acquired and used to fine-tune sophisticated computer models of future climate. We are now in a better position than ever to predict future warming of the atmosphere and its effects, including changes to the cryosphere, glacier hazards, sea-level rise, and changes in precipitation and severe storms. However, we still have a long way to go. Scientific information is being incorporated into the next assessment report of the Intergovernmental Panel on Climate Change, which will be released in 2014.
2. Risk associated with climate change can be determined if we know how much change will occur and how rapidly it will occur. For example, the impacts of sea-level rise can be calculated for a range of plausible future sea-level scenarios if the vulnerable population and infrastructure are known. The options and costs of different responses to these scenarios can then be determined and plans implemented to deal with the problem. If we are fairly confident that sea level will rise 70 cm by the end of this century, we can determine which coastal areas will be inundated or eroded. Risk associated with processes such as prolonged drought is more difficult to calculate because computer models do not yet accurately predict where and how severe droughts will be in the future. It is nonetheless an extremely important issue because widespread drought could lead to famine in areas where our food is grown. We do know from the geologic record that longer and more severe droughts than any of the historic period have occurred in some areas, including North America, in the recent geologic past.
3. Climate change is linked to many other natural hazards. As the oceans warm and sea level rises, some ocean shorelines will experience greater erosion than at present (Chapter 12). Researchers are now examining a possible link between ocean warming and the severity of hurricanes and extratropical storms. Also, as Earth's atmosphere warms, some places will become drier and others wetter, possibly leading to, respectively, periodic droughts and more frequent flooding.
4. Climate change could potentially have catastrophic effects. It is true that humans have repeatedly lived through, and successfully adapted to, major changes in climate in the past, but the human population was far smaller than it is today. With over 7 billion people living on Earth and with the complex symbiotic economic, technological, political, and social systems that characterize the modern world, the impacts of sea-level rise, drought, severe storms, and other hazardous phenomena related to climate change will be far greater in the future than ever before. As an example, a severe drought would have affected far fewer people 100 years ago than it would now.
5. The consequences of climate warming can be reduced, although not eliminated. Most scientists think that we need to address this issue sooner rather than later. Even though climate will warm gradually over the next several decades, that warming will have serious consequences. The best way to reduce the effects of climate warming is to attack the process itself and control emissions of greenhouse gases. However, this is far easier said than

(continued)

done—a reduction in greenhouse gas emissions is proving difficult for a variety of political, economic, and social reasons. Clearly we will have to adapt to some change. Some coastal areas will have to be defended, particularly those with large urban populations. Planning for prolonged drought and more severe storms can also reduce the impacts of these hazards. Governments will have to work more closely to provide relief to people affected by drought, heat waves, flooding, and other climate-related hazards.

Summary

An understanding of climate is essential for the study of natural hazards because the two are intimately linked. Climate is the characteristic atmospheric conditions within an area over a period of years to decades. Weather refers to atmospheric conditions over a much shorter timescale, typically days or weeks.

The atmosphere is the gaseous envelope that surrounds Earth and keeps it warm enough to support life. It comprises mainly nitrogen and oxygen, with smaller amounts of water vapour, argon, carbon dioxide, and other trace elements and compounds. Nitrogen, oxygen, and argon are permanent gases; their atmospheric concentrations do not vary over time. Carbon dioxide, water vapour, ozone, methane, and small droplets or particles known as aerosols are variable gases; their concentrations vary over time and they play an important role in the climate system.

Earth's atmosphere can be divided into several zones based on vertical changes in air temperature. The troposphere is the lowest of these zones and is where most weather occurs. Above the troposphere is the stratosphere, a zone of lower temperature with higher levels of ozone. Stratospheric ozone shields Earth's surface from most of the ultraviolet radiation that passes through the atmosphere. Ozone depletion has increased the incidence of skin cancer, but is not responsible for global warming.

Earth's climate has changed through geologic time. Several major glacial episodes have punctuated the past billion years of the planet's history. During each glacial episode, large ice sheets formed and decayed many times. Glacier advances were separated by warmer interglaciations, such as the one we are experiencing today. During peak glacial conditions, about 30 percent of Earth's land area was covered by glacier ice; present-day ice cover is only one-third this amount. Glaciation has many causes that operate on different timescales. On the longest timescale—tens of millions of years—glaciation is controlled by the position of the continents relative to the equator and the poles. On timescales of thousands to hundreds of thousands of years, variations in Earth's orbital pattern control the amount of solar energy reaching specific points on the planet's surface. Changes in oceanic and atmospheric circulation, volcanic activity, and energy output from the sun affect climate on the shortest timescales—that is, years to hundreds of years.

The most recent major glacial episode began early in the Cenozoic and intensified about 2.6 million years ago. Since then, Earth's mean annual temperature has fluctuated markedly. At present, Earth's temperature is rising because of a combination of natural and anthropogenic causes.

Trapping of heat by Earth's atmosphere is referred to as the greenhouse effect and occurs when water vapour and other gases in the troposphere absorb long-wave infrared radiation from the planet. Gases that contribute to this effect are called greenhouse gases. Carbon dioxide is an important greenhouse gas that has increased markedly in the atmosphere since the Industrial Revolution. The rise in global temperature since the end of the Little Ice Age in the late nineteenth century, and especially since 1980, appears to have been at least partly caused by large anthropogenic inputs of carbon dioxide and methane into the atmosphere.

Climate and natural hazards are intimately linked. Shifts in climate zones because of global warming may change the areas that are vulnerable to hurricanes, tornadoes, blizzards, drought, and wildfires. Global warming is a hazard because it can alter agricultural and weather patterns, accelerate the current rise in sea level, force plants and animals to migrate, and alter ecosystems. It could also lead to desertification of arid lands, drought, and an increase in frequency and intensity of wildfires.

Substantial scientific evidence demonstrates that climate is currently warming and that the warming is caused in part by human activities. Governments must work together to slow global warming by decreasing our dependence on fossil fuels. Weaning ourselves from hydrocarbons requires the adoption of a range of strategies, including conservation; large-scale development of alternative, non-carbon based sources of energy; and carbon sequestration.

Key Terms

adaptation (p. 418)
aerosol (p. 397)
anthropogenic (p. 395)
carbon sequestration (p. 417)
climate (p. 394)
cryosphere (p. 402)
desertification (p. 415)
drought (p. 414)
Earth system science (p. 394)
El Niño (p. 411)
forcing (p. 406)
glacier (p. 402)
global warming (p. 394)
greenhouse effect (p. 398)
greenhouse gas (p. 399)
Holocene Epoch (p. 403)
La Niña (p. 412)
Milankovitch cycle (p. 403)
ocean conveyor belt (p. 407)
Pleistocene Epoch (p. 402)
proxy data (p. 399)
weather (p. 394)

Did You Learn?

1. Explain the difference between climate and weather.
2. How does ozone differ from oxygen? Where is ozone found? Why is it important to humans?
3. What is methane? Why is it important to climate? What are the anthropogenic sources of methane?
4. What is an aerosol? How do aerosols influence climate?
5. Explain or draw the general global pattern of atmospheric circulation.
6. What are the causes of continental glaciation?
7. How can glaciers be hazardous?
8. Explain the greenhouse effect. Describe the types of radiation that are involved and the anthropogenic gases that contribute to global warming.
9. Describe how variations in Earth's orbit affect climate.
10. Explain solar, volcanic, and human forcing.
11. Name the natural hazards that might be affected by global warming.
12. What are the causes of sea-level change?
13. Describe how global warming will affect the biosphere.
14. What is desertification? How does it differ from drought?
15. Which parts of Earth are currently experiencing the greatest amounts of warming?
16. Explain how ozone depletion and climate change have been dealt with internationally.
17. What are the proposed methods for carbon sequestration? Which methods hold the greatest promise?
18. Explain El Niño and its effects.
19. What changes should individuals, the private sector, and government make to reduce greenhouse gas emissions?

Critical Thinking Questions

1. In this chapter, we discussed the possible effects of continued global warming. What do you think will be the effects of climate warming in the area where you live? Would you expect any change in risk from natural hazards because of climate warming? Think about the ways climate change might change the lives and lifestyles of people living in your area. Have you seen any evidence of climate change?
2. Assessing the cause and rates of change is important in many disciplines. Have a discussion with your parents or someone of their age and write down the major changes that have occurred in their lifetimes as well as in your lifetime. Characterize these changes as gradual, abrupt, surprising, or chaotic, or use other similar descriptive words of your choice. Were the changes local, regional, or global? Analyze the changes and consider which ones were important to you personally.
3. How do you think climate warming will affect you in the future? You can find information that will help you answer this question on the Environment Canada website (www.ec.gc.ca) and on websites of provincial, territorial, state, natural resource, environmental protection, public health, or emergency preparedness agencies. What adjustments will you have to make? What can you do to mitigate the effects of the warming?
4. Some people, for cultural, political, religious, or other reasons, do not accept the conclusion that global warming is happening and that it is primarily caused by people. What do you see as the basis for their opinions or beliefs? How do you think they might be convinced otherwise? If you share their opinions or beliefs, indicate why you do and what it would take to convince you that climate is warming due to human activities.

MasteringGeology

MasteringGeology **www.masteringgeology.com**. Looking for additional review and test prep materials? Visit the Study Area in MasteringGeology to enhance your understanding of this chapter's content by accessing a variety of resources, including **Self-Study Quizzes, Geoscience Animations, GEODe Tutorials, RSS feeds, flashcards,** web links and an optional **Pearson eText.**

CHAPTER 15

▶ **ARTIST'S RENDERING OF WILDFIRES THAT RAGED AFTER THE IMPACT OF AN ASTEROID OR COMET ON YUCATÁN PENINSULA 65 MILLION YEARS AGO** The impact of the asteroid and resulting wildfires contributed to the extinction of large dinosaurs, such as *Triceratops*, shown here. *(Alfred T. Kamajian/Scientific American)*

Impacts and Extinctions

Learning Objectives

Objects from space have bombarded our planet since its birth 4.6 billion years ago. The impact of the largest of these objects has been linked to a mass extinction of species, including the dinosaurs, 65 million years ago. We continue to face the risk of impacts from asteroids, comets, and meteoroids. Your goals in reading this chapter should be to

- Know the differences among asteroids, meteoroids, and comets
- Understand the physical processes associated with airbursts and meteorite impacts
- Understand the possible causes of mass extinction
- Know the evidence for the hypothesis that an impact produced a mass extinction at the end of the Cretaceous Period
- Know the likely physical, chemical, and biological consequences of impact from a large asteroid or comet
- Understand the risk of extraterrestrial impacts and airbursts

The Tunguska Event

Shortly before 7:00 A.M. on June 30, 1908, hundreds of people in Siberia reported seeing a blue-white fireball with a glowing tail descend from the sky (Figure 15.1). The fireball exploded above the Tunguska River valley in a sparsely populated, heavily forested area. It was later determined that the explosion had the force of 10 Mt (megatons) of TNT, equivalent to 10 hydrogen bombs. Few witnesses were close to the explosion, but its sound was heard hundreds of kilometres away and the air blast was recorded at meteorological stations throughout Europe. The blast flattened and burned more than 2000 km^2 of forest over an area more than twice the size of New York City.

A herdsman witnessed the devastation on the ground. His hut was completely flattened by the blast and its roof was blown away. Other witnesses a few dozen kilometres from the explosion reported being blown into the air and losing consciousness. They awoke to find a transformed landscape of smoke and burning toppled trees.

Russia was in the midst of political upheaval at the time of the event, and consequently there was no immediate investigation. In 1924, however, geologists working in the region interviewed surviving witnesses and determined that the blast from the explosion was heard over an area of at least 1 million km^2, equivalent in size to the province of British Columbia. Scientists returned to the area in 1927, expecting to find the impact crater of an asteroid. When they found no crater, they concluded that the devastation had been caused by an aerial explosion of a stony asteroid, probably at an altitude of about 7 km. The asteroid responsible for the explosion was later estimated to be 25 m to 50 m in diameter.[1,2]

Fortunately, the Tunguska event occurred in a sparsely populated region. If the asteroid had exploded over London, Paris, or Tokyo, millions of people would have died. Tunguska-type events are thought to occur about once every 1000 years.[3]

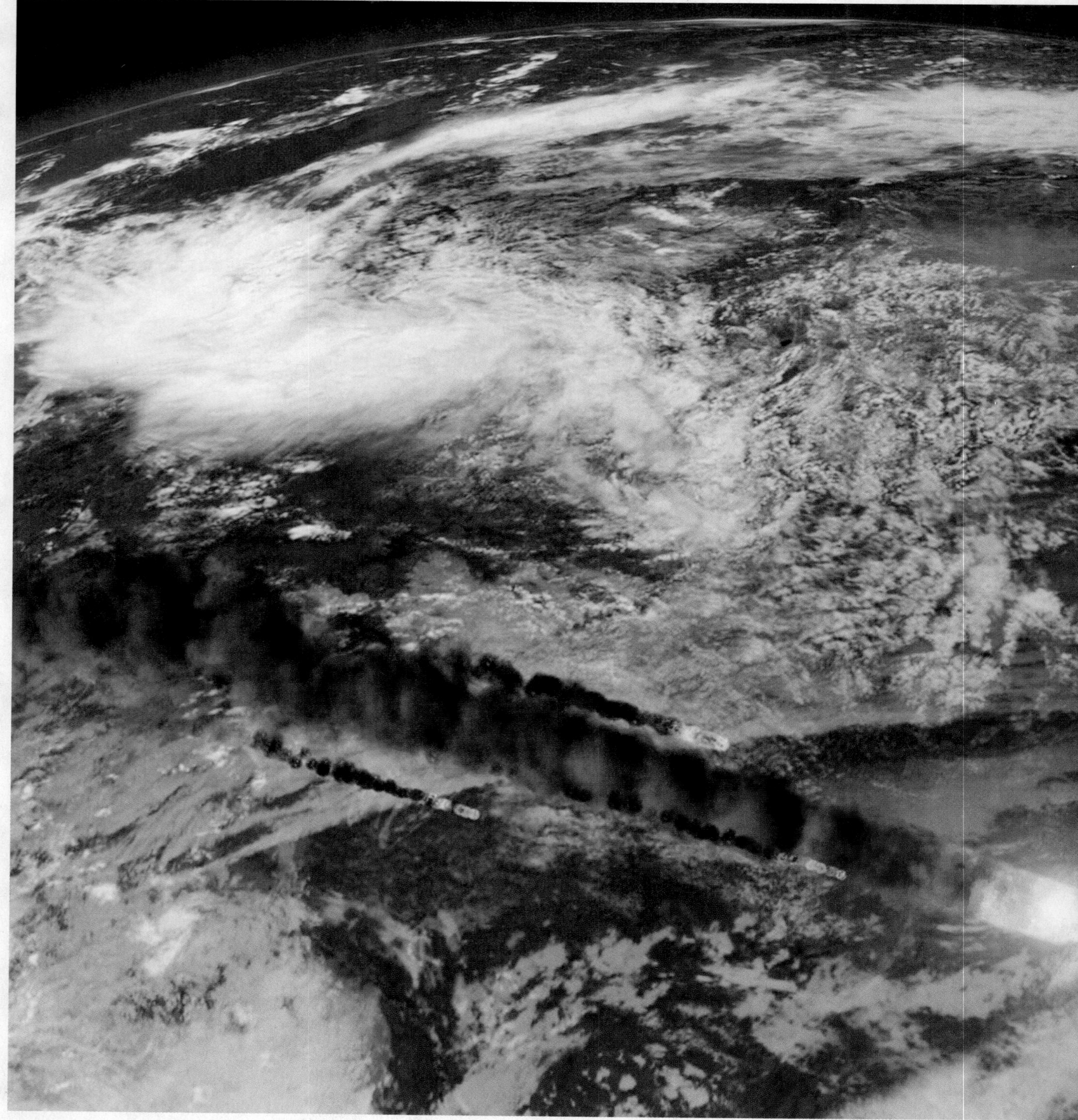

▲ **FIGURE 15.1 SIBERIAN AIRBURST** An artist's conception of a 25-m to 50-m asteroid approaching Earth just before exploding 7 km above Siberia with a force equal to 10 Mt (megatons) of TNT. The blast flattened and burned forest but left no crater. *(© Joe Tucciarone/Science Source)*

15.1 Earth's Place in Space

Preston Cloud, a famous geologist, wrote in 1978: "Born from the wreckage of stars, compressed to a solid state by the force of its own gravity, mobilized by heat of gravity and radioactivity, clothed in its filmy garments of air and water by the hot breath of volcanoes, shaped and mineralized by 4.6 billion years of crustal evolution, warmed and peopled by the sun, this resilient but finite globe is all our species has to sustain it forever."[4]

Cloud was referring to the birth of our planet about 4.6 billion years ago, but we can look much further back, some 14 billion years ago, to when the universe was created in an unimaginably large explosion known as the "Big Bang." This explosion produced the atomic particles that later formed galaxies, stars, and planets (Figure 15.2). The first stars probably date to about 1 billion years after the Big Bang, and stars continue to form today.

A star's life span depends on its mass—large stars have higher internal pressure and burn up more quickly than do small stars. Stars with a mass equal to that of our sun last around 10 billion years, whereas stars with masses 100 times that of the sun have life spans of only about 100 000 years. The sun is now a middle-aged star, about halfway through its life.

Stars die in a particularly spectacular way, releasing huge amounts of energy as *supernovas*. A supernova might have triggered gravitational collapse of a large molecular cloud from which the sun formed 5 billion years ago. The sun grew by accretion of matter from a flattened, pancake-like, rotating disk of hydrogen and helium dust called a *solar nebula* (Figure 15.3) and condensed under gravitational forces at the centre of the solar nebula. Other particles became trapped in orbits around the newly formed sun as rings, similar to the rings around the planet Saturn today. The gravitational forces of the largest, densest particles attracted other particles in the rings until they condensed to form the planets and other objects that orbit the sun (Figures 15.4 and 15.5). All planets in our solar system, including Earth, were bombarded early during their history by extraterrestrial objects ranging in size from dust particles to objects many kilometres in diameter. The bombardment occurred about 4.6 billion years ago[4,5] and is the part of the history that Cloud refers to when he states that Earth was "born from the wreckage of stars." Bombardment by asteroids and comets contributed to the growth of our planet. It did not cease to grow 4.6 billion years ago but has continued to the present, although now at a slow rate.

Asteroids, Meteoroids, and Comets

Trillions of particles remain in our solar system. Astronomers group them according to their size and composition (Table 15.1). The particles range in size from interplanetary dust a fraction of a millimetre in diameter to **asteroids** up to 1000 km across (Figure 15.6).[6] Asteroids consist of rock, metallic material, or mixtures of the two. Most asteroids are located in an *asteroid belt* between Mars and Jupiter (Figure 15.5) and would pose no threat if they remained there. Unfortunately, asteroids move around and collide with one another; some asteroids are in orbits that intersect Earth's orbit.

Smaller particles, ranging from dust to objects a few metres across, are termed **meteoroids** (Table 15.1). A **meteor** is a meteoroid that has entered Earth's atmosphere. As a meteor moves through the atmosphere, it becomes hot through frictional heating and gives off light. Spectacular *meteor showers* occur when large numbers of meteors streak across the night sky. If the object strikes Earth, we speak of it as a **meteorite**.

Comets are distinguished from meteoroids and asteroids by their glowing tail of gas and dust (Figure 15.7). They range from a few metres to several hundred kilometres in diameter and are thought to consist of a rocky core surrounded by ice and covered in carbon-rich dust. The ice includes frozen water, solid carbon dioxide (dry ice), carbon monoxide, and smaller amounts of other compounds. Astronomers believe that comets formed far outside the solar system and were thrown into an area called the **Oort Cloud** 50 000 AU (1.2 10^{14} km) from the sun (1 AU, or astronomical unit, equals 240 million km, which is the distance from Earth to the sun).[2] Comets also occur in the **Kuiper Belt** in the outer solar system (Figure 15.5).

Comets are best known for the beautiful light they create in the night sky from dust and gases released from their ice as it is heated by solar radiation. The released gas and dust form a spherical cloud around the comet. A tail of gas and dust is "blown" out from the cloud and away from the sun by the force of solar radiation, sometimes called the solar wind; thus the tail is always directed away from the sun.

The most famous of the comets that come close to Earth is Comet Halley, or "Halley's Comet." A spacecraft mission was conducted in 1986 to observe and study it. Sensors on the spacecraft found that the comet is a fluffy, porous body with very little strength. The nucleus of the comet consists of only about 20 percent ice; the remainder is empty space—a network of cracks and voids between loosely cemented materials.[1] Comet Halley crosses the space around Earth once every 75 to 76 years, giving every generation a chance to view it. It last visited us in 1986 and will next appear in mid-2061.

15.2 Airbursts and Impacts

Some asteroids and meteoroids contain carbonaceous material, whereas others are made of iron and nickel. Still others consist of silicate minerals, such as olivine and pyroxene, which are common in igneous rocks (Table 15.1).[7] Asteroids and meteoroids consisting of silicate minerals are referred to as *stony*. They are also said to be differentiated, meaning that they have experienced igneous and, in some cases, metamorphic processes during their histories.

Asteroids, comets, and meteoroids travel at velocities of 12–72 km/s (43 000–259 000 km/h) when they enter Earth's atmosphere.[1] They produce bright light as they heat up during their descent. A meteoroid will either explode in an **airburst** at an altitude between a few kilometres and 50 km, or collide with Earth as a meteorite (Figure 15.8; see Survivor Story). The Tunguska event mentioned at the beginning of this chapter was an airburst. Large numbers of meteorites have been collected from around the world, particularly from Antarctica, and more than 175 meteorite craters or small crater fields have been identified on Earth's surface.[6,8]

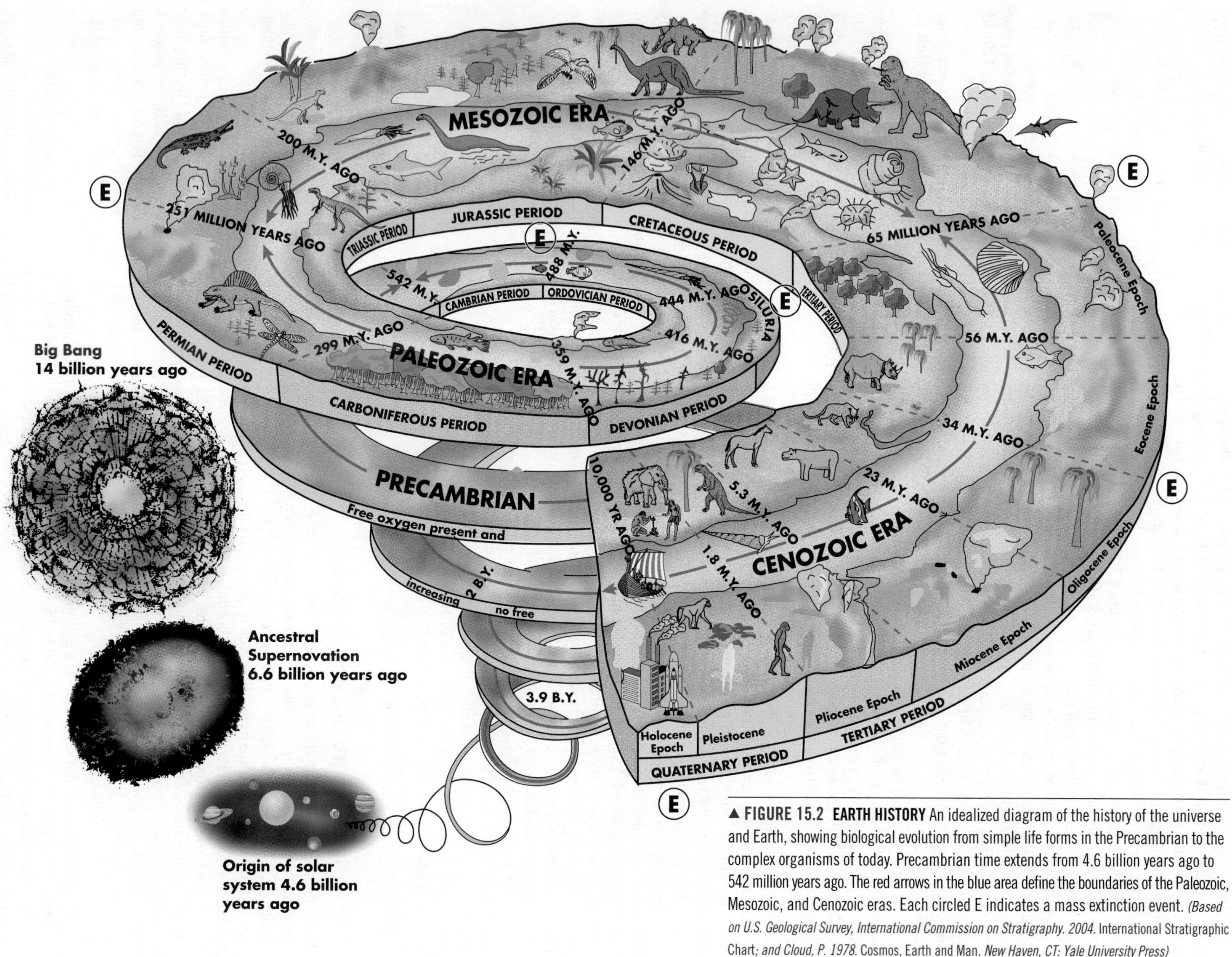

▲ **FIGURE 15.2 EARTH HISTORY** An idealized diagram of the history of the universe and Earth, showing biological evolution from simple life forms in the Precambrian to the complex organisms of today. Precambrian time extends from 4.6 billion years ago to 542 million years ago. The red arrows in the blue area define the boundaries of the Paleozoic, Mesozoic, and Cenozoic eras. Each circled E indicates a mass extinction event. *(Based on U.S. Geological Survey, International Commission on Stratigraphy. 2004.* International Stratigraphic Chart*; and Cloud, P. 1978.* Cosmos, Earth and Man. *New Haven, CT: Yale University Press)*

◄ **FIGURE 15.3 SOLAR NEBULA** Hubble telescope image of the Trifid Nebula within our own Milky Way Galaxy. This nebula consists of three bands of obscuring interstellar dust. Near the intersection of the dust banks is a group of recently formed, bright stars, which are easily seen in this image. The Trifid Nebula is about 9000 light-years from Earth. *(Science Source)*

◄ **FIGURE 15.4 THE PLANETS IN OUR SOLAR SYSTEM** Artist's depiction of the major objects in our solar system, showing the relative sizes of the sun, planets, and Pluto (a plantesmal). *(NASA)*

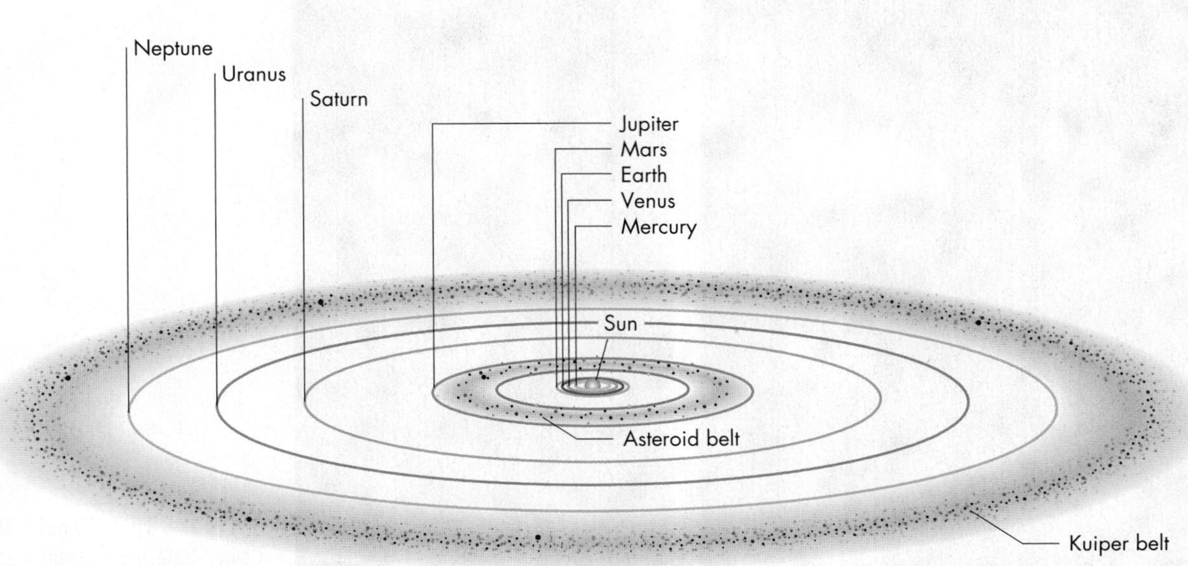

▲ **FIGURE 15.5 SOLAR SYSTEM** A diagram of our planetary system showing the asteroid and Kuiper belts. The Oort Cloud is too far from the sun to be seen in this figure. The orbits of the planets and belts are not to scale.

TABLE 15.1 Meteorites and Related Objects

Type	Diameter	Composition	Description
Asteroid	10 m–1000 km	Metallic or rocky	Metallic; some stony types are strong and hard, and may hit Earth. Weak, friable types likely will explode in the atmosphere at altitudes of several kilometres to hundreds of kilometres. Most asteroids originate in a belt between Mars and Jupiter.
Comet	A few metres to a few hundred kilometres	Frozen water, or carbon dioxide, or both with admixed small rock fragments and dust; like a dirty snowball	Weak and porous; comets generally explode in the atmosphere at altitudes of several kilometres to hundreds of kilometres. Most comets originate in the Kuiper Belt in the outer solar system and in the Oort Cloud, which is far outside the solar system, 50 000 AU[1] from the sun. The tail of a comet is produced as ices melt and gases and dust particles are shed from the object.
Meteoroid	Dust to 10 metres	Stony, metallic, or carbonaceous (carbon-bearing)	Most originate from collisions of asteroids or comets. They can be strong or weak.
Meteor	Dust to centimetres	Stony, metallic, carbonaceous, or icy	Most meteors are destroyed in Earth's atmosphere. Frictional heating produces light.
Meteorite	Dust to asteroid size	Stony or metallic	Meteorites are meteors that hit Earth's surface. The most abundant type of stony meteorite is called chondrite.[2]

[1]1 AU is the distance from Earth to the sun, about 240 million km.

[2]Chondrites contain chondrules, which are small (less than 1 mm) spheroidal glassy or crystalline inclusions.

Source: Data from Rubin, A. F. 2002. Disturbing the Solar System. *Princeton, NJ: Princeton University Press.*

◄ **FIGURE 15.6 ASTEROID 243 IDA** Numerous craters on the surface of asteroid 243 Ida show that collisions between objects in the asteroid belt are common. Ida is approximately 52 km long and orbits the sun once every five years at a distance of 441 million km. The photograph was taken by the NASA Galileo spacecraft in 1993. *(Jet Propulsion Laboratory/NASA Headquarters)*

◄ **FIGURE 15.7 COMET HALE-BOPP** Comet Hale-Bopp streaks across the night sky in 1997. Comets have a core of ice and dust and a tail of gases and dust grains. *(© Aaron Horowitz/Corbis)*

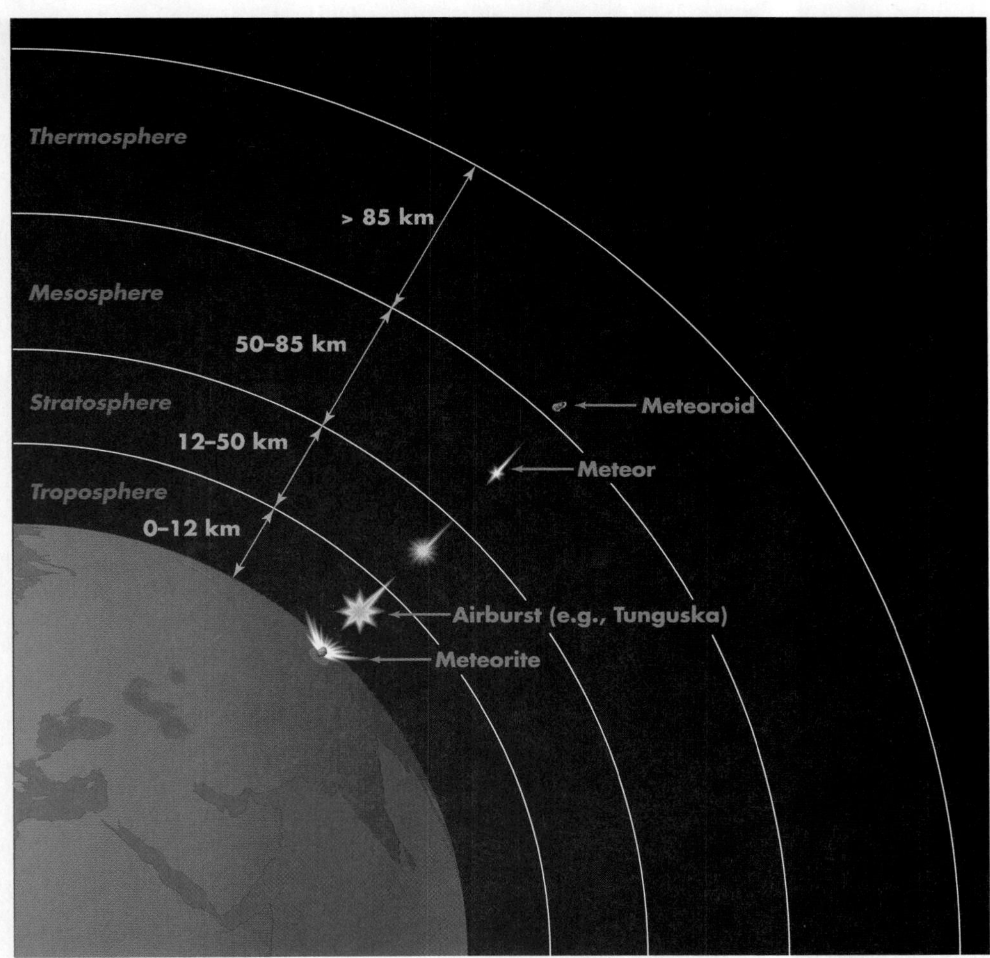

◄ **FIGURE 15.8 THE FATE OF A METEOROID IN EARTH'S ATMOSPHERE** An idealized diagram showing what happens to a meteoroid when it enters Earth's atmosphere. Meteors are small dust- to sand-sized meteoroids that emit light in the mesosphere and stratosphere. A large meteoroid may break apart in an airburst or crash into Earth as a meteorite. *(Based on R. Baldini, http://www-th.bo.infn.it/tunguska/impact/fig1_2.jpg)*

Impact Craters

The most direct and obvious evidence of the collision of extraterrestrial objects with Earth is the **impact craters** they produce (Figure 15.10).[6,8] The 50 000-year-old Barringer Crater, also known as Meteor Crater, in Arizona is an extremely well-preserved, bowl-shaped depression with a pronounced upraised rim (Figure 15.11a). The crater is 1.2 km in diameter and up to 180 m deep. The hummocky terrain surrounding the crater and rising up to 260 m above the surrounding flat Arizona desert is underlain by a layer of debris, referred to as an **ejecta blanket**. This debris layer comprises fragments of rock that were blown out of the crater on impact. The crater we see today is not nearly as deep as the initial impact crater because large amounts of fragmented rock fell back into it shortly after impact, forming a type of rock termed *breccia* (Figure 15.11b). Rocks below the crater floor were fractured, shocked, and locally melted by the impact and the extreme temperatures it generated.

The origin of Barringer Crater was debated in the late nineteenth century after its existence first became widely known. Ironically, G. K. Gilbert, the famous geologist who postulated that craters on the moon were formed by impacts, did not believe that Barringer Crater was formed in the same way. Careful field study later established that the crater was produced by the impact of a small asteroid, probably about 25 m to 100 m in diameter.[9]

Impact craters are easily distinguished from craters produced by other processes, notably volcanic activity. Impacts involve very high velocities and extreme pressures and temperatures that are not achieved with other geologic processes. Most of the energy of an impact is kinetic energy, or the energy of movement (see Chapter 10). This energy is transferred to Earth's surface through a shock wave that propagates into the uppermost part of the crust. The shock wave compresses, heats, melts, and excavates crustal rocks, producing a characteristic crater.[10] The shock wave can metamorphose rocks in the impact area, and melted material can mix with fragments of the impacting object itself. Most of the metamorphism involves high-pressure modification of minerals such as quartz. Such high-pressure metamorphism is produced only by meteorite impacts and thus is helpful in confirming an impact origin for a crater.

Impact craters can be grouped into two types: simple and complex. *Simple impact craters* are typically small—a few kilometres in diameter—and do not have an uplifted centre. Barringer Crater is a typical small impact crater (Figure 15.11). *Complex impact craters* form in the same way as simple craters, but are much larger (Figures 15.12 and 15.13). They can grow to more than 100 km across within seconds to several

SURVIVOR STORY

Meteorites in Illinois

When a meteorite shower lit up the sky on March 27, 2003, Pauline Zeilenga assumed the worst—her first thought was that it was nuclear war.

Zeilenga was in her living room with her husband, Chris, in their home in Park Forest, Illinois, when, shortly after midnight, the night sky lit up. "It wasn't like lightning," she says. "It was pitch-black outside, and then the night literally turned to day. I said, 'What the heck was that?'"

About a minute later came "the loudest thunder you've ever heard," she added. "But it kept going and going and going. And at a certain point, we realized it wasn't thunder, but an explosion."

It occurred to her that perhaps the Sears Tower had been hit, which was a reasonable thought given that the United States and its allies had shortly before begun an extended bombing campaign in Iraq.

Then, when the thunder finally subsided, the couple heard a sound like that of hail as small objects struck the exterior of their house. Chris went outside and quickly identified the culprit behind the disturbance—meteorites peppering the Park Forest region. But Pauline wasn't convinced: "I was telling my husband, find out what it was."

The area teemed with scientists and meteorite enthusiasts in the days after the event. Larger meteorites, including the ones that punched holes in the roofs of houses in the region, are worth thousands of dollars (Figure 15.9).

Overall, Zeilenga concludes, the experience was one of a kind, once the shock had subsided. "It was the coolest," she said.

—Chris Wilson

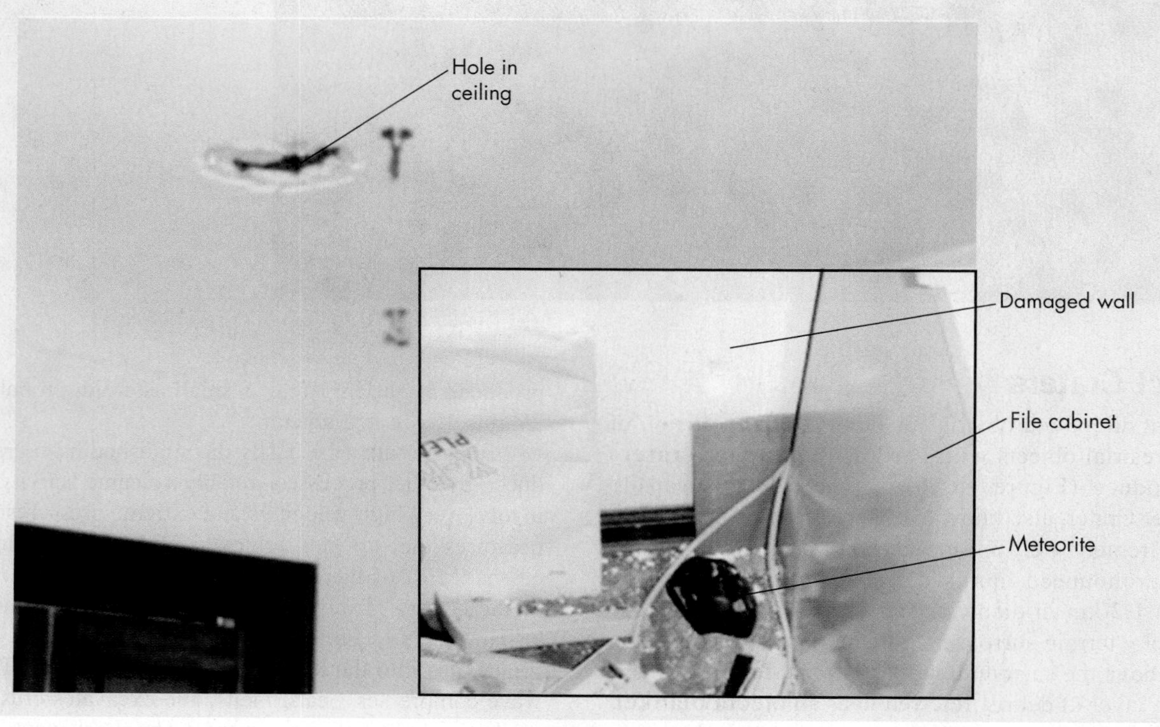

▲ **FIGURE 15.9 PARK FOREST METEORITE** A meteorite produced this hole in the house of Ivan and Colby Navarro in a suburb of Chicago on March 27, 2003. The meteorite lies on the floor in the inset picture. Colby was working at a computer when the meteorite crashed through the roof, struck the printer, banged off the wall, and came to rest next to the filing cabinet. No one was injured by any of the meteorites, but roofs, windows, walls, and cars were damaged. Park Forest is the most populated area to be hit by a meteorite shower in modern times. *(PARS International Corporation)*

minutes of the impact (see Case Study 15.1). The crater rim collapses and the centre of the crater floor rises following the impact. Most impact craters on Earth that are larger than about 6 km are complex, whereas smaller craters tend to be simple.

The Manicouagan impact structure, northeast of Quebec City, is a good example of an eroded, complex impact crater (Figure 15.10). A ring-shaped depression has been eroded in the impact breccia and is now occupied by a lake. The crater is one of the five largest impact craters in the world, with a diameter of about 100 km (Table 15.2).[8]

Old impact craters can be difficult to identify because they have been extensively eroded or filled with sediments

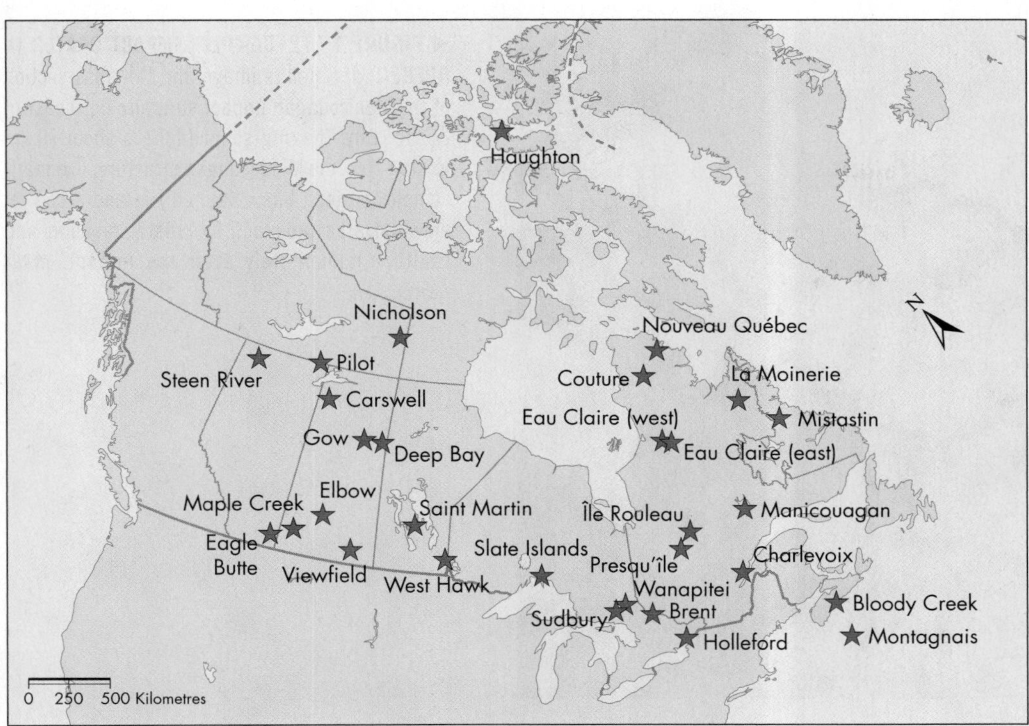

◄ **FIGURE 15.10 IMPACT CRATERS IN CANADA** This map shows locations of known large impact craters in Canada. All these craters are tens of millions to hundreds of millions of years old and thus have been extensively modified by erosion since they formed. Most of the craters are located on the Canadian Shield, an ancient part of Earth's crust. *(Reproduced or adapted with the permission of Natural Resources Canada 2013, courtesy of the Geological Survey of Canada (Bulletin 548))*

(a)

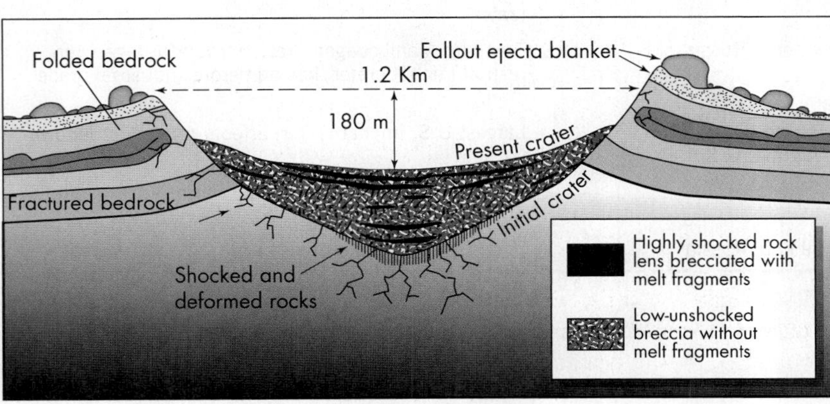

(b)

◄ **FIGURE 15.11 SIMPLE IMPACT CRATER IN ARIZONA** (a) The Barringer Crater is about 1.2 km wide and 180 m deep, and was produced by an asteroid impact about 50 000 years ago. (b) A generalized cross-section of Barringer Crater. Simple impact craters like this typically have raised rims and no uplifted centre. *((a) © Charles O'Rear/Corbis; (b) Based on Grieve, R., and M. Cintala. 1999. "Planetary impacts." In P. R. Weissman, L. McFadden, and T. V. Johnson (eds.).* Encyclopedia of the Solar System, *pp. 845–876. San Diego, CA: Academic Press)*

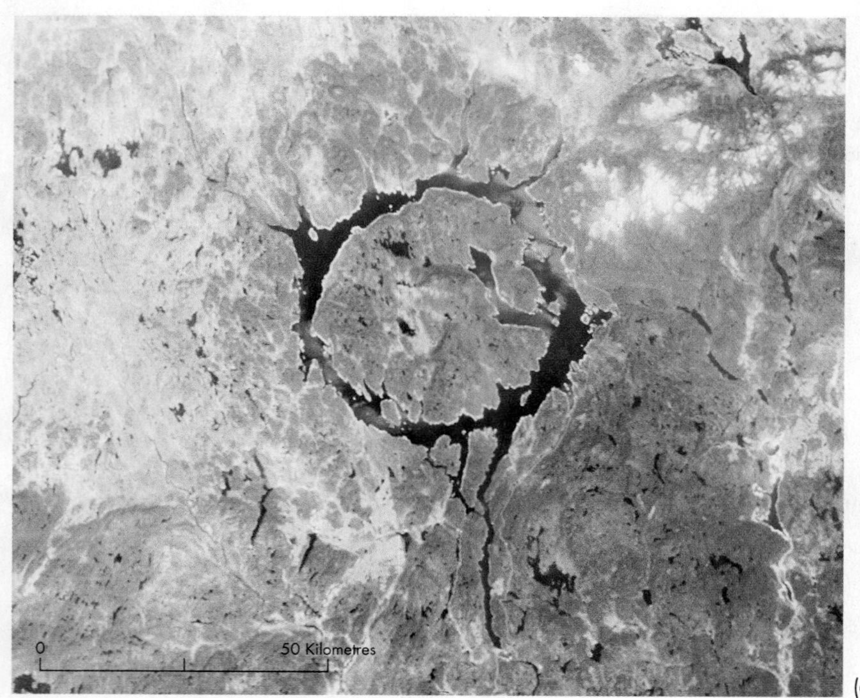

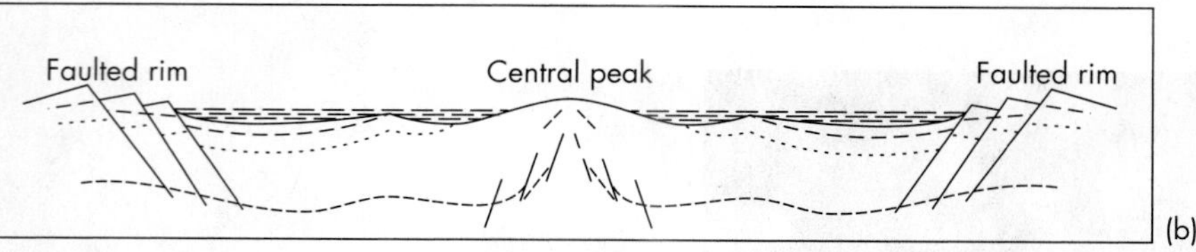

◄ **FIGURE 15.12 COMPLEX IMPACT CRATER IN QUEBEC** (a) Satellite image and (b) cross-section of the Manicouagan impact structure northeast of Quebec City. The ring-shaped lake is about 70 km in diameter. The Manicouagan structure, like many complex craters, has a faulted rim and a central elevated area produced by crustal rebound and faulting immediately after the impact. *(NASA Headquarters)*

TABLE 15.2 Notable Impacts and Airbursts of Extraterrestrial Objects

Age[1]	Feature/Event	Location	Significance
4 450 000 000	Birth of the moon	Earth	Mars-size planetary body collides with Earth to form the moon.
2 023 000 000	Vredefort Dome	South Africa	Earth's oldest and largest terrestrial impact crater.
1 850 000 000	Sudbury Crater	Ontario, Canada	Earth's second largest terrestrial impact crater; rich in nickel and copper ores.
214 000 000	Manicouagan Crater	Quebec, Canada	Tied with Popigai Crater for fourth largest terrestrial impact crater.
64 980 000	Chicxulub Crater	Below Yucatán Peninsula and Gulf of Mexico	Contributed to extinction of large dinosaurs; produced enormous tsunamis; third largest terrestrial impact crater.
35 700 000	Popigai Crater	Siberia, Russia	Tied with Manicouagan Crater for fourth largest terrestrial impact crater; has numerous industrial-grade diamonds.
35 500 000	Chesapeake Bay Crater	Buried below Virginia, U.S.A.	Largest U.S. impact crater; affects regional groundwater flow.
49 000	Barringer Crater (aka Meteor Crater)	Arizona, U.S.A.	First impact crater identified on Earth; training site for Apollo astronauts.
105	Tunguska Airburst	Siberia, Russia	Destroyed millions of trees over a 2200 km^2 area.

[1]Years before present.

Source: Based on Grieve, R. A. F. 2006. Impact Structures in Canada. *GEOtext 5. St. Johns, NF: Geological Association of Canada.*

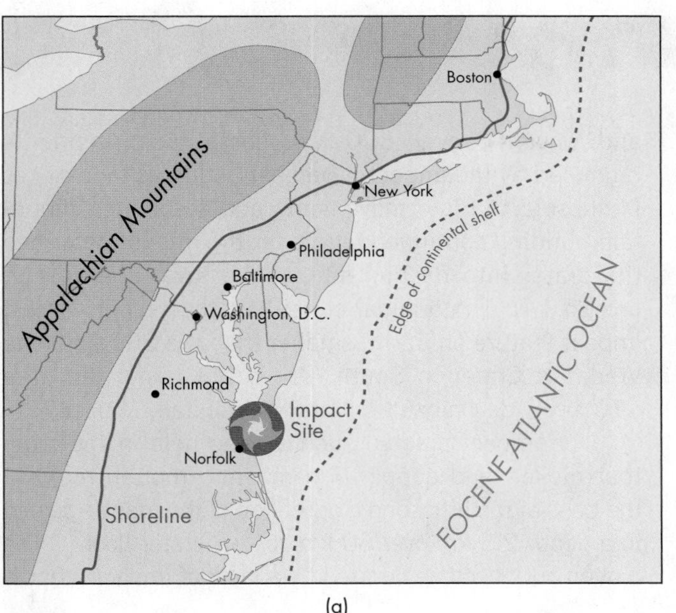

(a)

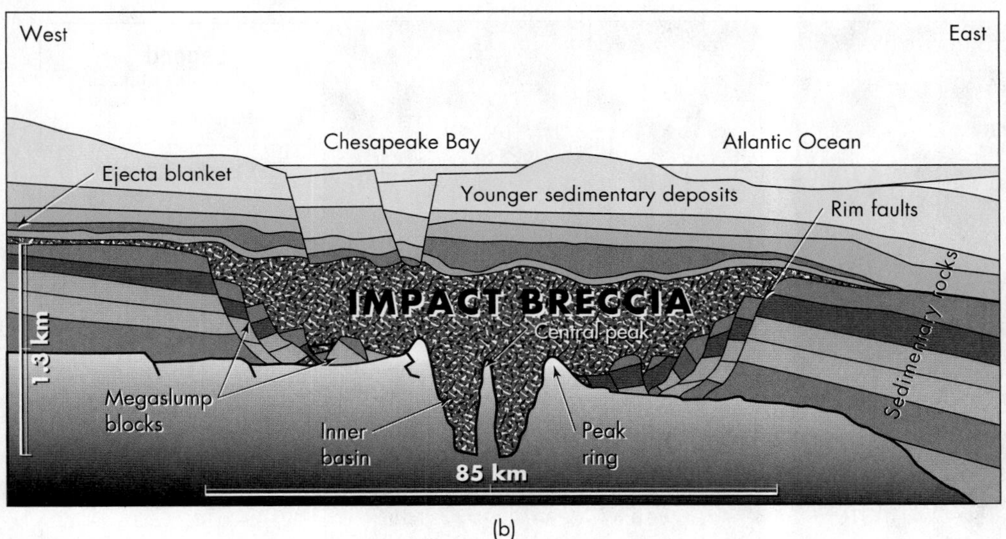

(b)

◄ **FIGURE 15.13 CHESAPEAKE BAY IMPACT CRATER** (a) A map showing the impact crater at the mouth of Chesapeake Bay. The crater formed about 35.5 million years ago when the shoreline of the Atlantic Ocean was located between the Appalachian Mountains and the U.S. cities labelled on the map (solid blue line). The dashed blue line is the seaward edge of the continental shelf 35 million years ago. (b) A cross-section of the Chesapeake Bay crater; note the 13 times vertical exaggeration of the cross-section. The crater is approximately 85 km wide and 1.3 km deep. Cemented angular rock fragments (breccia) cover most of the crater floor. U.S. Geological Survey personnel discovered this buried crater during a groundwater investigation. *(Williams, S., P. Barnes, and E. J. Prager. 2000. U.S. Geological Survey Circular 1199)*

that are younger than the impact event. Subsurface geophysical imaging and drilling in Chesapeake Bay off Virginia revealed a crater about 85 km in diameter buried beneath about 1 km of sediment (Figure 15.13). The crater was produced by the impact of a comet or an asteroid 3 to 5 km in diameter about 35.5 million years ago.[11] Compaction and faulting of sediments filling the crater may be partly responsible for the location of Chesapeake Bay.

Impact craters are much more common on the moon than on Earth for three reasons. First, most impact sites on Earth are in oceans where craters either were not produced or were buried by younger sediments and later destroyed by plate tectonic processes. Second, most impact craters on land have been eroded or buried and thus are more subtle features than those on the moon. Third, smaller meteoroids and comets burn up and disintegrate in Earth's atmosphere before striking the surface of Earth.

In 1993, Gene and Carolyn Shoemaker and David Levy discovered a comet circling Jupiter from photographs they had taken through a telescope in Southern California. Less than a year and a half later, this comet, which was named Shoemaker-Levy 9, exploded in one of the largest impacts ever witnessed. Shoemaker-Levy 9 was unusual in that it consisted of 21 fragments, many with bright tails. From telescopes on Earth and the Hubble Space Telescope in Earth's orbit, astronomers watched as the fragments of the comet entered Jupiter's atmosphere at speeds of 60 km/s. Each fragment exploded, releasing 10 000 Mt to 100 000 Mt of energy, depending on its size. More energy was released during this event than could be produced by the simultaneous detonation of all nuclear weapons on Earth. Hot compressed gases expanded violently upward from the lower part of Jupiter's atmosphere at velocities of up to 10 km/s. Gas plumes from the larger impacts rose more than 3000 km above the planet's surface, a height about 340 times the height of Mt. Everest. Extraordinarily large rings developed in Jupiter's atmosphere around the impact sites (Figure 15.15). The width of the rings

15.1 CASE STUDY

The Sudbury Impact Event

The world's richest nickel-copper mining district, and a huge contributor to Canada's economy in the twentieth century, is the Sudbury Basin in northern Ontario. The total value of metal production, from 1886 when the first mines in the Sudbury area opened until 2013, is about $350 billion.

Measuring 62 km long and 30 km wide, the Sudbury Basin owes its origin to the catastrophic impact of a 10-km meteorite 1.85 billion years ago at what is now Sudbury.[12] This impact was cataclysmic—it created a crater some 250 km across and up to 15 km deep. Ejecta was scattered over an area of 1.6 million km^2 and travelled over 800 km; some rock fragments produced by the impact have been found as far away as Minnesota.[12] Plate movements and associated mountain building continued long after the impact deformed the crater into its current, smaller oval shape (Figure 15.14). The original crater was the second largest impact feature on Earth, smaller than the 300-km-wide Vredefort Crater in South Africa, but larger than the 170-km-wide Chicxulub crater in Yucatán, Mexico.[13]

The impact created such a deep hole in the crust that nickel- and copper-rich magma (molten rock) at the base of the lithosphere or the top of the mantle pooled to a depth 2.5 km over 60 km^2 of the crater floor.[14] The molten rock became covered by a layer of impact breccia

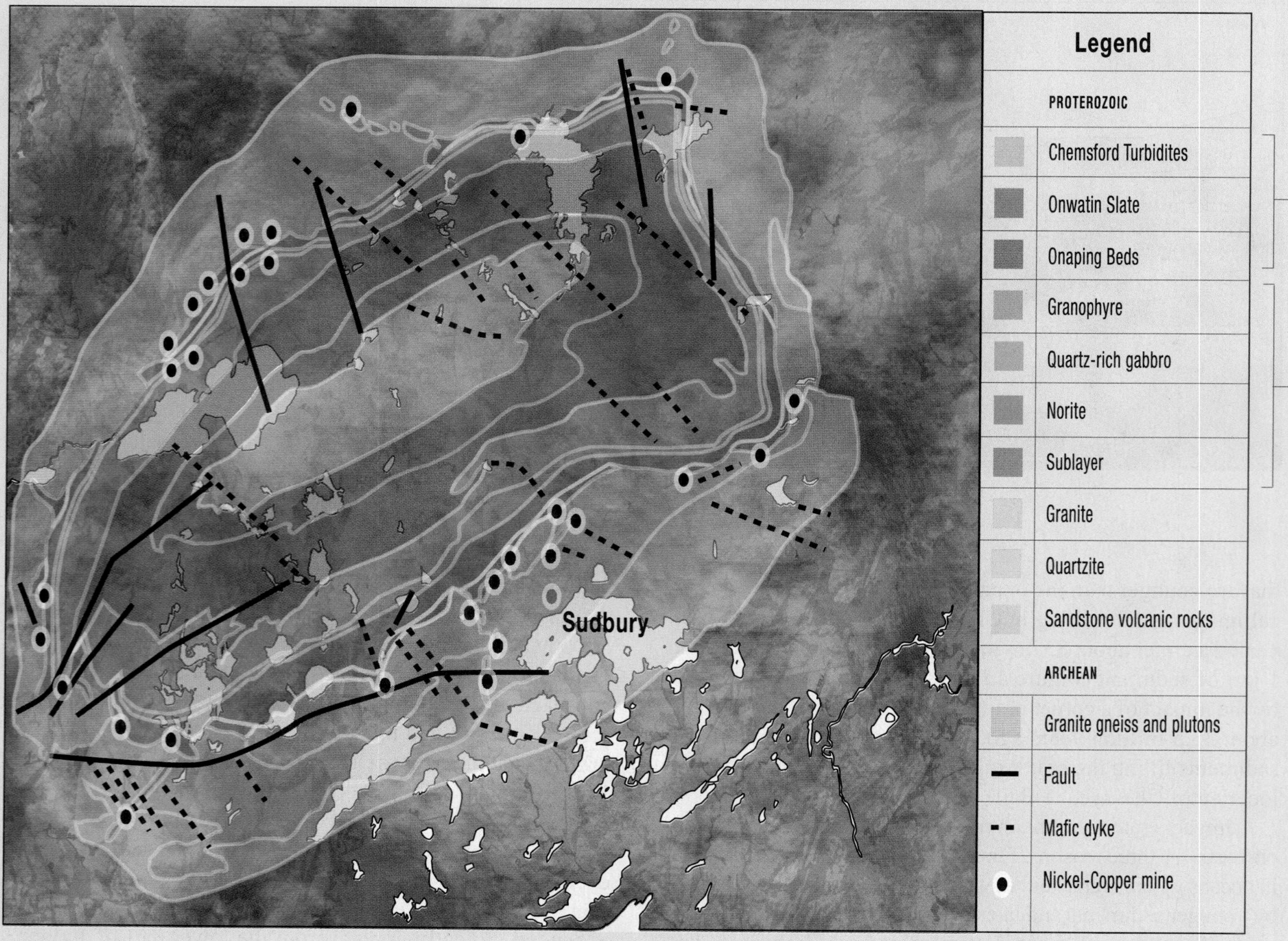

▲ **FIGURE 15.14 GEOLOGY OF THE SUDBURY BASIN** A geologic map of the Sudbury Basin, the largest nickel-copper mining district on Earth. The basin is the product of a huge meteorite impact about 1.85 billion years ago. The impact-produced rocks, which are known as the Sudbury Igneous Complex, host the most important metal deposits. They are overlain by the younger sedimentary rocks of the Whitewater Group. *(Based on Figure 5: Sudbury Igneous Complex: Geological Map (assembled from Naldrett, 1989; Naldrett et al., 1984; Pattison, 1979; and Shanks and Schwerdtner, 1991), Mineral Deposits of Canada: Magmatic Nickel-Copper-PGE deposits. http://gsc.nrcan.gc.ca/mindep/synth_dep/ni_cu_pge/index_e.php)*

about 3 km thick. The magma also intruded shattered host rocks, forming dykes that radiate up to 30 km from the crater and concentric dykes up to 14 km around it. The magma generated by the impact had an initial temperature of about 1700°C and took more than 300 000 years to solidify. During this time, heavier minerals settled to the bottom of the pool of magma, enriching the upper part of the pool in lighter minerals such as quartz and sodium feldspar. The lower layer, which is rich in heavy, calcium-rich feldspar, contains most of the nickel-copper deposits.[14]

The extreme pressures of the impact produced shatter cones, shock-metamorphosed minerals, micro-diamonds, and iridium anomalies. The impact possibly stirred up Earth's oceans, delivering a huge amount of oxygen to what was oxygen-deprived deep seawater.[15] The sudden oxygenation of the oceans instantly shut down deposition of marine sediments known as *banded iron formations*, which are massive deposits rich in iron oxides that accumulated at several times earlier in Earth's history when atmospheric concentrations of oxygen were low. The youngest banded iron formations are 1.85 billion years old, and, in northern Minnesota, they directly underlie a layer of Sudbury impact ejecta.[15] This association provides an intuitive connection between the Sudbury impact event and the shutdown of banded iron formation accumulation.

The mineral potential of the Sudbury Basin was first recognized in 1856, when Ontario provincial land surveyor Albert Salter located magnetic anomalies suggestive of mineral deposits while surveying a baseline west of Lake Nipissing.[14] Shortly afterward, Alexander Murray of the Geological Survey of Canada examined the area and confirmed "the presence of an immense mass of magnetic trap."[14] Due to the remoteness of the Sudbury area, Salter's discovery was initially ignored. Only with construction of the Canadian Pacific Railway through Sudbury in 1883 was the mineral potential of the area realized: Blasting connected with construction revealed a large deposit of nickel and copper ore that became the Murray Mine. Thus was triggered one of the largest mineral claim-staking rushes in Canadian history.

Ores in the Sudbury mining district are of two major types.[14] The first type, and economically by far the more important, is nickel-copper-platinum mineral deposits that formed in depressions at the base of the magma pool. The main ore minerals are pentlandite ([Fe, Ni]S) and chalcopyrite ($CuFeS_2$). Ores rich in zinc, copper, lead, silver, and gold occur in rocks overlying the impact sequence. They formed when hot fluids moved upward along faults from the melt sheet on the crater floor.

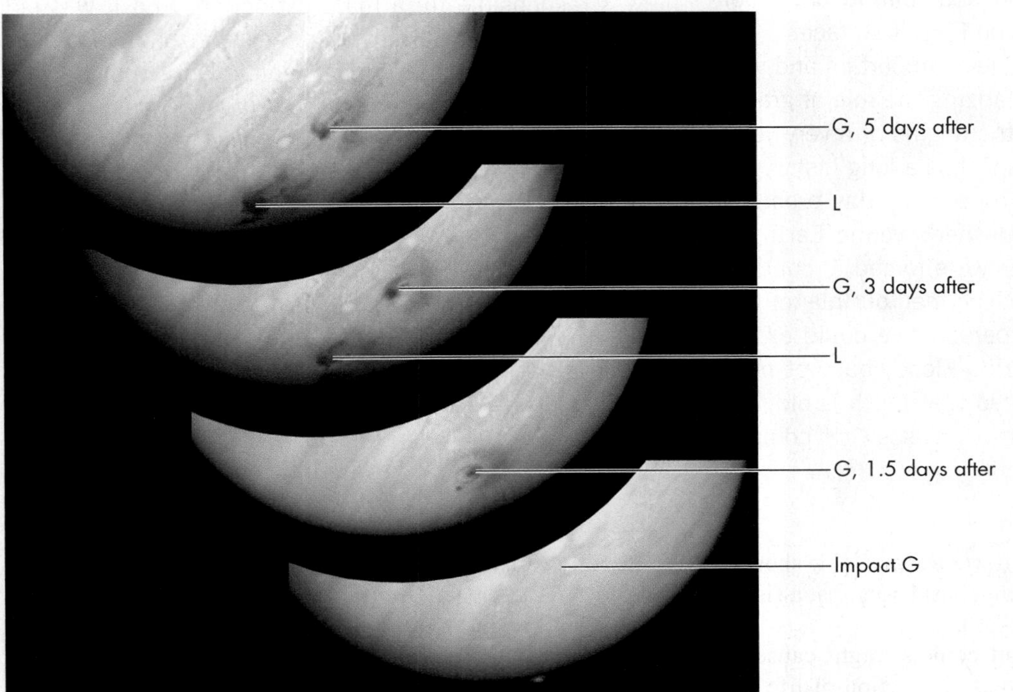

◄ **FIGURE 15.15 COMET HITS JUPITER** Spectacular impact of the comet Shoemaker-Levy 9G on Jupiter in 1994. The comet consisted of 21 fragments referred to as a "string of pearls." One fragment after another entered Jupiter's atmosphere and exploded. The four images shown here span a five-day period, from lower right to upper left. The reddish brown dots and rings mark the impact sites of fragments G and L. The width of the rings exceeds the diameter of Earth. *(R. Evans, J. Trauger, H. Hammel, and the HST Comet Science Team/NASA)*

15.2 CASE STUDY

Uniformitarianism, Gradualism, and Catastrophes

The idea that large objects from outer space could hit Earth was not widely accepted until the twentieth century, in spite of eyewitness accounts of meteorite impacts around the world. The first well-documented account may be a lethal meteorite that landed in Israel in B.C. 1420.[1] In the fifteenth century, Galileo invented the telescope and showed that the planets, including Earth, revolve around the sun. The religious establishment of the day did not accept this theory and dismissed the idea that extraterrestrial objects could strike Earth; in fact, Galileo was jailed for his beliefs.

In 1654, Irish Archbishop Ussher proclaimed that Earth was created in the year B.C. 4004. This belief was the dogma of the time and, given the power of the Church, was not to be disputed. However, people studying the formation of mountains and large river valleys could not understand how these features could form in only 6000 years. In 1785, Scottish doctor James Hutton wrote an influential book that introduced the concept of gradualism, or uniformitarianism. According to this concept, the same natural processes that operate today have produced Earth's features and rocks. The *principle of uniformitarianism* has since been paraphrased as "the present is the key to the past." Hutton also argued that Earth must be much older than 6000 years to allow the gradual processes of erosion, deposition, and uplift to form mountain ranges and other features on Earth's surface.

In 1830, Charles Lyell wrote an important and influential book on geology, popularizing the role of gradual processes and casting aside the dogma of a very young Earth. He proclaimed that Earth has a long history that can be understood by studying present-day processes and the rock record. When Ussher's young Earth was the accepted belief, scientists were forced to conclude that most of the processes that formed our planet were catastrophic in nature. This perspective could explain Biblical events, such as Noah's Flood, but not much else. Once scientists recognized that Earth is old, they reconstructed its history from processes that could be carefully observed, such as uplift and erosion. Charles Darwin was strongly influenced by Charles Lyell's book, and he applied the ideas of an ancient Earth and uniformitarianism to his concept of biological evolution. The concept of gradualism became widely accepted in the twentieth century, culminating in the 1960s with the discovery that lithospheric plates move with respect to one another. The theory of *plate tectonics* explains the origin and position of continents through the slow processes of seafloor spreading, faulting, subduction, and uplift (Chapter 2).

However, even as plate tectonics came into ascendancy, many scientists argued that catastrophic events have also played a role in shaping Earth. Some scientists pointed to large craters on Earth's surface that they believed were the result of asteroid impacts. Others argued that infrequent, relatively rapid extinctions over the past several hundred million years were difficult to explain using gradualism. The extinctions involved the disappearance of a large percentage (commonly half or more) of the species of plants and animals existing at the time. Five mass extinctions occurred in the distant geologic past; another is occurring today as a result of human activity (Figure 15.2). A mass extinction 65 million years ago marks the Cretaceous-Tertiary boundary and was probably caused by the impact of an asteroid approximately 10 km in diameter beneath what is now Yucatán Peninsula in Mexico. Even faced with evidence for the impact, scientists were initially skeptical that it was responsible for a mass extinction. And it was not until 1947 that Barringer Crater in Arizona was finally listed as a probable impact crater. Until that time, it was commonly referred to as a cryptovolcanic feature. Eventually, Barringer Crater was shown to be the result of an asteroid impact, and today more than 175 large impact features have been identified on Earth.

Scientific acceptance of the importance of catastrophic events in shaping Earth's history has led to a new concept, *punctuated uniformitarianism*. According to this concept, uniformitarianism explains the long geologic evolution of the lithosphere, but catastrophic events are responsible for mass extinctions and thus have played an important role in the evolution of life.

exceeded the diameter of Earth! It was a remarkable show for astronomers and a sobering event, considering that a similar impact might one day occur on Earth.[2]

The idea that asteroids and comets might cause catastrophes on Earth, and even mass extinction of life, was strongly resisted by scientists until very recently (see Case Study 15.2). Several bizarre ideas were suggested to explain the 1908 Tunguska event described at the beginning of this chapter, such as nuclear explosions and the explosion of an alien spaceship! This resistance is understandable, given that it wasn't until 1947 that Barringer Crater in Arizona was finally accepted as a probable impact feature. Until that time, it was commonly referred to as a *cryptovolcanic* structure, meaning that a "hidden" volcanic event had produced it. Eventually, Barringer Crater was shown to be the result of an asteroid impact, probably with an airburst. After the impact of Shoemaker-Levy 9 on Jupiter, the idea that a comet or an asteroid could strike Earth was nearly universally accepted.

15.3 Mass Extinctions

A **mass extinction** can be defined as the sudden loss of large numbers of species of plants and animals.[16] Mass extinctions coincide with boundaries of geologic periods or epochs because the geologic timescale was originally organized on the basis of the appearance and disappearance of groups of organisms (Figure 15.16). Most hypotheses to explain mass extinctions involve rapid climate change caused by movements of Earth's lithospheric plates, volcanic eruptions, or extraterrestrial impacts. Plate tectonic processes are slow, but on occasion, plate movements create new patterns of ocean circulation that have had major effects on climate. Extremely large volcanic eruptions can also cause significant climate change. Eruptions that produce huge volumes of *flood basalts*, such as the Miocene Columbia River basalts in Washington State, release large quantities of carbon dioxide into the atmosphere. Carbon dioxide is a greenhouse gas that warms climate (Chapter 14). In contrast, large explosive volcanic eruptions of more silica-rich magmas can inject tremendous quantities of volcanic ash and sulphur dioxide into the upper atmosphere, causing global cooling. Finally, climate change is one of several effects of extraterrestrial impacts or airbursts that can contribute to extinctions.

Geologists have documented five major mass extinctions during the past 550 million years, and a sixth is occurring today. The first two extinctions, which date to about 446 and 250 million years ago, could have been caused by global cooling followed by rapid warming.[17] Numerous large volcanic eruptions during the second event added huge amounts of ash and gases to the atmosphere, possibly contributing to the cooling. The third mass extinction occurred 202 million years ago and might also have been caused by volcanic activity. Eruptions at this time produced the largest amount of basaltic lavas ever released by volcanoes on land. Earth's temperatures, which were already 3°C warmer than today, increased another 3°C to 4°C due to the large release of carbon dioxide by the volcanoes.[17]

The fourth extinction coincides with the end of the Cretaceous Period and the so-called *K-T boundary* (see page 438) about 65 million years ago (the letter K is used because Cretaceous is spelled with a K in some languages). This event was sudden—abundant evidence suggests that it was caused by the impact of a large asteroid.[18] The impact brought an end to the large dinosaurs, which had been at the top of the food chain for 100 million years or more.[17] The fifth mass extinction occurred about 34 million years ago. Limited evidence suggests that an asteroid or comet could have hit Earth at that time, but most scientists link this extinction to cooling and glaciation that began about 40 million years ago.[17]

Finally, a mass extinction began near the end of the Pleistocene Epoch and has accelerated during the past two centuries due to rapid increases in human population, deforestation, agriculture, overfishing, and pollution.[16,19] North

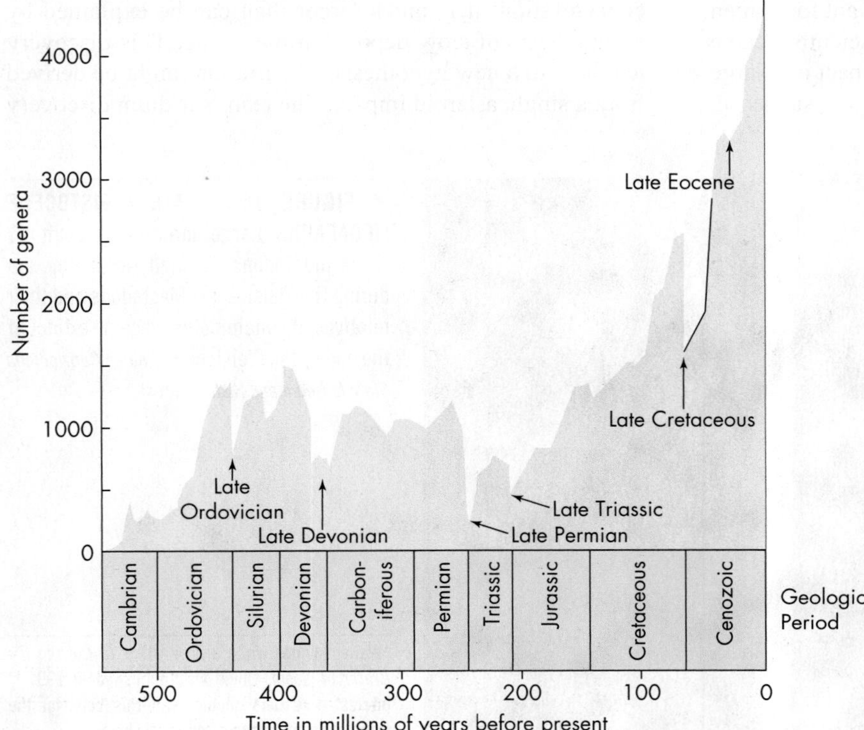

◄ **FIGURE 15.16 MASS EXTINCTION EVENTS** Diagram illustrating increasing diversification of life during the Phanerozoic Era, punctuated by mass extinction events.

and South America, Europe, Asia, and Australia lost 36 to 88 percent of all genera of large terrestrial animals near the end of the Pleistocene.[20] In North America, these animals included sabre-tooth cats, several species of ground sloths, the American lion and horse, a giant bear, and the mammoths and mastodons (Figure 15.17).[21]

Of the six extinctions mentioned above, the K-T extinction is the one most closely linked to an impact. It produced the third-largest crater on Earth, a blast shock wave, fall-out of ejecta, a large tsunami, global wildfires, atmospheric dust loading, and a collapse of food chains in the ocean and on land. Because of its importance, this event is explored in more detail in the next section.

K-T Boundary Mass Extinction

One of the great geologic detective stories of the past 50 years is the investigation of the K-T mass extinction. We now have evidence that, 65 million years ago, a comet or an asteroid smashed into Earth along the north shore of what is now Yucatán Peninsula. That event profoundly changed life and thus evolution. Dinosaurs disappeared, along with most species of plants and animals in the oceans and on land. Approximately 70 percent of all genera and their associated species died off. Some reptiles, including turtles, alligators, and crocodiles, and some birds, plants, and smaller mammals survived, due either to their habitat requirements or their location at the time of the impact. Not all areas experienced widespread wildfires, and some plants and animals were better adapted to the cooling that followed the impact.[22] The event set the stage for the evolution of mammals, which ultimately led to the appearance of primates and the genus *Homo*. What would life be like today if the K-T extinction had not happened? It is likely that humans would never have evolved!

Because the K-T extinction was so important for human evolution, we will look more closely at how scientists came to the conclusion that it was caused by the impact of a large comet or asteroid. The story is full of intrigue, suspense, rivalries, and co-operation, which is typical of many great scientific discoveries.[23] Walter Alvarez, a professor at the University of California, asked the question that started it all: What is the nature of the boundary between rocks of the Cretaceous and Tertiary periods?[1*]

In the 1970s, Alvarez developed an interest in rocks that span the Cretaceous-Tertiary boundary. With his father, who was a physicist, and some colleagues, he went to Italy to study a very thin layer of clay that marks the K-T boundary there (Figure 15.18). The extinction of many species coincided with the clay layer—fossils found in rocks below the clay were absent in the rocks above it. The scientists asked the question, How rapidly did these species disappear? Put another way, how much time did the clay layer span? Was it a few years, a few thousand years, or millions of years? The approach they took was to measure the amount of *iridium* in the clay. Iridium is a platinum-group metal found in very small concentrations in meteorites. The scientists chose this element for study because the global rate of accumulation of meteoritic dust and thus iridium on Earth is constant. Meteoric dust and iridium slowly accumulate on the deep ocean floor. Clay eroded from the continents and the remains of microscopic organisms slowly rain out on the ocean floor, diluting the iridium. The higher the rate of sedimentation, the more diluted the iridium becomes. Slow sedimentation allows time for more meteorite dust to accumulate, giving higher concentrations of iridium in deep ocean sediments.

What Alvarez and his colleagues found was entirely unexpected. They had expected to measure about 0.1 ppb (parts per billion) of iridium in the clay layer, a value consistent with slow accumulation through time. The concentration of iridium would be even less with rapid sedimentation. Instead, they found concentrations of about 9 ppb, 100 times more than they had expected. Although 9 ppb is a very small concentration, it is much larger than can be explained by a hypothesis of slow deposition over time. This discovery led them to a new hypothesis: The iridium might be derived from a single asteroid impact. The team's iridium discovery

◀ **FIGURE 15.17 LATE PLEISTOCENE MEGAFAUNA** Large mammals, such as these mastodons, roamed North America during the Pleistocene. Mastodons and their relatives, the mammoths, became extinct at the end of the Pleistocene. *(Karen Carr/Indiana State Museum and Historic Sites)*

*Walter Alvarez wrote a book titled *T. Rex and the Crater of Doom*, which was published in 1997.[23] Interested readers should read this book for the complete story summarized briefly here.

◀ **FIGURE 15.18 EVIDENCE FOR AN EXTRATERRESTRIAL IMPACT** The Cretaceous-Tertiary (K-T) boundary in Italy is delineated by a thin clay layer located below the left knee of the scientist on the left side of the photograph. The layer has an anomalously high concentration of the metal iridium. This metal occurs in high concentrations in extraterrestrial objects, such as meteorites. Its presence here is consistent with an asteroid impact 65 million years ago, at the end of the Cretaceous Period. *(Walter Alvarez)*

and their hypothesis of an extraterrestrial cause for the mass extinction at the K-T boundary were published in a journal in 1980.[24] In the same paper, they also reported elevated concentrations of iridium in deep-sea sediments at the K-T boundary in Denmark and New Zealand.

The discovery of the iridium anomaly at several places around the world added credence to the impact hypothesis, but the researchers had no impact crater and their hypothesis was criticized. The Alvarez paper, however, prompted other scientists to search for a 65-million-year-old crater. Could the crater even be found? The asteroid might have produced only a giant airburst, or it might have struck in the deep ocean and produced only a small crater. Or the crater might have been destroyed by subduction or filled with sediment and thus become obscure.

As it turned out, the K-T crater was found; it is covered by Tertiary sediments but was discovered in 1991 by geologists studying the structural geology of Yucatán Peninsula.[25] They found a nearly circular buried impact crater approximately 180 km in diameter. The boundary between nonfractured, younger rocks filling the crater and fractured rocks beneath and outside of the crater is clear. About half of the crater lies beneath the seafloor of the Gulf of Mexico; the other half underlies Tertiary sedimentary rocks on Yucatán Peninsula (Figure 15.19). A semicircular pattern of sinkholes,

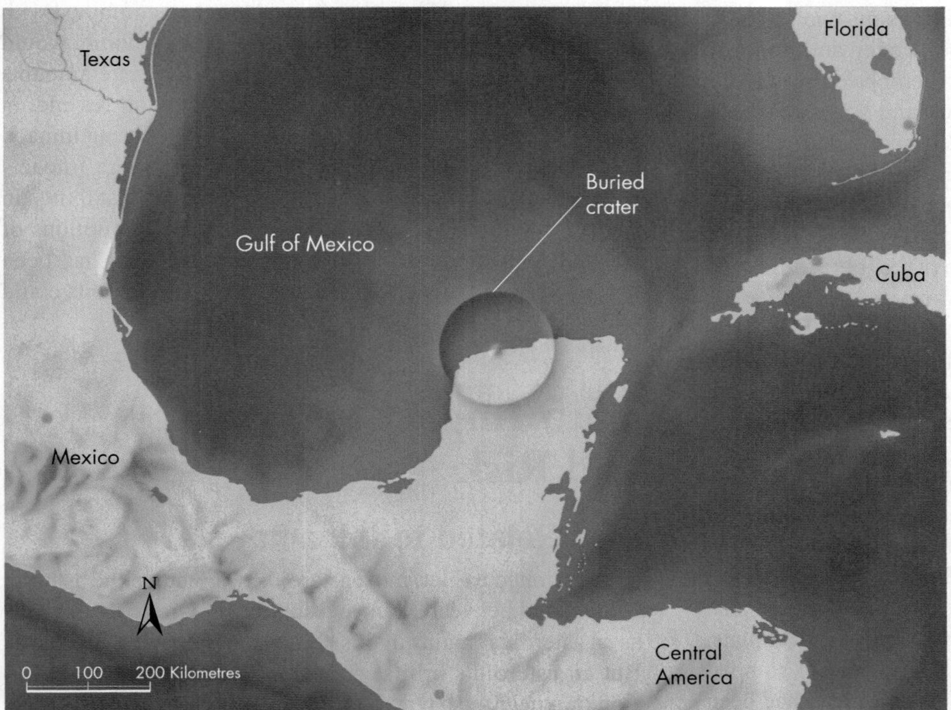

◀ **FIGURE 15.19 LARGE IMPACT CRATER IN MEXICO** A map showing the location of the Chicxulub impact crater on the north shore of Yucatán Peninsula. The crater underlies Tertiary sedimentary rocks on the peninsula and the adjacent seafloor of the Gulf of Mexico. It was produced by an asteroid impact 65 million years ago at the end of the Cretaceous Period. *(Photo Researchers, Inc./Science Source)*

known to the Mayan people as *cenotes*, delineates the edge of the impact crater on land. The cenotes are 50 m to 500 m in diameter and presumably formed by slow chemical weathering of fractured limestone at the edge of the crater. The crater was originally 30 to 40 km deep, but post-impact slumping of the crater walls filled much of the structure, and sedimentation over the past 65 million years completely buried it. While drilling within the crater, geologists found an impact breccia containing glassy melted rock at the base of the crater fill. It was clear that the force of the impact excavated the crater, fractured the rock outside it, and produced the breccia. The glassy melt rock indicates that sufficient heat was generated by the impact to melt rocks below the crater floor.[26] Another study of the crater found glass mixed with and overlain by the breccia, as well as evidence of shock metamorphism, which is commonly associated with impact structures.[27] The results of the studies of the Chicxulub crater have been published and accepted. Most scientists now believe that the asteroid or comet that struck the area 65 million years ago did, in fact, cause the K-T mass extinction.

Once the site of the crater had been identified, scientists naturally asked how such an event could cause a global mass extinction.[23] The asteroid that struck Yucatán Peninsula 65 million years ago was huge, perhaps larger than Mount Everest is high. It entered Earth's atmosphere at a speed of about 30 km/s, which is 150 times faster than a jet airliner travels. The amount of energy released during its encounter with Earth is estimated to have been about 100 million Mt, roughly 10 000 times the energy of the entire nuclear arsenal of the world.

Walter Alvarez and numerous other scientists have proposed a likely sequence of events for the impact and its aftermath (Figure 15.20). At an altitude of 10 km and moving at 30 km/s, the asteroid would have taken less than half a second to reach Earth's surface. It blasted a hole in the crust nearly 200 km across and 40 km deep. Within about two seconds, shock waves crushed the rocks in the crater and partially melted them. A huge ejecta blanket was emplaced around the crater, and a cloud of pulverized rock and gases rose in a gigantic fireball, which formed a mushroom cloud. The explosion ejected material far beyond Earth's surface, and the fireball ignited fires far from the impact site. Southern North America, all of Central America, and parts of South America, Africa, Asia, and Australia were ravaged by wildfires.[22] Vaporization of sulphur-bearing limestone and gypsum at the impact site produced sulphuric acid in the atmosphere. Nitric acid was added through oxidation of nitrogen in the atmosphere. Thus, following the impact, acid rain probably fell for a long time. Dust in the atmosphere encircled Earth, and for months little or no sunlight reached the lower atmosphere. Plants on the land and in the ocean stopped growing because of the lack of sunlight. Climate first cooled due to lack of sunlight, then warmed as aerosols and carbon dioxide from the pulverized limestone of Yucatán Peninsula enhanced the greenhouse effect. Acid rain was toxic to many organisms, particularly terrestrial and shallow marine plants and animals. As a result, the food chain was greatly disrupted or stopped functioning altogether because the base of the chain had been greatly damaged. Part of the impact occurred in the ocean and generated a tsunami with waves hundreds of metres high.[23,28] These waves raced across the Gulf of Mexico and inundated parts of North America. Geologists have found tsunami deposits in Texas and Europe.

15.4 Links with Other Natural Hazards

As we have seen, comet or asteroid impacts or airbursts trigger other hazardous natural processes, including tsunamis, wildfires, earthquakes, landslides, climate change, and possibly volcanic eruptions. Most impacts occur in the world's oceans, and large ones cause tsunamis, as was the case with the asteroid that hit Yucatán Peninsula 65 million years ago.

Wildfires of regional or global extent also resulted from this impact and from the 1908 Tunguska airburst. Superheated clouds of gas and debris reached temperatures needed to dry out and then ignite vegetation.[22] Computer simulations suggest that the pattern of wildfires following a large impact is complex, involving secondary impacts of rock blasted from the crater into space.[22] These secondary impacts occur over the entire surface of the planet, triggering fires everywhere. Fortunately, the simulations suggest that these wildfires did not burn the entire surface of the planet.

Seismic waves from a large impact cause numerous landslides both on land and on the seafloor. For example, the asteroid that formed the Chicxulub impact crater appears to have produced a magnitude 10 earthquake that triggered landslides on the continental slope of North America as far north as the Grand Banks off Newfoundland.[29]

Large asteroid impacts and airbursts cause global changes in climate. Impacts on land inject large quantities of dust and aerosols into the atmosphere that, combined with smoke from wildfires, cool climate for years. Cooling would be followed by prolonged warming from the large amounts of carbon dioxide produced by post-impact wildfires and, in some instances, by vaporization of limestone during impact.

Scientists have also hypothesized that large impacts trigger volcanic eruptions due to melting and instability in Earth's mantle.[30] Melting could result in huge eruptions of basalts, with perhaps 100 times more lava than has been produced by any historical eruption.[31] Such eruptions could alter climate and contribute to mass extinctions.

15.5 Impact Hazards and Risk

Risk Related to Impacts

As mentioned earlier, most asteroids originate in the asteroid belt located between Mars and Jupiter (Figure 15.5). As long as an asteroid remains in this belt, it poses no hazard to Earth. But an asteroid's orbital path can be disturbed by a collision or a near miss with another object. The path then might

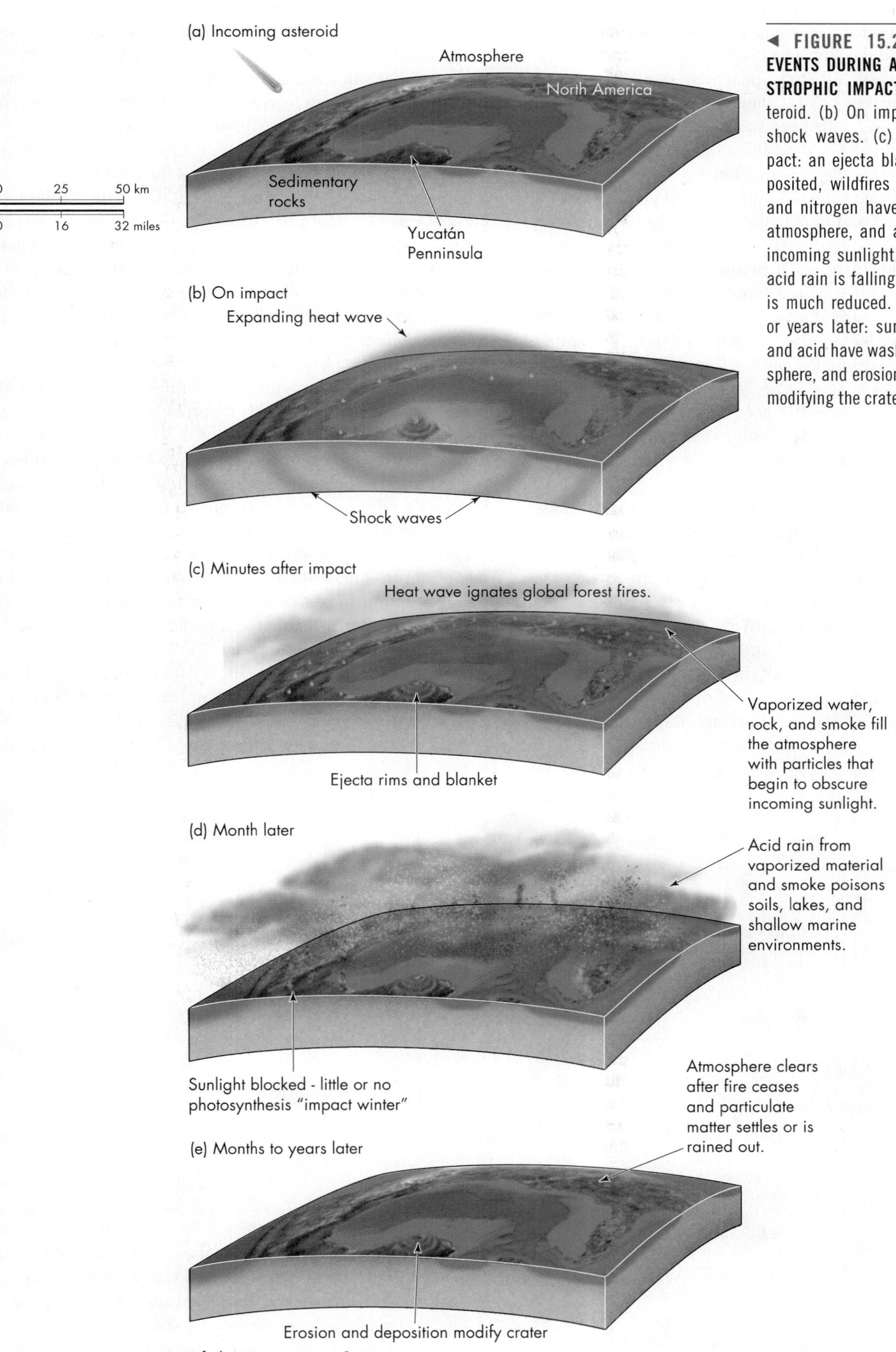

◄ **FIGURE 15.20 SEQUENCE OF EVENTS DURING AND AFTER A CATASTROPHIC IMPACT** (a) Incoming asteroid. (b) On impact: a fireball and shock waves. (c) Minutes after impact: an ejecta blanket has been deposited, wildfires are burning, water and nitrogen have been vaporized in atmosphere, and a dust cloud blocks incoming sunlight. (d) A month later: acid rain is falling and photosynthesis is much reduced. (e) Several months or years later: sunlight returns, dust and acid have washed out of the atmosphere, and erosion and deposition are modifying the crater.

become more elliptical, bringing the object into the space between Earth and the sun. The object might even regularly cross the orbital path of Earth, in which cases it is called a **near-Earth object** (NEO). The current estimate of the number of asteroids with near-Earth orbits and diameters larger than 100 m is 85 000; of these, about 1000 are larger than 1 km, and 30 are larger than 5 km.[32]

Near-Earth objects are currently defined on the basis of parameters that reflect the potential of the object to make a close approach to the Earth. An asteroid is considered "potentially hazardous" if it has a minimum orbit intersection distance from Earth of 0.05 astronomical unit (AU), equal to 7 500 000 km, and a size of more than about 150 m.[33] Objects that cannot approach closer to the Earth than 0.05 astronomical units (AU) or are smaller than 150 m are not considered potentially hazardous. The NASA Near Earth Object Catalog also includes the approach distances of asteroids and comets measured in lunar distances, which has become a common unit of measure used by the news media when discussing these objects.

Most asteroids are structurally weak, and if they entered Earth's atmosphere, they would probably explode at altitudes of about 30 km. These high-altitude explosions are spectacular but do not pose a significant hazard at the surface of Earth. The remaining objects are strong chondritic asteroids (Table 15.1). These relatively slow-moving objects could penetrate the atmosphere and explode at low altitudes or strike Earth's surface. Identifying all these objects will be extremely difficult, because most of them are difficult to identify and track.[1] We currently are able to track only objects larger than a few hundred metres in diameter.

Comets, another type of NEO, are generally a few kilometres in diameter and consist of dirty ice with a covering of rock particles and dust. Some comets meet a spectacular end by falling into the sun or colliding with a planet or moon in the solar system. Collisions between comets and planets or moons were common after the solar system formed; comets, for example, might have produced some of the craters on the moon. Likewise, many comets collided with Earth in its early stages. There are still many near-Earth comets, but an asteroid is more likely to collide with Earth than is a comet.

NEOs apparently have a relatively short life span, but meteoroids and asteroids are continuously ejected from the asteroid belt into near-Earth positions. Likewise, the orbits of some comets in the outer solar system could be perturbed by planets or other objects and become NEOs.

As we have seen with other natural hazards, the risk of an event is related to both its probability and its consequences should the event take place. The consequences of an airburst or direct impact from an extraterrestrial object several kilometres in diameter would be catastrophic. Such an impact might occur at sea, but its effects would be felt worldwide with a high potential for mass extinction. Impact events of this magnitude on Earth probably have return periods of tens to hundreds of millions of years (Figure 15.21). Impacts and airbursts from objects hundreds of metres in diameter are more frequent and could wreak havoc over a large area, causing great damage and loss of life. An object a few tens of metres in diameter would cause a regional catastrophe if it exploded in the atmosphere or struck land near a populated area. The size of the devastated area might be several thousand square kilometres. Even a small impact

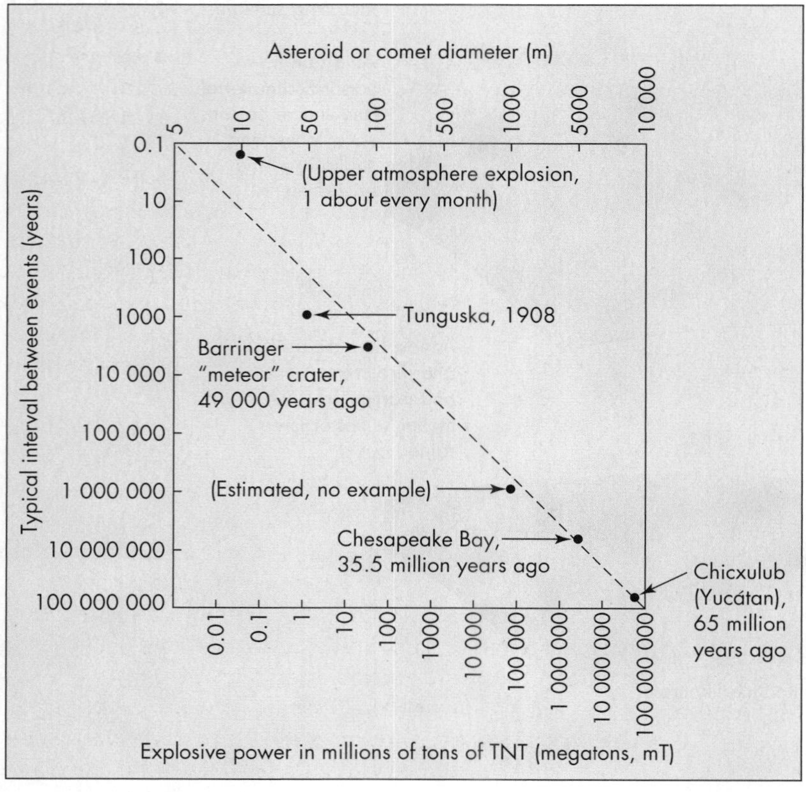

◀ **FIGURE 15.21 IMPACT FREQUENCY AND ENERGY** The relations among the size of asteroids or comets striking Earth, their frequency, and the energy released. *(Weissman, P. R., L.-A. McFadden, and T. V. Johnson, eds. 1999.* Encyclopedia of the Solar System. *San Diego, CA: Academic Press; and Brown, P., R. E. Spalding, D. O. ReVelle, E. Tagliaferri, and S. P. Worden. 2002. "The flux of near-Earth objects colliding with the Earth."* Nature *420: 294–296)*

could kill millions of people if it occurred over or in a large city. Scientists term these smaller impacts "Tunguska-type" events. An asteroid designated 2009 VA, which is about 7 m in diameter, whizzed past Earth at a distance of 14 000 km in November 2009. It was the third-closest known approach of a catalogued asteroid on record.

Recent research suggests that airbursts from asteroids 50 m to 100 m in diameter occur, on average, once every 1000 years or less (Figure 15.21). By using the Tunguska event as an example of what such a blast could do, scientists have estimated that an urban area is likely to be destroyed about once every 30 000 years by an airburst or the impact of a small asteroid. This conclusion is based on the distribution of 300 small asteroids known to have exploded in the atmosphere. Extrapolation of the data allows estimates to be made of the impacts of larger, even more damaging events.[3]

Predictions of the likelihood and type of future impacts have large statistical uncertainty. Deaths caused by an impact during a typical millennium could be as low as zero to as high as several hundred thousand people. A truly catastrophic event could kill millions of people; averaging this number over thousands of years produces a relatively high average annual death toll.

Overall, the risk from extraterrestrial impacts is higher than one might think. The probability that you will be killed by a catastrophic impact is approximately 0.01–0.1 percent. In comparison, the probability that you will be killed in a car accident is approximately 0.008 percent, and by drowning is about 0.001 percent. The risk of dying from the impact or airburst of a large comet or asteroid is comparable to that of many other risks we face in life. However, we emphasize again that this risk is spread out over thousands of years. Although the average death toll in any one year might appear high, it is just that—an *average*. Remember that such events are rare. Certainly there is a risk, but the interval between large impacts is so long that we shouldn't lose any sleep worrying about a global catastrophe caused by an extraterrestrial object.

Managing the Impact Hazard

Scientists and policymakers have only recently become aware of the risk that we face from asteroids and comets. Are we helpless, or is there something we can do to reduce this risk? We are not helpless. First, we can identify objects in our solar system that could threaten Earth. A program known as the *Spaceguard Program* was started in 1998 under the auspices of the U.S. National Aeronautics and Space Administration (NASA).[34] The objective of this program is to study the size distribution and dynamic processes associated with NEOs, specifically those objects with a diameter larger than 1 km. The program uses telescopes with digital imaging devices to identify and monitor fast-moving objects.[32] In 2007, NASA proposed to the U.S. Congress that existing surveys be extended to identify 90 percent of potentially hazardous NEOs larger than 140 m by 2020. When a potentially hazardous object is detected, it is catalogued at the International Astronomical Association's Minor Planet Center located at the Harvard-Smithsonian Center for Astrophysics. As of August 2012, 848 near-Earth asteroids larger than 1 km had been discovered, of which 154 are potentially hazardous. The estimate of near-Earth asteroids yet to be discovered is 70.[33] The five NASA-supported search facilities continue to improve their performance; thus good progress has been made toward eliminating the risk of a large, undetected impactor. Research to identify NEOs is likely to intensify in the future, with more and more objects catalogued and monitored. However, a complete evaluation will take a long time because many potentially dangerous objects have orbits that might not bring them close to Earth for decades. The good news is that most of the objects identified as potentially hazardous to Earth will not collide with our planet for at least several thousand years. Therefore, we have time to learn about them and develop appropriate technologies to reduce the risk.[1]

As surveys of NEOs continue, scientists will detect objects that are potentially hazardous to Earth. Communication of this hazard to the public and public officials must be both timely and authoritative. As with earthquake predictions, possible impacts must be evaluated by committees of scientists before action is taken. The IAU (International Astronomical Union) has established a technical review process for predictions that has had mixed results.[35] NASA's NEO Program Office maintains a website (http://neo.jpl.nasa.gov/) that classifies potentially hazardous NEOs using the *Torino Impact Hazard Scale* (Table 15.3).

The *Palermo Technical Hazard Impact Scale* is another scale used by astronomers, and is based on the probability of impact and the estimated kinetic energy release.[36] A rating of 0 means the hazard is as likely as the "background hazard," defined as the average risk posed by objects of the same size or larger over the years until the date of the potential impact. A value of +2 indicates the hazard is 100 times more likely than a random background event, and values less than −2 reflect events for which there are no likely consequences. Palermo Scale values between −2 and 0 indicate situations that merit monitoring.

Once a large NEO is known to be on a collision course with our planet, options to avoid or minimize the effects of an airburst or crater-forming impact are limited. In the case of smaller asteroids, people could be evacuated if authorities were provided several months' warning and if we could precisely predict the location of the impact. However, evacuating an area of several thousand square kilometres would be a tremendous, if not impossible, undertaking.[1]

No place on the planet is safe if a large asteroid strikes. Living things, including people, within the blast area would be killed instantly; those farther away would likely die in the ensuing months from the cold, acid rain, and destruction of the food chain. Little would be gained if we intercepted the object and blew it apart, because the smaller pieces would rain down on Earth and might cause more damage than a single larger body. And, in any case, we do not yet have the technology to destroy an incoming asteroid.

A better approach would be to try to divert the path of the asteroid so that it misses Earth. Let's assume we know that a 400-m asteroid will strike Earth in 100 years. The

TABLE 15.3 The Torino Impact Hazard Scale

No Hazard	0	Low collision hazard or object will burn up in atmosphere
Normal	1	Object will pass near Earth with collision extremely unlikely
	2	Somewhat close encounter; collision very unlikely and does not merit public attention
Merits Attention by Astronomers	3	Close encounter, with localized destruction possible; merits public attention if collision less than a decade away
	4	Close encounter with regional destruction possible; merits public attention if collision less than a decade away
	5	Close encounter with serious, but uncertain, regional destruction threat; merits contingency planning if less than a decade away
Threatening	6	Close encounter with serious, but uncertain, global catastrophe threat; merits contingency planning if less than three decades away
	7	Very close encounter with unprecedented, but still uncertain, global catastrophe threat; merits international contingency planning if less than a century away
	8	Collision will occur with object capable of localized destruction on land or tsunami if close off-shore; once every 50 to 100 years
Certain Collisions	9	Collision will occur with object capable of regional devastation or major tsunami; once every 10 000 to 100 000 years
	10	Collision will occur with object capable of global climatic catastrophe that threatens civilization; once every 100 000 years or more

Source: Based on Morrison, D., C. R. Chapman, D. Steel, and R. P. Binzel 2004. "Impacts and the public: Communicating the nature of the impact hazard." In M. J. S. Belton, T. H. Morgan, N. H. Samarasinha, and D. K. Yeomans (eds.). Mitigation of Hazardous Comets and Asteroids, *pp. 353–390. New York, NY: Cambridge University Press.*

asteroid has probably been crossing Earth's orbit for millions of years and would miss, rather than strike, Earth if we could slightly alter its orbit years before the impact. This scenario is possible because the probability that we would identify the object at least 100 years before impact is high, about 99 percent, and we have the technology to change the orbit of a threatening asteroid. Small nuclear explosions in the vicinity of the asteroid would alter its path without breaking it up. NASA has determined that such a "standoff" explosion would be 10 to 100 times more effective than non-nuclear alternatives, such as pulsed lasers or focused solar radiation, in changing the orbit of a potentially hazardous object.[37] The cost of such an expedition would likely exceed U.S.$1 billion and would require a coordinated effort involving NASA and the U.S. military. However, this expenditure would seem small if we are faced with the reality of a large asteroid on a collision course with Earth!

In summary, we continue to catalogue extraterrestrial objects that intersect Earth's orbit, and we are beginning to think about options to minimize the risk they pose. Because we have the ability to detect a large asteroid long before it reaches Earth, we may be able to intercept it and nudge it into a different orbit so that it misses our planet.

REVISITING THE FUNDAMENTAL CONCEPTS

Impacts and Extinctions

1. **Hazards can be understood through scientific investigation and analysis.**
2. **An understanding of hazardous processes is vital to evaluating risk.**
3. **Hazards are commonly linked to each other and to the environment in which they occur..**
4. **Population growth and socioeconomic changes increase the risk from natural hazards.**
5. **Damage and loss of life from natural disasters can be reduced.**

1. Studies of comets and asteroids and their interactions with planets allow us to predict the possibility and time that an object of a particular size could strike Earth. However, it is not yet possible to know whether a near-Earth object would actually collide with our planet or be a near miss. Nor is it possible to know the exact location that such an object would strike until a short time before it happened. Of course, 70 percent of impacts on Earth will be over or in oceans. Scientists also know the average return periods for impacts of different sizes. For example, an event of the size of Tunguska in 1908, which potentially could obliterate a large city, occurs on average about once every 1000 years. In contrast, an asteroid of the size of the one that struck Yucatán Peninsula 65 million years ago strikes Earth, on average, less frequently than once every 100 million years. This information is not particularly useful for, say, urban planning because strikes by large comets and asteroids are too rare to be of concern. Nevertheless, because the consequences of such impacts are so great, efforts are being made to track near-Earth objects.
2. Risk assessment is difficult for events with very long return periods, because considerable uncertainty exists about the frequency of the largest possible events. However, the risk—the product of the hazard, which is very small, and the consequences, which are very large—is still significant and something we need to think about. For example, a large urban area could be destroyed once every few tens of thousands of years. Given the growing size of cities and the billions of people who now live in them, the possible consequences are huge. However, the probability of such an impact is so low in any one year, or for that matter in a lifetime, that it is extraordinarily difficult to comprehend such an event ever happening.
3. Links exist between impacts of comets and asteroids with Earth and other natural hazards. For example, an impact in the ocean would trigger a large tsunami, and an impact on land would cause large-scale fires across the landscape. A large impact could result in strong cooling of the atmosphere, acid rain, and fallout of ash and other particulates. An impact from a large comet or asteroid could be what is sometimes termed a "planet cleaning" event, in that it would cause extinctions of plants and animals on a large scale.
4. The human population of Earth is now over 7 billion, compared to less than 2 billion in 1908 at the time of the Tunguska event. Comet and asteroid impacts pose a higher risk today than in 1908 because of the greater exposure of humans to the hazard. Even relatively small events could have serious ramifications.
5. It might be possible in the future to reduce the threat of a comet or asteroid impact. A large comet or asteroid that could impact Earth would probably be seen long before it actually arrived. We might have years to reduce or even eliminate the hazard from a large near-Earth object. Smaller objects capable of destroying a city are more difficult to identify; thus the lead-time for warning is much shorter. If we assume that we could see an object a month or so before impact, there might be time to evacuate the impact area if the area can be accurately identified. However, an orderly evacuation of a large city with a population of millions of people would be difficult. Efforts are being made to identify and track all near-Earth objects larger than 1 km. It has been recommended that smaller objects, with diameters of 100 m or less, also be identified. Even an object about 50 m in diameter could cause thousands of fatalities if it arrived without warning.

Summary

Near-Earth objects include asteroids, meteoroids, and comets. Some of these objects enter Earth's atmosphere and strike its surface. Small objects burn up in the atmosphere and are visible as meteors at night. Large objects, ranging from a few metres to hundreds of kilometres across, can explode in the atmosphere in an airburst or hit the surface of Earth. An impact by an asteroid as small as 5 km in diameter could cause a global catastrophe, including mass extinction of life. The largest and best-documented impact of such an object occurred 65 million years ago at the end of the Cretaceous Period—an asteroid about 10 km in diameter struck Earth, causing the extinction of 70 percent of plant and animal genera, including all of the dinosaurs. In addition to mass extinctions, impacts and airbursts of large asteroids and comets are likely to cause tsunamis, wildfires, earthquakes, landslides, climate change, and possibly volcanic eruptions.

The risk associated with an airburst or the direct impact of an extraterrestrial object is the product of its probability and the consequences should the event occur. Relatively small events like the 1908 Tunguska airburst occur somewhere on Earth every thousand years, on average. A Tunguska-like event would cause catastrophic damage if it happened over an urban area. Objects hundreds of metres across could cause catastrophic damage on a regional scale and can be expected to occur every few tens of thousands of years. Asteroids large enough to produce a catastrophe on a global scale probably strike Earth more rarely than every hundred million years. Programs such as *Spaceguard* identify asteroids larger than a few hundred metres in diameter long before they reach Earth. With adequate warning, it might be possible to intercept and divert an approaching asteroid using nuclear explosions in space. Millions of smaller, potential Tunguska-type objects could also produce catastrophic damage to cities and towns. Identifying all these objects will be extremely difficult.

Key Terms

airburst (p. 425)
asteroid (p. 425)
comet (p. 425)
ejecta blanket (p. 429)
impact crater (p. 429)
Kuiper Belt (p. 425)
mass extinction (p. 437)
meteor (p. 425)
meteorite (p. 425)
meteoroid (p. 425)
near-Earth object (p. 442)
Oort Cloud (p. 425)

Did You Learn?

1. Where is Tunguska? What happened there?
2. Explain how asteroids, meteors, comets, meteoroids, and meteorites differ.
3. Describe the composition of meteorites and comets.
4. Where do comets and asteroids originate?
5. Describe the characteristics of impact craters. How can impact craters be distinguished from other types of craters?
6. Explain the differences between simple and complex impact craters.
7. Why does Earth have fewer impact craters than the moon?
8. Explain the significance of Comet Shoemaker-Levy 9.
9. Summarize the hypotheses that have been proposed to explain mass extinctions.
10. What is the K-T boundary? Why is it significant in the study of natural hazards?
11. Name other natural hazards that are linked to extraterrestrial impacts and airbursts.
12. Is the risk of dying from an asteroid impact greater or less than the risk of dying in an automobile accident?
13. What can be done to avert an extraterrestrial impact that will happen in a year's time?

Critical Thinking Questions

1. What would be the likely effects of a Tunguska-type airburst over central Canada? If the event could be predicted 100 years in advance, what could be done to minimize damage and loss of life, assuming the object's orbit could not be changed? Outline a plan to minimize death and destruction.
2. How would the effects of an asteroid impact differ depending on whether it struck the ocean or land? Consider what would happen physically and chemically to the water and how the impact craters might differ.

MasteringGeology

MasteringGeology **www.masteringgeology.com**. Looking for additional review and test prep materials? Visit the Study Area in MasteringGeology to enhance your understanding of this chapter's content by accessing a variety of resources, including **Self-Study Quizzes, Geoscience Animations, GEODe Tutorials, RSS feeds, flashcards,** web links and an optional **Pearson eText.**

Glossary*

*Some words or terms in the Glossary have more than one meaning. The definitions that follow pertain to the usage in this book.

Aa A basaltic lava flow with a broken, blocky surface texture.

Abrupt climate change A large-scale change in the global climate system that takes place over a few decades or less.

Absorption The transfer of the energy from a wave, such as an electromagnetic wave, to matter through which the wave passes.

Acceptable risk The level of risk that an individual or community will tolerate before taking action to reduce its exposure to the hazardous process that is responsible for the risk.

Active fault A fault capable of rupturing during an earthquake. A common criterion for labelling a fault as active is that it has moved in the past 11 600 years, or several mappable displacements have occurred in the past 35 000 years. See *Inactive* and *Potentially active faults*.

Active layer The surface layer of sediment or rock in an area of permafrost that thaws during summer and is frozen at other times of the year. The active layer overlies material that is continuously frozen. See *Permafrost* and *Permafrost table*.

Active margin A continental margin characterized by earthquakes, volcanic activity, and mountain-building due to convergence of or horizontal displacement of tectonic plates.

Adaptation The collective adjustments that governments and individuals will have to make to deal with the changing climate of the twenty-first century.

Adit An opening driven horizontally into the side of a mountain or hill to provide access to a mineral deposit.

Advance An interval, commonly spanning years or decades, during which a glacier thickens and becomes longer. See *Retreat*.

Aerosol A suspension of microscopic liquid and solid particles, such as mineral dust and soot, in the atmosphere.

Aftershock One of the many earthquakes that occur a few minutes or up to about a year after a larger earthquake and in the same general area. See *Foreshock* and *Mainshock*.

Agricultural drought Drought that affects crop production. Agricultural drought can arise independently from any change in precipitation when soil conditions and erosion caused by poor agricultural practices reduce water available to the crops. It is also caused by *Meteorological drought.*

Airburst The explosion of a meteoroid in the atmosphere, generally at an altitude of 12–50 km.

Albedo The fraction of solar radiation reflected from Earth back into space. Albedo is a measure of the reflectivity of the surface; ice, for example, has a high albedo, whereas water's albedo is much lower.

Alberta Clipper An extratropical cyclone that forms east of the Rocky Mountains in Alberta and moves rapidly to the east or southeast across Canada and the northern United States. It is commonly associated with snowstorms, especially blizzards.

Alluvial Relating to unconsolidated sediment deposited by a stream or river.

Alluvial fan A fan-shaped body of coarse sediment deposited by a stream where it emerges from a mountain valley onto flatter terrain. Fans can be composed of stream deposits, debris-flow deposits, or a combination of the two.

Amplification Increase in ground motion during an earthquake; as P and S waves slow, some of their forward-directed energy is transferred to surface waves. See *P wave* and *S wave*.

Anastomosing river A river characterized by a small number of relatively stable channels separated by islands.

Andesite A grey to reddish-brown volcanic rock consisting mainly of plagioclase and feldspar, with little or no quartz.

Anthropogenic Produced or caused by humans Also called *human-induced.*

Anticline A fold in sedimentary rocks in which strata slope downward on both sides from a common crest.

Aphelion The point in a planet's orbit where it is farthest from the sun. At present, Earth's aphelion occurs about July 1, when the planet is 4.8 million km farther from the sun than at perihelion. The seasons in which aphelion and perihelion occur vary cyclically with a period of 21 000 years. See *Perihelion.*

Aquifer A water-bearing body of rock or sediment capable of providing useful amounts of groundwater to wells and springs.

Artificial bypass The human action that moves sand or other sediment around an obstruction. An example is dredging and pumping sand from the updrift to downdrift side of a jetty or harbour mouth.

Artificial control A human action or engineered structure that reduces or eliminates adverse effects of a hazardous surface process.

Ash fall The deposition of a layer of fine airborne rock and glass that have been erupted from a volcano.

Aspect The horizontal direction which a slope faces.

Asteroid A rocky or metallic body in outer space ranging from 10 m to 1000 km in diameter. See *Meteor, Meteorite,* and *Meteoroid.*

Asteroid belt The region between the orbits of Mars and Jupiter that contains most large asteroids in our solar system.

Asthenosphere The upper zone of Earth's mantle, located directly below the lithosphere; a hot, viscous, solid layer on which tectonic plates move.

Atmosphere A layer of gases surrounding a planet such as Earth.

Atmospheric pressure The force per unit area exerted by the gases that surround a planet. Also called *barometric pressure* because it is commonly measured with a barometer.

Atmospheric stability A condition of equilibrium in which a rising or sinking parcel of air returns to its original position. The atmosphere is stable when surface air and air aloft have similar temperatures.

Attenuation The decrease in the intensity of a seismic wave away from the earthquake source.

Avalanche The rapid downslope movement of snow under the influence of gravity.

Avalanche cord A red cord attached to a skier, snowshoer, or snowmobiler that allows a rescuer to locate the person in the event of a snow avalanche.

Avalanche dog A specially trained dog that works snowfields after an avalanche, searching for human scent through the snow.

Avalanche forecast An assessment of the likelihood and size of snow avalanches under existing or future conditions.

Avalanche shed A structure, generally made of concrete, that carries avalanching snow over a road or rail line.

Avalanche transceiver A portable device carried by skiers, snowshoers, and snowmobilers that emits and receives a radio signal. In receive mode, avalanche transceivers can locate a buried avalanche victim at distances up to 80 m in 5 to 15 minutes. Also called *beacons* or *beepers*.

Avaluator A rule-based decision support tool for backcountry skiers, snowboarders, and snowmobilers in Canada. The Avaluator comprises a pocket card and a companion booklet aimed at helping outdoor recreationists make critical decisions before and during backcountry trips during winter.

Avulse A sudden shift in the channel of a stream or river.

Banded iron formation A sedimentary rock consisting of alternating black layers rich in magnetite or hematite and reddish layers consisting of iron-poor shale and chert.

Bankfull discharge A water discharge that completely fills a stream channel. Most stream channels are formed and maintained by bankfull discharge.

Barometric pressure See *Atmospheric pressure*.

Barrier island A long, narrow, and low body of sand or gravel that is separated from the mainland by a bay or lagoon.

Basalt Volcanic rock consisting of calcium-rich feldspar and other silicate minerals rich in iron and magnesium. Basalt has a relatively low silica content and forms shield volcanoes.

Base level The theoretical lowest elevation to which a river can erode at a particular time.

Beach An accumulation of sand, gravel, or both along the shore of a lake or ocean. Beaches are produced by wave and current action.

Beach drift The movement of sand grains along the shore by wave run-up and backwash. See *Littoral transport.*

Beach face The sloping part of a shoreline affected by run-up and backwash of waves. See *Berm* and *Swash zone.*

Beach nourishment The artificial addition of sand to a shoreline to create or enhance a beach.

Beacon See *Avalanche transceiver.*

Bedding plane A planar surface separating two layers of sedimentary rock.

Bed load The sediment particles pushed or rolled along a river channel by flowing water. See *Dissolved load, Suspended load,* and *Total load.*

Beeper See *Avalanche transceiver.*

Bentonite The clay formed by the weathering of volcanic rock. Bentonite can absorb large amounts of water and expand to many times its dry volume.

Berm The relatively flat or landward-sloping part of a beach located just landward of the swash zone formed of wave-deposited sediment; the part of the beach where most people sunbathe. See *Beach face* and *Swash zone.*

Bermuda High A large, subtropical, semi-permanent area of high atmospheric pressure centred in the North Atlantic Ocean over the Azores. Clockwise circulation around the high brings hot humid air to the Atlantic coast of the United States and affects the weather of northern Africa and southern Europe. Also known as the *Azores High* or *North Atlantic High.*

Biogeochemical cycle The cycling of chemical elements or compounds through the atmosphere, hydrosphere, biosphere, and lithosphere.

Blind fault In tectonics, a fault whose rupture plane does not extend to Earth's surface.

Blizzard A severe winter storm during which large amounts of falling or blowing snow create low visibility for an extended period. The definition of *blizzard* differs slightly in Canada and the United States: In Canada winds must exceed 40 km/h with visibility less than 1 km for at least 4 hours; in the United States winds must exceed 56 km/h with visibilities of less than 0.4 km for at least 3 hours.

Block An angular rock fragment more than 64 mm across ejected from a volcano during an eruption. See *Lapilli.*

Blue zone Mapped areas in Switzerland and parts of the United States within which potentially dangerous snow avalanches can be expected every 30 to 300 years. Legislation might require that buildings within the blue zone have reinforced walls facing the avalanche path or that they be protected by deflection structures. See *Red zone.*

Bluff A high, steep bank or cliff along a lakeshore or river valley.

Body wave A seismic wave that travels outward from the focus of an earthquake through the interior of Earth. See *Surface wave.*

Body-wave scale An earthquake magnitude scale based on seismic waves that travel within Earth.

Bomb A smooth-surfaced block of volcanic rock more than 64 mm across that is ejected in a semi-molten state from a volcano during an eruption. A bomb acquires a streamlined shape as it cools while falling to the ground.

Bore See *Tidal bore.*

Braided A river channel pattern characterized by numerous intertwining, unstable channels separated by islands of sand and gravel.

Breaker zone That part of the beach and near-shore environment where waves build up, become unstable, and collapse toward the shore. See *Surf zone* and *Swash zone.*

Breaking wave A wave that curls over, crashes, and rushes ashore in a turbulent state.

Breakup The disintegration and downstream dispersion of an ice layer that forms on a river during winter. Breakup generally occurs in the spring.

Breakwater A wall protecting a beach or harbour from waves; it can be attached to the beach or located offshore.

Breccia A rock composed of angular fragments; breccia can have a sedimentary, volcanic, tectonic, or impact origin.

Caldera A large crater produced by a violent volcanic eruption or the collapse of the summit area of a volcano after an eruption.

Calve A separation of large blocks of ice from the toe of a glacier flowing into a lake or the sea; the calved blocks are icebergs.

Carbon dioxide A colourless, odourless gas composed of two oxygen atoms covalently bonded to a single carbon atom. It is formed during respiration and by the decomposition of organic matter, and is absorbed by plants through photosynthesis.

Carbon monoxide A colourless, odourless gas consisting of one carbon atom and one oxygen atom. It is produced by incomplete combustion of fossil fuels and is toxic to humans and other animals.

Carbon sequestration The capture and storage of carbon dioxide before it enters the atmosphere. Carbon dioxide can be stored below ground in geologic formations, at depth in the oceans, or above ground in plants or soils.

Carbon-14 method The method of dating fossil plant and animal material based on the slow radioactive decay of an unstable isotope of carbon. The method can be used to date fossils up to about 50 000 years old. See *Radiocarbon dating.*

Cascadia subduction zone The elongated zone within Earth's lithosphere located off the west coast of North America from northern California to central Vancouver Island where the oceanic Juan de Fuca plate slowly moves eastward beneath North America.

Catastrophe An event that causes damage to people and property on such a scale that recovery is long and complex. Natural processes that produce catastrophes include floods, hurricanes, earthquakes, tsunamis, volcanic eruptions, and large wildfires. See *Natural disaster.*

Catchment See *Drainage basin.*

Cause With reference to landslides, an internal or external factor that, over time, reduces the stability of a slope and brings it to the point of failure. Compare with *Trigger.*

Cave A natural subterranean cavity produced by slow solution of limestone or marble. Many caves consist of a series of chambers that are large enough for a person to enter.

Cave system See *Cave.*

Cenote A steep-walled collapse sinkhole that extends below the water table on the Yucatan Peninsula in Mexico.

Cenozoic Era The current and most recent of the three geological eras of the Phanerozoic (i.e., the past 541 million years). It follows the Mesozoic Era and spans approximately the past 65 million years.

Channel pattern The areal pattern of a flowing stream. Channel patterns include straight, meandering, anastomosing, and braided.

Channel restoration The process by which a stream channel is returned to a natural state.

Channelization The modification of a storm channel to permit more efficient conveyance of water and sediment. Channelization involves straightening, widening, and, in some instances, deepening and lining the channel.

Chlorine A toxic greenish-yellow halogen gas.

Chute A straight, steep gully formed and maintained by avalanches, rock falls, or debris flows.

Cinder cone A conical volcano consisting of pyroclastic deposits. Also called a *scoria cone.*

Clearcutting The forestry practice of harvesting all trees from a tract of land.

Climate The characteristic weather of a place or region over many years or decades. See *Weather.*

Coastal erosion The loss of rock or sediment from a shoreline because of wave action, landslides, wind abrasion, and runoff.

Coefficient of friction The ratio of the force necessary to initiate the movement of an object over an inclined surface and the frictional force that resists the motion of the object.

Cohesion The part of the shear strength of a rock or sediment that is independent of inter-particle friction.

Collapse sinkhole A depression in sediment or rock caused by the collapse of the roof of a cave. See *Sinkhole.*

Collapsible sediments (also referred to as collapsible soils) Sediments that compact and settle when they become wet.

Colluvium The broken rock or sediment deposited by mass movement processes, including creep and landslides.

Combustion The phase of a wildfire following ignition; it includes flaming, in which fine fuel and volatile gases are rapidly oxidized at high temperatures, and glowing or smouldering, which take place later at lower temperatures. See *Extinction* and *Pre-ignition.*

Comet An object that orbits the sun and is composed of a sponge-like rocky core surrounded by ice and covered by carbon-rich dust; a tail of gas and dust glows as a comet approaches the sun. Comets range from a few metres to a few hundred kilometres in diameter.

Community Internet Intensity Map An online map depicting the intensity of an earthquake. The map is based on email reports of the severity of shaking and structural damage in many different places.

Complex impact crater A large impact crater with an uplifted centre. Complex impact craters are commonly many tens of kilometres in diameter; the largest are more than 100 km in diameter.

Composite volcano A steep-sided volcanic cone produced by alternating layers of pyroclastic debris and lava flows. Also known as a *stratovolcano*.

Compression Forces acting on a rock mass that are directed toward one another.

Compression test The application of a vertical force with the back of a shovel blade on a column of undisturbed snow to evaluate the strength of the snowpack. Any weak layers will fracture along the exposed vertical face of the snow column.

Compressional wave See *P wave*.

Compressive strength The greatest compressive stress that a material can bear without failing.

Conduction The transfer of heat through a substance by molecular interactions. See *Convection* and *Radiation*.

Constructive interference The process by which two or more waves combine to form a larger wave; for example, on a lake or the sea.

Continental drift The slow movement of the continents over geological time, as explained by the theory of plate tectonics.

Continental shelf The shallow seafloor fringing a continent; it extends from the coastline to depths of 150 m to 200 m.

Continuous crown fire An intense wildfire that moves swiftly through forests, consuming the tops of most trees and releasing large amounts of heat. See *Intermittent crown fire*.

Continuous permafrost A region in which more than 90 percent of the ground is permanently frozen. See *Permafrost*, *Discontinuous permafrost*, and *Extensive discontinuous permafrost*.

Convection The transfer of heat by movement of particles. For example, hot water rises to the surface in a pot of boiling water and displaces cooler water, which sinks. Heat can be transferred vertically by convection within storm clouds, and convection is the main way that heat is transferred in a wildfire. See *Conduction* and *Radiation*.

Convection cell A circulation loop in which warmer fluid rises and cooler, denser fluid sinks. Convection cells operate on a large scale in Earth's atmosphere and mantle.

Convergence The movement of two or more objects toward one another.

Convergent boundary A boundary between two converging tectonic plates. The most common type of convergent plate boundary is a subduction zone, where one plate descends below another. Another type of convergent plate boundary is a collision zone, where a plate crumples and elevates the leading edge of another plate (e.g., the boundary between the Indo-Australia and Eurasia plates in the Himalayas).

Coping capacity The ability of a population to respond to and reduce the negative effects of a hazardous event.

Cordilleran Ice Sheet The large body of glacier ice that covered most of western Canada and adjacent areas of the United States at times during the Pleistocene Epoch. See *Laurentide Ice Sheet*.

Coriolis effect The apparent deflection of a moving object because of Earth's rotation. The deflection is to the right in the Northern Hemisphere and to the left in the Southern Hemisphere.

Crater A bowl-shaped depression at the top of a volcano that has formed by an explosion or collapse of the summit area.

Creep The slow downslope movement of soil, sediment, and highly fractured rock because of the force of gravity.

Critical facility An important building or other structure that must be located and built to survive an earthquake or other natural disaster.

Cross-loaded A slab of potentially unstable snow deposited by wind blowing parallel to a ridge crest. Cross-loaded slabs can collapse to produce snow avalanches. See *Top-loaded*.

Crown fire The flaming combustion of the upper canopy of a forest during a wildfire. Crown fires are commonly driven by high winds and are aided by steep slopes. See *Surface fire*.

Crust The outermost layer of Earth, composed of continental rocks rich in silicon and aluminum, and oceanic rocks rich in silicon and magnesium. The crust ranges in thickness from 8 km beneath the oceans to more than 60 km beneath the continents. See *Lithosphere*.

Cryosphere That part of Earth with water in its solid form; it includes snow, glaciers, ice sheets, lake ice, river ice, sea ice, and permafrost.

Cryptovolcanic A geologic feature that appears to be disrupted by volcanic activity but lacks associated volcanic rocks.

Crystallization The process by which molecules precipitate from solution to form a mineral.

Cumulonimbus A cumulus cloud that grows vertically until it extends into the lower stratosphere. Cumulonimbus clouds may have anvil-shaped tops. See *Cumulus* and *Cumulus stage*.

Cumulus Any cloud that develops vertically. Cumulus clouds generally have flat bases and cauliflower-like tops. See *Cumulonimbus* and *Cumulus stage*.

Cumulus stage The first phase of thunderstorm development, characterized by the growth of domes and towers that transform a cumulus cloud into a cumulonimbus cloud. See *Dissipative stage* and *Mature stage*.

Cutbank A steep slope eroded on the outside of a bend in a stream channel.

Cyclone General term applied to a large cell of moisture-laden air that rotates around an area of low pressure. As a result of the *Coriolis effect*, the winds in a cyclone rotate anticlockwise in the Northern Hemisphere and clockwise in the Southern Hemisphere. See *Extratropical cyclone* and *Hurricane*.

Dacite A grey to pale-brown volcanic rock with a silica content of 63 to 68 percent that consists of plagioclase, quartz, biotite, hornblende, and pyroxene.

Debris avalanche A very shallow slide of sediment or soil on a steep slope. The failure plane is generally either at the base of the organic soil or in colluvium.

Debris flow The rapid downslope movement of water-saturated sediment ranging in size from clay to boulders. Debris flows are triggered by heavy rain, rapid melting of snow and ice, or the sudden draining of a pond or lake.

Decompression melting Melting of deep rocks in Earth's crust when the pressure exerted on the rocks is reduced without a change in temperature. Decompression melting happens at divergent plate boundaries, continental rifts, and hot spots.

Delta A flat, triangular, or fan-shaped landform built by a stream where it enters a lake or the ocean. See *Delta plain*.

Delta plain The nearly flat surface of a delta. The delta plain is cut by distributary channels and may support extensive marshes, swamps, and other wetlands. See *Delta*.

Dendroclimatology The study of annual rings of living and dead trees to determine past climate.

Deposited See *Deposition*.

Deposition The accumulation of sediment on land and in lakes and the sea; deposition occurs by mechanical processes, chemical precipitation, or the build-up of dead plant matter.

Derecho A windstorm with straight-line winds exceeding 90 km/h over a distance of at least 400 km. The damage produced by a derecho may be equivalent to that of a tornado.

Desertification The conversion of land from a biologically productive state to one that resembles a desert.

Desiccation crack An open crack in mud produced by drying.

Differentiated In the context of meteoroids and asteroids, rock that has been affected by igneous or metamorphic processes.

Digital elevation model (DEM) A three-dimensional, computer-generated model of the land surface.

Dip-slip fault See *Normal fault*.

Direct effect A change directly induced by an event. Direct effects of a disaster or catastrophe include death, injury, and property damage. Also known as *Primary effect*. See *Indirect effect*.

Directivity In the context of an earthquake, the increased intensity of shaking in the direction of the fault rupture.

Disappearing stream A surface stream or river that enters a cave and flows underground.

Disaster A brief event that causes great damage or loss of life in a limited geographic area.

Disaster preparedness The actions of individuals, families, communities, states, provinces, or nations to minimize losses from a natural disaster before it occurs. See *Preparedness*.

Discharge The volume of water flowing past a point in a stream channel over a specific period, typically one second. Discharge is commonly measured in cubic metres per second (m^3/s).

Discontinuous permafrost A region in which 10 to 90 percent of the ground is permanently frozen. See *Permafrost*, *Continuous permafrost*, *Extensive discontinuous permafrost*, and *Sporadic discontinuous permafrost*.

Dissipative stage The final phase of a thunderstorm, when the supply of moist air is blocked by downdrafts in the lower levels of the cloud; precipitation decreases, and the storm begins to diminish. See *Cumulus stage* and *Mature stage*.

Dissolved load The part of a stream's load that is carried in solution. See *Bed load*, *Suspended load*, and *Total load*.

Distant tsunami A tsunami that travels thousands of kilometres across the open ocean and strikes remote shorelines with little loss of energy. Also called *tele-tsunami*. See also *Local tsunami*.

Distributary channel A stream channel that branches from the main channel on a delta or alluvial fan.

Divergence The movement of two or more objects away from one another.

Divergent plate boundary A boundary between two tectonic plates that are moving away from one another, where new crust is created. Divergent plate boundaries include mid-ocean ridges and some continental rift zones.

Doppler effect A change in wave frequency that occurs when the distance between a sound source and a receptor increases or decreases. For example, the pitch of an ambulance siren is higher when the ambulance approaches than it is when it moves away.

Doppler radar An instrument that emits electromagnetic waves to detect the velocity of precipitation that is falling toward or away from the receiver. Doppler radar is used to detect rotation of air within a thunderstorm that could lead to tornadoes. See *Doppler effect*.

Downburst A localized area of strong winds in a downdraft beneath a severe thunderstorm.

Downdrift The direction toward which a current of water, such as a longshore current, is flowing. See *Updrift*.

Downstream flood A flood in the lower part of a drainage basin because of high discharges of many tributary streams.

Drainage basin An area that contributes surface water to a stream. Also known as *catchment* and *watershed*.

Driving force A force that increases the downward-directed stress on a slope. See *Resisting force*.

Drought An extended period of unusually low precipitation that produces a shortage of water for people, other animals, and plants.

Dry slot A zone of dry, relatively cloud-free air that has sunk behind a cold front of an advancing low-pressure system.

Dryline An air mass boundary similar to a front, but in which the juxtaposed air masses differ in moisture content rather than temperature.

Dust storm A storm that transports large amounts of airborne silt and clay; visibility at eye level drops to less than 1 km for hours or days.

Dyke A linear embankment built to prevent flooding of an adjacent low-lying area by the sea or a river.

Dynamic equilibrium A state of balance between the work that a river does and the sediment load it receives. A stream in dynamic equilibrium maintains the gradient and cross-sectional shape it needs to move its sediment load.

Earth fissure A large, deep crack in loose sediment that forms when the water table falls and the sediment dries out.

Earth system science Science that transcends disciplinary boundaries and treats Earth as an integrated system comprising the atmosphere, lithosphere, hydrosphere, and biosphere, controlled by physical, chemical, biological, and human interactions.

Earthquake The sudden movement of rock on opposite sides of a fault; an earthquake releases strain that has accumulated within the rocks.

Earthquake cycle A hypothesis that explains successive earthquakes on a fault by a drop in elastic strain after an earthquake and the gradual accumulation of strain leading to the next quake.

Earth's energy balance The balance between incoming solar radiant energy and radiant energy leaving the planet.

Easterly wave An elongated trough of low atmospheric pressure that moves from east to west across the tropics, causing cloudiness and thunderstorms.

Eccentricity A measure of the departure of Earth's orbit from circularity.

Ecosystem A community of organisms that interact among themselves and in their physical environment.

Edge wave A surface wave that propagates along a rigid boundary and decreases exponentially in amplitude away from that boundary.

EF scale A scale with values ranging from EF0 to EF5 that is used to describe tornado intensity based on the maximum three-second wind velocity inferred from damage to buildings, towers, poles, and trees. The EF scale is a modification of the *F-scale* developed by T. Theodore Fujita.

Ejecta blanket The layer of breccia surrounding an impact crater, typically forming an irregular area of low mounds and shallow depressions.

El Niño A combined meteorological and oceanographic event in the equatorial Pacific Ocean during which trade winds weaken or even reverse, surface waters become anomalously warm, and the westward-moving equatorial ocean current weakens or reverses. These changes affect weather in Southeast Asia, Australia, New Zealand, and the Americas. See *La Niña*.

Elastic rebound The slow or rapid recovery of a strained body following the removal of a load.

Elastic strain The deformation that is reversed when stress is removed.

Electrical resistivity A measure of a material's ability to conduct an electrical current.

Electromagnetic energy The energy carried in waves by oscillations in electric and magnetic fields.

Electromagnetic spectrum The collection of electromagnetic waves of different wavelengths and frequencies.

Electromagnetic wave A wave that transfers energy in electric and magnetic fields.

E-line An imaginary line inland from the coast that marks the expected position of the shoreline after a particular number of years. See *E-zone*.

Enhanced Fujita scale See *EF scale*.

Epicentre The point on Earth's surface directly above the source, or focus, of an earthquake.

Eustasy A worldwide change in sea level, especially one caused by changes in the amount of glacier ice on land.

Evacuation The movement of people away from the site of a probable destructive natural event.

Expansive soil A type of clayey soil that, upon wetting and drying, expands and contracts, damaging foundations of buildings and other structures.

Exponential As used in this book, a type of growth characterized by a percentage increase in numbers or amount with each passing unit of time.

Extensive discontinuous permafrost A subzone of the discontinuous permafrost zone, in which 50 to 90 percent of the ground is permanently frozen. See *Permafrost*, *Discontinuous permafrost*, and *Sporadic discontinuous permafrost*.

Extinction The final phase of a wildfire, when all combustion, including smouldering, ceases because of insufficient heat, fuel, or oxygen. Also, the disappearance of a plant or animal species. See *Combustion* and *Pre-ignition*.

Extratropical cyclone A cyclone that develops over land or water in temperate regions, typically between 30° and 70° latitude.

Extreme cold Cold weather capable of causing injury or death. Extreme cold is associated with temperatures considerably below normal and moderate to strong winds. What constitutes extreme cold may vary across a country. Near-freezing temperatures are considered to be extreme cold in regions unaccustomed to cold winter weather.

Eye The centre of a hurricane, characterized by calm conditions and broken clouds, and ranging from 5 to over 60 km in diameter; surrounded by the *eyewall*.

Eyewall The portion of a hurricane bordering the eye; it is the region with the highest winds and heaviest rainfall. See *Eye*.

E-zone The area between the shore and an E-line that is expected to be lost because of erosion within a specified period. See *E-line*.

Factor of safety The ratio of resisting to driving forces on a slope. A slope with a factor of safety less than 1 might fail.

Fall A type of landslide in which rock or sediment bounds and rolls down a steep slope.

Fault A fracture along which adjacent rocks have moved relative to one another.

Fault creep The slow, continuous movement of rock or sediment along a fracture. Also called *tectonic creep*.

Fault scarp A linear escarpment at Earth's surface formed by movement along a fault during an earthquake.

Faulting The differential displacement of rocks on opposite sides of a fault. See *Fault*.

Fetch The distance that wind blows over a body of water. Fetch influences the height of wind-blown waves.

Fiord A glacially over-deepened valley, usually narrow and steep-sided, extending below sea level and filled with sea water. Also spelled *fjord*.

Fire break A strip of land that is cleared of trees and woody fuel to stop or control the spread of fire.

Fire management The control of wildfires to minimize loss of human life and property damage; accomplished through scientific research, public education, remote sensing of vegetation to determine fire potential, and prescribed burns.

Fire regime The potential for wildfire inferred from information about fuel types, terrain, and past fire behaviour.

Fissure In the context of volcanoes, a large fracture or crack in the upper crust that becomes a volcanic vent when lava or pyroclastic material erupts.

Flash flood The overbank flow resulting from a rapid increase in stream discharge; commonly occurs in upstream parts of drainage basins and in small tributaries downstream.

Flashy discharge Stream flow characterized by a rapid increase in discharge following precipitation.

Flood basalt An extensive series of basaltic lava flows.

Flood discharge The volume of water conveyed by a stream per unit time when a river is in flood and likely to damage property. See *Flood stage*.

Flood fringe The higher part of a floodplain where floodwaters are relatively shallow and slow. Some new development might be permitted on the flood fringe provided that it is adequately flood-proofed. See *Floodplain* and *Floodway*.

Flood stage The level of a stream at which water escapes the channel and threatens property. See *Flood discharge*.

Flooding The high water that inundates low-lying areas adjacent to a stream, lake, or the sea; includes stream overbank flows that deposit sediment on a floodplain and storm surges that raise water levels along a coast.

Floodplain The flat land adjacent to a stream produced by overbank flow and lateral channel migration.

Floodplain regulation Government restrictions on land use in areas subject to flooding by streams.

Floodway The portion of a floodplain with the deepest, fastest, and most destructive waters. See *Floodplain* and *Flood fringe*.

Flow A type of landslide in which material moves downslope in a viscous fluid state; includes debris flows and mudflows.

Flowstone A general term for an accumulation of calcium carbonate precipitated from water in a cave. See *Stalagmite* and *Stalagtite*.

Focus The point on a fault within Earth where rocks first rupture during an earthquake; seismic energy radiates out from this point. Also known as *hypocentre*. See *Epicentre*.

Fog A cloudlike mass of condensed water vapour lying close to the ground.

Foliation plane A planar surface in metamorphic rock resulting from the preferential alignment of platy minerals, mainly micas.

Footwall The rock directly below an inclined fault. See *Hanging wall*.

Forcing A factor that can cause climate to change; for example, solar radiation, aerosols, and greenhouse gases.

Forecast A public announcement that a flood, earthquake, volcanic eruption, or other event is likely to occur during a specified period, commonly made with a statement of probability.

Foreshock A small to moderate earthquake that occurs before and in the same general area as the main earthquake. See *Aftershock* and *Mainshock*.

Fractionation Progressive changes in the chemistry of a slowly cooling magma because of sequential crystallization of different minerals.

Front The boundary between two air masses that differ in temperature and, commonly, moisture content. See *Occluded front* and *Stationary front*.

Frost heaving The upward or lateral movement of land caused by the formation of ice below the surface.

Frostbite An injury to body tissues caused by exposure to extreme cold.

Frost-susceptible sediments Surficial sediments that expand when water within the sediments freezes. Also referred to as *frost-susceptible soils*.

Fumarole An opening in the ground, commonly in an active volcanic area, that emits steam and other gases such as carbon dioxide, sulphur dioxide, and hydrogen sulphide.

Funnel cloud A narrow, rotating column of air extending downward from a thunderstorm. A funnel cloud becomes a tornado if it reaches the ground.

Gabion A metal wire basket filled with rocks. Gabions are placed along stream banks to prevent erosion and on slopes to reduce the amount of sediment or rock falling onto roads, rail lines, and other structures.

Gallery A rigid structure built to protect a road or rail line from snow avalanches. A gallery can be made of steel, pre-stressed concrete, or timbers, and can be fully enclosed, like an artificial tunnel, or a lattice-like structure. See *Avalanche shed*.

Geographic information system (GIS) A three-dimensional, computer-generated model used to store, manipulate, retrieve, and display spatial data and to make maps using these data.

Geologic cycle The four associated sequences of Earth processes: the hydrologic, rock, tectonic, and geochemical cycles.

Geologic map A two-dimensional representation of surface Earth within a defined geographic area.

Geyser A spring that ejects hot water and steam at regular or irregular intervals. The most famous geyser is Old Faithful in Yellowstone National Park.

Glaciation A period in Earth history when climate was cooler and glaciers and sea ice were more extensive than they are today. See *Interglaciation*.

Glacier A large mass of ice that slowly flows under its own weight and persists from year to year. A glacier is formed by compaction and recrystallization of snow.

Glacio-isostatic uplift Upward movement of a large landmass caused by the decay of an ice sheet.

Global climate model (GCM) A computer program or programs that use mathematical equations to predict atmospheric and ocean circulation and climate in the future.

Global positioning system (GPS) A system of linked satellites and ground receivers that allows accurate location of points on Earth's surface. GPS can be used to detect changes in the land surface related to volcanic activity, earthquakes, and landslides.

Global warming The hypothesis that consumption of fossil fuels and land clearing, which produce carbon dioxide and other greenhouse gases, are increasing the mean temperature of Earth's lower atmosphere.

Gradient The average slope of a stream channel; that is, the ratio of the vertical drop of the stream over the horizontal distance of the drop; commonly expressed as a dimensionless number (m/m).

Great earthquake An earthquake of moment magnitude 8 or larger.

Greenhouse effect Trapping of heat in the lower atmosphere by water vapour, carbon dioxide, methane, and other gases.

Greenhouse gas Water vapour, carbon dioxide, methane, nitrogen oxides, and other gases that absorb infrared radiation and warm the lower atmosphere.

Groin A long, narrow rock, concrete, or other structure generally constructed perpendicular to the coast to protect the shoreline from erosion. Groins trap sediment in the zone of littoral drift. See *Groin field*.

Groin field A group of adjacent groins along a coast. See *Groin*.

Ground acceleration The rate at which the velocity of seismic waves increases per unit time. Ground acceleration is commonly expressed as a fraction or percentage of gravitational acceleration, which is 980 cm/s^2. Maximum ground acceleration is an important measure of the amount of structural damage an earthquake will cause.

Ground blizzard A type of blizzard caused by strong winds blowing snow from the ground. See *Blizzard*.

Ground penetrating radar (GPR) An electronic instrument that uses electromagnetic waves to image Earth's shallow subsurface. GPR is used to locate subsurface strata, caves, the water table, and buried pipes and tanks.

Groundwater mining The extraction of groundwater at an rate that is faster than the water is replenished.

Gust front The leading edge of strong and variable surface winds produced by downdraft during a thunderstorm.

Habitat An environment in which adapted plants and animals thrive.

Hailstone A rounded or irregular piece of ice that has increased in size while moving up and down within the clouds of a thunderstorm.

Halocarbon A synthetic chemical consisting of carbon, commonly hydrogen, and either chlorine, fluorine, bromine, or iodine. Most halocarbons contribute to ozone depletion in the stratosphere.

Hanging wall The rock directly above an inclined fault. See *Footwall.*

Hazard See *Natural hazard.*

Headland A peninsula on an irregularly shaped coastline that is more resistant to erosion than the surrounding shoreline.

Headwaters The tributaries of a stream or river near its source.

Heat energy The energy produced by random molecular motion. Heat energy may be transferred by conduction, convection, or radiation.

Heat index See *Humidex index.*

Heat island An area such as a city or industrial site that has consistently higher air temperatures than surrounding areas because of the greater retention of heat by buildings, concrete, and asphalt.

Heat wave Period of heat that is longer and hotter than normal; one of the deadliest natural hazards.

High-pressure centre An area of high atmospheric pressure characterized, in the Northern Hemisphere, by clockwise airflow.

Hoar The crystals of ice formed from snow by sublimation of water vapour.

Hoar frost The hoar that forms at the snowpack surface.

Holocene Epoch The interval of Earth history between 11 600 years ago and the present. The Holocene Epoch is commonly referred to as *post-glacial time.* See *Pleistocene Epoch.*

Hot spot A hypothesized, nearly stationary source of heat within the mantle that is the source of volcanic activity at specific, fixed points on Earth's surface.

Hot spring A natural discharge of groundwater at a temperature higher than that of the human body. See *Thermal spring.*

Hot tower A very tall cloud that develops within the eyewall of a hurricane before the storm intensifies. The towers release huge amounts of heat and contribute to a strengthening of winds in the hurricane.

Humidex index A numerical value that describes a person's perception of air temperature by taking into account relative humidity.

Humidity The amount of water vapour in the air.

100-year flood The largest flood likely to occur along a stream within any 100-year period; the 100-year flood has a 1 percent probability of occurring in any given year.

Hurricane A tropical cyclone with sustained winds of at least 118 km/h (119 km/h in the United States). Hurricanes are called "typhoons" in the western North Pacific Ocean and South China Sea, "tropical cyclones" in the South Pacific and southern Indian oceans, and "cyclonic storms" in the northern Indian Ocean. See *Tropical cyclone, Tropical depression, Tropical disturbance,* and *Tropical storm.*

Hurricane warning An alert issued for an area where hurricane conditions are expected within 24 hours. In Canada, this alert applies to areas that are likely to experience high waves and coastal flooding, but not necessarily hurricane-strength winds. See *Hurricane.*

Hurricane watch An alert for an area that could experience hurricane conditions in the near future. In the United States, the alert is restricted to a 36-hour period. See *Hurricane.*

Hydraulic fracturing A process by which a mixture of sand, water, and chemicals is injected under high pressure into hydrocarbon-bearing sedimentary rocks to create or widen fractures, thus allowing more oil and gas to flow out of the formation and into the well bore.

Hydrofluoric acid A corrosive aqueous solution of hydrogen fluoride.

Hydrogen sulphide A toxic flammable gas comprising molecules of hydrogen and sulphur. It is emitted by volcanoes and also is produced by the bacterial breakdown of organic matter in the absence of oxygen.

Hydrograph A graph showing the discharge of a stream over time.

Hydro-isostatic uplift Upward movement of the continental shelf caused by a lowering of sea level.

Hydrologic cycle A cyclic circulation of water between the oceans and atmosphere by means of evaporation, precipitation, surface runoff, and groundwater flow.

Hydrological drought Drought caused when water reserves in aquifers, lakes, reservoirs, and streams are below normal for an extended period. It commonly is caused by *meteorological drought.*

Hydrophobic layer A layer within soil that has accumulated water-repellent organic chemicals. Hydrophobic layers commonly develop in drier climates after very hot fires.

Hypocentre See *Focus.*

Hypothermia A medical emergency that occurs when the human body loses heat faster than it can produce heat. The result is an abnormally low and potentially dangerous body temperature.

Hypothesis A tentative explanation for an observation, a phenomenon, or a scientific problem that can be tested by further investigation.

Ice shelf The floating extension of an ice sheet.

Ice storm A period of freezing rain during which thick layers of ice accumulate on cold surfaces.

Iceberg A large block of floating ice that has broken, or calved, from the front of a glacier or ice shelf.

Igneous rock Rock formed by crystallization of magma. Extrusive igneous rock crystallizes on or very near Earth's surface; intrusive igneous rock crystallizes at depth.

Impact We use this word in two senses in this book: (1) the long-term consequences of a hazardous process; and (2) the collision of a comet or asteroid with Earth.

Impact crater The bowl-shaped depression produced by the impact of an asteroid.

Impervious cover A surface covered with concrete, asphalt, roofs, or other structures that impedes infiltration of water into the ground. The percentage of land with impervious cover increases as urbanization proceeds.

Inactive fault A fault that has not moved in the past 2 million years. See *Active* and *Potentially active faults.*

Indirect effect A secondary effect of an event. Indirect effects of a disaster or catastrophe include emotional distress; lost production and wages; donation of money, goods, and services; and payment of taxes to finance recovery. Also called a *secondary effect.* See *Direct effect.*

Injection well A drilled hole into which water, carbon dioxide, or hazardous waste is pumped.

Inner core The innermost part of Earth, with a radius of about 1200 km. The inner core is thought to consist primarily of an iron–nickel alloy and to have a temperature of about 5430°C, similar to the temperature of the surface of the sun. See *Outer core.*

Instrumental intensity Measurements of earthquake ground motion made with a seismograph. The data are used to produce maps showing perceived ground shaking and potential structural damage. See *Intensity.*

Insurance The guarantee of monetary compensation for a specified loss in return for payment of a premium.

Intensity A measure of the severity of shaking and damage caused by an earthquake at a specific place. The Modified Mercalli Scale provides a numerical estimate of an earthquake's effects on people and structures. See *Instrumental intensity, Magnitude,* and *Modified Mercalli Intensity Scale.*

Interglaciation A period in Earth history when climate was relatively warm and the extent of glacier and sea ice on the planet was similar to what it is today. See *Glaciation.*

Intermittent crown fire A wildfire that moves swiftly through forests, consuming the tops of some trees but not burning others. See *Continuous crown fire.*

Interplate earthquake An earthquake on a fault that bounds two lithospheric plates.

Intraplate earthquake An earthquake on a fault in the interior of a continent, far from a plate boundary.

Ion An electrically charged atom or molecule produced by the loss or gain of electrons.

Iridium A platinum-group metal that is more abundant in meteorites than in Earth's crust. High concentrations of the element in a clay layer separating Cretaceous and early Tertiary rocks have been attributed to an asteroid impact.

Isostasy Equilibrium in the lithosphere such that the forces tending to elevate the land balance the forces tending to depress it.

Jack-strawed The tilting of trees because of downslope movement of the rock or sediment in which they are rooted. Jack-strawed trees are a good indicator of recent or active landsliding.

Jet stream A concentrated flow of air near the top of the troposphere, characterized by strong winds.

Jetty A long, narrow engineered structure of rock, concrete, or other material, generally constructed perpendicular to the shore at the mouth of a river or an inlet to a lagoon, estuary, or bay. Jetties are commonly constructed in pairs and are designed to stabilize a channel, control sediment deposition, or deflect large waves.

Jökulhlaup An outburst flood from a glacier-dammed lake.

Karst An area of carbonate rock characterized by features produced by dissolution of calcite and dolomite, notably fissures, sinkholes, and caves. Surface drainage in karst areas is poor, and streams may flow underground.

Karst plain A generally flat land surface containing numerous sinkholes developed in limestone or marble. See *Karst*.

Karst topography A landscape characterized by sinkholes, caverns, and subterranean drainage. See *Karst*.

Kinetic energy The energy associated with the motion of an object. See *Potential energy*.

K-T boundary The boundary between the Cretaceous and Tertiary periods, dating to approximately 65 million years ago.

Kuiper Belt A ring-like zone of abundant comets in the outer solar system. See *Oort Cloud*.

Kyoto Protocol The international agreement to reduce emissions of greenhouse gases in an effort to manage global warming. The protocol is now legally binding in at least 128 countries, including Russia and most European nations. The United States is not a signatory.

La Niña A combined meteorological and oceanographic event during which surface ocean waters in the eastern tropical Pacific Ocean become anomalously cool. A La Niña event commonly begins during the summer and lasts one to three years. It is the counterpart to the El Niño "warm event," and, to a considerable degree, its development is the mirror image of El Niño, although La Niña events tend to be more irregular in their behaviour and duration. See *El Niño*.

Lag time The length of time between the peak rainfall during a storm and the ensuing maximum discharge of a stream. Urbanization generally decreases lag time.

Lahar A debris flow or mudflow on the slope of a volcano. Some scientists restrict the term to warm or hot debris flows triggered by eruptions, whereas other scientists apply it to any volcanic debris flow, whether it is eruptive or not.

Landslide The failure and downward and outward movement of a body of rock or sediment under the influence of gravity.

Landslide hazard map A two-dimensional representation of the land surface showing landslides and potentially unstable areas that might experience landslides in the future. See *Landslide risk map*.

Landslide risk map A two-dimensional representation of the land surface of an area showing different levels of risk that people face from landslides. See *Landslide hazard map*.

Land-use planning The development of a plan for future development of an area. The plan could recommend zoning restrictions and infrastructure that is appropriate for the community and its natural environment. Land-use planning is based on analysis of existing human activities and environmental conditions, including natural hazards.

Lapilli Gravel-size pyroclastic debris ranging in size from 2 mm to 64 mm. See *Block*.

Large igneous province A large accumulation of intrusive or extrusive igneous rocks, or both, formed over a period of 50 million years or less in an intraplate tectonic setting.

Latent heat The energy created by random molecular motion; it is absorbed or released when a substance undergoes a phase change, for example during evaporation or condensation. See *Sensible heat* and *Latent heat of vaporization*.

Latent heat of vaporization The energy absorbed by water molecules when they evaporate. See *Latent heat*.

Lateral blast A volcanic eruption characterized by an explosion directed away from the volcano more or less parallel to the ground surface; may occur when the side of a volcano collapses, as happened at Mount St. Helens in May 1980.

Lateral moraine A ridge consisting of loose glacial debris deposited at the margin of a valley glacier.

Laurentide Ice Sheet The large body of glacier ice that covered most of Canada and adjacent areas of the United States at times during the Pleistocene Epoch. At its maximum, the Laurentide Ice Sheet extended from the Rocky Mountains in the west to Atlantic Canada in the east. See *Cordilleran Ice Sheet*.

Lava The molten rock (magma) that flows from a volcano.

Lava flow The molten rock (magma) that flows downslope from a volcanic vent, cools, and solidifies.

Lava tube A natural conduit or tunnel through which lava has flowed. A lava tube is a type of cave.

Leda clay A highly sensitive clay deposited in the Champlain Sea in the St. Lawrence Lowland in southern Ontario and southern Quebec at the end of the last glaciation. Leda clay can change from a stiff state into a liquid mass when disturbed. See *Liquefaction*.

Levee A linear mound or embankment bordering a stream. Levees include natural features consisting of fine sediment deposited by overbank flood flows, and artificial embankments constructed by humans to protect adjacent land from flooding.

Lightning A natural, high-voltage electrical discharge between a cloud and the ground, between two clouds, or within a cloud. The discharge takes a few tenths of a second and emits a flash of light that is followed by thunder.

Linkage A physical or causal relationship between two objects or phenomena.

Liquefaction The transformation of water-saturated granular material from a solid state to a liquid state. Liquefaction commonly occurs during strong earthquakes.

Lithification The hardening of loose sediment into sedimentary rock by cementation and compaction.

Lithosphere The outermost layer of Earth, approximately 100 km thick, comprising the crust and the upper mantle. The lithosphere is broken into tectonic plates that move slowly with respect to one another. See *Crust*.

Lithospheric plate A slab of Earth's crust and upper mantle that forms part of the planet's hard outer shell. Earth's outer shell consists of about 16 major lithospheric plates.

Little Ice Age A recent period of variable, but generally cooler climate during which mountain glaciers around the world were larger than they are today. The beginning of the Little Ice Age is placed as early as the thirteenth century by some scientists and as late as the sixteenth century by others. The Little Ice Age terminated around the end of the nineteenth century.

Littoral cell A segment of coast that includes sediment sources and the littoral transport system that moves and deposits sand and gravel on the beach. See *Littoral transport*.

Littoral transport The movement of near-shore sediment parallel to the coast because of the return flow of water from waves. See *Longshore current* and *Beach drift*.

Local tsunami A tsunami that affects shorelines from a few kilometres to about 100 km from its source. See *Distant tsunami*.

Loess Silt deposited by wind.

Longitudinal profile With respect to streams, a graph showing the decrease in elevation of a streambed between its head and its mouth.

Longshore bar A submerged, elongated ridge of sand roughly parallel to the shore. Longitudinal bars are produced by wave action. See *Longshore trough*.

Longshore current A water flow parallel to the shore. Longshore currents develop in the surf zone where waves strike the coast at an angle. They are responsible for longshore drift. See *Littoral transport* and *Beach drift*.

Longshore drift The sand transported parallel to the shore by a longshore current. See *Longshore current* and *Beach drift*.

Longshore trough A submerged, elongated depression parallel to the shore and bordered by a longshore bar.

Love wave An earthquake-generated wave that travels along the surface and is characterized by a transverse, snake-like form.

Low-pressure centre An area of low atmospheric pressure characterized by anticlockwise airflow that forms along a stationary front in the Northern Hemisphere.

Maar A flat-bottomed, roughly circular volcanic crater produced by a single explosive eruption and commonly filled with water.

Magma The molten rock formed deep within Earth's crust or in the upper mantle.

Magnetic reversal A reversal in the polarity of Earth's magnetic field, such that magnetic north becomes magnetic south, or vice versa.

Magnetometer An instrument that measures the strength and, in some cases, the direction of Earth's magnetic field.

Magnitude The amount of energy released during an earthquake (symbol **M**). See *Intensity*.

Magnitude-frequency concept The concept that the size (intensity and extent) of an event is inversely proportional to its probability.

Mainshock The largest earthquake in a series of associated earthquakes. See *Aftershock* and *Foreshock*.

Major earthquake An earthquake of moment magnitude 7.0 to 7.9.

Mantle The hot viscous layer between Earth's crust and outer core. The mantle is about 2900 km thick and constitutes about 84 percent of the planet's volume.

Mass extinction The sudden disappearance of large numbers of plant and animal species.

Mass wasting A comprehensive term for any type of downslope movement of rock or sediment.

Material amplification An increase in the intensity of earthquake ground shaking because of the type and thickness of geologic material through which seismic waves pass.

Mature stage The second phase of a thunderstorm, when downdrafts move from the base of a cumulonimbus cloud and produce heavy precipitation. Tornadoes touch down during this phase of the storm. See *Cumulus stage* and *Dissipative stage*.

Meander An arcuate bend in the channel of a stream with a snake-like form. Sediment is deposited on a point bar on the inside of the meander bend and is eroded from a cutbank on the outside of the bend. See *Meandering river*.

Meander cutoff A crescent- or horseshoe-shaped stream channel that was abandoned when the stream established a shorter, more direct course across the narrow neck of a meander. The abandoned channel might be filled with water. See *Oxbow lake*.

Meander scroll A series of point bars formed by deposition of sediment on the inner bank of a meander. Migration of the meander loop produces a corrugated topography comprising sequentially deposited bars.

Meandering river A river with a single, sinuous channel that migrates back and forth across its floodplain over time. See *Meander*.

Mesocyclone An area 3 km to 10 km in diameter on the flank of a supercell storm and characterized by rotating clouds.

Mesoscale convective complex A large, circular, and long-lived group of thunderstorms that interact with one another.

Mesosphere The zone about 50 km to 80 km above Earth's surface where temperature decreases with altitude. The mesosphere is directly above the stratosphere. See *Stratosphere, Thermosphere,* and *Troposphere*.

Metamorphic rock Rock produced at depth in Earth's crust from pre-existing sedimentary, igneous, or metamorphic rocks through the action of heat, pressure, and chemically active fluids. Foliated metamorphic rocks consist of aligned mineral grains or bands of alternating light and dark minerals; non-foliated metamorphic rocks have neither of these characteristics.

Metamorphosed Altered through the action of heat, pressure, and the introduction of new chemical substances. See *Metamorphic rock*.

Meteor An extraterrestrial particle up to several centimetres in size that is consumed by frictional heat as it passes through Earth's atmosphere. Light emitted from the burning meteor forms a shooting star. See *Asteroid, Meteorite,* and *Meteoroid*.

Meteor shower A large number of meteoroids that generate light as they pass through Earth's atmosphere. Meteor showers commonly occur when Earth passes through the tail of a comet. See *Asteroid, Meteor*, *Meteorite*, and *Meteoroid*.

Meteorite An extraterrestrial particle that is dust- to asteroid-size and that hits Earth's surface. See *Asteroid*, *Meteor*, and *Meteoroid*.

Meteoroid An extraterrestrial particle in space that is smaller than 10 m in diameter and larger than dust-size; a meteoroid can form from the breakup of an asteroid. See *Asteroid, Meteor,* and *Meteorite*.

Meteorological drought Drought caused by a prolonged period with less-than-average precipitation.

Microburst A strong, localized downdraft beneath a thunderstorm that produces strong, potentially damaging winds.

Microearthquake An earthquake of moment magnitude less than 3. Also called a *very minor earthquake*.

Microzonation The detailed delineation of areas that are subject to different types and degrees of earthquake hazards. Microzonation is used in earthquake mitigation and land-use planning.

Mid-ocean ridge A long, relatively narrow mountain range on the ocean floor where new crust is formed by seafloor spreading. Mid-oceanic ridges are commonly found in the central parts of oceans; the Mid-Atlantic Ridge is an example.

Milankovitch cycles The cyclic variations in Earth's orbital path that produce changes in the amount of solar radiation reaching Earth's surface through time. The orbital cycles have periodicities of approximately 20 000, 40 000, and 100 000 years. See *Eccentricity, Obliquity,* and *Precession of the equinoxes*.

Mineral A naturally occurring, inorganic crystalline substance with an ordered atomic structure and a specific chemical composition.

Mitigation Any actions taken to offset the harmful effects of a hazardous natural event. Mitigation includes avoidance, construction to reduce the impact of the event, purchase of insurance, and education.

Modified Mercalli Intensity Scale An earthquake intensity scale with 12 categories of ground-shaking and structural damage.

Moho The boundary between Earth's crust and the underlying mantle. Also called the *Mohorovicic discontinuity*.

Moment magnitude A numerical measure of the amount of energy released by an earthquake. It is based on the seismic moment, which is defined as the product of the average amount of slip on the fault, the rupture area, and the shear modulus of the ruptured rocks. In simple terms, *shear modulus* is the measure of how hard a rock mass must be pushed to shear it. See *Magnitude*.

Moulin A vertical or near-vertical shaft in a glacier that carries surface water downward through the ice.

Mound A pile of rocks, gravel, or debris heaped in the run-out zone of an avalanche track to slow and break up avalanches. Mounds are commonly built in groups.

Natural disaster A natural event that causes serious injury, loss of life, and property damage within a specific geographic area. Natural processes that produce disasters include floods, hurricanes, earthquakes, tsunamis, volcanic eruptions, landslides, and wildfires.

Natural hazard A natural process that poses a potential threat to people and property.

Natural service function As used in this book, a benefit that arises from a natural event that might damage people or the environment.

Near-Earth Object (NEO) The asteroids that orbit between Earth and the sun or that occasionally enter the solar system and pass close to Earth.

Nor'easter A severe storm that tracks along the Atlantic coast of Canada and the United States and is immediately preceded by continuously blowing northeasterly winds. Nor'easters commonly produce hurricane-strength winds, intense precipitation, and large waves along the coast.

Normal fault A fault along which the hanging wall has moved down relative to the footwall. Also called a *dip-slip fault*. See *Reverse*, *Strike-slip*, and *Thrust faults*.

Nuées ardente See *Pyroclastic flow*.

Obliquity A change in the tilt of Earth's axis of rotation with respect to its plane of orbit.

Occluded front A composite front formed when a cold front overtakes a warm front, forcing the warm air mass aloft. See *Front* and *Stationary front*.

Ocean acidification An increase in the acidity of ocean waters due to the greater uptake of carbon dioxide from the atmosphere. Ocean acidification is a threat to many marine organisms, for example corals and other reef invertebrates.

Ocean conveyor belt The large-scale circulation pattern in the Atlantic, Indian, and southwest Pacific oceans driven by differences in water temperature and density; also called *thermohaline current*. Northward-flowing warm water maintains the moderate climate of northern Europe; if this flow were to slow, stop, or shift, glacial conditions might return to portions of the Northern Hemisphere.

Oort Cloud A zone of comets outside the solar system. See *Kuiper Belt*.

Organic sediment An *in situ* accumulation of partially decayed plant material, mainly peat. Organic sediment is common in swamps, marshes, bogs, and fens. See *Peat*.

Organizational stage The initial phase of a tornado, when a funnel cloud forms from rotating winds within a severe thunderstorm. See *Rope stage* and *Shrinking stage*.

Outer core A layer of liquid nickel and iron about 2260 km thick that lies between the solid inner core and the mantle of Earth. The temperature of the outer core ranges from about 4400°C to 6100°C. See *Inner core.*

Outflow boundary A line separating an air mass cooled by a thunderstorm and the surrounding air. It behaves like a cold front and is characterized by a shift in wind and drop in temperature. In large thunderstorms, the outflow boundary could persist for more than 24 hours and cause instability that leads to new storms.

Outlook An official statement providing information about expected river levels during spring thaw.

Overbank flow The floodwater that escapes the channel of a stream or river and spreads onto the surrounding floodplain.

Overwash The inundation of beaches, dunes, and, in some cases, an entire barrier island by a storm surge. Overwash commonly transports sand from the beach and dunes inland. See *Washover channel.*

Oxbow lake A crescent-shaped lake occupying an abandoned meander of a river or stream. See *Meander cutoff.*

Ozone A pungent, reactive form of oxygen that is most abundant in the stratosphere, where it protects Earth from high levels of ultraviolet radiation. Ozone is a pollutant in the lower atmosphere.

Ozone depletion The loss of stratospheric ozone because of releases of chlorofluorocarbons (CFCs) and other gases into the atmosphere.

P wave A seismic wave that travels from the hypocentre of an earthquake by compressing and extending rock and fluids along its path. P waves are the fastest of all earthquake waves. Also called a *primary* or *compressional wave.*

Pahoehoe A basaltic lava flow with a ropy surface texture; also called *ropy lava.*

Paleomagnetism The study of the magnetism of rocks at the time their magnetic signature was acquired.

Paleoseismicity The occurrence of earthquakes in the geologic past.

Paleoseismologist A scientist who documents prehistoric earthquakes and deformation by studying young sediments, rocks, and landforms.

Palermo Technical Hazard Impact Scale A logarithmic scale used by astronomers to rate the potential hazard of impact of a near-Earth object (NEO). The scale combines a measure of the probability of impact and the estimated kinetic yield.

Parcel A poorly defined body of air several hundred cubic metres in volume that acts independently of surrounding air.

Passive margin A continental margin bordering an ocean basin that expands by rifting.

Peat A sediment consisting mostly of fresh or decomposed plant remains, such as woody material, mosses, roots, and leaves. Peat is water-saturated in its natural state.

Perihelion The point in a planet's orbit where it is nearest the sun. At present, Earth's perihelion occurs about January 1, when the planet is 4.8 million km nearer the sun than at aphelion. The seasons in which aphelion and perihelion occur vary cyclically with a period of 21 000 years. See *Aphelion.*

Permafrost Rock or sediment that is colder than 0°C continuously for at least two years and contains disseminated or segregated ice. Permafrost underlies about 20 percent of Earth's land area, including about half of Canada, and is commonly dozens to hundreds of metres thick.

Permafrost table The upper surface of permanently frozen ground, directly below the active layer. See *Permafrost* and *Active layer.*

Permanent gas An atmospheric gas such as nitrogen or oxygen that cannot be compressed by pressure alone and is present in generally constant amounts. See *Variable gas.*

Photosynthesis The process by which green plants and some bacteria use light energy (from sunlight), carbon dioxide, and water to synthesize complex organic compounds.

Pile Long concrete, steel, or wooden beams driven into the ground to stabilize a slope.

Pillow A bulbous mass of basalt produced by extrusion of molten rock from the front of a cooling lava flow in a lake or the sea. Some lava flows consist largely of pillows and pillow fragments. See *Pillow breccia.*

Pillow breccia Basalt consisting of fragments of pillows. See *Pillow.*

Pineapple Express A strong and persistent flow of moist air and associated heavy rainfall originating in the waters adjacent to the Hawaiian Islands and extending to any location along the Pacific coast of North America.

Piping The slow subsurface removal of silt and fine sand from a sedimentary deposit by groundwater. Piping can produce subsurface cavities, tunnels, and sinkholes.

Plate tectonics A theory that explains the global distribution of earthquakes, active volcanoes, mountains, and other geologic features by the movement of large fragments of the Earth's crust on less rigid mantle rocks.

Plate-boundary earthquake An earthquake that occurs on a fault separating lithospheric plates; may be a *strike-slip*, *thrust*, or *dip-slip* earthquake.

Pleistocene Epoch The period of Earth history from 2.6 million years ago to 11 600 years ago, characterized by recurrent widespread continental glaciation. The Pleistocene is one of two epochs of the Quaternary Period and is commonly referred to as the *Ice Age*. See *Holocene Epoch.*

Plunging breaker A type of breaking wave in which the crest falls into the trough as the wave approaches the shoreline. Plunging breakers are associated with relatively steep beaches.

Point bar An accumulation of sand or gravel along the inside of a meander bend of a stream.

Point-release avalanche A type of snow avalanche in which a small mass of failed snow grows in size as it travels downslope, leaving a track with the shape of an inverted V.

Polar desert An arid, cold, high-latitude area that receives less than 250 mm of precipitation per year.

Polar jet stream A fast-flowing stream of air near the top of the troposphere that, in the Northern Hemisphere, typically crosses southern Canada or the northern conterminous United States and is characterized by strong winds blowing from west to east.

Pool A common erosional bedform in meandering and straight stream channels; characterized by slow-moving, deep water during low flow stages.

Pore water The water occupying pore space in sediment or sedimentary rock.

Pore water pressure The pressure of water held within the pores or fractures in sediments or rock.

Positive feedback Any form of feedback that magnifies or enhances a process or increases its output.

Potential energy The energy contained in an object by virtue of its position above a reference level. See *Kinetic energy.*

Potentially active fault A fault capable of producing an earthquake. See *Active* and *Inactive faults.*

Precession of the equinoxes A change in the direction of Earth's axis of rotation relative to the sun at the time of aphelion and perihelion. See *Aphelion* and *Perihelion.*

Precursor A physical, chemical, or biological phenomenon that immediately precedes an earthquake, volcanic eruption, landslide, or other hazardous event. See *Precursor event.*

Precursor event An event that precedes and signals an imminent earthquake, volcanic eruption, landslide, or other hazardous event.

Prediction Warning of a hazardous event such as an earthquake; involves specifying the date and size of the event.

Preheated The loss of water and other volatile chemical compounds during the pre-ignition phase of a wildfire. See *Pre-ignition.*

Pre-ignition The initial phase of a wildfire, during which fuel is brought to a temperature and water content that allow ignition; involves preheating and *pyrolysis*. See *Preheated*, *Combustion*, and *Extinction.*

Preparedness A state of readiness achieved by an individual in order to minimize damage and injury from a natural disaster, war, or pandemic. See *Disaster preparedness.*

Prescribed burn A fire purposely set and contained within a designated area to reduce the amount of fuel available for a wildfire. Also called a *controlled burn.*

Primary effect See *Direct effect.*

Primary wave See *P wave.*

Principle of uniformitarianism The scientific law stating that the geologic processes occurring today operated in the past and can therefore be used to explain past geologic events.

Proactive In this book, actions taken to prepare for disasters. See *Reactive.*

Process In the context of this book, ways in which volcanic eruptions, earthquakes, landslides, floods, and other hazards affect Earth's surface.

Punctuated uniformitarianism The concept that rare, catastrophic events have caused major changes to Earth's surface and biota.

Pyroclastic debris See *Pyroclastic deposit*.

Pyroclastic deposit The accumulation of volcanic debris blown from a volcano during an explosive eruption; the debris ranges from ash particles to blocks and bombs.

Pyroclastic flow A rapid flow of incandescent ash, blocks, and gas that have been explosively erupted from a volcano. Pyroclastic flows result from the collapse of an eruption column or lava dome. Also called *ash flow* or *nuées ardente*.

Pyroclastic rock Rock composed of volcanic glass, mineral crystals, and lithic fragments ejected during an explosive volcanic eruption.

Pyroclastic surge A hot, gaseous, turbulent flow of pyroclastic material that travels at high velocities down the flank of a volcano. Pyroclastic surges are less dense than pyroclastic flows.

Pyrolysis A group of chemical processes operating at high temperatures that split large fuel molecules into smaller ones; the products of these processes are volatile gases, mineral ash, tars, and carbonaceous char.

Quick clay Clayey sediment that may liquefy when disturbed, as during an earthquake.

Radiation The transfer of energy by electromagnetic waves or by moving subatomic particles. Radiation is the main mechanism by which heat is transferred from the sun to Earth, from Earth's surface to its atmosphere, and, to a lesser extent, from a wildfire to the atmosphere. See *Conduction* and *Convection*.

Radiocarbon dating A technique for determining the age of organic material, such as wood, charcoal, bone, and shell, by measuring the ratio of unstable carbon-14 to stable carbon-12. Radiocarbon dating can be used to date organic material up to about 50 000 years old. Also called *carbon dating* and *carbon-14 dating*. See *Carbon-14 method.*

Rain band Precipitation-producing clouds that spiral around a hurricane.

Rayleigh wave A seismic wave that travels at Earth's surface with a retrograde elliptical motion.

Reactive The action or response of a person or government to an accident, natural disaster, or catastrophe. See *Proactive*.

Recurrence interval The time between successive floods, earthquakes, or other disastrous events. Recurrence intervals are commonly expressed as average values (in years), based on a series of events.

Red zone Mapped areas in Switzerland and parts of the United States, within which snow avalanches capable of high impact pressures (30kPA or greater) occur on average once every 300 years. Buildings are generally not allowed in the red zone. See *Blue zone.*

Redirection Reflection or scattering of electromagnetic energy from the sun when it encounters Earth's atmosphere and surface.

Reflected Past tense of the verb *reflect*, which means that a surface or body returns waves—electromagnetic, light, heat, or sound—without changing or absorbing them.

Refracted Past tense of the verb *refract*, which means that the direction of a wave, for example a light or sound wave, changes as it passes from one medium to another. See *Refraction*.

Refraction The bending of surface waves as they enter shallow water. Waves move forward at a faster velocity in deep water than in shallow water.

Reinsurance The practice by which an insurer transfers part of its risk portfolio to another party in order to reduce the risk it faces from insurance claims.

Relative humidity The ratio of the amount of water vapour in air to a hypothetical amount that would saturate the air at a given temperature and pressure; commonly expressed as a percentage.

Relative sea level The local position of the sea relative to the land, which is influenced by global eustatic sea level and by local or regional uplift or subsidence.

Relief The difference in elevation between a high point, such as a mountain top, and an adjacent lower one, such as a valley floor.

Residence time The length of time that an element, compound, or other substance spends at one place or within a specific part of a natural system.

Resisting force A force that impedes the downslope movement of rock or sediment. See *Driving force*.

Resonance An increase in the amplitude of seismic waves when their frequency matches the natural vibrational frequency of an object.

Resurgent caldera The uplift of the central part of a giant volcanic crater that formed earlier during an explosion or collapse of the volcano's summit. See *Caldera.*

Retreat The interval of time during which a glacier thins and retreats. See *Advance*.

Retrofitting The renovation of engineered structures to withstand ground shaking and other, secondary effects of an earthquake.

Return period See *Recurrence interval*.

Return stroke The downward flow of electrons during a cloud-to-ground lightning discharge. This discharge produces the bright light of a lightning flash. See *Lightning* and *Stepped leader*.

Reverse fault A fault along which the hanging wall has moved upward relative to the footwall. See *Dip-slip*, *Normal*, *Strike-slip*, and *Thrust faults*.

Rhyolite A fine-grained, light-coloured, silica-rich volcanic rock consisting of feldspar, quartz, and ferromagnesian minerals.

Richter scale The range of earthquake magnitude values determined from trace deflections on a standard seismograph at a distance of 100 km from the epicentre. The Richter scale is logarithmic: an increase in magnitude of one (for example from 3 to 4) corresponds to a 10-fold increase in peak ground motion amplitude and a 30-fold increase in the total energy released by the earthquake. See *Magnitude*.

Ridge An elongated area of high atmospheric pressure.

Ridge push The gravitational force exerted by a plate as it moves away from the crest of a mid-ocean ridge toward a subduction zone.

Riffle A shallow section of stream channel where, at low flow, water moves rapidly over a gravel bed.

Rift A long, narrow trough bounded by normal faults. Rifts are produced when crustal rocks are pulled apart.

Ring of Fire A popular name given to the chain of active volcanoes bordering much of the Pacific Ocean.

Rip current The seaward flow of water in a narrow zone from the beach to beyond the breaker zone.

Riprap Large broken stones placed to protect a riverbank or shoreline from erosion.

Risk The product of the probability of a hazardous event and the expected damage if the event does occur. See *Risk analysis*.

Risk analysis The evaluation of the probability that a hazardous event will occur and its possible consequences, including death, injury, and property damage. See *Risk*.

River A large natural stream. See *Stream*.

Rock A solid, natural aggregate of one or more minerals, natural glass, or fossils.

Rock avalanche The downward and outward movement of fragmenting rock at high velocities. Rock avalanches can run out long distances on relatively gentle slopes.

Rock cycle A group of interrelated processes that produce igneous, metamorphic, and sedimentary rocks.

Rogue wave An abnormally large ocean wave produced by constructive interference of smaller waves.

Rope stage The final decaying phase of a tornado, during which upward-spiralling air in the funnel comes into contact with downdrafts from the thunderstorm. See *Organizational stage* and *Shrinking stage*.

Rotational A type of landslide in which material moves downward and outward on a well-defined, upward-curving slip surface.

Runoff Water that flows over the land surface; includes overland flow on slopes and channelized stream flow.

Run-out zone The lower part of an avalanche path where snow is deposited. See *Start zone* and *Track*.

Run-up The upward rush of a wave on a shore; also, the distance that a tsunami surges inland from the shoreline.

Rural–urban interface The area where forest, scrubland, or grassland borders developed, populated areas.

Rutschblock test A field test of snowpack stability in which a skier jumps on an excavated column of undisturbed snow. A score is applied based on the loading required to release a block from the upper part of the column.

S wave A seismic wave that travels in a snake-like fashion through solid material from the hypocentre of an earthquake. Also called a *secondary* or *shear wave*.

Sackung A German term for deep-seated rock creep; involves complex internal deformation of a failing rock mass along multiple shear planes. Surface features associated with sackung include ridge-top depressions, uphill-facing scarps, and bulging of the lower part of the slope.

Sand blow A cone or mound of sand that forms when liquefied, water-rich sand is ejected from a crack in the ground. Sand blows commonly result from strong ground shaking during earthquakes.

Sand storm A storm with high winds that transport sand particles to heights of up to 2 m above the ground. Most sand storms occur in arid areas.

Sand volcano See *Sand blow*.

Scientific method The principles and processes of discovery and demonstration that are characteristic of scientific investigation. The scientific method involves observation of phenomena, the formulation of a hypothesis, experimentation to demonstrate the truth or falseness of the hypothesis, and a conclusion that validates or modifies the hypothesis.

Scoria cone See *Cinder cone*.

Sea cliff A steep coastal bluff produced by wave erosion, landsliding, and, in some cases, groundwater seepage.

Seafloor spreading The process by which new oceanic crust is created at mid-ocean ridges. Convective upwelling of magma results in the continuous lateral movement of existing oceanic crust away from the axis of a mid-ocean ridge. See *Mid-ocean ridge* and *Spreading centre*.

Seamount An underwater mountain that rises from the ocean floor and has a peaked or flat-topped summit below the sea surface.

Seawall Engineered structure built along the shoreline to retard erosion and protect buildings from damage by waves and currents.

Secondary effect See *Indirect effect*.

Secondary wave See *S wave*.

Sector collapse A sudden collapse of the flank of a volcano, generally just before or during a volcanic eruption.

Sediment The fragments of inorganic or organic detritus transported and deposited by wind, water, gravity, or glacier ice.

Sediment budget The balance between changes in the volumes of sediment entering, leaving, and stored within a sedimentary system.

Sedimentary rock One of the three principal groups of rocks in Earth's crust. Sedimentary rocks consist of mineral grains and rock fragments deposited on land, in oceans, or in lakes.

Seismic gap The segment of an active fault that could produce a large earthquake, but has not done so recently.

Seismic source The part of a fault that ruptures to produce an earthquake.

Seismic wave A wave produced by sudden displacement of rocks along a fault. Seismic waves move through or along the surface of Earth.

Seismogram A written or electronic record of an earthquake made by a seismograph. See *Seismograph*.

Seismograph An instrument that records earthquakes. See *Seismogram*.

Seismologist A scientist who studies earthquakes.

Seismology The branch of science concerned with earthquakes and related phenomena.

Sensible heat The energy produced by random molecular motion. Sensible heat can be physically sensed and measured, such as with a thermometer. See *Latent heat*.

Set A group of waves of similar size and shape in the ocean or a large lake.

Setback The required minimum distance from the top of a bluff that a structure can be placed to ensure that it is not endangered by landsliding or fluvial erosion.

Shake map A computer-generated map of the shaking that people feel in different areas during an earthquake, produced in near-real time.

Shear strength The internal resistance of a body of rock or sediment to shear and failure.

Shear wave See *S wave*.

Shield volcano A gently sloping, broad, convex volcano consisting of basalt lava flows; the largest type of volcano. See *Stratovolcano*.

Shovel test A simple field test of snowpack stability in which the upper part of a column of undisturbed snow is pulled with a shovel.

Shrinking stage The late phase of a tornado, when the funnel thins and begins to tilt as its supply of warm moist air decreases. See *Organizational stage* and *Rope stage*.

Silicosis A respiratory disease caused by inhalation of crystalline silica dust.

Simple impact crater A small impact crater lacking an uplifted centre. Simple impact craters are generally smaller than a few kilometres in diameter.

Sinkhole A surface depression formed by solution of underlying limestone or by collapse of a cave. See *Collapse sinkhole*.

Slab A build-up of snow that can become unstable and avalanche.

Slab avalanche A type of avalanche in which large cohesive blocks of snow move rapidly downslope.

Slab pull The force that a subducting plate exerts on the oceanic lithosphere. This force results from the weight of the cold dense plate as it sinks into the mantle.

Slide The downslope movement of a coherent block of rock or sediment along a discrete failure plain.

Slip rate The long-term, average rate of displacement along a fault, generally expressed in millimetres or centimetres per year.

Slope An inclination of the land surface, expressed in degrees or percent.

Slope stability map A depiction of the land surface showing its susceptibility to landslides.

Slow earthquake A seismic event involving movement along a fault over a period of days to months.

Sluff A snow slide that is generally too small to bury a person.

Slump A type of landslide characterized by sliding along an upward-curved slip surface.

Slump block A mass of rock or sediment that has moved downward and outward along a curved slip surface. See *Slump*.

Smectite A group of clay minerals that can absorb large amounts of water; a common constituent of shrink–swell clay.

Snow avalanche See *Avalanche*.

Soil Surface rock or sediment that has been altered by mechanical, chemical, and biological processes and is a medium for plant growth. Engineers define *soil* as unconsolidated sediment that can be excavated or removed without blasting.

Soil slip The failure and rapid downslope movement of a thin layer of weathered rock or sediment. The slip surface is located in or at the base of weathered slope deposits lying on unweathered material.

Solar nebula A pancake-like, rotating disk of hydrogen and helium dust. The sun formed by accretion of hydrogen and helium within a solar nebula about 5 billion years ago.

Solution sinkhole A hole in the ground formed by solution of buried rocks, notably limestone, rock salt, or gypsum.

Spaceguard Program An initiative, stemming from the report of a 1992 U.S. Congressional study, to locate 90 percent of large near-Earth asteroids within a decade.

Spilling breaker A breaking wave that tumbles forward as it moves shoreward. Spilling breakers are associated with gently sloping beaches. See *Plunging breaker*.

Spit A long, narrow ridge of sand or gravel that extends parallel to the shore from a point of land on a coast.

Splitting wedge A masonry or concrete wedge-shaped structure designed to deflect avalanching snow away from a building.

Sporadic permafrost A sub-zone of the discontinuous permafrost zone, in which 10 to 50 percent of the ground is permanently frozen. See *Permafrost*, *Continuous permafrost*, *Discontinuous permafrost*, and *Extensive discontinuous permafrost*.

Spot fire A small fire ignited by wind-blown, burning embers ahead of the flaming front of a wildfire.

Spreading centre A linear ridge on the ocean floor along which basaltic magma is erupted and from which adjacent lithospheric plates move apart. See *Mid-ocean ridge* and *Seafloor spreading*.

Spring A continuous discharge of groundwater where the water table intersects the ground surface.

Squall line A line of thunderstorms accompanied by high winds and heavy rain. A squall line commonly forms in advance of a cold front.

Stage The level of water in a stream channel. Flood stage is the level at which the stream overflows its channel and spills onto its floodplain.

Stalactite A cylindrical or conical deposit of calcium carbonate that extends downward from the roof of a cave or an overhang. Stalactites form where carbonate-rich waters drip from the cave roof. See *Flowstone* and *Stalagmite*.

Stalagmite A cylindrical or conical deposit of calcium carbonate on the floor of a cave. Stalagmites form where carbonate-rich waters drip from the roof to the floor of a cave. See *Flowstone* and *Stalactite*.

Start zone The highest part of an avalanche path, where the snowpack fails. See *Run-out zone* and *Track*.

Stationary front A transition zone between two different air masses that are not moving. See *Front* and *Occluded front*.

Stepped leader A channel of ionized air that approaches the ground in a series of nearly invisible bursts; this channel becomes the path for the luminous return stroke of a cloud-to-ground lightning strike. See *Lightning* and *Return stroke*.

Stony meteorite A meteorite consisting of silicate minerals (75 to 90 percent) and metallic alloys of nickel and iron (10 to 25 percent). Stony meteorites are a diverse group, ranging from samples of matter that date to the beginning of the solar system to differentiated rocks derived from primitive planetary bodies. See *Differentiated*.

Stope An underground stepped excavation in a mine created to extract ore from steeply inclined or vertical veins.

Storm surge Wind-driven waves that flood low-lying coastal areas. Storm surges accompany hurricanes, nor'easters, and other severe storms.

Strain A change in the shape or size of a body because of application of a stress.

Stratosphere The zone in Earth's atmosphere above the troposphere, where temperature is either constant or increases with altitude. The stratosphere contains significant quantities of ozone, which protects life from ultraviolet radiation. See *Mesosphere, Thermosphere,* and *Troposphere*.

Stratovolcano A steep-sided, explosive volcano formed of pyroclastic deposits and lava flows. Also called a *composite volcano*. See *Shield volcano*.

Stream A ribbon-like body of water flowing in a channel; includes brooks, creeks, and rivers. See *River*.

Strength The ability of rock or sediment to resist deformation. Strength results from cohesive and frictional forces in the material.

Stress A force applied to an object.

Strike-slip fault A fault that displaces rocks laterally, with little or no vertical component of movement. See *Dip-slip*, *Normal*, *Reverse*, and *Thrust faults*.

Strong earthquake An earthquake of moment magnitude 6.0 to 6.9.

Subaqueous landslide A landslide that occurs on the floor of a lake or the sea.

Subduction The process by which one tectonic plate descends beneath another and eventually melts in the mantle.

Subduction earthquake An earthquake resulting from a sudden slip along the fault that separates two lithospheric plates at a subduction zone. The largest earthquakes on Earth are subduction earthquakes. See *Subduction* and *Subduction zone*.

Subduction zone An elongated zone, typically hundreds to more than 1000 km long, where two crustal plates converge, one moving slowly under the other. See *Subduction*.

Submarine canyon A steep-sided valley cut into the seafloor of the continental slope and commonly extending well onto the continental shelf.

Submarine trench A deep, long, and relatively narrow depression developed in oceanic crust on the ocean floor. Submarine trenches generally lie above subduction zones and are the deepest parts of the oceans.

Subsidence The lowering of Earth's surface because of sediment compaction, an earthquake, or other natural processes.

Subtropical jet stream A band of relatively strong winds concentrated between 20° and 40° latitude in the middle and upper troposphere. It can be present at any longitude but is generally strongest off the coast of Asia.

Suction vortices The small, intense, rotating wind cells that are responsible for much of the damage done by a tornado.

Sulphur dioxide A colourless, pungent gas with the formula SO_2. Sulphur dioxide is emitted into the atmosphere during many volcanic eruptions.

Supercell storm An unusually long-lived thunderstorm with a rotating updraft on the storm's flank.

Superelevation The difference in height of the inner and outer margins of a landslide or debris flow where it rounds a bend.

Supereruption A very large, explosive, caldera-forming eruption that ejects hundreds of cubic kilometres of pyroclastic debris. See *Supervolcano*.

Supernova An explosion of a star, accompanied by an extremely bright emission of vast amounts of energy.

Supervolcano A volcano that produces extremely large, but rare, explosive eruptions, with hemispheric or global impacts. See *Supereruption*.

Surf zone The nearshore area between the zone of breaking waves and the swash zone. See *Breaker zone* and *Swash zone*.

Surface fire A wildfire that burns primarily along the ground, consuming fuels such as grasses, shrubs, dead and downed limbs, and leaf litter. See *Crown fire*.

Surface wave A seismic wave that travels along the ground surface. Surface waves are generally strongest close to the epicentre, where they may cause much structural damage. See *Body wave*.

Surface-wave scale An earthquake magnitude scale based on measurements of Rayleigh waves, which travel along Earth's surface.

Surge A sudden rapid increase in the rate of flow of a glacier, producing a dramatic advance of the glacier terminus. The flow rate during a surge can be 100 times greater than the normal rate.

Suspended load The sediment particles transported in suspension within a river or stream. See *Bed load*, *Dissolved load*, and *Total load*.

Swash zone An area along the shoreline where waves run up and recede. See *Breaker zone* and *Surf zone*.

Swell A set of storm-produced waves with more or less uniform heights and lengths. Wave sets travel long distances with relatively little loss of energy.

Talus Fragments of rock that have fallen from a cliff or steep slope and accumulated at its base. Talus may form aprons and cones of blocky rubble.

Tectonic Pertaining to the structure or deformation of Earth's crust.

Tectonic creep See *Fault creep*.

Tectonic cycle The cyclic production and destruction of lithosphere through slow movement of tectonic plates; part of the geologic cycle.

Tectonic plate A very large, fault-bounded block of crust and upper mantle that slowly moves on top of the asthenosphere. Tectonic plates form at mid-oceanic ridges and are destroyed at subduction zones. Also called a *lithospheric plate*.

Tele-tsunami A tsunami that originates from a distant source, generally more than 1000 km from the area of interest. Also called a *distant* or *far-field tsunami*. See *Distant tsunami*.

Tensile strength The greatest extensional stress that a material can bear without failing.

Tephra A general term for fragmented volcanic material blown out of a volcano; includes ash, lapilli, blocks, and bombs.

Thaw flow slide A shallow, commonly fast-moving landslide resulting from failure of the active layer in an area of permafrost.

Thermal expansion The increase in the volume of seawater that occurs when the water warms.

Thermal spring A flow of heated groundwater onto land or into a river, lake, or ocean; includes warm springs and hot springs. See *Hot spring*.

Thermokarst An irregular terrain formed by thawing of permafrost.

Thermosphere The outermost zone of Earth's atmosphere (above 80 km in altitude), characterized by little gas and an increase in temperature upward. See *Mesophere, Stratosphere,* and *Troposphere*.

Thrust fault A low-angle reverse fault along which older rocks are displaced over younger rocks. See *Dip-slip*, *Normal*, *Reverse*, and *Strike-slip faults*.

Thunder The crashing or booming sound produced by rapidly expanding air along the path of the electrical discharge of lightning.

Tidal bore A high landward-flowing wave caused by the collision of tidal currents in an estuary.

Top-loaded A slab of potentially unstable snow deposited leeward of the crest of a slope or the top of a mountain by wind. See *Cross-loaded*.

Topography The shape and physical features of a part of Earth's surface.

Topple The pivoting of a rock mass about a point.

Torino Hazard Impact Scale A numerical scale for categorizing the impact hazard associated with asteroids and comets. A near-Earth object is assigned a numeral ranging from 1 to 10 based on its probability of collision and its kinetic energy measured in megatons of TNT.

Tornado A violently rotating, funnel-shaped column of air extending downward from a severe thunderstorm to the ground.

Total load The sum of the dissolved, suspended, and bed load that a stream or river carries.

Tower karst A landscape of steep-sided hills rising above a plain or above sinkholes that have formed by dissolution of limestone. See *Karst*.

Track The middle part of a snow avalanche path, where the avalanche accelerates and achieves its highest velocity. See *Run-out* and *Start zones*.

Transform boundary See *Transform fault*.

Transform fault A strike-slip fault that connects segments of mid-ocean ridges and forms the boundary between two plates; the San Andreas fault in California is an example. Also called a *transform boundary*.

Translational The downslope movement of rock or sediment along a well-defined, planar surface.

Transmission The unimpeded passage of electromagnetic energy from the sun through Earth's atmosphere.

Triangulation The act of locating an epicentre using distances from three seismographs.

Tributary A stream that flows into a larger stream.

Trigger With reference to mass wasting, a trigger is the event that sets off a landslide. Compare with *cause*.

Triple junction A point where three tectonic plates meet.

Troglobite An organism, such as a blind salamander, that spends its entire life in a cave.

Tropical cyclone A large thunderstorm complex rotating around an area of low pressure over warm water in the tropics or subtropics; includes tropical depressions, tropical storms, hurricanes, typhoons, severe tropical cyclones, and cyclonic storms. See *Hurricane*, *Tropical depression*, *Tropical storm*, and *Typhoon*.

Tropical depression A low-pressure centre in the tropics with winds less than 63 km/h, which is the threshold for a tropical storm. See *Hurricane*, *Tropical cyclone*, *Tropical disturbance*, and *Tropical storm*.

Tropical disturbance An area of disorganized but common thunderstorms associated with a low-pressure trough; the formative stage of a tropical depression, tropical storm, or hurricane. See *Hurricane*, *Tropical cyclone*, *Tropical depression*, and *Tropical storm*.

Tropical storm An organized system of thunderstorms centred on an area of low pressure. A tropical storm derives its energy from warm ocean waters and has sustained winds of between 63 km/h and 119 km/h. It is stronger than a tropical depression but weaker than a hurricane. See *Hurricane*, *Tropical cyclone*, *Tropical depression*, and *Tropical disturbance*.

Tropopause The boundary between the troposphere and the stratosphere.

Troposphere The lowermost layer of the atmosphere, characterized by a decrease in temperature with altitude. See *Mesophere, Stratosphere,* and *Thermosphere*.

Trough An elongated area of low atmospheric pressure.

Tsunameter An instrument placed on the seafloor to detect, measure, and report tsunamis in real time.

Tsunami The waves generated by a sudden upward or downward movement of a large area of the seafloor, an asteroid impact, subduction earthquake, collapse of the flank of a volcano, or a large landslide.

Tsunami warning An emergency notice from a government agency that a tsunami has been detected and is travelling toward the area receiving the alert. See *Tsunami watch*.

Tsunami watch A notice from a government agency that an earthquake, landslide, or volcanic explosion that could trigger a tsunami has occurred. See *Tsunami warning*.

Tsunami-ready High level of preparation by a community for a tsunami; involves a feasible strategy to alert the public in the event of a tsunami, a tsunami-preparedness plan, community awareness, and the ability to receive tsunami warnings.

Turbidity current A current of rapidly moving, sediment-laden water moving down a slope on the seafloor or the floor of a lake.

Tuya A flat-topped, steep-sided, extinct volcano that erupted into a lake beneath a former ice cap or ice sheet.

Typhoon A hurricane in the western Pacific Ocean; the term is derived from a Chinese word that means "scary wind" or "wind from four directions." See *Hurricane*, *Tropical cyclone*, *Tropical depression*, and *Tropical storm*.

Updrift The direction from which a current of water, such as a longshore current, is flowing. See *Downdrift*.

Uplift An increase in the elevation of an area because of tectonic or volcanic processes.

Variable gas A gas such as carbon dioxide or water vapour that is present in the atmosphere in variable amounts and cannot be compressed with pressure alone. See *Permanent gas*.

Vein A tabular body of rock that fills a fracture or fissure in an older rock. Veins form by precipitation of minerals from subsurface fluids.

Very minor earthquake See *Microearthquake*.

Viscosity A material's resistance to flow. Viscosity results from the internal friction of a material's molecules. Substances with a high viscosity do not flow readily; those with a low viscosity are more fluid.

Vog A volcanic fog or smog consisting of sulphur dioxide, other volcanic gases, sulphuric acid, and dust. On Hawaii, vog is an acrid blue haze produced by gases emitted from Kilauea volcano.

Volatile A chemical compound that exists in a gaseous state at Earth's surface and evaporates easily.

Volcanic crisis A circumstance in which a volcanic eruption, or the prospect of a volcanic eruption, presents a danger to a large number of people.

Volcanic dome A volcano formed from viscous magma with a high silica content; eruptive activity is generally explosive.

Volcanic vent A circular or elongated opening in the ground through which lava and pyroclastic debris are erupted.

Volcano A vent at Earth's surface through which molten rock (lava), rock fragments, ash, and associated gases are erupted; also the hill or mountain formed by this process.

Vortex A spinning column of air.

Vulnerability The susceptibility of people and property to a hazardous event. Vulnerability has technological and human dimensions.

Wadati-Benioff zone A dipping planar zone of earthquakes that delineates a downgoing oceanic crustal plate along a subduction zone. The earthquakes are produced by slip along the subduction thrust fault and by slip on faults within the downgoing plate due to bending and extension as the plate is pulled into the mantle.

Wall cloud A localized, persistent wall of condensed water vapour at the base of a severe thunderstorm. Wall clouds can rotate and are commonly associated with tornadoes.

Warning An announcement that a hazardous event, such as a hurricane or tornado, could happen in the near future.

Washover channel An erosional channel extending across a beach or through a coastal dune. Washover channels form during storm surges. See *Overwash*.

Watch An alert issued by a meteorological agency that weather conditions are favourable for severe weather such as tornadoes, a hurricane, a severe thunderstorm, or a blizzard.

Watershed See *Drainage basin*.

Waterspout A rapidly rotating column of air extending downward from a thunderstorm to an ocean, lake, or other water body.

Wave height The vertical distance between the crest and adjacent trough of a wave.

Wave normal The direction perpendicular to the crest of a wave.

Wave period The time that successive crests of a wave take to pass a reference point; the inverse of the frequency of the wave. See *Wavelength*.

Wave train A set of waves that constitutes a tsunami. See *Tsunami*.

Wavelength The horizontal distance between successive crests or troughs of a wave; applied to seismic, electromagnetic, water, and other waveforms. See *Wave period*.

Weather The atmospheric conditions, including air temperature, humidity, and wind speed, that characterize a particular place. See *Climate*.

Weathering Changes in the mineralogy, chemistry, and texture of rocks at or near Earth's surface because of physical, chemical, and biological activity.

Wildfire An uncontrolled fire in a forest, scrubland, or grassland.

Wind chill The additional cooling effect that wind has on humans in cold weather. Wind chill indexes quantify how humans lose heat in cold air that is moving.

Wind shear A sudden change in the speed or direction of the wind within a short distance.

Wind slab A thick, poorly bonded layer of snow deposited on a slope by wind. Wind slabs are the source of many snow avalanches.

References

CHAPTER 1

1. **Wikipedia.** 2010. "2010 Haiti earthquake." **http://en.wikipedia.org/wiki/2010_Haiti_earthquake.** Accessed September 11, 2013.

2. **Wikipedia.** 2013. "1989 Loma Prieta earthquake." **http://en.wikipedia.org/wiki/1989_Loma_Prieta_earthquake.** Accessed March 25, 2013.

3. **Roberts, N., Clague, J.,** and **Mora-Castro, S.** 2010. "A tale of two cities: Lessons for Vancouver from the Haiti earthquake." *Innovation* [Association of Professional Engineers and Geoscientists of the Province of British Columbia] 14(2):20–23.

4. **Wikipedia.** 2006. "Hurricane Katrina." **http://en.wikipedia.org/wiki/Hurricane_Katrina**. Accessed September 10, 2013.

5. **Wikipedia.** 2013. "Hurricane Sandy." **http://en.wikipedia.org/wiki/Hurricane_Sandy.** Accessed March 25, 2013.

6. **Natural Resources Canada.** 2009. "Natural hazards." **http://www.nrcan.gc.ca/earth-sciences/natural-hazard/natural-hazard/3612.** Accessed September 10, 2013.

7. **Etkin, D.** (ed.). 2010. "Canadians at risk: Our exposure to natural hazards." Institute for Catastrophic Loss Reduction, Toronto, ON. **http://www.iclr.org/images/Canadians_at_Risk_2010.pdf.** Accessed September 10, 2013.

8. **Clague, J. J.,** and **Roberts, N. J.** 2012. "Landslide hazard and risk." In *Landslides: Types, Mechanisms and Modeling,* eds. J. J. Clague and D. Stead, pp. 1–9. Cambridge, UK: Cambridge University Press.

9. **O'Keefe, P., Westgate, K.,** and **Wisner, B.** 1976. "Taking the naturalness out of natural disasters." *Nature* 260:566–567.

10. **United Nations Department of Humanitarian Affairs.** 1992. *Internationally Agreed Glossary of Basic Terms Related to Disaster Management.* Geneva, Switzerland: United Nations Department of Humanitarian Affairs, DNA/93/36.

11. **Villagrán de León, J. C.** 2006. *Vulnerability—A Conceptual and Methodological Review.* Bonn, Germany: United Nations University, Institute for Environment and Human Security.

12. **Advisory Committee on the International Decade for Natural Hazard Reduction.** 1989. *Reducing Disaster's Toll.* National Research Council. Washington, DC: National Academies Press.

13. **Abramovitz, J. N.** 2001. "Averting unnatural disasters." In *State of the World 2001,* ed. L. R. Brown et al., pp. 123–142. Washington, DC: World Watch Institute.

14. **Guha-Sapir, D., Hargitt, D.,** and **Hoyois, P.** 2004. *Thirty Years of Natural Disasters 1974-2003: The Numbers.* Brussels, Belgium: University of Louvain, Center for Research on the Epidemiology of Disasters (CRED).

15. **White, G. F.,** and **Haas, J. E.** 1975. *Assessment of Research on Natural Hazards.* Cambridge, MA: MIT Press.

16. **Abramovitz, J. N.,** and **Dunn, S.** 1998. *Record Year for Weather-related Disasters. Vital Signs Brief 98-5.* Washington, DC: World Watch Institute.

17. **Crowe, B. W.** 1986. "Volcanic hazard assessment for disposal of high-level radioactive waste." In *Active Tectonics,* ed. Geophysics Study Committee, pp. 247–260. National Research Council. Washington, DC: National Academies Press.

18. **Population Reference Bureau.** 2000. *World Population Data Sheet.* Washington, DC: Population Reference Bureau.

19. **Kates, R. W.,** and **Pijawka, D.** 1977. "From rubble to monument: The pace of reconstruction." In *Disaster and Reconstruction,* eds. J. E. Haas, R. W. Kates, and M. J. Bowden, pp. 1–23. Cambridge, MA: MIT Press.

20. **Costa, J. E.,** and **Baker, V. R.** 1981. *Surficial Geology: Building with the Earth.* New York, NY: John Wiley & Sons.

21. **Parry, M. L., Canziani, O. F., Palutikof, J. P., van der Linden, P. J.,** and **Hanson, C. E.,** eds. 2007. *Climate Change 2007: Impacts, Adaptation and Vulnerability; Contribution of Working Group II to the Fourth Assessment Report of the Intergovernmental Panel on Climate Change.* Cambridge, UK: Cambridge University Press.

22. **Trenberth, K. E., Jones, P. D., Ambenje, P., Bojariu, R., Easterling, D., Klein Tank, A., Parker, D., Rahimzadeh, F., Renwick, J. A., Rusticucci, M., Soden, B.,** and **Zhai, P.** 2007. "Observations: Surface and atmospheric climate change." In *Climate Change 2007: The Physical Science Basis; Contribution of Working Group I to the Fourth Assessment Report of the Intergovernmental Panel on Climate Change,* eds. S. Solomon, D. Qin, M. Manning, Z. Chen, M. Marquis, K. B Averyt, M. Tignor, and H. L. Miller. New York, NY: Cambridge University Press.

CHAPTER 2

1. **Wysession, M.** 1995. "The inner workings of Earth." *American Scientist* 83:134–147.

2. **Glatzmaier, G. A.** 2001. "The geodynamo." **http://es.ucsc.edu/~glatz/geodynamo.html.** Accessed March 27, 2013.

3. **Fowler, C. M. R.** 1990. *The Solid Earth.* Cambridge, UK: Cambridge University Press.

4. **Le Pichon, X.** 1968. "Sea-floor spreading and continental drift." *Journal of Geophysical Research* 73:3661–3697.

5. **Isacks, B. L., Oliver, J.,** and **Sykes, L. R.** 1968. "Seismology and the new global tectonics." *Journal of Geophysical Research* 73:5855–5899.

6. **Cox, A.,** and **Hart, R. B.** 1966. *Plate Tectonics.* Boston, MA: Blackwell Scientific Publications.

7. **Dewey, J. F.** 1972. "Plate tectonics." *Scientific American* 226:56–68.

8. **Mason, A. D.,** and **Raff, R. G.** 1961. "Magnetic survey off the west coast of North America, 40 degrees N. latitude to 52 degrees N. latitude." *Geological Society of America Bulletin* 72: 1267–1270.

9. **Cox, A., Dalrymple, G. B.,** and **Doell, R. R.** 1962. "Reversals of Earth's magnetic field." *Scientific American* 216(2):44–54.

10. **Clague, D. A., Dalymple, G. B.,** and **Moberly, R.** 1975. "Petrography and K-Ar ages of dredged volcanic rocks from the western Hawaiian Ridge and southern Emperor Seamount chain."

Geological Society of America Bulletin 86:991–998.

11. **Wilson, J. T.** 1963. "A possible origin of the Hawaiian Islands." *Canadian Journal of Physics* 41:863–870.

12. **Wilson, J. T.** 1965. "Transform faults, ocean ridges, and magnetic anomalies southwest of Vancouver Island." *Science* 150:482–485.

CHAPTER 3

1. **Wikipedia.** 2013. "2010 Haiti earthquake." **http://en.wikipedia.org/wiki/2010_Haiti_earthquake.** Accessed March 30, 2013.

2. **Wikipedia.** 2013. "1989 Loma Prieta earthquake." **http://en.wikipedia.org/wiki/1989_Loma_Prieta_earthquake.** Accessed March 30, 2013.

3. **U. S. Geological Survey.** 2006. *USGS Response to an Urban Earthquake, Northridge '94.* U.S. Geological Survey Open File Report 96-263.

4. **Clague, J. J.** 2001. "The Nisqually earthquake: A wakeup call from south of the border." *Innovation* [Association of Professional Engineers and Geoscientists of the Province of British Columbia] 5(5):14–17.

5. **Radbruch, D. H.,** and **Bonilla, M. G.** 1966. *Tectonic Creep in the Hayward Fault Zone, California.* U.S. Geological Survey Circular 525.

6. **Steinbrugge, K. V.,** and **Zacher, E. G.** 1960. "Creep on the San Andreas fault." In *Focus on Environmental Geology,* ed. R. W. Tank, pp. 132–137. New York, NY: Oxford University Press.

7. **Bolt, B. A.** 2004. *Earthquakes*, 5th ed. San Francisco: W. H. Freeman.

8. **Hough, S. E., Friberg, P. A., Busby, R., Field, E. F., Jacob, K. H.,** and **Borcherdt, R. D.** 1989. "Did mud cause freeway collapse?" *EOS, Transactions, American Geophysical Union* 70(47):1497, 1504.

9. **Yeats, R. S.** 2001. *Living with Earthquakes in California: A Survivor's Guide.* Corvallis, OR: Oregon State University Press.

10. **Jones, R. A.** 1986. "New lessons from quake in Mexico." *Los Angeles Times*, September 26, 1986.

11. **Hanks, T. C.** 1985. *The National Earthquake Hazards Reduction Program: Scientific Status.* U.S. Geological Survey Bulletin 1659.

12. **Camby, T. Y.** 1990. "California earthquake: Prelude to the big one." *National Geographic* 177(5): 76–105.

13. **Advisory Committee on the International Decade for Natural Hazard Reduction.** 1989. *Reducing Disaster's Toll.* National Research Council. Washington, DC: National Academies Press.

14. **Hough, S.** 2002. *Earthshaking Science: What We Know (and Don't Know) about Earthquakes.* Princeton, NJ: Princeton University Press.

15. **Atwater, B. F., Musumi-Rokkaku, S., Satake, K., Tsuji, Y., Ueda, K.,** and **Yamaguchi, D. K.** 2005. *The Orphan Tsunami of 1700: Japanese Clues to a Parent Earthquake in North America.* U.S. Geological Survey Professional Paper 1707.

16. **Clague, J. J.** 1997. "Evidence for large earthquakes at the Cascadia subduction zone." *Reviews of Geophysics* 35:439–460.

17. **Satake, K., Shimazaki, K., Tsuji, Y.,** and **Ueda, K.** 1996. "Time and size of a giant earthquake in Cascadia inferred from Japanese tsunami records of January 1700." *Nature* 378:246–249.

18. **Sieh, K.,** and **LeVay, S.** 1998. *The Earth in Turmoil: Earthquakes, Volcanoes and Their Impact on Humankind.* New York, NY: W. H. Freeman.

19. **Hamilton, R. M.** 1980. "Quakes along the Mississippi." *Natural History* 89:70–75.

20. **Mueller, K., Champion, J., Guccione, E. M.,** and **Kelson, K.** 1999. "Fault slip rates in the modern New Madrid Seismic Zone." *Science* 286:1135–1138.

21. **Adams, J. J.,** and **Basham, P. W.** 1989. "The seismicity and seismotectonics of Canada east of the Cordillera." *Geoscience Canada* 16:3–16.

22. **Adams, J. J.,** and **Basham, P. W.** 1991. "The seismicity and seismotectonics of Eastern Canada." In *Neotectonics of North America,* eds. D. B. Slemmons, E. R. Engdahl, M. D. Zobach, and D. D. Blackwell, pp. 261–276. Boulder, CO: Geological Society of America.

23. **Bent, A. L.** 2002. "The 1933 M_s = 7.3 Baffin Bay earthquake: Strike-slip faulting along the northeastern Canadian passive margin." *Geophysics Journal International* 150:724–736.

24. **Yeats, R. S., Sieh, K.,** and **Allen, C. R.** 1997. *The Geology of Earthquakes.* New York, NY: Oxford University Press.

25. **Youd, T. L., Nichols, D. R., Helley, E. J.,** and **Lajoie, K. R.** 1975. "Liquefaction potential." In *Studies for Seismic Zonation of the San Francisco Bay Region*, ed. R. D. Borcherdt, pp. 68–74. U.S. Geological Survey Professional Paper 941A.

26. **Atwater, B. F., Nelson, A. R., Clague, J. J., Carver, G. A., Yamaguchi, D. K., Bobrowsky, P. T., Bourgeois, J., Darienzo, M. E., Grant, W. C., Hemphill-Haley, E., Kelsey, H. M., Jacoby, G. C., Nishenko, S. P., Palmer, S. P., Peterson, C. D.,** and **Reinhart, M. A.** 1995. "Summary of coastal geologic evidence for past great earthquakes at the Cascadia subduction zone." *Earthquake Spectra* 11:1–18.

27. **Plafker, G.** 1965. "Tectonic deformation associated with the 1964 Alaska earthquake." *Science* 148:1675–1687.

28. **Bucknam, R. C., Hemphill-Haley, E.,** and **Leopold, E. B.** 1992. "Abrupt uplift within the past 1700 years at southern Puget Sound, Washington." *Science* 158:1611–1614.

29. **Plafker, G.,** and **Ericksen, G. E.** 1978. "Nevados Huascarán avalanches, Peru." In *Rockslides and Avalanches: 1, Natural Phenomena,* ed. B. Voight, pp. 277–314. New York, NY: Elsevier.

30. **Wikipedia.** 2010. "2008 Sichuan earthquake." **http://en.wikipedia.org/wiki/2008_Sichuan_earthquake.** Accessed September 10, 2013.

31. **U. S. Geological Survey.** 2010. "Magnitude 7.9—East Sichuan, China." **http://earthquake.usgs.gov/earthquakes/eqinthenews/2008/us2008ryan/.** Accessed September 10, 2013.

32. **Wang, X.-Y.,** and **Nie, G.-Z.** 2009. "Characteristics of landslides induced by Wenchuan M(S) 8.0 earthquake and preliminary analysis of their relations with ground motion parameters." *Chinese Journal of Geotechnical Engineering* 31:1378–1383.

33. **Hsu, Y.-S.,** and **Hsu, Y.-H.** 2009. "Impact of earthquake-induced dammed lakes on channel evolution and bed mobility; case study of the Tsaoling landslide-dammed lake." *Journal of Hydrology* 374:43–55.

34. **Wikipedia.** 2013. "2010-13 Haiti cholera outbreak." **http://en.wikipedia.org/wiki/2010_Haitian_cholera_outbreak.** Accessed March 31, 2013.

35. **Pakiser, L. C., Eaton, J. P., Healy, J. H.,** and **Raleigh, C. B.** 1969.

"Earthquake prediction and control." *Science* 166:1467–1474.

36. **Evans, D. M.** 1966. "Man-made earthquakes in Denver." *Geotimes* 10: 11–18.

37. **Reed, C.** 2002. "Triggering quakes with waste." *Geotimes* 47(3):7.

38. **Frohlich, C.,** and **Davis, S. D.** 2002. *Texas Earthquakes*. Austin, TX: University of Texas Press.

39. **Horner, R. B., Barclay, J. E.,** and **MacRae, J. M.** 1994. "Earthquakes and hydrocarbon production in the Fort St. John area of northeastern British Columbia." *Canadian Journal of Exploration Geophysics* 30:39–50.

40. **Wetmiller, R. J.** 1986. "Earthquakes near Rocky Mountain House, Alberta, and their relationship to gas production facilities." *Canadian Journal of Earth Sciences* 23:172–181.

41. **Baranova, V., Mustaqueem, A.,** and **Bell, S.** 1999. "A model for induced seismicity caused by hydrocarbon production in the Western Canada Sedimentary Basin." *Canadian Journal of Earth Sciences* 36:47–64.

42. **Montgomery, C. T.,** and **Smith, M. B.** 2010. "Hydraulic fracturing: History of an enduring technology." *Journal of Petroleum Technology* 62(12):26–32.

43. **BC Oil and Gas Commission.** 2012. *Investigation of Observed Seismicity in the Horn River Basin.* Victoria, BC: Province of BC, BC Oil and Gas Commission.

44. **Page, R. A., Boore, D. M., Bucknam, R. C.,** and **Thatcher, W. R.** 1992. *Goals, Opportunities, and Priorities for the USGS Earthquake Hazards Reduction Program.* U.S. Geological Survey Circular 1079.

45. **Eberhart-Phillips, D., Haeussler, P. J., Freymueller, J. T., Frankel, A. D., Rubin, C. M., Craw, P., Ratchkovski. N. A., Anderson, G., Carver, G. A., Crone, A. J., Dawson, T. E., Fletcher, H., Hansen, R., Harp, E. L., Harris, R. A., Hill, D. P., Hreinsdóttir, S., Jibson, R. W., Jones, L. M., Kayen, R., Keefer, D. K., Larsen, C. F., Moran, S. C., Personius, S. F., Plafker, G., Sherrod, B., Sieh, K., Sitar, N.,** and **Wallace, W. K.** 2003. "The 2002 Denali fault earthquake, Alaska: A large magnitude, slip-partitioned event." *Science* 300:1113–1118.

46. **Fuis, G. S.,** and **Wald, L. A.** 2003. *Rupture in South-central Alaska—The Denali Earthquake of 2002.* U.S. Geological Survey Fact Sheet 014-03.

47. **Basham, P. W., Weichert, D. H., Anglin, F. M.,** and **Berry, M. J.** 1985. "New probabilistic strong ground motion maps of Canada." *Bulletin of the Seismological Society of America* 75:563–595.

48. **California Geological Survey.** "Alquist-Priolo earthquake fault zones." **http://www.consrv.ca.gov/cgs/rghm/ap/Pages/index.aspx**. Accessed September 10, 2013.

49. **Committee on the Science of Earthquakes.** 2003. *Living on an Active Earth: Perspectives on Earthquake Science*. National Research Council. Washington, DC: National Academies Press.

50. **Scholz, C.** 1997. "Whatever happened to earthquake prediction?" *Geotimes* 42(3):16–19.

51. **Press, F.** 1975. "Earthquake prediction." *Scientific American* 232:14–23.

52. **Scholz, C. H.** 1990. *The Mechanics of Earthquakes and Faulting.* New York, NY: Cambridge University Press.

53. **Cervelli, P.** 2004. "The threat of silent earthquakes." *Scientific American* 290:86–91.

54. **Rogers, G.,** and **Dragert H.** 2003. "Episodic tremor and slip on the Cascadia subduction zone: The chatter of silent slip." *Science* 300:1942–1943.

55. **Rikitakr, T.** 1983. *Earthquake Forecasting and Warning*. London, UK: D. Reidel.

56. **Silver, P. G.,** and **Wakita, H.** 1996. "A search for earthquake precursors." *Science* 273:77–78.

57. **Gera, V.** 2009. "Investigation of building standards in quake zone." Associated Press Archives, **http://web.archive.org/web/20090414050139/http://www.google.com/hostednews/ap/article/ALeqM5jkcWIUobzfeODCXm1fJn_Xfj_QpgD97G7M700**. Accessed March 31, 2013.

58. **Aloisi, S.** 2009. "Italy quake exposes poor building standards." Reuters AlertNet. **http://web.archive.org/web/20090416073042/http://www.alertnet.org/thenews/newsdesk/L7932819.htm.** Accessed March 31, 2013.

59. **Lewis, A.** 2009. "Row over Italian quake 'forecast.'" BBC News. **http://news.bbc.co.uk/2/hi/europe/7986585.stm.** Accessed March 31, 2013.

60. **Allen, C. R.** 1983. "Earthquake prediction." *Geology* 11:682.

61. **Hait, M. H.** 1978. "Holocene faulting, Lost River Range, Idaho." *Geological Society of America Abstracts with Programs* 10(5):217.

62. **Gori, P. L.** 1993. "The social dynamics of a false earthquake prediction and the response by the public sector." *Bulletin of the Seismological Society of America* 83:963–980.

63. **Yeats, R. S.** 1998. *Living with Earthquakes in the Pacific Northwest.* Corvallis, OR: Oregon State University Press.

64. **Hickman, S. H.,** and **Langbein, J.** 2002. *The Parkfield Experiment: Capturing What Happens in an Earthquake.* U.S. Geological Survey Fact Sheet FS 0049–02.

65. **U. S. Geological Survey.** 2009. "San Andreas Fault Observatory at Depth." **http://earthquake.usgs.gov/research/parkfield/safod_pbo.php.** Accessed September 11, 2013.

66. **Holden, R., Lee, R.,** and **Reichle, M.** 1989. *Technical and Economic Feasibility of an Earthquake Warning System in California.* California Division of Mines and Geology Special Publication 101.

67. **Munich Reinsurance Company of Canada.** 1992. *Earthquake: Economic Impact Study.* Toronto, ON: Munich Reinsurance Company of Canada.

68. **Coburn, A.,** and **Spence, R.** 2002. *Earthquake Protection,* 2nd ed. Chichester, UK: John Wiley & Sons.

69. **Castell, G.** 2002. *Earthquake! Preparing for the Big One: British Columbia.* Vancouver, BC: Pacific Rim Earthquake Preparedness Program.

70. **Morgan, L.** 1993. *Earthquake Survival Manual.* Kenmore, WA: Epicenter Press.

CHAPTER 4

1. **Wikipedia.** 2013. "Tohoku earthquake and tsunami." **http://en.wikipedia.org/wiki/2011_Tōhoku_earthquake_and_tsunami.** Accessed April 28, 2013.

2. **WPRO.** 2012. *The Great East Japan Earthquake: A Story of a Devastating Natural Disaster, a Tale of Human Compassion*. Manila, Philippines: World Health Organization, Western Pacific Region.

3. **Mori, N., Takahashi, T., Yasuda, T.,** and **Yanagisawa, H.** 2011. "Survey of 2011 Tohoku earthquake tsunami inundation and run-up." *Geophysical Research Letters* 38, doi: 10.1029/2011GL049210.

4. **Bryant, E.** 2001. *Tsunami: The Underrated Hazard.* New York, NY: Cambridge University Press.

5. **Bolt, B. A.** 2006. *Earthquakes*, 5th ed. New York, NY: W. H. Freeman.

6. **U.S. Geological Survey.** 2010. "Tsunami and earthquake research at the USGS." West Coastal and Marine Geology. **http://walrus.wr.usgs.gov/tsunami/.** Accessed September 10, 2013.

7. **Shuto, N., Matsutomi, H., Tsuji, Y., Ito, H., Yamamoto, K.,** et al. 1993. "Tsunami devastates Japanese coastal region." *EOS, Transactions, American Geophysical Union* 74(37):417, 432.

8. **Subarya, C., Chlieh, M., Prawirodirdjo, L., Avouac, J.-P., Bock, Y., Sieh, K., Meltzner, A. J., Natawidjaja, D. H.,** and **McCaffrey, R.** 2006. "Plate-boundary deformation associated with the great Sumatra–Andaman earthquake." *Nature* 440:46–51.

9. **U.S. Geological Survey.** 2010. "Magnitude 9.1—Off the west coast of northern Sumatra, Sunday, December 26, 2004 at 00:58:53 UTC." Earthquake Hazards Program. **http://neic.usgs.gov/neis/eq_depot/2004/eq_041226/.** Accessed September 10, 2013.

10. **Owen, J.** 2005. "Tsunami family saved by schoolgirl's geography lesson." *National Geographic News.* **http://news.nationalgeographic.com/news/2005/01/0118_050118_tsunami_geography_lesson.html.** Accessed September 11, 2013.

11. **Watson, P.** 2005. "Swept into the world: Ancient knowledge saved endangered tribes from the tsunami, but the aid that poured in from outside could imperil their future." *Los Angeles Times.* **http://www.latimes.com/news/nationworld/world/la-fg-tribes3feb03,0,2790810.story?coll=la-home-headlines.** Accessed September 11, 2013.

12. **Misra, N.** 2005. "Nat Geo TV shows help tsunami islander save 1,500." *National Geographic News.* **http://news.nationalgeographic.com/news/2005/01/0107_050107_tsunami_natgeo.html.** Accessed September 11, 2013.

13. **Mott, M.** 2005. "Did animals sense tsunami was coming?" *National Geographic News.* **http://news.nationalgeographic.com/news/2005/01/0104_050104_tsunami_animals.html.** Accessed September 11, 2013.

14. **Atwater, B. F.,** and **Hemphill-Haley, E.** 1997. *Recurrence Intervals for Great Earthquakes of the Past 3500 Years at Northeastern Willapa Bay, Washington.* U.S. Geological Survey Professional Paper 1576.

15. **Kelsey, H. M., Nelson, A. R., Hemphill-Haley, E.,** and **Witter, R. C.** 2005. "Tsunami history of an Oregon coastal lake reveals a 4600 yr record of great earthquakes on the Cascadia subduction zone." *Geological Society of America Bulletin* 117:1009–1032.

16. **Tappin, D. R., Watts, P., McMurtry, G. M., Lafoy, Y.,** and **Matsumoto, T.** 2001. "The Sissano, Papau New Guinea tsunami of July 1998—Offshore evidence of the source animation." *Marine Geology* 175:1–23.

17. **McSaveney, M. J., Goff, J. R., Darby, D. J., Goldsmith, P., Barnett, A., Elliott, S.,** and **Nongkas, M.** 2000. "The 17 July 1998 tsunami, Papua New Guinea: Evidence and initial interpretation." *Marine Geology* 170:81–92.

18. **Heezen, B. C.,** and **Ewing, E. M.** 1952. "Turbidity currents and submarine slumps, and the 1929 Grand Banks [Newfoundland] earthquake." *American Journal of Science* 250: 849–873.

19. **Fine, I. V., Rabinovich, A. B., Bornhold, B. D., Thomson, R. E.,** and **Kulikov, E. A.** 2005. "The Grand Banks landslide-generated tsunami of November 18, 1929: Preliminary analysis and numerical modeling." *Marine Geology* 215:45–57.

20. **Miller, D. J.** 1960. *Giant Waves in Lituya Bay, Alaska.* U.S. Geological Survey Professional Paper P 0354-C:51–86.

21. **McMurtry, G. M., Fryer, G. J., Tappin, D. R., Wilkinson, I. P., Williams, M., Fietzke, J., Garbe-Schoenberg, D.,** and **Watts, P.** 2004. "Megatsunami deposits on Kohala Volcano, Hawaii, from flank collapse of Mauna Loa." *Geology* 32:741–744.

22. **Ward, S. N.,** and **Day, S.** 2001. "Cumbre Vieja Volcano: Potential collapse and tsunami at La Palma, Canary Islands." *Geophysical Research Letters* 28:3397–3400.

23. **Self, S.,** and **Rampino, M, R.** 1981. "The 1883 eruption of Krakatau." *Nature* 294:699–704.

24. **Winchester, S.** 2003. *Krakatoa: The Day the World Exploded: August 27, 1883.* New York, NY: Harper Perennial.

25. **U.S. Department of Commerce** and **U.S. Department of Interior.** 2005. "Fact sheet: Tsunami detection and warnings." **http://www.publicaffairs.noaa.gov/grounders/pdf/tsunami-factsheet2005.pdf.** Accessed September 11, 2013.

26. **Clague, J. J., Munro, A.,** and **Murty, T.** 2003. "Tsunami hazard and risk in Canada." *Natural Hazards* 28:433–461.

27. **Atwater, B. F., Musumi-Rokkatku, S., Satake, K., Tsuji, Y., Ueda, K.,** and **Yamaguichi, D. K.** 2005. *The Orphan Tsunami of 1700: Japanese Clues to a Parent Earthquake in North America.* U.S. Geological Survey Professional Paper P 1707.

28. **Satake, K., Wang, K.,** and **Atwater, B. F.** 2003. "Fault slip and seismic moment of the 1700 Cascadia earthquake inferred from Japanese tsunami descriptions." *Journal of Geophysical Research* 108(B11):148–227.

29. **Nelson, A. R., Atwater, B. F., Bobrowsky, P. T., Bradley, L.-A., Clague, J. J., Carver, G. A., Darienzo, M. E., Grant, W. C., Krueger, H. W., Sparks, R., Stafford, T. W., Jr.,** and **Stuiver, M.** 1995. "Radiocarbon evidence for extensive plate-boundary rupture about 300 years ago at the Cascadia subduction zone." *Nature* 378:371–374.

30. **Smith, K.,** and **Petley, D. N.** 2009. *Environmental Hazards: Assessing Risk and Reducing Disaster,* 5th ed. New York, NY: Routledge.

31. **Dudley, W. C.,** and **Lee, M.** 1998. *Tsunami!* 2nd ed. Honolulu, HI: University of Hawaii Press.

32. **Centers for Disease Control and Prevention.** 2004. "Fact sheet: Health effects of tsunamis." **http://www.bt.cdc.gov/disasters/tsunamis/healtheff.asp.** Accessed September 11, 2013.

33. **Clague, J., Yorath, C., Franklin, R.,** and **Turner, B.** 2006. *At Risk: Earthquakes and Tsunamis on the West Coast.* Vancouver, BC: Tricouni Press.

34. **California Seismic Safety Commission.** 2005. *The Tsunami Threat to California: Findings and Recommendations on Tsunami Hazards and Risk.* Report CSSC 05-03.

35. **Danielsen, F., Serensen, M. K., Olwig, M. F., Selvam, V., Parish F., Burgess, N. D., Hiraishi, T., Karunagaran, V. M., Rasmussen, M. S., Hansen, L. B., Quarto, A.,** and **Suryadiputra, N.** 2005. "The Asian tsunami: A protective role for coastal vegetation." *Science* 310:643.

36. **Geist, E. L.,** and **Parsons, T.** 2006. "Probabilistic analysis of tsunami hazards." *Natural Hazards* 37:277–314.

CHAPTER 5

1. **Wright, T. L.,** and **Pierson, T. C.** 1992. *Living with Volcanoes*. U.S. Geological Survey Circular 1073.

2. **Pendick, D.** 1994. "Under the volcano." *Earth* 3(3):34–39.

3. **Decker, R.,** and **Decker, B.** 1998. *Volcanoes,* 3rd ed. New York, NY: W. H. Freeman.

4. **Hickson, C. J.,** and **Edwards, B. R.** 2001. "Volcanoes and volcanic hazards." In *A Synthesis of Geological Hazards in Canada,* ed. G. R. Brooks, pp. 145–181. Geological Survey of Canada Bulletin 548.

5. **Hickson, C.** 2005. *Mt. St. Helens: Surviving the Stone Wind.* Vancouver, BC: Tricouni Press.

6. **Fisher, R. V., Heiken, G.,** and **Hulen, J. B.** 1997. *Volcanoes.* Princeton, NJ: Princeton University Press.

7. **Mathews, W. H.** 1956. "'Tuyas,' flat-topped volcanoes in northern British Columbia." *American Journal of Science,* 245:560–570.

8. **Mathews, W. H.** 1952. "Mount Garibaldi, a supraglacial Pleistocene volcano in southwestern British Columbia." *American Journal of Science,* 250:81–103.

9. **Osservatorio Vesuviano.** 2009. "Summary of the eruptive history of Mt. Vesuvius." Italian National Institute of Geophysics and Volcanology. http://web.archive.org/web/20061203041501/http://www.ov.ingv.it/inglese/vesuvio/storia/storia.htm. Accessed September 11, 2013.

10. **Wikipedia.** 2009. "Mount Vesuvius." http://en.wikipedia.org/wiki/Mount_Vesuvius. Accessed September 11, 2013.

11. **Kilburn, C.,** and **McGuire, B.** 2001. *Italian Volcanoes*. Hertsfordshire, UK: Terra Publishing.

12. **Gasparini, P., Barberi, F.,** and **Belli, A.** 2003. "Early warning of volcanic eruptions and earthquakes in the neapolitan area, Campania region, south Italy." *Proceedings, Second International Conference on Early Warning,* Bonn, Germany.

13. **Osservatorio Vesuviano.** 2009. "Vesuvius." http://www.ov.ingv.it/inglese/vesuvio/vesuvio.htm. Accessed September 11, 2013.

14. **Yellowstone Volcano Observatory.** 2012. "Questions about super volcanoes." U.S. Geological Survey. http://volcanoes.usgs.gov/volcanoes/yellowstone/yellowstone_sub_page_49.html. Accessed September 11, 2013.

15. **Francis, P.** 1983. "Giant volcanic calderas." *Scientific American* 248(6):60–70.

16. **Ambrose, S. H.** 1998. "Late Pleistocene human population bottlenecks, volcanic winter, and differentiation of modern humans." *Journal of Human Evolution* 34:623–651.

17. **Wikipedia.** 2009. "Supervolcano." http://en.wikipedia.org/wiki/Supervolcano. Accessed September 11, 2013.

18. **Schmincke, H.** 2004. *Volcanism*. New York, NY: Springer-Verlag.

19. **Clague, D. A.,** and **Dalrymple, G. B.** 1987. "The Hawaiian-Emperor volcanic chain: Part I, Geologic evolution." In *Volcanism in Hawaii,* eds. R. W. Decker, T. L. Wright, and P. H. Stauffer, pp. 5–73. U.S. Geological Paper 1350.

20. **Westgate, J. A.,** and **Naeser, N. D.** 1995. "Tephrochronology and fission-track dating." In *Dating Methods for Quaternary Deposits*, eds. N. W. Rutter and N. R. Catto, pp. 15–28. St. John's, NL: Geological Association of Canada.

21. **Clague, J. J., Evans, S. G., Rampton, V. N.,** and **Woodsworth, G. J.** 1995. "Improved age estimates for the White River and Bridge River tephras, western Canada." *Canadian Journal of Earth Sciences* 32:1172–1179.

22. **Pyne-O'Donnell, S. D. F., Hughes, P. D. M., Froese, D. G., Jensen, B. J. L., Kuehn, S. C.,** et al. 2012. "High-precision ultra-distal Holocene tephrochronology in North America." *Quaternary Science Reviews* 52: 6–11.

23. **Wikipedia.** 2013. "The Volcano (British Columbia)." http://en.wikipedia.org/wiki/The_Volcano_(British_Columbia). Accessed April 22, 2013.

24. **Brown, A. S.** 1969. "Aiyansh lava flow, British Columbia." *Canadian Journal of Earth Sciences* 6: 1460–1468.

25. **Rogers, G. C.,** and **Souther, J. G.** 1983. "Hotspots trace plate movements." *Geoscience* 12(2):10–13.

26. **Scott, K. M., Hildreth, W.,** and **Gardner, C. A.** 2000. *Mount Baker: Living with an Active Volcano.* U.S. Geological Survey Fact Sheet FS 0059–00.

27. **IAVCEE Subcommittee on Decade Volcanoes.** 1994. "Research at decade volcanoes aimed at disaster prevention." *EOS, Transactions, American Geophysical Union* 75(30):340, 350.

28. **Crandell, D. R.,** and **Waldron, H. H.** 1969. "Volcanic hazards in the Cascade Range." In *Geologic Hazards and Public Problems,* Conference Proceedings, eds. R. Olsen and M. Wallace, pp. 5–18. Office of Emergency Preparedness Region 7.

29. **Edwards, B. R.,** and **Russell, J. K.** 2000. "The distribution, nature, and origin of Neogene-Quaternary magmatism in the Northern Cordilleran Volcanic Province, Canada." *Geological Society of America Bulletin* 112:1280–1295.

30. **Simkin, T., Siebert, L.,** and **Blong, R.** 2001. "Volcano fatalities—Lesson from the historical record." *Science* 291:255.

31. **Fisher, R. V., Smith, A. L.,** and **Roobol, M. J.** 1980. "Destruction of St. Pierre, Martinique, by ash-cloud surges, May 8 and 20, 1902." *Geology* 8:472–476.

32. **Neal, C. A., Casadevall, T. J., Miller, T. P., Hendley II, J. W.,** and **Stauffer, P. H.** 1998. *Volcanic Ash—Danger to Aircraft in the North Pacific*. U.S. Geological Survey Fact Sheet 030–97.

33. **Holloway, M.** 2000. "The killing lakes." *Scientific American* 286(3): 90–99.

34. **Thorarinsson, S.,** and **Sigvaldason, G. E.** 1973. "The Hekla eruption of 1970." *Bulletin Volcanologique* 36:269–288.

35. **Scarth, A.** 1999. *Vulcan's Fury: Man against the Volcano*. New Haven, CT: Yale University Press.

36. **Sutton, J., Elias, T., Hendley II, J. W.,** and **Stauffer, P. H.** 2000. *Volcanic Air Pollution—A Hazard in Hawai'i* . U.S. Geological Survey Fact Sheet 169–97, version 1.1.

37. **Baxter, P. J.** 2005. *Human Impacts of Volcanoes*. New York, NY: Cambridge University Press.
38. **Crandell, D. R.,** and **Mullineaux, D. R.** 1969. *Volcanic Hazards at Mount Rainier, Washington*. U.S. Geological Survey Bulletin 1283.
39. **Hammond, P. E.** 1980. "Mt. St. Helens blasts 400 meters off its peak." *Geotimes* 25(8):14–15.
40. **Gardner, C.** 2005. "Monitoring a restless volcano: The 2004 eruption of Mount St. Helens." *Geotimes* 50(3):24–29.
41. **American Geophysical Union.** 1991. "Pinatubo cloud measured." *EOS, Transactions, American Geophysical Union* 72(29):305–306.
42. **Kilburn, C. R. J.,** and **Sammonds, P. R.** 2005. "Maximum warning times for imminent volcanic eruptions." *Geophysical Research Letters* 32:L24313, doi: 10.1029/2005GLO24184.
43. **McGuire, B.** 2006. *Hazard and Risk Review 2006*. London, UK: University College London, Benfield Hazard Research Centre.
44. **Francis, P.** 1976. *Volcanoes*. New York, NY: Pelican Books.
45. **Richter, D. H., Eaton, J. P., Murata, K. J., Ault, W. U.,** and **Krivoy, H. L.** 1970. *Chronological Narrative of the 1959–60 Eruption of Kilauea Volcano, Hawaii*. U.S. Geological Survey Professional Paper 537E.
46. **Tilling, R. I.** 2000. "Volcano notes." *Geotimes* 45(5):19.
47. **Gardner, C. A.,** and **Guffanti, M. C.** 2006. *U.S. Geological Survey's Alert Notification System for Volcanic Activity*. U.S. Geological Survey Fact Sheet 2006-3139.
48. **Murton, B. J.,** and **Shimabukuro, S.** 1974. "Human response to volcanic hazard in Puna District, Hawaii." In *Natural Hazards*, ed. G. F. White, pp. 151–159. New York, NY: Oxford University Press.
49. **Mason, A. C.,** and **Foster, H. L.** 1953. "Diversion of lava flows at Oshima, Japan." *American Journal of Science* 251:249–258.
50. **Williams, R. S., Jr.,** and **Moore, J. G.** 1973. "Iceland chills a lava flow." *Geotimes* 18(8):14–18.

CHAPTER 6

1. **Bergman, B.** 2003. "100th anniversary of Frank Slide disaster." *Macleans*, April 23, 2003.
2. **Wikipedia.** 2009. "Frank Slide." **http://en.wikipedia.org/wiki/Frank_Slide.** Accessed September 11, 2013.
3. **Frank Slide Interpretive Centre.** 2012. "FAQ: Frequently asked questions." **http://www.history.alberta.ca/frankslide/faq/faq.aspx.** Accessed September 10, 2013.
4. **Alberta Geological Survey.** 2009. "Turtle Mountain monitoring project & field laboratory." **http://www.ags.gov.ab.ca/geohazards/turtle_mountain/turtle_mountain.html.** Accessed September 11, 2013.
5. **Field, M.,** and **MacIntyre, D.** "Big boulder crosses Canada." Frank Slide Interpretive Centre. **http://whaton.uwaterloo.ca/waton/s004.html.** Accessed September 10, 2013.
6. **Varnes, D. J.** 1978. "Slope movement types and processes." In *Landslides: Analysis and Control*, eds. R. L. Schuster and R. J. Krizek, pp. 1–33. National Research Council, Transportation Research Board Special Report 176.
7. **Pestrong, R.** 1974. *Slope Stability*. New York, NY: McGraw-Hill.
8. **Mathews, W. H.,** and **McTaggart, K. C.** 1978. "Hope rockslides, British Columbia Canada." In *Rockslides and Avalanches: 1, Natural Phenomena,* ed. B. Voight, pp. 259–275. New York, NY: Elsevier.
9. **Plafker, G.,** and **Ericksen, G. E.** 1978. "Nevados Huascarán avalanches, Peru." In *Rockslides and Avalanches: 1, Natural Phenomena,* ed. B. Voight, pp. 277–314. New York, NY: Elsevier.
10. **Heezen, B. C.,** and **Ewing, M.** 1952. "Turbidity currents and submarine slumps, and the 1929 Grand Banks [Newfoundland] earthquake." *American Journal of Science* 250:849–873.
11. **Evans, S. G., Clague, J. J., Woodsworth, G. J.,** and **Hungr, O.** 1989. "The Pandemonium Creek rock avalanche, British Columbia." *Canadian Geotechnical Journal* 26:427–446.
12. **Chow, V. T.** 1959. *Open-channel Hydraulics*. New York, NY: McGraw Hill.
13. **Apmann, R. P.** 1973. "Estimating discharge from superelevation in bends." *ASCE Journal of the Hydraulics Division* 99:65–79.
14. **Nilsen, T. H., Taylor, F. A.,** and **Dean, R. M.** 1976. *Natural Conditions that Control Landsliding in the San Francisco Bay Region*. U.S. Geological Survey Bulletin 1424.
15. **Burroughs, E. R., Jr.,** and **Thomas, B. R.** 1977. *Declining Root Strength in Douglas Fir after Felling as a Factor in Slope Stability*. USDA Forest Service Research Paper INT-190.
16. **Terzaghi, K.** 1950. "Mechanism of landslides." In *Application of Geology to Engineering Practice,* ed. S. Paige, pp. 83–123. Geological Society of America Berkey Volume.
17. **McCulloch, D. S.,** and **Bonilla, M. G.** 1970. *Effects of the Earthquake of March 27, 1964, on the Alaska Railroad.* U.S. Geological Survey Professional Paper 545–D.
18. **Leggett, R. F.** 1973. *Cities and Geology*. New York, NY: McGraw-Hill.
19. **Aylsworth, J. M., Lawrence, D. E.,** and **Evans, S. G.** 1997. *Landslide and Settlement Problems in Sensitive Marine Clay, Ottawa Valley*. Geological Association of Canada, Mineralogical Association of Canada, Field Trip Guidebook B1.
20. **Peckover, F. L.,** and **Kerr, J. W. G.** 1977. "Treatment and maintenance of rock slopes on transportation routes." *Canadian Geotechnical Journal* 4: 487–507.
21. **Lewkowicz, A. G.,** and **Harris, C.** 2005. "Frequency and magnitude of active-layer detachment failures in discontinuous and continuous permafrost, northern Canada." *Permafrost and Periglacial Processes* 16:115–130.
22. **Schuster, R. L.** 1996. "Socioeconomic significance of landslides." In *Landslides: Investigation and Mitigation,* eds. A. K. Turner and R. L. Schuster, pp. 12–35. Transportation Research Board Special Report 247, National Research Council. Washington, DC: National Academies Press.
23. **Flemming, R. W.,** and **Taylor, F. A.** 1980. *Estimating the Cost of Landslide Damage in the United States*. U.S. Geological Survey Circular 832.
24. **Clague, J. J., Brooks, G. R., Evans, S. G.,** and **VanDine, D. F.** 2000. "Quaternary and engineering geology of the Fraser and Thompson River valleys, southwestern British Columbia." In *Guidebook for Geological Field Trips in Southwestern British Columbia and Northern Washington*, eds. G. J. Woodsworth, L. E. Jackson, Jr., J. L. Nelson, and B. C. Ward, pp. 49–86. Vancouver, BC: Geological Association of Canada, Cordilleran Section.
25. **Kiersch, G. A.** 1965. "Vaiont reservoir disaster." *Geotimes* 9(9):9–12.

26. **Miller, D. J.** 1960. "The Alaska earthquake of July 10, 1958—Giant wave in Lituya Bay." *Bulletin of the Seismological Society of America* 50: 253–266.

27. **Friele, P. A.**, and **Clague, J. J.** 2004. "Large Holocene landslides from Pylon Peak, southwestern British Columbia." *Canadian Journal of Earth Sciences* 41:165–182.

28. **Guthrie, R. H., Friele, P., Allstadt, K., Roberts, N., Evans, S. G., Delaney, K. B., Roche, D., Clague, J. J.,** and **Jakob, M.** 2012. "The 6 August 2010 Mount Meager rock slide-debris flow, Coast Mountains, British Columbia: Characteristics, dynamics, and implications for hazard and risk assessment." *Natural Hazard and Earth System Science* 12:1–18.

29. **Clague, J. J., Munro, A.,** and **Murty, T.** 2003. "Tsunami hazard and risk in Canada." *Natural Hazards* 28: 433–461.

30. **Erman, D., and the SNEP Team.** 1997. "Sierra Nevada ecosystems." In *Status of the Sierra Nevada: The Sierra Nevada Ecosystem Project, Final Report to Congress, Vol. 1.* U.S. Geological Survey Digital Data Series DDS-43. **http://pubs.usgs.gov/dds/dds-43/.** Accessed September 11, 2013.

31. **Swanson, F. J.,** and **Dryness, C. T.** 1975. "Impact of clear-cutting and road construction on soil erosion by landslides in the Western Cascade Range, Oregon." *Geology* 7:393–396.

32. **Jones, F. O.** 1973. *Landslides of Rio de Janeiro and the Sierra das Araras Escarpment, Brazil.* U.S. Geological Survey Professional Paper 697.

33. **Leighton, F. B.** 1966. "Landslides and urban development." In *Engineering Geology in Southern California,* eds. R. Lung and R. Proctor, pp. 149–197, Special Publication, Los Angeles Section, Association of Engineering Geologists.

34. **Committee on the Review of the National Landslide Hazards Mitigation Strategy.** 2004. *Partnerships for Reducing Landslide Risk.* National Research Council. Washington, DC: The National Academies Press.

35. **Eisbacher, G. H.,** and **Clague, J. J.** 1981. "Urban landslides in the vicinity of Vancouver, British Columbia, with special reference to the December 1979 rainstorm." *Canadian Geotechnical Journal* 18:205–216.

36. **Jones, D. K. C.** 1992. "Landslide hazard assessment in the context of development." In *Geohazards,* eds. G. J. McCall, D. J. Laming, and S. C. Scott, pp. 117–141. New York, NY: Chapman & Hall.

37. **Spiker, E. C.,** and **Gori, P. L.** 2003. *National Landslides Mitigation Strategy—A Framework for Loss Reduction.* U.S. Geological Survey Circular 1244.

38. **Slosson, J. E., Yoakum, D. E.,** and **Shuiran, G.** 1986. "Thistle, Utah, landslide: Could it have been prevented?" In *Proceedings of the 22nd Symposium on Engineering Geology and Soils Engineering,* ed. S. H. Wood, pp. 281–303. Boise, ID: 22nd Symposium on Engineering Geology and Soils Engineering.

39. **Piteau, D. R.,** and **Peckover, F. L.** 1978. "Engineering of rock slopes." In *Landslides*, eds. R. Schuster and R. J. Krizek, pp. 192–228. Transportation Research Board Special Report 176, National Research Council. Washington, DC: National Academies Press.

40. **Pierce County Department of Emergency Management.** 2008. "Mount Rainier Volcanic Hazards Plan, working draft." **http://www.co.pierce.wa.us/DocumentCenter/View/3499.** Accessed September 11, 2013.

CHAPTER 7

1. **Jamieson, B.** 2001. "Snow avalanches." In *A Synthesis of Geological Hazards in Canada*, ed. G. R. Brooks, pp. 81–100. Geological Survey of Canada Bulletin 548.

2. **Perla, R. I.** 1980. "Avalanche release, motion and impact." In *Dynamics of Snow and Ice Masses*, ed. S. C. Colbeck, pp. 397–462. New York, NY: Academic Press.

3. **Armstrong, B. R.,** and **Williams, K.** 1992. *The Avalanche Handbook.* Golden, CO: Fulcrum Publishing.

4. **Schaerer, P. A.** 1981. "Avalanches." In *Handbook of Snow: Principles, Processes, Management and Use*, eds. D. M. Gray and D. H. Male, pp. 475–516. Toronto, ON: Pergamon Press.

5. **Colbeck, S. C.** 1991. "The layered character of snow covers." *Reviews of Geophysics* 29:81–96.

6. **McClung, D. M.,** and **Schaerer, P. A.** 1993. *The Avalanche Handbook.* Seattle, WA: The Mountaineers.

7. **Abromeit, D., Deveraux, A. M.,** and **Overby, B.** 2004. "Avalanche basics." U.S. Forest Service, National Avalanche Center. **http://www.fsavalanche.org.** Accessed September 11, 2013.

8. **Daffern, T.** 1993. *Avalanche Safety for Skiers and Climbers,* 2nd ed. Calgary, AB: Rocky Mountain Books.

9. **Jamieson, B.** 1997. *Backcountry Avalanche Awareness,* 6th ed. Revelstoke, BC: Canadian Avalanche Association.

10. **Woods, J.** 1962. *Snow War: An Illustrated History of Rogers Pass, Glacier National Park,* 3rd ed. Toronto, ON: Canadian Parks and Wilderness Society.

11. **Morrall, J. F.,** and **Abdelwahab, W. M.** 1992. "Estimating traffic delays and the economic cost of recurrent road closures on rural highways." *Logistics and Transportation Review* 29: 159–177.

12. **Schweizer, J.,** and **Föhn, P. M. B.** 1996. "Avalanche forecasting—and expert system approach." *Journal of Glaciology* 42:318–332.

13. **Fredston, J.,** and **Fesler, D.** 1994. *Snow Sense: A Guide to Evaluating Snow Avalanche Hazard.* Anchorage, AK: Alaska Mountain Safety Center.

14. **Parks Canada.** 2010. "Avalanche terrain ratings for backcountry touring in the mountain national parks (2nd ed.)." **http://www.pc.gc.ca/pn-np/inc/pm-mp/visit/visit7a1_e.pdf.** Accessed September 11, 2013.

15. **Falk, M., Brugger, H.,** and **Adler-Kastner, L.** 1994. "Avalanche survival chances." *Nature*, 368:21.

16. **Tremper, B.** 2001. *Staying Alive in Avalanche Terrain*. Seattle, WA: The Mountaineers.

17. **Gilmore, K.** 2002. "Introduction to avalanche rescue dogs." 1st Specialist Response Group. **http://www.1srg.org/Contributed-Materials/SAR%20Dog%20Avalanche%20promo.htm.** Accessed September 11, 2013.

CHAPTER 8

1. **Brambati, A., Carbognin, L., Quaia, T., Teatini, P.,** and **Tosi, L.** 2003. "The lagoon of Venice: Geological setting, evolution and land subsidence." *Episodes*, 26:264–268.

2. **Waltham, T.** 2002. "Sinking cities." *Geology Today* 18(3):95–100.

3. **Fletcher, C.,** and **Da Mosta, J.** 2004. *The Science of Saving Venice.* New York, NY: Umberto Allemandi & C.

4. **Nosengo, N.** 2003. "Save our city!" *Nature* 424:608–609.

5. **Galloway, D., Jones, D. R.,** and **Ingebritsen, S. E.** 1999. "Introduction." In *Land Subsidence in the United States*, ed. D. Galloway, D. R. Jones, and S. E. Ingebritsen, pp. 1–6. U.S. Geological Survey Circular 1182.

6. **Waltham, T., Bell, F.,** and **Culshaw, M.** 2005. *Sinkholes and Subsidence: Karst and Cavernous Rocks in Engineering and Construction*. New York, NY: Springer-Verlag.

7. **Bloom, A. L.** 1991. *Geomorphology: A Systematic Analysis of Late Cenozoic Landforms*. Englewood Cliffs, NJ: Prentice Hall.

8. **Wikipedia.** 2013. "Permafrost." **https://en.wikipedia.org/wiki/Permafrost.** Accessed September 10, 2013.

9. **Nelson, F. E., Anisimov, O. A.,** and **Shiklomanov, N. I.** 2001. "Subsidence risk from thawing permafrost." *Nature* 410:669–890.

10. **Goldman, E.** 2002. "Even in the high Arctic nothing is permanent." *Science* 297:1493–1494.

11. **Hodek, R. J., Johnson, A. M.,** and **Sandri, D. B.** 1984. "Soil cavities formed by piping." In *Sinkholes: Their Geology, Engineering and Environmental Impact*, ed. B. F. Beck, pp. 249–254. Rotterdam , Netherlands: A.A. Balkema.

12. **Cockfield, W. E.,** and **Buckham, A. E.** 1946. "Sink-hole erosion in the white silts at Kamloops, British Columbia." *Transactions of the Royal Society of Canada* 40:1–10.

13. **Penvenne, L.** 1996. "The disappearing delta." *Earth* 5(4):16–17.

14. **Scheffe, K. F.** 2005. "Collapsible soils in the Rio Grande Valley of central New Mexico." *Geological Society of America Abstracts with Programs* 37(7):327.

15. **Scheffe, K. F.,** and **Lacy, S. L.** 2004. "Hydro-compactible soils." In *Understanding Soil Risks and Hazards: Using Soil Survey to Identify Areas with Risks and Hazards to Human Life and Property*, ed. G. Muckel, pp. 60–64. Lincoln, NE: U.S. Department of Agriculture, Natural Resources Conservation Service, National Soil Survey Center.

16. **Fischetti, M.** 2001. "Drowning New Orleans." *Scientific American* 287(10):76–85.

17. **Ingebritsen, S. E., McVoy, C., Glaz, B.,** and **Park, W.** 1999. "Florida Everglades: Subsidence threatens agriculture and complicates ecosystem restoration." In *Land Subsidence in the United States,* eds. D. Galloway, D. R. Jones, and S. E. Ingebritsen, pp. 95–106. U.S. Geological Survey Circular 1182.

18. **Wilding, L. P.,** and **Tessier, D.** 1988. "Genesis of Vertisols: Shrink-swell phenomena." In *Vertisols: Their Distribution, Properties, Classification and Management,* eds. L. P. Wilding and R. Puentes, pp. 55–81. College Station, TX: Texas A&M University Soil Management Support Services Technical Monograph No. 18.

19. **Noe, D. C., Jochim, C. L.,** and **Rogers, W. P.** 1999. *A Guide to Swelling Soils for Colorado Homebuyers and Homeowners.* Colorado Geological Survey Special Publication 43.

20. **Clague, J. J.** 1997. "Evidence for large earthquakes at the Cascadia subduction zone." *Reviews of Geophysics* 35:439–460.

21. **Schmidt, W.** 2001. "Sinkholes in Florida." *Geotimes* 46(1):18.

22. **Comiso, J. C.,** and **Parkinson, C. L.** 2004. "Satellite-observed changes in the Arctic." *Physics Today* 57(8): 38–44.

23. **Dixon, T. H., Amelung, F., Ferretti, A., Novali, F., Rocca, F., Dokka, R., Sella, G., Kim, S.-W., Wdowinski, S.,** and **Whitman, D.** 2006. "Subsidence and flooding in New Orleans: A subsidence map of the city offers insight into the failure of the levees during Hurricane Katrina." *Nature* 441:587–588.

24. **Jones, D. E.,** and **Holtz, W. C.** 1973. "Expansive soils—The hidden disaster." *Civil Engineering* 43(6):49–51.

25. **Ritter, D. F., Kochel, R. C.,** and **Miller, J. R.** 2002. *Process Geomorphology*. New York, NY: McGraw-Hill.

26. **Rankin, A.** 2010. "Expert says long-term effects of permafrost melt unknown." Northern News Service online. **http://www.nnsl.com/frames/newspapers/2010-04/apr29_10cli.html.** Accessed May 2, 2013.

27. **Wright, L.** 2012. "Thawing permafrost sinks buildings, hikes costs in North." CBC News North. **http://www.cbc.ca/news/canada/north/story/2011/11/16/north-big-fix-permafrost.html.** Accessed May 2, 2013.

28. **McDonald, T.** 2012. "UN warns permafrost is melting." ABC News. **http://www.abc.net.au/news/2012-11-28/un-warns-permafrost-is-melting/4396340.** Accessed May 2, 2013.

29. **National Research Council, Joint Academies Committee on the Mexico City Water Supply.** 1995. *Mexico City's Water Supply: Improving the Outlook for Sustainability.* Washington, DC: National Academies Press.

30. **Harris, R. C.** 2004. "Giant desiccation cracks in Arizona." *Arizona Geology* 34(2):1–4.

31. **Slaff, S.** 1993. *Land Subsidence and Earth Fissures in Arizona.* Arizona Geological Survey Down-to-Earth Series 3.

32. **Roberts, H. H.** 1997. "Dynamic changes of the Holocene Mississippi River delta plain: The delta cycle." *Journal of Coastal Research* 13: 605–627.

33. **Morton, R. A., Bernier, J. C.,** and **Barras, J. A.** 2006. "Evidence of regional subsidence and associated interior wetland loss induced by hydrocarbon production, Gulf Coast region, USA." *Environmental Geology* 50:261–274.

34. **Louisiana Coastal Wetlands Conservation and Restoration Task Force.** 1998. *Coast 2050: Toward a Sustainable Coastal Louisiana.* Baton Rouge, LA: Louisiana Department of Natural Resources.

35. **Penland, S., Wayne, L. D., Britsch, L. D., Williams, S. L., Beall, A. D.,** and **Butterworth, V. C.** 2000. *Process Classification of Coastal Land Loss between 1932 and 1990 in the Mississippi River Delta Plain, Southeastern Louisiana.* U.S. Geological Survey Open-File Report 00-0418.

36. **Karst Waters Institute.** 2002. "What is karst (and why is it important)?" **http://www.karstwaters.org/aboutkarst/index.php.** Accessed September 10, 2013.

37. **Holtzer, T. L.,** and **Galloway, D. L.** 2005. "Impacts of land subsidence caused by withdrawal of underground fluids in the United States." In *Humans as Geologic Agents*, ed. J. Ehlen, W. C. Haneberg, and R. A. Larson, pp. 87–99. Geological Society of America, Reviews in Engineering Geology 16.

38. **Bull, W. B.** 1974. "Geologic factors affecting compaction of deposits in a land subsidence area." *Geological Society of America Bulletin* 84:3783–3802.

39. **Kenny, R.** 1992. "Fissures." *Earth* 1(3):34–41.

40. **Craig, J. R., Vaughan, D. J.,** and **Skinner, B. J.** 1996. *Resources of the Earth,* 2nd ed. Upper Saddle River, NJ: Prentice Hall.

41. **Rahn, P. H.** 1996. *Engineering Geology*, 2nd ed. Upper Saddle River, NJ: Prentice Hall.

42. **Kappel, W. M., Yager, R. M.,** and **Miller, T. S.** 1999. "The Retsof Salt Mine collapse." In *Land Subsidence in the United States,* eds. D. Galloway, D. R. Jones, and S. E. Ingebritsen, pp. 111–120. U.S. Geological Survey Circular 1182.

43. **Péwé, T. L.** 1982. *Geologic Hazards of the Fairbanks Area, Alaska.* Alaska Division of Geological and Geophysical Surveys Special Report 15.

44. **Coplin, K. S.,** and **Galloway, D.** 1999. "Houston-Galveston, Texas: Managing coastal subsidence." In *Land Subsidence in the United States,* eds. D. Galloway, D. R. Jones, and S. E. Ingebritsen, pp. 35–48. U.S. Geological Survey Circular 1182.

45. **McCarthy, D. F.** 2002. *Essentials of Soil Mechanics and Foundations: Basic Geotechnics,* 6th ed. Upper Saddle River, NJ: Prentice Hall.

46. **Liquori, A., Maple, J. A.,** and **Heuer, C. E.** 1983. The design and construction of the Alyeska Pipeline. *International Conference on Permafrost, Proceedings* 3(2):151–157.

47. **Geological Survey of Canada**. 2009. "Permafrost." **http://www.nrcan.gc.ca/earth-sciences/node/433.** Accessed September 10, 2013.

48. **Galloway, D., Jones, D. R.,** and **Ingebritsen, S. E.** 1999. "Mining ground water." In *Land Subsidence in the United States,* eds. D. Galloway, D. R. Jones, and S. E. Ingebritsen, pp. 7–13. U.S. Geological Survey Circular 1182.

CHAPTER 9

1. **Wikipedia.** 2013. "2013 Alberta floods." **http://en.wikipedia.org/wiki/2013_Alberta_floods.** Accessed August 2, 2013.

2. **Osborn, G. D.** 1987. "Geologic and hydrologic hazards in Calgary." Chapter 11 in *Geology of the Calgary Area,* ed. L. E. Jackson, Jr. and M. Wilson, pp. 121–127. Calgary, AB: Canadian Society of Petroleum Geologists.

3. **Keller, E. A.,** and **Florsheim, J. L.** 1993. "Velocity reversal hypothesis: A model approach." *Earth Surface Processes and Landforms* 18:733–748.

4. **Clague, J. J.,** and **Evans, S. G.** 1994. *Formation and Failure of Natural Dams in the Canadian Cordillera.* Geological Survey of Canada Bulletin 464.

5. **Rahn, R. P.** 1984. "Flood-plain management program in Rapid City, South Dakota." *Geological Society of America Bulletin* 95:838–843.

6. **U. S. Department of Commerce.** 1973. *Climatological Data, National Summary* 24(13).

7. **Anonymous.** 1993. "The flood of 93." *Earth Observation Magazine* September:22–23.

8. **Mairson, A.** 1994. "The great flood of 1993." *National Geographic* 185(1):42–81.

9. **Bell, G. D.** 1993. "The great Midwestern flood of 1993." *EOS, Transactions, American Geophysical Union* 74(43):60–61.

10. **Anonymous.** 1993. "Flood rebuilding prompts new wetlands debate." *U.S. Water News* November:10.

11. **Pinter, N., Thomas, R.,** and **Wollsinski, J. H.** 2001. "Assessing flood hazard on dynamic rivers." *EOS, Transactions, American Geophysical Union* 82:333–339.

12. **Saulny, S.** 2007. "Development rises on St. Louis area flood plains." *New York Times*, May 15, 2007.

13. **Pinter, N.** 2005. "One step forward, two steps back on U.S. floodplains." *Science* 398:207–208.

14. **Beyer, J. L.** 1974. "Global response to natural hazards: Floods." In *Natural Hazards,* ed. G. F. White, pp. 265–274. New York, NY: Oxford University Press.

15. **McCain, J. F., Hoxit, L. R., Maddox, R. A., Chappell, C. F.,** and **Caracena, F.** 1979. "Meteorology and hydrology in Big Thompson River and Cache la Poudre River basins." In *Storm and Flood of July 31–August 1, 1976, in the Big Thompson River and Cache la Poudre River Basins, Larimer and Weld Counties, Colorado.* U.S. Geological Survey Professional Paper 1115A.

16. **Shroba, R. R., Schmidt, P. W., Crosby, E. J.,** and **Hansen, W. R.** 1979. "Geologic and geomorphic effects in the Big Thompson Canyon area, Larimer County." In *Storm and Flood of July 31–August 1, 1976, in the Big Thompson River and Cache la Poudre River Basins, Larimer and Weld Counties, Colorado.* U.S. Geological Survey Professional Paper 1115B.

17. **Bradley, W. C.,** and **Mears, A. I.** 1980. "Calculations of flows needed to transport coarse fraction of Boulder Creek alluvium at Boulder, Colorado." *Geological Society of America Bulletin,* Part II, 91:1057–1090.

18. **Wikipedia.** "Lynmouth Flood." **http://en.wikipedia.org/wiki/Lynmouth_Flood.** Accessed May 4, 2013.

19. **Wikipedia.** "Yellow River." **http://en.wikipedia.org/wiki/Huang_He.** Accessed September 11, 2013.

20. **Agricultural Research Service.** 1969. *Water Intake by Soils*. U.S. Department of Agriculture, Miscellaneous Publication No. 925.

21. **Strahler, A. N.,** and **Strahler, A. H.** 1973. *Environmental Geoscience*. Santa Barbara, CA: Hamilton Publishing.

22. **Brooks, G. R., Evans, S. G.,** and **Clague, J. J.** 2001. "Floods." In *A Synthesis of Geological Hazards in Canada*, ed. G. R. Brooks, pp. 101–143. Geological Survey of Canada Bulletin 548.

23. **Brooks, G. R.,** and **Lawrence, D. E.** 2000. "Geomorphic effects of flooding along reaches of selected rivers in the Saguenay region, Québec." *Géographie physique et Quaternaire* 54:281–299.

24. **Wikipedia.** "Saguenay Flood." **http://en.wikipedia.org/wiki/Saguenay_Flood.** Accessed May 4, 2013.

25. **Linsley, R. K., Jr., Kohler, M. A.,** and **Paulhus, J. L.** 1958. *Hydrology for Engineers*. New York, NY: McGraw-Hill.

26. **Leopold, L. B.** 1968. *Hydrology for Urban Land Planning*. U.S. Geological Survey Circular 554.

27. **Seaburn, G. E.** 1969. *Effects of Urban Development on Direct Runoff to East Meadow Brook, Nassau County, Long Island, New York.* U.S. Geological Survey Professional Paper 627B.

28. **Lliboutry, L., Arnao, B. M., Pautre, A.,** and **Schneider, B.** 1977. "Glaciological problems set by the control of dangerous lakes in Cordillera Blanca, Peru." *Journal of Glaciology* 18: 239–254.

29. **Mason, K.** 1929. "Indus floods and Shyok glaciers." *Himalayan Journal* 1:10–20.

30. **Wikipedia.** "2010 Pakistan floods." **http://en.wikipedia.org/wiki/2010_**

Pakistan_floods. Accessed May 4, 2013.

31. **Mackin, J. H.** 1948. "Concept of the graded river." *Geological Society of America Bulletin* 59:463–512.
32. **The Canadian Encyclopedia.** 2010. "Hydroelectricity." **http://www.thecanadianencyclopedia.com/index.cfm?PgNm=TCE&Params=a1ARTA0003932**. Accessed September 11, 2013.
33. **Clague, J.**, and **Turner, B.** 2003. *Vancouver, City on the Edge: Living with a Dynamic Geological Landscape.* Vancouver, BC: Tricouni Press.
34. **U.S. Congress.** 1973. *Stream Channelization: What Federally Financed Draglines and Bulldozers Do to Our Nation's Streams.* House Report No. 93–530. Washington, DC: U.S. Government Printing Office.
35. **Wikipedia.** "Red River Floodway." **http://en.wikipedia.org/wiki/Red_River_Floodway**. Accessed September 10, 2013.
36. **Rosgen, D.** 1996. *Applied River Morphology*. Lakewood, CO: Wildland Hydrology.
37. **Pilkey, O. H.**, and **Dixon, K. L.** 1996. *The Corps and the Shore*. Washington, DC: Island Press.
38. **Baker, V. R.** 1976. "Hydrogeomorphic methods for the regional evaluation of flood hazards." *Environmental Geology* 1:261–281.
39. **Bue, C. D.** 1967. *Flood Information for Floodplain Planning*. U.S. Geological Survey Circular 539.
40. **Schaeffer, J. R., Ellis, D. W.**, and **Spieker, A. M.** 1970. *Flood-hazard Mapping in Metropolitan Chicago*. U.S. Geological Survey Circular 539.
41. **Smith, K.**, and **Ward, R.** 1998. *Floods*. New York, NY: John Wiley and Sons.
42. **Canada Department of Environment.** 2013. *"Flood Damage Reduction Program."* **http://www.ec.gc.ca/eau-water/default.asp?lang=En&n=0365F5C2-1**. Accessed September 11, 2013.

CHAPTER 10

1. **Lecomte, E. L., Pang, A. W.**, and **Russell, J. W.** 1998. *Ice Storm '98*. Toronto, ON: Institute for Catastrophic Loss Reduction Research Paper Series No. 1.
2. **Wikipedia.** 2009. "North American ice storm of 1998." **http://en.wikipedia.org/wiki/North_American_ice_storm_of_1998**. Accessed September 10, 2013.
3. **Changnon, S. A.**, and **Changnon, J. M.** 2002. "Major ice storms in the United States, 1949–2000." *Environmental Hazards* 4:105–111.
4. **Government of Quebec.** 1999. *Facing the Unforeseeable: Lessons from the Ice Storm of '98*. Montreal, QC: Publications du Québec, Montreal.
5. **DeGaetano, A. T.** 2000. "Climatic perspectives and impacts of the 1998 northern New York and New England ice storm." *Bulletin of the American Meteorological Society* 81:237–254.
6. **National Climatic Data Center.** 2006. "Eastern U.S. flooding and ice storm." National Oceanic and Atmospheric Administration. **http://www.ncdc.noaa.gov/oa/reports/janstorm/janstorm.html**. Accessed September 11, 2013.
7. **Smith, G. A.**, and **Pun, A.** 2006. *How Does Earth Work? Physical Geology and the Process of Science*. Upper Saddle River, NJ: Pearson Prentice Hall.
8. **Smith, J.**, ed. 2001. *The Facts on File Dictionary of Weather and Climate*. New York, NY: Checkmark Books.
9. **Lutgens, F. K.**, and **Tarbuck, E. J.** 2007. *The Atmosphere: An Introduction to Meteorology,* 10th ed. Upper Saddle River, NJ: Pearson Prentice Hall.
10. **Christopherson, R. W.** 2006. *Geosystems: An Introduction to Physical Geography,* 6th ed. Upper Saddle River, NJ: Prentice Hall.
11. **Aguado, E.**, and **Burt, J. E.** 2007. *Understanding Weather and Climate,* 4th ed. Upper Saddle River, NJ: Pearson Prentice Hall.
12. **American Meteorological Society.** 2002. "Updated recommendations for lightning safety—2002." **http://www.ametsoc.org/POLICY/Lightning_Safety_Article.pdf**. Accessed September 11, 2013.
13. **Rauber, R. M., Walsh, J. F.**, and **Charlevoix, D. J.** 2005. *Severe and Hazardous Weather: An Introduction to High Impact Meteorology,* 2nd ed. Dubuque, IA: Kendall/Hunt.
14. **National Weather Service.** 2004. "Lightning—The underrated killer." **http://www.lightningsafety.noaa.gov/overview.htm**. Accessed September 11, 2013.
15. **Environment Canada.** 2002. "Lightning activity." **http://weatheroffice.ec.gc.ca/lightning/index_e.html**. Accessed September 11, 2013.
16. **National Weather Service.** 2004. "Lightning risk reduction outdoors: When thunder roars, go indoors!" **http://www.lightningsafety.noaa.gov/outdoors.htm**. Accessed December 29, 2009.
17. **Ackerman, S. A.**, and **Knox, J. A.** 2003. *Meteorology: Understanding the Atmosphere*. Pacific Grove, CA: Thomson Brooks/Cole.
18. **Ashley, W. S.**, and **Mote, T. L.** 2005. "Derecho hazards in the United States." *Bulletin of the American Meteorological Society* 86:1577–1592.
19. **National Weather Service.** 2011. "Weather spotter's field guide." National Oceanic and Atmospheric Administration, National Weather Service. **http://www.wrh.noaa.gov/pqr/Skywarn/SpotterGuide-2011.pdf**. Accessed September 11, 2013.
20. **Burt, C. C.** 2004. *Extreme Weather: A Guide and Record Book*. New York, NY: W. W. Norton.
21. **Grazulis, T. P.** 2001. *The Tornado: Nature's Ultimate Windstorm*. Norman, OK: University of Oklahoma Press.
22. **National Weather Service.** 2006. "The Enhanced Fujita Scale (EF Scale)." Norman, OK: NOAA/National Weather Service, Storm Prediction Center. **www.spc.noaa.gov/efscale/**. Accessed September 10, 2013.
23. **Wikipedia.** "Edmonton tornado." **http://en.wikipedia.org/wiki/Edmonton_Tornado**. Accessed September 11, 2013.
24. **Wikipedia.** "2011 Joplin tornado." **http://en.wikipedia.org/wiki/2011_Joplin_tornado**. Accessed May 9, 2013.
25. **Environment Canada.** 2013. "Tornadoes." **http://www.getprepared.gc.ca/cnt/hzd/trnds-eng.aspx**. Accessed September 10, 2013.
26. **Wikipedia.** "Regina Cyclone." **http://en.wikipedia.org/wiki/Regina_Cyclone**. Accessed May 9, 2013.
27. **Wikipedia.** "2011 Goderich Ontario tornado." **http://en.wikipedia.org/**

wiki/2011_Goderich_Ontario_tornado. Accessed May 9, 2013.

28. **Wikipedia.** 2010. "Blizzard." http://en.wikipedia.org/wiki/Blizzard. Accessed May 9, 2013.

29. **National Weather Service.** 2001. "Winter storms: The deceptive killers." http://www.nws.noaa.gov/om/winterstorm/winterstorms.pdf. Accessed September 11, 2013.

30. **Environment Canada.** 2004. "Winter hazards." http://www.ec.gc.ca/meteo-weather/default.asp?lang=En&n=46FBA88B-1. Accessed September 10, 2013.

31. **The National Snow and Ice Data Center.** 2004. "The blizzards of 1996." Boulder, CO: University of Colorado. http://nsidc.org/snow/blizzard/plains.html. Accessed September 10, 2013.

32. **Whittow, J.** 1980. *Disasters: The Anatomy of Environmental Hazards.* London, UK: Penguin Books.

33. **Jones, R. L.** 2005. "Canadian disasters—An historical survey." http://web.ncf.ca/jonesb/DisasterPaper/disasterpaper.html. Accessed September 10, 2013.

34. **NOAA.** 2013. "Drought." http://www.nws.noaa.gov/om/brochures/climate/Drought.pdf. Accessed May 8, 2013.

35. **Wikipedia.** 2013. "Drought." http://en.wikipedia.org/wiki/Drought. Accessed May 9, 2013.

36. **Subcommittee on Disaster Reduction.** 2003. *Reducing Disaster Vulnerability through Science and Technology: An Interim Report of the Subcommittee on Disaster Reduction.* Washington, DC: Committee on the Environment and Natural Resources, National Science and Technology Council, Executive Office of the President of the United States.

37. **U. S. Environmental Protection Agency.** 2013. "Heat island effect." http://www.epa.gov/hiri/. Accessed September 10, 2013.

38. **De Bono, A., Peduzzi, P., Giuliani, G.**, and **Kluser, S.** 2004. "Impacts of summer 2003 heat wave in Europe." In *Early Warning on Emerging Environmental Threats.* United Nations Environment Programme, Division of Early Warning and Assessment—Europe. http://www.grid.unep.ch/products/3_Reports/ew_heat_wave.en.pdf. Accessed September 10, 2013.

39. **Houghton, J.** 2004. *Global Warming: The Complete Briefing,* 3rd ed. Cambridge, UK: Cambridge University Press.

40. **Goldenberg, S., Landsea, C., Mestas-Nunez, A. M.,** and **Gray, W. M.** 2001. "The recent increase in hurricane activity: Causes and implications." *Science* 293:474–479.

41. **Pyne, S. J., Andrews, P. I.,** and **Laven, R. D.** 1996. *Introduction to Wildland Fire.* New York, NY: John Wiley & Sons.

42. **Golden, J. H.**, and **Adams, C. R.** 2000. "The tornado problem: Forecast, warning, and response." *Natural Hazard Review* 1(2):107–118.

43. **Roberts, R. D., Burgess, D.,** and **Meister, M.** 2006. "Developing tools for nowcasting storm severity." *Weather and Forecasting* 21:540–558.

44. **Godschalk, D. R., Brower, D. J.,** and **Beatly, T.** 1989. *Catastrophic Coastal Storms: Hazard Mitigation and Development Management.* Durham, NC: Duke University Press.

45. **Albey, M.** 1998. *The Ice Storm: An Historic Record in Photographs of January, 1998.* Toronto, ON: McClelland & Stewart.

46. **Auerbach, P. S.** 2003. *Medicine for the Outdoors.* Guilford, CT: The Lyons Press.

47. **Gill Jr., P. G.** 2002. *Wilderness First Aid: A Pocket Guide.* Camden, ME: Ragged Mountain Press.

CHAPTER 11

1. **Brooks, D.** 2005. "The best-laid plan: Too bad it flopped." *The New York Times*, September 11, 2005.

2. **van Heerden, I.,** and **Bryan, M.** 2006. *The Storm: What Went Wrong and Why during Hurricane Katrina—The Inside Story from One Louisiana Scientist.* New York, NY: Viking Penguin.

3. **Knabb, R. D., Rhome, J. R.,** and **Brown, D. P.** 2006. "Tropical cyclone report: Hurricane Katrina 23–30 August 2005." National Oceanic and Atmospheric Administration, National Weather Service, National Hurricane Center. http://www.nhc.noaa.gov/2005atlan.shtml. Accessed September 11, 2013.

4. **Seed, R. B., Bea, R. G., Abdelmalak, R. I., Athanasopoulos, A. G., Boutwell, G. P., Bray, J. D., Briaud, J.-L., Cheung, C., Cobos-Roa, D., Cohen-Waeber, J., Collins, B. D., Ehrensing, L., Farber, D., Hanemann, M., Harder, L. F., Inkabi, K. S., Kammerer, A. M., Karadeniz, D., Kayen, R. E., Moss, R. E. S., Nicks, J., Nimmala, S., Pestana, J. M., Porter, J., Rhee, K., Riemer, M. F., Roberts, K., Rogers, J. D., Storesund, R., Govindasamy, A. V., Vera-Grunauer, X., Wartman, J. E., Watkins, C. M., Wenk Jr., E.,** and **Yim, S. C.** 2006. *Investigation of the Performance of the New Orleans Flood Protection Systems in Hurricane Katrina on August 29, 2005. Volume I: Main Text and Executive Summary.* Berkeley, CA: University of California at Berkeley.

5. **Wolshon, B., Urbina, E., Wilmot, C.,** and **Levitan, M.** 2005. "Review of policies and practices for hurricane evacuation. 1: Transportation planning, preparedness, and response." *Natural Hazards Review* 6:129–142.

6. **Johns, C.,** ed. 2005. "Katrina: Why it became a man-made disaster; Where it could happen next." *National Geographic* Special Edition.

7. **Fitzpatrick, P. J.** 1999. *Natural Disasters: Hurricanes; A Reference Handbook.* Santa Barbara, CA: ABC-CLIO.

8. **Stewart, S. R.** 2005. "Tropical cyclone report: Hurricane Ivan 2–24 September 2004." National Oceanic and Atmospheric Administration, National Weather Service, National Hurricane Center. http://www.nhc.noaa.gov/2004ivan.shtml. Accessed September 11, 2013.

9. **Waugh, W. L., Jr.** 2006. "Shelter from the storm: Repairing the national emergency management system after Hurricane Katrina." *The Annals of the American Academy of Political and Social Science* 806:288–332.

10. **Emanuel, K.** 2006. *Divine Wind: The History and Science of Hurricanes.* New York, NY: Oxford University Press.

11. **Pasch, R. J., Blake, E. S., Cobb III, H. D.,** and **Roberts, D. P.** 2006. "Tropical cyclone report: Hurricane Wilma 15–25 October 2005." National Oceanic and Atmospheric Administration, National Weather Service, National Hurricane Center. http://www.nhc.noaa.gov/pdf/TCR-AL252005_Wilma.pdf. Accessed September 10, 2013.

12. **National Weather Service.** 2006. "Tropical cyclone introduction.

Jetstream—Online school for weather." **http://www.srh.noaa.gov/jetstream/tropics/tc.htm.** Accessed September 11, 2013.

13. **Ahrens, C. D.** 2005. *Essentials of Meteorology: An Invitation to the Atmosphere.* Belmont, CA: Thomson Brooks/Cole.

14. **Aguado, E.,** and **Burt, J. E.** 2007. *Understanding Weather and Climate,* 4th ed. Upper Saddle River, NJ: Pearson Prentice Hall.

15. **Simpson, R. H.,** and **Riehl, H.** 1981. *The Hurricane and Its Impact.* Baton Rouge, LA: Louisiana State University Press.

16. **Lutgens, F. K.,** and **Tarbuck, E. J.** 2007. *The Atmosphere: An Introduction to Meteorology,* 10th ed. Upper Saddle River, NJ: Pearson Prentice Hall.

17. **Ackerman, S. A.,** and **Knox, J. A.** 2003. *Meteorology: Understanding the Atmosphere.* Pacific Grove, CA: Thomson Brooks/Cole.

18. **Flanagan, R.** 1993. "Beaches on the brink." *Earth* 2(6):24–33.

19. **Burt, C. C.** 2004. *Extreme Weather: A Guide and Record Book.* New York, NY: W. W. Norton.

20. **Wikipedia.** 2013. "Hurricane Sandy." **http://en.wikipedia.org/wiki/Hurricane_Sandy.** Accessed May 10, 2013.

21. **Stewart, S. R.** 2002. "Tropical cyclone report: Tropical Storm Allison 5–17 June 2001." National Oceanic and Atmospheric Administration, National Weather Service, National Hurricane Center. **http://www.nhc.noaa.gov/2001allison.html.** Accessed September 11, 2013.

22. **National Weather Service Southern Regional Headquarters.** 2002. *Tropical Cyclones & Inland Flooding for the Southern States.* Fort Worth, TX: National Oceanic and Atmospheric Administration.

23. **Negri, A. J., Burkardt, N., Golden, J. H., Halverson, J. B., Huffman, G. J., Larsen, M. C., McGinley, J. A., Updike, R. G., Verdin, J. P.,** and **Wieczorek, G. F.** 2005. "The hurricane-flood-landslide continuum." *Bulletin of the American Meteorological Society* 86:1241–1247.

24. **Edwards, R.** 1998. "Tornado production by exiting tropical cyclones." Preprints, *23rd Conference on Hurricanes and Tropical Meteorology*, Dallas, TX: American Meteorological Society, pp. 485–488.

25. **Crossett, K. M., Culliton, T. J., Wiley, P. C.,** and **Goodspeed, T. R.** 2004. *Population Trends along the Coastal United States.* National Oceanic and Atmospheric Administration, National Ocean Service.

26. **Houghton, J.** 2004. *Global Warming: The Complete Briefing,* 3rd ed. Cambridge, UK: Cambridge University Press.

27. **National Hurricane Center.** 2013. "National Hurricane Center." National Oceanic and Atmospheric Administration, National Weather Service. **http://www.nhc.noaa.gov/.** Accessed September 10, 2013.

28. **Environment Canada.** 2013. "Canadian Hurricane Centre." Meteorological Service of Canada. **http://www.ec.gc.ca/ouragans-hurricanes/.** Accessed September 10, 2013.

29. **National Hurricane Center.** 2004. "Hurricane preparedness: Disaster supply kit." National Oceanic and Atmospheric Administration, National Weather Service. **www.nhc.noaa.gov/HAW2/english/prepare/supply_kit.shtml.** Accessed September 11, 2013.

CHAPTER 12

1. **Solomon, S., Qin, D., Manning, M., Chen, Z., Marquis, M., Averyt, K. B., Tignor, M.,** and **Miller, H. L.,** eds. 2007. *Contribution of Working Group I to the Fourth Assessment Report of the Intergovernmental Panel on Climate Change.* New York, NY: Cambridge University Press.

2. **Minkel, J. R.** 2004. "Surf's up—Way up." *Scientific American* 291(4):38.

3. **Davis, R. E.,** and **Dolan, R.** 1993. "Nor'easters." *American Scientist* 81:428–439.

4. **Bartsch-Winkler, S. R.,** and **Winkler, D. K.** 1988. *Catalogue of Worldwide Tidal Bore Occurrences and Characteristics.* U.S. Geological Survey Circular C 1022.

5. **Komar, P. D.** 1998. *Beach Processes and Sedimentation,* 2nd ed. Upper Saddle River, NJ: Prentice Hall.

6. **Johnson, J. W.** 1956. "Dynamics of nearshore sediment movement." *Bulletin of the American Association of Petroleum Geologists* 40:2211–2232.

7. **Garrison, T.** 2005. *Oceanography: An Invitation to Marine Science.* Belmont, CA: Thomson Brooks/Cole.

8. **Plafker, G.** 1965. "Tectonic deformation associated with the 1964 Alaska earthquake." *Science* 148: 1675–1687.

9. **National Oceanic and Atmospheric Administration (NOAA).** 1998. *Population at Risk from Natural Hazards by Sandy Ward and Catherine Main.* NOAA's State of the Coast Report. Silver Spring, MD: NOAA. **http://oceanservice.noaa.gov/websites/retiredsites/sotc_pdf/PAR.PDF.** Accessed September 11, 2013.

10. **Norris, R. M.** 1977. "Erosion of sea cliffs." In *Geologic Hazards in San Diego,* ed. P. L. Abbott and J. K. Victoris. San Diego, CA: San Diego Society of Natural History.

11. **Clague, J. J.,** and **Bornhold, B. D.** 1980. "Morphology and littoral processes of the Pacific coast of Canada." In *The Coastline of Canada, Littoral Processes and Shore Morphology*, ed. S. B. McCann, pp. 339–380. Geological Survey of Canada Paper 80-10.

12. **Williams, S. J., Dodd, K.,** and **Gohn, K. K.** 1991. *Coasts in Crisis.* U.S. Geological Survey Circular 1075.

13. **Morton, R. A., Pilkey Jr., O. H., Pilkey Sr., O. H.,** and **Neal, W. J.** 1983. *Living with the Texas Shore.* Durham, NC: Duke University Press.

14. **Komar, P. D.** 1997. *The Pacific Northwest Coast: Living with the Coasts of Oregon and Washington.* Durham, NC: Duke University Press.

15. **Larsen, J. I.** 1973. *Geology for Planning in Lake County, Illinois.* Illinois State Geological Survey Circular 481.

16. **Buckler, W. R.,** and **Winters, H. A.** 1983. "Lake Michigan bluff recession." *Annals of the Association of American Geographers* 73(1):89–110.

17. **Shaw, J., Taylor, R. B., Forbes, D. L., Solomon, S.,** and **Ruz, M.-H.** 1998. *Sensitivity of the Coasts of Canada to Sea-level Rise.* Geological Survey of Canada Bulletin 505.

18. **Thieler, E. R., Gayes, P. T., Schwab, W. C.,** and **Harris, M. S.** 1999. "Tracing sediment dispersal on nourished beaches: Two case studies. In *Coastal Sediments '99*, ed. N. C. Kraus and W. G. McDougal, pp. 2118–2136. Reston, VA: American Society of Civil Engineers.

19. **Forbes, D. L., Taylor, R. B.,** and **Shaw, J.** 1989. "Shorelines and rising sea levels in eastern Canada." *Episodes* 12:23–28.

20. **U. S. Department of Commerce.** 1978. *State of Maryland Coastal*

Management Program and Final Environmental Impact Statement. Washington, DC: U.S. Department of Commerce.

21. **Leatherman, S. P.** 1984. "Shoreline evolution of North Assateague Island, Maryland." *Shore and Beach*, July:3–10.
22. **Wilkinson, B. H.**, and **McGowen, J. H.** 1977. "Geologic approaches to the determination of long-term coastal recession rates, Matagorda Peninsula, Texas." *Environmental Geology* 1:359–365.
23. **Clague, J. J.** 1989. "Sea levels on Canada's Pacific coast: Past and future trends." *Episodes* 12(1):29–33.
24. **Committee on Coastal Erosion Zone Management.** 1990. *Managing Coastal Erosion*. National Research Council. Washington, DC: National Academies Press.
25. **Carter, R. W. G.**, and **Oxford, J. D.** 1982. "When hurricanes sweep Miami Beach." *Geographical Magazine* 54:442–448.
26. **Flanagan, R.** 1993. "Beaches on the brink." *Earth* 2(6):24–33.
27. **Pilkey, O. H.**, and **Dixon, K. L.** 1996. *The Corps and the Shore*. Washington, DC: Island Press.
28. **Ioannidis, D.**, and **Karambas, Th. V.** 2007. "Soft shore protection methods: Beach drain system." In *10th International Conference on Environmental Science and Technology, CEST2007*, Kos Island, Greece, pp. A-528–535.
29. **Rowntree, R. A.** 1974. "Coastal erosion: The meaning of a natural hazard in the cultural and ecological context." In *Natural Hazards: Local, National, Global*, ed. G. F. White, pp. 70–79. New York, NY: Oxford University Press.
30. **Leatherman, S. P.** 2003. *Dr. Beach's Survival Guide: What You Need to Know about Sharks, Rip Currents, and More before Going in the Water*. New Haven, CT: Yale University Press.

CHAPTER 13

1. **Wikipedia.** 2009. "Okanagan Mountain Park fire." **http://en.wikipedia.org/wiki/2003_Okanagan_Mountain_Park_Fire.** Accessed September 11, 2013.
2. **Ruddiman, W. F.** 2001. *Earth's Climate: Past and Future*. New York, NY: W. H. Freeman.
3. **Ruddiman, W. F.** 2005. *Plows, Plagues, and Petroleum: How Humans Took Control of Climate*. Princeton, NJ: Princeton University Press.
4. **Rossotti, H.** 1993. *Fire*. Oxford, UK: Oxford University Press.
5. **Pyne, S. J.**, **Andrews, P. L.**, and **Laven, R. D.** 1996. *Introduction to Wildland Fire*, 2nd ed. New York, NY: John Wiley & Sons.
6. **Ward, D.** 2001. "Combustion chemistry and smoke." In *Forest Fires: Behavior and Ecological Effects*, eds. E. A. Johnson and K. Miyanishi, pp. 57–77. San Diego, CA: Academic Press.
7. **Arno, S. F.**, and **Allison-Bunnell, S.** 2002. *Flames in Our Forests*. Washington, DC: Island Press.
8. **Verdi, R. J.**, **Tomlinson, S. A.**, and **Marella, R. L.** 2006. *The Drought of 1998–2002: Impacts on Florida's Hydrology and Landscapes*. U.S. Geological Survey Circular 1295.
9. **Florida Department of Community Affairs.** 2004. "Wildfire mitigation in Florida: Land use planning strategies and best development practices." **http://www.fl-dof.com/publications/index.html.** Accessed September 11, 2013.
10. **Atlas of Canada.** 2013. "Forest fires." **http://atlas.nrcan.gc.ca/site/english/maps/forestry.html.** Accessed September 10, 2013.
11. **Earth Policy Institute.** 2009. "Wildfires by region: Observations and future prospects." **http://www.earthpolicy.org/images/uploads/graphs_tables/fire.htm.** Accessed May 11, 2013.
12. **Canadian Forest Service.** 2009. "Historical analysis: Large fire database." **http://cwfis.cfs.nrcan.gc.ca/en_CA/lfdb.** Accessed September 10, 2013.
13. **Natural Resources Canada.** 2013. "Canadian Wildland Fire Information System." **http://cwfis.cfs.nrcan.gc.ca/home.** Accessed September 10, 2013.
14. **Wikipedia.** 2013. "2011 Slave Lake Fire." **http://en.wikipedia.org/wiki/2011_Slave_Lake_wildfire.** Accessed May 11, 2013.
15. **National Interagency Fire Center.** 2013. "Statistics." **http://www.nifc.gov/fireInfo/fireInfo_statistics.html.** Accessed September 10, 2013.
16. **Mackes, K. H.**, **Lynch, D. L.**, **Kelly, S. K.**, and **Eckhoff, M.** 2007. "Missionary Ridge fire assessment." *Journal of Testing and Evaluation*, doi: 10.1520/JTE100044.
17. **Wikipedia.** 2013. "Waldo Canyon Fire." **http://en.wikipedia.org/wiki/Waldo_Canyon_fire.** Accessed May 11, 2013.
18. **Chandler, C.**, **Cheney, P.**, **Thomas, P.**, **Trabaud, L.**, and **Williams, D.** 1983. *Fire in Forestry: Vol. I—Forest Fire Behavior and Effects*. New York, NY: John Wiley & Sons.
19. **DeBano, L. F.** 1981. *Water Repellent Soils: A State-of-the-art*. U.S. Forest Service General Technical Report INT-79.
20. **U. S. Geological Survey.** 1998. *USGS Wildland Fire Research*. U.S. Geological Survey Fact Sheet 125-98.
21. **Jordan, P.**, and **Covert, S. A.** 2009. "Debris flows and floods following the 2003 wildfires in southern British Columbia." *Environmental and Engineering Geoscience* 15:217–234.
22. **Westerling, A. L.**, **Hidalgo, H. G.**, **Cayan, D. R.**, and **Swetnam, T. W.** 2006. "Warming and earlier spring increase western U.S. forest wildfire activity." *Science* 313:940–943.
23. **Kasischke, E. S.**, and **Turetsky, M. R.** 2006. "Recent changes in the fire regime across the North American boreal region—Spatial and temporal patterns of burning across Canada and Alaska." *Geophysical Research Letters* 33:L09703.
24. **Westerling, A. L.**, **Cayan, D. R.**, **Brown, T. J.**, **Hall, B. L.**, and **Riddle, L. G.** 2004. "Climate, Santa Ana winds and autumn wildfires in southern California." *EOS, Transactions, American Geophysical Union* 85(31):294, 296.
25. **Tolmé, P.** 2004. "Will global warming cause more wildfires?" *National Wildlife* 42(5):14–16.
26. **Fried, J. S.**, **Torn, M. S.**, and **Mills, E.** 2004. "The impact of climate change on wildfire severity: A regional forecast for northern California." *Climatic Change* 64:169–191.
27. **Canadian Forestry Service.** 2012. "Mountain pine beetle." **http://tidcf.nrcan.gc.ca/insects/factsheet/2816.** Accessed September 10, 2013.
28. **British Columbia Forest Service.** 2007. "Mountain pine beetles in British Columbia." **http://www.for.gov.bc.ca/hfp/mountain%5Fpine%5Fbeetle/.** Accessed September 10, 2013.
29. **British Columbia Wildlife Management Branch.** No date. "Very large wildfires." **http://bcwildfire.ca/History/LargeFires.htm.** Accessed September 10, 2013.

30. **Fuller, M.** 1991. *Forest Fires: An Introduction to Wildland Fire Behavior, Management, Firefighting and Prevention*. New York, NY: John Wiley.

31. **Jones, S. R.**, and **Cushman, R. C.** 2004. *A Field Guide to the North American Prairie*. New York, NY: Houghton Mifflin.

32. **Canadian Council of Forest Ministers.** 2009. "National Forestry Database: Forest fires—Background." **http://nfdp.ccfm.org/fires/background_e.php**. Accessed September 10, 2013.

33. **Minnich, R. A.** 1983. "Fire mosaics in southern California and northern Baja California." *Science* 219:1287–1294.

34. **Minnich, R. A. Garbour, M. G., Burk, J. H.**, and **Sosa-Ramiriz, J.** 2000. "California mixed-conifer forests under unchanged fire regimes in Sierra San Pedro Martir, Baja California, Mexico." *Journal of Biogeography* 27:105–129.

35. **Burgan, R. E., Klaver, R. W.**, and **Klaver, J. M.** 1998. "Fuel models and fire potential from satellite and surface observations." *International Journal of Wildland Fire* 8:159–170.

36. **Partners in Protection.** 2005. "Partners in Protection: Working together for safer communities in the wildland urban interface." **http://www.partnersinprotection.ab.ca**. Accessed September 10, 2013.

37. **Arrowood, J. C.** 2003. *Living with Wildfires: Prevention, Preparation, and Recovery*. Denver, CO: Bradford Publishing Company.

CHAPTER 14

1. **NASA Earth Observatory.** 2009. "Arctic sea ice." **http://earthobservatory.nasa.gov/Features/WorldOfChange/sea_ice.php**. Accessed September 10, 2013.

2. **NASA.** 2009. "Satellites show Arctic literally on thin ice." **http://www.nasa.gov/topics/earth/features/arctic_thinice.html**. Accessed September 10, 2013.

3. **Rignot, E.**, and **Kanagaratnam, P.** 2006. "Changes in the velocity structure of the Greenland ice sheet." *Science* 311:986–990.

4. **BBC News.** 2006. "Greenland melt speeding up." **http://news.bbc.co.uk/2/hi/science/nature/4783199.stm**. Accessed September 11, 2013.

5. **Wikipedia.** 2009. "Greenland ice sheet." **http://en.wikipedia.org/wiki/Greenland_ice_sheet**. Accessed September 10, 2013.

6. **Shepherd, A.**, and **Wingham, D.** 2007. "Antarctic and Greenland ice sheets." *Science* 315:1529–1332.

7. **Jeffries, M. O.** 1986. "Ice island calvings and ice shelf changes, Milne Ice Shelf and Ayles Ice Shelf, Ellesmere Island, N.W.T." *Arctic* 39:15–19.

8. **Hattersley-Smith, G.** 1963. "The Ward Hunt Ice Shelf: Recent changes of the ice front." *Journal of Glaciology* 4:415–424.

9. **NASA Earth Observatory.** 2004. "Breakup of the Ward Hunt Ice Shelf." **http://earthobservatory.nasa.gov/Features/wardhunt/**. Accessed September 11, 2013.

10. **European Space Agency.** 2009. "Satellite imagery shows fragile Wilkins Ice Shelf destabilised." **http://www.esa.int/esaCP/SEMRAVANJTF_index_0.html**. Accessed September 11, 2013.

11. **Intergovernmental Panel on Climate Change.** 2007. "Summary for policy makers." In *Climate Change 2007: The Physical Science Basis. Contribution of Working Group I to the Fourth Assessment Report of the Intergovernmental Panel on Climate Change*, ed. S. Solomon, D. Qin, M. Manning, Z. Chen, M. Marquis, K. B. Averyt, M. Tignor, and H. L. Miller. Cambridge, UK: Cambridge University Press.

12. **Hansen, J., Sato, M., Pushker, K., Russell, G., Lea, D. W.**, and **Siddall, M.** 2007. "Climate change and trace gases." *Philosophical Transactions of the Royal Society, Mathematical, Physical and Engineering Sciences* 365:1925–1954.

13. **Cloud, P.** 1990. Written communication.

14. **National Aeronautics and Space Administration (NASA).** 1990. *EOS: A Mission to Planet Earth*. Washington, DC: NASA.

15. **United States Environmental Protection Agency.** 2013. "Overview of greenhouse gases; methane emissions." **http://epa.gov/climatechange/ghgemissions/gases/ch4.html**. Accessed September 10, 2013.

16. **Oldfield, F.** 2005. *Environmental Change: Key Issues and Alternative Approaches*. New York, NY: Cambridge University Press.

17. **Houghton, J.** 2004. *Global Warming: The Complete Briefing*, 3rd ed. New York, NY: Cambridge University Press.

18. **Goudie, A.** 1984. *The Nature of the Environment*, 3rd ed. Oxford, UK: Blackwell Scientific.

19. **Grainger, A.** 1990. *The Threatening Desert*. London, UK: Earthscan Publications Ltd.

20. **Jansen, E., Overpeck, J., Briffa, K. R., Duplessy, J.-C., Joos, F., Masson-Deltte, V., Olago, D., Otto-Bliesner, B., Peltier, W. R., Rahmstorf, S., Ramesh, R., Raynaud, D., Rind, D., Solomina, O., Villalba, R.**, and **Zhang, D.** 2007. "Paleoclimate." In *Climate Change 2007: The Physical Science Basis. Contribution of Working Group I to the Fourth Assessment Report of the Intergovernmental Panel on Climate Change,* ed. S. Solomon, D. Qin, M. Manning, Z. Chen, M. Marquis, K. B. Averyt, M. Tignor, and H. L. Miller. New York, NY: Cambridge University Press.

21. **Petit, J. R., Jouzel, J., Raynaud, D., Barkov, N. I., Barnola, J. M., Basile, I., Bender, M., Chappellaz, J., Davis, M., Delaygue, G., Delmotte, M., Kotlyakov, V. M., Legrand, M., Lipenkov, V. Y., Lorius, C., Pepin, L., Ritz, C., Saltzmann, E.**, and **Stievenard, M.** 1999. "Climate and atmospheric history of the past 420,000 years from the Vostok ice core, Antarctica." *Nature* 399:429–436.

22. **NOAA.** 2002. "Paleo proxy data." In *Introduction to Paleoclimatology*. National Oceanic and Atmospheric Administration. **http://www.ncdc.noaa.gov/paleo/primer_proxy.html**. Accessed September 10, 2013.

23. **Weart, S. R.** 2003. *The Discovery of Global Warming*. Cambridge, MA: Harvard University Press.

24. **Aguado, E.**, and **Burt, J. E.** 2007. *Understanding Weather and Climate,* 4th ed. Upper Saddle River, NJ: Prentice Hall.

25. **Rahmstorf, S., Cazenave, A., Church, J. A., Hansen, J. E., Keeling, R. F., Parker, D. E.**, and **Somerville, R. C. J.** 2007. "Recent climate observations compared to projections." *Science* 316:709.

26. **Bennett, M. R.**, and **Glasser, N. F.** 1996. *Glacial Geology: Ice Sheets and Landforms*. Chichester, UK: John Wiley and Sons.

27. **Dawson, A. G.** 1992. *Ice Age Earth: Late Quaternary Geology and Climate*. New York, NY: Routledge.

28. **Lamb, H. H.** 1977. *Climate: Present, Past and Future. Vol. 2, Climatic History and the Future*. New York, NY: Barnes & Noble Books.

29. **Grove, J. M.** 2004. *Little Ice Ages: Ancient and Modern*. London, UK: Routledge.

30. **Crowley, T. J.** 2000. "Causes of climate change over the past 1000 years." *Science* 289:270–277.

31. **Trenberth, K. E., Jones, P. D., Ambenje, P., Bojariu, R., Easterling, D., Klein Tank, A., Parker, D., Rahimzadeh, R., Renwick, J. A., Rustiucci, M., Soden, B.,** and **Zhai, P.** 2007. "Observations: Surface and atmospheric climate change." In *Climate Change 2007: The Physical Science Basis. Contribution of Working Group I to the Fourth Assessment Report of the Intergovernmental Panel on Climate Change*, ed. S. Solomon, D. Qin, M. Manning, Z. Chen, M. Marquis, K. B. Averyt, M. Tignor, and H. L. Miller. New York, NY: Cambridge University Press.

32. **Karl, T. R.** 1995. "Trends in U.S. climate during the twentieth century." *Consequences* 1(1):3–12.

33. **Seager, R.** 2006. "The source of Europe's mild climate." *American Scientist* 94:334–341.

34. **Evans, R.** 2002. "Blast from the past." *Smithsonian* 33(4):52–57.

35. **Rignot, E.,** and **Kanagaratnam, P.** 2006. "Changes in the velocity structure of the Greenland Ice Sheet." *Science* 311:986–990.

36. **National Snow & Ice Data Center.** 2013. "The contribution of the cryosphere to changes in sea level." **http://nsidc.org/crysophere/sotc/sea_level.html.** Accessed May 13, 2013.

37. **Arendt, A. A., Echelmeyer, K. A., Harrison, W. D., Lingle, C. S.,** and **Valentine, V. B.** 2002. "Rapid wastage of Alaska glaciers and their contribution to rising sea level." *Science* 297:382–386.

38. **Hall, M. H. P.,** and **Fagre, D. B.** 2003. "Modeled climate-induced glacier change in Glacier National Park, 1850–2100." *BioScience* 53:131–140.

39. **U. S. Geological Survey.** 2012. "USGS repeat photography project documents retreating glaciers in Glacier National Park." **http://nrmsc.usgs.gov/repeatphoto/.** Accessed June 2, 2013.

40. **Appenzeller, T.** 2007. "The big thaw." *National Geographic* 211(6):56–71.

41. **Steig, E. J., Schneider, D. P., Rutherford, S. D., Mann, M. E., Comiso, J. C., et al.** 2009. "Warming of the Antarctic ice-sheet surface since the 1957 International Geophysical Year." *Nature* 457:459–462.

42. **Davis, C. H., Li, Y., McConnell, J. R., Frey, M. M.,** and **Hanna, E.** 2005. "Snowfall-driven growth in East Antarctic ice sheet mitigates recent sea-level rise." *Science* 308:1898–1901.

43. **Monagham, A. J., Bromwich, D. H., Fogt, R. L., Wang, S.-H., Mayewski, P. A., et al.** 2006. "Insignificant change in Antarctic snowfall since the International Geophysical Year." *Science* 313: 827–831.

44. **Dickinson, W. R.** 2009. "Pacific atoll living—How long already and until when." *GSA Today* 19(3):4–10.

45. **Delcan.** 2012. *Cost of Adaptation—Sea Dykes & Alternative Strategies; Final Report.* Victoria, BC: B. C. Ministry of Forests, Lands and Natural Resource Operations.

46. **Tufnell, L.** 1984. *Glacier Hazards.* New York, NY: Longman.

47. **Costa, J. E.,** and **Schuster, R. L.** 1988. "The formation and failure of natural dams." *Geological Society of America Bulletin* 100:1054–1088.

48. **Clague, J. J.,** and **Evans, S. G.** 1994. *Formation and Failure of Natural Dams in the Canadian Cordillera.* Geological Survey of Canada Bulletin 464.

49. **Geertsema, M.,** and **Clague, J. J.** 2005. "Jökulhlaups at Tulsequah Glacier, northwestern British Columbia, Canada." *The Holocene* 15:310–315.

50. **Evans, S. G., Tutubalina, O. V., Drobyshev, V. N., Chernomorets, S. S., McDougall, S., Petrakov, D. A.,** and **Hungr, O.** 2009. "Catastrophic detachment and high-velocity long-runout flow of Kolka Glacier, Caucasus Mountains, Russia in 2002." *Geomorphology* 105:314–321.

51. **Solomon, S., Qin, D., Manning, M., Alley, R. B., Berntsen, T., Bindoff, N. L., Chen, Z., Chidthaisong, A., Gregory, J. M., Hegerl, G. C., Heimann, M., Hewitson, B., Hoskins, B. J., Joos, F., Jouzel, J., Kattsov, V., Lohmann, U., Matsuno, T., Molina, M., Nicholls, N., Overpeck, J., Raga, G., Ramaswamy, V., Ren, J., Rustiucucci, M., Somerville, R., Stocker, T. F., Whetton, P., Wood, R. A.,** and **Wratt, D.** 2007. "Technical summary." In *Climate Change 2007: The Physical Science Basis. Contribution of Working Group I to the Fourth Assessment Report of the Intergovernmental Panel on Climate Change*, eds. S. Solomon, D. Qin, M. Manning, Z. Chen, M. Marquis, K. B. Averyt, M. Tignor, and H. L. Miller. Cambridge, UK: Cambridge University Press.

52. **University Corporation for Atmospheric Research.** 1994. *El Niño and Climate Prediction.* Washington, DC: NOAA Office of Global Programs.

53. **Dennis, R. E.** 1984. "A revised assessment of worldwide economic impacts: 1982–1984 El Niño/Southern Oscillation event." *EOS, Transactions, American Geophysical Union* 65(45):910.

54. **Canby, T. Y.** 1984. "El Niño's ill winds." *National Geographic* 165:144–181.

55. **Philander, S. G.** 1998. "Who Is El Niño?" *EOS, Transactions, American Geophysical Union* 79(13):170.

56. **U. S. Climate Change Program.** 2008. *Climate Change and Ecosystems.* Washington, DC.

57. **Union of Concerned Scientists.** 2013. "Early warning signs of global warming: Coral reef bleaching." **http://www.ucsusa.org/global_warming/science_and_impacts/impacts/early-warning-signs-of-global-2.html.** Accessed September 10, 2013.

58. **Stone, R.** 2007. "A world without corals?" *Science* 316:678–681.

59. **Appell, D.** 2009. "Can 'assisted migration' save species from global warming?" *Scientific American* 300:78–80.

60. **Botkin, D. B., et al.** 2007. "Forecasting effects of global warming on biodiversity." *BioScience* 57:227–236.

61. **Mainguet, M.** 1994. *Desertification*, 2nd ed. Berlin, Germany: Springer-Verlag.

62. **Wheaton, E. E.,** and **Arthur, L. M.** 1989. "Executive summary." In *Environmental and Economic Impacts of the 1988 Drought: With Emphasis on Saskatchewan and Manitoba*, eds. E. E. Wheaton and L. M. Arthur, pp. iii–xxiv. Saskatchewan Research Council Publication No. 3-2330-4-E-89.

63. **Karl, T. R., Heim, R. R., Jr.,** and **Qualye, R. G.** 1991. "The greenhouse effect in central North America: If not now, when?" *Science* 251:1058–1061.

64. **Lemmen, D. S., Vance, R. E., Wolfe, S. A.,** and **Last, W. M.** 1997. "Impacts of future climate change on the southern Canadian prairies:

A paleoenvironmental perspective." *Geoscience Canada* 24:121–133.

65. **Vance, R. E., Mathewes, R. W.,** and **Clague, J. J.** 1992. "A 7000-year record of lake-level change on the northern Great Plains: A high resolution proxy of past climate." *Geology* 20:879–882.

66. **Wolfe, S. A.,** and **Lemmen, D. S.** 1999. "Monitoring of dune activity in the Great Sand Hills region, Saskatchewan." In *Holocene Climate and Environmental Change in the Palliser Triangle: A Geoscientific Context for Evaluation of the Impacts of Climate Change on the Southern Canadian Prairies*, eds. D. S. Lemmen and R. E. Vance, pp. 199–210. Geological Survey of Canada Bulletin 534.

67. **Stocks, B. J., Fosberg, M. A., Lynham, T. J., Mearns, L., Wotton, B. M., Yang, Q., Jin, J.-Z., Lawrence, K., Hartley, G. R., Mason, J. A.,** and **McKenney, D. W.** 1998. "Climate change and forest fire potential in Russian and Canadian boreal forests." *Climate Change* 38:1–13.

68. **Nameroff, T.** 1997. "The climate change debate is warming up." *GSA Today* 7(12):11–13.

69. **Fahey, D. W.** 2006. "Twenty questions and answers about the ozone layer: 2006 update." Nairobi, Kenya: United Nations Environmental Programme, Ozone Secretariat. **http://ozone.unep.org/Frequently_Asked_Questions.** Accessed September 10, 2013.

70. **Schag, D. P.** 2007. "Preparing to capture carbon." *Science* 315:812–813.

71. **Watson, A. J., Bakker, D. C. E., Ridgwell, A. J., Boyd, P. W.,** and **Law, C. S.** 2000. "Effect of iron supply on Southern Ocean CO_2 uptake, and implications for glacial atmospheric CO_2." *Nature* 407:730–733.

72. **Chisholm, S. W.** 2000. "Oceanography: Stirring times in Southern Ocean." *Nature* 407:685–687.

73. **White, D. J., Burrowes, G., Davis, T., Hajnal, Z., Hirsche, K., Hutcheon, I., Majer, E., Rostron, B.,** and **Whittaker, S.** 2004. "Greenhouse gas sequestration in abandoned oil reservoirs: The International Energy Agency Weyburn Pilot Project." *GSA Today* 14(7):4–10.

74. **Friedman, S. J.** 2003. "Storing carbon in Earth." *Geotimes* 48(3):16–21.

75. **Service, R. F.** 2004. "The carbon conundrum." *Science* 305:962–963.

76. **Bartlett, K., Pinsker, L. M.,** and **Reed, C.** 2003. "Demonstrating carbon sequestration." *Geotimes* 48(3):22–25.

77. **McGeehin, J. P., Barron, J. A., Anderson, D. M.,** and **Verardo, D. J.** 2008. *Abrupt Climate Change. Final Report, Synthesis and Assessment Product 3.4.* Washington, DC: U.S. Climate Change Science Program.

78. **Alley, R. B.** 2004. "Abrupt climate change." *Scientific American* 291(5):62–69.

79. **Broecker, W.** 1997. "Will our ride into the greenhouse future be a smooth one?" *GSA Today* 7(5):1–7.

CHAPTER 15

1. **Lewis, J. S.** 1996. *Rain of Iron and Ice*. Redding, MA: Addison-Wesley.

2. **Rubin, A. F.** 2002. *Disturbing the Solar System*. Princeton, NJ: Princeton University Press.

3. **Brown, P., Spalding, R. E., ReVelle, D. O., Tagliaferri, E.,** and **Worden, S. P.** 2002. "The flux of small near-Earth objects colliding with the Earth." *Nature* 420:294–296.

4. **Cloud, P.** 1978. *Cosmos, Earth and Man*. New Haven, CT: Yale University Press.

5. **Davidson, J. P., Reed, W. E.,** and **Davis, P. M.** 1997. *Exploring Earth*. Upper Saddle River, NJ: Prentice Hall.

6. **Grieve, R. A. F.** 2001. "Impact cratering on Earth." In *A Synthesis of Geological Hazards in Canada,* ed. G. R. Brooks, pp. 207–224. Geological Survey of Canada Bulletin 548.

7. **Gehrels, T.,** ed. 1995. *Hazards Due to Comets and Asteroids*. Tucson, AZ: University of Arizona Press.

8. **Grieve, R. A. F.** 2006. *Impact Craters in Canada*. St. John's, NF: Geological Association of Canada GEOtext 5.

9. **Lipschutz, M. E.,** and **Schultz, L.** 2007. "Meteorites." In *Encyclopedia of the Solar System*, 2nd ed., eds. L.-A. McFadden, P. R. Weissman, and T. V. Johnson, pp. 251–282. New York, NY: Elsevier.

10. **Grieve, R. A. F., Cintala, M. J.,** and **Tagle, R.** 2007. "Planetary impacts." In *Encyclopedia of the Solar System*, 2nd ed., eds. L.-A. McFadden, P. R. Weissman, and T. V. Johnson, pp. 813–828. New York, NY: Elsevier.

11. **Williams, S. J., Barnes, P.,** and **Prager, E. J.** 2000. *U.S. Geological Survey Coastal and Marine Geology Research—Recent Highlights and Achievements*. U.S. Geological Survey Circular 1199.

12. **Wikipedia.** 2009. "Sudbury Basin." **http://en.wikipedia.org/wiki/Sudbury_Basin.** Accessed September 11, 2013.

13. **Planetary and Space Science Centre.** 2009. "Earth Impact Database." **http://www.unb.ca/passc/ImpactDatabase/.** Accessed September 10, 2013.

14. **Pye, E. G., Naldrett, A. J.,** and **Giblin, P. E.** 1984. *The Geology and Ore Deposits of the Sudbury Structure*. Ontario Ministry of Natural Resources, Ontario Geological Survey Special Volume 1.

15. **Slack, J. F.,** and **Cannon, W. F.** 2009. "Extraterrestrial demise of banded iron formations 1.85 billion years ago." *Geology* 37:1011–1014.

16. **Dott Jr., R. H.,** and **Prothero, D. R.** 1994. *Evolution of the Earth,* 5th ed. New York, NY: McGraw-Hill.

17. **Stanley, S. M.** 2005. *Earth System History*. New York, NY: W. H. Freeman.

18. **Hildebrand, A. R.** 1993. "The Cretaceous/Tertiary boundary impact (or the dinosaurs didn't have a chance)." *Journal of the Royal Astronomical Society of Canada* 87:7–118.

19. **Hallam, T.** 2004. *Catastrophes and Lesser Calamities: The Causes of Mass Extinctions*. New York, NY: Oxford University Press.

20. **Barnosky, A. D., Koch, P. L., Feranec, R. S., Wing, S. L.,** and **Shabel, A. B.** 2004. "Assessing the causes of late Pleistocene extinctions on the continents." *Science* 306:70–75.

21. **Lange, I. M.** 2002. *Ice Age Mammals of North America: A Guide to the Big, the Hairy, and the Bizarre*. Missoula, MT: Mountain Press Publishing Company.

22. **Kring, D. A.,** and **Durda, D. D.** 2003. "The day the world burned." *Scientific American* 289(6):98–105.

23. **Alvarez, W.** 1997. *T. Rex and the Crater of Doom*. New York, NY: Vintage Books.

24. **Alvarez, L. W., Alvarez, W., Asaro, F.,** and **Michel, H. V.** 1980. "Extraterrestrial cause for Cretaceous-Tertiary extinction." *Science* 208:1095–1108.

25. **Pope, K. O., Ocampo, A. C.,** and **Duller, C. E.** 1991. "Mexican site for the K/T impact crater?" *Nature* 351:105.

26. **Swisher III, C. C., Grajales-Nishimura, J. N., Montanari, A., Margolis, S. V., Claeys, P., Alvarez, W., Ranne, P., Cedillo-Pardo, E., Maurrasse, F. J-N. R., Curtis, G. H., Smit, J.,** and **McWilliams, M. O.** 1992. "Coeval $^{40}Ar/^{39}Ar$ ages of 65.0 million years ago from Chicxulub crater melt rocks and Cretaceous-Tertiary boundary tektites." *Science* 257:954–958.

27. **Hildebrand, A. R., Penfield, G. T., Kring, D. A., Pilkington, N., Camargo, Z. A., Jacobsen, S. B.,** and **Boynton, W. V.** 1991. "Chicxulub crater: A possible Cretaceous/Tertiary boundary impact crater on the Yucatan peninsula, Mexico." *Geology* 19: 867–871.

28. **Matsui, T., Imamura, F., Tajika, E., Nakano, Y.,** and **Fujisawa, Y.** 2002. "Generation and propagation of a tsunami from the Cretaceous-Tertiary impact event." In *Catastrophic Events and Mass Extinctions*, eds. C. Koeberl and K. G. MacLeod, pp. 69–77. Geological Society of America Special Paper 356.

29. **Norris, R. E.,** and **Firth, J. V.** 2002. "Mass wasting of Atlantic continental margins following the Chicxulub impact event." In *Catastrophic Events and Mass Extinctions*, eds. C. Koeberl and K. G. MacLeod, pp. 79–95. Geological Society of America Special Paper 356.

30. **Jones, A.** 2005. "Meteorite impacts as triggers to large igneous provinces." *Elements* 1:277–281.

31. **Kerr, R. A.** 2007. "Humongous eruptions linked to dramatic environmental changes." *Science* 316:527.

32. **McFadden, L.-A.,** and **Blinzel, R. P.** 2007. "Near-Earth objects." In *Encyclopedia of the Solar System,* 2nd ed., eds. L.-A. McFadden, P. R. Weissman, and T. V. Johnson, pp. 283–300. New York, NY: Elsevier.

33. **Wikipedia.** 2013. "Near-Earth object." **http://en.wikipedia.org/wiki/Near-Earth_object.** Accessed May 15, 2013.

34. **National Aeronautics and Space Administration.** 2003. "Near Earth Object Program." **http://neo.jpl.nasa.gov/neo/report.html.** Accessed May 15, 2013.

35. **Morrison, D., Chapman, C. R., Steel, D.,** and **Binzel, R. P.** 2004. "Impacts and the public: Communicating the nature of the impact hazard." In *Mitigation of Hazardous Comets and Asteroids*, eds. M. J. S. Belton, T. H. Morgan, N. H. Smarasinha, and D. K. Yeomans, pp. 353–390. New York, NY: Cambridge University Press.

36. **Wikipedia.** 2013. "Torino Technical Hazard Impact Scale." **http://en.wikipedia.org/wiki/Palermo_Technical_Impact_Hazard_Scale.** Accessed May 15, 2013.

37. **National Aeronautics and Space Administration.** 2008. *Near-Earth Object Survey and Deflection Analysis of Alternatives: Report to Congress.* Washington, DC: National Aeronautics and Space Administration.

Index

A

aa, 135
abrupt climate change, 418
absorption, 277
acceptable risk, 15, 448
acid rain, 138
acidification, 210
active, 55
active layer, 169, 214
active margin, 337
adaptation, 418. *see also* specific hazards
adits, 228
aerosols, 397, 448
aesthetic resources, 225
aftershock, 65
agricultural drought, 297
air pressure, 280–281
airbursts, 425–437, 443, 448
Alaska
  earthquakes, 68, 70, 94, 194, 216
  landslides, 172
  subsidence, 216, 220
  tsunamis, 94, 101, 102, 103
  wildfires, 383
albedo, 278
Alberta, 234–236
Alberta Clippers, 296
Aleutian Islands, 36, 94
Aleutian Trench, 36
alluvial fan, 238, 239
Alpine fault, 36
Alvarez, Walter, 438
Alyeska pipeline, 76
American Avalanche Association, 201
amplification, 61, 62, 448
anastomosing, 238
Anatolian fault, 79
Anchorage earthquake, 19
Andes mountains, 35, 36
andesite, 118
Antarctic Ice Sheet, 410
anthropogenic sources, 395, 448
Appalachian Mountains, 36
aquifers, 229, 448
Arctic
  avalanches, 193
  climate change, 390–394
  coastal erosion, 350
  coastal hazards, 346
  earthquakes, 67
  ice sheets, 392–393
  ice shelves, 393
  permafrost thaw, 218, 221–222, 302
  sea ice, 390, 392
Arizona, 431, 436, 437
Armenia, 84
artificial bypass, 358
artificial control, 21, 448
ash falls, 129, 131, 135–136, 448
ash flows, 135
asteroid belt, 425
asteroids, 425, 428, 442–443, 448
asthenosphere, 27, 448
Atkinson, Gail, 82
atmosphere
  atmospheric pressure, 280–281
  circulation in, 397–398
  composition of, 278–279, 394–397
  Coriolis effect, 284–285
  defined, 278, 448
  effects of wildfires on, 376
  and energy, 274, 276
  fundamental concepts, 303–304
  structure of, 279–280, 397
  summary, 304–305
  vertical stability of, 282
atmospheric pressure, 280–281
atmospheric river, 316
atmospheric stability, 282
attenuation, 60
avalanche cords, 204, 448
avalanche dogs, 204, 448
avalanche forecasts, 198, 448
avalanche sheds, 196, 199
Avalanche Studies and Model Validation in Europe (SATSIE), 200
Avalanche Terrain Exposure Scale (ATES), 201
avalanche transceivers, 198, 204, 448
avalanches. *see* landslides; snow avalanches
Avaluator, 201, 202, 448
avulse, 238

B

Baffin Island, 67
banded iron formations, 435
Bangladesh, 5
bankfull discharge, 245
barometric pressure, 280–281
barrier islands, 343, 448
Barringer Crater, 429, 436, 437
basalt, 118
base level, 237
beach budget, 347–348
beach drainage, 359
beach drift, 342
beach face, 342
beach nourishment, 358–359, 449
beaches, 339, 342–343, 346, 358, 449
bed load, 238
bedding planes, 165
bentonite, 216
Berkeley, 384
berms, 196, 342
Bermuda High, 315
Bhuji, 83
Big Bang, 425
Big Blow of 1962, 326
Big Thompson Canyon, 246, 249, 266
biogeochemical cycle, 12–13, 449
biosphere, 412–414
blind, 55
blizzards, 294–296, 449
blue zone, 196
bluffs, 342, 346, 449
body-wave scale, 49
body waves, 57, 449
Borah Peak earthquake, 79
bores, 95
Borrero, Jose, 110
Bowen's Reaction Series, 122
braided, 238, 240
break up, 244
breaker zone, 342, 449
breaking waves, 340–341, 449
breakwaters, 356–358, 449
breccia, 429
Britannia, 259, 260
British Columbia
  earthquakes in, 50, 65–66, 68
  flooding, 259, 260
  floods, 240, 241, 248–249
  landslides, 161, 164, 173–174, 179–180
  slip events, 79
  snow avalanches in, 194, 197–198
  volcanoes, 120, 121, 124
  volcanoes in, 126, 127, 131–132
  wildfires, 364–366
British Columbia Oil and Gas Commission, 73
building codes, 84, 384–385

C

calderas, 124, 127, 128, 131, 449
Calgary, 234–236
California
  earthquakes, 46, 48, 81–82, 84
  hazard reduction programs, 75
  landslides, 176
  subsidence, 226, 227
  wildfires, 376–377, 384, 386
Canada
  droughts, 298
  earthquake risk, 24
  earthquakes, 51, 65–66, 67, 68, 101
  floods, 249, 253, 267
  ice storms, 272–274, 275
  impact craters, 431
  karst in, 218
  landslides, 156–159, 167, 169, 171, 173–174, 177
  natural hazards, effects of, 9
  permafrost, 214, 221–222

snow avalanches, 193, 197–198, 286
tornadoes, 295
volcanoes, 120, 121, 131–132
wildfires, 372–374, 381
Canada Water Act, 265–266
Canada Water Conservation Assistance Act, 265
Canada–British Columbia Flood Damage Reduction Agreement, 267
Canadian Avalanche Association, 201, 202
Canadian Avalanche Centre, 201, 202
Canadian Forest Service, 374
Canadian National Seismograph Network, 75
Canadian Ski Patrol System, 202
carbon-14 method, 10, 449
carbon dioxide
in atmosphere, 395, 398, 399, 408
and ocean acidification, 413
and subsidence, 210–211
and volcanoes, 138
and wildfires, 377
carbon monoxide, 137
carbon sequestration, 417, 449
Carrizo Plain, 37
Cascade volcanoes, 35
Cascadia subduction zone, 24, 26, 35, 100–101, 130
catastrophes, 7, 93, 449
catchment, 237
cause, 159
cave systems, 212, 226, 449
Cedar fire, 377, 386
cenotes, 440
Cenozoic Era, 402–403
channel patterns, 238, 449
channel restoration, 263–264, 449
channelization, 261–263, 449
chemical phenomena, 79
Chernobyl, 7
Chesapeake Bay impact crater, 433
Chile, 46, 65, 94, 123
Chilkoot Pass, 186
China
earthquakes, 5, 46, 48
floods, 10, 246
landslides, 71
population, 16
chutes, 159, 191
cinder cones, 120, 122–124, 126, 449
clearcutting, 175
cliffs, 342, 346, 349
climate, 396
components of. *see also* climate change
defined, 394, 449
and landslides, 166–167
modelling, 400–402
and snow avalanches, 189
stabilization of, 419
and wildfires, 376–379
of the world, 395
climate change. *see also* global warming
abrupt climate change, 418
adaptation to, 418
Arctic threat, 390–394
causes of, 406–409
and coastal hazards, 350
consequences of, 419–420
effects of, 409–415, 419
evidence of, 404–406
forcings, 406–409
and fossil fuels, 417
fundamental concepts, 419
and glacier ice, 410
long timescales, 402–404
and natural hazards, 21, 320
and other natural hazards, 419
risk, 419
short timescales, 404–409
study of, 399–402, 419
and subsidence, 222
summary, 420
temperature, 399, 405, 406, 409
and volcanoes, 145
Cloud, Preston, 424
clouds, 280
coastal erosion, 326, 345–349, 350, 351, 352, 358, 360, 362
coastal hazards
coastal zone management, 361
erosion, 345–349
flooding, 220
fundamental concepts, 361–362
hard stabilization, 356–358
and human activity, 352–356, 362
introduction to, 337
minimizing the hazard, 356–359, 360, 362
natural service functions of, 352
and other natural hazards, 350–352, 362
perception of hazard, 359
rip currents, 344–345
sea-level change, 343–344
sea-level rise, 349–350
soft solutions, 358–359
summary, 362
tsunamis, 112
waves, 337–341
coastal processes, 337–343, 350–352, 361–362
collapsible sediments, 216, 449
colluvium, 166, 449
Colombia, 139
Colorado, 172, 188, 375
Colorado Front Range, 246, 249, 266
combustion, 367–368, 449
comets, 425, 428, 435, 442, 449
community internet intensity maps, 50
complex landslides, 160
composite volcanoes, 120, 121
compression, 35
compression test, 200
compressional waves, 57
compressive strength, 192
conduction, 276, 368, 450
constructive interference, 338
continent-to-continent collisions, 36
continental drift, 29, 450
continental shelf, 163
continents, 28
convection, 28, 276, 368, 450
convection cell, 276
convergence, 280, 339
convergent boundaries, 35, 43, 450
coping capacity, 7, 450
Cordilleran ice sheet, 124
Coriolis effect, 284–285, 314, 450
coseismic uplift, 70
Crater Lake, 127, 131
craters, 124, 450
creep, 160, 161, 450
critical facilities, 83–84
cross-loaded, 190
crown fires, 371, 450
crust, 26, 450
cryosphere, 402, 450
cryptovolcanic structure, 437
crystallization, 11, 450
cumulonimbus clouds, 280
cumulus clouds, 280, 283
Curry, Dave, 213
cutbank, 238
cyclones. *see also* hurricanes
adjustments to, 331
Bangladesh (1970), 5
classification of, 311
defined, 311, 450
development of, 312–318
effects of, 322–323, 325–326
extratropical cyclones, 315–318
forecasting, 328–330, 331
fundamental concepts, 331–332
introduction to, 311
minimizing the hazard, 328–330, 332
naming, 311
as natural service function, 21
and other natural hazards, 326–327, 331
paths of, 320
perception of hazard, 331
risk, 318–320, 322
structure of, 318
summary, 332
tropical cyclones, 312–314

D

dacite, 118
Dafoe, Eric, 197
dams, 256–257
Darwin, Charles, 436
data collection, 381
debris avalanches, 166, 450
debris flows, 139–140, 159, 160, 450
decompression melting, 117–119, 450
deep waste disposal, 73
Deepwater Horizon, 352
delta, 238, 239
delta plain, 216, 229
deltaic sedimentation, 229
Denali, 76
dendroclimatology proxy record, 399
deposit/deposition, 11, 450

derechos, 288
desertification, 414, 450
desiccation cracks, 216, 222
Dietz, Robert, 41
dip-slip faults, 54, 450
direct effects, 17, 451
directivity, 60
disappearing streams, 213
Disaster Financial Arrangements, 267
disaster preparedness, 21, 450
disasters, 7, 8, 17, 19–20, 450
discharge, 238, 245, 451
discharge-frequency curve, 252
disease, 72, 105
displacement, 52, 56
dissipative stage, 286
dissolved load, 238
distant tsunami, 95, 451
distributary channels, 223
divergence, 280, 339
divergent boundaries, 31, 451
dogs, 204
Doppler effect, 302
Doppler radar, 302, 328
downbursts, 327
downdrift, 342
Downie slide, 179–180
downstream floods, 246, 248–249
drainage basin, 237, 451
drainage control, 179–180
driving forces, 163, 451
drought, 297–299, 414, 451
drumlins, 334
dry slot, 316
drylines, 288
Drynoch landslide, 179
Dust Bowl, 298–299
dust storms, 299, 451
dykes, 216, 260–261
dynamic equilibrium, 256–257

E

E-lines, 357, 451
E-zones, 357, 451
Earth
  history of, 426
  magnetic field, 38–40
  place in space, 424–425
  structure of, 26–28, 29, 43–44
earth fissures, 222
Earth system science, 394, 451
Earth's energy balance, 276, 277
earthflows, 162
earthquake cycle, 63–65, 451
earthquakes
  adjustments to, 83–86
  Alaska (1964), 65, 68, 70, 216
  Alaska (2002), 194–195
  Anchorage (1964), 19
  Armenia (1998), 84
  Baffin Island (1933), 67
  Bishop (1986), 15
  Borah Peak, 79
  California (1989), 46, 48
  California (1994), 46, 48
  Canada, 50, 51, 65–66, 67, 68, 101
  Chile (1960), 65
  Chile (2010), 46
  China (1556), 46
  China (1976), 5, 46
  China (2008), 46, 48
  death and damage from, 46–49
  defined, 451
  Denali (2002), 76
  education about, 84
  effects of, 68–71
  energy from, 6, 51
  forecasting, 77–83
  Haiti (2010), 2–5, 46, 72
  human interaction with, 73–75
  India (2001), 83
  Indian Ocean (2004), 5, 46
  Indus Valley, 46
  intensity, 50–53
  introduction to, 49
  Iran (2003), 46
  Italy (2009), 80–81
  Japan (1964), 70, 78
  Japan (1995), 5, 79, 83
  Japan (2011), 5, 37, 46, 90–94
  Landers (1992), 82
  and landslides, 172
  Loma Prieta (1989), 61, 62, 65, 68, 82
  Los Angeles (1971), 75
  Mexico (1985), 62, 63
  minimizing hazard, 75–83
  Mississippi Valley (1811–1812), 67
  natural service functions of, 21, 72–73
  New Madrid (1811–1812), 67
  New Zealand (2011), 36, 60, 70
  Newfoundland (1929), 101, 163, 172
  Niigata (1964), 78
  nomenclature, 49
  Northridge (1994), 8, 19–20, 60, 84
  Pacific Northwest (1700), 24, 65, 100–101
  Pakistan (2005), 5, 46
  perception of hazard, 83
  and plate tectonics, 43
  prediction, 77–83
  preparation for, 5, 83–86
  processes, 53457
  Quebec (2010), 67
  Quebec (2013), 67
  Queen Charlotte Islands (2012), 50, 65–66, 68
  recent catastrophic, 6
  risk, 65–67
  Roatan (2009), 69
  rupture, direction of, 60
  San Francisco (1906), 68, 72
  shaking from, 57–62
  slow earthquakes, 55
  and snow avalanches, 194–195
  and subsidence, 216, 219
  Sumatra (2004), 46, 96–97
  summary, 87
  Sylmar (1971), 50
  and tsunamis, 94–96
  Turkey (1999), 83
  in the US, 51
  and volcanoes, 145
  warning systems, 83
  Washington state (2001), 46, 48, 58, 60
  worldwide, 52, 54
eccentricity, 403
ecosystems, 226, 256, 412
edge waves, 95
edifice collapse, 139
Edmonton, 293
education
  for earthquake preparedness, 84
  for tsunami preparedness, 108–109, 111
  on wildfires, 384
EF Scale, 292, 451
Egypt, 256
ejecta blanket, 429, 451
El Niño, 5, 352, 411–412, 413, 451
El Salvador, 71
elastic rebound, 63, 64
elastic strain, 63, 64
electrical resistivity, 79
electromagnetic energy, 277
electromagnetic spectrum, 277, 278
electromagnetic waves, 276
emergency relief, 84
energy
  and Arctic ice, 392
  and asteroid or comet impacts, 442
  and atmospheric processes, 274
  behaviour of, 277–278
  from earthquakes, 6, 51, 94
  at Earth's surface, 276–278
  heat transfer, 276
  and magnitude and displacement, 52
  sources of, 6
  types of, 276
  from volcanoes, 146
  of wildfires, 372
England, 246
Enhanced Fujita Scale, 292
environment, 16
Environment Canada, 303
epicentre, 49, 50, 57, 59, 60, 451
episodic tremors, 79
erosion, 167, 238, 350. *see also* coastal erosion
European heat wave (2003), 299, 300
eustasy, 344, 451
evacuation, 21, 385, 451
expansive soils, 216, 217, 229, 451
exponential, 17, 451
exposure, 7
extinction, 368, 451
extinctions, mass, 436
extratropical cyclones, 311, 315–318, 451. *see also* cyclones
extreme cold, 296
eye, 313, 451

eyewall, 313, 451
Eyjafjallajokull, 43

F

factor of safety (FS), 163, 451
falls, 159, 160, 451
fault activity, 55
fault creep, 55
fault scarp, 68
faulting, 54
faults, 49, 54, 56, 451
Federal Emergency Management Agency (FEMA), 67, 267, 357
fertile land, 256
fetch, 322, 337
fire management, 381–383
Fire Potential Index, 381
fire regime, 381, 452
fire suppression, 381–383
fire triangle, 367
fires
  and earthquakes, 72
  and lightning, 302
  and tsunamis, 105
  and volcanoes, 144
fissures, 127
flash floods, 19, 246, 452
flashy discharge, 259
flood basalts, 127, 437
Flood Damage Reduction Program, 266–267
flood discharge, 245
flood fringes, 267
flood-proofing, 265
flood stage, 245
flooding
  adjustments to, 265–269
  Alberta (2013), 234–236
  Big Thompson Canyon (1976), 266
  Britannia (1921), 259, 260
  in Canada, 249, 253, 254
  channel restoration, 263–264
  China (1887, 1931), 246
  China (1998), 10
  Colorado (1976), 249
  Colorado Front Range (1976), 246
  and cyclones, 326–327
  and dams, 257–258
  death and damage from, 255–256
  defined, 241, 452
  England (1952), 246
  explained, 241
  flash floods, 246
  forecasting, 264
  Fraser River (1894), 240
  Fraser River (1972), 248–249
  frequency of, 251–252
  fundamental concepts, 269
  Himalayas (1929), 253
  and human activity, 256–259
  India (2005), 246
  and land use, 256–257, 269
  magnitude of, 251–252
  minimizing the hazard, 260–264
  Mississippi River (1973, 1993), 242–244
  natural service functions of, 21, 256
  and other natural hazards, 269
  outburst floods, 250, 253
  Pakistan (2010), 253, 255
  Pennsylvania (2006), 264
  perception of hazard, 264
  Peru (1941), 253
  recent catastrophic, 5
  relocation of people, 268–269
  risk, 253, 269
  Saguenay flood (1996), 250
  structural prevention of, 260–263
  and subsidence, 222
  summary, 270
  and thunderstorms, 301
  in United States, 254
  and urbanization, 258–259
floodplain regulation, 265, 452
floodplains, 236, 243, 269, 452
floodways, 267
Florida, 213, 216, 263
flows, 159, 160
flowstone, 212
fluorine, 138
flux, 13
focal depth, 57, 60
focus, 49, 50, 452
fog, 297, 452
foliation planes, 165
footwall, 54
forces on slopes, 163–170
forcing factors, 406, 452
forecasting
  of cyclones, 328–330
  defined, 452
  of earthquakes, 77–83
  of flooding, 264
  of hurricanes, 306, 328–330
  of impacts and airbursts, 443, 445
  of natural processes, 14
  of snow avalanches, 198, 200
  of storm surges, 328–329
  of volcanoes, 147–150
  of weather, 302
foreshocks, 13, 452
forest fires. *see* wildfires
fossil fuels, 417
fracking, 73
fractionation, 120, 122
France, 188
Frank Slide, 156–159, 171
Fraser River, 240, 248–249
fronts, 282
frost heaving, 215, 220, 452
frost-susceptible sediments, 214, 452
frostbite, 296
Fujita, T. Theodore, 292
funnel clouds, 289

G

gabions, 180, 263
Galileo, 436
galleries, 196
Galveston, 330
gases
  greenhouse gases, 399
  permanent gases, 394
  variable gases, 395–396
  volcanic gases, 137–139, 149
  of wildfires, 367
Geological Survey of Canada, 50, 75, 229, 435
geology
  of a catastrophe, 93
  and earthquakes, 62
  effects of wildfires on, 376
  and flooding, 246
  and landslides, 159
  and subsidence, 218
  of the Sudbury basin, 434
  and tsunamis, 100–101
  and volcanoes, 149–150
geophysical surveys, 231
Geostationary Operational Environmental Satellite (GOES), 105
geothermal power, 146
geysers, 128
Gilbert, G. K., 429
glaciation, 159, 402–404
Glacier National Park, 197–198
glaciers, 402–403, 410–411, 452
glacio-isostatic uplift, 344
global climate models, 400–402
Global Positioning System (GPS) stations, 79
global warming. *see also* climate change
  adaptation, 418
  and biosphere changes, 412–414
  carbon sequestration, 417
  and climate patterns, 411–412
  and cyclones, 327
  defined, 394, 452
  and desertification, 414
  and drought, 414
  and Earth system science, 394
  and human activity, 407
  international agreements, 416–417
  minimizing the effects of, 415–418
  and other natural hazards, 21, 409–415
  and permafrost thaw, 222
  and sea-level rise, 222
  and weather events, 301–302
  and wildfires, 415
Goderich, 294
gradient, 237
grading, 180
gradualism, 436
Grand Coulee Dam, 258
gravity, 6
Great Hurricane, 319
greenhouse effect, 398–399, 452
greenhouse gases, 399, 415–416, 416–417, 452
Greenland Ice Sheet, 392–393, 410
groins, 356, 358, 452
Gross, Richard, 92

ground acceleration, 68
ground penetrating radar (GPR), 231
ground rupture, 68, 76
groundwater, 72, 219, 226
groundwater mining, 229, 453
Gruntfest, Eve, 266
gust front, 288

H

H1N1, 7
habitats, 239, 412
Haida Gwaii, 36, 65–66, 68
hail, 288
Haiti, 2–5, 49, 72
halocarbons, 396
Hamilton, Don, 151
hanging wall, 54
Hansen, James, 394
Harappa, 46
hard stabilization, 356–358
Harvey Pass, 194
Hawaii, 20, 40, 41, 101, 130–131, 138
Hawaiian-Emperor chain, 40, 41
hazard reduction programs, 75
hazardous areas, 18
hazardous weather. *see* weather
hazards, 7, 179. *see also* natural hazards
headlands, 339
headwaters, 237
Healthy Forests Restoration Act, 383
Healy, Ingrid, 198
heat energy, 6, 276
heat index, 300, 453
heat islands, 299
heat transfer, 275
heat waves, 299
heavy rains, 326
Hekla, 138
Hess, Harry, 40, 41
Hickson, Catherine, 145
high-pressure centres, 280–281
high winds, 325–326
Hill, Philip, 345
Himalayan range, 36, 253
historical period, 399
history, 10–11
hoar, 191, 453
Holocene Epoch, 55, 403, 453
Honduras, 10
Hoover Dam, 73
Hope Slide, 161, 171
hot spots, 37, 40, 41, 129–130, 453
hot springs, 128
hot towers, 313, 314
human activity
  and coastal hazards, 352–356, 362
  and earthquakes, 73–75
  and flooding, 256–259
  and global warming, 407
  and landslides, 159, 175–177
  and natural hazards, 10
  and snow avalanches, 195
  and subsidence, 226–229, 231
  and weather, 301, 327–328
human forcing, 407
humidex index, 300, 301, 453
humidity, 278
Hurricane Betsy, 309
Hurricane Camille, 309
Hurricane Frieda, 320
Hurricane Hazel, 247–248, 320
Hurricane Ivan, 309
Hurricane Juan, 320–322
Hurricane Katrina
  damage from, 9
  deaths, 5
  described, 306–309
  effects of, 17
  and land-use practices, 10
  survivor story, 310
Hurricane Mitch, 10, 311
Hurricane Sandy, 324–325
hurricane warnings, 328, 453
hurricane watches, 328, 453
Hurricane Wilma, 311
hurricanes. *see also* cyclones; specific hurricane names
  categories of, 312
  and coastal hazards, 350, 351
  cross-section of, 313
  damage from, 247, 320, 324, 330
  death from, 247, 319, 320, 324
  defined, 311, 453
  forecasting, 328–330
  fundamental concepts, 331–332
  and global warming, 411
  and human activity, 328
  naming, 311
  past 50 years, 309
  paths of, 314
  recent catastrophic, 6
  risk, 331
  summary, 332
  and tornadoes, 327
Hutton, James, 436
hydraulic fracturing, 73
hydro-isostatic uplift, 344
hydrocarbon extraction, 73–74
hydrocarbon trap, 22
hydrograph, 245, 246, 251
hydrologic cycle, 6, 12, 237, 453
hydrologic monitoring, 148
hydrological drought, 297
hydrophobic layer, 376
hypocentre, 49
hypothermia, 303
hypothesis, 13, 453

I

ice-contact volcanoes, 124, 127
ice cores, 400
ice jams, 241, 244–246, 245
ice sheets, 392–393, 402
ice shelves, 392–393
ice storms
  Canada (1998), 17
  defined, 453
  described, 296
  Quebec and Ontario (1998), 272–274, 275
  recent catastrophic, 6
icebergs, 411
Iceland, 34, 43, 137, 138, 152
igneous rock, 11, 453
Ilgachuz Range, 120
impact, 9–10, 453
impact craters, 429–437, 453
impacts, of comets or asteroids
  airbursts, 425–437
  Earth's place in space, 424–425
  forecasting, 443, 445
  fundamental concepts, 445
  impact craters, 429–437, 453
  management of the hazard, 443–444
  and other natural hazards, 440, 445
  risk, 440–443, 445
  sequence of events, 441
  summary, 446
  Torino Impact Hazard Scale, 444
  warnings, 443
impervious cover, 258
inactive, 55
India
  earthquakes, 83
  floods, 246
  population, 16
Indian Ocean, 5, 9, 10, 46, 93, 96–99, 109
indirect effects, 17, 453
Indus Valley, 46
injection wells, 229
inner core, 26, 453
insect infestations, 378
instrumental intensity, 52, 53
instrumental period, 399
insurance
  defined, 453
  for earthquakes, 84
  for flooding, 267
  for natural hazards, 20–21
  for wildfires, 385
intensity, 49, 453
international agreements, 416–417
International Decade for Natural Hazards Reduction, 7
intraplate earthquakes, 66–67, 453
Inuvik, 221–222
ions, 238
Iran, 46
iridium, 438–439
island arc, 36
isostasy, 344, 453
Italy, 80–81, 171

J

jack-strawed, 177, 454
Jakob, Matthias, 178
Japan
  earthquakes, 5, 46, 70, 78, 79, 90–94
  tsunamis, 5, 90–94, 96
  volcanoes, 114–116
jet streams, 315, 454
jetties, 356–358, 454

jökulhlaups, 124, 145
Juan de Fuca plate, 35, 37, 55
Juan de Fuca Ridge, 31
Jupiter, 435

K

K-T boundary, 437–440
karst, 210–213, 218, 225, 454
karst plain, 211, 454
Kauai, 40
Kelowna fire, 364
Kennedy, Betty, 247–248
kinetic energy, 276
Kissimmee River restoration, 263
Kobe, 79, 83
Krakatoa, 94, 102
Krans, Bob, 377
Kuiper Belt, 425, 454
Kyoto Protocol, 416

L

La Niña, 412, 454
lag time, 259
lahars, 16, 117, 139–140, 454
Lake Mead, 73
Lake Michigan, 346
Lake Nyos, 137–138
land-level changes, 70, 78–79
land surface monitoring, 148
land use, 10, 20, 107, 256–257, 269, 454
Landers, 82
landform development, 72–73
landscaping, 229
landslide hazard map, 177, 179
landslides
  adjustments to, 182–183
  in California, 176
  in Canada, 156–159, 161, 167, 169, 171, 173–174, 176–177
  China (2008), 71
  and coastal hazards, 350
  Colorado (2004), 172
  defined, 159, 454
  and earthquakes, 71, 172
  effects of, 171
  El Salvador (2001), 71
  forces on slopes, 163–170
  forecasting, 183
  Frank Slide (1903), 156–159, 171
  fundamental concepts, 183
  Hope Slide (1965), 161, 171
  and human activity, 159
  and human interaction, 175–177
  introduction to, 159
  minimizing the hazard, 177, 179–182, 183
  Mount Meager (2010), 173–174
  natural service functions, 175
  and other natural hazards, 171–172, 183
  Pandemonium Creek (1959), 164, 165
  perception of hazard, 182–183
  Peru (1970), 162, 163
  prevention of, 179–181
  reactivation of, 168
  Rio de Janeiro (1988), 175–176
  risk, 170–171
  risk analysis, 177
  solutions to, 174
  summary, 184
  and tsunamis, 101–102, 172
  types of, 159–163
  variables, 159
  velocity of, 164
  and volcanoes, 145
  warning systems, 174, 182
Lang, Christine, 99
lapilli, 118
L'Aquila earthquake, 80–81
large igneous provinces, 128–129
latent heat, 276
lateral blasts, 135, 454
lateral moraines, 164
Laurentide ice sheet, 349
lava, 114, 454
lava flows, 133, 135, 454
lava fountain, 121
lava tubes, 120, 122
Leda clay, 167, 169
Leighton, F. B., 176
levees, 216, 225, 229, 242, 243, 244, 260, 454
Levy, David, 433
Light Detection and Ranging (LIDAR), 328
lightning, 6, 287–288, 290, 302, 454
limestone dissolution, 211
liquefaction, 68, 69, 167, 454
lithification, 11, 454
lithosphere, 27, 454
lithospheric plates, 11, 28, 454
Little Ice Age, 404
littoral cell, 347
littoral transport, 323, 342–343, 454
Lituya Bay, 101, 102
load, 238
local tsunami, 96, 454
location, 14, 59
loess, 216
Loma Prieta, 61, 62, 65, 68, 82
Long Valley, 128, 129
longitudinal profile, 237
longshore bar, 342
longshore currents, 342
longshore drift, 342
longshore trough, 342
Los Angeles, 75, 176
Love wave, 57
low-pressure centres, 280–281
Lyell, Charles, 436

M

M, 49–50
maars, 124, 454
Mackenzie River delta, 239
Macoun, John, 414
Macoun, Tony, 198
magma, 28, 114, 117–119, 122, 454
magma chambers, 216, 219
magnetic field, 38–40
magnetic monitoring, 148
magnetic reversal, 38, 454
magnitude
  comparisons, 2
  defined, 49, 454
  and displacement and energy, 52
  explained, 49–50
  worldwide, 52
magnitude-frequency concept, 9–10, 455
mainshock, 65
Maldives, 351
Mammoth Cave National Park, 225
Manicouagan impact structure, 430
mantle, 26, 455
mapping
  avalanche paths, 196
  community internet intensity maps, 50
  earthquake hazard map of Seattle area, 78
  flood hazard maps, 267
  floodplan zoning, 265
  hurricane track map, 329
  for landslide risk, 177, 179
  seismic ground acceleration map, 75, 77
  shake map, 52, 53
  for subsidence, 229
  tsunamis, 103, 107
  weather map, 283
Marianas Trench, 36
Marshall, Rich, 197
Martinique, 136
Maryland, 354–355
mass extinctions, 436, 437–440, 455
mass wasting, 159, 455
Matthews, Drummond, 41
mature stage, 286, 291
Mayfield, Max, 306
meander cutoff, 239
meandering, 238, 240
mega-fault, 24
mega-quake, 24
Meisner, Harris, 334
Meisner Island, 334–337
mesocyclone, 289
mesoscale convective complexes, 288
metamorphic rock, 11, 455
metamorphosed, 11, 455
meteor showers, 425
meteorites, 425, 428, 430, 455
meteoroids, 425, 429, 455
meteorological drought, 297
Meteorological Service of Canada, 303
meteors, 425
methane, 396
Mexico, 62, 63, 123–124, 126, 439–440
Miama Beach, 358, 360, 361
microbursts, 288
microearthquakes, 77
microzonation, 84
Mid-Atlantic Ridge, 29, 32, 43
mid-ocean ridges, 29, 38, 129, 455
Milankovitch cycles, 403–404, 455
minerals, 11, 72, 455
mining, 159, 228, 229

Missionary Ridge, 375
Mississippi Delta, 220, 223–225, 256
Mississippi River floods, 242–244
Mississippi Valley, 67
mitigation, 7, 303, 455
Modified Mercalli Intensity Scale, 50, 52, 455
Moho, 27, 455
moment magnitude (M_w), 49, 455
Mono Craters, 126
Monte Carlo simulation, 108
Montreal Protocol on Substances that Deplete the Ozone Layer, 416
Moon, John and Jane, 69
moulins, 410
mounds, 196
Mount Baker, 35, 132–133
Mount Edziza, 120, 126
Mount Fuji, 121
Mount Garibaldi, 121, 124
Mount Hood, 35
Mount Lassen, 35
Mount Loihi, 40
Mount Mazama, 131
Mount Meager, 121, 131, 173–174
Mount Rainier, 35, 121, 139
Mount St. Helens, 35, 121, 123, 131, 133, 141–144, 151
Mount Unzen, 114–116
Mount Vesuvius, 121, 123, 124, 125
mountain pine beetle, 378
Murray, Alexander, 435

N

Naples, 125
NASA, 392, 443
National Center for Atmospheric Research, 320
National Earthquake Prediction Evaluation Council, 81
National Flood Insurance Program, 267
National Flood Insurance Reform Act, 268
National Oceanic and Atmospheric Administration (NOAA), 105, 303, 381
National Research Council (NCR), 357
National Weather Service, 303
natural disaster, 7
natural hazards
  and climate change, 21
  damage reduction, 17–21
  death and damage from, 5, 8–9
  effects of in Canada and the US, 9
  fundamental concepts, 13–21
  and geologic cycle, 11–13
  glacier hazards, 411
  history of, 10–11
  and human activity, 10
  magnitude and frequency of, 9–10
  natural service function, 21
  and plate tectonics, 43
  proactive response, 20–21
  reactive response, 17–20
  recent catastrophic, 5–6
  and science, 13
  study of, 5–9
  summary, 22
natural service functions
  defined, 455
  of earthquakes, 21, 72–73
  of landslides, 175
  of natural hazards, 21
  of snow avalanches, 195
  of subsidence, 225–226, 225–226
  of volcanoes, 21, 146
  of wildfires, 379, 381
near-Earth object (NEO), 442, 455
New Madrid, 67, 81
New Orleans, 220
New Zealand, 36, 60, 70
Newfoundland, 101, 102, 163, 172
Nicholson, Andrew, 197, 198
Nile River, 256
nitrogen oxides, 396
nor'easters, 296, 311, 316
normal fault, 55, 56, 455
normal fault earthquakes, 66
North American plate, 35, 55
North Anatolian fault, 36
North Pacific tsunami, 9, 95
Northridge earthquake, 8, 19, 60, 84
Nostetuko Lake, 241
Nova Scotia, 334–337
nuclear explosions, 74
nuclear reactors, 90, 93
nuées ardentes, 135

O

Oakland, 384
obliquity, 403
occluded fronts, 282, 316
ocean acidification, 413
ocean basins, 28
ocean circulation, 406–407
Ocean City, 354–355
ocean conveyor belt, 407, 455
ocean floor. *see* seafloor spreading
oceanic crust, 28
oceanic-to-oceanic collision, 36
oil and gas industry, 72, 73, 76
Okanagan Mountain fire, 364, 366
Okefenokee Swamp, 375
Ontario, 67, 272–274, 275
Oort Cloud, 425, 455
organic sediments, 216, 455
organizational stage, 289
Ottawa River, 169
outburst floods, 250, 253
outer core, 26, 456
outflow boundaries, 288
outlooks, 264
overbank flow, 239
overwash, 323, 456
oxbow lakes, 239
ozone, 396

P

P waves, 57, 456
Pacific Northwest, 24, 65, 104, 216
Pacific Tsunami Warning Center, 98, 105
pahoehoe, 135
pahoehoe lava flow, 133, 135
Pakistan, 5, 46, 253, 255
paleo-proxy record, 399
paleomagnetism, 38–40, 456
paleoseismicity, 55
paleoseismologists, 75, 456
Palermo Technical Hazard Impact Scale, 443
Palliser, John, 414
Palliser Triangle, 414–415
Pandemonium Creek, 164, 165
Papua New Guinea, 94, 101
parcels, 282
Park Forest meteorite, 430
Parkfield Experiment, 81–83
Parks Canada, 201, 203
passive margin, 337
peat, 216, 218
Pemberton, 20
Pennsylvania, 264
permafrost
  defined, 213, 455
  explained, 213–215
  and geophysical surveys, 231
  and human activity, 228–229
  and landslides, 169
  and subsidence, 218
  thawing, 219–220, 221–222, 228–229, 411
permafrost table, 214
permanent gases, 394
Peru, 162, 163, 253
Peru–Chile Trench, 36
Philippine plate, 36
Philippines, 171
photosynthesis, 13, 456
physical barriers, 260–261
physical phenomena, 79
piles, 180
pillow breccias, 124
pillows, 124
Pineapple Express, 316
piping, 215, 456
plate boundaries, 29, 31, 33–36
plate-boundary earthquakes, 65–66, 456
plate motion, 36–37
plate tectonics
  defined, 456
  explained, 28–31
  fundamental concepts, 43–44
  and gradualism, 436
  and hazards, 43
  hot spots, 37, 40
  mechanisms of, 42–43
  model of, 33
  plate motion, 36–37
  summary, 44
  theory of, 6
  and volcanoes, 129–131
Pleistocene Epoch, 55, 402–403, 456
plunging breakers, 340
point bar, 238
point-release avalanches, 189, 456
poisonous gases, 137–139

polar deserts, 398
polar jet streams, 315–316
pollen, 400
Pompeii, 121, 123, 124, 125
Pontes, Lisza, 386
pools, 239, 240
population growth, 16–17, 19
pore water, 215
Portugal, 94
positive feedback, 392
potential energy, 276
Prairie Farm Rehabilitation Administration, 414
pre-ignition phase, 367, 456
precession of the equinoxes, 403
precursor events, 14, 456
precursors, 78, 456
prediction, 14, 456
predictions, 456. *see also* forecasting; specific natural hazards
preheating, 367
preparedness, 5, 83–86, 111, 303, 304, 456
prescribed burns, 382, 456
primary effects, 133
primary waves, 57
principle of uniformitarianism, 436
proactive response, 17, 20–21, 456
probability, 14, 107–108
probes, 204
processes
- coastal processes, 337–343
- defined, 6
- faulting, 54
- nature of, 13–14
- understanding, 15–16
- of wildfires, 367–371

Professional Profile
- Atkinson, Gail, 110
- Borrero, Jose, 110
- Gruntfest, Eve, 266
- Hickson, Catherine, 145
- Hill, Philip, 345
- Jakob, Matthias, 178
- Krans, Bob, 377
- Statham, Grant, 202

proxy data, 399
Public Safety and Emergency Preparedness Canada (PSEPC), 267
punctuated uniformitarianism, 436
pyroclastic debris, 118
pyroclastic deposits, 118, 457
pyroclastic flows, 116, 135, 136, 457
pyroclastic surges, 135, 457
pyrolysis, 367, 457

Q

Quebec, 17, 67, 167, 272–274, 275, 432
Queen Charlotte fault, 36, 65–66
Queen Charlotte Islands, 50, 65–66, 68
quick clay failures, 168, 169

R

radiation, 276, 368, 454
radiocarbon dating, 75
rain, 326
rain bands, 313, 457
Rapid City flash flood, 19
Rayleigh wave, 57
Razzak, Abdul, 98
recovery, stages of, 19–20
recreation, 146
recurrence interval, 67, 196, 245, 457
Red River Floodway, 263
red zone, 196
redirection, 277
refract, 339
reinsurance, 21, 457
relative humidity, 279, 457
relative sea level, 344, 457
relief, 166
relocation, 268–269
reservoirs, 73
residence time, 12, 457
resisting forces, 163, 457
resonance, 323, 457
retaining walls, 180
retrofitting, 84
return stroke, 287
reverse fault, 54, 457
rhyolite, 118
Richter, Charles F., 49
Richter scale, 49, 457
ridge push, 42
ridges, 299
riffles, 239, 240
Ring of Fire, 114, 117
Rio de Janeiro, 175–176
rip currents, 344–345, 457
riprap, 263, 356
risk
- and climate change, 419
- of cyclones, 318–320, 322
- defined, 7, 457
- of earthquakes, 65–67
- of flooding, 253, 269
- versus hazard, 179
- of hurricanes, 331
- of impacts, 440–443, 445
- of landslides, 170–171, 177, 179
- and population and socio-economic changes, 16–17
- seismic risk estimation, 75, 76
- of snow avalanches, 193, 195–196, 205
- of subsidence, 218–219, 231
- of tsunamis, 103–104, 112
- of volcanoes, 150–151, 153
- and weather hazards, 303
- of wildfires, 371–375, 387

risk analysis, 13, 15–16, 457
river flood. *see* flooding
rivers
- channel patterns, 238
- defined, 237, 457
- discharge, 238
- erosion, 238
- introduction to, 237
- material transported, 238
- sediment deposition, 238
- velocity, 238

Roatan, 69
rock avalanches, 158, 161, 162, 457
rock cycle, 11, 457
rocks, 11, 457
rockslides, 160
Rocky Mountain Arsenal, 73
Rogers Pass, 194, 197
rogue waves, 338, 457
Roland, Floyd, 221–222
rope stage, 291
rotational movement, 165
Roth, Dale, 197
run-out zone, 191, 196, 457
run-up, 95, 108, 164, 457
runoff, 237
rutschblock test, 200

S

S waves, 57, 457
sackung, 161, 458
sacrificial sand, 357
Saffir, Herbert, 311
Saffir-Simpson Scale, 311, 312
Saguenay flood, 250
San Andreas fault, 36, 37, 41, 79, 81–83
San Andreas Fault Observatory at Depth (SAFOD), 82
San Francisco, 68, 72
sand blows, 68
sand storms, 299
sand transport, 342
sand volcanoes, 68
Santa Ana winds, 377
Saskatchewan, 294–296
science, 13
scientific method, 13, 458
scientific research, 381
scientific resources, 225
scoria cones, 122–124
Scott Lake, 213
sea cliffs, 342, 458
sea ice, 392
sea-level change, 343–344
sea-level rise, 349–350, 410
seafloor displacement, 94
seafloor spreading, 29, 38–40, 458
seamounts, 40
seawalls, 356, 458
secondary effects, 133
secondary waves, 57
sector collapse, 139, 458
sediment
- and climate change, 399
- collapsible sediments, 216, 449
- compaction of, 215–216
- defined, 458
- deltaic sedimentation, 229
- and flooding, 238, 256, 257, 258
- frost-susceptible sediments, 214, 452
- organic sediments, 216, 455
- and permafrost, 214
- and the rock cycle, 11

sediment (*continued*)
  and subsidence, 218, 224
  transport of, 343
sediment budget, 346, 458
sedimentary rock, 11, 458
seismic gaps, 79
seismic ground acceleration map, 75, 77
seismic sources, 54
seismic waves, 55, 57, 458
seismograms, 57, 58
seismographs, 28, 58, 458
seismologists, 49, 458
seismology, 28, 147–148, 458
sensible heat, 276
setbacks, 356
sets, 338
severe weather. *see* weather
shafts, 228
shake map, 52, 53
shaking, 68
shear strength, 163, 458
shear waves, 57
shield volcanoes, 120–121, 458
Shoemaker, Gene and Carolyn, 433
shovel test, 200
shovels, 204
shrinking stage, 291
Siberia, 422, 424
silicosis, 136, 458
Simpson, Robert, 311
sinkholes, 211–212, 219, 458
sinking plates, 29, 31
slab avalanches, 190, 458
slab pull, 42
Slave Lake, 374–375
slides, 458. *see also* landslides; rockslides
slip events, 79
slip rate, 75
slope stability map, 179
slopes
  and avalanches, 192
  and flooding, 257
  forces on, 163–170
  supports for, 180–181
slow earthquakes, 55, 458
sluffs, 189, 192, 458
slump blocks, 163
slumps, 159, 160, 458
smectite, 216
Smith, Tilly, 98
snow avalanches
  in Canada, 193
  Chilkoot (1896), 186
  climatology, 189
  control through explosives, 196
  danger scale, 201
  death and damage from, 194
  defined, 188, 458
  forecasting, 198, 200
  France (1999), 188
  fundamental concepts, 205
  Glacier National Park (2003), 197–198
  Harvey Pass (2008), 194
  and human activity, 195
  and infrastructure, 195–196, 205
  initiation of, 189–190
  introduction to, 188–189
  mapping, 195–196
  minimizing the hazard, 195–196,
    198–199, 198–201
  modelling, 200
  motion of, 191
  natural service functions, 195
  and other natural hazards, 194–195
  rescue, 203–205
  risk, 193, 195–196, 205
  Rogers Pass (1910), 194
  safety, 201–202
  size of, 189
  and slope, 192
  summary, 205–206
  survival, 204
  terrain factors, 191–192
  triggers, 191
  weak layers, 190–191
  and weather, 205
snowpack observations, 200
socio-economic changes, 16–17
soil
  defined, 210, 458
  expansion and contraction, 210, 220, 222
  expansive soils, 216, 217, 229
soil and rock conditions, 62
soil slips, 167
solar forcing, 407
solar nebula, 425, 427
solar system, 427
space, 424–425
Spaceguard Program, 443
spilling breakers, 340
spits, 343, 458
splitting wedges, 196
spot fires, 370
spreading centres, 29, 31, 458
springs, 213, 458
spruce budworm, 378
squall lines, 288
St. Lawrence River valley, 67, 167
St. Pierre, 136
stage, 245
stalactites, 212
stalagmites, 212
start zone, 191, 196, 459
Statham, Grant, 203
stationary fronts, 282
Statoil, 417
stepped leader, 287
stony, 425
storm surges, 322–323, 328–329, 459
strain, 54
stratigraphy, 67
stratovolcanoes, 120, 121, 123, 129, 459
streams, 237, 459
strength, 54
strike-slip faults, 54, 56, 65, 459
structural control, 106–107
structural protection, 84, 385
subaqueous landslides, 162, 459
subduction earthquakes, 65, 92, 459
subduction zones, 29, 129, 459
submarine trench, 36
subsidence
  adjustments to, 229–230
  and climate change, 222
  coastal flooding, 220
  defined, 210, 459
  and earthquakes, 78–79, 216, 219
  effects of, 219–220
  Everglades, 216
  and flooding, 222
  forecasting, 231
  fundamental concepts, 231
  and groundwater extraction, 226
  and human activity, 226–229, 231
  introduction to, 210
  and landscaping, 229
  magma chamber deflation, 216, 219
  minimizing the hazard, 229–230
  and mining, 228, 229
  Mississippi Delta, 223–225
  natural service functions, 225–226
  and other natural hazards, 222, 231
  perception of hazard, 229
  and permafrost, 213–215, 219–220, 222,
    228–229, 228–229, 231
  risk, 218–219, 231
  Scott Lake (2006), 213
  sinkholes, 211–212, 219
  soil volume changes, 220
  summary, 232
  Venice, 208
  and wetlands, 220, 229
  and withdrawal of fluids, 226–227, 229
subtropical jet streams, 315
suction vortices, 291
Sudbury impact event, 434–435
sulphur dioxide, 138
sulphuric acid, 138
Sumatra, 46, 94, 128
sun energy, 6
super eruptions, 128, 459
supercell storms, 288
superelevation, 164
supernovas, 425
superstorm Sandy, 5, 21
superstorms, 5
supervolcanoes, 128, 459
surf zone, 342, 459
surface fires, 370–371, 459
surface rupture, 50
surface-wave scale, 49
surface waves, 57, 459
surge, 411
Survivor Story
  Curry, Dave, 213
  Hamilton, Don, 151
  Kennedy, Betty, 247–248
  Lang, Christine, 99
  Moon, John and Jane, 69

Ogg, Danny, 172
Pontes, Lisza, 386
Utley, Michael, 290
Zeilenga, Pauline, 430
Zeitoun, Abdulrahman, 310
suspended load, 238
suture zone, 36
swash zone, 342, 459
swell, 339
swine flu, 7
Sylmar, 50
synclines, 62

T

talus slopes, 167, 459
tectonic, 11, 28, 459
tectonic creep, 55, 459
tectonic cycle, 11, 12, 459
tectonic effects, 344
tectonic plates, 24, 30, 55. *see also* plate tectonics
tele-tsunami, 95
tensile strength, 192
tephra, 118, 459
Texas, 229
thaw flow slides, 169, 170, 459
thermal expansion, 344
thermal monitoring, 148
thermal springs, 128
thermokarst, 214, 215
Thistle landslide, 181
thrust faults, 54, 56, 65, 92, 459
thunder, 287
thunderstorms
  and flooding, 301
  life cycle of, 286
  in North America, 286
  processes of, 282–283, 286
  severe, 288–289
Tibetan Plateau, 36
tidal bores, 340, 459
timber harvesting, 175
time, and landslides, 169
Toba eruption, 128
tombolos, 343
top-loaded, 190
topographic relief, 166
topography, 166, 369
topples, 161, 459
Torino Impact Hazard Scale, 444
tornadoes
  classification of, 292
  deaths from, 294
  defined, 289
  Edmonton (1987), 293
  energy of, 6
  Goderich (2011), 294
  and hurricanes, 327
  occurrence of, 293–294
  stages of, 289, 291
  watches and warnings, 302
total load, 238
tower karst, 212–213
track, 191, 460
transform boundaries, 36, 41, 460
transform fault, 36, 460
translational movement, 165
transmission, 277
tree rings, 399, 400
Trenberth, Kevin, 320
triangulation, 60
tributaries, 237
triggers, 159
triple junctions, 36, 460
troglobites, 226
tropical cyclones, 311, 312–314, 460
tropical depressions, 312, 460
tropical disturbances, 312, 460
tropical storms, 313, 460
tropopause, 286
troposphere, 280, 460
troughs, 312
tsunameters, 105
tsunami-ready, 111, 460
tsunami warning, 109, 460
tsunami watch, 109, 460
tsunamis
  adjustments, 111
  Alaska (1958), 101
  Alaska (1964), 103
  deaths from, 9
  defined, 94
  detection and warning, 105–106, 111
  and earthquakes, 94–96
  education, 111
  education about, 108–109
  effects of, 104–105, 109
  examples of, 94
  formation of, 95
  fundamental concepts, 111–112
  geologic evidence for, 100–101
  Indian Ocean (2004), 5, 9, 10, 46, 96–99, 109
  introduction to, 94
  Japan (1700), 24
  Japan (1993), 94, 96
  Japan (2011), 5, 90–94
  Krakatoa, 102
  and land use, 107
  and landslides, 101–102, 172
  local, 96, 454
  mapping, 103, 107
  minimizing the hazard, 105–109
  Newfoundland (1929), 101, 102, 172
  North Pacific (1964), 9, 95
  and other natural hazards, 105, 112
  Pacific Northwest (1700), 104
  preparedness, 93, 111
  probability analysis, 107–108
  risk, 103–104, 112
  structural control, 106–107
  summary, 112
  and volcanoes, 102
Tuktoyaktuk, 220
Tunguska event, 422, 436
turbidity current, 162
Turkey, 36, 83
tuyas, 124
typhoons, 311, 411, 460

U

undertow, 345
uniformitarianism, 436
United States
  earthquakes, 51, 79
  floods, 254
  natural hazards, effects of, 9
  thunderstorms in, 286
updrift, 342
uplift, 78–79
urbanization, 175–177, 258–259
U.S. Federal Emergency Management Agency (FEMA), 303
U.S. Federal Fire Policy, 382
U.S. Geological Survey (USGS), 50, 75, 81, 381
U.S. National Flood Insurance Program, 267
U.S. National Weather Service, 306
Ussher, Archbishop, 436
Utah, 181
Utley, Michael, 290

V

Vancouver, 24
variable gases, 395–396
vegetation, 167
veins, 72
velocity, 164, 238
Venice, 208
vents, 120, 124, 127
Victoria, BC, 24
Vine, Frederick, 41
viscosity, 118
Voellmy, A., 200
vog, 138
volatiles, 118, 460
Volcan Osorno, 121, 123
volcanic crisis, 150–151, 460
volcanic domes, 120, 122, 126, 460
Volcanic Explosivity Index (VEI), 133
volcanic forcing, 407
volcanic soils, 146
volcanic vents, 120, 124, 127, 460
volcanoes
  adjustments to, 151–152
  Aleutian volcanoes, 36
  Cascade volcanoes, 35
  characteristics of, 120
  Chile (1869), 123
  and climate, 145
  Colombia (1985), 139
  damage from, 117
  and earthquakes, 145
  energy from, 6
  Eyjafjallajokull (2010), 43
  features of, 124–129
  forecasting, 147–150, 153
  fundamental concepts, 153
  gas emissions, 137–139, 149

volcanoes (*continued*)
geologic history, 149–150
Hawaiian-Emperor chain, 40, 41
hazards of, 133–139
historic eruptions, 134
Iceland (1845, 1970), 138
Iceland (2010), 137
introduction to, 114
Japan, 114
Kauai, 40
Krakatoa, 102
and landslides, 145
Mammoth Lakes (1982), 15
Martinique (1902), 136
Mexico (1943), 123–124
minimizing the hazard, 147–150, 153
Mount Baker (1843), 132–133
Mount Loihi, 40
Mount Rainier, 139–140
Mount St. Helens (1980), 123, 133, 141–144, 151
Mount Unzen, 114–116
Mount Vesuvius, 121, 123, 124, 125
natural service functions of, 21, 146
and other natural hazards, 144–145, 153
and plate tectonics, 129–131
regions with active volcanoes, 131–133
risk, 153
risk perception, 150–151
Sumatra, 128
summary, 153–154
and tsunamis, 102
types of, 120–125
warnings, 150
vortex, 289
vulnerability, 7, 460

W

Wadati-Benioff zone, 31, 33, 460
Wadati, Kiyoo, 49
Waldo Canyon, 375
wall cloud, 289
warning, 14–15, 105–106, 460
warning systems
for earthquakes, 82
for hurricanes, 328
for landslides, 174, 182
for tornadoes, 302
for tsunamis, 106
for volcanoes, 150
for wildfires, 384
warnings, 460
Washington State, 46, 48, 58, 60, 67, 70, 121
washover channels, 323
waste disposal, 73
watches, 460
water, 167–169, 225, 395–396
water cycle. *see* hydrologic cycle
water supply, 12
watershed, 237
waterspouts, 292
Watkins, Abby, 197
wave height, 338
wave normals, 339
wave period, 338
wavelength, 277, 338
waves, 337–341
weather
adjustments to, 303
and atmosphere, 278–282
blizzards, 294–296
and climate, 394
consequences of, 304
defined, 394, 460
drought, 297–299
dust storms, 299
extreme cold, 296
fog, 297
forecasting, 302
fundamental concepts, 303–304
hail, 288
heat waves, 299–300
and human activity, 301, 327–328
ice storms. *see* ice storms
and landslides, 159
lightning, 6, 287–288, 290
map, 283
minimizing the hazard, 302
natural service functions of, 302
and other natural hazards, 301–302, 304
processes of, 280–282
risk, 303
and rock falls, 169
sand storms, 299
and snow avalanches, 200
summary, 304–305
thunderstorms. *see* thunderstorms
tornadoes. *see* tornadoes
and wildfires, 370
wind chill, 296
weathering, 11, 460
Wegener, Alfred, 29, 38
West Coast and Alaska Tsunami Warning System, 92, 105
wetlands, 219, 220, 223–225, 229
wildfires
in Alaska, 383
in British Columbia, 364–366
in California, 376–377, 384, 386
in Canada, 372–374, 381
causes of, 372
Cedar fire (2003), 377, 386
and climate, 376–379
damage from, 372, 375, 387
data collection, 381
deaths from, 375
defined, 366–367
effects of, 376, 387
environment of, 368–371
fuel, 368–369
fundamental concepts, 387
global map, 382
and global warming, 415
impacts on plants and animals, 379
intensity of, 372
introduction to, 366–367
and lightning, 302
management of, 381–383
minimizing the hazard, 384–385
Missionary Ridge (2012), 375
natural service functions of, 21, 379, 381
Oakland and Berkley (1991), 384
Okefenokee Swamp (2002), 375
and other natural hazards, 387
perception of hazard, 383–384
phases of, 367–368
processes, 367–371
risk, 371–375, 387
scientific research, 381, 387
Slave Lake (2011), 374–375
and solar energy, 6
summary, 388
suppression of, 381–383
and topography, 369
types of, 370–371
Waldo Canyon (2012), 375
and weather, 370
and wind, 369
Yellowstone National Park (1988), 380
Yukon, 383
Wilson, J. Tuzo, 40–42
wind, 190
wind chill, 296, 460
wind patterns, 285
wind shear, 288
wind slab, 190, 460
Woods Hole Research Center, 297
World Agency of Planetary Monitoring and Earthquake Risk Reduction, 81
World Meteorological Organization, 311
Wyss, Max, 81

Y

Yangtze River, 10
Yellow River, 246
Yellowstone National Park, 40, 128, 129, 133, 380
Yucatán Peninsula, 439–440
Yukon, 131–132, 383

Z

Zeilenga, Pauline, 430